Geomorphology

This textbook provides a modern, quantitative, and process-oriented approach to equip students with the tools to understand geomorphology. Insight into the interpretation of landscapes is developed from basic principles and simple models, and by stepping through the equations that capture the essence of the mechanics and chemistry of landscapes. Boxed worked examples and real-world applications bring the subject to life for students, allowing them to apply the theory to their own experience. The book covers cutting-edge topics, including the revolutionary cosmogenic nuclide dating methods and modeling, highlights links to other Earth sciences through up-to-date summaries of current research, and illustrates the importance of geomorphology in understanding environmental changes. Setting up problems as a conservation of mass, ice, soil, or heat, this book arms students with tools to fully explore processes, understand landscapes, and participate in this rapidly evolving field.

BOB ANDERSON has taught geomorphology since 1988, first at University of California, Santa Cruz, and now at University of Colorado, Boulder. Bob has now studied most parts of landscapes, from their glaciated tips to their coastal toes, with significant attention to sediment transport mechanics, the interaction of the geophysical and geomorphic processes that shape mountain ranges, and the evolution of bedrock canyons and glaciated landscapes. He has participated in the development of a new tool kit that employs cosmogenic radionuclides to establish timing in the landscape. He develops numerical models of landscapes that honor both field observations and the first principles of conservation; these models in turn have served to hone his field efforts. In the course of this academic adventure he has been founding editor of the *Journal of Geophysical Research – Earth Surface*, co-authored the textbook *Tectonic Geomorphology* (2000, Wiley-Blackwell) with Doug Burbank, and has been honored by election as a Fellow of the American Geophysical Union.

SUZANNE ANDERSON has been on the faculty at University of Colorado, Boulder, since 2004, where she teaches courses on geomorphology, Earth's Critical Zone, landscapes and water, and glaciers and permafrost. Her awards include an Outstanding Graduate Student Instructor award at University of California, Berkeley, a NASA Graduate Student Fellowship in Global Change Research, and an NSF Earth Sciences Post-doctoral Fellowship. Suzanne's research has taken her to Svalbard, Alaska, Oregon, and Nepal, and has focused on interactions between chemical weathering, hydrology, and physical erosion mechanisms. She currently directs the Boulder Creek Critical Zone Observatory, an NSF environmental observatory based at the University of Colorado that involves researchers from four institutions and agencies. Suzanne was editor of *Arctic, Antarctic, and Alpine Research* from 2004–2006, and served as an associate editor of the *Journal of Geophysical Research – Earth Surface* from 2002–2006.

Praise for this textbook

"This book is terrific! Anderson and Anderson have hit it just right on all the main points: their book is engaging and informal; thorough but not pedantic; and shot through with the sheer pleasure of understanding how things work. It's packed with physical insight, useful information, and interesting problems; and it is simply a pleasure to read. This is a model of what a textbook should be, and it's also the first place I'd send a student or colleague to get them excited about landscapes and how we study them."

CHRIS PAOLA – *Professor of Geology and Geophysics, St. Anthony Falls Laboratory, Minneapolis*

"This much needed, skilfully crafted text will be welcomed by the geomorphology community. ... I applaud Bob's and Suzanne's approach of focusing on "how geomorphic things work" independently of where and when ... From this perspective the text is aptly titled, and it will have a long, healthy lifespan ... The text offers a systematic coverage of essential ingredients ... the presentation of various topics spans a range of sophistication ... so that the text can be used for an introductory course, or as part of a more advanced course. The writing is clear, sometimes playful, and possesses personality. The overall reaction of my students using a draft version has been very positive."

DAVID JON FURBISH – *Professor and Chair, Department of Earth and Environmental Sciences, Vanderbilt University*

"Geomorphology has entered a new era. Building on decades of research on the mechanisms of Earth surface processes and driven by stunning new tools that provide both the age and elevation of the landscape, geomorphologists now endeavor to truly predict the form of the Earth. The Anderson's new book is the first to pull this information together in a consistent framework. Its synthesis will be used to date the arrival of geomorphology as a mature, coherent, predictive science. The book is both authoritative and accessible, encouraging students (and instructors) to think creatively and precisely about how the landscape evolves. Unlike previous geomorphology texts, it provides a consistent approach for defining and solving models for the full range of features found on the surface of the Earth."

PETER R. WILCOCK – *Professor and Associate Chair, Department of Geography and Environmental Engineering, Johns Hopkins University*

"A wonderful, wide ranging review of the modern science of geomorphology."

NIELS HOVIUS – *Lecturer, Department of Earth Sciences, University of Cambridge*

Geomorphology

THE MECHANICS AND CHEMISTRY OF LANDSCAPES

Robert S. Anderson

AND

Suzanne P. Anderson

University of Colorado, Boulder, USA

CAMBRIDGE
UNIVERSITY PRESS

CAMBRIDGE UNIVERSITY PRESS
Cambridge, New York, Melbourne, Madrid, Cape Town, Singapore,
São Paulo, Delhi, Dubai, Tokyo, Mexico City

Cambridge University Press
The Edinburgh Building, Cambridge CB2 8RU, UK

Published in the United Kingdom by
Cambridge University Press, UK

www.cambridge.org
Information on this title: www.cambridge.org/9780521519786

First published 2010
Reprinted with corrections 2011

Printed in the United Kingdom at the University Press, Cambridge

A catalog record for this publication is available from the British Library

Library of Congress Cataloging-in-Publication data

Anderson, Robert S. (Robert Stewart), 1952–
  Geomorphology : the mechanics and chemistry of landscapes /
Robert S. Anderson and Suzanne P. Anderson.
    p. cm.
  Includes bibliographical references and index.
  ISBN 978-0-521-51978-6 (pbk.)
1. Geomorphology. I. Anderson, Suzanne P. II. Title.
  GB401.5.A43 2010
  551.41–dc22   2010004400

ISBN 978-0-521-51978-6 Paperback

Additional resources for this publication at
www.cambridge.org/9780521519786

The Blue Hills badlands in central Utah comprise a landscape of
diffusive hillslopes developed in the shales of the Cretaceous interior
seaway, bounded by incising channels. Downcutting of the sinuous
channel here is accomplished by a series of headward-migrating
knickpoints, and reflects baselevel control by the Fremont River.
In the middle distance is a silhouette of South Caineville Plateau,
capped with 60 m of massive sandstone. The snow-capped laccolithic
Henry Mountains in the distance were the subject of Grove Karl
Gilbert's 1877 "Report on the Geology of the Henry Mountains",
which laid the foundation for modern geomorphology.

CONTENTS

PREFACE

Geomorphology is the study of the shape of the Earth. In this book we take this quite literally, and address the shape of the Earth at many scales. We ask why it is spherical, or not quite spherical, why it has a distribution of elevations that is bimodal, one mode characterizing a quite well-organized set of ocean basins, another the terrestrial landscape. At smaller scales, we address why hilltops are convex, why glacial troughs are U-shaped, why rivers are concave up. At yet smaller scales, sand is rippled, beaches are cusped, hillslopes are striped, and mud is cracked. These are some of nature's most remarkable and visible examples of self-organizing systems. Each cries out for both explanation and appreciation.

Goals

We wrote this textbook to provide modern teachers and students of geomorphology with a formal treatment of geomorphic processes that acknowledges the blossoming of this field within the last two decades. It brings together between two covers the background that serves to attach our field with those of geophysics, atmospheric sciences, geochemistry, and geochronology. It honors the heightened importance of geomorphology in understanding the environment and its changes, with an attendant need to pose these problems more formally.

The book is intended to be used in an introductory geomorphology course in which the attention is more on the processes that shape landscapes than on the cataloging of landforms. Most likely such a course will fit into a third and fourth year undergraduate or an introductory graduate curriculum. The students must be comfortable with or be accepting of the challenge of a mathematical treatment of the topic. We have tried to be friendly by providing steps in the derivations, by providing a comprehensive math backdrop in the appendix, and by setting a conversational tone, as if we were in the room teaching.

The long gestation of the book (we began this book a decade ago) is in part due to the breadth of the territory we have tried to cover. But it also reflects the high productivity of the community of scientists for whom this book is intended. The last decade has seen the emergence of new journals in which to publish, new methods to employ in the field, and, of course, continued growth of computational capacity available to the field. These new papers serve as a distraction at the very least, and as new material to try to synthesize or incorporate in some fashion. The field is therefore a moving target, as it should be in any burgeoning field of science. We have tried to capture it in motion, and to give a sense that it is ever-broadening through incorporating the latest material.

Our goal is to allow the reader of this book to view landscapes in a more systematic way. We focus on the formal treatment of geomorphic processes that allows the student to see the connective tissue between sub-disciplines in geomorphology. We show how one can set up problems by employing the concept of continuity, or of conservation of some quantity, in, for example, hillslopes, glaciers, alluvial rivers, and dating methods. The word picture for all of these problems is: the rate of change of storage of some quantity = the rate of inputs minus the rate of loss of that quantity. Setting up the problem in this way then demands that we understand quantitatively how material (or energy) moves in the environment, and what the sources or sinks of that material might be. This then motivates both theoretical work on fluxes, and field experiments designed to constrain such theory. The student is encouraged to gain an appreciation of this approach by sheer repetition, from application to application, from chapter to chapter. If by the end of the book, or of the course based upon it, the student is heard to groan "not again . . .," we will have succeeded.

The practice of modern geomorphology often includes the generation of numerical models of landscapes or of key landforms. This exercise absolutely requires the formal problem set-up we advocate. The computer demands that we think in concrete, careful, and logical terms. In this textbook we honor that demand and demonstrate through repeated use of this approach how to set up quantitative problems in geomorphology. In this sense this textbook therefore connects more directly to similar approaches in our sister sciences of physics and chemistry.

So that the student need not scurry off to find another math or physics textbook, we have both provided detailed derivations within the textbook, and have supported the steps with reference to an extensive math appendix meant to serve as a refresher for all math from algebra through differential equations and probability density functions.

Novelties

We cover explicitly several topics that are not broken out in most geomorphology textbooks. These include several of the first chapters in the book:

- The whole Earth shape (Chapter 2). We ask why the Earth is a sphere, or really not quite a sphere,

and what governs the largest features on the Earth. This introduces isostasy.
- Large-scale forms attributable to large-scale geophysical processes (in the mantle) (Chapter 3).
- Tectonic geomorphology (Chapter 4). Here we discuss the geophysical processes responsible for the growth of individual mountain ranges. As most of these involve faults, this requires addressing slip rates and how we know them, which verges on paleoseismology.
- Establishing timing in the landscape (Chapter 6). Here we dwell on the developments in the use of cosmogenic radionuclides, and break out a section on thermochronometry as it has become so useful in constraining long-term exhumation patterns.

The end of the book is ornamented with two novel chapters:

- The geomorphology of big floods (Chapter 17). We could not help but assemble in one place all those stories we hear about in different corners of the literature about the biggest of the geomorphic events – the big floods: Bonneville, Spokane, Lake Agassiz, and so on. These are the stories we all tell around the campfire, discussing when we would like to have lived, what events we would like to have witnessed. The evidence for these is writ large on some landscapes, for there has not been the power in any subsequent event to erase them from the landscape.
- Whole landscapes (Chapter 18). In this chapter we assemble information from all quarters on the evolution of the Santa Cruz landscape as an illustration of how all of the parts of the book are useful in compiling a more comprehensive understanding of one landscape.

Geomorphology is indeed the most visible of the Earth science disciplines. It is the study of the scenery that inspires photography. We launch each chapter with a photograph meant to capture the beauty of the topic, accompanied by a quote or a poem similarly inspired.

Arrangement of the book

We have organized the book to proceed from large scale to small scale. Treatment of the large scale requires an acknowledgement of the various roles of geophysics in generating and in accommodating topography. We augment these precursor chapters with one on dating (Chapter 6) and one on the roles of the atmosphere in geomorphology (Chapter 5). Armed with these tools, we then tackle the more classical topics within geomorphology – those that tear down and attack the geophysically generated topography. We treat first the processes and forms that characterize cold environments. We admit these are topics of particularly strong interest for both of us. But these lie a little outside the organization that naturally arises in the remainder of the book. After treatment of cold environments (in Chapters 8 and 9), we have organized the remainder of the topics according to what one needs to know first: we need to produce regolith before we can transport it. We need to know how material moves on hillslopes before it gets to the rivers. We need to know how water moves on hillslopes and in rivers before we can address how water transports sediment. Finally, we need to know all of these pieces before we can fully understand a particular landscape. We employ the Santa Cruz landscape in coastal California as our chief example. Tectonics matters, sea level variation matters, orographic precipitation matters, and so on.

How to use the book

One may teach a course based on the material in this book in many ways. The more common approaches to teaching geomorphology would skip the large-scale material in the first few chapters and begin with the small scale, e.g., sediment transport, hillslopes, or wind. After all, it is often these topics that have attracted the student into a class on geomorphology. As the book is designed such that all chapters can stand alone, one may order the course however one wishes. If the students have been exposed to the large-scale backdrop material in other classes, then begin with glaciers, or sediment transport. We recommend, however, that the course designer sweep through the text to locate where we have introduced certain topics. The table of contents is a good place to start. For example, fluid mechanics is introduced in earnest in the chapter on rivers (Chapter 12), the development of the full Navier–Stokes equation being tucked in an appendix to that chapter. Heat transfer is covered in the chapter on the effects of large-scale geophysics (Chapter 3), as this is where we first encounter conduction and diffusion in studying the bathymetry of ocean basins. Settling speeds are introduced in the hillslopes chapter (Chapter 10), as it is here that we need them first to calculate the kinetic energies of raindrops. The student will need this backdrop on settling speeds again in studying sediment transport mechanics; we spare the space by not reproducing the development in that chapter (Chapter 14).

Student and teacher support

We have included material in boxes scattered throughout the book. These boxes serve several purposes: to allow us the occasional historical aside, to illustrate a topic with an example, or to develop an analogy with another field altogether. For example, corduroy roads are analogous to eolian ripples; the common day grilling of a cheese sandwich develops insight into thermal problems.

We have posed several student problems at the end of each chapter in order to challenge the student to use the material and the approaches presented. Some of these exercises simply promote paying close attention to one or another illustration in the text. Others involve more complicated calculations. We also pose a couple of thought questions, which are more qualitative, open-ended questions meant to inspire review of the chapter or connection with other chapters.

We also point the reader to a smaller text in which the guiding principles of this larger book are illustrated. In this *Little Book of Geomorphology*, available on the web since January 2008 at http://instaar. colorado.edu/~andersrs/The_little_book_010708_web. pdf, many of the geomorphic examples we discuss in this larger book are sketched and briefly discussed. The little book is subtitled "exercises in continuity." Its brevity places the analyses more cheek by jowl to allow more immediate appreciation of this theme. The little book will continue to be available on the website.

Finally, we have included a very thorough and up-to-date reference list, so that the book is tightly attached to the modern literature. Each chapter ends with a list of suggested reading. These are usually key books in the field covered in that chapter, to which the reader should turn for a more extensive discussion of the literature.

All the figures in the book will be available on a long-lived website so that professors may use them to illustrate lectures based on the material. This site will also have other photographs to support the material.

What we do not cover

In writing any textbook one must choose what to cover and what to omit. We have not covered karst landscapes. We have not surrounded the growing literature on submarine landscapes. And we have stuck to our own planet Earth. While the examples that we cover are overwhelmingly terrestrial, the general principles and the approach to posing geomorphic problems more formally can be applied to the surface of any object in the solar system (or beyond) if the appropriate environmental conditions are considered. In this new century, in which we have already marveled at how several landers have crawled around the surface of Mars, have launched a mission to Mercury, and have watched as a spacecraft slipped through the rings of Saturn to begin a several-year exploration of the Saturnian system, it is relevant to ask how well our understanding of surface processes here on Earth translates into an ability to understand the features of other bodies in our solar system. These extra-terrestrial landscapes serve as ultimate tests of our knowledge, as they represent natural experiments in which the controlling variables have been significantly tweaked from those on Earth: gravity, wind speed, atmospheric composition, solar radiation, tectonic rates, the mechanical and chemical properties of the materials comprising the surface, and so on. It is indeed an exciting time to be a student of not only our planet but of planets in general.

ACKNOWLEDGEMENTS

We were initially inspired to write a textbook during a conversation with Tom Dunne many years ago. He challenged us to articulate the fundamental principles of geomorphology. We have tried to take up that challenge, pulling most strongly on the theme of conservation. We thank Roger Hooke for carefully reading a draft of the book. Dave Furbish has been a strong supporter of our effort, including sponsoring an altogether too brief writing visit for Suzanne. Pete Adams, Greg Hancock, Eric Kirby, Kirsten Menking, Noah Snyder, and perhaps a few others have used the book in draft form in classes, and their feedback and encouragement has been very helpful.

We also acknowledge deeply those who have inspired us through their teaching, both formal and informal: among them, Bernard Hallet, Tom Dunne, Bill Dietrich, Peter Haff, Ron Shreve, and Bob Sharp have set the highest of standards. As professors, we also learn through the eyes, ears, legs, minds, and hearts of our students. To those students and post-docs at UC Santa Cruz and at Colorado, we offer our heartfelt thanks for the challenges you accepted, and the adventures in which you shared.

We dedicate this work to our parents,
John and Florence Anderson and
Ken and Lois Prestrud, who first introduced
us to mountain landscapes and spawned
our love of science.

We also dedicate this work to our children,
Hannah and Grace Anderson, who have never
known a time when mom and dad were not
working on the book, who will help carry their
generation forward, and who we hope will
strive to understand their surroundings and
sustain their environment.

Suzanne dedicates her contributions to the
memory of her brother, Kris.

CHAPTER 1

Introduction to the study of surface processes

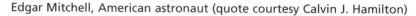

Suddenly, from behind the rim of the moon, in long, slow-motion moments of immense majesty, there emerges a sparkling blue and white jewel, a light, delicate sky-blue sphere laced with slowly swirling veils of white, rising gradually like a small pearl in a thick sea of black mystery. It takes more than a moment to fully realize this is Earth . . . home.

Edgar Mitchell, American astronaut (quote courtesy Calvin J. Hamilton)

In this chapter

The Earth is a blue and white sphere, spotted with green and brown, spinning through space. Geomorphology is the study of the shape of the Earth's surface and of the processes that are responsible for its evolution. Viewed from space, this may seem a trivial pursuit, given that Earth is nearly a perfect sphere with a very smooth surface. (If we ignore the slight oblateness of Earth, the ratio of surface roughness to radius is very close to the tolerance for roughness of an American billiard ball.) Yet the scale of surface roughness – of mountains and plains, of sand dunes and glaciers – is significant for living organisms. The processes occurring in the surficial zone where rock, water, air, and life interact do more than shape the surface of the Earth; they also drive chemical cycling between rock and atmosphere, influence tectonic motions, and support living systems. This book focuses on landscape evolution processes, but does not neglect connections to large-scale and long-period geologic processes and climate. In this chapter, we briefly outline the specific processes that are responsible for generating topography. We introduce several broad themes that cross-cut the discipline, and that serve as the weft to the process-specific warp in this book. Principal among these themes is the notion that the study of each of these landscapes can be formalized by conserving one or another quantity: sand, ice, water, energy . . .

The Blue Marble image of planet Earth, the most detailed true-color image of Earth to date. Using a collection of satellite-based observations, largely from the MODIS sensor, scientists have generated a true-color mosaic of the entire planet. (From NASA Goddard Space Flight Center. Image by Reto Stöckli (land surface, shallow water, clouds). Enhancements by Robert Simmon (ocean color, compositing, 3D globes, animation). Data and technical support: MODIS Land Group; MODIS Science Data Support Team; MODIS Atmosphere Group; MODIS Ocean Group Additional data: USGS EROS Data Center (topography); USGS Terrestrial Remote Sensing Flagstaff Field Center (Antarctica); Defense Meteorological Satellite Program (city lights).)

While the principles we address in this book are universal in that they can be applied to the study of any rocky planetary surface, we focus on our own planet and therefore must address why this planet is special. That the surface temperatures of Earth span the phase boundaries of H_2O has not only promoted the evolution of life on the planet, but has led to the diversity of surface processes involved in the evolution of the planet's surface. Ice sculpts landscapes in one way, water as rivers another, water as raindrops another. This diversity is mirrored in the landscapes.

But identifying processes and even formulating mathematical statements about how landscape might evolve in the face of this or that suite of processes is insufficient to capture the essence of the Earth's surface. We must also determine how rapidly or how efficiently these processes are acting. This boils down to the need to document rates – erosion rates, transport rates and the like. While some processes operate at speeds that are measurable over a single field season, or more importantly over the period of a PhD dissertation, most geomorphic rates are very slow. We have needed new tools. In this chapter we touch upon how new tools have emerged to aid the geomorphologist in establishing timing in the landscape.

We introduce the two chief drivers of geomorphic processes. The Earth's surface is the boundary between rock put in motion by deep geophysical processes and the atmosphere put in motion by uneven solar heating. Neither motion is steady. Earthquakes punctuate the motion of the rock. Storms punctuate the motion of the atmosphere. We discuss here the reasons why we must embrace this complexity, and introduce attempts of modern geomorphologists to treat the non-uniformity and non-steadiness of these processes.

The role of the atmosphere is yet more complicated. Even the statistical mean of the weather – the climate – changes on timescales over which landscapes evolve. Any study of the Earth's surface must acknowledge the role of climate history reaching back tens of millions of years. Here we quickly summarize this climate context, focusing on the last few tens of millions of years over which time the majority of the landscapes of the Earth have been rewritten. Discussion of the importance of the climate context sets the stage for the present challenge of addressing how the Earth's surface will respond to climate change induced by anthropogenic changes in the gas content of the atmosphere.

The global context

We live on a blue, white, brown, and green, nearly spherical, spinning, canted planet, 150 million kilometers from a medium-sized 4.5-billion-year-old star (Figure 1.0). One moon adorns the sky and tugs the ocean of its parent planet into a giant moving permanent wave. The moon was born early of a massive collision. That event set the planet spinning on an axis tilted with respect to the plane of the ecliptic, yielding daily and seasonal variations in radiation reaching the surface. The Earth is cooling down. Heat moved efficiently toward the surface by convection of the mantle is more slowly conducted through the outermost, coolest layer, which behaves as a solid on geological timescales, and which is broken into a small number of tectonic plates. The descent of old, cold, thickened plates from the surface also drives a creeping circulation of the mantle, at speeds of several cm per year, and establishes the relative motions of the plates. These motions crinkle the margins of the plates, generating belts of mountains, and drive volcanism that dots the topography with volcanoes where plates descend.

But this topography is subject to attack. That the Earth is both blue and white reflects the fact that water can be found in all three phases at the surface of the planet – blue liquid water, white water vapor (clouds), and ice. This unique aspect of Earth is allowed by being the right distance from the Sun, having an atmosphere that contains gases capable of absorbing long-wavelength radiation, and being large enough to retain these gases. The atmosphere and ocean of the planet are in motion as well; unlike the mantle, motion of fluids of the hydrosphere and atmosphere is turbulent, at speeds up to many meters per second, driven by both the uneven solar heating of the planet and its spin. Water evaporated from lakes and oceans, and transpired by land plants, is transported by storms spawned within the atmosphere, and then precipitates as either rain or snow. It is the motion of these substances, rain immediately and snow more slowly, where it accumulates sufficiently to become a glacier, moving down slopes ultimately generated by crustal processes, that leads to the dissection and sculpting of the land surface. None of these phenomena are steady on geologic timescales. Wind, water, and ice erode, transport and deposit sediment in discrete episodes of activity. Movement of continents on the surface of the planet slowly changes the circulation of atmosphere and oceans. The celestial mechanics of our planet's motion, which includes interactions with other planets in the solar system, leads to variations in the Earth's orbit, which in turn drive variation in the delivery of energy to the Earth. In the last couple of million years, this has resulted in numerous major swings in the climate on the Earth, leading to the growth and demise of huge ice sheets on northern continents. These set the climatic context within which human civilization has arisen and have greatly influenced the landscapes with which we interact.

Our planet supports a tremendous diversity of living organisms, whose activities fundamentally impact the chemical and physical properties of the surface. The green patches on the Earth's surface are evidence of photosynthetic organisms that harvest solar radiation and produce oxygen. Yellow to brown colors are indicative of organic detritus of living systems. Organisms have been found in environments ranging from cold seeps in the dark depths of the ocean, to hot springs on the land surface, encased in rocks in the cold Antarctic Dry Valleys and encrusting deep hot mine shafts. Life harnesses energetic chemical reactions, and the accumulation of products and consumption of reactants of these reactions are capable of changing the chemistry of the surroundings. Living systems build edifices and strengthen soils, and conversely churn soil and break rock. The furtive activities of animals scurrying and digging near the surface, the labor of worms mining the subsurface, the slower actions of roots and plant exudates, the growth of corals and black smoker towers, the protection of the surface by dense canopies of leaves, all of these living systems shape the environment so fundamentally that we struggle to consider what the Earth's surface must have looked like and how it must have behaved before significant life had evolved.

It is on this one planet, whose rocky plates are driven about by an internal heat engine, whose surface is irradiated by a distant Sun, and protected by a cocoon of gases, that now six billion humans strive to extract a living. The landscape serves as the scenic backdrop to this toil, and as a challenge to some who would seek the peaks of the terrestrial landscape and the depths of the oceans. It serves as the generator of catastrophes in the form of floods and landslides. It serves as the interface where atmospheric gases, circulating waters, and rocks of the lithosphere interact. Soil, a consequence of these interactions, both supports

the terrestrial biosphere and in turn is modified by that biosphere. The landscape serves as witness to the history of civilization, the fragments of past cultures preserved in deposits. It serves as the signature for any particular civilization, the topography and the climate conspiring to determine the types of plants that can be grown, the transportation corridors, and the landmarks of the culture.

Overview of geomorphology

Earth's landscapes are sculpted by a suite of geomorphic processes that vary with position on the Earth and with time due to changes in the climate. Landscapes contain signatures of the principal active processes that we would like to learn to read.

Terrestrial landscapes consist of hillslopes bounded by channels. Most hilltops are convex upward – they are rounded, not pointy – and are mantled by a layer of soil or mobile regolith. When close enough to the surface, mechanical and chemical weathering processes work on the rock, creating a zone in which the rock is significantly altered, weakened, and broken. When the bits become small enough, they can be transported downhill by hillslope processes. The motion can be either slow (creep) or fast (landsliding), and can involve motion of individual grains or of large masses. The specific processes involved are diverse, and include cold region (periglacial) processes of frost creep and solifluction, the action of burrowing animals, rainsplash, flow of water over the surface, and so on.

Hillslopes deliver water and sediment to streams at their base. Streams naturally and inevitably form a dendritic network that breaks the landscape into drainage basins – these are the quanta of the geomorphic system. Water discharge generally increases downstream, and the size and shape of channels accommodating that discharge change in an orderly fashion. Stream profiles tend to be concave up – they decline in slope with distance downstream. Streams can erode bedrock where tectonic processes are driving rock upward relative to sea level. Outside of these regions, streams are alluvial – sediment mantles the channel floors throughout – and transport the sediment delivered to the river from hillslopes toward the ocean. While natural stream channels can handle the most common annual peak water runoff from rain and snowmelt, every few years the water discharge exceeds the capacity of the channel, and a flood occurs. This delivers water and sediment to the floodplain, which should therefore be considered a part of the river.

Just as streams bound hillslopes, coastlines bound the entire terrestrial landscape. This line of intersection between land and sea varies in time because global sea level changes, because local tectonics raises or subsides rock, and because the coastline either can be eroded by waves or can accrete by deposition. Waves move sediment and erode rock, releasing energy transferred to the water by storms far out at sea.

At high altitudes and latitudes, where winter snow accumulates enough to outpace summer melting, glaciers grow to fill valleys or to overwhelm entire landscapes. A glacier transports ice from a high altitude zone of net accumulation to a zone of net ablation. It accomplishes this transport by movement as a very viscous fluid, and by sliding at the bed. Glaciers can erode the rock over which they slide, sculpting such characteristic forms as cirques, U-shaped valleys, string-of-pearl lakes, and fjords. In the late Cenozoic ice ages, massive ice sheets covered the northern continents, greatly modifying the landscapes of northern North America and Eurasia.

Wind also generates recognizable landforms, some from erosion of rock by sediment but most from deposition of sediment. The sandy deserts are ornamented by ripples on the backs of dunes on the backs of yet larger dunes – these are the poster children of eolian processes. All of these landforms reflect the process of downwind bouncing of sand that we call saltation. Dust, on the other hand, travels different more wiggly paths in a process we call suspension, and can be wafted up to thousands of kilometers by the atmosphere. Loess deposits of far-flung dust mantle a surprisingly large fraction of the Earth's surface.

Guiding principles

How should we organize our thoughts about how the Earth's surface works? What are the guiding principles? What is the connective tissue between the topics or sub-disciplines within geomorphology?

Conservation

One of the strongest organizing principles upon which we found our study of surface processes is the rule of

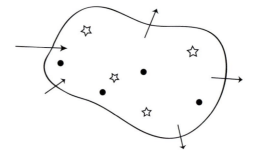

Figure 1.1 Schematic of the concept of continuity we will employ throughout this book. We wish to craft a statement of conservation of some quantity within the control volume represented by the irregular shape. In general, this quantity can be created within the volume (circles), or decay within the volume (stars), or cross into or out of the volume through its boundaries (arrows).

conservation. In most instances one may cast a problem as the conservation of some quantity. The word statement would go something like: "the rate of change of [fill in the blank] within a definable volume equals the rate at which it is produced within that space, plus the rate at which it is transported into the volume across the boundaries, minus the rate at which it is lost across the boundaries." The corresponding generic diagram is shown as Figure 1.1. We will learn to translate this word statement into a mathematical one. The quantity of concern might be heat, or it might be mass of regolith, or the volume of sediment, ice, or water; it might be concentration of a radioactive isotope, or it might be momentum. We will use each of these in developing the major equations in this book.

This approach serves to help us parse up a problem, take it apart into pieces we can address individually. For example, in the problem of conservation of heat, heat energy can be produced within a volume by the decay of radioactive elements. We must then know how to constrain how much heat will be generated by these decay reactions, and this will require knowledge of the concentration of such elements in the volume – something that is measurable. In the case of conservation of ice in a parcel of a glacier, we must know how ice moves across the edges of the parcel – we must know the physics that determines the transport rate of ice. This in turn finds us chasing down the physics of deformation of ice, and of sliding of ice against its bed.

To be concrete about it, we catalog here a number of examples in which we employ this approach:

Conservation of heat in the lithosphere thickness problem

Conservation of heat in the planetary temperature problem

Conservation of radionuclides in dating methods

Conservation of ice in a glacier

Conservation of heat in permafrost temperature profiles

Conservation of immobile elements in weathering profiles

Conservation of regolith in hillslopes

Conservation of water in overland flow

Conservation of water in a groundwater field

Conservation of sediment in bedform profiles

Conservation of momentum in development of the Navier–Stokes equation

Conservation of sediment in littoral cells

Conservation of water in flood discharge calculations

When transformed into mathematical statements, these all look similar. The rates of change are governed by spatial gradients in transport rates, and by any sources or sinks of the quantity of concern.

Transport rules

Material, be it water, ice, or air is moved from one place to another on the Earth's surface by the action of forces that include body forces (usually gravity) and surface tractions, or stresses. The rate of motion is set by the material properties, in particular how a material responds to stresses. The relationship between rate of motion, or strain rate, and an applied stress is called the rheology of the material. In geomorphology, we run into materials with widely differing rheologies. Take the Greenland Ice Sheet for instance. Air cooled near the surface of the ice sheet slides down the surface slope, generating katabatic winds with velocities that can exceed 100 km/hr, while water produced by melting flows down the same surface slopes at velocities rarely greater than 1 m/s. The ice also flows downslope, but at velocities of no more than a few tens of meters per year. Finally, the ice sheet is thick and extensive enough to invoke displacements of the mantle underneath; the mantle is capable of flowing at rates of the order of 1–10 cm/yr. We must become conversant in linear viscous, nonlinear viscous, and coulomb rheologies. More broadly speaking, however, we will find that the transport of a substance is often

proportional to the gradients in some quantity. Water and soil move down topographic gradients; chemicals move down chemical concentration gradients; heat moves down thermal gradients. This basic realization serves as connective tissue among many problems; it makes the solutions to many problems conceptually and even mathematically analogous.

Event size and frequency

One of the long-standing discussions in geomorphology is how to accommodate the fact that the geomorphic systems are forced by highly variable environmental conditions. We will find that many of the processes important in the transport of material on the Earth's surface are dependent upon the weather in a complex way. Some of these processes are naturally "thresholded"; for instance, the transport of sand does not occur until the velocity of water or of wind rises above some value. Once above this threshold, the transport increases dramatically with further increase in the flow. We will explore what determines the thresholds, and why the dependence is nonlinear above the threshold. This general characteristic of many transport processes is fundamental to geomorphology in that it bears heavily upon the relative importance of rare, large events. One might ask the question "What is the relative importance of every-day events versus those that occur once every decade, or every century, or every thousand years?" In order to address such a question, we need to know two things: what is the distribution of the sizes of events, and how does the system respond as a function of the size of the event. Weather events are typically distributed such that small events (small wind, small precipitation, etc.) are much more common than large events. We will characterize this more formally, but this is the essence. Note that if the process in which we are interested is thresholded, the little events do nothing – even if they are statistically the most frequent. Conversely, large events will perform a lot of geomorphic work when they occur, given the nonlinear character of transport processes, but they are rare. In essence, this means that there is an inevitable tradeoff between the rarity of an event and the geomorphic work that is accomplished by it. Big events do count, and the more nonlinear the process, the more important they become.

It is in this context that we must assess the principle of uniformitarianism on which much of geology is founded. We acknowledge that the physics and chemistry of the processes acting are indeed immutable; understanding of the present processes is the key to unlocking the past. But we also acknowledge that the external forces acting on the geomorphic system change on a wide variety of timescales. Rates of processes change on timescales of seconds to millions of years. Even the dominance of one process over another has changed. Acknowledgement of the role of major events capable of writing their signature so boldly on the landscape that thousands to millions of years of subsequent Earth history has been incapable of erasing it is part of modern geomorphology. We revel in the stories of these large events. In some sense this leads to something of a neocatastrophist view of the world, one in which the roles of large floods unleashed by glacially dammed lakes, of tsunamis generated by volcano collapse or magnitude 9 earthquakes, and of impacts of 10 km-diameter bolides are acknowledged for the work they can perform in sculpting the Earth's surface.

Establishing timing: rates of processes and ages of landscapes

One of the major advances in our science, one that has allowed us to make significant progress within the last 50 years, is the ability to establish timing in the landscape. The fathers of our field, such greats as Grove Karl Gilbert, Walther Penck, John Wesley Powell, William Morris Davis, and the like, had few tools to employ in dating geomorphic features. There was little constraint on determinations of how fast a landscape evolved, and for that matter on how old the Earth was. We therefore focus in an early chapter in the book on dating methods. While we do not present an exhaustive review, we try to give the reader a sense of the newer techniques. It will not be a surprise to those who know our research that we focus on the utility of cosmogenic nuclides, those rare isotopes of elements formed only by interaction of cosmic rays with atoms in near-surface materials. These have a very short history of use within geomorphology, beginning only in the late 1980s, and evolving rapidly ever since. This method has opened up to quantitative dating the Plio-Pleistocene (essentially the last five million years) during which most of the modern landscape has been developed. We can now date moraines, marine and fluvial terraces, and even caves. Using twists on the same methods, we can now determine the rate of erosion of a point on the landscape, or of a basin, over timescales that are much longer than humans have been around to measure them. While establishment of timing in the landscape is important in telling more precise stories of landscape evolution, its importance goes well beyond this. It has allowed us to test quantitatively models of landscape evolution. We can no longer be satisfied with models that get the shapes of the landscape correct. The models must also evolve at the right rate.

What drives geomorphic processes?

The Earth's surface responds to processes driven from both below and above the surface. We discuss in Chapter 2 how the deep Earth works, and how these processes impact the Earth's surface. The Earth is cooling by both conduction and convection, the former dominating in the lithosphere (and in the inner core), the latter in the lower mantle and outer core. These processes move heat from the interior to the surface of the Earth, arriving at a rate of roughly $40\,mW/m^2$. This is not much compared to the rate at which solar energy is delivered to the Earth's surface by radiation; in fact it is lower by a factor of about a million! More importantly to geomorphology, however, is the internal engine of the Earth that drives plate tectonics. This has established the context within which we understand the broadest features of the Earth's surface – its ocean basins – and the locations of the different styles of deformation of the Earth's surface. On a smaller scale, the collisions and movements of lithospheric plates drive crustal deformation that results in both faults and folds, and produces earthquakes as these structural elements grow. These processes serve to move rock up or down relative to sea level, delivering rock into harm's way for geomorphic erosion processes driven by the solar engine. Variation in uplift of rock from place to place alters the slope of the land surface. We will see that slopes, both of rivers and of hillslopes, are primary determinants of the level of geomorphic activity in a landscape. This broad topic is the realm of tectonic geomorphology, which is itself the subject of textbooks. Our coverage of this topic is therefore again not exhaustive. While plate tectonics and its crustal manifestation are important in forcing geomorphic processes, it is often the study of geomorphic features that constrains the patterns and the rates of tectonic processes.

The majority of the energy driving the surface system is supplied by the Sun. On average, the incident radiation supplies $1370\,W/m^2$ of energy to the outer reaches of the Earth's atmosphere, most of which makes its way through the atmosphere to the surface. Spatial variation in solar energy input, and the fact that the Earth is spinning, drives the circulation of the atmosphere. In part, this drives oceanic circulation as well. Solar energy also governs evaporation of water from the surface, which is moved elsewhere by the atmosphere, and precipitates as either rain or snow. This precipitation then moves down gradients, percolates into the ground, runs down hillslopes, and down rivers, or if it becomes ice, flows or slides downhill. It is the motion of these fluids – air, water and ice – that powers many surface processes. While the uplift of rock into harm's way is accomplished largely by tectonic processes, it is the transfer of water from low areas (mostly oceans and lakes) to high ones, that supplies these heavy fluids to the tops of the landscape so that they can do geomorphic work as they travel back down the landscape to the sea under the force of gravity. Tectonics allows local conversion of heat energy to potential energy to raise rock relative to the sea. The atmosphere similarly converts thermal energy into potential energy to raise water above sea level, in a hydrologic cycle with a much shorter timescale.

The surface temperature of the Earth

Ice, clouds, and oceans make the Earth seen from space largely blue and white. This is why the astronauts marveled at the Earth from their trips around the Moon in the 1960s and 1970s. It is made so in large part because the water on or near the Earth's surface exists in solid, liquid, and vapor states. This is unique in the solar system. Why is Earth the "water planet"? Liquid water is stable in only a narrow range of pressure and temperature conditions (Figure 1.2). The mean surface temperature of the Earth is $15\,°C$ above the freezing point of water. This is governed by a radiation balance that we explore in detail in Chapter 5. If the Earth had no atmosphere at all, the radiation balance suggests that the mean temperature of the Earth's surface should be about $255\,K$, or $-18\,°C$ (Figure 5.7).

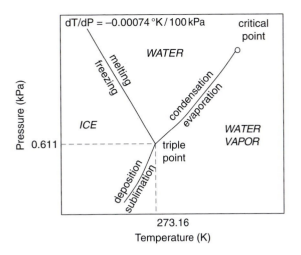

Figure 1.2 Phase diagram $P(T°)$ for the molecule H_2O. Note the negative slope of the melting curve separating water and ice (redrawn from Lock, 1990, Fig. 2.4, with permission of Cambridge University Press).

Clearly this is not the case. If it were, the Earth's surface would be frozen solid, and water would not exist in liquid and vaporous forms. This calculation does not take into account the chemistry of the Earth's atmosphere, which contains gases, some of them in trace amounts, that are excitable by particular slices of the spectrum of radiation emitted by the Sun and the Earth. The excitation of the gases leads to absorption of some of the energy and alters the simple radiation balance we have just calculated. Gases like H_2O, CO_2, CH_4, O_2, and O_3 absorb energy in specific wavelength bands (Figure 5.8), many of which are around $10\,microns$, in what we call the infrared band. Light in wavelengths around $0.5\,microns$, the visible spectrum, is not absorbed by these same gases, meaning that little of the radiation in the visible spectrum is altered by the atmosphere. These so-called greenhouse gases, transparent in the visible (from the Sun) and absorptive in the infrared, are therefore very important for allowing the surface temperature of the Earth to be above $0\,°C$, rather than well below it. And it is presently the human alteration of the concentrations of these same gases that is forcing the global change of climate in the last hundred years.

The climate context

While the climate varies over all timescales, there is some order to the system. It responds to changes in the Sun's luminosity, variations in the orbit of the Earth around the Sun, variations in the tilt of the Earth's axis of rotation relative to the plane of the ecliptic, and the distribution of continents on the globe. Understanding climate and landscapes is a double-edged sword. On one side, we need a means of deciphering what climate history has been; on the other side, we must know the climate history to understand how a landscape has responded to climate change. Most of the landscapes we see, and most of the geomorphic features we study, are young. This is partly because recent events remove the record of the older history. It is also in part due to the fact that we happen to live in extraordinary times, in the midst of a cycle of ice ages. It is obvious that growth of glaciers and ice sheets affects the areas they override. The climate changes that drove the growth of ice sheets, however, also wrought changes in non-glacial landscapes: increased wetness filled "pluvial" lake basins, withdrawal of water lowered the sea level, regions surrounding ice sheets were locked in permafrost. Because of these events, we generally turn to continuous marine records to provide a history of the climate, and use the knowledge of climate history derived from the marine records to interpret terrestrial landscapes. This approach has its drawbacks, of course. Marine records, while continuous, tend to provide a measure of conditions averaged over a large area – the globe or an ocean basin. To understand a particular landscape, we want to know the local climate, which may be quite different from the global mean or even from the nearby ocean area.

The most commonly used marine climate record is derived from the oxygen isotopic composition of the shells, or tests, of benthic, or bottom dwelling, deep-sea foraminifera. The carbonate shells of these organisms form in isotopic equilibrium with the seawater in which they grew. The ratio of ^{18}O to ^{16}O, expressed as $\delta^{18}O$, in the foraminifera depends on the temperature and the isotopic composition of the seawater. Changes in the temperature of the seawater cause a slight, but known, change in the fractionation of ^{18}O between the carbonate shell and the seawater so that warming reduces $\delta^{18}O$ and cooling raises $\delta^{18}O$ in the foraminifera. The other factor that affects the ^{18}O record of the benthic foraminifera is the growth of ice sheets. Water evaporated from the oceans is much lighter isotopically (−15 to −40 permil (parts per thousand, also written ‰) VSMOW) than mean ocean water (0 permil VSMOW), resulting in a shift in the composition of the ocean. Full Northern hemispheric

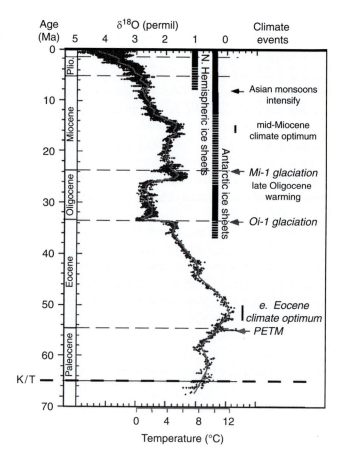

Figure 1.3 Oxygen isotope histories over the Cenozoic derived from deep-sea cores. Climate, tectonic, and biotic events are labeled, as are the occurrences of Antarctic and Northern hemispheric ice sheets. PETM = Paleocene–Eocene Thermal Maximum (after Zachos et al., 2001, Figure 2, with permission from the American Association for the Advancement of Sciences).

glaciation and Antarctic glaciation increases the $\delta^{18}O$ of seawater by about 2.4 permil, and hence increases the $\delta^{18}O$ of the foraminifera during glaciations. Many schemes have been devised to separate these two effects in the marine isotopic record.

A compilation of benthic foraminifera $\delta^{18}O$ composition reproduced in Figure 1.3 shows that Earth's climate has cooled over most of the Cenozoic, following the early Eocene climatic optimum. Several shifts in climate are evident in this general cooling trend, such as a period of full-scale Antarctic glaciation in the Oligocene, from ∼34 Ma to ∼26 Ma, and the re-initiation of full-scale Antarctic glaciation in the mid-Miocene. The onset of Northern Hemispheric glaciation in the Pliocene, at ∼3 Ma, coincides with the deepest cold plunge in the Cenozoic record. These events result from several processes. The slow reorganization

(a)

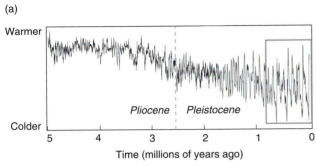

(b)

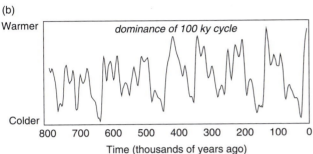

Figure 1.4 Climate history over last 4 Ma from marine isotopic records. (a) 4 Ma to present, showing slow cooling trend starting at roughly 3.5 Ma. The 40 ka (ka = thousands of years, kilo-annum) cycle dominates climate swings until roughly 700 ka, after which the 100 ka cycle dominates. (b) Detail of last 800 ka, showing strong 100 ka cycle. Note also how anomalous the present state of the climate is relative to the mean over the Pleistocene ((a) after δ^{18}O data in Zachos *et al*., 2001, with permission from the American Association for the Advancement of Science; (b) after Imbrie *et al*. (1984) stacked δ^{18}O record, with permission from Reidel Press).

of oceanic or atmospheric circulation due to tectonic rearrangements of plates is an important trigger for the onset of glaciations. Antarctic glaciations began when the Southern Ocean opened, isolating the Antarctic plate at high latitudes, and allowing unimpeded circulation of ocean water and atmosphere around it. The onset of Northern Hemispheric glaciation is closely associated with the closing of the Panama seaway. Another strong influence is the variation in the orbital parameters that control the amount, distribution, and seasonal timing of solar radiation on the Earth's surface. These orbital parameters oscillate in predictable ways on timescales of 10^4–10^5 yr, too short to show up in the record shown in Figure 1.3. An examination of the deep-sea foraminifera δ^{18}O record shown in Figure 1.4 over the last 4 Ma, the period when Northern Hemisphere glaciation began, and the period most relevant to modern landscapes, shows these short-term oscillations in climate clearly. Over most of the record, swings in δ^{18}O occur at

about 41 ky, the period associated with variations in obliquity, or the tilt of the Earth's axis of rotation. Over the last 800–900 ky, however, the period of the dominant oscillations shifts to 100 ky, that associated with variations in eccentricity of the orbit, or the deviation of the Earth's orbit about the Sun from circular. This 100 ky oscillation has dominated the late Quaternary climate, which is marked by glacial–interglacial cycles with this frequency.

Several aberrations also mark the Cenozoic climate record depicted in Figure 1.3. These brief, odd climate intervals are named: the Paleocene-Eocene Thermal Maximum (PETM, also called the Late Paleocene Thermal Maximum, LPTM), the Oi-1 Glaciation, and the Mi-1 Glaciation. Each of these events developed rapidly (10^3–10^5 yr), and cannot be explained by either tectonic plate rearrangements or variations in orbital parameters alone. For instance the PETM may have been triggered by a rapid release of methane to the atmosphere from the vast reservoir of methane stored in clathrates (open-structure ices) on continental margins. These climatic aberrations serve as notice that the climate system is capable of changing rapidly and dramatically in response to triggers that we do not fully understand.

Quite remarkably, a major increase in sedimentation to the world's oceans occurred within the last several million years, coinciding with the onset of the Plio-Pleistocene glacial cycles (Figure 1.5). That this is not a local phenomenon but is rather a global signal has been recently confirmed by analysis of records from many sedimentary basins. One such study documents the 30 Ma history of erosion within the eastern Alps, which we show as the inset to Figure 1.5. Within the Plio-Pleistocene, the geomorphic system appears to have woken up from a mid-Cenozoic stupor. This is a remarkable event, and one that is still crying out for an explanation beyond the simple one that "glaciers are very efficient at eroding the landscapes"; this record includes landscapes not subjected to late Cenozoic glaciation.

While we now know the last few millions of years of global ice volume well from the marine deep sea records, it would have been difficult to deduce this history from the land record alone. Save for a few continuous terrestrial records, such as the massive deposits of silt in China's loess plateau, the terrestrial record is woefully incomplete. In glaciated regions, younger alpine glaciers and ice sheets generally remove the record of older ones by overprinting or

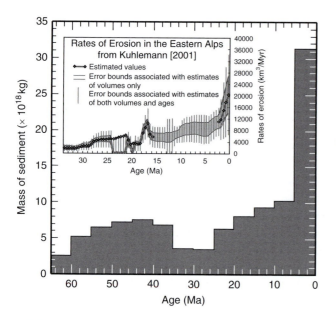

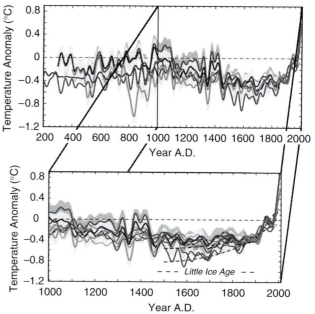

Figure 1.5 Global terriginous sediment yield, binned into 5 Ma intervals (after Hay *et al.*, 1988, after Molnar, 2004, Figure 1). Inset: rates of erosion in the eastern Alps. In both histories, the very late Cenozoic increase in sediment accumulation rate, and in erosion rate deduced therefrom, is dramatic (from Kuhlemann, 2001, after Molnar, 2004, Figure 4, reprinted with permission from the *Annual Review of Earth and Planetary Sciences*, Vol. 32 © 2004 Annual Reviews, www.annualreviews.org).

Figure 1.6 Proxy records of Northern Hemisphere climate history over the last 1–2 millennia. All records are smoothed with a 40-year low-pass filter; shading surrounding each record represents 95% confidence interval (see Mann *et al.*, 2009 for descriptions of each record). Dashed curves are based upon borehole temperature profiles. Bold curve beginning in 1850 is composite instrumental record. All proxy records are scaled to have the same mean as the instrumental record over the overlap interval. Note the "hockey stick" geometry of the curves. The Little Ice Age (LIA) is clearly visible as the low temperatures between about 1500 and 1900 (After Mann *et al.*, 2009, Figure 3, with permission from the National Academy of Sciences, © 2009 National Academy of Sciences, USA).

eroding the older features. Landslides and the inevitable operation of hillslope and channel processes also erase the odd but telling (often flat) elements of the landscape that record sea levels (marine terraces) and river levels (fluvial terraces) that provide clues of older landscapes. We must somehow see through this, and learn to read the incomplete record in the terrestrial landscape with the full knowledge of the climate history found in more complete records elsewhere.

While the comings and goings of glaciers are the most obvious manifestations of climate change in the Quaternary, many other aspects of the climate and of the landscape also varied. Glaciers can have a far-flung impact on landscapes elsewhere; glaciers serve as a major source of sediment, which is passed to the fluvial system down-valley. As both temperature and precipitation varied, the fauna and flora responded, which in turn affects the efficiency of geomorphic processes. In non-glacial environments, it may be the strength of the monsoon that dictates the level of activity of the geomorphic system. Monsoon strength has varied as well, driven by the insolation of landmasses that sets up pressure gradients that in turn drive water-laden atmosphere over the continents.

Finally, we acknowledge that we live in a time in which the climate is being strongly influenced by the activities of humans. One of the most important graphic displays of this phenomenon is the "hockey stick" curve of Jones and Mann (2004) reproduced in Figure 1.6. This shows the departure of the global temperature trend from "natural" and is coincident in time with the increase in greenhouse gases depicted in the now-famous "Keeling curve" of CO_2 (Figure 1.7). There is little doubt that the increase in CO_2 and other greenhouse gases that has been so carefully documented over the last 50 years is being manifested in the climate of the world (Mann and Jones, 2003; Jones and Mann, 2004; see commentary by Crowley in *Eos Forum*, 12 July 2005 and references therein). This figure and the data behind it form a prominent pillar in *The Fourth Assessment Report of the Intergovernmental Panel on Climate Change* (IPCC) (Solomon *et al.*, 2007). While the landscapes most sensitive to change are the high latitudes, all

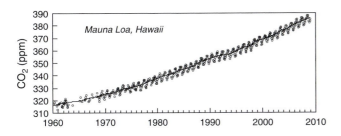

Figure 1.7 The "Keeling curve" from Mauna Loa, Hawaii. Longest record of atmospheric CO_2 concentrations, initiated by Charles Keeling of Scripps Oceanographic Institute. Seasonal variations ride atop the long-term rising trend (data from up-to-date Scripps CO_2 site, http://scrippsco2.ucsd.edu/data/atmospheric_co2.html (see Keeling *et al.*, **2001**)).

landscapes are expected to be impacted; several feedbacks lead to globalization of the changes, not the least of which is the rise in sea level.

We live at a time when the world is responding to major perturbations in its climate system. Much of this response is ultimately geomorphic in nature: changing ice volumes raise sea level, which alters the location of wave attack; changing sea-surface temperatures produce more frequent and/or more intense hurricanes; changing sea ice area results in stronger wave attack of Arctic coasts and may result in alteration of the thermohaline circulation system; changing air temperatures change the nature of storms, increasing the likelihood of convective storms, which have larger raindrops; shifts in the proportion of precipitation that falls as rain as opposed to snow will alter the hydrographs of rivers; change in surface temperatures can change plant cover, in some arid areas diminishing it to the point that the soil is exposed to entrainment by wind. The list could go on. While we do not focus on these changes in this book, we have tried to provide the student with the mechanics and chemistry of the Earth's surface processes necessary to address these issues quantitatively.

Summary

Our home planet is special. The orbit of this third planet is just the right distance from the Sun that water can exist in all its phases somewhere or another on the surface of the Earth. Solar energy dominates in providing energy to the geomorphic engine, so that it can attack rock that was brought upward toward the surface by tectonic processes that are ultimately driven by an Earth attempting to cool down. That attack occurs largely due to the degradation of rocks in the near-surface conditions, and the motion of water, ice, and air across a land surface on which biological actors serve both to bind material and to aid in its motion.

We can organize our thoughts about how many components of the geomorphic system work by appeal to continuity. This basic accounting procedure forces us to acknowledge the need to understand the processes that transport material, momentum and heat around on the surface of the Earth. This quest to understand process is the focus of both the book, and of much of the geomorphic discipline.

We will find that these geomorphic processes are governed by several components of the climate. These are both spatially and temporally variable on all scales. Some of this variability is deterministic, owing for example to the spherical nature of the Earth. Much of it is stochastic, owing to the turbulence of the atmosphere that serves to deliver the precipitation to the surface of the Earth in storms. We must acknowledge the role of weather, representing the temporal variability of the climate, as many geomorphic processes are nonlinear in their response to precipitation or wind. Climate history that is relevant to geomorphology includes the gradual cooling of climate in the Cenozoic, which culminates in the huge glacial–interglacial swings that have characterized the Plio-Pleistocene. The sedimentation records of basins around the world document a dramatic several-fold increase in sedimentation rate that is coincident with the advent of major northern hemispheric glaciations.

That we live in an interglacial, indeed that civilization has evolved in this interglacial, is important to recognize. Our present climate is nowhere close to the average climate of the last three million years. The most recent climate changes are human-induced, most importantly resulting in the warming of climate captured in the instrumental record, and deduced by comparison of many proxy records of climate that reach back one to two millennia.

Problems

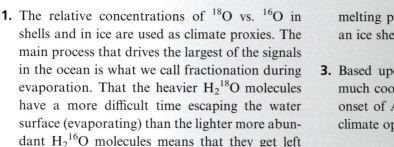

1. The relative concentrations of ^{18}O vs. ^{16}O in shells and in ice are used as climate proxies. The main process that drives the largest of the signals in the ocean is what we call fractionation during evaporation. That the heavier $H_2^{18}O$ molecules have a more difficult time escaping the water surface (evaporating) than the lighter more abundant $H_2^{16}O$ molecules means that they get left preferentially behind in the water. This question relates to why this happens. The temperature of a liquid measures the mean kinetic energy of the molecules in that liquid. Kinetic energy is given by $E = \frac{1}{2}Mv^2$, where M is the mass of the object and v its speed.

   If the kinetic energies of the two isotopes of water mentioned above are equal, but they have different masses, how different are their speeds? Report your answer in terms of percentage difference.

   (*Hint*: write out the formula for the kinetic energy of each, and set them equal to each other. Things like the $\frac{1}{2}$ will then cancel out, and it will also become independent of the units you use for the mass (so you can just use the atomic weight of each). Notice that the effect is subtle. It gives the light isotope just a little edge in escaping more efficiently.)

2. Given the slightly negative slope of the phase boundary between liquid water and ice in Figure 1.2, how much would the pressure-melting point of ice be depressed at the base of an ice sheet 1 km thick?

3. Based upon the $\delta^{18}O$ curve in Figure 1.3, how much cooler, in °C, was the Earth surface at the onset of Antarctic glaciation than at the Eocene climate optimum? And in °F?

4. Calculate the global average terrestrial erosion rate over the last 5 Ma based upon the 31×10^{18} kg of sediment delivered to the oceans in the most recent 5 Ma shown in Figure 1.5.

5. Compare the total energy available to the Earth surface from below and from above. At this distance from the Sun, the solar energy flux is roughly 1370 W/m^2. The global mean heat flux from Earth's interior is 41 mW/m^2. Sum these over the appropriate surface areas, and give the answer as a ratio of solar to deep-Earth derived energy.

6. *Thought question.* Catalog the changes in the geomorphic system that you might imagine could be induced by a warming climate. (We hope you will have a longer list after reading the entire book!)

7. *Thought question.* Describe an everyday activity in which you employ an accounting procedure of the sort we list in this chapter.

Further reading

Anderson, R. S., 2008, *The Little Book of Geomorphology: Exercising the Principle of Conservation*, available at: http://instaar.colorado.edu/~andersrs/publications.html#littlebook.

This self-published book serves as something of an abstract for this larger textbook. Because it is smaller, the analyses that demonstrate the utility of the principle of conservation in working geomorphic problems are more immediately juxtaposed.

Solomon, S., D. Qin, M. Manning, *et al.* (eds.), 2007, *Fourth Assessment Report of the Intergovernmental Panel on Climate Change*, Cambridge, UK and New York, USA: Cambridge University Press.
This is the scientific summary of the fourth IPCC report. It establishes the consensus of the scientific community on the climate context for the next century.

Zachos, J. C., M. Pagani, L. Sloan, E. Thomas, and K. Billups, 2001, Trends, rhythms, and aberrations in global climate 65 Ma to Present, *Science* **292**: 686.
In this important review, the authors have surrounded the Cenozoic record of climate, noting major events and trends in the data. This paper best sets the climatic context within which the landforms of the world have formed.

Whole Earth morphology

> The Earth was absolutely round. I believe I never knew what the word round meant until I saw Earth from space.
>
> Aleksei Leonov, USSR cosmonaut (quote courtesy Calvin J. Hamilton)

In this chapter

The roundness of the Earth is a feature it holds in common with other planets in the solar system. In this chapter we seek an explanation for this sphericity, and the largest scale departures from it. These departures include the oblateness of the Earth, and the existence of ocean basins and land masses with distinctly different elevations. These most general features of the morphology of Earth can be understood from a few simple physical principles: the pressure field within the Earth associated with both self-gravitation and spin, the response of deep Earth material to gradients in pressure, and the requirement for forces to balance when in a state of hydrostatic equilibrium.

In this chapter we first address the shape of the Earth as a whole, asking first why it is round, and then why it is not quite perfectly round. We then present the statistics of the Earth's topography, its hypsometry, and address quantitatively why the continents ride high above the ocean basins. This serves to introduce the principle of isostasy.

The Earth rising over the Moon; color photo taken from the Apollo 11 spacecraft. Note that the Moon really is this colorless (image courtesy NASA).

Why an oblate spheroid?

First let us assemble some facts about the Earth's shape. To within less than a third of one percent, the Earth is a sphere (Figure 2.1), with a mean radius of about 6367 km (see Box 2.1). While the terrestrial planets

Figure 2.1 Size comparison of Earth and Moon accomplished by juxtaposing images from Galileo spacecraft on its way to Jupiter in 1992 (image from http://solarviews.com/cap/earth/earthmo3.htm).

(Mercury, Venus, Earth, and Mars) are all similarly spherical, other objects such as asteroids and the moons of Mars, Phobos, and Deimos, are decidedly not round (Figure 2.2). Some look more like pockmarked potatoes. As these non-spherical objects are significantly smaller than even the smallest planets, size apparently plays a role. But the figure of the Earth is not perfectly spherical. It is slightly flattened such that the distance to the center of the Earth from the pole is 1/300th shorter than from a point on the equator.

We now ask why. The degree to which a planetary object approaches a sphere has apparently something to do with its size. Size dictates the pressure to which material inside the object is subjected. In addition, the thermal state of the interior depends upon size. It takes a certain time to cool from the initial formation process in which the kinetic energy of incoming planetisimals was converted to heat, and for any major radiogenic isotopes to decay. The larger the planet is, the longer it will take to cool off, or to transport internal heat to the surface to be lost as radiation to space. And material behavior, or rheology, depends sensitively upon both temperature and pressure. Over the roughly 4 billion years since planetary formation by collisions of planetesimals, the materials composing these larger objects have flowed at the pressures and temperatures found in their interiors, and have reached states of hydrostatic equilibrium. Consider the pressures within a planet with a hypothetical bump and dimple on its surface (Figure 2.3). At a given distance from the center of the planet, the pressures will be highest beneath the bump, lowest beneath the dimple, and intermediate elsewhere.

Box 2.1 What is the radius of the Earth?

If you do not happen to carry this number around in your head, how might you remember it? An efficient way to remember this number is to memorize instead the original definition of a kilometer: 1/10 000th of the distance between the equator and the North pole along a meridian through Paris (this turns out to be off slightly because the first standard meter, scribed on a bar of platinum-iridium alloy, was 0.2 mm short, owing to a miscalculation of the flattening of the Earth due to rotation; the modern System International (SI) definition of a meter is the length of the path traveled by light in vacuum during a time interval of 1/299 792 458 of a second, but the difference is immaterial for this calculation). This being the case, the Earth is 40 000 km in circumference, and the radius is easily calculated from the formula for the circumference of a circle: $C = 2\pi R$, or $R = C/(2\pi)$. Call π about 3, and you have a back-of-the-envelope estimate of the radius: 40 000/6 or about 6600 km. You won't be far off. And if you happen to have a calculator in the cafe in which this discussion is taking place, you can calculate it exactly: 6367 km.

(a)

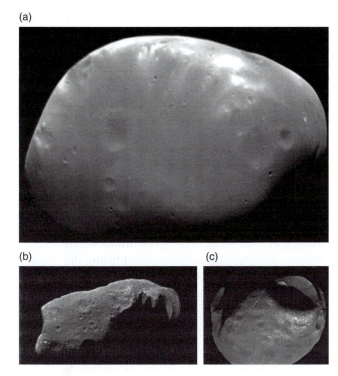

(b)   (c)

Figure 2.2 (a) Deimos, moon of Mars (image courtesy NASA). (b) The asteroid Ida, in the main asteroid belt between Mars and Jupiter, is $56 \times 24 \times 21$ kilometers in size (image courtesy NASA). (c) Phobos, moon of Mars (27 km across; orbits 5700 km above Martian surface). (Viking Project, JPL, NASA, image mosaic by Edwin V. Bell II, NSSDC/Raytheon ITSS).

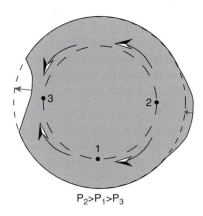

$P_2 > P_1 > P_3$

Figure 2.3 Sketch of a bulge and a dimple on the surface of a planet-sized object, and the pressures that they exert on material in the interior. High pressures under the bulge, and low pressures under the dimple, result in pressure gradients that drive flow of material (arrows) away from the bump and toward the dimple. Convergence of material beneath the dimple will raise the surface there, while divergence of material beneath the bump will lower the amplitude of the bump. A condition of hydrostatic equilibrium, in which flow no longer occurs, demands a spherical surface, shown by the thinly dashed line.

This variation in pressure results in pressure gradients, which drive a fluid from one place to another. The material, here the mantle of the planet, will move in the direction of the lower pressure; in other words, it will move down the pressure gradient. The result is a net loss of mantle from beneath the bump, and a net accumulation of material beneath the dimple. This in turn reduces the height of the bump and the depth of the dimple, leading to a removal of these topographic features from the planet. The pressure gradients driving the flow in turn decline, leading ultimately to a state in which all pressure gradients have been removed. This end-state is called hydrostatic equilibrium, and it is found in the interior of a sphere-shaped object. Seen at the timescale of the solar system, then, planet-sized objects can be viewed as self-gravitating fluids that seek and can achieve states of hydrostatic equilibrium. Asteroids of one to a few tens of kilometers in diameter have not deformed internally over this timescale, as the pressures and temperatures are too low to incite such flow; the pressure gradients exist, but the pressures are too low and the temperatures too small to allow the deformation needed to attain hydrostatic equilibrium. Instead, asteroids and small moons behave more like rigid solids whose strength is sufficient to counteract the non-uniform interior pressures resulting from a complex potato-like shape.

What about the flattening of the Earth? Note that while a 1/300 difference in the Earth's radius from the pole vs. the equator does not sound like much, this difference translates into about 21 km; this is slightly greater than the difference between the deepest oceanic trench (the Marianas trench: 11.04 km) and the tallest terrestrial mountain (Everest: 8.84 km)! This oblateness of the Earth is therefore the largest topographic feature of the planet, and ought to be treated in a "geo-morphology" or Earth-shape class!

We have found that even the rocky interior planets are spherical because they behave as fluids. But they are not perfectly spherical. Planets are slightly flattened because they are rotating about a spin axis. This introduces another force acting upon any mass within the planet. This is the centrifugal force, which acts outward normal to the axis of rotation, and is proportional to both the spin rate and the distance from the spin axis (Figure 2.4). This means that a mass on the surface of the Earth at low latitudes is acted upon by a stronger outward-directed centrifugal force than a

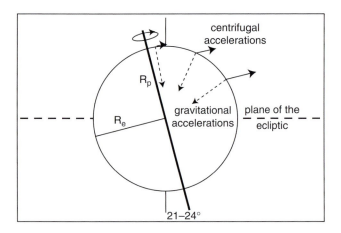

Figure 2.4 Slight flattening of the Earth results in a difference between the radius measured to the poles vs. that to a point on the equator. $R_e/R_p = 1.0033$. Given that the Earth is spinning, the centrifugal accelerations are greatest at the equator and vanish at the poles.

mass nearer the poles. This counters some of the center-directed gravitational force, and therefore reduces the pressure at the base of a column of rock at low latitudes. Over geologic time this results in a slow flow of mass toward the equator until the system is in hydrostatic balance. The problem is actually considerably more complex than this, as the rearrangement of mass itself creates higher order complexity to the shape of the rotating object. Astronomers call these shape descriptors J_2 for the flattening, and J_4, J_6, etc. for the higher order harmonics. The bottom line is that at long timescales the Earth behaves as a rotating fluid, a balance of forces being achieved by slight flattening of the sphere in the plane of the equator. It is also worth noting that the atmosphere and the ocean, which are much lower viscosity fluids, are acted upon by these same (and other) forces, resulting in equatorial bulges in these fluids as well. While this is

---

Box 2.2 Mass of the Earth and its speed in orbit

Although somewhat tangential to this course, these quantities are of interest for the understanding of several large-scale characteristics of the Earth. The speed of the Earth in its orbit is easily calculated from knowledge of the radius of the orbit ($R_o = 150$ million km), and the period of the orbit (1 year). The velocity is given by $v = C/P$, where the circumference of the orbit is $C = 2\pi R_o$, and P is the period. For a back-of-the-envelope calculation, note that the number of seconds in a year may be very closely approximated by $\pi \times 10^7$. The velocity is therefore nearly $v = 2R_o/10^7 = 30$ km/s. One geomorphic consequence, and really one of the only manifestations of the speed of our motion around the sun, is that any bolide impacts with Earth will occur at relative speeds of this order. Indeed, this corresponds well with estimates of bolide impact speeds (i.e., tens of kilometers/second) made using other methods.

The mass of the Earth is worth knowing, as it will constrain the mean density of the Earth (the mass over the volume is the mean density). This allows us to assess how representative (or not) Earth surface materials are of the planetary mean. The total mass of the Earth can be estimated from the orbit of a satellite around it. The centripetal (literally center-directed) acceleration necessary to maintain a circular orbit of radius r is v^2/r. Noting that $F = ma$, this implies that a center-directed force of mv^2/r must be acting on a satellite of mass m, velocity v in a circular orbit of radius r. The only force available is the gravitational attraction of the Earth, $F_g = GMm/r^2$. Here, M is the mass of the Earth, and m is the mass of the satellite, but note that by equating these forces the mass of the satellite drops out. The resulting equation for the mass of the Earth is $M = rv^2/G$. The most convenient satellite to use is obviously our Moon, which is 3.84×10^5 km from Earth (a little shy of a quarter of a million miles), and travels in its orbit with a period of 27.33 days. Using $G = 6.6732 \times 10^{-11}$ m^3 kg^{-1} s^{-2} and converting days to seconds, we find that the Earth has a mass of roughly 6×10^{24} kg. Dividing by the volume of the Earth ($V = \pi D^3/6 = 4\pi R^3/3$), we find that the mean density of the Earth is 5500 kg/m^3. This is roughly twice that of common continental crustal rocks (granite), and is still a lot more dense than even the ultramafic rocks that are rarely exposed from the lower crust and upper mantle. This is one of the first hints that the density structure of the Earth is non-uniform. Most of the mass is concentrated in a metallic Fe-rich core with density of over 12 000 kg/m^3, and a radius of roughly 3500 km – over half that of the Earth.

unlikely to be noticed by the typical human, it can be very apparent to those who climb tall mountains at high latitudes. The atmosphere at the same altitude on Mt. Denali (65°N; 6194 m) is decidedly thinner than that on a comparable mountain near the equator, Mt. Kilimanjaro (3.07°S; 5895 m).

Topographic statistics: Earth's hypsometry

Although the Earth looks as smooth as a billiard ball when seen from space, we inhabitants of the Earth's surface know that it has topography that is significant to us. Viewed as a whole planet, the ~20 km of relief on the Earth's surface represents only 0.31% variation in the radius of the planet. But surface roughness can also be measured as the scatter of elevations around the mean surface elevation. The distribution of elevations on the Earth viewed as a histogram (Figure 2.5) shows a great deal of scatter around the mean elevation of −2.1 km, a good thing for us air-breathing land dwellers. The distribution is distinctly bimodal, as seen in the two humps in the histogram; the lower hump reflects the distribution of elevations in the oceans, while the upper hump reflects subaerial elevations.

This statistical distribution of topography is often presented in another way, in integral form, in something we call the hypsometric curve. Formally, the hypsometry is the surface area as a function of elevation, or in other words, the integral of the area with respect to elevation, beginning at the lowest elevation. The bimodal distribution of elevations translates into two flat places in the hypsometric curve (Figure 2.5). From this curve it is easy to determine the total area or percentage of the Earth that is above sea level. The intersection of the hypsometric curve with present sea level suggests roughly 29% of the Earth's surface is above sea level. Note, however, that a slight drop in sea level, such as occurred in the last glacial maximum (120 m), will expose the broad shallow-sloping continental shelves (Figure 2.5). At such times the percentage of the Earth that is land rises to 36%.

Taking the means of the areas below and above present sea level, we find that the mean depth of the ocean basins is 3.8 km, while the mean elevation above sea level of the continents is 0.8 km. This mean elevation of the land surface above sea level is called

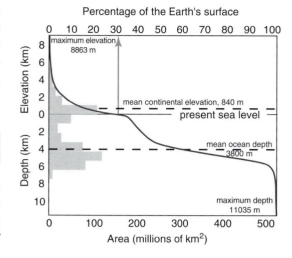

Figure 2.5 Hypsometry of Earth topography, showing both terrestrial and oceanic realms. Data are shown as a histogram (gray boxes) with 1 km resolution, and as a cumulative "hypsometric curve" (solid line). This is plotted integrating from the top, and therefore may be read as the area (bottom axis) or the percentage of the Earth's surface (top axis) that lies above a given elevation (data from Sverdrup et al., 1942; plot after Anikouchine and Sternberg, 1973; adapted from Allen, 1997, Figure 1.37, with permission from Wiley-Blackwell Publishing).

the continental freeboard. We will explore what causes this fundamental bimodality of elevations on the Earth, in other words what sets the quantitative difference between the mean elevations of land and sea surfaces. We seek here to understand the second most important large-scale feature of the planet's topography, after the centrifugal bulge.

Fundamental to this understanding is the following pair of observations: the crust associated with continents is lower in density and is thicker than that associated with ocean basins, and, as we have already seen, at large time and length scales, the deep Earth behaves as a fluid. Continental crust is lower in density because it is andesitic to granitic in composition (from partial melting of basalt), while oceanic crust is basaltic (derived from partial melting of peridotite). Both types of crust can be thought of as floating in the mantle, the lighter continental crust floating higher than the denser oceanic crust. Consider this more quantitatively. What do we mean by floating? This implies that a force balance has been established in which the downward forces associated with the weight of the object are balanced by the buoyant forces associated with the density difference between the object and the fluid in which it is immersed

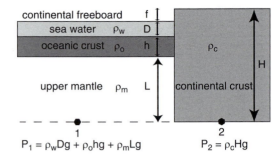

Figure 2.6 Schematic for calculation of continental freeboard, *f*. When pressures resulting from material columns at points 1 and 2, P_1 and P_2, are equal (isostatic = equal pressure), flow is prevented and a condition of dynamic equilibrium is achieved. Continental crust rides higher than oceanic crust because it is both lighter and thicker.

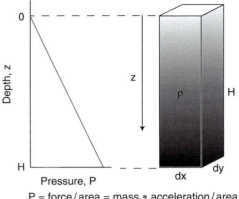

$$P = \text{force}/\text{area} = \text{mass} * \text{acceleration}/\text{area}$$
$$= [\rho H dx dy]g/dx dy = \rho g H$$

Figure 2.7 Schematic for the pressure in a column of material of density ρ. The pressure is the weight of the overlying material (a force) divided by the cross-sectional area of the base of the column, $dx dy$. If the density is uniform, the pressure linearly increases with depth, and is $\rho g H$ at the base of the column.

(Figure 2.6). Formally, this means that the pressures exerted at some arbitrarily large depth beneath both the continental and oceanic crust must be equal. Why is this? Just as in the explanation for the removal of bumps and dimples from planetary surfaces, this is because fluids respond to pressure differences by flowing from regions of high pressure to regions of low pressure. (Prove this to yourself by clapping your hands just in front of your face. As your hands get closer together the air pressure between them is ever so slightly higher than in the surrounding air, and the air responds by flowing toward the regions of low pressure. You feel this as a wind on your face with each clap.) The same happens if you place an ice cube in a glass of water. The water responds by flowing out from under the cube, where it momentarily experienced higher pressures than in the surrounding column of water. When the cube comes to rest with a certain portion of its thickness exposed above water, all blobs of the fluid within the glass are in a state of hydrostatic equilibrium; the forces on each blob of fluid are balanced. Now consider the continental and oceanic crusts floating in the mantle (we will ignore differences in the mantle structure beneath continents and ocean crust for the moment; the density contrast is largest between the two types of crust). We show a force balance beneath two columns, one with a continent of thickness H, the other with an ocean crust of thickness h, itself covered by water of depth D. We ask in particular what sets the freeboard of the continent, f, i.e., the mean elevation of the continental mass above sea level?

In order to write the pressure balance between the two columns, we must know how to assess the pressure at a specific depth within a fluid (Figure 2.7). Pressure is force per unit area. The force acting at the base of a column of material (fluid or crust or a pile of fish) is simply the weight of the overlying column. At the base of the column of height H, this is $\rho g H dx dy$, where dx and dy are the horizontal dimensions of the column, ρ is the density of the material, and g is the acceleration due to gravity. Dividing by the cross-sectional area of the column then yields the pressure at the base as simply $\rho g H$. For materials of uniform density, the pressure increases linearly with depth into the material, as shown in the plot of pressure vs. depth in Figure 2.7. Putting the calculation of pressure to use in the specific case illustrated in Figure 2.6, we find that isostatic equilibrium (equality of pressures) between the points labeled 1 and 2 may be written as:

$$P_1 = P_2 \Rightarrow \rho_w g D + \rho_o g h + \rho_m g L = \rho_c g H \qquad (2.1)$$

The acceleration due to gravity, g, cancels out from each term. We can solve for the unknown excess depth to which the continental crust protrudes in the mantle, L:

$$L = \frac{\rho_c H - \rho_w D - \rho_o h}{\rho_m} \qquad (2.2)$$

The continental freeboard then arises from noting how all the thicknesses are related:

$$f = H - (D + h + L) \qquad (2.3)$$

In order to evaluate f, we must now assume certain values for the densities and thicknesses involved. The densities are easily determined. Samples of continental and oceanic crust are readily available, and water of course we know to have a density of $1000\,kg/m^3$ (actually $1003\,kg/m^3$ for seawater charged with 3000 ppm solutes). Taking granite to be representative of continental crust, we assign $\rho_c = 2700\,kg/m^3$. Taking basalt to be representative of oceanic crust, we assign $\rho_o = 3000\,kg/m^3$. The density of upper mantle material is more difficult to establish. Some ultramafic xenoliths (literally, foreign rocks) embedded in erupted lavas are derived from great depths, and suggest peridotite, with a density of $3300\,kg/m^3$, is representative. But how do we determine the thicknesses of the ocean, and the two types of crust? Happily, the density contrasts between seawater and the seafloor, and between the crust and the upper mantle, allow seismic reflection techniques to measure these thicknesses. Ocean depths have been measured using acoustic echo-sounders on ships for many decades, and we now know the mean ocean depth to be 3.8 km. Similarly, crustal thickness can be evaluated from seismic wave reflection, owing to the large contrast in the seismic velocity between the crust and upper mantle. The depth of this Mohorovičić (Moho) discontinuity has been measured well, allowing us to state with confidence that mean oceanic crustal thickness is of the order of 5 km, while continental thickness is about 23 km, although its thickness is much more variable than oceanic crust.

When we use these values, we obtain $L = 13.1\,km$, and $f = 1.1\,km$. This compares favorably with the 0.84 km mean elevation of the land surface shown on the hypsometry plot. We can therefore explain this important observation about the Earth's topography by appealing to simple physics of a pressure balance. This physical process of floating of materials in a denser fluid is called isostasy, iso- meaning "equal," and -stasy derived from Greek stasis, or standstill. The crust, when considered at this continental scale, is said to have achieved an isostatic balance.

We can therefore explain the continental freeboard by a combination of crustal densities and thicknesses that are quite reasonable. The continents ride higher in the mantle by virtue of both lower density and greater thickness than their oceanic counterparts. This explains the most striking feature of the statistics in the world's topography: the bimodality of the probability distribution, and the quantitative difference between the two maxima.

Box 2.3 The mean depth of the ocean basins

Well before sonar was developed in World War II, we knew the mean depth of the ocean basins, especially the Pacific. The tool was tsunamis, large wavelength, low-amplitude waves in the ocean surface that can be generated by large earthquakes or other disturbances of the seafloor that can travel across the ocean basin. (Note that the Pacific is essentially ringed with subducting margins capable of generating tsunamis.) Knowing the time and location of tsunami generation and its arrival time on a distant shore constrains the mean speed of the wave across the ocean. The speed of the wave is in turn dictated by the mean water depth. Although it sounds perhaps ridiculous, the theory employed here is shallow water wave theory, in which $v = \sqrt{gD}$, where D is the mean depth of the fluid. How can this be considered shallow water, if it is several kilometers deep? All that matters is the ratio of the water depth to the wavelength of the wave; when this ratio is very small, the wave behaves as a shallow water wave. Tsunamis have wavelengths of tens to hundreds of kilometers, set by the length of the seafloor disturbance, which in major earthquakes can be up to 1000 km. Let us work an example. The 1960 M9 Chilean earthquake took 14 hours to impact Hawaii, 10 600 km away. The average speed of the wave was therefore 210 m/s. Solving the equation for depth, $D = v^2/g = 4500\,m$. This is a reasonable mean depth for an ocean basin, as we will see in the next chapter.

Summary

The Earth is round because it is large enough to behave as a self-gravitating fluid on long timescales. The slight flattening of this sphere is attributed to the fact that it is spinning, which accounts for a bulge that is of the order of 20 km. This is roughly the magnitude of the roughness of the Earth's surface, as measured by the full range of its topography. Neither the bulge nor the roughness is significant when seen from the Moon – the Earth looks very round and very smooth. The histogram of the Earth's surface topography is bimodal. This reflects the two distinct compositions and thicknesses of crustal material: thin, dense oceanic crust and thick lower density continental crust. Both the sphericity of the Earth, and the quantitative difference between the ocean depths and the continental freeboard can be attributed to the fluid-like behavior of the mantle at long time and length scales. Stresses associated with large-scale topography force the Earth's mantle to flow away from bumps such that at some depth within the mantle the pressures are equal. Isostasy reigns. We begin the next chapter with a discussion of the bathymetry in the oceans. We will see that there is distinctive regularity to the oceanic bathymetry that results from a combination of the maturation of the lithospheric plates and these same isostatic principles.

Problems

1. What is the mean density of the Earth?

2. How fast is the Earth moving in its orbit? (Note that this scales the speed at which a piece of debris would hit the Earth.) How fast is Mars moving?

3. How much new water must be added to the ocean each year to sustain the measured 2 mm/yr rate of rise attributable to new water? Give your answer in both cubic km and gigatonnes.

4. Estimate the sea level rise we can expect from a warming of the ocean column by 1 °C. Assume that the whole ocean column heats uniformly and that the thermal expansion coefficient is $1 \times 10^{-5}/°C$.

5. How thick a column of rock is needed to exert a pressure of 1 bar (or 10^5 Pa) at the Earth's surface? How thick a column of water is needed? How thick a column of air, if its density is that of air at the Earth's surface, $1.22\,\mathrm{kg/m^3}$?

6. Given the present hypsometry of the Earth's surface, what fraction of the surface would be terrestrial if sea level were to drop by 500 m?

7. *Thought question*. Describe the conditions that would allow a protoplanet to retain a geometry that is well out-of-round.

8. *Thought question*. Explain in words why the world has a bimodal hypsometry. Then discuss how this might differ on a planet that either lacks plate tectonics, or on which the density difference between its crust and mantle is significantly different from Earth, or on which there is no ocean.

Further reading

Beatty, J. K., C. C. Petersen, and A. Chaikin (eds.), 1998, *The New Solar System*, 4th edition, Cambridge: Cambridge University Press.
This is a recent review of the planetary system, with discussions of each planet and recent findings about it based upon planetary missions.

Turcotte, D. L., and G. Schubert, 2002, *Geodynamics*, 2nd edition, Cambridge: Cambridge University Press, 456 pp.
This textbook, now in its second edition, serves as a quantitative introduction to the dynamics of the earth. It is highly mathematical, but also very complete. The treatment of heat and of flexure are particularly well done.

CHAPTER 3

Large-scale topography

If the fit between South America and Africa is not genetic, surely it is a device of Satan for our frustration.

Chester Longwell (1958), cited in Denmark Kommissionen, *Meddelelser om Grønland* (1983)

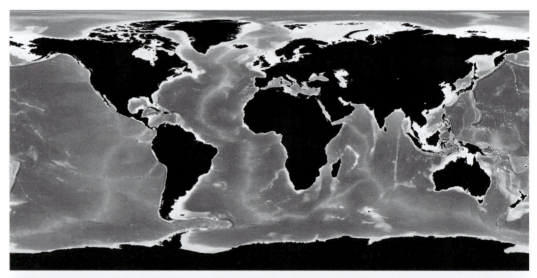

In this chapter

We focus on the roles of tectonics and, more broadly, geophysics in the generation of topography at the largest length scales. At these scales the mantle is involved. Its density dictates the scale of deflection of the surface by topographic loads. Its viscosity dictates the timescales for responding to changes in loads. Its convection drives the lithospheric plates about, and indeed the plates themselves are but the cooled tops of these convecting cells. In this chapter, in addition to revisiting the isostasy problem, we will introduce the conservation of heat, conduction of heat, and convection. We defer until the next chapter a discussion of the geophysical generation of topography at smaller scales – that at which individual mountain ranges occur in association with brittle behavior of the upper crust, and in which loads are supported at least partially by flexure.

Digital world bathymetry (top) and topography (bottom) (images courtesy of http://veimages.gsfc.nasa. gov/8392/gebco_bathy (top), and http://veimages.gsfc.nasa.gov/8391/srtm_ramp2.world.5400x2700. jpg (bottom)).

In this chapter we must come to grips with how tectonics works. We must get specific about what a plate is, and why it moves at rates of the order of centimeters per year. This in turn will serve to introduce the two major means by which heat moves about within the Earth – conduction (to create the lithosphere) and convection (to move the lithospheric plates about). We will also address the mechanisms by which large-scale topographic loads (continents, ice sheets, large lakes) are supported – by buoyancy. Indeed, while the support of such large static loads is associated with the density contrast of the materials involved, changes in loads such as ice sheets and lakes can incite a mantle response that determines the timescale for the response of the Earth. It is in this way that the study of geomorphology, and in particular documentation of the deformation rates one can deduce from deformed geomorphic markers such as shorelines, can provide some of the strongest constraints on mantle rheology.

We cannot begin to discuss the topography of the Earth and the processes responsible for modifying it before outlining first how plate tectonics works. The broadest outlines of the Earth's topography are all connected to plate tectonics: the mountain chains that reflect both the spreading centers and the subduction zones, the organized bathymetry of the ocean basins separating the ridges from the subduction zones, the highest topography associated with continent–continent collisions, and even the volcanic hot spot tracks all owe their patterns to the motions of lithospheric plates that comprise the exterior carapace of the Earth. We first discuss what the plates are: the lithosphere is simply the conductively cooled thermal boundary layer at the top of the mantle. This fact has tremendous leverage in allowing us to understand the bathymetry of ocean basins. We then describe briefly the process by which plates move, with the target of understanding quantitatively the plate speeds of roughly 10 cm/yr.

Ocean basins: the marriage of conduction and isostasy

Beyond the basic oblate spheroidal shape of the Earth discussed in Chapter 2, the second largest topographic features on Earth are its ocean basins. Since beginning to collect very detailed bathymetric information in the 1930s, we have learned that the ocean basins are remarkably well organized. In particular, each major basin has a topographic high, away from which the elevations of the sea floor fall off with considerable regularity. The pattern is quite visible in Figure 3.1. This gross structure of the ocean floor has since been explained by recognition of these topographic highs, or mid-ocean ridges, as spreading centers, where mantle is welling up, partially melted by pressure-release to produce oceanic crust consisting largely of basalts. This thin (roughly 5 km thick) scum of crust is wafted away from the spreading center at a rate of a few centimeters per year. On closer inspection, the decay from the heights of the mid-ocean ridges to the depths of the abyssal plains is remarkably similar from place to place. Why does the bathymetry display this regular pattern? First, what do the data say?

As shown in Figure 3.2, the bathymetry looks similar from basin to basin when plotted not against distance from the ridge, but as elevation vs. *age* of crust. It falls off first rapidly, then more and more slowly with age. We can investigate what sorts of mathematical functions best fit this pattern. If we expect that this pattern is a power law, we might plot the elevation vs. distance on log–log paper, while if we expect that it is exponential, we would choose semi-log paper (see Math appendix). When this is done, it appears that the pattern is well fit with a power of about 1/2. The formula for the bathymetry shown in either of these two basins is therefore

$$D = D_0 + a\sqrt{t} \tag{3.1}$$

where D is ocean depth, D_0 is the depth at the spreading center, t is the age of the plate at some distance x, and a is a constant. Clearly, we can substitute $x/(u/2)$ for the age as long as we know the half-spreading rate, $u/2$. Why should ocean depth increase as the square root of the age of the lithosphere? That this phenomenon so clearly displays a square root of time dependence cries out for a physical explanation. The answer lies in the thermal evolution of the oceanic lithosphere.

As the plate moves away from the spreading center, it thickens. But what defines the "plate" and why does it thicken? There are several ways to think about it: (1) it is all that material at depth that moves with the surface, (2) it is all that material that behaves as a solid on geological timescales (millions of years). In either instance, note that the definition is one based not on compositional differences (as for instance the Moho at the base of the oceanic crust is defined),

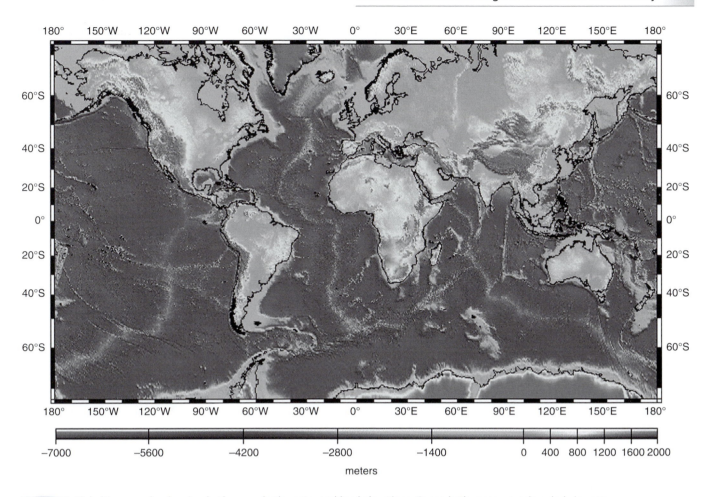

Figure 3.1 Global topography showing both ocean bathymetry and land elevations. Ocean bathymetry reveals orderly increase in depth away from major spreading centers, e.g., the mid-Atlantic Ridge. Note also the continental shelves.
(source: http://www.ldeo.columbia.edu/~small/GDEM.html, 2' Digital Elevation Model generated by D. T. Sandwell, W. H. F. Smith, and C. Small. Continental elevations derived from EROS GTOPO30 DEM. Submarine elevations based on Scripps/NOAA predicted bathymetry derived from Geosat, ERS-1, and Topex/Poseidon Satellite altimetry).

but on material behavior differences. *The lithosphere is a rheological boundary layer*. The rheology of a material, derived from the Greek rhein for "to flow," describes the rate and style of deformation of that material under stress. The main determinant of the rheology of the mantle is its temperature. This means that we can think of the lithosphere as a thermal boundary layer as well as a rheological one. But what is meant by this fancy phrase "boundary layer"? We will see it several times in geomorphology. Most generally, it is a region close to the surface of some object in which some property changes dramatically. Here the surface is the bottom of the ocean, the object is the Earth, and the property undergoing rapid or large-scale change is the temperature. As another example, we also live in the atmospheric boundary layer, the base of the atmosphere where one of several properties

changing rapidly is the wind velocity. It is zero at the ground surface and changes rapidly within the bottom kilometer, even in the bottom few meters. The other edge of a boundary layer is often arbitrary. In the case of the lithosphere, we chose to take the bottom boundary of the lithosphere to be where the mantle has reached a temperature of about 1200 °C. This choice is based upon laboratory experiments that show that mantle materials above this temperature deform sufficiently rapidly to behave as fluids on geological timescales. So the base of the lithosphere is defined by an isotherm. We reiterate that the lithosphere, unlike the crust, is a layer defined by rheology, and not by composition.

Given the above discussion of the lithosphere, we see that the problem of the gross pattern of topography on the ocean floor is therefore transformed into a problem in heat flow. Here we introduce the

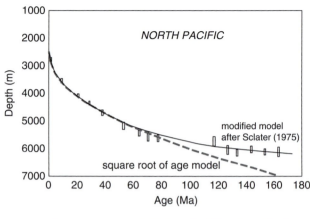

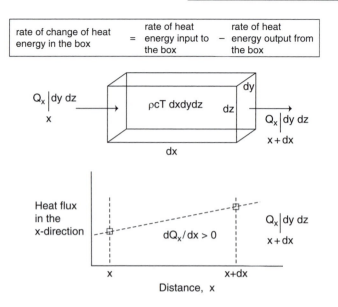

Figure 3.3 Word picture and box diagram required for derivation of heat flow equation.

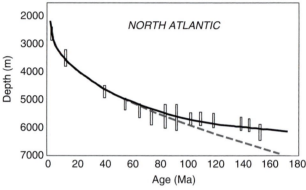

Figure 3.2 Bathymetry (vertical bars) of (a) North Pacific and (b) North Atlantic oceans as a function of lithospheric age, in transects perpendicular to the ridge crest. Dashed line is a model based upon the square root of lithospheric age, solid line is a modification of this from Sclater et al. (1975). Square root of age model works well out to roughly 80 Ma (redrawn from Parsons and Sclater, 1977, with permission from American Geophysical Union).

concepts of heat conservation and conduction to treat this problem formally. The upwelling hot mantle material is roughly 1600 °C, while the ocean floor is very close to 0 °C. Ignoring for the moment the region very close to the spreading center, the principal means by which the near-surface rock is cooled is by conduction. Vibrational energy is traded off between adjacent atoms in such a way as to even out the energy, meaning that it flows from regions of high energy (temperature) to regions of low temperature. The result is a relationship that has become known as Fourier's law, where the flux of heat in a particular direction, say along the x-axis, is proportional to the local gradient of temperature in that direction, through a constant called the thermal conductivity, k:

$$Q_x = -k_x \frac{\partial T}{\partial x} \tag{3.2}$$

This heat flux, Q, is defined as having units of energy per unit area per unit time, E/L^2T (in this book all fluxes will have units of something per unit area per unit time). We can then determine what the units of thermal conductivity must be to render the equation dimensionally correct. As the temperature gradient has units of degrees per unit distance (°T/L), k must have units of $(E/L^2T)/(°T/L) = E/°TLT$ or $(E/T)/°TL$. In SI units, in which lengths are in meters, masses in kilograms, and times in seconds (hence the old MKS system), this is W/m °C. The minus sign in Equation 3.2 assures that the heat travels *down* thermal gradients. We need to formalize one more concept, the conservation of energy. Consider a box of material depicted in Figure 3.3, and think about the quantity of thermal energy (heat) in the box. The conservation of heat can be written, in words:

the rate of change of heat in the box = rate of input of heat into the box minus the rate of heat output from the box

To turn this word picture into a mathematical statement, we need expressions for each of the terms. The heat in the box is simply the temperature of the box (which we can think of as the concentration of heat per unit mass), times the mass per unit volume (the density) times the thermal heat capacity (energy per unit mass per degree of temperature): $H = \rho c T dx dy dz$.

Box 3.1 Using the Taylor series

We could arrive at the same differential equation using a different method. Consider again the fluxes of heat through the edges of the box. While we can assert the flux of heat at a position x, $Q(x)$, we required an expression for the flux of heat at the other side of the box, $Q(x + dx)$. Recall from Appendix B the Taylor series representation for the value of a function at a new position given its value and its derivatives at a known position. Applying this to our situation, we have

$$Q(x + dx) = Q(x) + \frac{1}{1!}\frac{dQ}{dx}dx + \frac{1}{2!}\frac{d^2Q}{dx^2}dx^2 + \cdots$$

In the one-dimensional heat balance, and ignoring sources and sinks of heat, we now have

$$\frac{\partial(\rho cT dx dy dz)}{\partial t} = Q(x)dydz - Q(x + dx)dydz$$

Representing $Q(x + dx)$ with the Taylor series, and ignoring higher order terms, those with derivatives of second and higher order, which are multiplied by higher and higher powers of the small increment dx, this becomes

$$\frac{\partial(\rho cT dx dy dz)}{\partial t} = Q(x)dydz - \left[Q(x) + \frac{dQ}{dx}dx\right]dydz = -\frac{\partial Q}{\partial x}dx dy dz$$

Finally, dividing by $\rho c dx dy dz$, we arrive at an equation for the evolution of temperature:

$$\frac{\partial T}{\partial t} = -\frac{1}{\rho c}\frac{\partial Q}{\partial x}$$

This is identical to Equation 3.5.

The rate of change of this quantity with time is therefore the left-hand side of the equation: $\frac{\partial(\rho cT dx dy dz)}{\partial t}$. The rate at which heat is coming across the left-hand side of the box we denote $Q_x(x)dydz$ (which reads "heat flux in the x-direction, evaluated at the location x") and that going out the right-hand side of the box is similarly $Q_x(x+dx)dydz$, or the heat flux evaluated at $x + dx$. Our equation then becomes

$$\frac{\partial(\rho cT dx dy dz)}{\partial t} = Q_x(x)dydz - Q_x(x + dx)dydz \qquad (3.3)$$

Up to this point we have made very few assumptions. The only thing missing is any heat that is either produced or consumed within the box, which could happen by radioactive decay of elements, or strain heating of a fluid, or the change of phase of the material. We will ignore all of these for the present. The oceanic lithosphere is relatively poor in radioactive elements, meaning that we can safely ignore a radioactive heat source. For now we will also ignore the other potential heat sources. Now let's simplify this a little. We can hold the volume of material in this solid constant through time, allowing us to pull the dx, dy, and dz out of the partial derivative with respect to time. Let's also assume that the density and the heat capacity of the material don't change over the temperature range we are concerned with. When we divide both sides by $\rho c dx dy dz$, the equation simplifies to

$$\frac{\partial T}{\partial t} = -\left(\frac{1}{\rho c}\right)\frac{[Q_x(x + dx) - Q_x(x)]}{dx} \qquad (3.4)$$

The last step involves the simple recognition that if we were to shrink the size of our box so that dx tends toward zero, the term in brackets on the right-hand side is the spatial derivative of the heat flux, $\partial Q_x/\partial x$ and the final equation for the conservation of heat becomes

$$\frac{\partial T}{\partial t} = -\left(\frac{1}{\rho c}\right)\frac{\partial Q_x}{\partial x} \qquad (3.5)$$

This equation says that the temperature in a region will rise if there is a negative gradient in the flux of heat across it, and vice versa. We have shown this in Figure 3.3. If there is a positive gradient in the x-direction, then more heat is leaving out the right-hand side of the box than is arriving through the left-hand side of the box, and the temperature in the box ought to decline.

Note that we have not yet specified the mechanism by which heat is transported across the edges of the box. In the case at hand, the cooling of the oceanic lithosphere, in which conduction is the primary mechanism of heat flow, we turn to the physics of conduction, and apply Equation 3.2, Fourier's law for conduction of heat, for the heat flux, Q_x. This results in

$$\frac{\partial T}{\partial t} = -\left(\frac{1}{\rho c}\right)\frac{\partial\left(-k\frac{\partial T}{\partial x}\right)}{\partial x} \qquad (3.6)$$

Making the final assumption that the conductivity is uniform within the material and thus does not depend upon x, the conductivity can be removed from the x-derivative and the heat equation becomes

$$\frac{\partial T}{\partial t} = \kappa \frac{\partial^2 T}{\partial x^2} \qquad (3.7)$$

where $\kappa = k/c\rho$ and is called the thermal diffusivity and has the simple units of length^2/time. Equation 3.7 is called the diffusion equation. We will see the diffusion equation in several contexts within this book. You have heard the adage that nature abhors vacuums. It also hates sharp corners in things. Diffusive processes destroy sharp corners by removing sharp peaks and filling in valleys. Equation 3.7 describes how diffusion smoothes the sharp step in the temperature profile of the oceanic lithosphere imposed by placing hot new material against cold seawater. We will also encounter the diffusion equation when we talk about the decay of sharp topography through time. Diffusion also describes the gradual decay of spikes in concentration of a chemical in a fluid, such as ink in water. In each of these examples, the transport of heat or soil or a chemical is driven by a conduction-like process, one in which flux of something is proportional to the gradient in that quantity. Peaks and valleys in concentration (of mass, heat, topography, ink ...) can be recognized by their curvature, or second derivative, peaks being high negative curvature, valleys positive. The rate of change of the temperature is dictated by the local curvature of the temperature profile, and by the thermal diffusivity.

Now let's return to the cooling of the oceanic lithosphere. Armed with this mathematical description of how conduction works, and mathematical statements that capture this physics, consider now the problem of how the oceanic lithosphere thickens with time. We simplify the problem to one in which a column of new crustal material of roughly uniform temperature is brought suddenly into contact along its top surface with the cold base of the ocean. The starting profile is uniform except at its very top surface. The ocean acts as a "boundary condition" in the problem. We also assume that the ocean so efficiently mixes that it does not warm up in the process (as might rock at the edge of a magma chamber, say). So, to a good approximation, the top boundary condition is held at a constant temperature, of say 4 °C. Given the word picture of how conduction works, think about what will happen to the temperatures in the column of rock. Consider a single parcel of rock just below the surface or box, and think about the fluxes of heat into and out of it. No heat flows into the base of it, as there is no thermal gradient driving the flux. This is definitely not the case at the top, where the top feels a huge thermal gradient, a very large temperature difference across a short distance. Heat therefore flows rapidly out the top of the box, and none flows in the base. By conservation of heat, the heat content of the box, and therefore the temperature of the box, declines. Now take a later snapshot of the thermal drama. The temperature of the box is now distinctly below its original temperature. There is therefore a thermal gradient driving flow toward it from the bottom. As well, there continues to be a large gradient driving flow out the top, although note that this gradient has now declined because the temperature of the box is now lower, and therefore closer to that of the top boundary, the ocean floor. So it ought to continue to cool, but at a lower rate. The temperature profiles through time will therefore evolve in the way we have depicted them in Figure 3.4. The depth to a particular isotherm increases rapidly at first, and more slowly thereafter. The formal solution for the depth, L, to a chosen temperature with time is

$$L = \eta\sqrt{\kappa t} \qquad (3.8)$$

where the dimensionless factor η is of order unity (close to 1), and t is the time since the temperature at the surface was lowered. The depth in fact varies as *the square root of time*! This is a principal and very

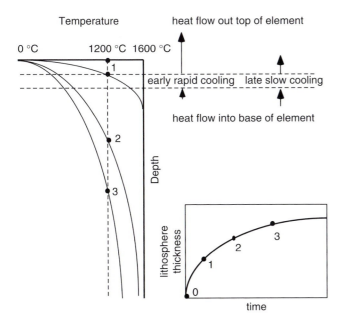

Figure 3.4 Evolution of thermal structure in a cooling column of rock. Initial temperature profile (bold line) of column at uniform temperature of 1600 °C. Subsequent to exposure at spreading ridge, top surface is held at 0 °C ocean bottom temperature. Temperature cools at all depths from time 1 through time 3. The 1200 °C isotherm, reflecting the rheological definition of the base of the lithosphere, deepens through time, from time 0 through time 3, as shown by dots at intersection with the temperature profile, and in lower right box. Heat flux into an element delineated by dashed lines shows large mismatch between heat flowing out top and into base at early times, changing to much lower mismatch at later times. The large mismatch in temperature causes rapid cooling in early times, decaying to slower cooling rates at later times.

useful result of diffusion problems: length scales and timescales are related through the simple formula $\delta \sim \sqrt{\kappa t}$ or $t \sim \delta^2/\kappa$, where δ is a length scale (say the distance from a boundary to a particular isotherm) and t is a timescale. The exact values depend upon the particular isotherm chosen, and the particular temperatures of the boundaries.

The bottom line, then, is that the physics of conductive cooling of a column of material subjected to a new and constant top temperature dictates that the oceanic lithosphere (remember, defined by an isotherm) thickens at a rate proportional to the square root of time. But why does the lithosphere droop into the underlying cooler mantle as it moves away from the spreading ridge? The reason is that it is denser than the underlying mantle because its mean temperature is lower than that of the underlying mantle. It wants to sink into the mantle, and the thicker it is the lower it sinks. Note that the lithosphere ultimately does founder entirely, at subduction zones. If we know what the mean density of the lithosphere is, then we can do another isostatic column balance to determine how low lithosphere of a given thickness ought to ride in the underlying asthenosphere. The density is related to the temperature through a constant that is characteristic of the material called the coefficient of volumetric thermal expansion, α_v. For mantle materials this is roughly 3×10^{-5} per degree centigrade, meaning that for every 1 °C drop in temperature the volume

Box 3.2 The cheese sandwich

While in geological problems it is worth committing to memory the rule of thumb that Earth materials have a thermal diffusivity of about $1\,\text{mm}^2/\text{s}$, consider the following experiment for determining the thermal diffusivity of another common substance. When we grill a cheese sandwich, we are counting on conduction of heat to allow a thermal wave to propagate through the bread and into the cheese. The sandwich is "done" when the cheese has obtained a particular temperature, namely that of its melting point. This will be some fairly large fraction of the temperature of the base of the skillet you are frying this thing in. The cheese is usually thin relative to the bread surrounding it. So the length scale of concern is the thickness of the bread slice. Call it 10 mm. We also know from experience that it usually takes about 5 minutes to do the job. Using the above equation for the timescale, setting this to 300 seconds, and plugging in 10 mm for the length scale, we find that the diffusivity of bread is about $0.3\,\text{mm}^2/\text{s}$. More to the point, especially if you tend to be prone to culinary impatience, you can see how much faster you would be sitting down to eat if you used thinner bread. Sawing off a pair of 7 mm slices instead of 10 mm slices will save you $7^2/10^2$ or roughly 50% in cooking time.

Now think about how much time and energy you would save by dicing up that potato you are trying to boil!

decreases by 3×10^{-5} of the original volume. Expressed mathematically, this is

$$\frac{V}{V_o} = 1 + \alpha_v \Delta T \tag{3.9}$$

Consider a chunk of mantle material of given mass, m, and initial density, ρ_o, at a particular temperature. The new density associated with a different temperature is therefore the same mass divided by its new volume. The ratio of the final to the initial density is

$$\frac{\rho}{\rho_o} = \frac{1}{1 + \alpha_v \Delta T} \tag{3.10}$$

We can calculate the density once we know the temperature change to which the material has been subjected. Consider the thermal profile through the lithosphere. At the top of the lithosphere the temperature is that of the ocean floor (about 4 °C), and by definition the temperature at the base is roughly 1200 °C. The thermal profile has a slight curvature throughout, as in Figure 3.4,) so that any parcel of lithosphere continues to cool through time. But to first order the profile is about linear, meaning that the mean temperature in the lithosphere is roughly (1200–0)/2, or 600 °C. The mean temperature of the column of material has therefore declined by 1200–600 or 600 °C. The resulting density change is simply calculated from the above equation to be about +2% (i.e., $\rho/\rho_o = 1.02$). We can now perform a column balance on two columns, one at the ridge, the other some distance $x(= (u/2)T)$ away from the ridge, as shown in Figure 3.5. We want to know if this can explain, quantitatively, the differing depths of water over the two columns. The ridge top is at a water depth D_o. The other column is under a water depth of $D_o + D$. We want to know how the additional water depth, D, is related to the lithospheric thickness of the column, L. The balance, shown in the figure, results in a ratio of additional ocean depth to lithospheric thickness of roughly 2%. This is very close to the measured relationship.

To summarize, the lithosphere thickens at a rate proportional to the square root of time. The mean density of the oceanic lithosphere is roughly 2% greater than that of the surrounding hotter mantle, and this mean density is approximately the same from place to place. The depth of the droop of the denser lithosphere into the underlying mantle asthenosphere, and hence the depth of the ocean away from the ridge, is therefore controlled solely by the thickness of the

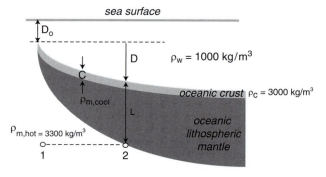

$$\rho_w D_o g + \rho_{m,hot}(D + L)g = \rho_w D_o g + \rho_w D g + \rho_{m,cool} L g$$
$$\rho_{m,hot} D + \rho_{m,hot} L = \rho_{m,cool} L + \rho_w D$$
$$D/L = (\rho_{m,cool} - \rho_{m,hot}) / (\rho_{m,hot} - \rho_w)$$

Figure 3.5 Schematic of cooling oceanic lithosphere. Crust achieves and then maintains a 5 km thickness, while lithosphere continues to thicken with time and hence with distance from the spreading center. Depth of the ocean is dictated by isostatic balance: pressures at points 1 and 2 must be equal. Note very large vertical exaggeration.

lithosphere, and therefore increases as the square root of distance from the ridge. This is in full accord with the observed relationship, which, recall, we determined to be well fit by a square root relationship. We emphasize that the important factor setting the gross topography of the ocean floor is the thickness of the lithosphere and not its temperature. The *mean* temperature of the lithosphere is everywhere the same, as we have defined it by an isotherm on the base, and the top is held at the temperature of the deep ocean, but the thickness of the lithosphere is dictated by the slow downward propagation of the 1200 °C isotherm.

There are a few interesting wrinkles in this story. As noted when inspecting the bathymetric data, there are systematic departures of the data from this simple square root relation (recall the pattern in Figure 3.2). In particular, both very close to the ridge and very far from it the relation does not appear to describe the full data set. This has prompted considerable research, as departures from simple theories ought to do. The near-ridge anomaly has been attributed to the action of another heat flow mechanism that cools the lithosphere more rapidly than could conduction. The high thermal gradients can drive hydrothermal convection cells within the crust and thin overlying sediment near the ridge, which are very efficient at moving heat about. At greater distance from the ridge, the gradients driving hydrothermal convection diminish, and

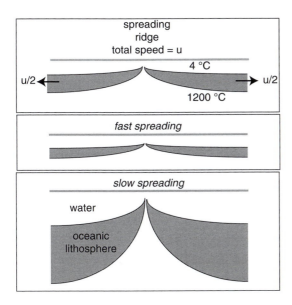

Figure 3.6 Schematic illustration of the role of spreading rate on the bathymetry of an ocean basin. Slow spreading results in a deeper oceanic basin at the same distance from the spreading center, a corollary to the dependence of lithospheric thickness on age.

a layer of sediment acts to seal the fracture permeability of the sea floor, leaving only conduction to perform the task of removing heat from the lithosphere. At very great distances from the ridge (lithospheric ages of more than 80 Ma) it is thought that there might be other sources of heat at the base of the lithosphere, perhaps associated with straining of the underlying mantle materials.

Let us now ponder some large-scale geologic consequences of this model. Consider the age structure of the sea floor. It is straightforward to see that the distribution of ages results in a distribution of ocean depths using the simple model of square root age dependence described above and knowledge of the spreading rates. The present mean age of the oceanic lithosphere is 60 Ma. Note that in the Phanerozoic there is thought to have been little change in the total volume of sea water (save the 150 m-worth that is occasionally sequestered on land in huge ice sheets ... but we shall talk about that later). This means that if we change the mean depth of the ocean floor, and in particular if we diminish it, we will force this water out over the continents, as shown in Figure 3.6. Although there is some current debate about this, some think that this happened in the Cretaceous. The age distribution of the sea floor can be changed

by swallowing a spreading ridge, for instance at a subduction zone, or by increasing the speed of ocean spreading at all or some of the ridges. If the mean ocean depth changes by only a few hundred meters, large portions of continents will be submerged. (Remember, the mean freeboard of the continents, including the Tibetan Plateau and other such high places, is only 800 m!) In this hypothesis, faster mean seafloor spreading rates could be responsible for the generation of the huge mid-Cretaceous seaway in the center of the North American continent (Heller *et al.*, 1996), which was itself so important in generating the massive coal deposits along the shores of the western USA. There are other ideas, for example, the thought that the lithosphere is pulled down by the sinking Farallon plate beneath the center of North America, but we will touch upon them later.

Note that we have not dealt here, nor will we deal elsewhere, with the shorter wavelength topography of the ocean basins. This includes seamounts, the abyssal hills (themselves of great interest lately), transform ridges, etc. Two larger scale features bear mention, however. The first are the continental shelves. As one can see from the hypsometric curve, and from the bathymetry of any continental margin, these shelves are confined to water depths of about 0–150 m. They range in width from essentially non-existent on some actively subducting margins, to more than 100 km on passive margins. This raises the question of what controls the width, and why the 150 m maximum depth? It is at least intriguing that the scale of swing of the sea level within the Pleistocene is about 150 m. Highstands, such as the present one associated with low global ice volume, alternate with lowstands as low as 150 m that are associated with large glacial maxima. The latest of these was only 20 thousand years ago. Another aspect of the shelves is that they are in places dissected by very impressive submarine canyons that are at least as deep as the Grand Canyon. The origins of these canyons, when they are active, and what the processes are that etch them into bedrock, remain actively debated and studied problems.

A second major feature of ocean basins is the many chains of volcanic mountains that reflect the passage of a hot spot. At least as interesting as the volcanic mountains themselves are the broad topographic swells that they ornament. We will treat these in the next chapter when discussing flexural support of loads.

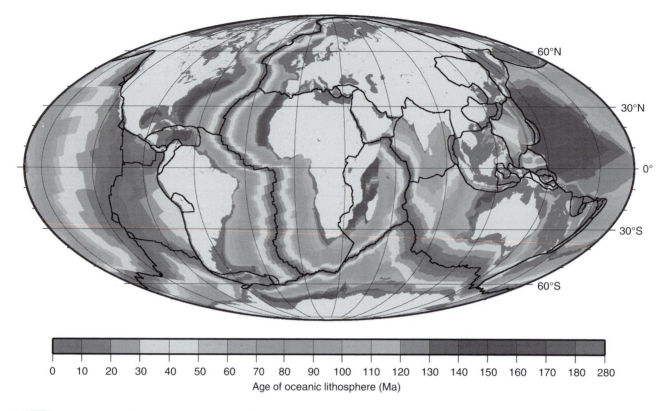

0 10 20 30 40 50 60 70 80 90 100 110 120 130 140 150 160 170 180 280

Age of oceanic lithosphere (Ma)

Figure 3.7 Lithospheric plates in their present configuration, with age of oceanic lithosphere. Continents are light gray, shelves dark gray (after Müller *et al.*, 2008, Figure 1a, with permission from American Geophysical Union).

Plate tectonics overview

In this brief overview we lay out first the basic kinematics of plate tectonics (where the plates are, how fast they move, and so on), then turn to a conceptual picture of how this system works, and end with a simple yet illuminating calculation of what sets the speed of plate motion. The major lithospheric plates of the present globe are shown in Figure 3.7. Most are many thousand kilometers in horizontal extent. The plate speeds vary, but most are of the order of 5–15 cm/yr, as can be deduced from the ages of the seafloor.

The motion of plates

The motion of lithospheric plates in plate tectonics is a manifestation of the convection of the Earth's mantle. This outer half of the planet is being both cooled from above, and heated from below. Just as a pot of soup heated from below will ultimately convect, moving hot soup up from below and cold soup from the surface downward, so too will a body of

rock heated and cooled under the proper conditions. Likewise, just as a lake cooled from above in the fall and winter will ultimately "turn over," moving cold surface water down to the bottom, so too will the mantle. Both bottom-up and top-down forcing can drive convection. The cooling occurs ultimately through radiation of heat to outer space, and is dictated essentially by the great difference in temperature between the Earth's surface and outer space. The heating at the base of the mantle is caused by its contact with the very rapidly convecting liquid iron of the outer core. Heat moves into the mantle from the core by conduction; it is lost from the mantle to the atmosphere and the ocean by conduction through the lithosphere. The regions within which conduction reigns as the heat transport mechanism are called conductive boundary layers. They are each of the order of a few tens to 100 km thick; the energy transport in the remaining part of the 2900 km-thick mantle is dominated not by conduction but by convection.

Convection can occur whenever the forces driving overturn of a fluid are greater than those resisting such motion. In most general terms, the force promoting convection is buoyancy, which is scaled by the density difference between the one portion of the fluid and another, and by how big this blob of anomalous fluid is. The density difference can arise from either compositional or thermal effects; in the case of mantle convection, thermal effects dominate. The force resisting this motion scales with the viscosity of the fluid, and with the surface area of the anomalous blob. It is common to derive a ratio of driving to resisting forces that may be used to characterize whether or not convection should occur, and how vigorous the convection might be. This ratio, which is dimensionless, is called the Rayleigh number, denoted Ra, and is one of many non-dimensional numbers we will encounter in this book. When this Rayleigh number is above about 2000, convection driven by thermal buoyancy should occur. For the Earth's mantle, the Rayleigh number is presently of the order of 10^5, meaning that it ought to be convecting quite vigorously.

Plate speeds

Given this physics, how fast ought plates to be moving in the Earth mantle system? Can we, from first principles, predict the rates at which the plates are moving about on the Earth's surface? Given that plate tectonic movement rates determine the pace of mountain building events, which drive rock above sea level where terrestrial erosive surface processes can attack it, and generate topographic gradients that dictate the rates of geomorphic processes, this is a crucial number. In addition, note that if we have a theory that allows calculation of these rates, we will be in a position to evaluate quantitatively how these rates might have been different in the past, when the Earth was younger and perhaps even more vigorously convecting.

The problem may be reduced to a force balance. If we assume that the plates are not accelerating, in other words that they are moving at the same rate today as they were last year as they were hundreds of years ago, which is probably a safe assumption, then the forces that are driving them must be in balance. This is a simple restatement of Newton's second law: $F = ma$, where F is the sum of the forces operating to cause motion, m is the mass of the object, and a is the acceleration. The object of concern is a lithospheric plate. We simply have to identify the forces acting on

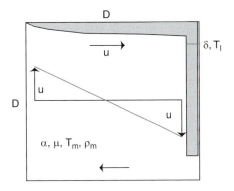

Figure 3.8 Sketch of mantle "convection in a box" of width and depth D at a rate u. Lithosphere (gray) thickens to δ at time of subduction. Balance of negative buoyant force driving convection, and viscous drag resisting it, yields an estimate of the speed of convection, u (see text for details) (redrawn from Davies, 1999, Figure 8.1, with permission from Cambridge University Press).

the plate, and equate them, as shown in Figure 3.8. The problem is akin to a settling problem with which we are more familiar, and which we will turn to in discussing sediment transport in fluids, except that the object that is settling is not a sphere, but a slab. The forces operating are F_b, the (negative, downward) buoyant force of the slab, caused by its having a density that is slightly higher than the remainder of the mantle, and F_r, the (positive, upward) resisting force, caused by viscous drag along the surface of the slab as it descends through the very viscous mantle. Our task is to estimate the magnitude of each of these terms and set them equal (so that their vector sum is zero). First, we need an equation for the buoyancy force. Let's consider a unit thickness of mantle (into the page), in which case we need expressions for the force per unit width. If the thickness of the mantle is D, and the width of the convecting cell is also D (a crude approximation), then the buoyant force per unit width of slab pulling it downward is

$$F_b = D\delta g(\rho_l - \rho_m) \tag{3.11}$$

where δ is the thickness of the lithosphere upon subduction, g is the acceleration due to gravity, and ρ_l and ρ_m are the lithospheric and mantle densities, respectively. The force increases linearly with the thickness of the slab upon subduction.

The resisting force is a drag force, or a surface traction, that operates on the surfaces of the slab. This requires knowledge of the shear stress (a force per unit area) and the area over which it operates. Given that we are doing a balance on only a 1 m sliver, the area really translates to the length of the boundary between mantle and lithosphere, or $2D$. The shear stress is the product of the viscosity of the deforming material, μ, here the mantle, and the rate at which it is straining, or the shear strain rate, $\partial u/\partial r$. In other words,

$$F_r = 2D\tau = 2D\mu\frac{\partial u}{\partial r} \qquad (3.12)$$

Here the shear rate is the rate at which the velocity within the mantle changes with distance away from the lithosphere (see Figure 3.8). Using knowledge that the whole mantle cell is turning over with a speed of u at its perimeter, meaning that while the right-hand side is going down at u, the left-hand side is coming up at $-u$. The difference, $2u$, occurs over a distance of D, meaning that $\partial u/\partial r = 2u/D$. Equating the buoyant driving and the resisting forces per unit length, and solving for the unknown velocity, results in

$$u = \frac{D\delta g(\rho_l - \rho_m)}{4\mu} \qquad (3.13)$$

The thicker and longer the lithospheric slab, δ, the faster it will settle into the mantle, while the higher the viscosity the lower its speed will be. It remains to calculate the thickness of the lithosphere, and the density difference between lithosphere and mantle. The lithosphere thickens by conduction (as we have discussed above), which is controlled by the thermal diffusivity, κ, and the time since cooling at the surface began. In particular, $\delta = \sqrt{kT} = \sqrt{k(D/u)}$, where T is the time it takes for the lithosphere to move from spreading center to subduction zone. Note that the faster the mantle turnover rate, the thinner the lithosphere will be by the time it subducts. We have also discussed the density difference between lithosphere and mantle, as we needed this number in order to evaluate the isostatic balance. The density difference is simply

$$\Delta\rho = \alpha\Delta T\rho_m \qquad (3.14)$$

where ΔT is the difference in temperature between the lithosphere and the surrounding mantle. This leaves us with the following equation for the speed of the plate:

$$u = \frac{Dg\alpha\Delta T\rho_m\sqrt{\kappa D/u}}{4\mu} \qquad (3.15)$$

or, noting that u is involved in the thickness of the lithosphere,

$$u = \left(\frac{Dg\alpha\Delta T\rho_m\sqrt{\kappa D}}{4\mu}\right)^{2/3} \qquad (3.16)$$

We now need to estimate all of these variables. What is the temperature difference between mean mantle and mean lithosphere? Given that the surrounding mantle is vigorously convecting and hence does not vary greatly from that at the base of the lithosphere, we can take its temperature to be that of the base of the lithosphere. The mean temperature of the lithosphere is well approximated by the average of its top and its base, given that the thermal profile within the lithosphere is crudely linear. Hence, $\Delta T = 1200 - ((1200 - 0)/2) = 600\,°C$. The other variables may be taken to be: $D = 2900\,km = 2.9\times10^6\,m$, $g = 10\,m^2/s$, $\rho_m = 3.3\times10^3\,kg/m^3$, $\kappa = 1\,mm^2/s = 10^{-6}\,m^2/s$, $\mu = 10^{21}\,Pa\text{-}s$, and $\alpha = 3\times10^{-5}/°C$. The resulting estimate of plate speed (really the half-spreading rate) is $3.6\times10^{-9}\,m/s$, or $0.11\,m/yr$. We have already documented actual plate speeds, which are a little slower than this, typical convergence rates being more like 5–10 cm/yr, and hence 2.5–5 cm/yr for expected half-spreading rates. Nonetheless, given the approximations made in setting up the problem, and the necessarily crude estimates of the values of the parameters, we have come astonishingly close to reality!

It is also instructive, before we leave this calculation, to inspect the final equation for the velocity, and assess how this might have changed through geologic time. Among all these variables, what is likely to have changed? Physical constants like g and α and κ certainly will not have changed. The variable that is most susceptible to change is the viscosity of the mantle. Viscosity is a temperature-dependent property of a material. In particular, it declines greatly with increasing temperature, exponentially so. If the mantle in early times were significantly warmer than it is now, then the viscosity would have been much lower. Note that $u \sim \mu^{-2/3}$. If the viscosity were an order of magnitude (ten times) lower in the early Earth, then the plate speed would have been $(1/10)^{-2/3}$ or 4.6-fold faster then. It is a good thought experiment to explore the implications of this for the topography of the early Earth.

Large-scale mountain ranges: orogens

The support of the world's largest mountain ranges is similar to that of the ocean basins. Isostasy

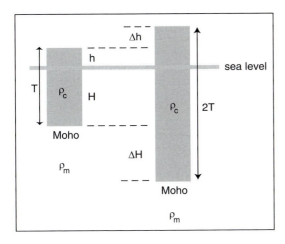

Figure 3.9 Isostatic response to doubling of the crustal thickness. Deepening of the Moho greatly exceeds the increase in the freeboard of the continent (its height above sea level).

reigns. The support of the topography is largely through buoyancy. The topography reflects the thickness of low-density crustal material beneath it. In this section we will explore the geophysical consequences of erosion on a large scale, and explore issues related to the response of continental land-masses to the thickening of the crust, taking the Tibetan Plateau as our example. We will ask the questions: how does the elevation of a mountain range respond to the removal of material from the range by erosion? Does the pattern in which the material is removed play a role? We will see that we need to be careful in exactly how we pose the questions, and where within the landscape one might seek the answers to the questions.

Effects of thickening the crust

In Figure 3.9 we show a column of continental crustal material of thickness T, in isostatic balance such that it floats with a mean elevation of height h above sea level, and a depth to the Moho of H below sea level. If by some mechanism the crustal material were greatly thickened, say doubled, by how much would the mean elevation of the range and the mean depth to the Moho change? The problem is perfectly analogous to exploring the change in freeboard of a slab of ice if you place another slab on top of the first. The pressure at an arbitrary depth within the mantle beneath the base of the crust must be the same in

the two cases, allowing us to set P_1 and P_2 equal again, as in our previous column balances:

$$P_1 = P_2 \Rightarrow \rho_c T g + \rho_m \Delta H g = \rho_c (2T) g \qquad (3.17)$$

Again dividing through by the gravitational acceleration, g, and solving for ΔH yields

$$\Delta H = \frac{\rho_c}{\rho_m} T \qquad (3.18)$$

Note from the figure that the change in mean elevation, Δh, is simply obtained from $2T = \Delta h + T + \Delta H$. Using our expression for ΔH, we arrive at the expression for the change in mean elevation:

$$\Delta h = T - \Delta H = T \left[1 - \frac{\rho_c}{\rho_m} \right] \qquad (3.19)$$

Given typical values for densities of crustal and mantle material, of 2700 and 3300 kg/m^3, respectively, the term in the brackets becomes 2/11, or roughly 20%. If no erosion were to occur during the thickening of the crust, doubling of the crust would drive a change of elevation of roughly 20% of the change in crustal thickness, while the rest, or 80%, results in lowering of the Moho.

Let's explore an example. The Tibetan Plateau in central Asia has resulted from the collision of India with Asia, a collision that began roughly 40 Ma, and continues today at plate convergence rates of about 6 cm/yr. This is the world's greatest example of continent–continent collision, and results in the largest welt of high topography (Figure 3.10). Mean elevations of the plateau are close to 5 km over a 1200 km N–S transect. Let's work in the reference frame of Asia. Sitting on Asia, it looks like the Indian subcontinent, with its continental crust of thickness T, is arriving at the plate boundary at a rate u. What happens at depth at this boundary is hotly debated. But it appears that a large fraction of the Indian continental crust is somehow tucked underneath the Asian continental crust. If the crusts of the two continents were roughly 25 km thick to begin with, this would result in a 50 km-thick crust beneath the Plateau. The change in thickness of 25 km should result in a change in elevation of roughly 2/11 of 25, or roughly 5 km. This is remarkably close to the mean elevation of the Plateau, supporting the notion that the crust has been roughly doubled.

Can this explain the width of the Plateau as well? We wish to craft a volume balance on the crustal

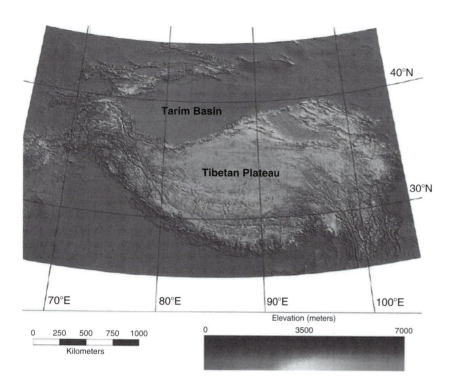

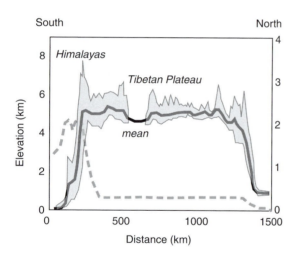

Figure 3.10 (a) Topography of Tibet using SRTM 90 m data. (b) N–S topographic swath profile across the plateau. Gray shaded band encloses all topography; black line is the mean topography in the 100 km-wide swath. Note the abrupt edges and the flat top of the plateau, with mean elevation ~5 km. The precipitation profile (dashed) illustrates the strong orographic effect: precipitation is effectively milked from the clouds as they encounter the Himalayan front from the south. Highest relief is at the edges of the plateau, most impressively at the southern edge in the Himalayas (after Fielding *et al.*, 1994, Figures 2 and 3).

material, depicted in Figure 3.11. (Actually, this is a cross-sectional area balance.) If the entire Indian continental crust has been tucked beneath the Plateau, then the total cross-sectional area of crust ought to be the original thickness T times the width W, plus that contributed since the collision. The crust added since collision is simply the thickness of Indian continental crust, T, times the rate at which it is arriving at the plate boundary, u, times the total time since collision began, t. The measured depth to the Moho is roughly 50 km, or $2T$. Setting the measured cross-sectional area $2TW$ equal to the original TW plus the newly arrived crust Tut, results in an expected width $W = ut$. Given the convergence rate of $6\,\mathrm{cm/yr} = 6\times10^{-5}\,\mathrm{km/yr}$ and a time since collision of $40\,\mathrm{Ma} = 40\times10^{6}$ years, we predict the width of the plateau to be 2400 km – too wide by almost a factor of two. The error could be due to any of several assumptions we have made: the thickness of Indian crust could be less than that of Asia; the plate rate we have used might not be representative of the mean rate since 40 Ma; the time of collision could be more recent; or

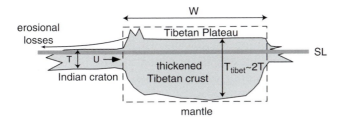

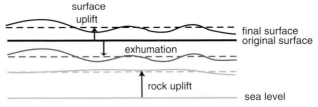

surface uplift = spatial average of (rock uplift − exhumation)

Figure 3.11 Application of conservation of mass to the evolution of the Tibetan Plateau and the thickened Tibetan crust that supports it. Conservation of mass within the dashed box requires accounting for gains of mass due to arrival of Indian crust and losses of mass through erosion. The crust beneath Tibet is roughly double that of the Indian craton.

Figure 3.12 Definition sketch for various terms related to motion of the surface. Exhumation removes rock and therefore would lower the topography if nothing else were happening. The exhumation pattern can be quite variable in space as depicted by the departures from the mean exhumation (dashed). Rock uplift moves rock upward relative to a datum, here sea level. The rock uplift pattern is usually relatively smooth compared to the pattern of exhumation. The surface of the topography moves up or down depending upon the local rock uplift and exhumation rate. The surface uplift relevant to the geophysicist interested in changes in the potential energy of the system is the mean of the surface uplift pattern (dashed).

some of the Indian crust might be getting swallowed into the mantle.

Effects of erosion on the isostatic balance

In the above discussion, we have ignored erosion. Consider now how removal of some mean thickness ΔT will change the elevation of the crustal column, as shown in Figure 3.11. We can still employ Equations 3.18 and 3.19, but with $-\Delta T$ replacing T. While some mean thickness ΔT was eroded, the elevation change is only a small fraction (2/11) of this. Most of the erosion goes into pulling up the base of the Moho. Even more interestingly, the erosion has resulted in uplift of rock relative to sea level. Note that no tectonic forces were required to drive this uplift of rock. This is a point made by Molnar and England (1990) in a very stimulating paper aimed at the broad geologic audience involved in both tectonic and climatic studies. Among the many contributions of the paper, they attempted to clarify the meaning of the word uplift, which has been used by different subsets of geologists to mean different things. The problem has been largely one of reference frame. They define *uplift of rock* to mean uplift of a rock mass relative to sea level (a convenient reference frame, and one that doesn't move around too much). They define relative motion of rock and the Earth's local surface to be *exhumation*. Finally, they define uplift of the surface, or surface uplift, to be the motion of the Earth's surface relative to sea level. These quantities are related through a simple equation, and illustrated in Figure 3.12:

surface uplift = uplift of rock − exhumation

The "surface" of interest to the geophysicist is that averaged over an area large enough to reflect changes of crustal thickness. The length scale is therefore roughly that of the crustal thickness, meaning that the area of interest is roughly 30×30 km or about 1000 km^2. So the columns we have been drawing would have to be at least 30 km on a side. The geomorphologist, on the other hand, is often interested in smaller scale topography. Ridges and valleys are more typically one to a few kilometers across. What do mountain peak heights and valley bottom heights do in the face of erosion and its isostatic consequences? Molnar and England pointed out the following scenario. Say erosion is not uniform in the column we have been considering. For instance, what if the same volume of material was eroded, but it was all removed from valleys, leaving peaks untouched, as in Figure 3.13? Since it is the same mean thickness of material removed (ΔT), the Moho comes up by the same height ΔH we have already calculated, and rock uplift throughout the column is thus the same. But what happens to the mean elevation of the column, and to the elevation of the peaks? Since the tops of the peaks we have left untouched by erosion, they remain a height T above the Moho, meaning that they are now $h + (9/11)h = 1.8h$ tall! They have increased in elevation dramatically due to the isostatic consequences of non-uniform erosion, or "relief production." The mean elevation is half of this, given the

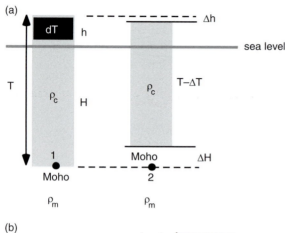

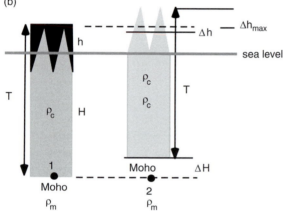

Figure 3.13 Effects of non-uniform erosion on peak elevation. While mean elevation declines due to removal of crustal mass, the greater erosion in valleys during times of increasing relief can result in the uplift of peaks (adapted from Molnar and England 1990, Figure 2, with permission from *Nature*).

crude geometry of the triangular valleys we have chopped into the crustal block, or $0.9h$. Note that no matter whether we are eroding uniformly or non-uniformly, the mean elevation of a range declines during erosion – by a small fraction of the mean thickness removed, but a decline nonetheless.

The possibility of raising the elevations of peaks within a range by simply deepening the valleys within it, enhancing the relief of the range, is intriguing. As again noted by Molnar and England in their provocative paper, this linkage can set up some interesting feedbacks in the climate–topography–geomorphology–isostasy system. As we will see in our discussion of glaciers, the health of a glacier (called its mass balance) depends upon how much of the mountain valley within which it exists lies above the snowline, an elevation (the equilibrium line altitude, or ELA) above which some of

a winter's snow remains unmelted at the end of the next summer. The higher a mountain mass, the more healthy the glacier. If in addition, the glaciers, which occupy the valleys of the range, are efficient erosional agents, as we will see they are, then the possibility exists that further erosion of the valleys makes the glaciers more healthy, further deepening the valleys, and so on. A positive feedback loop has been established.

Another feedback has to do with precipitation as it interacts with a mountain mass. Local precipitation is milked out of an air parcel as it is forced to rise over a mountain range in something we call an orographic effect (oro- for mountain related, like orogeny). On a local scale, the air mass does not care about the mean elevation of a range, but instead must go over the peaks in the range. So peak elevations matter to the atmosphere. The higher they get, the more precipitation will be milked from the clouds locally on the windward side of the range, leaving less for the downwind side. This orographic effect implies that relief production within a range might intensify rain shadows in the lee of a range. Let us explore a couple of examples.

The Tibetan Plateau again serves as an example. Note that the south side of the Plateau is bounded by the huge Himalayan chain of mountains, with many of the world's 8000 m peaks. As shown in the cross section of the topography in Figure 3.10, compiled from a large digital elevation data set, the mean elevation of the Himalayas is roughly that of the Tibetan Plateau. In contrast, the relief (the differences between highs and lows in the topography) of the Plateau is minuscule in comparison to that within the Himalayas. Note the pattern of precipitation rates. The precipitation arrives at the Himalayas from the south, delivered by the Asian monsoon from the Indian Ocean. Tibet is extremely dry, living in the rain shadow of the Himalayas. To first order, we can think of the Himalayas as simply an eroding edge of the plateau. The non-uniformity of this erosion, resulting from efficient rivers and glaciers, has generated huge valleys, and high relief. The isostatic response to this relief production has raised the highest of the peaks to roughly 1.8 times the mean height of the low-relief plateau, or 8000 m.

Many questions remain in this landscape. When within the collision was this relief produced? Has it always been there? Has the global cooling in the late Cenozoic produced glaciers, which in turn have generated relief, producing rock uplift? There exists a strong possibility of misinterpretation of the rock record, as again pointed out by Molnar and England (1990), and their principal motivation for the paper. Is it a pulse of tectonics in the late Cenozoic that raised the Himalayas and produced a pulse of sediment from the mountains, or was it climate change? (That both can lead to a pulse of sedimentation in basins that bound ranges makes the interpretation of such basin sediments ambiguous – leading to the title of their paper asking whether it is the chicken or the egg.) Another potential consequence of the raising of Tibet is the aridification of central Asia. At present, the monsoon rains are trapped at the leading edge of the plateau, the Himalayas, and cannot penetrate into the interior (see the precipitation pattern in cross section of the plateau in Figure 3.10). One of the key attributes of central Asia is its deserts, the Gobi, the Taklamakan, and so on. Downwind of these, throughout what is now China, is the most extensive deposit of dust in the world, the Loess Plateau, with loess thicknesses of hundreds of meters. The age of the base of the loess deposit is roughly 8 Ma, which corresponds well with the age of the inception of the monsoon as deduced from paleo-oceanographic records in the Arabian Sea. Perhaps the two events are linked: the growth of the Tibetan plateau triggers the monsoonal pattern of precipitation, which in turn aridifies the region downwind of the plateau.

Mantle response times: geomorphology as a probe of mantle rheology

We now ask what determines the timescales of response to the emplacement or removal of crustal loads. Formally, the elastic response is essentially instantaneous, traveling at the speed of sound within the material. Stand on a diving board and it responds immediately; jump off from it and it rebounds instantly to its old position. The fluid response, however, which becomes increasingly important as the scale of the load increases (as the degree of compensation C increases toward 1.0), is dictated by the time it takes for the deep crustal or upper mantle material to move out of the way of a new load in response to the pressure gradients established by that load. This timescale is determined by the viscosity of the fluid, and hence serves as a probe of the mantle viscosity. The higher the viscosity, the longer it takes to respond

to an applied load. For mantle materials, the timescale is of the order of several thousand years. Importantly, we know this timescale largely from the deflection of datable geomorphic markers in the landscape. The large-scale ice sheets and lakes that have relatively recently met their demise constitute natural experiments that provide us with major constraints on mantle viscosity. The wavelength of the load has to have been large enough (of the order of several hundred kilometers) to involve the buoyant response in order to allow the load to serve as a probe of mantle viscosity. The longer the wavelength of the load, the greater the volume of mantle is involved in the response to the loading; larger loads are therefore better probes of deeper mantle viscosity structure.

The first discussion of the use of this sort of geomorphic deep-Earth probe can be found in G. K. Gilbert's (1890) classic monograph on Lake Bonneville (the first USGS monograph). Gilbert used the variation in the heights of shorelines that were all formed in a very short period of time, called the Bonneville shoreline, to map the deflection field associated with the Bonneville Lake load. Lake Bonneville flooded the eastern portion of the Basin and range province, which is studded with north–south trending mountain ranges bounded by normal faults. Because each of these ranges has shorelines corresponding to the Bonneville highstand, Gilbert was able to map the subsequent deflection of the shorelines simply by noting their elevations. The resulting map reproduced in Figure 3.14 looks like a bulls eye, with a maximum upward deflection of the shorelines of over 50 m. We now know from ^{14}C dating that the Bonneville shoreline was formed about 14.5 ka. Of course, at the time it was formed, this shoreline was everywhere the same elevation. That it is now warped up in the center implies that at the time it was formed the load of the lake deflected it downwards by at least this much. Importantly, there are also other younger shorelines that show a pattern of deflection as well. These can be used to document the temporal evolution of the Bonneville area to the removal of the load. It is from the timescale of this response that the viscosity of the upper mantle can be read (see Bills et al., 1994).

Other sites that have been used as probes of mantle behavior include the rebounded shorelines around the edge of Hudson's Bay and in Fennoscandia, reflecting the isostatic rebound from the removal of the Laurentide and Fennoscandian ice sheets, respectively

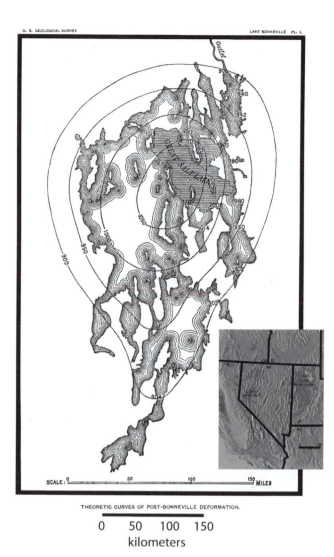

THEORETIC CURVES OF POST-BONNEVILLE DEFORMATION.

0   50   100   150
kilometers

Figure 3.14 Pattern of isostatic rebound associated with loss of Lake Bonneville (G. K. Gilbert, 1890, Plate L). Present remnant of Lake Bonneville is the Great Salt Lake (horizontal lines). Contours of Bonneville shoreline elevations (Gilbert's original is in feet above the Great Salt Lake elevation) show warping of at least 150 ft, or about 46 m.

(e.g., Walcott, 1972; Andrews, 1968; see summaries in Cathles, 1975, and Watts, 2001). Lake Ayre in Australia and Lake Uyuni in the altiplano of the Andes have been used as well.

Ice sheet and ocean loading and the response of the Earth surface to it

We have discussed the load associated with Lake Bonneville. While this is locally important, the volume of water involved is not globally significant. In great contrast, the volume of ice tied up in ice sheets at present and in the past is significant. The sum of the present volumes of ice on Greenland and in Antarctica is worth about 80 m of sea level; if they were to melt, sea level would rise 80 m. And at the last glacial maximum (LGM), enough additional ice was tied up in the great ice sheets covering northern North America and Fennoscandia that sea level was depressed by 120–150 m. In other words, the loading of the northern continents by ice coincided with the unloading of all the world's oceans, and conversely, during deglaciation, the unloading of the northern continents coincided with the reloading of the ocean basins. We illustrate this in Figure 3.15. Here we summarize the response of the Earth's surface to these large-scale changes in the pattern of loading. It is important to acknowledge that the Earth is still responding to these changes, as the timescale for the response, dictated by the mantle viscosity, is a significant fraction of the time since deglaciation. We cannot understand the pattern of the modern rates of relative sea level change, as documented by tide gages, for example, at points along the world's coasts without addressing this component of surface uplift or subsidence. What is clear from both the data and from models of the full mantle response is that the signal of growth and decay of ice loads on the surface of the Earth has incited a truly global response, as seen in Figure 3.15, that must be acknowledged when trying to interpret modern gravity (GRACE), tide gage, tilt, and GPS records. While it is clear that a river like the Mississippi, whose headwaters were covered by the southern margin of the Laurentide ice sheet, will have been directly and greatly affected by the ice loading, it is less clear how a river like the Amazon will have been affected. We will discuss rivers at greater length elsewhere, but it should be intuitive that the larger a river is, the smaller will be its slope. We will see that the broad tilting of the continental margin driven by the loading and unloading of the adjacent ocean basins is of the same scale as the slopes of these large rivers.

Consider the simple representation of a pattern of loading that is triangular, tapering from a peak at the load center to zero at the margin. This is depicted in Figure 3.16. For the moment, imagine that this load is emplaced instantaneously. For our purposes, this means it is placed more rapidly than the underlying mantle can respond. We ask what determines the

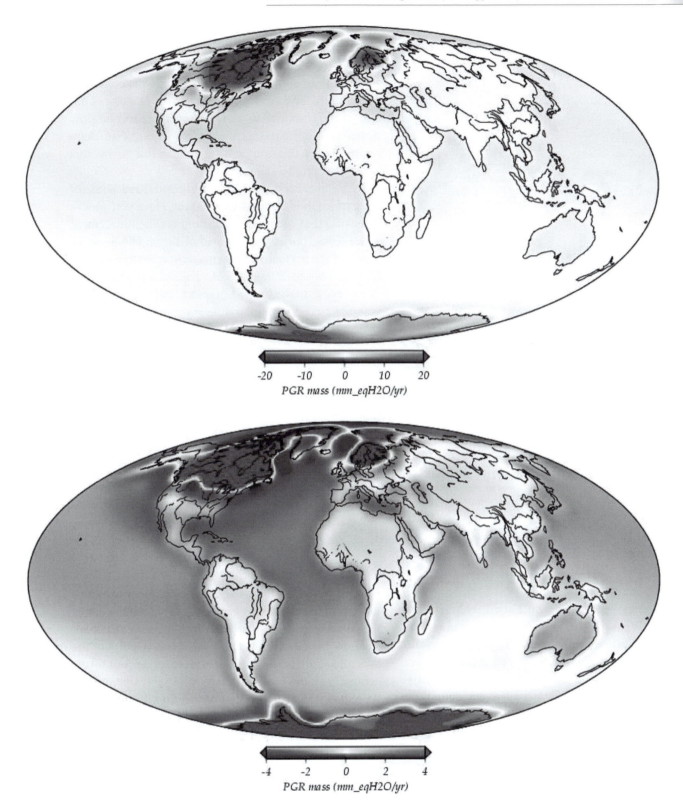

Figure 3.15 Predicted global distribution of post-glacial rebound (PGR), calculated using a prescribed glacial history (ICE-5G from Peltier (2004)), 90 km-thick lithosphere, 1170 km-thick upper mantle with a viscosity of 0.9×10^{21} Pa-s, and lower mantle viscosity of 3.6×10^{21} Pa-s. Scales reported as surface changes in mm of water equivalent per year. The panels differ only in the color scale. Top: wide range allows details to be seen in the region of large load changes over the ice sheet sites. Bottom: reduced scale allows details to be seen in the remainder of the world far from ice sheet influences. While collapse of the forebulges associated with the ice sheets produces the greatest subsidence rates, the reloading of the ocean basins causes subsidence that is greatest immediately outboard of the continental margins (figure in personal communication from A. Paulson, with permission to reproduce).

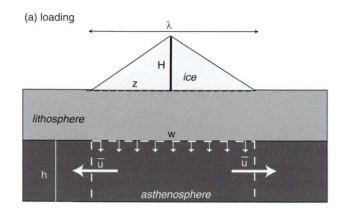

(a) loading

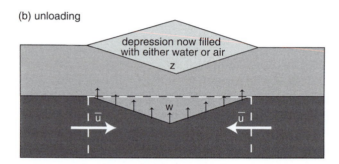

(b) unloading

Figure 3.16 Problem set-up for glacial loading (a) and unloading ((b): rebound) by movement of asthenosphere in a low-viscosity channel. After addition (or removal) of the ice sheet, here treated as a triangular ice load of density ρ_i, maximum height H and width λ, the pressure gradient set up in the asthenospheric channel of density ρ_m, and thickness h, serves to push mantle away from (back under) the site of the load. Conservation of volume in the asthenosphere requires deflation (inflation) of the channel at a rate w, and hence decrease (increase) in the elevation of the surface, z.

and hence decreasing the pressure gradient. This will continue until the horizontal pressure gradient driving the mantle away declines to zero, at which point in time no more flow can occur. This condition is formally one of isostatic balance. We therefore expect the response to be fastest at first, and to slow down through time as the pressures equalize. You could sketch this out; it is a good habit to do this before turning to the math.

In the discussion above, we considered the response of the surface to loading by an ice sheet. However, we live in a world that is still responding not to the emplacement of ice loads, but to the removal of them. The great ice sheets of the northern continents most recently met their demise between 22 ka and 7 ka. The unloading problem is analogous to the loading problem, but with the gradients reversed: the mantle flow is now directed toward the centers of the now-vanished ice sheets, and the convergence of flow there in turn results in thickening of the asthenosphere and uplift of the landscape.

Conservation of volume in the asthenosphere demands that the flow into the region beneath the load, across the "gates" shown as dashed vertical lines in Figure 3.16, demands that the change in volume between the gates equals the gain or loss across the edges. This results in the equation for the vertical velocity of the surface, w:

$$\lambda w = 2h\bar{u}$$

or

$$w = \frac{2h\bar{u}}{\lambda} \tag{3.20}$$

where λ is the width of the load, $\bar{u}$ the mean speed of the asthenosphere in the channel of thickness h, and z the mean elevation of the surface beneath the load. We now need an equation for the mean velocity of the asthenosphere across the gate. Here we appeal to the geometry of the problem as we have set it up, and use an expression for the flow between two rigid plates (although of course this is a simplification). Assuming that the asthenosphere can be approximated as a linear viscous substance, this results in

$$\bar{u} = -\frac{1}{12\mu}h^2\frac{dP}{dx} \tag{3.21}$$

The velocity is inversely proportional to the viscosity, and is sensitive to the thickness of the channel. The negative sign assures that the flow moves down the gradient, dP/dx. The final piece of the puzzle is then

temporal response, which will have a characteristic timescale. We want to know what determines this timescale. The load generates a horizontal pressure gradient in the mantle; the pressure beneath the load is larger than the pressure at the same depth relative to the geoid at the edge of the load. We will see more formally elsewhere (Chapter 10) that fluids respond to pressure gradients by flowing down the pressure gradient, from regions of high pressure toward sites of low pressure. The mantle will therefore flow from beneath the load toward its edge. The second part of the problem is conservation of mass in the asthenosphere. Loss of mass from asthenosphere beneath the load leads to thinning of the asthenosphere. As the asthenosphere thins, the load settles downward, decreasing the pressure beneath the center of the load,

the estimation of the horizontal pressure gradient. This is scaled by the difference in pressure at the load center and edge, divided by the distance between these. In the case of unloading, this pressure gradient can be approximated by

$$\frac{\mathrm{d}P}{\mathrm{d}x} = \frac{P_{\text{edge}} - P_{\text{center}}}{\lambda/2} = \frac{2\Delta\rho g\Delta z}{\lambda} \tag{3.22}$$

where $\Delta\rho$ corresponds to the density difference between asthenospheric mantle and the infilling material (water or air). We can now assemble these pieces to arrive at a differential equation for the rate of uplift:

$$w = \frac{\mathrm{d}(\Delta z)}{\mathrm{d}t} = -\frac{h^3 \Delta\rho g}{3\lambda^2 \mu} \Delta z \tag{3.23}$$

This linear first-order ordinary differential equation may be solved to yield

$$w = w_\mathrm{o} e^{-\dfrac{t}{3\mu\lambda^2/\Delta\rho g h^3}} \tag{3.24}$$

This exponential equation conforms to our expectation of a rapid initial response, which then decays through time. Note that the collection of constants in the numerator must have units of a timescale. Indeed, this is the characteristic timescale, τ, for the response of the mantle: $\tau = (3\lambda^2\mu)/(\Delta\rho g h^3)$. The equation may therefore be written more simply as

$$w = w_\mathrm{o} e^{-t/\tau} \tag{3.25}$$

The history of uplift rate, $w(t)$, is shown in Figure 3.17. The surface elevation rises rapidly at first and then more slowly through time. The timescale, τ, corresponds to the time it takes for the rate of rise of the surface to decline to $(1/e)$ of its original rate. The lithosphere is therefore still on the move for times of a few times this characteristic time, or several τ. We can see now that the timescale is determined by the mantle viscosity. This calculation lies at the core of the utility of geomorphic markers as probes of mantle viscosity.

By integrating the equation for the rate of rise of the surface, we derive an equation for uplift of the surface through time:

$$\Delta z = \Delta z_\mathrm{o} \left[1 - e^{-t/\tau} \right] \tag{3.26}$$

where Δz_o is the expected final deflection of the surface at very long times $t \gg \tau$. This is asymptotically

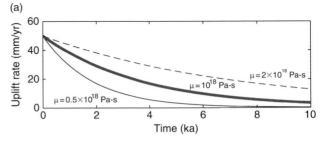

(a)

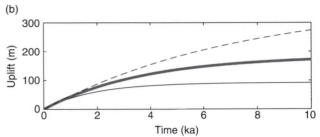

(b)

Figure 3.17 Theoretical rebound histories for three mantle viscosities with identical prescribed initial rebound rates. (a) Rebound rate histories displaying different characteristic timescales, shorter times associated with lower viscosities. Asthenospheric channel thickness 200 km, 1000 km-wide load, density contrast between mantle and infilling material = 3300 kg/m³. (b) Deflection histories obtained by integrating the rebound rate histories.

approached at long times, $t \gg \tau$, after the removal of the load, as seen in Figure 3.17.

It has long been recognized that glacial rebound is the cause of the raised beaches in the Hudson Bay area, and in Sweden (Figure 3.18). These beaches can be dated using ^{14}C in shells they contain. These dates and the elevation of the shorelines comprise a data set that has been used to constrain local mantle viscosities. In Figure 3.19 we show the time series of rebound from dated beaches near the Angerman River, in Sweden. The best-fit curve depicted has a timescale τ of 3.8 ka: within 3.8 ka of the removal of the load, 1/e or roughly 37% of the rebound had occurred. The implied mantle viscosity may be calculated by solving the equation for the timescale for viscosity: $\mu = \Delta\rho g R\tau$. For $R = 1000$ km, $g = 9.8$ m²/s, and the density difference of 3300–917, or roughly 2300 kg/m³, the implied viscosity is $\mu = 3 \times 10^{21}$ Pa-s (see Andrews, 1968; Walcott, 1972; Cathles, 1975; Peltier and Andrews, 1976; Peltier et al., 1986). In Figure 3.20 we show results from Paulson et al. (2005) in which the structure of mantle viscosity is inferred from data sets of this sort.

This analysis is quite simplified. The glacial rebound problem has been repeatedly revisited as more data

Figure 3.18 Flight of raised beaches tucked in the northwest sector of the Richmond Gulf Peninsula, Quebec, on the southeastern coast of Hudson Bay, near the site of maximum ongoing uplift of land due to post-glacial rebound (photograph by Claude Hillaire-Marcel, Université du Quebec à Montreal, with permission to reproduce).

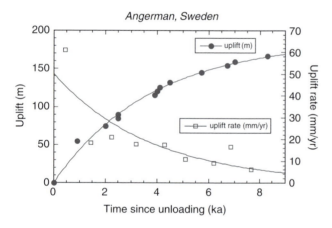

Figure 3.19 Rebound of Angerman River area, Sweden, in the aftermath of the demise of the Fennoscandian ice sheet. Data from dated shorelines. Boxes: uplift rates derived from differencing elevation and age data. The curves shown are the expected asymptotic exponential solution derived in the text for a site within the footprint of removed load. Same best-fit timescale for elevation data (3.77 ka) is used to fit rebound rate data. Note that the modern rate of uplift is about 5 mm/yr.

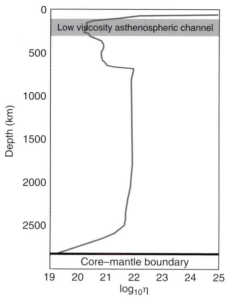

Figure 3.20 Profile of viscosity of the Earth's mantle showing low-viscosity "channel" in the upper mantle beneath the very high viscosity (rigid) lithosphere, relatively uniform viscosity in the lower mantle, and decline in viscosity at the core–mantle boundary. Note the logarithmic scale for the viscosity. Structure is constrained by numerous observations, including glacio-eustatic rebound histories around the globe (redrawn from Paulson et al., 2005, Figure 3, with permission from Wiley-Blackwell Publishing).

have been assembled. Most recent work (e.g., Peltier, 2004; Paulson et al., 2005; Mitrovica and Forte, 2004) acknowledges the complexity of the spatial and temporal pattern of loading, and the potential for both horizontal and vertical structure in the viscosity of the mantle to alter the simple patterns. The data now used to constrain this viscosity structure include the raised beaches already discussed, the modern tide gage records of the world's coastlines both inside and outside the loads, the tilting of the Great Lakes (deduced from lake level records on the north and south sides of the lake), gravity data from the GRACE satellite that documents the uplift rate over the entire region, and so on.

Note the two-way nature of the problem we have just treated. The elevations of the geomorphic features, here raised beaches, have allowed a calculation of an important geophysical number, the viscosity of the mantle, demonstrating their utility as probes of mantle viscosity. Just as importantly, the geomorphologist must be aware that the rearrangement of loads on the surface of the Earth incites a deep Earth response resulting in a deflection of the surface whose amplitude may be important for certain processes, which will be delayed relative to the loading by a time constant of several thousand years, determined by mantle viscosity.

Mantle flow and its influence on topography

While we commonly attribute much of the detail of the world's topography to the action of plate tectonics, and focus specifically on the margins of plates, where orogeny and volcanoes produce topography to be attacked by geomorphic processes, the effects of mantle convection are felt even in the interiors of continents. The details of the mantle-driven processes and their manifestation on the surface of the Earth are presently being debated. This is healthy, and will result in a rapid evolution of our thought about how the deep Earth influences the Earth's surface. The focus over the last half-century has been on horizontal tectonics, and its role in the formation of the mountain ranges of the world. The plate tectonic theory so pervaded the literature that the evidence for very broad scale vertical motions of the surface has been largely neglected. Pure vertical motion did not fit the new paradigm of horizontal plate motions translating into vertical motion. Only in the last decade has this problem re-emerged as a reasonable target for geodynamicists. In the interests of covering in this chapter the full panoply of geophysical processes of importance to geomorphology, here we briefly summarize an array of connections between deep Earth (mantle) and surface processes that should serve as catalysts for future study. These include the flow field set up within the mantle by motion of slabs in the mantle, and the tractions this can set up on the base of the lithosphere (called "dynamic topography"); the foundering of dense crustal roots from beneath mountain ranges; the gooshings of the asthenosphere over large distances in response to changing ice and water loads on the Earth's surface; and the motion of the lower crust in response to pressure gradients set up by the topography itself.

Dynamic topography

Recall that the timescale for a lithospheric slab to reach the core–mantle boundary, at 3000 km depth, is of the order of 100 million years if the vertical speed of the slab is 3 cm/yr (3×10^6 m/3×10^{-2} m/yr $= 10^8$ years). These long timescales result in a very long legacy for the influence of lithospheric slabs on the Earth's surface. As we have discussed, a subducting slab of lithosphere sinks within the surrounding mantle

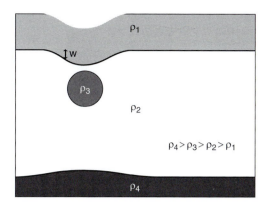

Figure 3.21 The essence of dynamic topography. Mantle density anomaly, here spherical, imposes flow in mantle interior that in turn deforms other density interfaces in the system (inspired by Davies, 1999, Figure 8.5).

because it is negatively buoyant – its mean density is higher than that of the surrounding mantle. The motion of this slab through the mantle incites a flow within the mantle, the detailed geometry of which depends upon the dip of the slab, the viscosity of the mantle, and so on. This flow inevitably affects the top surface, or the overlying lithosphere, by exerting a suction on it – a distribution of downward normal forces. This downward pull generates a huge dimple in the lithosphere, as depicted in Figure 3.21. The time and length scales of this deflection of the lithosphere are important. As you might guess, the horizontal scales are continental in size – they are dictated by the horizontal dimensions of the downgoing slab. The timescales are also set by the activity of the downgoing slab, but importantly involve as well the viscosity of the mantle (see for example the simulations in Figure 3.22). The higher the viscosity of a fluid, the longer it will take to respond to changes in the forces acting upon it. In the case of the Earth's upper mantle, the viscosity is so high that the timescales are of the order of tens of millions of years. This means that it might take 30 million years for the dimple to be produced as a subducting plate slides beneath a continental mass, and likewise it will take 30 million years for the dimple to decay away, or rebound, after the back edge of the slab passes by. But how deep might this dimple be? It is the vertical scales of the deflection that are surprising; they are of the order of a kilometer. This scale alone makes this phenomenon worthy of consideration by geomorphologists, as the mean

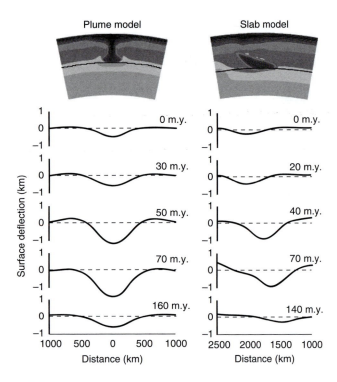

Plume model   **Slab model**

Figure 3.22 Dynamic topography over a rising plume (left) and a sinking slab (right), shown at different times in the calculation. Top cross section shows initial conditions for the models (= 0 Ma). Note the vertical scale of the deflection is of the order of 1 km (from Pysklywek and Mitrovica, 1998, Figures 2 and 3).

elevation of a continent (the so-called continental free-board) is only about 800 m. During subduction of a plate beneath a continent, its interior could potentially be pulled beneath sea level, inciting deposition of marine sediments there, and upon cessation of subduction, this region ought to pop back up above sea level. In the absence of any knowledge of this process, one would falsely attribute such continental flooding to a rise in global sea level.

But how do we know this happens? How do we untangle the records of eustatic changes in sea level from those of dynamic topography associated with subduction? Close attention to the rock record is the key; where subtle tiltings of rock packages on sub-continental scales can be demonstrated (Mitrovica *et al.*, 1989), and evidence of synchroneity of eustatic sea level highstands with the deposition of these sediments is lacking, the finger can be pointed at large-scale mantle dynamics. Note that the tilts we are talking about here are small. They may be roughly scaled by the maximum vertical deflection (1 km) divided by the horizontal scale of the deflection

(several thousand kilometers), meaning tilts of a few times 10^{-4}. Examples of continental flooding that may be attributed to dynamic topography can be seen in the rock records of the western interior seaway of North America, and in Australia. Present examples may include Indonesia (e.g., Gurnis, 2001).

The impact on geomorphic processes is several-fold. First, as we will see, rivers run across the land surface until they reach a stable water surface, either a lake or the ocean, which serves as a baselevel. Dynamic topography clearly can affect greatly the location of the coastline on a continent. Second, rivers transport sediment and incise into bedrock at rates that are dictated in part by the local slope of the river. River slopes can be quite subtle, ranging from a few meters per kilometer (0.001 for mountain streams) to a few centimeters per kilometer (10^{-5}, e.g., the Amazon). For big rivers, then, the scale of the tilting can be comparable to the river's slope, meaning that it could be either enhanced significantly, or reduced significantly (even reversed!) by this continental-scale process, depending upon the direction of the river flow, the orientation of the downgoing slab, and whether the subducting slab is just beginning to pass beneath the river, or just finishing. Finally, given that we are talking about a kilometer of vertical motion, one can imagine that the geomorphic processes active on existing topography will change as it passes from one elevation band to another. Both the precipitation forcing many geomorphic processes, and the vegetation that acts as a strong determinant of the type of geomorphic process change dramatically with elevation.

Topographic oozing of the Tibetan Plateau margin

In a landmark article, Marin Clark and Wicki Royden proposed that the smooth topographic ramp leading from southeastern China up to the Tibetan Plateau resulted from the oozing of hot lower crustal material from beneath the Plateau (Clark and Royden, 2000). This ramp contrasts sharply with the abrupt topographic front of the Himalayas that bounds the plateau to the south, visible in Figure 3.23. The physics of the problem is identical to that we have just introduced in the glacial rebound problem.

The analysis of Clark and Royden follows and simplifies an earlier model of the deformation of the

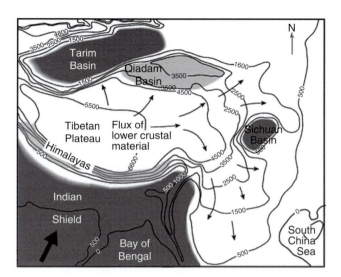

Figure 3.23 Map of Tibetan Plateau and surrounding lowlands, showing proposed direction of transport of lower crustal material from beneath plateau. Note the broad topographic ramp to the SE, and its contrast with the abrupt topographic front imposed by the Himalayas to the south of the plateau (from Clark and Royden, 2000, Figure 5).

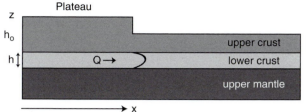

Figure 3.24 Set-up for flow of lower crustal material from beneath Plateau. Change of thickness of the low-viscosity channel, h, will result from gradients in the discharge of lower crustal material. Pressure driving the flow is tied to the thickness of the upper crust, h_o, which is non-uniform due to thickening of crust to form the Tibetan Plateau.

Tibetan Plateau as a whole (Shen *et al.*, 2000). Viscous lower crust (in this case) is being driven down a channel by a pressure gradient (Figure 3.24). The chief difference is that the channel through which the lower crust is being driven is unbounded – material is not being gooshed into or out of a cylinder but into a slot of effectively infinite length. We emphasize that whether the hypothesis is right or wrong (and there is indeed some support for it in river incision along this ramp (Schoenbohm *et al.*, 2006)), their treatment can be understood using the same physics we have been exploring in this section.

We depict their proposed system in Figure 3.24. The rate of change in thickness of the lower crustal channel is determined by the gradient in the discharge of lower crustal material down the channel. In other words,

$$\frac{\partial z}{\partial t} = \frac{\partial h}{\partial t} = -\frac{\partial Q}{\partial x} \tag{3.27}$$

where h is the thickness of the channel, and Q is the discharge of viscous material along it. Again, the discharge, Q, is the volume of material per unit time per unit width of channel [$= L^3/LT$ or L^2/T]. Since the thickness of upper crust above the viscous channel, h_o, does not change with time, the rate of change

of elevation of the surface, z, simply follows the rate of change of thickness in the channel. It is, however, important that the thickness of crust h_o does vary in space. It is the thickened crust beneath the Tibetan Plateau that leads ultimately to its heating, which in turn leads to its reduction in viscosity and hence tendency to flow.

As in the rebound case, the discharge in the viscous channel is calculated from the integral of the velocity profile,

$$Q = \int_0^h U(z)\mathrm{d}z = -\frac{1}{12\mu}\frac{\partial p}{\partial x}h^3 \tag{3.28}$$

While at first the pressure gradient is set by the gradient in thickness of upper crust, gradients in lower crustal thickness will begin to play a role as it evolves:

$$\frac{\partial p}{\partial x} = \rho_c g \frac{\partial\,[h_o + h]}{\mathrm{d}x} \tag{3.29}$$

Inserting this expression into the statement of conservation of volume in the channel results in

$$\frac{\partial h}{\partial t} = \frac{\rho_c g}{12}\frac{\partial\left[\dfrac{h^3}{\mu}\dfrac{\partial(h_o + h)}{\partial x}\right]}{\partial x} \tag{3.30}$$

For now let's assume that the viscosity of the lower crust is uniform, so that we may pull it out of the derivative. Taking the derivative of the remaining product results in two terms:

$$\frac{\partial h}{\partial t} = \frac{\rho_c g}{12\mu}\left[h^3\left(\frac{\partial^2 h_o}{\partial x^2} + \frac{\partial^2 h}{\partial x^2}\right) + 3h^2\left\{\left(\frac{\partial h}{\partial x}\right)^2 + \frac{\partial h_o}{\partial x}\frac{\partial h}{\partial x}\right\}\right] \tag{3.31}$$

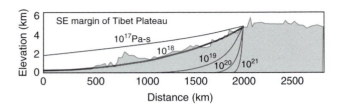

Figure 3.25 Topographic profile of ramp leading from Tibetan Plateau to the SE. Lines show solution for expected profile given an assumed viscosity of the lower crust channel (redrawn from Clark and Royden, 2000, Figure 4a).

Taking the leading term in this expression, dominated by h^3, leaves us with the simpler

$$\frac{\partial h}{\partial t} = \frac{\rho_c g}{12} \frac{h^3}{\mu} \left[\left(\frac{\partial^2 h_o}{\partial x^2} \right) + \left(\frac{\partial^2 h}{\partial x^2} \right) \right]$$

$$= \kappa \left[\frac{\partial^2 h_o}{\partial x^2} + \frac{\partial^2 h}{\partial x^2} \right] \qquad (3.32)$$

The first term corresponds to a couplet of source–sink associated with the non-changing but non-uniform thickness distribution of the upper crust. The second term then acts to smear out this constant source, leading to the thickening of the lower crust (and associated inflation of topography) on the lengthening ramp, and simultaneous deflation of the lower crust below the edge of the Plateau. This is a diffusion equation, analogous to the one derived in our treatment of thermal problems except for this spatially distributed source term. The analogy is made explicit here by assigning an effective diffusivity, κ, to the collection of terms in front of the curvature. This constant reflects the efficiency with which changes in thickness occur. As expected, this efficiency is low when the viscosity is high, and is greatly enhanced when the channel is thick; the collection h^3/μ is difficult to disentangle.

Clark and Royden present results reproduced in Figure 3.25, based upon calculations assuming a channel thickness h of 15 km. Given this choice of channel thickness, the best-fitting viscosity of the lower crustal material is 10^{18} Pa-s.

Gooshing of mantle across the continental edge

We have talked about gooshing the mantle about by ice loads, and gooshing the lower crust by topographic loads. There is at least one other case worthy of our attention. When a huge ice sheet is constructed on a continent, that water comes from somewhere and that somewhere is the ocean. It gets there circuitously, delivered as snow by storms, but its ultimate source is evaporation from the ocean. So when the ice sheet volume is large the ocean volume is small. To give this a scale, at the Last Glacial Maximum some 20 thousand years ago, sea level was drawn down about 120 m relative to today. Just when the land is pushed down by the giant plunger of an ice sheet or two in the northern hemisphere, forcing mantle to flow outward away from the load, the ocean basins of the world are being unloaded. The converse is also true: when the ice sheets dwindle, mantle rushes back in, while at the same time the ocean basins are being loaded up, and mantle should be pushed away from them. Consider then a continental margin well away from any ice load. The variations in the ocean load adjacent to the continental margin should drive a gooshing of the mantle back and forth across the continental margin. Let's marry the two examples we have discussed so far, the ice load and the topographic ooze, and sprinkle in a periodic variation in the system to construct a model of this situation. Our goal is to assess the amplitude of the effect on the topography of the edge of the continent, and the distance inland over which this signal declines.

Assume a viscous channel in the upper mantle, beneath a uniform crust. Consider the simplest case of a very steep continental margin, in which the full 120 m swing of sea level does not cause significant lateral migration of a coastline. This is shown in Figure 3.26. Now allow sea level to vary sinusoidally with a fixed amplitude, Δh_{sea} and period, P:

$$h_{sea} = h_{sea} + \Delta h_{sea}(\sin(2\pi t/P)) \qquad (3.33)$$

This variation in sea level translates into a variable pressure at the depth of the mantle channel, which in turn drives variation in the pressure gradient in the channel that causes transport of mantle. In addition, the thickness variation in the mantle then can drive flow as well, as the mean pressure in the thicker channel is greater than that in the thinner portion of the channel.

The conservation of fluid in the mantle channel is identical to the one we wrote for the Tibetan lower crustal ooze (Equation 3.10), as is the equation for lateral discharge of mantle. What differs is the pressure field driving flow, which now includes that of the dynamic oceanic load. The pressure is therefore

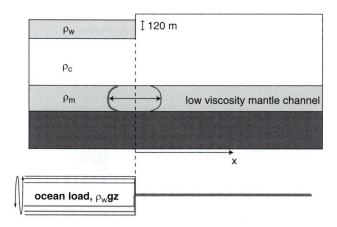

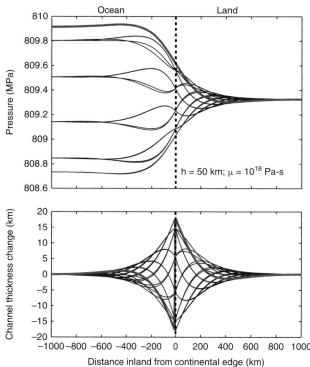

Figure 3.26 Schematic diagram illustrating the possible transport of mantle in a low-viscosity channel in response to oscillating sea level. The ocean load varies with sea level, which in the Pleistocene has varied by 120 m between glacial lowstands and interglacial highstands with periods of 20–100 ka. At highstands (light lines), mantle is forced from beneath the ocean load toward the interior of the continents (light velocity profile), while at lowstands (bold lines), mantle will be pushed toward the ocean basins (bold velocity profile). Our goal is to predict the resulting pattern of thickening of the channel, and the consequent raising or lowering of the land surface.

Figure 3.27 Elevation changes associated with periodic variations in sea level. (a) Pressure field calculated in upper-mantle channel at 10 times within a sinusoidal variation in sea level with period 20 ka, full amplitude 120 m (two oscillations shown). (b) Mantle channel thickness change driven by gradients in discharge of viscous channel material. This should result in elevation changes in the overlying landscape that extends hundreds of kilometers from the continental margin. Last time stamp (bold line) corresponds to sea level highstand.

$$P = \rho_w g h_{sea} + \rho_c g h_c + \rho_m g(h/2) \qquad (3.34)$$

When we acknowledge these contributions from all components of the load, the discharge becomes

$$Q = -\frac{1}{12\mu}\frac{dp}{dx}h^3$$

$$= -\frac{h^3}{12\mu}\left[\frac{d(\rho_w g h_{sea} + \rho_c g h_c + \rho_m g(h/2))}{dx}\right] \qquad (3.35)$$

Because the thickness of the crust does not vary in time, and it is largely uniform in space in this problem, we drop this middle term. The statement for conservation of volume in the mantle channel then becomes

$$\frac{\partial h}{\partial t} = \frac{\rho_c g}{12\mu}\left\{h^3\left[\left(\frac{\partial^2 h_o}{\partial x^2}\right) + \left(\frac{\partial^2 h}{\partial x^2}\right)\right] + 3h^2\left[\left(\frac{\partial h}{\partial x}\right)^2 + \frac{\partial h}{\partial x}\frac{\partial h_o}{\partial x}\right]\right\} \qquad (3.36)$$

This formulation explicitly acknowledges that the thickness of the channel, h, varies, which in turn causes variations in the discharge through the strong nonlinear h^3 dependence.

We have used a numerical model to explore the behavior of the system when forced with a 120 m oscillation in sea level. Results are shown in Figure 3.27. The system displays a diffusive behavior. It displays both an exponential decay of amplitude with distance from the margin, and a phase lag, much like the solution for temperature within a half space when forced by oscillation of the surface temperature (e.g., Gold and Lachenbruch, 1973; Turcotte and Schubert, 2002). In the thermal case, the length scale that dictates both the decay rate and the time lag is set by the square root of the thermal diffusivity and the period of the oscillation:

$$L = \sqrt{\frac{\kappa P}{\pi}} \qquad (3.37)$$

The same appears to be the case in this system. Numerical experiments show that the penetration of the effect into the continent indeed depends upon

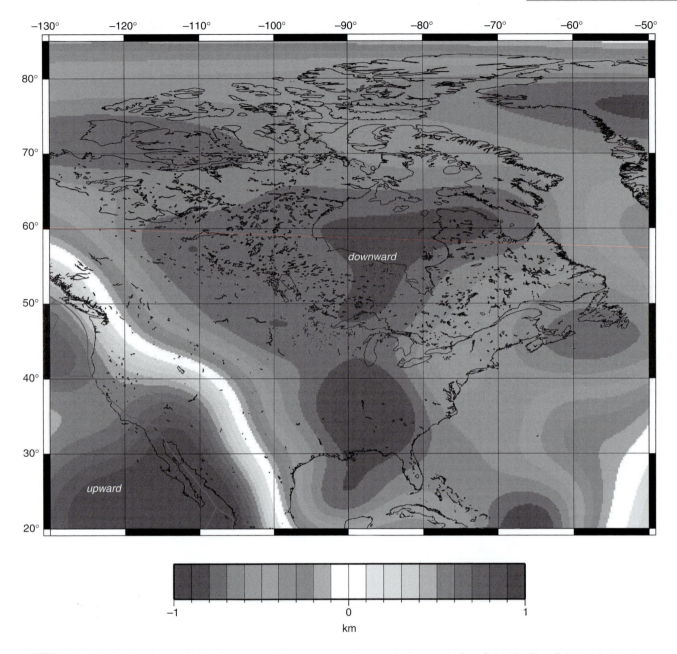

Figure 3.28 Vertical deflection of the North American lithosphere associated with the mantle flow field. The flow field is calculated from buoyancy forces set up by the density structure of the mantle, which is in turn constrained by seismic velocity structure (tomography). Note the amplitude is of the order of 1 km over central North America (reproduced from Forte *et al.*, 2007, Figure 2, with permission from the American Geophysical Union).

both the period of the oscillation and those variables that take the place of the diffusivity:

$$\kappa = \frac{\rho g h^3}{12\mu} \qquad (3.38)$$

While this problem has yet to be fully exploited, we challenge the reader to look for hints that this gooshing of the mantle across the continental edge occurs. It seems to the authors that the geomorphic signal of this phenomenon will be best displayed where large rivers approach the coastline. Large rivers have small slopes. For example, the slope of the Amazon is around 1 cm/km, or 10^{-5}, while that of the Mississippi is perhaps a few times this

$(2 \times 10^{-5}$ on the delta itself; see Syvitski and Saito (2007)). The smaller the slope, the more likely it will be tweaked by the small tilts of the continental edge imposed by the movement of mantle across the margin. The maximum tilts in the case we have illustrated are about $15\,\mathrm{m}/300\,\mathrm{km}$, or 5×10^{-5}; given the slopes of major rivers as they approach the coast, this tilt is worthy of discussion. Note also that a topographic sag comes and goes well inland of the margin – 500 km inland in the case we have illustrated – and that the sagging is out of phase with adverse tilt at the margin.

In working this problem (more or less as a teaser) we have not accounted for the rigidity of the lithosphere above the channel, which will serve to smooth any sharp gradients in the predicted thickness of the mantle channel. These occur at the continental margin. We also note that the larger effect on a river draining the continent is the more obvious and dramatic oscillation of the baselevel. During glacial times these major rivers seek to join an ocean (baselevel) that is 120 m below present, and should incise their margins as they seek that level. The greater hope to find the effect we have illustrated comes instead from its signal far inland from the margin, and in any temporal lag of the signal. After all, sea level reached its present highstand roughly 6000 years ago, while the high viscosity of the mantle channel should result in continued, ongoing warping of the margin.

In this section we have explored a few impacts of mantle physics on the overlying topography. These examples serve not only to illustrate another application of the principles of conservation, but to alert the geomorphologist that what happens deep in the Earth does indeed matter to its surface. These deep gooshings constitute the largest length scale processes to which the Earth's surface is subjected.

Calculation of the effects of large-scale mantle flow field on the surface of the Earth is now a growing subfield within geophysics. Here it is the motions of the cold negatively buoyant plates, and the upwellings of warmer mantle that exert tractions on the base of the lithosphere. One such calculation is shown in Figure 3.28 over North America. The magnitude of the effect is of the order of a kilometer, while the wavelengths over which these surface deflections occur are thousands of kilometers, and the timescales over which they change are tens of millions of years. These are surely subtle signals, but could have profound effects on the inundation history of continents, and on long-term evolution of drainage systems. They should not be overlooked in any attempt to understand long-term evolution of continental scale topography.

Summary

In this chapter we have introduced topographic features and geologic processes that could influence the deflection of the Earth's surface at scales so large that they involve the mantle in some fashion. Large wavelength loads on the Earth's surface (topography in the crust, ice, ocean) are supported not by the strength of the lithosphere but by buoyancy provided by the high density of the mantle. The timescales for response to changes in loads, and that set the rate at which mantle moves about beneath topography, are determined by the viscosity of that portion of the mantle involved.

The lithospheric plates themselves should be perceived as the rigid lids to a convecting mantle. They represent a rheological boundary layer. Given the strong dependence of rheology on temperature, the problem of plate thickness and its evolution translates into a thermal problem. We introduced the conservation of heat equation and the process of conduction by which heat moves in the lithosphere. The resulting diffusion equation could be solved for the thickness of the cooled boundary layer as a function of time since the top was exposed to the $4\,^{\circ}\mathrm{C}$ base of the ocean at the spreading center. The lithosphere thickens as the square root of time. As any material cools it contracts and becomes more dense. The mean density of the lithosphere is therefore higher – by about 3% – than the underlying asthenosphere. Given this density contrast and the pattern of plate thickness, we could quantitatively explain the bathymetry of the world's oceans using an isostatic

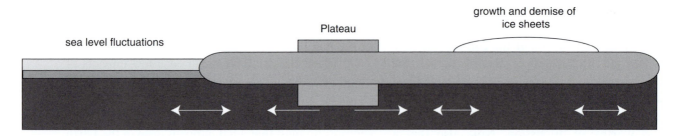

Figure 3.29 Summary diagram of the response of the mantle to a variety of Earth surface loads, each of which produces a pressure gradient in the underlying material. If the loads are emplaced for long enough, and the viscosity of the underlying material is weak enough to flow on these timescales, they will result in flow that in turn raises or lowers the surrounding landscape. Long-term growth of plateaus induces mantle or upper crustal flow away from the high topography. Other loads such as lakes and ice sheets are transient; they come and go, producing flow first away from the emplaced load, and then back under when it vanishes.

balance. As the ocean floors make up roughly two-thirds of the surface of the Earth, this calculation takes us a long way toward understanding the topography of the Earth.

We introduced briefly a simple model for the motion of the lithospheric plates. The roughly 10 cm/yr speeds of the plates could be amazingly well predicted by a theory in which the dense plates are falling downward through a dense, viscous mantle. While quite simplistic, the calculation nonetheless hints at what material properties of the mantle control these rates – the thermal diffusivity that controls the thickening rate of a plate, the coefficient of thermal contraction that controls the associated densification of the plate, and the mantle viscosity that controls the resistance of the mantle to shearing motions, and hence the drag on the plate as it descends.

Despite its importance not only in governing plate rates but also setting the response times to changes in surface loads, the viscosity of the mantle is difficult to measure. We have introduced several large-scale experiments Nature has performed for us in the form of the growth and demise of ice sheets and large lakes that provide a constraint on mantle viscosity. These are summarized in Figure 3.29. In most of these cases, the experiment involves deflection of some geomorphic marker on the Earth's surface that we know should have been horizontal at the time of formation. In most cases these are shorelines. The subsequent deflection of these horizontal markers, and

their ages, has been a target of geomorphic study for more than a century.

We also discussed several examples of 1000 km scale deflections of the Earth's surface that are currently thought to have resulted from mantle (or deep crustal) dynamics. These are also shown in Figure 3.29, and include (1) continental-scale deflection of the surface by tractions on the base of the lithosphere associated with the flow field induced by slabs in the upper mantle, (2) the "oozing" of deep crustal material from beneath the edge of the Tibetan Plateau, driven by the pressure gradient set up by the great contrast in topography, and (3) the pumping of mantle back and forth across the continental margin associated with sea level oscillations in the ice ages.

Note that in treating this material we have employed the principle of conservation in at least two ways. We first encounter it in developing an equation for the conservation of heat, and second in developing the equation for conservation of mantle in a viscous channel. In both cases we find that the rate of change of the quantity of concern (heat, mantle volume) depends upon spatial gradients in the flux of that quantity. This forces us to acknowledge what governs the fluxes. In the case of heat in the lithosphere, this is conduction, represented by Fourier's law, and in the case of the mantle it is viscous fluid flow. We will encounter both of these again in this book, but more importantly we will see further examples of the utility of the conservation equation.

Problems

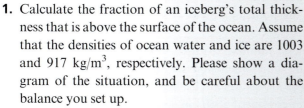

1. Calculate the fraction of an iceberg's total thickness that is above the surface of the ocean. Assume that the densities of ocean water and ice are 1003 and 917 kg/m³, respectively. Please show a diagram of the situation, and be careful about the balance you set up.

   Now observe an ice cube in a glass of water as it melts, and plot both the root and the freeboard as functions of time. Plot their ratio as a function of time.

2. Calculate the expected deflection of the lithosphere beneath a thick ice sheet. Assume that the ice sheet is 4000 m thick, and that ice has a density of 917 kg/m³. (The Antarctic Ice Sheet is roughly this thick at its maximum thickness.) The thickness of the crust (of density 2700 kg/m³) beneath the center of the ice sheet and the region outside of it is the same. The density of the upper mantle that gooshes out of the way to allow this deflection of the surface is 3300 kg/m³. How far down is the rock depressed beneath the load of the ice?

3. How much does the exposed land area change (a) during a sea level drop of 150 m and (b) during a sea level highstand of 500 m? Report your answers as fractions of the Earth's surface area that are land.

4. Assuming a continental crustal thickness of 25 km, an oceanic crustal thickness of 5 km, and an ocean depth of 4 km, calculate the continental freeboard using reasonable assumptions for continental crustal, oceanic crustal, mantle, and ocean densities.

5. Now consider what will happen to the maximum elevations of the land when V-shaped valleys are chopped throughout the landmass, extending down to sea level. Assume that the valleys take up the entire landscape, and that the tips of the ridges between them are not eroded. How much will the ridges rise? (*Hint:* scale this by the depth of the valleys or, equivalently, the relief of the landscape.)

6. *Oceanic lithospheric thickness.* In this chapter, we discussed the square root of age dependence of oceanic lithospheric thickness, and how this controls the bathymetric profile across an ocean basin. As we have seen, this collapses to a simple equation for ocean depth:

   $$D = D_o + A\sqrt{\kappa t}$$

   where D is the ocean depth in m, D_o the depth at the spreading ridge in m, t is the age of the lithosphere in years, and κ is the thermal diffusivity of the lithosphere.

   Given this equation, and the plot of ocean depth vs. age for the first 80 Ma for the North Pacific from Sclater's work (Figure 3.2), calculate a value for the dimensionless constant, A. Assume a reasonable thermal diffusivity of 1 mm²/s. (*Hint:* watch out for units here, as the age of the lithosphere is given on the plot in millions of years (Ma). To make the answer for

A dimensionless, you will have to convert the diffusivity into m^2/yr and the age into years.)

7. *Heat flow in the Basin and Range.* In the Basin and Range province of the western USA the heat flow at the Earth's surface is considerably enhanced. In a 300 m-deep borehole in Owens Lake in the eastern California portion of the Basin and Range, the temperature increases by 18 °C from the surface to the base of the borehole. The conductivity of the materials is 4 W/(m K).

What is the heat flux (also sometimes called the heat flow) at the surface of the Earth? Express this first in W/m^2, and then in heat flow units (HFU, where $1 \ HFU = 41.84 \ mW/m^2$). Note that the worldwide average heat flow is 1.67 HFU.

8. *Thermal profile in and beneath an ice sheet.* Consider a portion of the East Antarctic Ice Sheet that is 2 km thick (in places it is significantly more than this). The mean annual surface temperature is –55 °C. The heat flux is $54 \ mW/m^2$, which is a decent average for Antarctica. The thermal conductivity of ice is 2.2 W/(m K), and that of the underlying bedrock is 3.5 W/(m K).

(i) Calculate and then plot the steady-state geotherm for this location, taking the temperature profile down into the underlying rock by 2 km. Assume no heat production from radioactivity takes place.

(ii) At what depth into the bedrock does the temperature of the rock rise above the freezing point of water?

9. *Thought question.* Consider a hypothetical world of the same dimensions as Earth, in which plate speeds are steady at 10 cm/yr, and on which one mega-continent 6000 km across exists. At time 0 the continent splits in half along a N–S rift extending from the north pole to the south pole, while a subduction zone is simultaneously born along the western edge of the western half of the continent. How long would it take before a major continent–continent collision occurs? (This timescale governs a very long cycle of mega-continent assemblies.)

10. *Thought question.* How would the world differ if the plate rates were twice as fast? (*Hint*: focus first at least on the shapes and depths of ocean basins.) How much might sea level change due to such a speed-up?

11. *Thought question.* Review the means by which information from geomorphology has been used to evaluate the character of the mantle (for example, its viscosity).

Further reading

Cathles, L. M., 1975, *The Viscosity of the Earth's Mantle*, Princeton, NJ: Princeton University Press, 386 pp.
A classic treatise on what we know about the viscosity of the Earth's mantle. While we now know much more than then, the book still illustrates well the types of data one assembles in these problems, and provides a snapshot of our knowledge in the early 1970s.

Davies, G., 1999, *Dynamic Earth: Plates, Plumes and Mantle Convection*, Cambridge: Cambridge University Press, 458 pp.
This textbook addresses the physics necessary to understand quantitatively the dynamics of the Earth with particular focus on the mantle. Davies lays out the thermal and fluid mechanics pieces at several levels of complexity, making his arguments accessible to a wide array of students.

Turcotte, D. L. and G. Schubert, 2002, *Geodynamics*, 2nd edition, Cambridge: Cambridge University Press, 456 pp.
This textbook on geophysics serves as an excellent introduction to quantitative geophysics, with tendrils that reach into geomorphology and planetary science.

Tectonic geomorphology

In many places, earthquakes with similar characteristics have been shown to recur. If this is common, then relatively small deformations associated with individual earthquake cycles should accumulate over time to create geological structures.

King *et al.*, 1988a, p. 13 307

In this chapter

In Chapter 3 we discussed motions of the surface of the Earth that could directly be attributed to the underlying ductile mantle. In this chapter we address smaller scale mountain ranges resulting from the very different behavior of the Earth's crust and lithosphere. At crustal-scale wavelengths the lithosphere is strong and can support the weight of the topography by bending beneath it in a process we call flexure. In effect, this strength serves to hand off the duties of supporting surface loads, from being purely buoyancy (isostatic) at long wavelengths, as we discussed in the last chapter, to being purely flexural at short wavelengths. In addition, the upper crust is brittle, and responds to far-field tectonic forcing (ultimately resulting from the motion of lithospheric plates) by both folding and faulting.

Normal-fault bounded Toiyabe Range, Basin and Range Province. The active normal fault brings the adjacent valley floor down relative to the mountain range. Faceted spurs abound on the upthrown footwall block, while alluvial fans accumulate on the hanging wall to smooth the valley floor (photograph by R. S. Anderson).

In order to study the geomorphology of mountain ranges we must understand the underlying geophysical processes and rates responsible for motion of the crust. Armed with this information, one can then study the geomorphic attack of the tectonically generated topography. In this chapter we introduce the physics of the crustal and lithospheric processes responsible for the growth of mountain ranges, and illustrate these with examples of well-studied ranges around the world. We begin with consideration of individual fault structures and describe the expected pattern of surface deflection associated with motion on the faults. We then acknowledge that larger mountain ranges and plate boundaries require motion on one or more faults, or on a fault that is so large that it does not fail in individual events. The linkages of fault segments become important, and the cross-talk between them. We introduce paleoseismology, the study of the history of motion on individual faults and the sizes of earthquakes that this motion causes. Many of the tools used in this field are geomorphic, or involve displacements of geomorphic markers. Finally, we acknowledge that the long-term slip on faults can result in significant overlapping of crust if reverse motion occurs, or stretching and thinning of crust if normal fault motion occurs. This alters the distribution of load on the lithosphere, which in turn flexes. We therefore discuss the physics of flexure, which governs the magnitude and wavelength of the deflection. In the end, the long-term growth of mountain ranges therefore requires both the displacement of the surface by motion on faults, and flexing of the lithosphere. Of course, geomorphic processes tangle with this growing topography, altering the shape of the topography as it grows tectonically ... but that is left for later chapters.

Deformation associated with individual faults

Many mountain ranges are formed by slip along one or more crustal faults. In western North America we have excellent examples of both normal-fault bounded ranges within the Basin and Range province, and the older thrust-fault bounded Laramide ranges further east. While these sets of ranges were born of very different tectonic regimes, they illustrate the types of mountain ranges that can form by motion

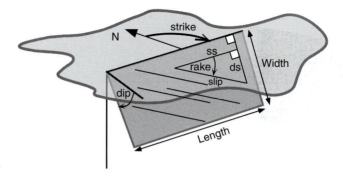

Figure 4.1 Definitions of features of a dipping fault (after Burbank and Anderson, 2000, Figure 11.1, with permission from Blackwell Publishing).

on such crustal structures. We wish to know the vertical deformation field associated with such structures, as this exerts the pattern of tectonically driven rock uplift and subsidence that will drive the surface processes that ultimately modify this geologic structure into a topographic mountain range.

We first define the geometry of the fault itself, and of the slip pattern on that fault, as sketched in Figure 4.1. Defining a specific fault, or plane across which there is shear displacement of rock, requires identification of the strike and dip of the plane, the length along strike, the width down-dip, and, if the fault is buried, the depth from the surface of the Earth to the top of the fault. The pattern of slip on the fault is described by a vector, the direction of which might be scribed on the fault surface as slickensides. This vector can be defined, equivalently, as the magnitude (slip) and direction (rake) of the vector, or by the components of dip-slip and strike-slip. To be more precise, rake is defined to be the angle between the horizontal and the lines of slip that would be obtained by laying a protractor down on the fault plane.

In Figure 4.2 we present the expected vertical deformation pattern when an elastic sheet is cut by a dislocation and then subjected to shear strain. Note that the geometry of the vertical displacement is more complex than the simple sliding blocks one sees in introductory geology texts. This reflects the fact that these faults end within the subsurface. The Earth is not a pair of wooden blocks that can be slid one across the other. Faults are finite; they have tips, edges. By definition, the pattern of vertical displacement of the surface from motion on a normal fault is such that the hanging wall is pulled down relative to the footwall. The mapview shape of the vertical deflection pattern is

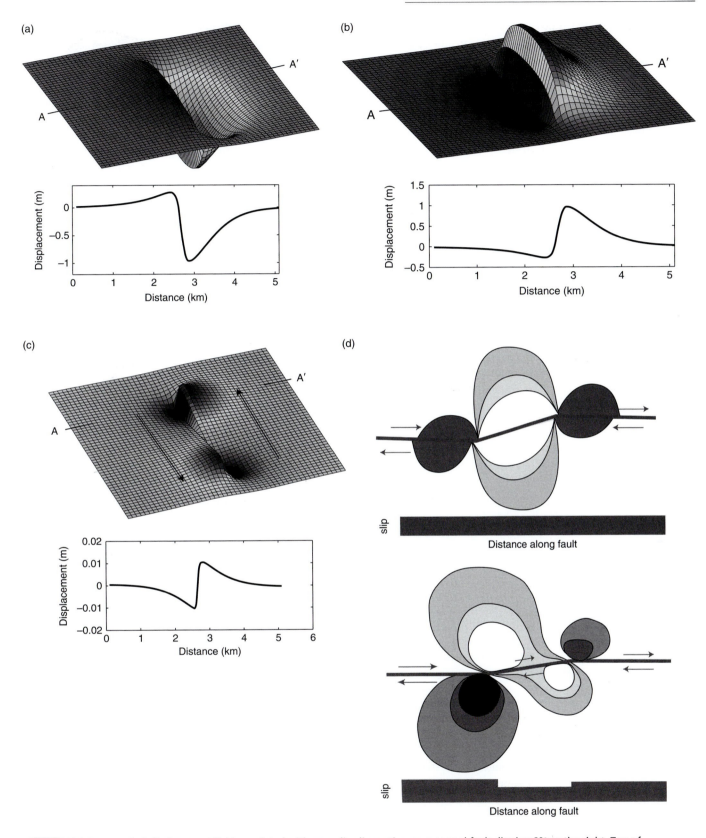

Figure 4.2(a) Top: vertical displacement field associated with pure dip-slip motion on a normal fault dipping 60° to the right. Top of the fault is slightly buried, making it a blind normal fault. Note greatest throw is in the center of the fault. Bottom: profile of vertical displacement along the line A–A', showing greater down-throw of hanging wall than uplift of footwall block.

Figure 4.2(b) Top: vertical displacement field associated with pure dip-slip motion on a reverse fault dipping 60° to the right. Note greatest throw is in the center of the fault. Bottom: profile of vertical displacement along the line A–A', showing greater

shown in Figure 4.2(a). The displacement is such that the slip on the fault and hence the magnitude of the vertical displacement reaches a maximum in the center of the fault. The hanging wall is actually warped downward near the fault, while the footwall is thrown slightly upward by a small amount relative to the down-throw of the hanging wall. It too flexes to grade smoothly into the far field, non-deformed rock.

The vertical displacement pattern is exactly the opposite in a thrust fault: the hanging wall is thrown up more than the footwall is shoved downward. This is illustrated in Figure 4.2(b). If the fault does not daylight, meaning that the fault tip is somewhere below the surface of the Earth, we call it a blind fault. We are blind to it. The deformation field is then a blurred version of the pattern in which the fault cuts the surface. The rounded top of the deformation field implies that if the surface is composed of layered deposits of some sort, these deposits will be warped into this pattern. The closer to the surface the fault tip comes, the sharper this fold. It is often by the presence of such surface geological folds that we can infer the presence of blind thrust faults at depth. This is especially important in the assessment of the seismic hazard in the Los Angeles basin, which is underlain by numerous blind thrust faults.

It may come as a surprise that strike-slip faults can also generate vertical displacement. This arises from two aspects of faults that generate related effects. Uplift can be generated both at bends in strike-slip faults, and where there are gradients in the horizontal slip on the fault, as shown in Figure 4.2(c). Bends come in two flavors: restraining bends and releasing bends. Restraining bends (left stepping bends in right lateral strike-slip faults, or right steps in left lateral faults) restrict motion on the fault, and cause the crustal material to pile up at the bend. This both restricts the motion on the fault (hence the name) and causes thickening of the local crust at the bend. Some of this thickening causes uplift of the surface,

while some inevitably deflects the base of the crust downward. Depending upon the magnitude of the bend, the deformation will be accommodated by elastic deformation of the adjacent rock, or may cause failure of the adjacent rock to generate ancillary thrust faults. Releasing bends are the opposite (right bends in right lateral and left bends in left lateral faults). They generate a thinning of the crust, and subsidence, in some instances producing a discrete graben.

The vertical deflection resulting from variations in slip is a little more subtle. The most obvious examples are at the tips of a particular segment of a strike-slip fault. Beyond the tip, the crust is not moving, while near the tip the crust is moving either away from or toward the far-field crust, depending upon which side of the fault we inspect. In the case of crustal material trying to move toward the non-moving crust, the crust must pile up, in the same way it does across a restraining bend. It must thicken, and hence uplift. Conversely, the crust on the other side must get stretched, and therefore subside. This results in the most fundamental pattern of uplift associated with slip on a perfectly straight strike-slip fault, which we call the dog bone pattern, depicted in Figure 4.2(d). While this tip-effect is the most obvious and perhaps the largest amplitude gradient in horizontal displacement, any variation in slip along the fault can generate uplift or subsidence. As a cautionary note, recall that we are talking about vertical deformation. This effect is distinct from the slip of topographic features, such as a slice of a hill, along the fault that could give the false sense that this side of the fault is rising. Translation of topography is different from its generation in place. Both happen in strike-slip settings.

Now return to thinking about the strike-slip case, defining x to be parallel to the strike of the fault, y to be the other horizontal dimension normal to the fault, and z vertical. A finite amount of slip parallel to the fault, D, results in a displacement gradient at

Caption for Figure 4.2 (*cont.*)
uplift of hanging wall than subsidence of footwall block. Calculation made using the program 3DDEF (see Gomberg and Ellis, 1994) by assessing strains throughout the elastic material in which a dislocation has been inserted, while imposing a convergent displacement parallel to the line A–A'.

Figure 4.2(c) Top: vertical deformation field due to left-lateral motion on a vertical strike-slip fault. Uplift and subsidence occur at the fault tips where the material converges and diverges, respectively, due to gradients of slip of opposite sign. Bottom: profile of vertical displacement taken normal to the strike of the fault, along the line A–A'. Note the small amplitude of the vertical displacement.

Figure 4.2(d) Strain field associated with slip on a restraining bend in a right-lateral strike-slip fault. Top: uniform slip through a 16° bend in the fault. Bottom: non-uniform slip (bottom plot) through a 4° bend. Light tones = compression resulting in thickening and uplift; dark tones = extension resulting in thinning and subsidence. Contours are 2.5, 5, and 7.5% strain (after Bilham and King, 1995, Figure 13, with permission from American Geophysical Union).

Box 4.1 How does horizontal slip translate into vertical deformation?

Without delving into the gory details, we can understand this phenomenon using the principle of mass conservation. For near-surface crustal materials, we may safely assume that the materials are incompressible. Under these conditions, conservation of mass requires that

$$\frac{\partial u}{\partial x} + \frac{\partial v}{\partial y} + \frac{\partial w}{\partial z} = 0 \tag{4.1}$$

where u is the displacement in the x-direction, v is the displacement in the y-direction, and w is the displacement in the z-direction. This equation states that any gradient in displacement in one direction causes gradients of the opposite sign in the other dimensions, as shown graphically in Figure 4.3. Take for example a rubber band. Note its original length and its original thickness. Now stretch it along the band, from left to right, tying one end to a pencil. All points have moved to the right, but each pair of points has moved further apart. We define this to be a positive strain; it results from a displacement gradient in the left–right direction. What has happened in the other dimension? The band has thinned. It has contracted, moving points in this dimension closer together. It has experienced a strain of the opposite sign.

the tips of the fault, where, over some distance, L, the displacement must die off to 0. This displacement gradient is therefore roughly $\partial u/\partial x = -D/L$. It is negative, reflecting that it decays along the positive x-axis. What happens in the other dimensions? The arrival of new crust can be accommodated by stretching in either the fault-normal or the vertical directions. Because there is less resistance in the vertical dimension (air is easier to move out of the way than rock), most of what happens is stretching in the vertical, meaning thickening of the crust. This results in uplift. Exactly the opposite takes place on the other side of the fault, where the displacement gradient along the fault is now of the opposite sign: $\partial u/\partial x = D/L$. Conservation of mass requires that this region of the crust thins, and therefore subsides.

Fault scaling and fault interaction

The study of complex orogens, in which many faults chop up the terrain, has led to an understanding of how faults grow, and how they interact. In extending regions, for example, it appears that once-isolated faults can link to become larger faults on which the throw (or slip) can be even larger. The central Apennines of Italy, the normal faults of Greece, and the rift faults so well preserved and so well imaged seismically in the North Sea basin serve as primary large-scale playgrounds for this sort of work. Smaller scale analogs are to be found in the volcanic tablelands of the Bishop area in the eastern Sierras, where the

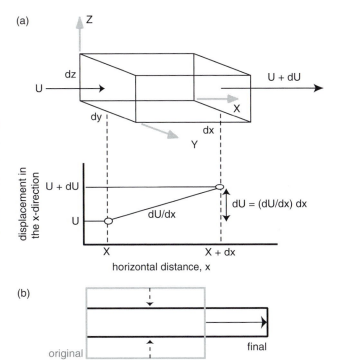

Figure 4.3 (a) Strain results from gradients in displacement; here U is the displacement in the x-direction. (b) Strain in an incompressible medium requires that lengthening in one dimension (full arrow) results in thinning in others (dashed arrows).

Bishop Tuff is caught up in the westernmost Basin and Range extension, and in the grabens of the Canyonlands National Park, in the Utah desert, where blocks are gliding on Paleozoic salt toward an incising Colorado River (Trudgill, 2002).

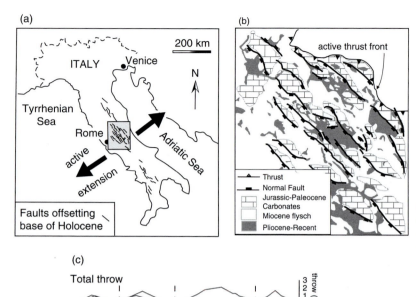

Figure 4.4 (a) Map of central Apennines, Italy, a region of active extension. Faults offsetting base of Holocene are shown. (b) The swarm of normal faults depicted accomplish the extension, generating basins filled with Plio-Pleistocene sediments (courtesy of Greg Tucker). (c) Pattern of vertical throw on several of the faults as a function of distance along strike from NW to SE. Segment boundaries are shown as vertical dashed lines. Throw rates are deduced from dated offset markers, or from total throw assuming that the scarp is 18 ka (after Roberts and Michetti, 2004, Figures 2 and 8, *J. Structural Geology*, with permission from Elsevier).

The key here is that there is a predictable scaling between fault throw, S, and fault length, L: $S = \gamma L$, where γ is a dimensionless constant. Each fault segment displays a preferred ratio of slip to length. As the fault lengthens, the center of the fault slips further. If a fault links with another to become a longer fault, the slip at the link point can then grow rapidly. This scaling can be tested in the field by obtaining slip rates on faults at many points along their length.

Interactions among faults are very well documented in the Italian Apennines, where active extension of the range that dominates the boot of Italy is accommodated by numerous normal faults mapped in Figure 4.4. Each fault shows maximum throw at its center, and declines toward each fault tip. The slip rates deduced by dating of offset markers are of the order of 2–3 mm/yr in the center of the fault. The faults are separated by complex fault ramps that can in turn be broken through by relay faults to allow the now connected faults to act as a single longer fault.

The physics behind these linkages and the evolution of complexly faulted regions lies in the evolution of stresses within the upper crust. We have employed the theory of elastic dislocations to illustrate the vertical displacement fields at the surface of the crust

(e.g., Chinnery, 1961; Mansinha and Smylie, 1971; Gomberg and Ellis, 1994; Okada, 1992). The theory allows calculation of the displacement in all three directions at all points within the material caused by a prescribed displacement on a fault of given geometry (as we have seen in Figure 4.2(a)–(c)). Alternatively, one may drive the deformation by imposing a far-field displacement on the boundaries, and allowing the fault or faults to slip. The calculation is akin to taking a sheet of foam rubber, slitting it in locations meant to represent faults, and then displacing one or another edge relative to the opposite one. These displacements also rearrange the stress field within the material. Imagine that the stresses are uniform in a uniform, flaw-free sheet. But when a discontinuity is imposed, across which no stresses can be transmitted, say, the stresses must be enhanced in the material adjacent to the slit. Stresses are indeed highly altered near the fault tips. We can also assess whether the altered stresses either promote or delay failure on any other plane in the region (e.g., King *et al.*, 1994). Assuming a Mohr–Coulomb failure criterion as the condition in which a fault will slip, one can calculate the "coulomb stress change" on all faults. Research on fault hazards and "stress-triggering" of earthquakes has recently blossomed; each new earthquake, corresponding to a displacement on a fault, alters the likelihood of activity on all surrounding faults (e.g., Stein, 2003; Toda *et al.*, 1998). This method has been employed most dramatically to analyze the 1990s earthquake sequence in the Mojave Desert, and has been used to assess the apparent zippering of the North Anatolia fault in Turkey.

One can also employ these physics on longer time-scales to assess how fields of faults mature through time. In the case of normal faulted regions such as that shown in Figure 4.5, initial faults aligned along strike of one another are the most likely to link through time. Such linkage allows some initial faults in less favorable locations to cease slipping while the motion becomes concentrated in a smaller set of master faults.

Coulomb stress changes

At the core of modern research on both short-term triggering of one earthquake by another (so-called stress-triggering) and on long-term linkage between existing faults, is the calculation of the change in stresses within the crust caused by slip on a fault. In particular, we calculate what has been called the changes in the coulomb failure stress in the crust. But let us explore why this is done. The slip (or failure) on a fault is taken to be analogous to the failure of a granular material. This is captured as the Coulomb failure criterion (see Chapter 10) relating shear stress and normal stress on a plane in the material. This failure criterion traces a straight line on a plot of shear stress vs. normal stress, with an intercept of S (cohesion) and a slope of $\tan(\phi)$ (the coefficient of internal friction)

$$\tau \geq S + \tan(\phi)(\sigma - P) = S + \tan(\phi)\sigma' \qquad (4.2)$$

Recall from structural geology how we can visualize the stresses on an arbitrary plane in the material using the Mohr circle on the same plot, as shown in Figure 4.6.

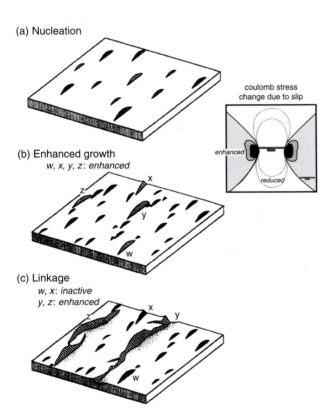

(a) Nucleation

(b) Enhanced growth
w, x, y, z: enhanced

(c) Linkage
w, x: inactive
y, z: enhanced

coulomb stress
change due to slip

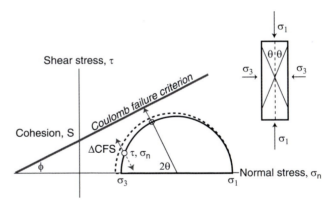

Figure 4.5 Schematic of linkage of normal faults as the crust slowly extends. (a) Initial faults are small, each with elliptical throw profiles. Interaction of faults is governed by the stress changes associated with faulting (inset to right), promoting failure on faults in the grayed zones that are along strike, while reducing it in shadow zones. (b) Linkage extends favorably oriented faults and suppresses growth of others. (c) A few master faults remain once the pattern has matured, while a few faults are now abandoned and cease to move (after Cowie, 1998, Figures 1 and 9, *J. Structural Geology*, with permission from Elsevier).

Figure 4.6 Mohr circle representation of stresses in a rock mass. The Coulomb failure criterion is shown as a straight line with intercept S and slope $\tan(\phi)$. The state of stress on all possible planes with principal stresses σ_1 and σ_3 is shown by the circle. Optimally oriented plane is oriented with angle θ from the maximum principal stress (shown in inset on right). An increase in pore pressure on the fault lowers the effective normal stress, moving the circle to the dashed circle, promoting failure. The change in Coulomb failure stress, ΔCFS, used in assessing stress-triggering of earthquakes, is depicted on an arbitrary plane with original stresses τ, σ_n, and is taken toward failure in one case (solid arrows, moving its state of stress toward the failure criterion), and away from it in another (dashed, moving it into a stress shadow). Such a stress change on an optimally (most susceptible) plane depicted by the polygon would cause slip or failure on that plane.

For given values of the principal stresses σ_1 and σ_3, the shear and normal stresses can be calculated on all possible planes in the material oriented at angles ϕ relative to the maximum principal stress (see insert). These planes trace a circle on the Mohr diagram. If this circle just touches the failure envelope, the plane oriented at ϕ will fail (slip). At any state of stress, represented by σ_1 and σ_3, the plane most susceptible to shear failure (hence, point on the circle) is that closest to the failure envelope.

Stress triggering of earthquakes

Now consider how stresses might change through time in the material, causing changes in the location and/or size of the Mohr circle. Many events can cause changes in stress. Far-field stresses caused by far-field plate tectonic motion will incrementally ratchet up the stresses. Pore pressures in the rock can change, lowering the normal stress on the fault (moving σ_3 to the left). As we will see in the study of soil mechanics and landslides (Chapter 10), we should in fact be using not the normal stress but effective normal stress, in this diagram: $\sigma' = \sigma_n - P$. Finally, slip on an existing fault can alter the stress field in the surrounding material. In general, both the normal and shear stresses are changed within the surrounding material, depending on the sense of slip on the fault. In some regions this combination of changes in stresses ought to favor failure on a given plane, while in other regions it will reduce the likelihood of failure. The calculation of "changes in coulomb failure stresses" is meant to capture this effect. We define the Coulomb failure stress (also called the Coulomb failure function, see Reasenberg and Simpson, 1992) to be

$$\text{CFS} = \tau - \tan(\phi)(\sigma - P) - S \qquad (4.3)$$

It is generally assumed that neither of the material properties S and $\tan(\phi)$ change due to an earthquake. This implies that changes in CFS can be written:

$$\Delta \text{CFS} = \Delta \tau - \tan(\phi)(\Delta \sigma - \Delta P) \qquad (4.4)$$

which is usually simplified further by assuming that changes in P are either negligible or unknown. Graphically, on Figure 4.6, we may visualize this as motion in stress space that is orthogonal to the failure criterion. Movement of the state of stress on a given plane (point on the circle) toward the failure envelope promotes failure, or advances it toward failure from

where it would otherwise have been given its principal stresses. And conversely, movement of the state of stress away from the failure envelope will reduce its likelihood of failure, or retard its time to failure through incremental growth of far-field stresses.

This calculation can be applied in two ways: (1) calculation of stress changes on a specific fault, and hence of a given orientation, and (2) on arbitrary faults in the region. The first calculation can be used to assess whether existing mapped faults are taken toward or away from failure by a given earthquake. For example, Reasenberg and Simpson (1992) assessed changes in coulomb stress on the San Francisco Bay area set of strike-slip faults by the Loma Prieta earthquake of 1989. Such assessments have now become part of the toolbox in earthquake seismology and hazard assessment (e.g., Stein, 2003; Harris *et al.*, 2002). The second calculation is used to analyze aftershocks in the region surrounding an earthquake. Assuming that these are caused by small slip on existing arbitrarily oriented cracks in the crust, and that those that are most favorably oriented are most likely to slip, we simply calculate the change in coulomb failure stress on such hypothetical planes. It was this sort of calculation performed in the aftermath of the 1992 Landers earthquake that convinced the seismological community of the validity of the approach. As we discuss elsewhere, the distribution of slip in the strike-slip Landers earthquake was very well documented. It occurred in the desert. Many cultural and geomorphic markers could be used to assess the slip distribution, which in turn allowed calculation of the expected pattern of stress changes in the region surrounding the earthquake. In addition, the Mojave desert was immediately instrumented with seismometers to listen for patterns of aftershocks. And they occurred in a pattern that beautifully matched that of the calculated changes in Coulomb failure stress (Stein *et al.*, 1992). They occurred dominantly where the ΔCFS was positive and should therefore promote failure, and they were absent where ΔCFS was negative, where the region was thrown into a stress shadow by the slip in the Landers event. One of the surprising aspects of such calculations is that the calculated stress changes are truly small. They are about 1 bar (0.1 MPa), which is very small compared to the stress drop during an earthquake, implying that the Earth's crust is poised at a state of stress that is very close to failure.

Linkage of faults

We are now armed to understand the strategy employed to address how a field of faults matures through time. The changes in the state of stress can also be used to assess how a fault (or crack) will grow. As we illustrate in Figure 4.5, Cowie calculates the change in stress due to slip in a given normal fault, which promotes extension of the fault along its strike, promotes slip on some favorably oriented faults, and throws other faults into a stress shadow. This leads to growth of master faults and cessation of motion on minor faults that effectively lose the battle for access to the proper flavor of stress.

Determination of offsets from modern earthquakes

In the aftermath of an earthquake, a variety of survey tools can be employed to document the deformation pattern. Traditionally, one measures the new horizontal and vertical locations of monuments or markers of some sort whose pre-seismic positions are already known. Simple mapping of offset cultural features remains a rapid, cheap means of documenting slip. Note the offset vehicle tracks in Figure 4.7. The most accurate tools to document

vertical deformation are leveling surveys in which very well calibrated segmented rods are used. In a first-order survey, in order to minimize false signals of uplift or subsidence, rods are used that have the property of not expanding significantly due to temperature changes. The material used is "invar," short for invariant with temperature. Horizontal changes are documented using classic survey methods. The transit of old has been replaced with modern electronic distance measurement methods in a total station that measures both distance and angle. More recently, in one of the more interesting civilian applications of the global positioning satellites (GPS), the vertical and horizontal changes in position of a monument on the ground surface can be documented down to a resolution of 1 mm in the horizontal and less than 1 cm in the vertical.

Both GPS and leveling surveys were employed in the aftermath of the 1989 Loma Prieta earthquake in northern California to document the uplift pattern. The resulting pattern (seen in Figure 18.4) is close to that one might expect from oblique (right lateral plus thrust with west side up) slip on the blind thrust fault that generated the earthquake. The maximum uplift on the hanging wall is offset from the center of the seismically determined fault in the sense that one might expect from adding in a little dose of dog bone. The dog bone pattern is best seen in the slight subsidence zones, which

Figure 4.7 Right-lateral offset of tire tracks in the Mojave Desert across one of the faults (dashed) involved in the 1992 Landers earthquake. It is by mapping such offset features that the slip distribution along the fault can be documented (photograph by R. S. Anderson).

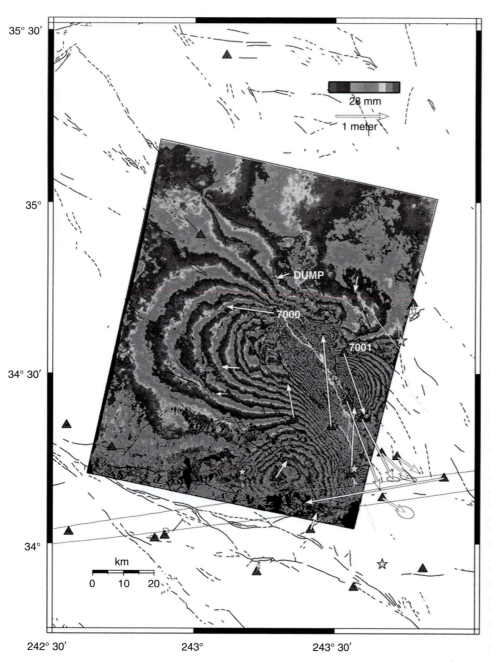

Figure 4.8 InSAR and GPS data covering the 1992 Landers earthquake. Stars mark the locations of the Joshua Tree, Landers, Big Bear, and Hector Mine epicenters. The red triangles denote GPS sites (displacements from Freymueller *et al.*, 1994). The SAR frame covering the Landers earthquake is outlined by a black rectangle. A full color cycle represents 28 mm of displacement. White lines depict traces of the Landers and Hector Mine surface ruptures. Note broad similarity of the inSAR pattern with expected vertical displacements associated with slip on a right lateral strike-slip fault (after Price and Bürgmann, 2002, Figure 2, with permission from BSSA).

occur diagonally from each other in the manner to be expected from a component of right lateral slip.

More recently, satellite imagery has allowed us to document the full displacement field rather than that along pre-existing lines or points of deformation. The method employs two satellite radar images that can be used to construct an interferogram such as that shown in Figure 4.8. The interference fringes correspond to small changes in ground surface elevation (in reality, distances (or "ranges") to the satellite, which does not look vertically). The resolution is such that each fringe is worth only a few mm of uplift or subsidence. As seen by inspection of the deformation field generated by the 1992 Landers earthquake in the Mojave Desert of southern California, this pattern can be quite complex. While technologically complex, this new tool is extremely useful both in remote terrain, where pre-existing survey monuments are unlikely to exist, and in urban terrain where the sharp corners from the many buildings provide good radar reflectors.

But this method requires modern imagery from satellites flying radar wavelength radiation. Most recently, algorithms have been developed that employ optical imagery – Landsat, for example – to assess rates of motion of features on the land (e.g., Michel *et al.*, 1999a,b; Leprince *et al.*, 2008). This promotes use of archived images that reach back several decades. The method requires careful geospatial registration of two images from two times (say prior to and after an earthquake). Once the images are well referenced to many ground control points, the images are compared in great detail. At any point (pixel) in the later image, a patch of surrounding pixels is moved until it best correlates with the earlier image. The motion can be only a fraction of a pixel, and in any direction. The displacement that yields the highest correlation is assumed to represent the motion of the land between the times of the two photographs. Because this process is automated, one can perform this correlation at every pixel in the image, resulting in a displacement map. The method can be employed to document a really distributed displacement of any topographic surface (Leprince *et al.*, 2008), and has therefore begun to be used to study landslides, glaciers (e.g., Scherler *et al.*, 2008), and other features of geomorphic interest.

Paleoseismology

At present, there is no way to predict the timing of an earthquake with enough resolution or confidence to use prediction as a means of reducing the hazards of tectonically active environments. Instead, the main means of limiting the trauma associated with earthquakes is to build well against them. In order to assess the degree of seismic hazard at a site, and hence the standards to which buildings must be built, one must know the history of recent earthquakes at that site. When was the last rupture? At what interval does a fault rupture, regularly or not, i.e., what is its recurrence interval? How much slip occurs in an event and what reach of a fault breaks in individual earthquakes? These are the ingredients for the assessment of seismic hazards, as they go into the calculation of the expected magnitude of an earthquake, and the amplitudes and durations of shaking in the nearby landscape. This is the realm of paleoseismology, and has been summarized

well in recent books (e.g., Yeats *et al.*, 1997). We merely touch upon some of the techniques here.

Strike-slip faults

One of the more obvious markers of slip on a fault is the offset of linear features by strike-slip motion on a fault. The right lateral offsets of fences across the San Andreas Fault (SAF) in the aftermath of the 1906 San Francisco earthquake were famously photographed by G. K. Gilbert. But what about slip on faults that occurred prior to human roads and railroads and fence lines? Here we must use geomorphic markers. Many geomorphic markers are indeed linear – stream channels, edges of terraces, coastlines. Or at least their original geometry can be deduced from their remains. An example of this is well documented in the Carrizo Plain along the San Andreas Fault as it passes west of the southern San Joaquin Valley. Here the bed of a small ephemeral creek running from a subtle topographic high on the east side of the valley, and informally named Wallace Creek after the geologist Robert Wallace, is offset by the fault. We know that the slip of the SAF along this reach of the fault was 11 m in the 1857 Fort Tejon earthquake (Sieh, 1978). But one can also find on the west side of the fault numerous dry channel features that appear to be beheaded former extensions of Wallace Creek, as shown schematically in Figure 4.9. This site and nearby sites along the fault have been trenched to

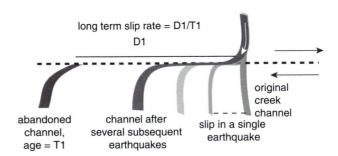

Figure 4.9 Schematic of a stream channel affected by a right lateral strike-slip fault. The original channel crosses the fault and is etched into a datable deposit of age T1. Subsequent earthquakes offset the channel right-laterally, but remain attached to the main channel until the segment along the fault is too long to maintain. An older abandoned beheaded channel allows the long-term slip rate to be determined. At Wallace Creek along the San Andreas Fault, Sieh and Jahns (1984) showed $D1 = 130$ m, and $T1 = 3800$ years, yielding a long-term slip rate of 34 mm/year.

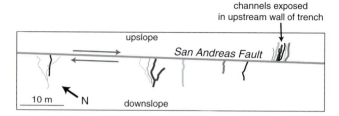

Figure 4.10 Offset channels in Carrizo Plain, California, showing magnitude of offsets on the San Andreas Fault (shaded bar) during major earthquakes. While channels are tightly clustered in a modern channel on the upslope wall of the scarp, buried channels on the lower scarp unearthed in excavations reveal offsets averaging ~7 m per event (after Liu *et al.*, 2004, Figure 2).

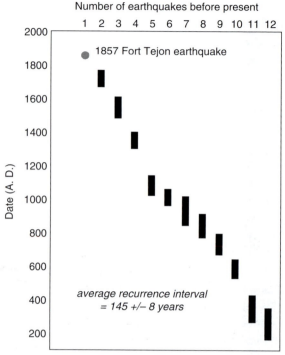

Figure 4.11 Paleoseismic record derived from Pallett Creek section along the San Andreas Fault, southern California. Well-dated vegetation provides the timescale for each of these events. The last event to rupture the site was the Fort Tejon earthquake of 1857 (after Sieh and Jahns, 1984, Figure 16, with permission from the American Geophysical Union).

uncover the ages of these slip events. It was from ^{14}C dating of the deposits into which the creek is incised, and those exposed by trenching the channel floors of the abandoned channels, that the slip history of this segment of the San Andreas fault was first documented (Sieh and Jahns, 1984). They found that the long-term slip rate was 34 mm/yr. This was one of the first cases in which a fault was trenched in order to deduce its slip history. This is now a primary tool in paleoseismology. The field of paleoseismology was effectively founded by Kerry Sieh working on both the Wallace Creek site and that at Pallett creek further south, where even more datable events were recorded (Sieh, 1978). In Figure 4.10 we diagram a recent set of excavations that reveals the distribution of slip magnitudes for 13 past events. The average slip per event here is 7 m. At Pallett Creek, the stream is temporarily ponded by the motion of the fault, and vegetation associated with these ponding events allows dating of the deposits exposed by trenching. This revealed the history of slip on an event-by-event basis, as reproduced in Figure 4.11. The history shows not only the long-term slip rates, but the variability of slip per event. This remains one of the best-documented slip histories of any fault.

The ingredients needed to determine the slip rate of a strike-slip fault are therefore the offset of a geomorphic marker, and dating of the surface associated with that marker. The art is finding such features in the environment, and carefully employing the best available dating method. Many subtle problems lurk below the surface. Geomorphic markers are often not straight but wiggle in map view, introducing error in the documentation of total slip. As faults often lie at the base of mountain ranges, it is features that run

normal to the range that are cut by the fault. These include both moraines (if the climate is such that glaciers once poked out from the mountain range), but more often river channels. In some ways the Wallace Creek example is ideal. It is a small stream, the individual slip events are huge, and the long-term rate fast. Cowgill (2007) has addressed many of the issues one faces in dealing with offset fluvial forms in less ideal cases. He and coworkers have revisited sites that involve rivers crossing the Altyn Tagh Fault on the Tibetan Plateau, a left lateral strike-slip fault that is responsible for far-field deformation of the plateau as the Indian subcontinent continues to ram into Asia. This study suggests that the slip rate is 9.4 mm/yr, which is in agreement within error with recently measured rates using GPS.

Normal faults

The recurrence interval of normal fault-generated earthquakes is often documented from logging of

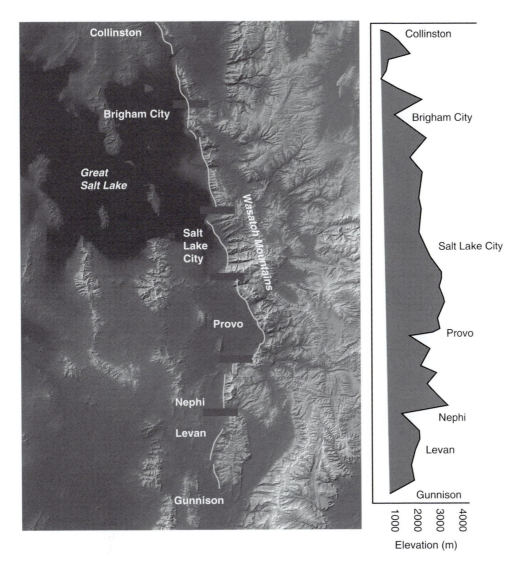

Figure 4.12 The segmentation of the normal faulting along the normal faulted Wasatch Mountain front, eastern Utah. Bars represent proposed segment boundaries. Right: topographic profile along crest of the Wasatch Range shows minimum heights at segment boundaries (after Schwartz and Coppersmith, 1984, Figures 1 and 10, with permission from the American Geophysical Union).

trenches cut across fault scarps. The dip-slip component of coseismic offset is deduced from the magnitude of offset of strata within the deposit, and timing is documented by dating buried soil horizons that bracket these displacements. Such work in the normal-fault bounded Wasatch Range in Utah shown in Figure 4.12, at the eastern edge of the Basin and Range province, has had strong implications for the generation of mountain ranges by numerous earthquakes. Detailed and exhaustive trenching studies in the 1980s along the Wasatch front near Salt Lake City (Machete *et al.*, 1991) suggest that this faulted front appears to consist of a set of roughly 10 km long segments that have ruptured independently, with their own slip pattern and their own recurrence interval. This gave rise to the notion of a characteristic earthquake (Schwartz and Coppersmith, 1984). These authors also noted that the skyline of the Wasatch Range along the fault showed significant dips at their proposed segment boundaries, and peaked in mid-segment (Figure 4.12). In dip-slip dominated systems it may well be that the segments of which the faulted mountain front consists fire off with a particular pattern of slip that varies relatively little from one seismic event to the next. By repeating such earthquakes, the topography of the resulting mountain range associated with each segment should mimic the slip pattern.

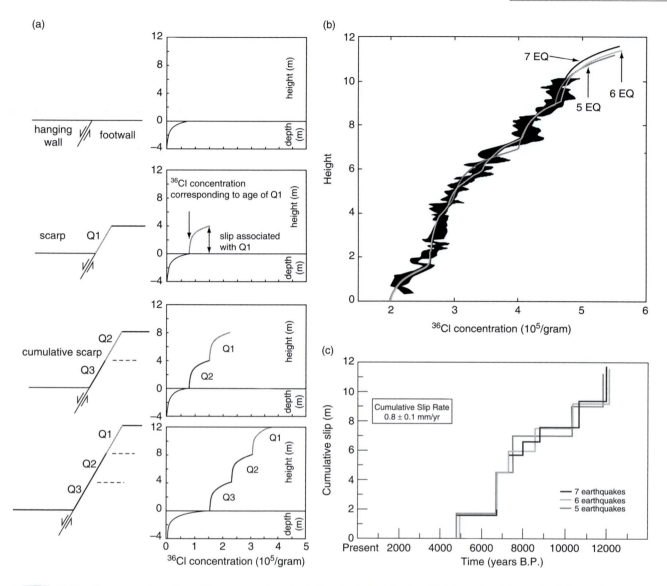

Figure 4.13 Use of cosmogenic radionuclide concentrations to deduce fault slip histories. (a) Schematic of the method, in which a slip event on the normal fault exposes a new segment of the rock surface on the footwall to cosmic ray bombardment. Curvature of the CRN concentration on the newly exposed surface reflects production in the subsurface under increasingly thick hanging wall edge. Subsequent increase in concentration is linear with time. Multiple slip events will result in multiple curved segments. (b) ^{36}Cl concentration profile on footwall of the a fault block in Italian Apennines. (c) Slip history implied by best-fitting models of the ^{36}Cl profile, showing five-, six-, and seven-event scenarios (after Palumbo *et al.*, 2004, Figures 4 and 5).

Here the branching of the master fault at depth into individual segments can promote the segmented behavior at the surface. The size and geometric nature of the relay zones between fault segments governs their ability to form barriers to linkage at the surface (e.g., Soliva *et al.*, 2008). If linkage occurs, as in the strike-slip Landers earthquake of 1992, the total slip during the event can be much higher, and the magnitude of the earthquake will be much greater. The generality of the notion of a characteristic earthquake to all fault geometries has been questioned, as some segments of

major strike-slip systems have yielded data on slip patterns and timing of earthquakes that fail to show consistently repeated events (see Ward and Goes, 1993).

In the central Apennines of Italy yet another method can be employed to deduce the timing and sizes of recent slip events. The aridity of the climate allows great preservation of the fault scarps on the upthrown limestone footwalls of the normal faults responsible for these mountain ranges. As summarized in Figure 4.13, using concentrations of cosmogenic ^{36}Cl on the rock of the fault scarp, researchers have documented a record of 5–7 slip events in the Holocene scarp. During interseismic intervals, the exposed scarp accumulates ^{36}Cl at a rate dictated by the elevation and slope of the site. In the shallow sub-surface, however, rock on the footwall accumulates ^{36}Cl at a rate that declines exponentially with burial depth. Upon the next slip event, this exponential profile is instantaneously raised to become subaerial, and the process repeats. The resulting series of linked exponential segments of the concentration profile with height up the scarp, clearly visible in Figure 4.13, can then be inverted for the best-fitting series of slip events. These in turn can be used to assess recurrence intervals, time since the last event, and so on; all such information is important in assessing seismic risk of an area.

Megathrust faults

The largest earthquakes on Earth are related to the megathrust faults between the downgoing subducting plate and the overlying plate in subduction zones. No one is going to trench that fault! Instead, we must rely upon a cycle of vertical motion generated by the subduction-related stick-slip motion at the plate interface. The data indicate that vertical strain along the boundary can be explained by the following cycle: during interseismic periods, the plate interface is locked, and the overlying plate is dragged downward, bending as it does. This bending can produce uplift at inland sites. As sketched in Figure 4.14, during the coseismic slip, the locked patch slips and the overlying plate snaps back upward near the trench and downward at inland sites that had been arched. The sign of the vertical cycle depends upon how far the site is from the position of the locked patch at depth. Two types of vertical motion indicators have been employed to detect this cycle, each championed by imaginative researchers.

In the Pacific Northwest of North America, along the coastlines of Oregon and Washington, Brian Atwater of the USGS was the first to demonstrate the utility of the stratigraphic record in coastal estuaries. These low-energy environments protected from major wave action are dominated by muds and by vegetation that is subtly adjusted to elevation above or below sea level. What he and other scientists see in bay after bay is a sequence of salt-water tolerant peat that slowly gives way, upward in the stratigraphy, to fresh-water peats, which are then capped at an abrupt transition by salt-water tolerant species. Sometimes a

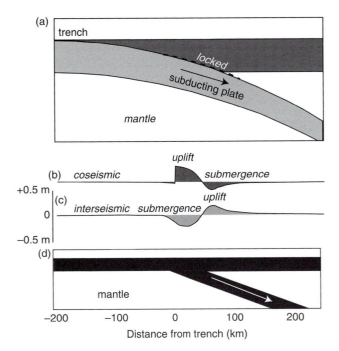

Figure 4.14 Expected patterns of coseismic and interseismic vertical motion associated with motion on the megathrust fault of the subduction zone. (a) Schematic of the locked zone that generates stick-slip motion. (d) Abstraction of this schematic into simpler geometry on which a slip dislocation is used to calculate coseismic (b) and interseismic (c) vertical displacement. Scale of motion is of the order of 1 m, and the patterns of motion are nearly the inverse of one another so that long-term displacement is minor (after Natawidjaja *et al.*, 2004, Figure 18, with permission from the American Geophysical Union).

sand layer is preserved at this interface. In places, the sequence is repeated many times within the stratigraphy one can exhume by trenching coastal deposits. How do we interpret such records? The Pacific Northwest coastline is on the hanging wall of the subduction zone as the Juan de Fuca plate dives beneath the North American Plate. Given the distance of the Pacific Northwest coastline from the trench, locking of the plate interface in the seismic zone will cause an upward bending of the hanging wall and the coastline above it, while the interface is locked, all of which is undone when the slip occurs. The resulting geologic record is summarized in Figure 4.15. The interseismic signal is then slow emergence of the land relative to the ocean, and the coseismic signal is rapid dropping of the land relative to the sea. Because peats can be dated using radiocarbon, we can date the seismic events. The same vertical cycle is documented in the diatom assemblages found in the same deposits: slow land emergence is followed by rapid subsidence.

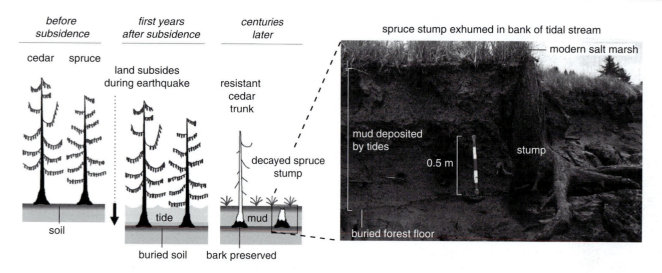

Figure 4.15 Evidence of past megathrust earthquakes associated with subduction zones. Drowning of coastal forests by coseismic subsidence results in rapid death of trees, creating ghost forests. Subsequent slow burial by fine sediment preserves roots and outer bark of these trees (seen at right, from Naselle River, Willapa Bay, Washington), allowing precise dating of the earthquake. Accumulation of the salt marsh can be tracked using ^{14}C on peats, which can switch from salt-water tolerant species to fresh water species as interseismic deformation proceeds (after Atwater *et al.*, 2005, p.17, USGS Professional Paper 1707, Chapter 1).

In addition, trees growing close to sea level during the interseismic emergence of the land are abruptly killed when the land subsides, taking their roots into brackish water. Trees so killed in the last big earthquake to hit the Pacific Northwest can be dated to the year using dendrochronology, indicating an event in 1700. But the story is even more interesting, and even more international (Atwater *et al.*, 2005). This record of a large subduction-related earthquake in the Pacific Northwest is confirmed by written records of a tsunami hitting the east coast of Japan on the evening of January 26, 1700. That it arrived with no seismic warning implies that the event occurred elsewhere – in this case across the entire Pacific Ocean basin. A snapshot of a recent simulation of the seismically generated wave field is reproduced in Figure 4.16, just as it impacts the Japanese shoreline.

In historic time, one of the two largest earthquakes to occur was the great 1964 Alaskan earthquake, which ruptured a large segment of the Aleutian arc shown in the map of Figure 4.17. Dating of buried peat sequences from coastal settings around the Aleutian arc has revealed the timing of many past earthquakes on this subduction margin. The work of many researchers is summarized in Carver and Plafker (2008). For example, as revealed in the inset to Figure 4.17, the sections in the Prince William Sound region lying on the eastern margin of the 1964 rupture reveal the timing of nine

past earthquakes over the last 5000 years, the most detailed record coming from the subsiding Copper River delta.

In the aftermath of the 1964 Alaskan earthquake, tide gage records have been used to document not only the coseismic pattern of rock uplift, but the subsequent post-seismic pattern (Gilpin, 1995). As you can see from the map in Figure 4.17, Kodiak Island lies astride the hinge line between uplift and subsidence. Its complex coastline provides many sites at which one can monitor tides. The coseismic pattern was one of dramatic northward tilt of the island that aided in constraining the position of the locked patch at depth, and the magnitude of the slip on it (20 m!). The post-seismic tidal pattern appears to support later aseismic slip on a patch further down the plate interface (Gilpin *et al.*, 1994).

In Japan, both tide gage records and repeat leveling surveys have been used to document the pattern of both interseismic and coseismic deformation (Hyndman *et al.*, 1995). In lines running normal to the trench, and crossing the island of Shikoku, leveling data support the concept that strain in the interseismic period can be attributed to a locked patch at depth on the plate interface whose lower downdip extent is controlled by temperature. Above a certain temperature, total locking transforms to a transitional zone and then to free slip. Given the great

Figure 4.16 Simulation of the tsunami generated by the January 26, 1700, megathrust earthquake on the Cascadia margin, shown as it reaches Japan (simulation by Kenji Satake; cover image in Atwater *et al.*, 2005, USGS Professional Paper 1707).

earthquakes of 1944–1946 in this region, the period immediately before the quakes can be assessed as "late interseismic" while that immediately after the quakes is "early interseismic." The two patterns of deformation differ in detail but not in basic form, implying evolution of the extent of the locked patch through time. The coseismic pattern of vertical motion almost perfectly undoes the interseismic motion, as can be seen well on Figure 4.18.

Working on a subduction margin a long way from the Pacific Northwest, Kerry Sieh has shown that subtle vertical movements of the land relative to the ocean can also be detected using corals. As discussed below, corals are restricted to a latitudinal band near the equator. In particular, individual coral heads that are more or less elliptical in plan view, and that are a meter or more in height, can be used to document the vertical movement along the Sumatran coastline mapped in Figure 4.19. These coral heads, also called micro-atolls, display annual growth bands that allow great temporal resolution in the record. In stable water levels, the coral heads will grow only outwards. If relative sea level rises, they will grow upward, and if it drops, subaerially

exposed coral will die. Sieh and coworkers (e.g., Natawidjaja *et al.*, 2004) have documented the seismic cycle in micro-atolls along the Sumatran arc by sawing the coral heads along vertical planes, noting the pattern of growth, and carefully dating the growth bands corresponding to changes in growth pattern. The slow interseismic submergence and rapid coseismic emergence are well recorded in the coral heads, an example of which is reproduced in Figure 4.20. Much like the bay muds employed first by Brian Atwater along the Pacific Northwest of North America, by constraining the timing of earthquakes along the margin, they were able to document the pattern of megathrust earthquakes, including the time since the last great earthquake. These define the several patches of the megathrust shown in Figure 4.19. The great Sumatran earthquake of 2004 that spawned the tsunami responsible for the deaths of more than 200 000 people was therefore of no great surprise. Considerable effort is now being spent to document further the spatial and temporal pattern of great megathrust earthquakes on the Sumatran plate boundary, and to put in place both educational materials and warning systems that can

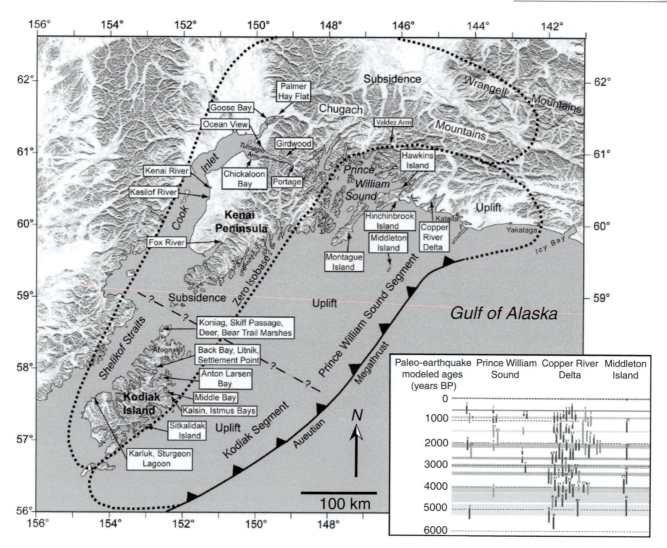

Figure 4.17 Location map of paleoseismic studies and regions of coseismic uplift and subsidence of the great 1964 Alaska earthquake that ruptured the eastern Aleutian subduction zone (after Carver and Plafker 2008, Figure 1). Inset shows plots of radiocarbon ages for evidence of paleosubduction earthquakes at paleoseismic study sites in the Prince William segment of the eastern Aleutian subduction zone. Vertical dark bars show 2-sigma ranges for minimum limiting ages, lighter bars denote maximum limiting ages. The number above each age bar reflects the position in the paleoseismic sequence of the dated sample based on the local stratigraphy, with 1 being the most recent paleoearthquake. The horizontal bars are overlap ages for paleosubduction earthquakes from sequence age models for Kodiak, Girdwood, and the Copper River delta sites (after Carver and Plafker, 2008, Plate 2).

avert the deaths associated with the inevitable repeat of great earthquakes (Sieh, 2006).

Long-term deformation: cumulative displacement deduced from offsets of geomorphic markers

On most subduction margins, the net displacement over a seismic cycle is not great. The land bobs up and down or down and up, and does not accumulate elevation. This is not the case in other settings, where repeated earthquakes result in significant topography. The deformation averaged over many seismic events can be deduced by appeal to deformation of geomorphic features whose original geometry one can constrain. These therefore serve as markers against which deformation can be measured. If the age of these features can be constrained as well, we can divide the total displacement of the feature by the age to obtain the deformation rate field. The most obvious and most useful of these features are shorelines, either of lakes or of oceans. These are generated by waves impacting the coastlines, leaving either beaches, or notches in bedrock, and in low-latitude coastlines, the generation of coral platforms. In such cases, these features record the sea level at the time of formation of the feature, and were originally horizontal. Subsequent warping of these features therefore reflects non-uniform rock uplift. In the ideal case, if the age of the geomorphic feature is known, then one may infer the deformation rate, and hence slip rate on the responsible fault or faults. Marine terraces have been used as tectonic markers for many years. Unfortunately, not all active structures we might wish to explore happen to be adjacent to lakes or oceans. We must therefore use rivers as our markers, whose original courses have some knowable profile or plan view form. We will illustrate with examples of both coastal and fluvial worlds.

Consider a simple case of a fault outcropping on an alluvial fan shown in Figure 4.22. Total throw on the fault can be deduced by assessing the offset of the fan surface, and the time since the fan began to be offset can be documented by dating the fan surface. Geomorphic processes act to diffuse the scarp form, taking material from the convex upper corner and depositing it in the concave lower corner. How round the scarp profile looks (formally, its curvature) will therefore depend upon the time since the fault broke the fan surface, and the efficiency of the geomorphic processes in moving the materials of which the fan is composed. The resulting curvature has been used to estimate ages of fault scarps and of offset shorelines in the Basin and Range province of the US (e.g., Hanks *et al.*, 1984; Hanks and Wallace, 1985), a calculation that can only be accomplished if the geomorphic efficiency, captured by the topographic diffusivity, can be assigned.

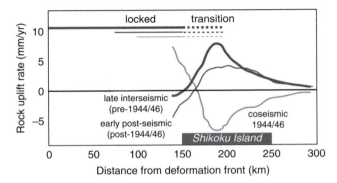

Figure 4.18 Leveling data from Nankaido transect, Shikoku Island, Japan. Interseismic and coseismic vertical motions are opposite in sign, reflecting locking and release of a locked patch on the plate interface. Coseismic data from the 1944/1946 earthquakes. Data from the post-earthquake interval reveal the pattern of early interseismic deformation, while data from the pre-earthquake reveal the pattern of late interseismic deformation. Location of base of locked patch coincides with modeled depth of 350 °C at the plate interface (redrawn from Hydnman *et al.*, 1995, Figure 13, with permission from the American Geophysical Union).

Marine platforms

Long-term uplift of portions of the California coastline results from emergence of the Coast Ranges. Here the rock acts as a tape recorder of sea level changes wrought by glacial–interglacial cycles of global ice volume (hence, the global ocean volume). The recording is literally etched into the rocky margin by wave attack on the edge of the continent, producing the abrasional platforms sketched in Figure 4.23. This conceptual picture of the generation of marine terraces by long-term steady uplift of the rock and oscillation of sea level was first suggested by

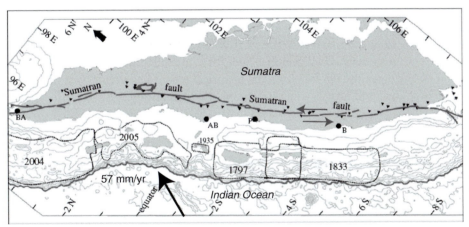

Figure 4.19 Map of Sumatra and patches of the underlying megathrust fault that have failed in recent centuries, as deduced from microatolls on the islands outboard of the mainland. The 1833 event was likely between 8.7 and 8.9 in magnitude. Sunda trench between subducting Indian-Australian plate and overlying Eurasian plate is shown as a red line. The lateral Sumatran fault takes up the strikeslip component of the plate convergence (after Sieh, 2006, Figure 4, with permission from The Royal Society).

(a)

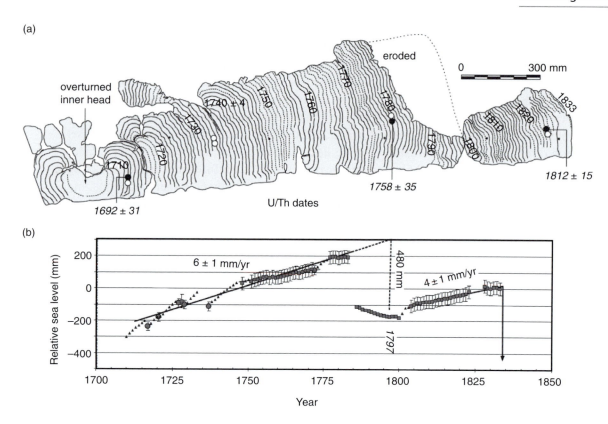

(b)

Figure 4.20 (a) Detailed cross section of microatoll of Porites coral showing relative sea level changes before final emergence in 1833 seismic event. (b) Steady upward growth of the coral head documents relative sea level drop associated with interseismic submergence of the land of 6 mm/year before and 4 mm/year after a late 1700s (1797?) event. Three U/Th dates constrain the ages of the corals and confirm that annual band counting is accurate (after Natawidjaja *et al.*, 2006, Figure 16, with permission from the American Geophysical Union).

Box 4.2 The Himalayan front

One other tectonic setting deserves attention: continent–continent collisions. The Himalayan orogen reflects the collision of the Indian subcontinent with Asia. Initial collision is thought to have initiated around 40 Ma. The current rate of convergence, as deduced from GPS instruments across the orogen, is roughly 25 mm/yr. Some fraction of this convergence is taken up on discrete crustal structures, as revealed by growth of anticlines associated with the main frontal thrust (see Figure 4.26 below). The earthquakes produced by slip on these structures can be devastating because the region is home to many millions of people, and in general they cannot afford to build strongly against the shaking. The historical record has been compiled from British colonial and Indian accounts, and reveals a smattering of M6–M7 earthquakes, many of which killed tens of thousands of people. But when one sums the slip associated with these events, most of the convergence is missing, unaccounted for, as summarized in Figure 4.21 (Bilham *et al.*, 2001; Bilham and Ambraseys, 2004). Recent trenching of the thrust faults at the very southern margin of the orogen reveals that much larger earthquakes than have occurred historically must occasionally rock the sub-Himalayan landscape. These are likely magnitude 8–9 events, the most recent of which was in 1505.

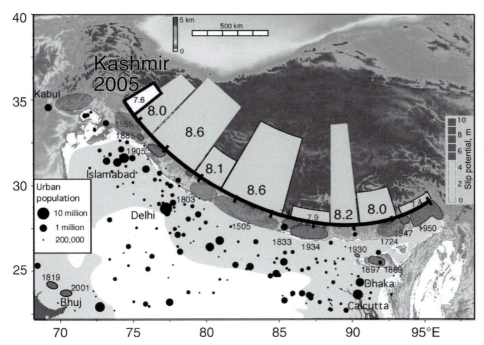

Figure 4.21 Earthquakes along the Himalayan arc. Historic earthquakes are shown in shaded ellipses with dates, and the subsequent accumulated slip potential on the segment translated into predicted earthquake magnitude. Note the proximity of large population centers on the Indian subcontinent, many with populations exceeding one million people. The Kashmir earthquake of 2005 is shown, which was responsible for over 80 000 deaths (after Roger Bilham website, 2008, with permission: http://cires. colorado.edu/~bilham/).

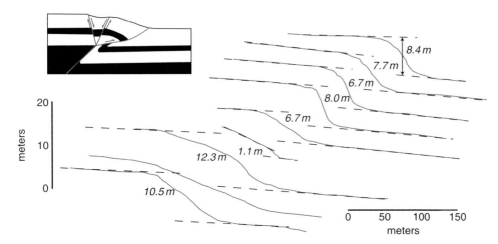

Figure 4.22 Topographic profiles across a thrust fault scarp (inset) crossing the northern piedmont of the Tien Shan mountains. Vertical offset of the terrace surface is deduced from offset of projected planes from the surveyed surfaces. Roughly 10 m vertical offsets of the 10 ka surface suggest vertical offset rates of roughly 1 mm/yr (after Avouac *et al.*, 1993, Figure 7a, with permission from the American Geophysical Union).

C. S. Alexander (1953), working in the Santa Cruz edge of Monterey Bay, California (see Chapter 18). While at the time he knew neither the origin of the sea level fluctuations nor the source of rock uplift, the conceptual picture remains valid. Documentation of the pattern of long-term uplift rate has since been accomplished along many coastlines. In northern California, the Kings Range is emerging in association with complex tectonic setting of the migration of the Mendocino triple junction. The marine platforms along this coast show great variation in height above modern sea level, suggesting a welt of uplift that

moves northward with the triple junction. It is quite common that these platforms are difficult to date. Although there are a few exceptions, one is generally happy to be able to date one of the platforms, and assign ages to the others simply by appeal to steady uplift. Modern platforms consist of an abraded rocky bench, called the abrasional platform, backed by a seacliff, the base of which is called the inner edge, or the wavecut angle. As sea level drops relative to the land, reflecting the onset of another ice age, this platform is mantled by beach sands. The trick is to date the beach sands marking this abandonment. In order

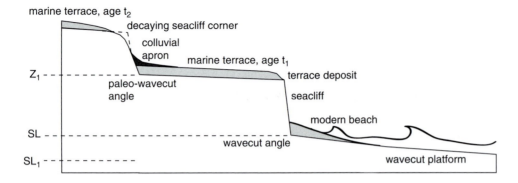

Figure 4.23 Anatomy of a modern rocky coastline. The intersection of the modern seacliff with the wavecut platform, often mantled with modern beach sands, defines the wavecut angle. This is a proxy for modern mean sea level, just as wavecut angles on paleo-seacliffs are proxies for past sea levels. The elevation of ancient wavecut angles must be determined by exposures through the colluvial apron at the base of the decaying paleo-seacliffs and marine terrace sands, or by extrapolating the platform inland from surveys of the terrace. The rock uplift rate can be calculated from the difference in elevation between the paleo-wavecut angle (here Z_1) and the sea level at the time of its formation (here, SL_1), divided by the age, t_1.

to assign a long-term rock uplift rate, one needs to know (1) the modern elevation of the platform inner edge (as illustrated in Figure 4.23, this is sometimes difficult to find beneath both beach sands and colluvium at the base of the ancient seacliff), (2) the age of the platform, and (3) the sea level at the time of erosion of the platform. The first is the easiest, and can come from river gashes cut through the marine terraced landscape, from digging trenches at the base of the cliff, or most commonly by projection of the platform tops to the base of the paleo-cliff, and subtraction of the cover deposit thickness. Dating the platform requires finding something to date. As we see below, coral platforms can be dated using U/Th. Solitary corals, such as *Belanophilia elegans*, can be dated on higher latitude sites (although their range does not extend much above 40°N). More recently we have used [10]Be profiles to date the terrace deposit sands themselves (see Chapter 18). But how would we know the sea level at the time of platform formation? We must know this independently. It is commonly the case that we know the age of one terrace (at best), in which case we proceed by assuming that the rock uplift has been steady. As summarized in Figure 4.24, this is well illustrated in the Mendocino coastline of northern California, where the uplift is caused by the migration of the Mendocino triple junction. In some special cases, it turns out that we can invert the problem to solve for sea level if we know well both the uplift rate and the ages of the platforms. This is best done on coral platforms, as discussed

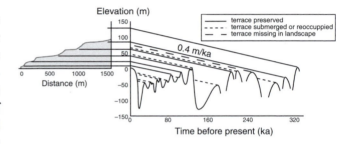

Figure 4.24 Terrace prediction diagram. Left: surveyed elevation profile of marine terraces near Mendocino triple junction, Bruhel point, California, with inner edge elevations noted as horizontal lines. Right: sea level history. If rock uplift rate has been steady, each sea level highstand, corresponding to an interglacial period, will result in generation of a marine terrace. Parallel solid lines represent predicted steady uplift histories of inner edges. Dashed lines represent highstands that generated terraces that are either submerged, or have been reoccupied by later terraces. Implied steady uplift rate at this location on the coast is 0.4 mm/yr (after Merritts and Bull, 1989, Figure 2).

below, from which we have extracted a sea level curve going back many glacial cycles.

In southern California, slip on a buried reverse fault has been shown to result in uplift of the Palos Verdes peninsula, which juts into Santa Monica Bay. The blind fault generates an anticline that eventually emerges above sea level. As shown in cross section in Figure 4.25, as the land emerges, it is notched at each sea level highstand to produce a set of 13 marine terraces that appear as bathtub rings around the peninsula. The pattern of elevations of these terraces can be used to constrain both the geometry of the

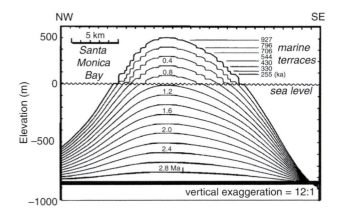

Figure 4.25 Simulation of growth of the Palos Verdes anticline by slip on the blind Palos Verdes thrust fault, and the formation of marine terraces as it emerges above sea level. Here the view is along the strike of the fault, parallel to the anticline axis, and the curves represent the anticline at 200 ka intervals. Marine terraces are all assigned 500 m widths, and form bathtub rings around the peninsula (after Ward and Valensise, 1994, Figure 5, with permission from the American Geophysical Union).

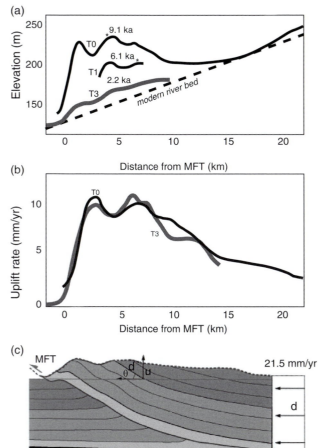

Figure 4.26 Deformation of alluvial terraces along the Bagmati River, Nepal. (a) Deformed terrace profiles and ages of terraces deduced from radiocarbon. Note that the patterns up-deformation look similar (after Lave and Avouac, 2001, Figure 4a). (b) Rock uplift pattern obtained by dividing terrace height above modern river by the terrace age. Uplift rate appears to have been steady in the Holocene. (c) Interpretation of deformation pattern as being caused by horizontal shortening at a rate $d = 21.5$ mm/yr, transformed through the dip of the structure into the uplift rate, u (from Avouac, 2003, Figure 19(a), and Figure 21(b and c), with permission from the American Geophysical Union).

fault (dipping 67°, and slipping from 6–12 km depths), and the 3–3.7 mm/yr rate of oblique dextral-reverse slip on it.

River profiles

In settings that lack baselevel markers such as shorelines and coastal terraces, one must rely instead upon river profiles as the metric against which deformation is measured. Here we rely upon the fact that in the absence of variations in rock uplift, most river profiles display a monotonically declining slope with area (see Chapter 13). If a river encounters a region in which rock is being uplifted beneath it, the river profiles etched into the rock walls of the valley will become warped as the rock into which they were etched rises above the incising river. The formation of the fluvial terraces themselves is caused by lateral erosion of the channel floor, typically in times of alluvial aggradation, which are then abandoned as the river again narrows and incises downward. The terraces so formed are called strath terraces. High on valley walls, the scraps of these strath terraces can be either barren rocky platforms displaying the markings of fluvial erosion, or river gravel-capped bedrock platforms. In either case, the form can be dated using [10]Be or some other dating method. The pattern of river erosion can be documented by noting the pattern of elevation difference between the modern river profiles and the ancient strath. This pattern is commonly assumed to reflect the pattern of rock uplift along that reach of channel. This pattern of warping of terraces can be used to document the rock uplift pattern. As summarized in Figure 4.26, along the frontal thrust of the Himalayas, Lave and Avouac (2001) have shown that the warping produces a growing anticline that implies a long-term convergence of 21.5 mm/yr.

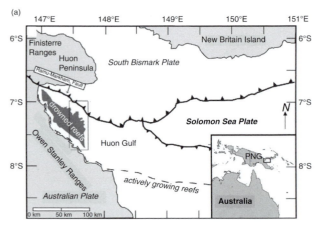

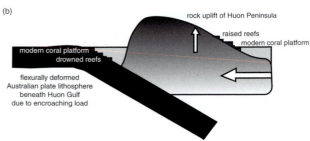

Figure 4.27 (a) Map of setting of Huon Peninsula of Papua New Guinea, showing sites of both raised and drowned coral reefs (after Webster *et al.*, 2004b, Figure 1, AGU). (b) Origin of both emergent (raised) and submergent (drowned) coral reefs associated with growth of the Finisterre Range, Papua New Guinea. Rock uplift associated with suture of an old island arc with the edge of the Australian plate (not shown) causes rock uplift of the Huon Peninsula, and abandonment of coral reefs. Conversely, reefs are drowned on the loaded plate beneath the Huon Gulf whenever the sum of the rate of submergence of the plate and the rate of rise of sea level exceeds the rate at which corals can grow (redrawn from Galewsky, 1998, Figure 1, with permission from Wiley-Blackwell Publishing).

The special case of corals

We have already seen how individual modern coral heads can be used as paleoseismological tools. Reef-growing corals grow within a few meters of sea level. Their occurrence in past deposits is therefore a proxy for that portion of the landscape having been very near sea level at the time of deposition. Those that grow closest to sea level are the best markers, just as those taxa that live for a short period of time and go extinct abruptly are the best index fossils. While reef-forming corals dominantly grow in a latitudinal band of 25°N to 25°S (with exceptions up to 35°N where influenced by warm ocean currents), there is indeed much of tectonic interest that occurs in these

latitudes. The entire complex of island arcs in the western Pacific lies in this band. We illustrate here their use as both macroscopic benches, as reefs as a whole – the equivalent of marine abrasional platforms – and as individual coral heads that preserve annual banding akin to tree rings. The former are used as long-term geomorphic markers to constrain the rates and patterns of rock uplift and submergence, while we have already discussed the latter used as paleoseismic sensors.

As we have seen it is usually very difficult to date marine terraces, especially if they are older than the ^{14}C tool will allow us to date. The real strength of the coral platform is that it can be dated using U/Th methods, back to several hundred thousand years ago, meaning that we can date coral platforms that are many sea level cycles old.

Perhaps the most famous site for the study of coral terraces is the Huon Peninsula of Papua New Guinea, which is mapped in Figure 4.27. This peninsula is rapidly rising as an old island arc is being sutured onto the northeastern margin of the Australian plate (which includes most of New Guinea) to create the Finisterre Range (Abbott *et al.*, 1997). The suture is progressing eastward, extending the range and the peninsula that marks its easternmost end as it emerges from the sea. The corals are beautifully expressed as a series of discrete benches. Early mapping of terrace elevations, shown in Figure 4.28, revealed that the elevation of any particular bench varies systematically along the peninsula, implying a pattern of uplift that is not uniform. The rate of uplift can then be deduced from dating the coral, and from knowing independently what the sea level was at that time. The rate of rise on the Huon peninsula is so high that many minor sea level highstands are recorded. It is as if the system behaves as a strip chart recorder and the strip chart motor is running at a high speed, allowing a high fidelity resolution of the record. It is for this reason that the Huon record has often been turned inside out to be interrogated as a record of sea level rather than of uplift. The logic is that if we can assume that at any particular site along the coast the uplift rate has been steady in time, then we can deduce from the elevation and the age of any other terrace what the sea level must have been at that time. We know the uplift rate from dating, say, the MIS 5e terrace (120 ka) whose elevation we can measure, z_{5e}, and we know from independent information elsewhere

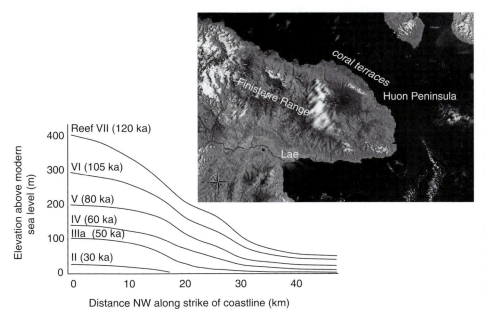

Figure 4.28 Huon Peninsula, Papua New Guinea, and its famous coral terraces. Terraces on the northern shore of the peninsula are being uplifted at roughly steady but non-uniform rates as the Finisterre Range rises due to collision of an island arc with the edge of the Australian continent. The coral terraces are datable using U/Th methods, and have been used to reconstruct sea level histories over the last several glacial cycles. Their rapid rate of rise provides great detail in the sea level history, much as a rapidly moving strip-chart recorder (satellite image from NASA, August 7, 2006; profiles after Chappell, 1974, Figure 2, with permission of the American Geophysical Union).

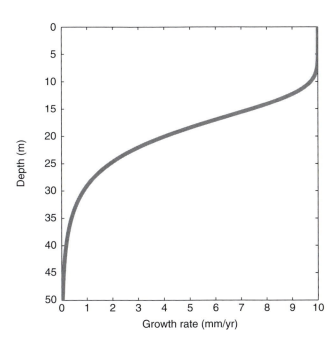

Figure 4.29 Rate of growth of corals as a function of water depth (see Galewsky, 1998). As the available light declines, growth rate declines. Below roughly 10 m water depths, growth rates decline precipitously. If the rate of relative sea level rise (sum of eustatic rise and land subsidence rates) exceeds the coral growth rate, corals will be drowned.

terrace i. The sea level at the time terrace i coral bench was formed is then $z_{SLi} = z_i - UT_i$. The sea level curve constructed from the Huon terraces has for years been one of only a few long dated records.

Interestingly, submerged or drowned coral terraces have been discovered in the same region. While it is straightforward to understand why a coral reef dies when it emerges above sea level, the process of drowning a reef is a little more subtle. Consider the following combination of processes (Galewsky et al., 1996; Galewsky, 1998). Corals can grow at a rate that is dictated by the amount of sunlight that reaches them. This in turn declines with ocean depth, as light is extinguished with depth. A typical growth rate curve is shown in Figure 4.29. Importantly, it shows a maximum growth rate of about 10 mm/yr. Let sea level oscillate with a period P and an amplitude Δz. Finally, now imagine that the entire rock platform to which the coral reef is attached is submerging at a given rate. Whenever the rate of relative sea level rise, the sum of submergence rate and sea level rise rate, exceeds the maximum rate at which a coral can grow, the coral will be drowned. If, by the next time sea level drops, the corals are at such great depths that they can no longer grow upward into the light, they are forever drowned. Such reef complexes were imaged in swath bathymetry in the Huon Gulf on the other side of the Huon peninsula noted on the map in Figure 4.27(a). Here, the edge of the plate onto which

that the sea level at that time, z_{SL5e}, was +6 m above present. The rate is simply $U = (z_{5e} - z_{SL5e})/T_{5e} = (z_{5e} - 6)/120$ ka. Armed with this rate, we can date and measure the elevation of any other terrace; call it

the arc complex causing growth of the Finisterre Range is being sutured is bending under the load of the encroaching mountain range, causing submergence. These reefs have been sampled by dredging. The dates of the reefs obtained by U/Th methods indeed correspond to times of very rapid sea level rise caused by terminations of major glaciations (Galewsky *et al.*, 1996; Galewsky, 1998; Webster *et al.*, 2004a,b). Dating of the reefs, and knowing the water depth in which they presently are found, reveals that the average rate at which the platform is subsiding is 5.7 mm/yr. This is indeed a rapid rate of submergence.

Flexure

We now consider the response of the Earth to the rearrangement of the topographic load on the Earth surface by long-term slip on faults. So far we have only dealt with situations in which the material at depth responds to the load by flow. Now we introduce flexure. Have you ever thought about the forces acting upon you as you stand on the surface of the Earth? We need to reconcile the fact that if I let go of a piece of chalk it will fall to the ground, while I can simply stand here without accelerating. Somehow the forces are balanced on me ($F = ma$) while they are not on the piece of chalk, at least when I let it go from my hand. We have focused so far on only one force, gravity. The (vector) sum of these forces is unbalanced on the chalk, and balanced on me. Why? The difference is that I am in contact with the ground, and somehow the ground is pushing back at me with exactly enough force to cancel out my weight, making the vector sum equal zero. How is this happening, in general and in detail? The material I am standing on is behaving like a spring. I am loading it up with a certain force (my weight), which compresses it to just the right level, so that it pushes back just enough to balance my weight. Of course it is more complex than this. It isn't macroscopic springs involved, but rather atomic ones. At human time and length scales, the Earth behaves as an elastic solid. We push on it temporarily and it compresses; we move one step over and it flexes back. How does this flexure work? (See a detailed discussion in Turcotte and Schubert (2002), and an entire book on it by A. B. Watts (2001).) We summarize the physics in Figure 4.30. Atoms within

minerals have a preferred spacing, which is dictated by the sizes and charges of the atoms of which they are made. The atoms sit in potential energy wells in which the atomic energy is minimized. Any closer together and the nuclei will repel each other; any further apart and the bonds established by sharing of electrons will be unhappy. What happens when we stand on a rock surface is we exert a net downward force on the solid, which compresses the atoms in their lattices immediately beneath us, and stretches them in lattices within the rock mass at greater depths and distances. Both of these result in the movement of the atoms from their preferred or equilibrium (minimum energy) positions within the minerals. This motion is resisted by forces experienced by each atom as it is shoved up one or the other wall of its potential energy well. The sum of all these atomic forces, which results in something we call "fiber stresses" within the underlying plate, exactly balances our weight. The larger the weight, the larger the displacements of the atoms from their preferred spacings, and the larger the volume of rock is involved in the resistance. This microscopic response to forces dictates the elastic properties of the rock mass, which in turn dictate the degree to which it resists compression for an applied pressure (Young's modulus) and the degree to which a stretching in one dimension results in a contraction in another dimension (Poisson's ratio). Other derivative properties you might have heard of include the rigidity of the material, the compressibility of the material, and so on. When considering an object resting upon the Earth, be it a human, a skyscraper, or a mountain range, the bottom line is that the downward force that we call an object's weight is balanced by the sum of a lot of atoms being a little unhappy.

The pattern of deflection associated with a complex load is the sum of the deflection associated with a set of point loads. The simpler case, for which there is an analytic solution, is that in which the loads may be considered long with respect to their width, in other words in which the loads look like lines. The solution for the deflection beneath a line load (see for example Turcotte and Schubert, 2002, and Watts, 2001) is

$$w = w_0 e^{-x/\alpha}[\cos(x/\alpha) + \sin(x/\alpha)] \qquad (4.5)$$

where w_0 is the vertical deflection directly beneath the load, and the length scale α is called the flexural parameter:

(a)   the macroscopic view

(b)   the microscopic view

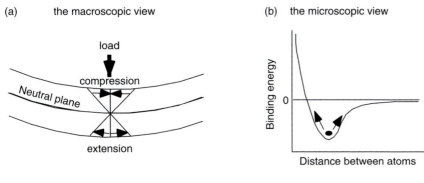

(c)   the flexural pattern

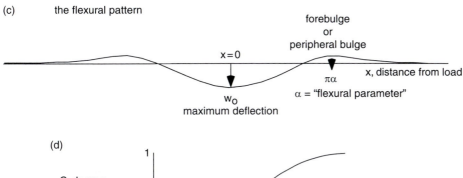

forebulge
or
peripheral bulge

$x = 0$

$\pi\alpha$ x, distance from load

w_O
maximum deflection

α = "flexural parameter"

(d)

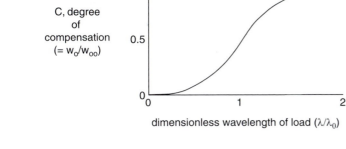

C, degree
of
compensation
$(= w_o/w_{oo})$

dimensionless wavelength of load (λ/λ_0)

Figure 4.30 Flexural primer.
(a) Macroscopic view. Compression
on the inside of a bent plate, and
extension on the outside, with a
plane of no strain (neutral plane)
in the middle. (b) Microscopic view.
Atoms forced from their equilibrium
position by strain gain
energy (storing strain energy).
(c) Large-scale flexural pattern
with maximum deflection, w,
beneath line load, falling off with
a length scale set by the flexural
parameter, α, and subtle forebulge
a distance of $\pi\alpha$ away from the
load. (d) Degree of compensation
by Airy (buoyancy-driven)
support. Buoyant support grows
in importance as the width of
the load increases. Bottom axis
is a non-dimensional length
scale (see text).

$$\alpha = \left[\frac{4D}{(\rho_{\text{mantle}} - \rho_{\text{infill}})g}\right]^{1/4} \quad (4.6)$$

This dictates the horizontal length scale of the response, and is related, as you might expect, to the elastic properties of the crust, and to the thickness of the plate being bent. Note that the densities involved are the densities of the material that must move beneath the load in response to it, and that of the material that fills the deflection (air or water or sediment, depending upon the geologic setting). The quantity that encapsulates the plate properties is D, the flexural rigidity:

$$D = \frac{ET_e^3}{12(1 - \nu^2)} \quad (4.7)$$

where E is Young's modulus, ν is Poisson's ratio, and T_e is the effective elastic thickness of the plate being bent. Using typical values of the elastic constants, the

flexural parameter α is usually of the order of several tens of to a couple of hundred kilometers.

First let us inspect the form of the deflection (Equation 4.5). All the sine and cosine functions do is bounce between –1 and 1. The amplitude of the deflection is controlled entirely by the exponential factor, which decays monotonically away from the load, located at $x = 0$. As plotted in Figure 4.31, you can see that the response is symmetric about the line load. The largest deflection is directly beneath the load, while at very great distances (many times α) it decays to zero. The surprising aspect of the solution is that at intermediate distances there exists a slight bulge, centered at $\pi\alpha$, and a small fraction of the amplitude of the deflection beneath the load. (To prove that this is where the bulge should be, set the first derivative of the solution to zero and solve for x. You would find that it is zero at $x = 0$, $x = +$ infinity, and $x = \pi\alpha$.) This bulge is dubbed the forebulge, the flexural bulge, or the peripheral bulge. The amplitude of the bulge may now be calculated by evaluating the deflection equation at $x = \pi\alpha$. The value is $e^{-\pi}$, or roughly 4% of w_o, the deflection below the load itself. While subtle, this bulge is important, for example, in controlling the pattern of deflection around the periphery of the Pleistocene ice sheets.

But what determines the amplitude of the flexure, the maximum flexure, w_o? This ought to increase with the magnitude of the load, and to decrease with the strength of the plate. In addition, and importantly, this maximum deflection varies with the horizontal length scale of the load, λ. The interplay of all these variables is best represented by a non-dimensional number, C, called the degree of compensation, graphed in Figure 4.32:

$$C = \frac{w_o}{w_\infty} = \frac{\rho_m - \rho_c}{\rho_m - \rho_c + \dfrac{D}{g}\left(\dfrac{2\pi}{\lambda}\right)^4} \quad (4.8)$$

where w_∞ is the deflection if the load were supported entirely by buoyancy. This expression represents the proportion of the weight of the load that is supported by fluid buoyancy, what we have been treating as a pure isostatic response, and that supported by the elastic response – fiber stresses. A degree of compensation, C, of 1.0 corresponds to a load that is supported entirely by fluid buoyancy, while $C = 0.0$ corresponds to a load being supported entirely by flexure, by an infinitely strong plate. Inspection of the x-axis on Figure 4.32 reveals the importance of the length scale, λ, of the load. At a specific wavelength of $\lambda_o = 2\pi D/(g(\rho_m - \rho_c))^{1/4}$ the compensation is 50:50 flexural:buoyancy. All else being equal (in other words for a given thickness and strength of the plate, and given densities of crust and mantle), it is the length scale (width) of the load that dictates whether it is supported flexurally or buoyantly, and dictates the magnitude of the deflection beneath the load.

In fact, for reasonable values of the elastic properties of the Earth's near-surface materials, we find that

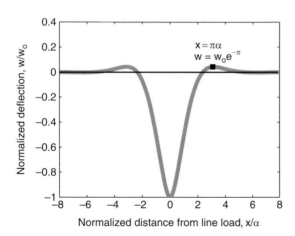

Figure 4.31 Deflection associated with a line load imposed at $x = 0$. Deflection normalized by that immediately beneath the load. Distances normalized by the flexural parameter.

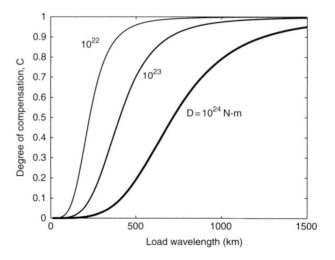

Figure 4.32 Dependence of the degree of isostatic compensation on load wavelength, shown for three flexural rigidities, D. The greater the degree of compensation, the greater the role of buoyancy (pure Airy isostasy) in supporting the load.

Box 4.3 Your load

Consider, for example, the load *you* exert while standing on the Earth's crust. What supports your weight? If it were buoyancy, given that your density is roughly that of water, while that of the mantle is about 3.3 times this, the deflection beneath you would be about (1/3.3) of your height. The simple observation that we are not wallowing around knee-deep in crust implies that the support comes instead from the elastic response of near-surface materials to our weight.

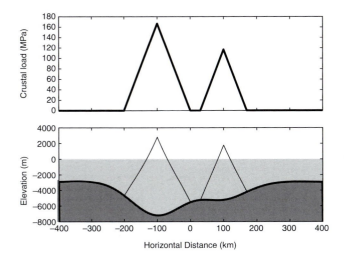

Figure 4.33 Calculation of deflection associated with two triangular volcanic chains, assumed to be two line loads 200 km apart. (a) Load distribution in force per unit area. (b) Topography resulting from deflection of the base of the ocean floor, flexing to accommodate the loads. Far-field ocean floor depth is 3 km. The peak subaerial topography is roughly 3 km. Effective elastic thickness = 25 km.

loads up to about 400 km in width are supported largely by elastic stresses. It was indeed appropriate to treat the Tibetan Plateau load (1300 km wide at its narrowest) as one being supported by buoyancy, justifying our simple treatment as an Airy isostatic problem. Smaller mountain ranges, however, require treatment of the flexural response. A classic and interesting example of a line load is the Hawaiian chain of volcanoes. The problem is sketched in Figure 4.33. The width of these basaltic piles is about 150 km at their base. Inspection of the bathymetry around the Hawaiian volcanoes shows a prominent "moat" corresponding to the deflection of the oceanic lithosphere associated with these large loads. One can also see the subtle peripheral bulge in the bathymetry.

An interesting corollary to this story is the expected bathymetric history of a parcel of seafloor near the present Hawaiian chain. As has been documented by Wessell and Keating (1994), a seamount riding along on this seafloor, just off the line connecting the volcanoes, first sees the arrival of the swell associated with the hotspot, which raises the oceanic lithosphere in a bulge 100 km wide. As this swell leaks out basaltic products, building the volcanoes, and these loads get closer and closer to the seamount, they bend down the lithosphere creating the flexural moat. The seamount therefore rides down into the flexural moat.

The vertical scale of the movements is not trivial. The Cross seamount has been shown to have been raised to at least sea level, resulting in its decapitation by coastal processes about 3.2 Ma, and has subsequently been drowned by at least 375 m.

Many loads fail to display the symmetry of a line. Volcanoes and sedimentary deltas, for example, have geometries better characterized as points or cones. Happily, the differential equation that lies behind flexure problems is a linear ordinary differential equation (it is fourth order, but it is linear). This linearity allows us to calculate the deflection associated with a complicated load by summing that due to a set of independent loads of differing amplitude, each at a different distance from our location of concern. With the solution for a line load, or a point load, or a disk load in hand, one can simply construct the complicated load out of a set of lines or points or disks. We illustrate this below with a few examples.

The solution for the deflection associated with a point load (e.g., see Watts, 2001) is

$$w = \left[\frac{P\beta^2}{2\pi D} \right] \text{kei} \left(\frac{r}{\beta} \right) \tag{4.9}$$

where P is the load, β is the characteristic length scale for decay of the amplitude of the deflection with distance from the load, r. The function kei is one of a class of functions called Bessel–Kelvin functions, in this case a Bessel–Kelvin function of order zero. These arise in radially symmetric problems involving membranes, such as the deflection of a drum skin upon being tapped. The function looks very similar to the exponentially decaying sinusoid of the line load solution. Happily, many numerical packages now incorporate Bessel functions as simple one-line calls.

Consider now the characteristic length scale for the decay of deflection away from the point load:

$$\beta = \left[\frac{D}{(\rho_{\text{mantle}} - \rho_{\text{infill}})g} \right]^{1/4} \tag{4.10}$$

The length scale β differs from α, the characteristic distance for a line load, by the lack of the 4 in the numerator; the length scale for a point load is therefore smaller by a factor or $4^{1/4} = 1.414$, or about 40%. The difference between these two length scales reflects the symmetry of the problem: in the case of the point load, the load is being supported from all sides rather than just two; for a load of the same magnitude, we should therefore expect the deflection to be smaller.

We note that there must still be an isostatic balance on the large scale. One can still appeal to the isostatic balance, set up by the relative densities of the materials. The balance is simply no longer "local," but is instead regional. This translates into an "integral constraint" on the deflection, or a requirement that when the deflection field is summed over the entire pattern, the force balance is enforced:

$$\int_0^R P(x,y)\mathrm{d}x\mathrm{d}y = \int_0^R w(x,y)\rho_\mathrm{m}g\mathrm{d}x\mathrm{d}y \qquad (4.11)$$

where $P(x,y)$ is the field of point loads (e.g., $P = \rho_\mathrm{c}gH(x,y)$), w the deflection field, and R a distance very large relative to β, the flexural parameter for a point load.

Let us consider some examples. Consider the load represented by the cone of sediments that comprise the Amazon sediment fan, shown in the map in Figure 4.34. Here the density of the load itself is the bulk density of sediment, and the infill material is yet more sediment. The load can be represented by a cone with prescribed slope, diameter, and density. The total volume of the load is so huge that the deflection beneath it is of the order of 1–2 km, and a significant forebulge is produced. Whereas the bulge to the east, offshore, plays no geomorphic role, it has been suggested that the bulge to the west, onshore, is sufficient to deflect local drainages (Driscoll and Karner, 1994). The bulge corresponds spatially to the Garupa Arch, upstream of which there are numerous lakes. As we discussed in the last chapter, this is one of several examples of the role played by geophysical processes in tweaking large rivers. That these rivers have such low gradients (the Amazon has a slope of order cm/km, or 10^{-5}) makes them particularly susceptible to the subtle changes in slope that these geophysical processes can produce.

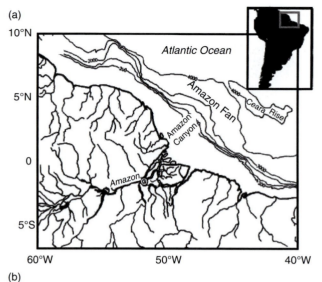

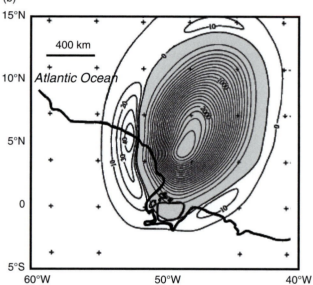

Figure 4.34 Flexure associated with the load of the Amazon fan. (a) Map of mouth of Amazon River adjacent to continental shelf, and depositional fan. (b) Deflection field associated with the load of sediment. Downward deflections are contoured at 100 m intervals, while upward deflections in the peripheral bulge are contoured at 10 m intervals (Driscoll and Karner, 1994, Figures 1 and 3, with permission from the American Geophysical Union).

Unloading

Loads can come in many flavors. When erosion acts to remove rock, which is replaced with air, the deflection of the remaining rock is upward. We speak of this as "unloading," the erosion pattern representing a field of "negative loads." In this case, the upper mantle moves back under the removed load, while only air fills the deflection. Erosion typically occurs on small length scales. Widths of individual valleys are always small relative to the flexural length scale, meaning that we must assess the pattern of expected response to their erosion using a flexural model rather than a pure Airy isostatic model. We have evaluated the role of glacial erosion in causing rock uplift of

(a)

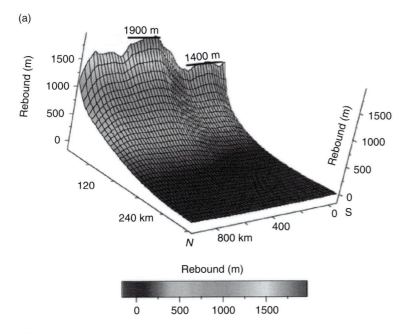

Rebound (m)

(b)

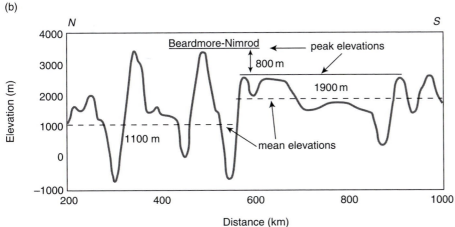

Figure 4.35 Rebound of Transantarctic Mountains range crest due to glacial erosion of fjords. (a) The expected deflection of the lithosphere, driven by the two-dimensional pattern of unloading by glacial erosion. (b) Major glacial troughs sourced in the East Antarctic Ice Sheet punctuate the coast-parallel profile of the crest of the range (after Stern et al., 2005, Figure 5).

mountain ranges. Here again the erosion is very localized, confined to glacial valleys while the interfluves are effectively untouched. In narrow ranges such as the Wind River range of Wyoming, the erosional response to removal of many hundreds of meters of rock in these isolated valleys is minor – a hundred meters or so (Small and Anderson, 1998). In the Alps, on the other hand, the massive erosional unloading in the center of the range, as deduced from long-term sediment accumulations in basins adjacent to the orogen, may be large enough to account for many hundreds of meters of rock uplift in the Plio-Pleistocene (Champagnac et al., 2007).

Stern and colleagues (2005) have calculated the flexural isostatic rebound associated with the erosional unloading of the edge of the rift flank associated with the Transantarctic Mountains in Antarctica. The margin is punctuated by huge glacial troughs that deliver ice from the East Antarctic Ice Sheet to the continental shelf. The resulting isostatic rebound is shown in Figure 4.35. The response is significant, and helps to generate the back-tilting of the entire range toward the interior of the continent.

Generation of mountain ranges by repeated earthquakes

The pieces are now in place to consider longer-term growth of mountain ranges and associated geologic

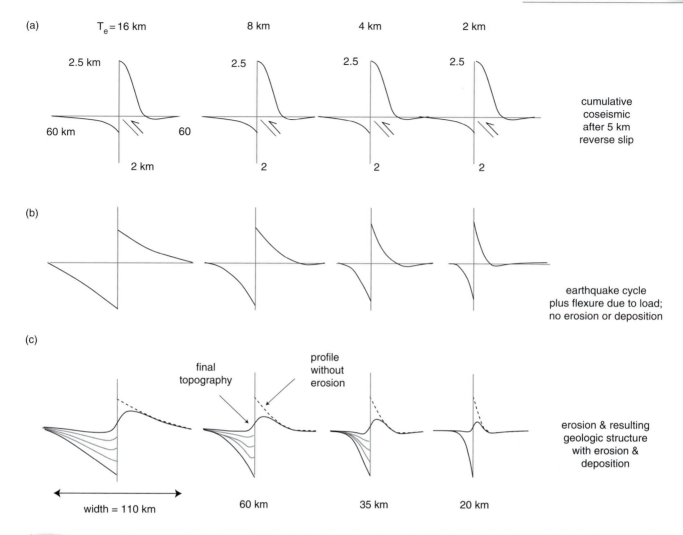

Figure 4.36 Mountain range growth as modeled by repeated earthquakes on a reverse fault. (a) Cumulative coseismic displacement only, after 5 km of reverse slip, (b) cumulative displacement, and flexure associated with changes in the load, (c) coseismic displacement plus erosion of peaks and deposition on the downthrown block, plus flexural support of the changed load. Note that the width of the structure (mountain range and basin) decreases as the effective elastic thickness, T_e, decreases from left to right (after Burbank and Anderson, 2000, Figure 11.10; after King *et al.*, 1988a, Figure 7, with permission from the American Geophysical Union).

structures in adjacent basins. This notion of the characteristic quake with a characteristic vertical deformation or rock uplift pattern was embedded in a simple model for the generation of crustal scale mountain ranges in a pair of papers in the late 1980s by King *et al.* (1988a,b). The various components of the problem are summarized in Figure 4.36. They combined three physical elements to generate the topography and the associated geological structures to be expected from motion on a thrust fault. They first calculate the expected vertical displacement field associated with slip on the fault, treating the problem as an elastic dislocation. They then allowed the resulting topography to flex under the load of the new tectonically generated topography. This served to make the

deformation field more symmetrical, to spread the effect in accord with the effective elastic thickness (or rigidity) of the local lithosphere, and deepened the hole in the downthrown block. They then attacked the topography with the simplest of all geomorphic rules – they diffused the tips of the mountain range, and threw the resulting products into the subsidence basin adjacent to the fault. In comparing the general patterns of the resulting geological structures in the American west (dips of the strata at depth, combined width of the range and adjacent basin), they found that in order to match the model with the available observations, they had to appeal to a very floppy crust, with an effective elastic thickness of only a few kilometers, as shown in Figure 4.36.

Summary

When strained sufficiently by far-field tectonic stresses, the brittle upper crust of the Earth breaks. Faults are planes across which shear motion of the adjacent walls takes place when the frictional strength of the materials is exceeded. The resulting earthquakes are major hazards for humans on the Earth's surface. Tectonic geomorphology is the study of the deformation of the Earth's crust that results in topography that is then attacked by a variety of geomorphic processes that we will study in the remainder of this book. The crustal processes of interest occur on several timescales, and their study utilizes the offsets of geomorphic markers to deduce patterns and rates of crustal motion.

The deformation pattern associated with slip on any single fault generates topography. In dip-slip faults the maximum throw is in the center of the fault segment and tapers toward the tips. Even strike-slip faults can at least temporarily generate topography as the displacement leads to displacement gradients that bunch up or stretch the local crust. Faults occur not as single planes, but commonly as complex networks of faults. The stresses generated by motion on one fault produce changes in the static stresses in the surrounding crust. When resolved on other existing faults, these stresses can lead to either enhancement or reduction of the likelihood of failure on that fault. Fault networks evolve through these stress interactions, leading to mature faulting patterns in which one or more master faults survive to accommodate the tectonically driven strains, while other initial faults may cease to move.

The deformation field associated with individual earthquakes (coseismic deformation) and the straining of the Earth during the interseismic interval (interseismic deformation) can be documented by repeat surveys of leveling lines, by records of tide gages, by satellite interferometry and by the growth patterns of individual coral heads.

Paleoseismology is the study of past earthquakes. Here we often must look into the stratigraphic record exposed in trenches dug across the fault. We must identify strata that are offset by motion on the exposed fault plane, and must date both the material being offset and the material that covers it in order to bracket the age of the earthquake. In tropical settings past motion on the megathrust fault plate boundaries can be deduced using the pattern of coral head growth.

Long-term deformation associated with motion on faults is perhaps of greatest interest to the geomorphologist wishing to know the tectonic origin of topography. Here it is often geomorphic features such as edges of streams, marine terraces and shorelines, and river channel profiles that serve as the datums against which deformation is measured. We can use any feature whose original shape can be known.

The growth of significant topography alters the distribution of the loads carried by the lithosphere. The altered topographic load is supported by bending, or flexing of the lithosphere if the load is short wavelength, and by buoyancy if the load is long wavelength. In the last chapter we dealt with the long wavelength end-member in our discussion of isostasy. In this chapter we acknowledge the handoff between the two processes at intermediate wavelengths, and provide line load and point load solutions for calculating flexure at short wavelengths. An important length scale emerges, called the flexural parameter, which governs the distance over which a load is supported. This is commonly many tens to hundreds of kilometers. Loads can be negative as well, causing upward bowing of the Earth's surface. These are caused by erosion, in which a rock load is effectively replaced by air. Local erosion of a short wavelength mountain range can cause rock uplift over a broader region.

Mountain ranges are therefore born of motion on faults that generates rock uplift, which in turn causes flexing of the lithosphere as the topography grows. Given that the atmosphere delivers moisture to the growing range, the resulting mountain range will be attacked by geomorphic processes as it grows. While the simplest model of geomorphic processes simply rounds the growing mountain tops and places that debris in the adjacent basins, our task in the remainder of the book is to acknowledge the details of the geomorphic processes so that we can more realistically address how tectonically generated topography is altered.

Problems

1. Given the distance between the site of the 1700 Cascadia earthquake rupture and Japan where the tsunami arrived (see Figure 4.16), and the mean ocean depth of the Pacific, how much time is likely to have elapsed between the earthquake and the arrival of the tsunami in Japan?

2. If the elevation of the inner edge of a wave-cut platform is 35 m above present sea level, and we date it to be roughly 120 ka, or marine isotope stage 5e, what is the rate of rock uplift at this site?

3. Discuss why the sea level highstand of about 170–180 ka shown in Figure 4.24 is not recorded as a marine terrace on this coastline. Roughly, what rate of rock uplift would be required to record this sea level highstand as a terrace?

4. On the classic Huon Peninsula coral terraces depicted in Figure 4.28, consider the variation in uplift rate along reef VII. Report both the maximum and minimum rates of rock uplift.

5. On a subsiding platform on which coral reefs are growing, calculate the rate of sea level rise that is required to drown the reef. The subsidence rate of the platform is 3 mm/yr. The growth rate of corals is well captured by the curve shown in Figure 4.29. Contrast this with the sea level histories reported in Chapter 16, and answer whether there is a time within the last deglaciation that this rate was exceeded.

6. If the maximum deflection of the Earth's surface beneath the Laurentide ice sheet is 1 km, calculate the height of the forebulge that is expected to have existed around the perimeter of the ice sheet. Assuming that the flexural parameter is 200 km, at what distance from the edge of the ice sheet would this bulge have reached its maximum height?

7. Consider the isostatic rebound after removal of an ice load. Given the response time, τ, of 4 ka (as found for the Fennoscandian ice sheet), a time since unloading of 9 ka, an asymptotic maximum rebound w_o of 180 m, and a response function that looks like:

$$w = w_o*(1 - e^{-t/\tau})$$

where t is the time of removal of the load, answer the following questions:
   (i) How much rebound has yet to occur in the future? What is the present rate of rebound?
   (ii) Discuss what the 180 m implies in terms of the ice load, and in terms of the duration of the loading.

8. Starting with Equation 4.5, formally analyze the expected height of the forebulge relative to the depth of the deflection beneath a line load. Show all your steps in the analysis.

9. Given the plot of the predicted flexure associated with the load of the Amazon Fan in Figure 4.34b, assess the maximum outward (westward) tilt beyond the forebulge displayed on land west of the mouth of the river. Report your answer in m/km.

10. *Thought question*. What are two means of supporting mountain ranges? Name them, but then discuss why one or the other mechanism is used in particular cases. (*Hint*: what supports the Sears Towers in Chicago vs. what supports Tibet?)

11. *Thought question*. How would the topography of a mountain range or of an orogen differ if the upper mantle of the Earth were significantly hotter (as it might well have been early in Earth's history)?

Further reading

Atwater, B. F., S. Musumi-Rokkaku, K. Satake, Y. Tsuji, K. Ueda, and D. K. Yamaguchi, 2005, *The Orphan Tsunami of 1700: Japanese Clues to a Parent Earthquake in North America*, US Geological Survey Professional Paper 1707, Reston, VA/Seattle, WA: US Geological Survey/University of Washington Press, 133 pp.
This is one of the wonderful mystery stories in the Earth Sciences. It is extremely well illustrated, and connects events in the Pacific Northwest with those in Japan.

Burbank, D. W. and R. S. Anderson, 2000, *Tectonic Geomorphology*, Oxford: Blackwell Science, 274 pp.
These authors review the state of the topic of this chapter as of 2000, targeting upper undergraduate and graduate students.

Keller, E. A. and N. Pinter, 2002, *Active Tectonics (Earthquakes, Uplift and Landscape)*, New Jersey: Prentice Hall, 362 pp.
This is another compilation of the literature relevant to how modern earthquakes can be documented and how they affect the Earth's surface, targeted for a more introductory audience.

Watts, A. B., 2001, *Isostasy and Flexure of the Lithosphere*, Cambridge: Cambridge University Press.
This is a key reference for a more in-depth treatment of how the Earth responds to the loads placed on its surface. The author nicely summarizes the history of thought as well.

Yeats, R. S., K. Sieh, and C. R. Allen, 1997, *The Geology of Earthquakes*, New York: Oxford University Press, 568 pp.
Written by luminaries in the science of paleoseismology, this book is a synthesis of how we have come to know about earthquakes in all types of tectonic settings.

CHAPTER 5

Atmospheric processes and geomorphology

Blow, winds, and crack your cheeks! rage! blow!
You cataracts and hurricanes, spout
Till you have drench'd our steeples, drown'd the cocks!

William Shakespeare (King Lear)

In this chapter

We review those aspects of the climate system that are relevant to the study of geomorphology. The primary driver of geomorphic processes is the atmosphere. This is remarkable given its thickness and density. The atmosphere delivers the winds and precipitation in storms, it determines the temperatures, and it sets the pace of the hydrologic cycle that in turn dictates the vegetation of a region. All of these vary dramatically with location on the Earth's surface, largely because the solar radiation that drives the climate system is delivered non-uniformly on the surface. In addition, the climate varies strongly in time. The most obvious temporal variation, and the most relevant to the current state of the landscape, is that associated with the ice ages of the last two to three million years.

Clearing of storm above Black Glacier, at crest of Mt Olympus, Washington. Snowfall is high here as Pacific storms collide with the high topography of the Olympic Mountains (photograph by R. S. Anderson).

Table 5.1 **Gas concentrations in the atmosphere (after Marshall and Plumb (2008) Table 1.2)**

Gas	Molecular weight (g/mole)	Proportion by volume
N_2	28.01	78%
O_2	32.00	21%
Ar	39.95	0.93%
H_2O vapor (water vapor)	18.02	0.5%
CO_2 (carbon dioxide)	44.01	380 ppm
CH_4 (methane)	16.04	1.7 ppm
O_3 (ozone)	48	500 ppb

ppm = parts per million
ppb = parts per billion

Climate is a synthesis of the weather at a point or in a region of the Earth's surface (Hartmann, 1994). The climate at any point on the Earth must be described in a statistical way: at the very least we must describe both the mean and the variation about that mean. The short-term variations are what we call weather, which is driven by the storms embedded in the general circulation pattern of the atmosphere, and it is these storms that deliver the precipitation to the Earth's surface. We will review the basic physics and chemistry of the atmosphere, and will discuss the climate and how it varies over the Earth.

The atmosphere is dominantly a mixture of molecular nitrogen (N_2) and oxygen (O_2), with many minor constituents (Table 5.1). Several of these trace gases play strong roles in controlling the passage of radiation through the atmosphere. This gaseous envelope is indeed thin. If collapsed to have a uniform density of that at the Earth's surface, $1.23\,kg/m^3$, it would be less than 10 km thick. And if compressed to have a density of that of water, it would be only 10 m thick; the atmospheric pressure at sea level is roughly $10^5\,Pa$, or 1 bar, the same pressure one would experience beneath 10 m of water at a density of $1000\,kg/m^3$.

The Sun

Our local star, the Sun, is about 26 000 light years from the center of the Milky Way galaxy, and one of roughly 10^{11} stars in the galaxy. It is huge,

three-quarters of a million kilometers in radius, and contains 99% of the mass of the solar system. The fine structures within the corona of the Sun, within its atmosphere, are much larger than the Earth, as seen in the artificial juxtaposition of images in Figure 5.1. The Sun is at least a second generation star, and probably third generation. Although the radiation from the Sun is largely produced by fusion of H (75% of the Sun's mass) to He (24%) (8.9×10^{37} H nuclei are converted to He nuclei per second), it is not made solely of these two elements. Rather, as seen in Figure 5.2, the remaining roughly 1% consists of the full periodic table of elements, in a cascade of elemental abundances that are identical (with a couple of exceptions) to that found in the Earth. That these higher atomic weight elements are synthesized only in stars is proof that there must have been earlier generations of stars that lived and then died cataclysmically to generate the makings of our Sun and the objects that constitute the remainder of the solar system. This is also reflected in the difference between the age of the Earth (and hence Sun) of around 4.55 Ga, and the age of the universe, which is thought to be about 15 Ga. This is the root of the statement that you are the stuff of stars. It is meant quite literally.

The Sun is in the middle of a typical 11 billion year life for a main-sequence star of this size. It has not always been this bright. This has led to the faint young sun paradox: while astronomers theorize that in the first billion years the sun should have been less radiative, which in turn should have dropped the mean annual temperatures of the Earth well below freezing, there is geological evidence, in the form of water-lain sediments, for running water on the surface of the Earth even then. This is usually explained by appeal to an even stronger greenhouse effect back in those early years, with very high concentrations of CO_2. The Sun's output of radiation also varies on smaller timescales. Most famously, there is the 11-year sunspot cycle, which is accompanied by a variation in solar radiation of roughly 0.1% over the 11-year cycle; magnetic storms associated with the sunspots are thought to inhibit local convection in the outer Sun, reducing radiation. Radiation from the Sun has also varied on longer timescales and less regularly; this is reflected in century to several-decade variations in sunspot activity illustrated in Figure 5.3.

Figure 5.1 Earth shown at proper scale, dwarfed by coronal flares from the Sun (from NASA, TRACE).

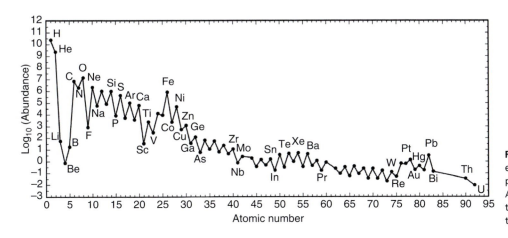

Figure 5.2 Abundances of elements in the solar system, plotted on a logarithmic scale. Abundances are normalized to Si, which is normalized to 10^6 atoms.

Climate and weather processes

The atmosphere is a low-density, low-viscosity fluid, and the forces acting on it result in speeds that dictate that it is most often in a state of turbulence. The whorls of turbulence range from the little eddies we can visualize as snow blows through trees, to the huge thousand-kilometer diameter storms best seen from space. It is the passage of these disturbances in the atmosphere, which cannot be predicted but which dominate the everyday world, that constitute what we call weather. And in most geomorphic systems, it is the weather that matters. Individual atmospheric events that deliver the wind, or that deliver the precipitation that runs off the landscape and governs the discharge of rivers, are what do the work. Weather, in other words, is stochastic; while it cannot be predicted, one can describe it statistically, for example with its mean and its standard deviation, or with a probability distribution of events (wind speeds, precipitation events). Averaged over a long enough period of time, these statistics become stationary, as shown in Figure 5.4 – they no longer change with increasing averaging times. It is these averages, this statistical representation of the stochastic process of weather, that we call the climate of a place. As an example of a climatology product, we show in Figure 5.5 the annual average map of rainfall in the world.

In order to proceed toward an understanding of climate and of weather, we must understand the physics behind the basic distribution of climates around

the world, we must know how and why the atmosphere is in motion, and about how its motion governs the distribution of precipitation over the Earth.

Why is Earth the "water planet"?

Because this is such an important question, in this section we delve into the detail needed to understand the answer at a quantitative level. Liquid water is stable in only a narrow range of pressure and temperature conditions (recall Figure 1.3). If we accept that atmospheric pressure is fixed by the exhalation of gases from the planetary interior and the pull of gravity preventing them from escaping to space, then we must understand the surface temperature of the Earth. This can be cast as a radiation problem. The mean surface temperature of the Earth is 15 °C above the freezing point of water. The temperature is dictated by a radiation balance shown in Figure 5.6. The incoming radiation from the very hot Sun a long way away is balanced by the outgoing radiation of the much cooler Earth. Planet Earth is just the right distance from the Sun. Let us illustrate this with a simple calculation. If the planetary temperature is roughly steady averaged over a year, then the solar radiation intercepted at this distance from the Sun must equal that radiated from the planet:

$$Q_o A_e = Q_e A_{es} \qquad (5.1)$$

where Q_o is the heat flux from the Sun at a distance equal to the Earth's orbit, A_e is the area of the disk presented to the incoming radiation (a disk of diameter equal to the diameter of the Earth), Q_e is the

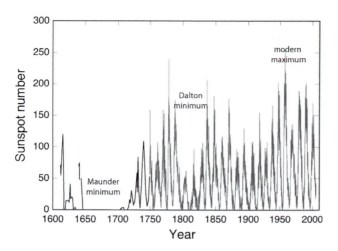

Figure 5.3 Four centuries of regular sunspot number observations. Since ~1749, continuous monthly averages of sunspot activity began to be collected in Zurich, and are plotted here from data available from NASA's Marshall Solar Physics site, where they report the International Sunspot number. Prior to 1749, only sporadic observations of sunspots are available, but show a distinct minimum in activity called the Maunder minimum (lasting roughly from 1645–1715).

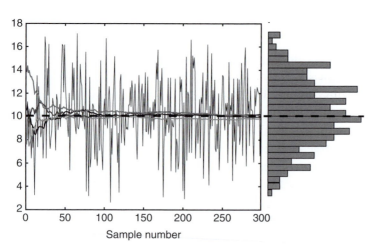

Figure 5.4 Random process and its statistical representation. Thin line: realization of a series of numbers selected from a Gaussian (normal) distribution with mean = 10, standard deviation = 3. Right: histogram of numbers drawn from the Gaussian population in the series displayed. Bold lines: means of this and three other realizations (not shown), which change as the number in the sample increases. These show asymptotic convergence of the mean with that of the full population. It typically takes 30–60 numbers to represent the mean faithfully.

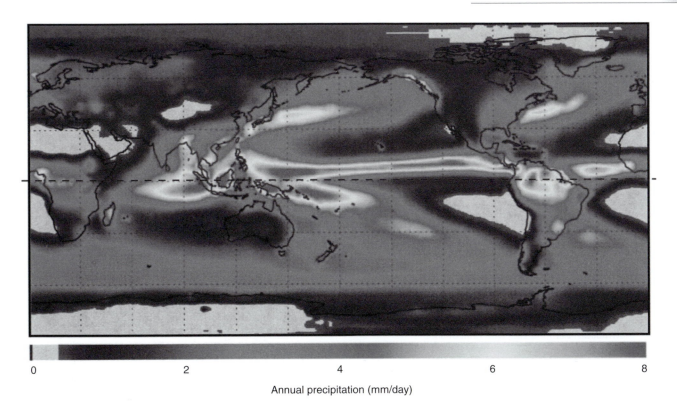

0   2   4   6   8

Annual precipitation (mm/day)

Figure 5.5 Global mean annual precipitation measured over 23-year period from 1979–2001, reported in mm/day. Band of highest precipitation centers on the equator, while lowest precipitation centers on 30 °N and 30 °S and on the poles (after Adler *et al.*, Figure 4, copyright 2003 American Meteorological Society (AMS)).

outgoing heat flux from the Earth, and A_{es} is the surface area of the entire sphere of the Earth. Note that the units of heat flux are energy per unit time per unit area. The statement above therefore equates the total heat arriving and the total heat lost per unit time.

We need to be able to calculate each of the quantities in this equation. We will need to know the size of the Earth, and the distance from the Sun. (We have listed many important numbers in Appendix B to this book; some are worth committing to memory.) The portion of the beam of radiation from the Sun that the Earth intercepts is a circle whose radius is R, or $A_e = \pi R^2$. In contrast, the area over which heat is lost to space by radiation from the Earth is the whole surface area of the Earth, or the surface area of a sphere: $A_{es} = 4\pi R^2$. But how do we calculate the heat fluxes, and how do we get to temperature from the equation above? Radiation obeys the following formula:

$$Q_s = \sigma T_s^4 \tag{5.2}$$

where σ is the Stefan–Boltzmann constant (also supplied in the appendix), and T is the surface temperature of the body emitting the radiation. Here we have written this general expression for the specific case of heat flux from the Sun, Q_s, and the surface temperature of the Sun, T_s. In a transparent (non-absorbing) medium, the same total heat is passed through spherical shells of increasing radius, or

$$QA = Q_s A_s \tag{5.3}$$

where A is the surface area and Q is the heat flux through a sphere at some distance centered around the Sun. At increasing distances from the Sun, the radiative heat flux Q must go down as the area of the sphere through which it is transmitted increases. Solving this for the heat flux at a specific distance from the object (here the Sun, with radius R_s), we get

$$Q = Q_s \frac{A_s}{A} = Q_s \frac{\pi R_s^2}{\pi R^2} = Q_s \left[\frac{R_s}{R}\right]^2 \tag{5.4}$$

The heat flux falls off with the inverse of the square of the distance from the object (Figure 5.7). This is known as the inverse square law. If the arbitrary R in the above equation is replaced with R_o, the radius

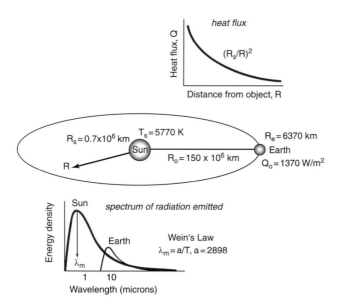

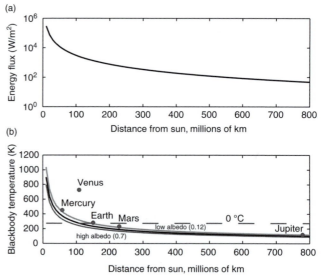

Figure 5.6 Elements of the radiation balance for the Earth. Middle: Earth shown in orbit around Sun, with radii of Sun, Earth, and orbit, temperature of the Sun's surface, and the energy flux Q_o arriving at the outer edge of the Earth's atmosphere. Top: inverse-square law for energy flux as a function of distance between source and the object. Bottom: blackbody radiation spectra for both Sun and Earth, showing both the dependence of the peak wavelength on surface temperature, and of the total energy emitted (integral beneath the curves). Note logarithmic scale for wavelengths.

Figure 5.7 Calculated solar radiation arriving at the orbits of planets out to Jupiter (a) and the expected surface temperatures of these planets (b) based upon the assumed albedos (labeled). Dots correspond to measured blackbody surface temperatures. The intermediate albedo case (0.3) corresponds roughly to the Earth's present measured albedo. That the Earth lies above the expected temperature associated with its albedo indicates the role of the greenhouse gases in its atmosphere. Only Venus lies outside the expected envelope, owing to its extreme greenhouse gas atmosphere.

of the Earth's orbit around the Sun, then we can calculate the heat flux arriving at the top of the Earth's atmosphere. The distance R_o is roughly 150 million kilometers (or 93 million miles, as you may have memorized in elementary school). We can now assemble the bits of the problem in one final equation. The temperature we would expect for the Earth's surface may be calculated from

$$T_e = T_s \left(\frac{1 - \alpha}{4} \right)^{1/4} \left(\frac{R_s}{R_o} \right)^{1/2} \qquad (5.5)$$

The one additional element in this equation is α, the albedo of the planet. The albedo (see discussion later in the chapter) is a measure of the reflectivity of the planet, 1 being perfectly reflective, 0 being perfectly absorbing. Only that portion of the Sun's beam that is not reflected contributes to heating the Earth. The albedo of the Earth is roughly 0.3. We can now see that the mean temperature of any planet is dictated by the temperature of the Sun, by the radius of the orbit around the Sun, and by the albedo of the planet when seen from space. The temperature of the Sun is roughly 5770 K. How do

we know this? We know it from measurement of the peak wavelength, λ_{max}, of the light emitted, and the relationship between this and the surface temperature, T, known as Wien's law: $\lambda_{max} = a/T$. The constant $a = 2898$ if the temperature is in kelvins, and the wavelength is in microns (10^{-6} m). As the temperature of the radiating body increases, the peak wavelength of the radiation emitted declines. Given these numbers, we calculate that the mean temperature of the Earth's surface should be about 255 K, or –18 °C. This is shown in Figure 5.7 along with those of other terrestrial planets. Clearly this is not the case. If it were, the Earth's surface would be frozen solid, and water would not exist in liquid and vaporous forms. What have we done wrong in the calculation, or what is missing?

We have not taken into account the chemistry of the Earth's atmosphere, which contains gases, some of them in trace amounts, that are excitable by a particular slice of the spectrum of radiation emitted by the Sun and the Earth. The upward shift in surface temperature attributable to greenhouse gases in the atmosphere is about $T = 33$ °C (from –18 °C predicted from radiative equilibrium in the absence of an

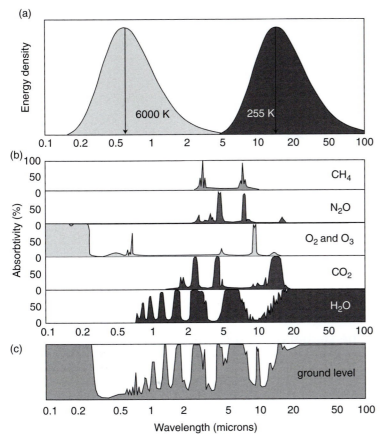

Figure 5.8 (a) Blackbody radiation from the Sun and Earth, assuming surface temperatures of 6000 K and 255 K, respectively. Vertical scale is normalized to the peak in each case. (b and c) Absorptivity spectra of major atmospheric trace gases, and the resulting spectrum of radiation received at the Earth's surface (after Piexoto and Oort, 1992, Figure 6.2, with permission from the American Institute of Physics).

atmosphere to $+15\ ^{\circ}C$ observed). As this is all-important for the state of the Earth, and as the green-house gas content of the atmosphere is changing rapidly with the activity of humans (recall the Keeling curve in Figure 1.7), we will describe in more detail how this works. The excitation of the gases leads to absorption of some of the energy and alters the simple radiation balance we have just calculated. One can see from Figure 5.8 how gases like H_2O, CO_2, CH_4, O_2, and O_3 absorb energy in specific wavelength bands. Many of these wavelengths are around 10 microns, in what we call the infrared band. Note also that light in wavelengths around 0.5 microns, the visible spectrum, is not absorbed by these same gases. Little of the radiation in the visible spectrum is altered by the atmosphere. Now recall Wien's law also illustrated in Figure 5.6, which implies that a body with a temperature of around $0\ ^{\circ}C$ will radiate at about 10 microns, while a body with the Sun's surface temperature will radiate at about 0.5 microns. First, it is not a coincidence that wavelengths around 0.5 microns are called visible light – humans and most other animals have evolved to utilize

this portion of the spectrum. Second, it is in the longer (infrared) wavelength band, that being emitted by an Earth with a surface temperature around $15\ ^{\circ}C$, that is most efficiently absorbed by the molecules listed above. It is therefore this outgoing radiation that is intercepted by the atmosphere and used to excite atmospheric gases. The higher the concentration of these gases, the more of these wavelengths of radiation will be absorbed, and prevented from getting to outer space to balance the incoming radiation. Therefore, the Earth will have to be yet hotter to produce enough outgoing radiation to balance the incoming solar radiation. Hence the surface of the Earth is warmer than we calculated when we ignored the atmosphere. These so-called greenhouse gases are therefore very important in that they allow the surface temperature of the Earth to be above $0\ ^{\circ}C$, rather than well below it. And it is presently the human alteration of the concentrations of these same gases that is forcing the global change of climate in the last hundred years.

But just how does this work? How do these gases absorb energy? Many of the gases of concern are

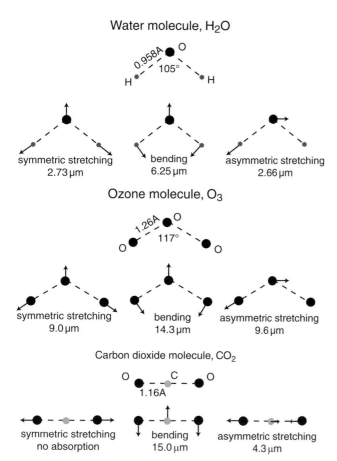

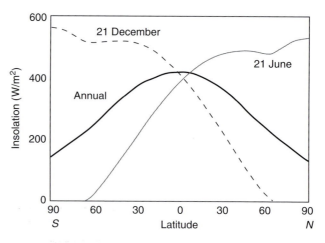

Figure 5.10 Latitudinal distribution of insolation at winter and summer solstices, and mean annual (after Hartmann, 1994, Figure 2.7, with permission from Elsevier).

Figure 5.9 Molecular structure of three principal trace gases in the Earth's atmosphere. Their vibrational modes are noted, and the wavelength of absorption associated with each, are denoted (after Piexoto and Oort, 1992, Figure 6.5 (from Houghton, 1985), with permission from the American Institute of Physics).

The spatial pattern of radiation

As we have seen above, the temperature of the surface of the Earth is dictated by a radiation balance between incoming shortwave radiation emitted by the Sun and outgoing longwave radiation emitted back to space from the Earth. We calculated that in the absence of greenhouse gases the surface temperature would be well below 0 °C, roughly 33 °C colder than it is. But that was a calculation for the mean annual surface temperature for the Earth as a whole. The radiation balance is not everywhere the same; it is non-uniform because the Earth is a sphere, and it varies in time annually because the Earth is spinning and its axis of rotation is tilted at 23° relative to the plane of the orbit. As shown in Figure 5.10, the annually averaged radiation is greatest at the equator and least at the poles. It varies with season. This can be easily understood by recognizing that the energy delivery we are describing is a flux, or energy per unit area. At the poles the flux is low because the same unit of energy in the solar beam is distributed over a broader area (Figure 5.11). In fact, the radiative flux varies as the cosine of (latitude + tilt), where the tilt varies from −23.45 to +23.45° from summer to winter solstice. We discuss later the effects of the longer term variability of the tilt. It is currently decreasing within its 41 ka cycle.

As we saw earlier, the albedo of a surface is a measure of its whiteness (think albino, or albite, a

molecules consisting of two or more atoms bound by bonds of particular lengths and strengths. It is the vibrations and rotations of atoms in these molecules that are excited at specific wavelengths, thereby using some fraction of the energy in that wavelength band of the radiation. Individual atoms absorb energy in high frequency bands associated with electronic orbital jumps. Molecules absorb at lower energies and longer wavelengths because these frequencies excite rotational and vibrational resonance of the bonds between the atoms of the molecule. We show this in Figure 5.9 for several key atmospheric molecules. It is pure coincidence that many of these absorption bands occur within the wavelength range of outgoing infrared radiation, allowing these trace gases to mask big swaths of the outgoing long wavelength radiation.

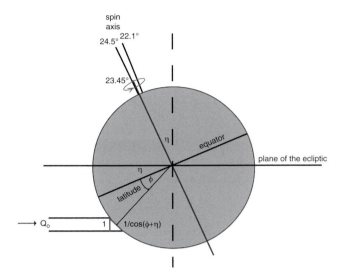

Figure 5.11 Range of tilt of the Earth's rotational axis relative to the plane of the ecliptic. The tilt varies between the two extremes depicted, on a 41 ka period. The present tilt is 23.45°, and is increasing. Incoming solar beam with flux Q_o is spread over an area that depends on the tilt and the latitude.

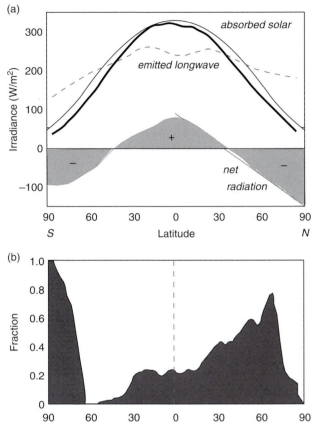

Figure 5.12 (a) Absorbed, emitted, and net radiation averaged in latitudinal bands. Also shown is a calculation of the average incoming solar radiation, as discussed in the text (smooth gray line). Note the net gain of heat from radiation in low latitudes, and the net loss at high latitudes (after Hartmann, 1994, Figure 2.12, with permission from Academic Press). Dashed line behind the northern hemisphere net radiation curve shows simplified pattern used in calculation of the poleward transport needed to balance the heat, depicted in Figure 5.13. (b) Fraction of land area at each latitudinal band on the Earth. Note the strong concentration of land area in the northern hemisphere, and Antarctic in deep southern latitudes (after Hartmann, 1994, Figure 1.12, with permission from Elsevier).

white feldspar; the Latin for white is *albus*). A surface with an albedo of 1 is perfectly reflective, while a surface with an albedo of 0 is perfectly absorbing. The average albedo of the whole Earth is about 0.3, but, as with the distribution of incoming radiation, it too is highly non-uniform, largely due to the distribution of oceans and ice. Water has an albedo of about 0.03, deserts and grasslands 0.2, ice 0.7, snow and cloud tops 0.8; very pure snow can have an albedo greater than 0.9. These values also depend upon the angle of incidence of the illumination. The poles, where most of the large bodies of snow-covered ice occur either as sea ice or as ice sheets, therefore have consistently very high albedos relative to lower latitudinal bands. This exacerbates the discrepancy between the radiation reaching the Earth's surface at the poles and equator; not only does the equator receive more solar radiation at the outer edge of the atmosphere, but the poles suffer from having a large fraction of the incoming radiation reflected back to space. This further enhances the need for poleward (meridional) circulation to move heat from the equator toward the poles.

The pattern of energy transport within the climate system is dictated by the pattern of net radiation. The sum of incoming shortwave solar radiation and outgoing longwave infrared radiation dictates the pattern of net radiation on the Earth. It is clear from Figure 5.12 that there exists a net positive radiation input in low latitudes, and a net negative radiative loss at the poles. If this were all that were happening, the poles would cool due to this continued loss of heat by radiation, and would continue to do so until the local outgoing radiation declined enough to come into local balance with the incoming radiation. The feedback would be through the dependence of radiative flux on temperature (Equation 5.2). The converse is true in the equatorial latitudes: these would warm until the outgoing radiation increased sufficiently to balance the high incoming radiation. That the Earth's mean annual temperature is roughly constant (yes, we are warming, but only slowly and due to other processes) implies that the misbalance in radiative energy input must be offset by other energy transport processes. This is indeed the role of the climate system: the oceans and atmosphere must accomplish the transport of heat by advection, the transport of heat by a fluid in motion. Before exploring in detail the means by which the heat is transported, we can calculate just how much heat *must* be transported for a heat balance to be assured. The basic equation is one of conservation of heat energy. We will see this in several other contexts, but for now consider a system in which both radiative and advective transport occurs. The basic statement of conservation of heat energy is: the rate of change of heat in a latitudinal band on the Earth is equal to the inputs of heat to that band and the losses of heat from that band. Put mathematically, this is

$$\frac{\partial E}{\partial t} = q_{in} - q_{out}$$
$$= Q_{radnet}A(\phi) + Q_{o+a}(\phi) - Q_{o+a}(\phi + d\phi)$$

(5.6)

where the first term on the right-hand side represents the net radiative input or output from a latitudinal band with area A, and the other terms represent the horizontal meridional (N–S) poleward transport of heat by ocean and atmosphere across latitude ϕ and across $\phi + d\phi$. We can then evaluate the steady case in which $\partial E/\partial t = 0$. Setting the left-hand side of the equation to zero shows that the gradient of the oceanic and atmospheric heat flux is equal to the net radiation input. This is illustrated in Figure 5.13 for examples in both equatorial and polar latitudinal bands. We can therefore visualize, before we actually perform the calculation, what the shape of the transport profile must look like: it must increase with latitude in the equatorial zone; it will reach its maximum where the net balance is zero, and will then decline toward the poles. Formally, the transport of heat required for the climate system (the oceans and atmosphere) is therefore the integral of the net radiation balance. This integration is performed and shown in Figure 5.14. Indeed, it shows the features we expected. In addition, measured transport by the

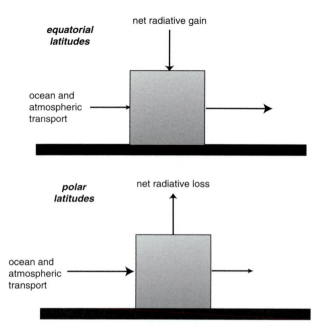

Figure 5.13 Cartoon of the ocean transport rates needed to balance the net radiative gain or loss. (a) Equatorial latitude case in which the excess radiation must be balanced by a positive gradient in oceanic and atmospheric transport. (b) Polar latitude case in which the net radiative loss must be compensated by a net gain through oceanic and atmospheric loss, accomplished by a negative gradient in poleward transport.

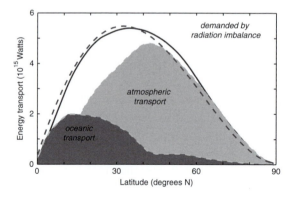

Figure 5.14 Transport of energy in petawatts demanded by the imbalance of net radiation, and the patterns of transport accomplished by ocean and atmosphere components of the climate system. The smooth dashed curve is poleward transport calculated assuming a simple linear decline in net radiative balance depicted in Figure 5.12 (observed transport components after Trenberth and Caron, 2001, American Meteorological Society (AMS)).

oceans and the atmosphere is plotted, revealing that these fluids share the duties roughly equally.

The majority of the transport in the atmosphere is actually zonal, or E–W. The mean flow of air from the equator to the poles is in fact small, and the majority of the transport of heat occurs by eddies – storms in the atmosphere and major currents in the oceans.

Vertical structure of the atmosphere

Why might we want to know the vertical structure of the atmosphere as geomorphologists? For one, we must know the temperature structure in the atmosphere, $T(z)$, in order to describe the mass balance of a glacier, as the rate of melting of snow and ice is well correlated with the temperature, and the phase of precipitation (rain vs. snow) depends upon temperature. We need to know the density structure of the atmosphere in order to calculate the dependence of production rate of cosmogenic nuclides either in the atmosphere (^{14}C or on surfaces at different elevations, e.g., ^{10}Be and ^{26}Al). As we all know from experience in climbing mountains, it gets colder as we climb higher, and it gets harder to breathe. But at one level the temperature structure doesn't make sense. After all, we are getting closer to the sun as we walk or fly higher. Isn't this what melted the wax-glued feathers to the wings of Icarus as he and Daedalus attempted to escape the Cretan tower of King Minos? Some other phenomenon must overpower the increasing strength of the Sun's radiation.

In fact, it is the basic law of gases that dictates that as the pressure goes down, the temperature goes down. Recall the ideal gas law: $PV = nRT$. This can be rewritten as

$$P = (\rho/m_g)RT \qquad (5.7)$$

where ρ is the density of the gas, and m_g is the molecular weight of the gas. Consider an atmosphere in hydrostatic equilibrium. The balance between gravitational force and the pressure gradient force is simply

$$\frac{dP}{dz} = -\rho g \qquad (5.8)$$

Note that this equation can be derived by appeal to a force balance on an element of fluid that results in the Navier–Stokes equation we present in the appendix to Chapter 12. Considering the z-component of this equation, and assuming that the vertical speed of the atmosphere may be neglected, we arrive at the simple Equation 5.8 above. We can anticipate that by

integrating this equation we would find that the pressure at any level in the atmosphere is proportional to the integral of the density structure above it. If the density is uniform with height, as it is to a close approximation in either the ocean or in crustal rock, the pressure increases linearly with depth. The assumption of uniform density is definitely not valid in the case of the atmosphere, which by contrast is composed of a compressible gas rather than a liquid or a solid. We can combine the gas law and force balance expressions above to rewrite the expected pressure gradient:

$$\frac{dP}{dz} = -\frac{Pm_g g}{RT} \qquad (5.9)$$

For an isothermal atmosphere, which we acknowledge is a simplifying assumption for now, we can define a length scale, H,

$$H = \frac{RT}{m_g g} \qquad (5.10)$$

Integration of Equation 5.10 then reveals that the pressure should decay exponentially with height above the surface:

$$P = P_s e^{-z/H} \qquad (5.11)$$

where P_s is the surface pressure at the base of the atmosphere. The measured surface pressure is approximately 1×10^5 Pa, or 1 bar. (Actually, it is 1.013×10^5 Pa.) The molecular weight of atmospheric gas is about 29 g/mole (see Problem 1 in this chapter). The length scale that dictates the rate of decline with height, called the scale height of the atmosphere, H, is roughly 8 km, at least near the Earth's surface. Therefore, by a height of 8 km above the surface, the pressure of the air is $1/e$ ($e = 2.71828$, or roughly 3) of that at the surface. By three scale heights, the pressure is $1/e^3$, or about 3% of that at the surface. Given this pressure distribution, we can differentiate it and solve for the expected distribution of density. For an isothermal atmosphere, the expected distribution would be exponential, with the same scale height, H:

$$\rho = \rho_s e^{-z/H} \qquad (5.12)$$

Most of the atmospheric mass is therefore within the bottom 20 km or so of the atmosphere. On top of Mt Everest, at an elevation of 8.8 km, the density of the atmosphere should be $e^{-8.8/8}$, or about 33% of that at sea level. No wonder it is hard to breathe up there.

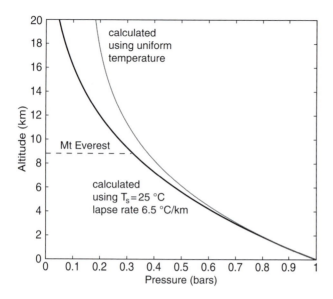

Figure 5.15 Calculated air pressure profiles for uniform temperature, and hence scale height, H_o, and variable scale height, calculated using a surface temperature of 25 °C and a lapse rate of 6.5 °C/km. Elevation of Mt Everest is shown, with corresponding pressure roughly a third of that at sea level.

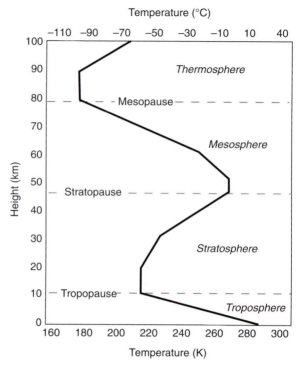

Figure 5.16 Temperature structure of the atmosphere (after US Standard atmosphere (1976)) (adapted from Piexoto and Oort, 1992, Figure 2.3, with permission from the American Institute of Physics).

In fact the atmosphere is not isothermal, but temperature declines with height, as discussed below. Taking this into account, the density profile, shown in Figure 5.15, departs from the simple exponential, although not markedly over the region of concern nearest the Earth's surface.

The full vertical thermal structure of the atmosphere (Figure 5.16), which defines the major named layers within the atmosphere, reveals the complexity of radiative transfer in a thinning atmosphere. The temperature drops steadily with height within the troposphere, the bottom 10–12 km of the atmosphere. The globally averaged temperature gradient within the troposphere, called the lapse rate, Γ, is roughly 6.5 °C/km, or 0.065 °C/m.

Above the tropopause, the temperature gradient reverses and temperatures increase with further height within the stratosphere. The zone of warm temperatures that defines the stratosphere is caused by efficient absorption of UV radiation by the trace gas ozone, O_3. (Note from Table 5.1 how efficient this absorption must be, given the low ($<$1 ppm) concentrations of ozone.) One must therefore ask why ozone dominates at this height. It is here that the photo-dissociation (photolysis) of O_2 occurs. The free atomic O produced by photolysis then links with molecular O_2 to create O_3.

Wind and atmospheric circulation

Wind is the motion of air. This fluid, like any other, moves when there is a net force on it, typically either a gravitational body force, or a potential gradient (here a difference of pressure at the same elevation from one place to another). The atmosphere is driven about by variations in pressure, the highs and the lows being the H and L on the weather maps. Air moves from high pressure toward low pressure. Because the Earth is spinning, the Coriolis effect causes the apparent trajectory of the air mass as seen by an observer on the Earth to bend to the right in the northern hemisphere and to the left in the southern hemisphere. The wind system is therefore established by the patterns of highs and lows. We discuss here both the major wind system on the Earth, and several specific winds set up by local but persistent pressure fields: monsoons, sea breezes, and katabatic winds.

Hadley cells

Let us compare the vertical pressure profile in a column of warm air vs. that in cold air. At any level

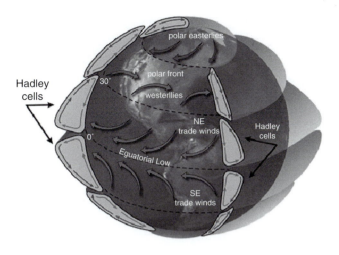

Figure 5.17 General global atmospheric circulation system on Earth. The Hadley cells (circulation) are a major part of the climate system (Image from NASA Earth Observatory, *Fewer Clouds Found In Tropics*).

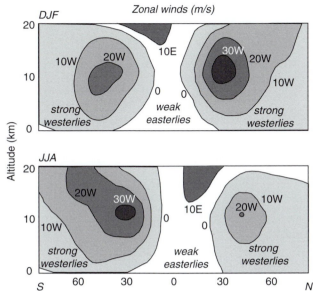

Figure 5.18 Cross-section of the zonal average wind speeds for DJF (top) and JJA (bottom) periods. Strong westerlies dominate, bordering weak easterlies. Note that the pattern swings to the south in the southern hemisphere summer (after Hartmann (1994), Figure 6.4, with permission from Elsevier; data from Oort (1983)).

in the atmosphere, the pressure in the cold air column will be higher than that in the warm air column. This sets up a horizontal pressure gradient that will drive air near the Earth's surface from the cold toward the warm. Indeed, at low latitudes, air moves toward the equator, toward the semi-permanent low-pressure zone there (Figure 5.17). It converges there, and of course cannot simply accumulate. The air at the equator, or really at the inter-tropical convergence zone (ITCZ), where these N-directed and S-directed air masses collide, must rise. This air is being heated from below, which serves as well to evaporate the oceanic water, raising the humidity and decreasing the density of the air mass. The heating also makes it buoyantly unstable, and allows it to rise. One can see this from space: a cloud band encircles the globe, marking the zone of condensation of the water embedded in the rising air masses. The poleward flow of this upper air completes the circuit, descending at roughly 30°N and S. The twisting of the S-directed surface winds to the right in the northern hemisphere, and the left twist of the N-directed winds in the southern hemisphere, generate the northeastern and southeastern trade winds, respectively. The semi-permanence of the equatorial low that attracts these winds makes for a very regular, steady wind field near the equator. The sinking flow of the top limb to these Hadley cells promotes drying and warming of the air. These "horse latitudes," so named because the horses onboard the sailing ships were

either eaten or thrown overboard to lighten the loads and allow the ships to sail better in the light breezes (see for example Boorstin, *The Discoverers*, 1983), are regions of high pressure, the subtropical high-pressure zone. Surface winds directed away from these highs toward the poles are also deflected by the Coriolis effect, and generate the strongest of the zonal winds, the westerlies. In Figure 5.18 we see that these winds average 20–30 m/s (45–70 miles/hr), and shift latitudinally with season.

It is clear to see that the strongest winds are E–W, or zonal, winds that in and of themselves will not transport heat meridionally. Much of the N–S transport is accomplished not by the average flow, or the mean atmospheric circulation, but by eddies in it. These are disturbances and storms, which, by their N–S component of motion, advect heat meridionally.

We have talked about why the Earth has the general circulation pattern it does, and it includes the facts that the Earth is a sphere (dictating non-uniform net radiation), and that it is spinning (which twists the flow attempting to transport heat from surfeit toward deficit). Now we must come to grips with the fact that the fluid in motion is of low enough viscosity, is thin enough, and is moving fast enough, to be turbulent.

Figure 5.19 Tracks of major historical hurricanes born in the Atlantic (1851–present) and eastern Pacific (1949–present). Note lack of cyclogenesis in near-equatorial latitudes, and clockwise paths of the storms (from NOAA/National Hurricane Center).

This is not ice in motion, it is air. A manifestation of turbulence is the instability of flows to disturbances, such that they evolve toward chaotic motion that includes many scales of eddies.

Take hurricanes as one end-member of such eddies. These storms feed on the heat from warm sea surfaces. This produces the low pressure (warm humid air being less dense), which, once it is warm enough to rise buoyantly, in turn carries the evaporated water vapor with it. The flow of air to replace this rising air sets up the circulation about the central core as the air flow is bent to the right in the northern hemisphere (generating counter clockwise or cyclonic circulation), and to the left in the southern hemisphere (generating clockwise or anti-cyclonic circulation in the southern equivalent, typhoons). The tracks of these tropical storms in general depart from purely zonal. As seen in Figure 5.19, the tracks of many hurricanes in the Atlantic are northward. Hurricanes, then, take a big gulp of energy from low latitudes, where the sea surfaces are warmest, and move it toward the poles.

While the geomorphic relevance of these and other storms will be discussed later, it is worth noting that the winds, the waves, and the precipitation generated by these systems are far from steady. The forcing of the geomorphic system is therefore not well characterized by the mean flow of the atmosphere, the mean annual temperature, the mean annual precipitation, and the mean waves that might well be used to characterize the "climate" of a region. Rather, we must acknowledge the reality that these quantities are stochastic and can be only statistically described. Just as they are important in the N–S transport of energy by the climate system, it is these stochastic features of the climate system – its weather, really – that drive the geomorphic system.

Monsoons

A monsoon refers to a wind pattern that changes direction with the season. It was originally used to describe the winds of the Indian subcontinent. The great Indian monsoon, and its less intense cousin, the southwest monsoon in North America, deliver the majority of the precipitation to the landmass. They are driven by the heating and cooling of the landmass, which as we have seen can change temperature much more readily than the ocean. In the case of the Indian monsoon, Tibet, which is roughly 1000×1000 km in area, heats significantly in the summer. The air column therefore becomes less dense, and the surface pressure drops. As the temperature above the ocean changes much less, a major horizontal pressure gradient is established from high pressure over the ocean to low pressure over the Tibetan Plateau. The resulting winds bring water-laden air over the

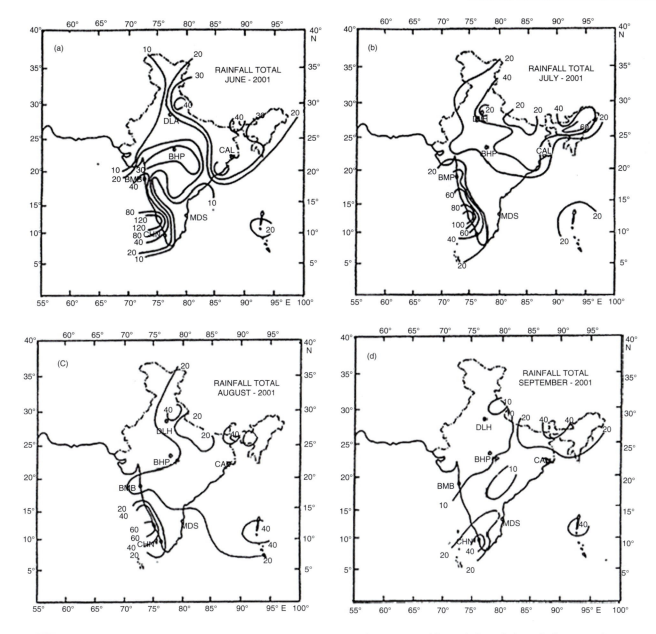

Figure 5.20 Measured precipitation in the 2001 Indian monsoon season, showing monthly totals (in cm) through the 4-month period. Rainfall is greatest on the west coast as the winds encounter the Western Ghats, sustaining more than 1 m of rain in each of June and July (after Mitra *et al.*, 2003, Figure 3, copyright 2003 American Meteorological Society (AMS)).

edge of the continent, moving it northward. As it encounters the slope of the land, uplift of the air causes precipitation (see the discussion of the orographic effect below). The Indian monsoonal precipitation, which is mapped in Figure 5.20, is intense, sustaining more than 1 m of rainfall in each of the two wettest months, June and July. While the onset of the monsoon, the day it starts, varies from year to year, one can nonetheless count on strong precipitation. The rainfall is greatest on the western coast of India, where the winds encounter the great step in topography called the Western Ghats – an example of orographic precipitation, as discussed below. From a geomorphic perspective, these rains drive major increases in the flow of the rivers

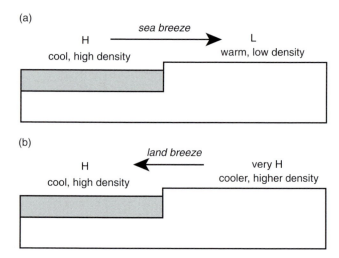

Figure 5.21 Sea breeze system. (a) Mid-afternoon warming of land surface results in low density of surface air, and low pressure at the surface, relative to the cool, higher density column of air above the sea. The resulting pressure gradient promotes landward-directed winds, called the sea breeze. (b) The opposite occurs at night, when the low heat capacity land surface cools off more quickly than the sea surface. The land breeze generated is typically less strong than the afternoon sea breeze.

draining the continent, with associated sediment transport and river incision. Rainfall associated with the southwest monsoon in the arid south-west of the USA is responsible for the majority of the precipitation in that region.

Sea breezes

The sea breeze is a classic example of pressure-driven local winds. We illustrate this schematically in Figure 5.21. While these are rarely major players in a geomorphic sense, they nonetheless can dominate the climate of a region. In mid-summer, when the land surface heats during the day, the air column above the land becomes less dense and hence near-surface pressures decline. In contrast, the density of the air column over the adjacent ocean remains high as the ocean's high thermal inertia prevents it from warming much during daytime. The resulting pressure gradient promotes the flow of air towards the land, from high to low pressure. In coastal California this dependable breeze supports the wind-aided sports of hang-gliding and surf-sailing along the coast. The breezes can be strong enough and sustained enough to drive significant sand transport inland from the beaches, creating near-coast dune fields.

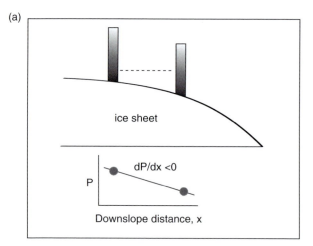

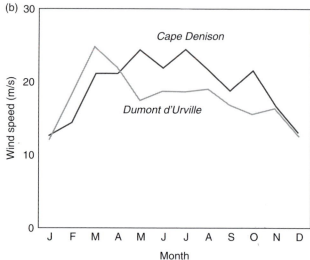

Figure 5.22 Physics of katabatic winds. (a) Intense cooling of the surface, here depicted above an ice sheet, cools adjacent air, generating a strong density contrast with altitude. Adjacent air columns on a slope will therefore display a strong horizontal pressure difference, dictating a negative pressure gradient in the downslope direction. This drives the drainage winds. (b) Mean monthly wind speeds at two nearby sites in Adélie Land, Antarctica. Cape Denison, where Mawson over-wintered, is likely the windiest place on Earth near sea level. Dumont d'Urville is the site of the French manned station (part (b) after Wendler *et al.*, 1997, Figure 3, with permission from American Geophysical Union).

Katabatic winds

Cooling of air against a very cold sloping surface can generate a persistent wind called a katabatic wind (from the Greek word *katabatikos* meaning "going downhill"). As the air cools it becomes more dense in a layer nearest the surface. Given that the air at a similar level in the atmosphere downslope of the surface is less dense because it is further from the cold surface, a pressure difference is established. As seen in Figure 5.22, the

Figure 5.23 Portrait of Sir Douglas Mawson, Australasian Antarctic Expedition 1930–1931, taken by Frank Hurley (nla. pic-an23478526) (reproduction with permission of the National Library of Australia).

pressure gradient is always in the downslope direction, and drives a downslope wind. The primary examples of cool surfaces are glaciers and ice caps. Anyone who has tried to climb up valleys occupied by glaciers knows they will be climbing against the wind. The most famous katabatic wind in Europe is the Mistral, which blows down the Rhône valley in southern France and out into the Mediterranean. But the strongest and most sustained is related to the polar ice caps. Those who have lived in Antarctica or have tried to go to the South Pole know that the winds there are constantly draining down toward the ocean. Winds at the edge of the continent can be maintained at 80 mph, with gusts well above 100 mph. Douglas Mawson, the great geologist and Australian explorer (Figure 5.23), writes of these winds in his *Home of the Blizzard* (1930), his account of over-wintering at Cape Dennison in 1913 after returning to the coastal hut only a few hours too late to catch the

ship back north. They measured wind speeds that averaged 80 mph over the winter! Recent re-occupation of this and adjacent sites in Adélie Land has shown that hurricane wind speeds (>32 m/s) occur 20% of the time, and 30% of the time in the winter months (Wendler *et al.*, 1997). These drainage winds serve to keep landscapes downwind much cooler than they would be in the absence of the advected cold air. In the case of Antarctica, they serve both to transport significant quantities of snow, and to generate sea ice-free areas near the coast (polynyas).

The williwaws of both Tierra del Fuego and Alaska, generated by the presence of near-coast glaciers, are other examples of katabatic winds. But glaciers are not required: the Santa Ana winds of southern California are perhaps even more familiar, as they descend from the cold winter interior plateau of the western USA.

Orographic effects

Here we briefly discuss the interaction of the atmosphere with topography. When moisture-laden winds encounter a change in slope of the land, they are made to rise. The rising air cools, leading to condensation of the moisture and ultimately to the formation of rain or snow. Once the drops are large enough to settle against the upward component of the air speed, they will fall, wafting downwind as they do so. The net result is a pattern of precipitation that is tightly tied to the pattern of topography. It should be obvious that this sets up a strong potential for feedback in the climate-topography system, in which precipitation is milked from the atmosphere on the windward sides of mountain ranges, inciting rapid erosion there, which in turn could alter the pattern of rock uplift.

Many land masses lie athwart the major wind systems of the world. These include the Southern Alps of New Zealand, the Cascade Mountains of the Pacific Northwest in the USA (Anders *et al.*, 2007), and the Himalayas (e.g., Anders *et al.*, 2006; Bookhagen and Burbank, 2006). The Cascades and Olympics form a topographic barrier to the westerlies that deliver the famous rains of the Pacific Northwest. In the case of the Himalayas, they form the ramp between the Indian continent and the Tibetan Plateau, toward which the monsoonal winds blow in the northern hemispheric summer.

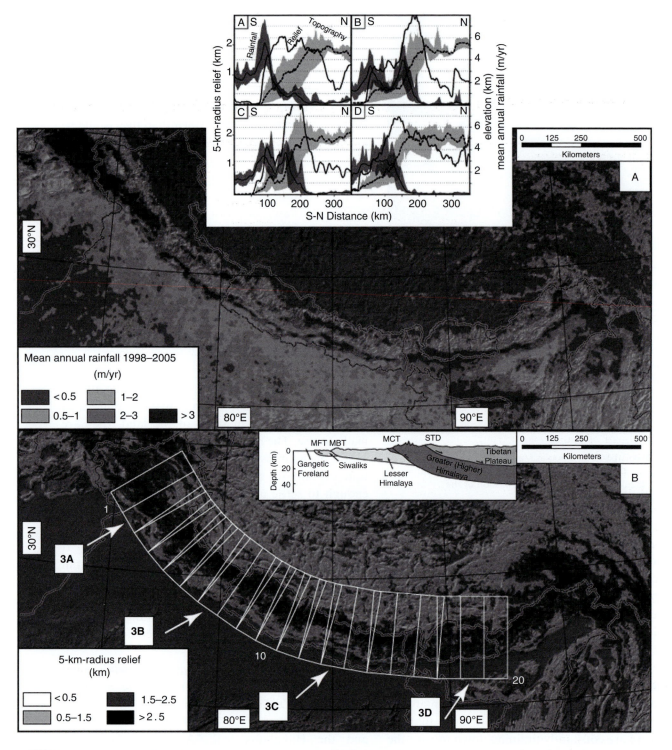

Figure 5.24 Dramatic orographic effect on precipitation pattern along the front of the Himalayas. (a) Map of precipitation in m/yr. (b) Relief calculated in 5 km circles, showing high Himalayas flanked to the south by foothills of the lesser Himalayas (cross section in inset) and to the north by the low-relief Tibetan Plateau. Top: Cross sections of the precipitation pattern showing that the most intense precipitation rates are not at the peak of the topography but on the windward side of the Himalayas at the most prominent break in slope (from Bookhagen and Burbank, 2006, Figures 1 and 2, with permission from American Geophysical Union).

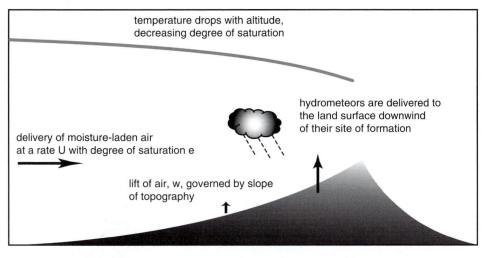

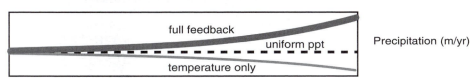

Figure 5.25 The essence of orographic precipitation. In this schematic illustration, the effects of temperature drop with altitude are overwhelmed by the lifting of the atmosphere dictated by the topographic slope. The locus of highest precipitation will be downwind of the greatest slope in the topography because hydrometeors formed in the atmosphere drift downwind as they fall, with a cant governed by the ratio of the fall velocity of the precipitation to the wind speed (after Roe *et al.*, 2002, Figure 1).

This problem of orographic precipitation was first reviewed comprehensively by Smith (1979), and more recently by Roe (2005). Roe *et al.* (2002, 2003) have addressed the expected changes in the shapes of steady-state stream profiles if the interaction of the precipitation field with the topography is incorporated formally (see Chapter 13 on bedrock river incision). As we see there, it is required to know the distribution of precipitation, P, so that we can estimate the distribution of river discharge that in turn governs the transport of sediment and erosion of bedrock.

What distribution of precipitation is appropriate for simple landscapes, and in what phase will this precipitation fall (rain or snow)? The orographic effect can now be documented using satellite-based measurements of precipitation (the Tropical Rainfall Mission, or TRMM satellite). In swaths taken perpendicular to the Himalayan front, the pattern of precipitation is dramatically enhanced at major changes in slope of the topography. One can easily see in Figure 5.24 the double band of monsoonal rain along the Himalayan front, the southernmost tripped by the foothills governed by the first set of thrusts accommodating the Indian collision, the northern

band coinciding with the topographic break to the Higher Himalayas.

The physics of the process can be summarized as follows, as depicted in Figure 5.25, and as discussed in Roe (2005) and in Roe and Baker (2006). Assume for simplicity that the wind is steady in both speed and direction, approaching a mountain front at speed U. There are three effects (at least) that must be captured. First, droplets of water must grow from the airmass through condensation to sizes large enough to fall at rates that exceed the rate of rise of the airmass. The formation of droplets depends upon the degree of saturation of the airmass. This in turn depends upon the temperature of the air, T, and upon the rate of lifting of the air, w ($= US$). Second, once droplets begin to fall (at which time they are dubbed by the atmospheric science community to be "hydrometeors"), they will drift further downwind before they reach the ground. The effects of temperature and rate of lift have opposing influences (Figure 5.25). The result is a pattern in which the maximum precipitation is shifted downwind from the maximum upwind-facing slope of the topography. For short hills, this can yield a precipitation maximum downwind of the crest, but more commonly, at the

(a)

$Nh/U \ll 1$

U

h

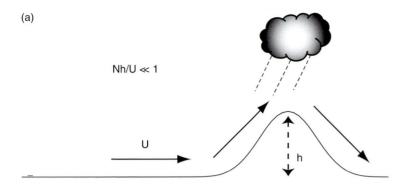

(b)

$Nh/U \gg 1$

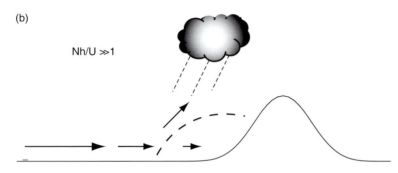

Figure 5.26 Role of the non-dimensional number *Nh/U* in determining the pattern of orographic precipitation. (a) Condition in which precipitation maximum lies just upwind of the summit of the mountain range. (b) Condition in which significant blocking of the flow occurs, moving the precipitation maximum well upwind of the mountain range (after Galewsky, 2009, Figure 1, with permission from American Geophysical Union).

(a) $Nh/U \ll 1; \beta \gg 1$; no rotation

(b) $Nh/U \gg 1; \beta \gg 1$; no rotation

(c) $\beta = 1$

(d) $Nh/U \gg 1; \beta \gg 1$; with rotation

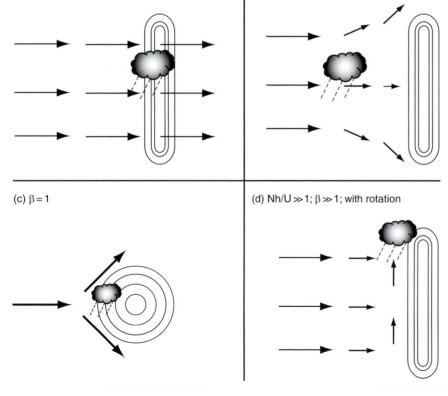

Figure 5.27 Diagrams depicting the roles of the non-dimensional number *Nh/U*, range symmetry, *β*, and rotation, on the pattern of precipitation associated with a mountain range. Changes in *Nh/U* (compare a vs. b) are reflected in the degree of blocking of the flow. Changes in range symmetry (a vs. c) govern the likelihood of flow splitting. Rotation (b vs. d) moves the maximum of precipitation off the axis of topographic symmetry (after Galewsky, 2009, Figure 2, with permission from American Geophysical Union).

range scale, this results in a precipitation maximum upwind of the crest of the range. More recent summaries of orographic precipitation have explored the roles of the dimensions of the mountain range involved. Galewsky (2008, 2009) has advocated the use of two dimensionless numbers that capture the essence of the variability in the orographic impact of a range. As illustrated in Figure 5.26, these are the aspect ratio of the range (its length to width), and the number Nh/U. When the wind speeds are low and the range high (and/or the stability criterion N is high), the range effectively blocks the flow of the atmosphere, creating an up-range sector of low air speed. The air effectively rises over this bubble rather than over the range itself, causing the lifting of the atmosphere well upwind of the surface slope on the range. The precipitation likewise will fall well upwind. In Figure 5.27 we illustrate further geometric effects. When the range is short in the cross-wind direction (small β), winds will diverge around the range, diminishing the orographic enhancement of precipitation. Finally, rotation of the Earth causes the Coriolis effect, which promotes the preferential passage of winds in one direction around the range, localizing precipitation.

We will see that these orographic effects come strongly into play when we address quantitatively the evolution of mountainous landscapes. As two examples, these effects govern both the pattern of precipitation relevant to the positive mass balance on glaciers (the snowfall pattern; see Chapter 8 and the discussion of the glaciers of Kings Canyon, Sierras), and the pattern of precipitation and hence discharge in a river (see Chapter 13).

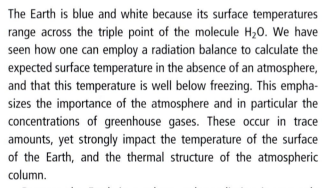

Summary

The Earth is blue and white because its surface temperatures range across the triple point of the molecule H_2O. We have seen how one can employ a radiation balance to calculate the expected surface temperature in the absence of an atmosphere, and that this temperature is well below freezing. This emphasizes the importance of the atmosphere and in particular the concentrations of greenhouse gases. These occur in trace amounts, yet strongly impact the temperature of the surface of the Earth, and the thermal structure of the atmospheric column.

Because the Earth is a sphere, solar radiation is unevenly supplied to it. This produces a global radiation budget in which the equatorial latitudes are net positive while the polar latitudes are net negative. Yet in the long term the energy is balanced in every latitude band. The unbalanced radiation budget generates temperature gradients and pressure gradients in the fluid envelopes of ocean and atmosphere that then accomplish the balancing of the energy budget. These fluids equally participate in moving excess heat from the equator toward the poles.

The atmosphere is turbulent at most scales. While there is a mean flow of the atmosphere that can be well described by the Hadley cells at low latitudes, for example, the transport of heat and other quantities takes place largely in turbulent structures. The largest of these are synoptic storms. It is these storms, and the wind and precipitation associated with them, that drive the geomorphic modification of the planet. Any description of the climate at a site must acknowledge the full distribution of weather events that comprise the climate.

At smaller scales, the atmosphere interacts in many ways with the topography of the surface of the Earth. Topography acts as roughness elements for the atmosphere, steering it. The temperature of the land surface can change dramatically, especially in contrast with that above the ocean, setting up gradients that promote local to regional flow of the atmosphere. These latter effects include monsoons, katabatic winds, and sea breezes. The local deflection of flow by mountains can also excite significant enhancement of precipitation upwind and depletion of precipitation downwind of mountains. These effects imply that there ought to be significant co-evolution of topography and precipitation, providing an intellectual link between the geomorphic and atmospheric science communities.

Problems

1. Given the chemical species in the atmosphere and their proportions in Table 5.1, what is the mean atomic weight of atmospheric gas, in kg/mole? Use only the three dominant species: nitrogen, oxygen, and argon. Then use the atomic weight to calculate the scale height in the atmosphere, H.

2. How would the mean annual temperature of the Earth change if the average albedo of the Earth were (a) raised to 0.5 from its present value of about 0.3, or (b) lowered to 0.1?

3. Calculate the annual variation in solar radiation arriving at the Earth in epihelion vs. perihelion if the Earth's orbit is 3% eccentric.

4. Determine the expected emission temperature of Venus, which is 0.72 AU from the Sun, and whose planetary albedo is 0.77. The observed temperature on the surface of Venus is about 750 K. Discuss why your calculation might differ from this.

5. Calculate the expected scale height of the atmosphere on Mars.

6. In calculating the distribution of pressure within the atmosphere, we assumed that the acceleration due to gravity was independent of height in the atmosphere. Evaluate this assumption, making use of the fact that the acceleration due to gravity decays with the square of the distance from the center of the Earth, over the bottom 50 km of the atmosphere. Express your answer in percent change of g from that at the base of the atmosphere.

7. (a) What is the residence time of water in the atmosphere? Two facts that are relevant: the annual total global precipitation, over both ocean and land, is 2.61 mm/day. The mass of water in the atmosphere is 15.5×10^{15} kg. (As an additional calculation, you should be able to demonstrate that, if condensed to a uniform layer, the atmosphere contains the equivalent of 25 mm of water.) (b) Perform the same calculation for the residence time of water in the world's oceans.

8. *Thought question.* Australia has moved northward at a rate of several cm/year for many millions of years. Its climate is therefore likely to have changed as it entered the band of latitudes centered on 30°S. Determine its northward rate of motion from exploration of the literature, and using its N–S dimension calculate how long Australia will stay in the arid band of latitudes. We are looking for a timescale here. Then discuss how this climate change might be manifested in the landscapes of Australia, and what evidence you might look for to document either the climate history or the landscape response to it.

Further reading

Boorstein, D., 1983, *The Discoverers*, London: Random House, 745 pp.
This remains a gateway into the history of some of the great discoveries. Written by a Librarian of Congress with a flare for storytelling, we learn about the history of our thoughts about time, about the discovery of the New World, to mention only a couple of topics, all assembled in an accessible way.

Hartmann, D. L., 1994, *Global Physical Climatology*, San Diago, CA: Academic Press, 411 pp.
This is an undergraduate textbook on climate with emphasis on the global scale and on the physics rather than chemistry and biology.

Marshall, J. and R. A. Plumb, 2008, *Atmosphere, Ocean and Climate Dynamics, an Introductory Text*, Burlington, MA: Elsevier Academic Press, 319 pp.
Very well illustrated with simple graphics that get to the core of the physics, this provides a new entrance point for the student of the fluid envelopes of the Earth.

Mawson, D., 1930, *The Home of the Blizzard, Being the Story of the Australasian Antarctic Expedition, 1911–1914*, 3rd edition, Birlinn Ltd., 440 pp.
This is a wonderful entrance into the world of Antarctic exploration. It is not only a great adventure story, but is studded with his astute observations of the side of the continent he experienced.

CHAPTER 6

Dating methods, and establishing timing in the landscape

It is of great use to the sailor to know the length of his line, though he cannot with it fathom all the depths of the ocean.

John Locke

In this chapter

Within the last two decades, a major revolution has occurred in our ability to establish timing in the landscape. The presentation of dating methods in this chapter will be significantly skewed toward the new methods that have driven this revolution, the new lines with which we plumb time. While the exercise of exploring the processes involved in a landscape is largely one of physics and chemistry, the documentation of the rates at which these processes are acting in the landscape requires obtaining dates of surfaces and of horizons within deposits. New dating techniques have often triggered surges of research activity. This is especially the case if the new technique allows use of materials thought to be barren of timing information, or if the timescale over which such techniques will yield dates is a new one. While we will very briefly review older or classical relative dating techniques, in which we can tell that one surface is older or younger than another surface, we will focus on the newer techniques that yield absolute ages. See more comprehensive reviews of dating methods in Pierce (1986), and in Burbank and Anderson (2000). Several of these methods are quite young at present. One must therefore be aware of the pitfalls of each method, the sources of uncertainty. We will once again encounter the utility of writing out a balance equation.

In addition, we describe briefly methods used to establish timing and process rates at both very short and very long scales. The first entails the use of cosmogenic radionuclides, with a focus on ^{10}Be and ^{26}Al. The latter employs proxies for how long a rock parcel has spent below a particular temperature. These thermochronometric methods include the counting of fission tracks and documentation of the quantity of trace gases in a mineral that reflect the decay of radioactive elements in the mineral lattice.

Jason Briner and Aaron Bini sampling granitic bedrock outcrop in Sam Ford Fjord, eastern Baffin Island, for cosmogenic exposure dating of the retreat of the Laurentide Ice Sheet from the fjord (photo by R. S. Anderson).

Relative dating methods

The classic tools available to the geomorphologist include an array of relative dating methods. Perhaps the most rudimentary of these are associated with basic stratigraphic methods (the oldest layers are at the bottom of a pile), and structural geologic principles such as cross-cutting relationships. The peat at the bottom of a moraine-ponded lake is the oldest peat in the stratigraphic column in the lake deposit. The moraine with continuous margins that cross-cuts or truncates another moraine is the younger of the two.

Long before radiometric dating tools were developed, the geomorphologist was employing the degree of weathering of a surface as an indicator of its age (see Pierce, 1986). Among the various means by which degree of weathering has been quantified are the carbonate and/or clay content of an arid-region soil (see Birkeland, 1999), the heights of weathering posts or depths of weathering pits on surfaces of boulders, and the thicknesses of weathering rinds on boulders, and of hydration rinds on obsidian. Where quantified in settings where the age of the surface is known independently, for example on lava flows, these weathering rinds appear to increase in thickness as the square root of time (e.g., Pierce *et al.*, 1976; Colman and Pierce, 1986). Such behavior smacks of a diffusive system, here involving the diffusion of weathering products through the outer skin of the rock. As long as the reaction rates involved depend upon the gradient of the concentrations, then as the rind thickens the growth rate of the rind will decline. This is another example of a growing boundary layer.

As geomorphology has evolved to demand more quantitative ages, some have turned to the roughness of a surface as a proxy for the age of the surface. This has been most successfully applied in desert alluvial or debris flow fan dating. The idea is that such surfaces begin with high local roughness due to channels and levees, and due to the large grain sizes involved in these flows, and that these features decay through time. As we see in the hillslope chapter (Chapter 10), many processes tend to smooth or diffuse topography. In addition, boulders break down to smaller clasts as they break apart by weathering. If these processes are diffusive (dependent on the local topographic slope), then the surface ought to decline in roughness rapidly at first and more slowly thereafter; theory suggests that

roughness ought to decline as the inverse square root of time. Given this, quantifying the roughness of a surface became the challenge. This can be done using topographic profiles collected either on the ground using classical survey methods, albeit at small spacing, or from airborne methods. Tom Farr at JPL in Pasadena employed this latter method on the fans of the eastern California desert, using radar (Farr, 1992; Evans *et al.*, 1992). While this is an expensive and rather blunt tool to quantify the many surfaces in these fans, the effort served as both a method development opportunity and an inspiration for the important shuttle radar topography mission (SRTM) that in the late 1990s collected the topography of the Earth from 60°N to 60°S (Farr *et al.*, 2007).

Absolute dating methods

The most easily understood absolute dating methods involve materials that are annually layered. These include tree rings and varves (laminated muds). While no individual tree lives more than a few thousand years, longer chronologies can be constructed by patching together the living tree record with that from older now-dead logs. This has been done in part by matching sections of tree ring width time series in two logs, the pattern of which is dictated by climatic conditions that are at least regional. These chronologies, which contain important paleoclimate information, now extend through and beyond the Holocene. As we will see below, a principal use of these chronologies is now in the calibration of other methods, such as radiocarbon dating, which can be performed on the same materials.

Varves are annually laminated sediments. The layers are usually one to a few millimeters in thickness and like tree rings vary in thickness through time. In this case, the thickness of a varve reflects a complicated function of distance from the shoreline of the water body, and the sediment supplied to it in that year. In using varves, one must demonstrate that they are indeed annual layers. This can be done either by independently dating material in a particular varve (say a radiocarbon date on a piece of plant material or charcoal), or by documenting a pattern of some other nuclide whose history is independently known. Two candidates for this are ^{210}Pb and bomb-derived ^{137}Cs (Figure 6.1). The peak of ^{137}Cs concentration

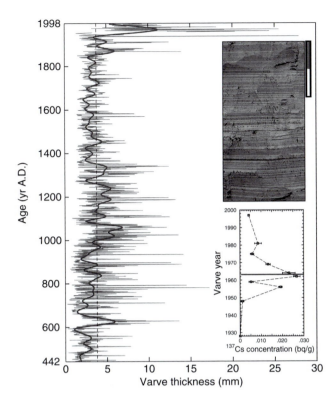

Figure 6.1 Time series of varve thickness from sedimentary section in freshly drained Iceberg Lake, Alaska. Record extends more than 1500 years. Inset: photograph of varves with 1 cm scale bar. Lower inset: profile of ^{137}Cs, which is expected to peak in the year of maximum atmospheric atomic bomb testing (1963; horizontal band). That the peak corresponds to the year that layer counting suggests is 1963 strongly supports the interpretation of the layers as being annual layers, i.e., varves (after Loso *et al.*, 2004, Figures 3, 5, and 7).

should occur in a varve that corresponds to 1963, the spike in atmospheric testing of nuclear weapons.

Paleomagnetic dating

On much longer timescales, one may employ the reversals of the Earth's magnetic field as a means of dating stratigraphic sequences. Various compilations of the reversal chronology have been published. In general they are limited by the age of the oldest ocean floor from which the magnetic field can be assessed. Cande and Kent (1992, revised in 1995) have produced the most complete tables of the reversal dates. Most recently, unmanned airborne surveys have generated magnetic profiles across ocean basins (Gee *et al.*, 2008) that may serve to increase our knowledge of the timescale, shown in Figure 6.2.

The most recent reversal, from the Matayama reversed to the Bruhnes normal polarity epoch, occurred at roughly 700 ka. Happily for geomorphologists interested in the latest Cenozoic, reversals have occurred roughly every million years in this time. This contrasts greatly with the extended normal epoch called the Cretaceous Quiet Zone during which the magnetic field appears not to have reversed for some 20 million years. The Earth's magnetic field not only performs these dramatic flips: over shorter timescales, the field wanders (polar wander), and occasionally undergoes short-lived reversals.

Optically stimulated luminescence (OSL)

Two luminescence techniques, thermal luminescence (TL) and optically stimulated luminescence (OSL), rely on a solid state property of common minerals quartz and feldspar that allows them to record the time they have been sitting in a deposit (e.g., see Berger, 1995; Aitken, 1998). The property is that ionizing radiation, most of it from decay of radioactive elements (U, Th, K, Rb) in nearby sedimentary grains, can create free electrons that become trapped in defects in the mineral lattice. They are then released as luminescence upon exposure to radiation of sufficient intensity. The amount released is proportional to the duration of the exposure to radiation (age of the deposit) and the local intensity of the radiation, as depicted in Figure 6.3. This means that one must measure both the luminescence in the lab, called the equivalent dose or paleodose, D_e, and local radiation in the field setting, called the dose rate, D_r; the age is then simply

$$T = \frac{D_e}{D_r} \qquad (6.1)$$

Operationally, one must sample the deposit in the dark, or beneath a cover that limits the sunlight, so that the sample is not zeroed in the sampling process. Stored in a light-tight container, it is then measured in the lab. The local radiation is either measured in the field, or a sample of the nearby sediment is collected to be analyzed for the concentrations of radioactive elements in the lab. OSL has become the preferred method, as the measurement of the optically stimulated luminescence can be done in small pulses, allowing multiple measurements on a single sample, and the time needed to extract the luminescence signal is short. The radiation used to stimulate

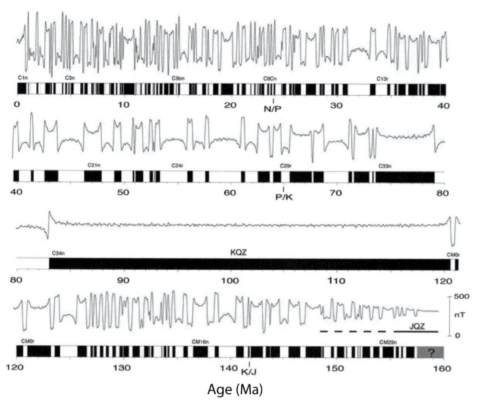

Figure 6.2 Paleomagnetic timescale through mid-Mesozoic. Dark bands = normal polarity, white = reversed. Note the long Cretaceous quiet zone (KQZ) from 121–83 Ma (after Gee *et al.*, 2008, Supplementary Figure 1, with permission from the American Geophysical Union).

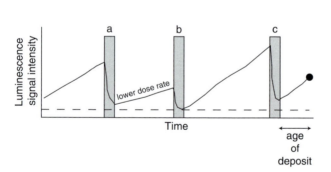

Figure 6.3 Hypothetical history of luminescence signal in a representative grain in a sedimentary deposit whose age we wish to determine. Three transport events (a–c, shaded) expose the grain to light, releasing the electrons trapped in crystal defects. The time represented by these transport events can be as short as a few hours, while the time between events can be thousands to tens of thousands of years. In the case shown, only event b fully "zeros" the signal. The luminescence measured in the lab will therefore yield an age estimate that is too large. The radiation dose rate provided by the nearby sediment in the deposit in the interval between transport events a and b is smaller than those during its other times of repose.

release from the most sensitive sites is commonly blue, green, or infrared.

This method has been applied to both fine-grained sediments (e.g., silt in loess), and more coarse sediments (e.g., fluvial packages, see Wallinga, 2002). The range of reliable ages extends to 200–300 ka, although some ages up to almost 1 Ma have been reported. One must assume that the grains being dated have been "reset" or zeroed during transport to the site of the deposit. In bright sunlight, the exposure time is of the order of 100–10 000 seconds, or much less than a day. The silt in loess, which travels to the depositional site in suspension in the air, over distances that can be up to thousands of kilometers, will certainly be reset. Coarser sediment that travels in bedload, bouncing along the bed at the base of a flow of water, will take longer to be zeroed.

Amino acid racemization

While ^{14}C dating is the better known method for determining the age of biological specimens, we will see that it is restricted in its age range to about 40–50 ka.

Another method employs amino acids in biological specimens and has a greater range, perhaps viable back to several hundred thousand years. Amino acids are the building blocks of life. In live tissues and bones and shells, the amino acids are twisted or coiled in a certain direction and are said to be racemized. Upon death, the molecules begin to flip back toward being randomly racemized, half to the left, half to the right. The degree to which the amino acids in a specimen have departed from purely left-coiling is a clock. The rate at which this transformation takes place, and hence the rate at which this clock ticks, depends upon temperature in an orderly fashion, as in any chemical reaction (Figure 6.4). It is an Arrhenius process with a definable activation energy. The different amino acids have different activation energies, so that in a single specimen there may be several amino acid clocks ticking.

Given the thermal dependence of the process, one can turn this dating method on its head and use it as a thermometer if we know the age independently. This has been done on the shorelines of Lake Bonneville, which are independently dated using ^{14}C on shells (see Kauffman, 2003; see also Miller *et al.*, 1997, for a southern hemisphere story). The reaction progress in the shells is well below what we would expect if the shells had been maintained at the present day temperatures. The mean annual temperatures during the LGM in the interior of western North America must have been at least 10 °C below present temperatures, confirming inferences made from periglacial features such as frost wedge casts from nearby Wyoming.

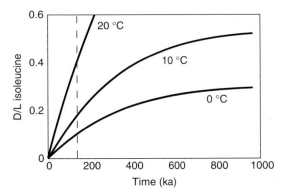

Figure 6.4 Evolution of the D/L ratio in the amino acid isoleucine at three steady temperatures. Soon after death of the organism, the L–D reaction dominates and the D/L ratio climbs rapidly from zero. At later times, the forward L–D reaction and backward D–L reaction rates are roughly equal, and the D/L ratio ceases to evolve. The warmer the temperature, the greater the rate of racemization, such that at 125 ka, for example, the D/L ratios are 0.1, 0.2, and 0.4, for the average temperatures of 0, 10, and 20 °C (after Kaufman and Miller, 1992, Figure 1 and temperature dependence after Hearty *et al.*, 1986, Figure 2).

Box 6.1 Extinctions in Australia

The amino acid racemization dating of shells of two large flightless birds in Australia (the emu and a now-extinct variety known as Genyornis) has allowed the telling of an amazing tale of ecological change (Miller *et al.*, 1999, 2005; Magee *et al.*, 2004) (Figure 6.5). The ostrich-sized Genyornis, as well as 85% of the Australian megafauna, disappeared about 55 000 years ago. The emu and Genyornis co-existed until then. This timing coincides with the arrival of humans on the Australian continent, raising the question of the mechanism of extinction. The stable isotopes of C and O in the shells tell some part of this story. Because "you are what you eat" (meaning the stable isotopes of what you eat are recorded in your tissues and bones), we can deduce the diets of animals from their bones, or in this case the eggs that they laid. It appears that while the Genyornis continued to eat the same plants, the emu switched its diet from C3 to the C4 plants that now dominate (Miller *et al.*, 2005). This in turn raises the question of why the change in plant types appears to coincide with the arrival of humans. The use of fire is hypothesized to have driven this change in vegetation. The switch to spinifex plants in northern Australia, which are far less effective at evapotranspiration, is now thought to alter the ability of moisture to penetrate into the interior of the continent; the shorelines of the large interior Lake Eyre in monsoon cycles that post-date the arrival of humans are lower than those during monsoon intervals prior to human arrival.

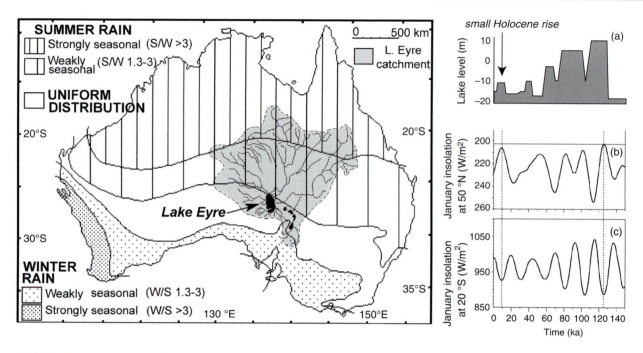

Figure 6.5 Map of Australia highlighting the closed interior Lake Eyre basin. The history of Lake Eyre levels has been documented using 14C, U/Th, and amino acid racemization of shells from huge birds (a). January insolation histories at 50°N (b) and at 20°S (c) suggest control of monsoonal delivery of precipitation into the Australian interior at 125 ka and 9 ka (dashed vertical lines) by northern hemisphere insolation. The small size of the Phase I lake compared to the Phase IV and V lakes has been interpreted to reflect a significant change in vegetation in northern Australia that began at 50 ka associated with the arrival of humans on the continent (after Magee et al., 2004, Figures 1, 2 and 3).

Oxygen isotopes and the marine isotope stages

Any student of geomorphology should be keenly aware of the record of oxygen isotopes derived from sediment cores from the deep sea floor. Our ability to measure stable isotopes and to drill deep cores evolved in the aftermath of World War II. The chief stable isotopes of concern are those of hydrogen and of oxygen found in water: H, ^{2}H and ^{3}H, ^{16}O, ^{17}O, ^{18}O. These lead to different masses of water molecules, which in turn influence their behavior in the physics of phase changes, and in biological processes. These isotopes are now routinely measured in small mass spectrometers, and are reported as the ratio of the isotopes in a sample relative to the ratio in a standard, known as delta values. For example, δ^{18}O:

$$\delta^{18}O \equiv 1000 \left[\frac{\left(^{18}O/^{16}O\right)_{sample} - \left(^{18}O/^{16}O\right)_{standard}}{\left(^{18}O/^{16}O\right)_{standard}} \right]$$

(6.2)

where the subscript "standard" refers to an accepted value of this ratio in an aliquot of a world-accepted standard. For water samples this is commonly SMOW, standard mean ocean water, and in carbonate samples this is commonly PDB, Peedee Belemnite, a marine carbonate shell. As the departures of the ratios from the standards are generally small, we multiply the numbers by 1000, yielding units of permil (‰).

By the early 1970s, records of the depth series of δ^{18}O from foraminifera (forams) in cores were being produced. These revealed complicated patterns that were interpreted to represent some combination of temperatures of the water in which the forams grew, and the isotopic concentration in the ocean above the core site. A breakthrough came when the picking of forams (the selection of individual tests from the sediment sample) became more selective, in particular isolating those forams that were known to grow in the deep-water column or in the sediments themselves – benthic forams. (As these are much rarer than the planktic forams, picking enough tests to run a sample on the mass spectrometers of the time was quite painstaking.) The signal and the interpretation of it were simplified because surface water temperature was effectively eliminated as a variable. In a series of now classic papers, Shackleton and Opdyke (1973, 1977) showed that the $\delta^{18}O$ signal varies strongly on 100 ka, 40 ka, and 20 ka cycles over the last million years, and that the signal was found in numerous cores. This is shown in Figure 6.6. They argued that this signal revealed a global story that supported the Milankovitch theory that global climate varied with a beat set by the

variations in the Earth's orbit. They defined what have become known as marine isotope stages (MIS).

Such records from benthic forams have now been extracted from many sites around the global oceans. The story is indeed a global one. The spectra of the time series consistently show distinct peaks at 100 ka, 40 ka, and 19–23 ka, as expected from control of climate by ellipticity of the orbit, tilt of the spin axis (also called obliquity), and precession of the equinox, respectively (Figure 6.7). At least in the Quaternary, the strongest component of the benthic foram signal is global ice volume. The argument is as follows. Water containing lighter isotopes of H and of O is easier to evaporate than that with heavier isotopes. Their smaller mass translates into higher speeds for the same temperature (see Problem 1 in Chapter 1). If (and only if) the water that evaporates is sequestered on land as ice (in the great ice sheets, for example), the water remaining in the ocean will become enriched in the heavy isotopes. The degree of enrichment is proportional to the amount of water sequestered as ice sheets, and hence is a proxy for global ice volume. Forams grown in the ocean will take on the isotopic concentration of the water, and will therefore faithfully record this proxy.

While forams are ubiquitous, and a continuous depth series can be obtained by closely sampling the deep sea cores, the absolute timing of these records is less well known than one might think. Absolute dates

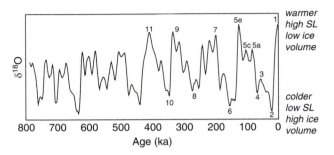

Figure 6.6 Marine oxygen isotopic signal over last 800 ka, based upon benthic forams. The dominant 100 ka period ice volume signal is clear, with 40 ka and 20 ka period smaller amplitude fluctuations. The last few marine isotope stages (MIS) are numbered. Of greatest importance to the geomorphic story written on the landscape are the last glacial maximum (LGM) at ~20 ka, and last major interglacial (MIS 5e).

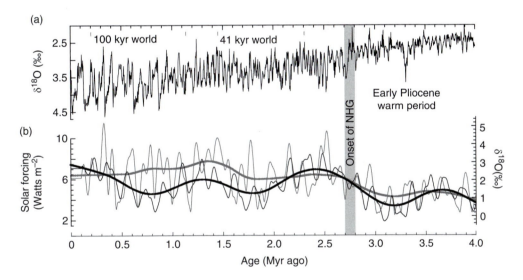

Figure 6.7 (a) Oxygen isotope record covering the majority of Pliocene–Pleistocene time showing early warm period of the Pliocene, onset of northern hemispheric glaciation at roughly 2.7 Ma, and transition from 41 ka world to 100 ka world at 1 Ma. (b) Solar radiative heat flux (after after Ravelo *et al.*, 2004, Figure 1, reproduced with permission of Nature Publishing Group).

come largely from magnetic stratigraphy. Cores extracted from the deep sea are sampled densely and assessed in sensitive magnetometers. Flips in the magnetic field are documented, and counted backwards down through the core. The last time the magnetic field was reversed was roughly 700 ka (recall Figure 6.2). This Bruhnes-Matayama transition from reversed (Matayama) to normal (Bruhnes) is therefore often the first major tie-point in the isotopic record. One can imagine that if such tie points to absolute time are so rare (roughly every million years), there is a lot of wiggle room in the true dates of events between them.

The isotopic records have now been measured back through the entire Cenozoic – basically as deep in time as the oldest ocean floor (see review in Zachos *et al.*, 2001; and Figures 1.6 and 6.6). As we hinted in the introductory chapter, several features of this record are worth becoming familiar with. Times of light water, low in ^{18}O, correspond to low ice volume, high sea level interglacials. These are labeled with odd marine isotope stages (MIS). We are presently in MIS 1. Times of heavy ocean water, high in ^{18}O, correspond to high ice volume, low sea level glacials, and are even isotope stages. The last glacial maximum (LGM) is MIS 2, and is centered around 22 ka. Inspection of the marine isotopic record reveals the complicated beat of the climate first documented by Shackleton and Opdyke. The ice volume history leading to the LGM was punctuated by smaller scale reversals in ice volume that have been subdivided using letters. The last major glacial was in MIS 6, and was followed by MIS 5e, the last time sea level exceeded present sea level (by about 6 m). Letters a, c, and e correspond to low ice volume, b and d to high ice volume. MIS 5e, 5c, and 5a, for example, all have corresponding sea level highstands recorded in coral terraces in Papua New Guinea.

Taking a little broader view, we can see in Figure 6.6 that the oscillations in ^{18}O in the last million years have been dominated by the 100 ka cycle. The larger glacials are separated by 100 thousand years. This was not the case in the early Pleistocene (Figure 6.7), when the signal is dominated by 40 ka cycles; variations in the tilt of the Earth's axis ruled the climate then. The transition around 1 Ma remains a target of study, with many hypotheses about its cause. At an even larger scale (Figure 1.6), we can see the trend toward a cooler Earth through the Cenozoic, a trend that is punctuated with abrupt events such as that at the Paleocene–Eocene boundary (the PE thermal maximum, or PETM). One can also pick out the onset of major Northern hemisphere ice sheets at around 2.4 Ma. All of these signals, so remarkably recorded in deep sea cores, not only pose challenges to the paleoclimate community for understanding of the climate system. Many of the explanations put forth for the major events in the record involve geomorphic systems, including variations in ice sheets or weathering. In turn, the variation in climate that is implied by these and other proxies derived from deep-sea cores must be acknowledged in the study of the Earth's surface, as they are the best continuous proxy we have of how the climate that intimately influences all geomorphic systems has evolved over the timescales of landscape evolution.

Radiometric dating methods

The most common absolute dating method is the ^{14}C or radiocarbon method. In this method we take advantage of the decay of ^{14}C atoms to ^{14}N, which occurs with a half-life of roughly 5730 years. ^{14}C atoms are produced in the atmosphere by the collision of cosmic ray particles (see next section) with gases in the atmosphere. These atoms then become incorporated in the trace gases CO_2 and CO, the former of which is incorporated in plants upon photosynthesis. Some small fraction of the CO_2 (about 10^{-10} percent) used by the plant has this ^{14}C atom, as long as the plant is alive, it continues to incorporate these anomalous atoms in the ratio in which they occur in the atmosphere. Upon death, the ^{14}C atoms in the tissues of the plant will begin to decay to their daughter products. We can measure the ratio of the ^{14}C to ^{12}C in the organic material of the plant as a clock. While the decay of any one atom of ^{14}C is a stochastic or probabilistic event, we do know that the probability of the decay of any one of the ^{14}C atoms to its daughter in any unit of time (say a year) is a constant, set by the decay constant, λ:

$$\frac{dN}{dt} = -\lambda N \tag{6.3}$$

where N is the number of parent atoms. The rate of decay of the concentration of parent atoms, N, depends simply upon the concentration. The resulting equation for the total concentration of parent ^{14}C through time is then

$$N = N_o e^{-\lambda t} \qquad (6.4)$$

where N_o is the initial concentration at $t = 0$, the time of death of the plant. Let us make sure we understand this most fundamental equation for radiometric dating. You are probably familiar with the notion of a half-life. This is the time, $t_{1/2}$, that it takes for the concentration to fall to one half, or 50%, of the original concentration, as seen in Figure 6.8. Given a graph of the concentration history, you could read off this time on the time-axis. At two half-lives, the concentration will have fallen to half of a half, or a quarter, of its original concentration, and so on. How do we connect this to the equation shown above? We must find the relationship between the decay constant and the half-life. $N = N_o/2$ should occur at the half-life. Inserting this into the equation we find that

$$\frac{N_o}{2} = N_o e^{-\lambda t_{1/2}}$$

$$\frac{1}{2} = e^{-\lambda t_{1/2}} \qquad (6.5)$$

$$t_{1/2} = -\frac{1}{\lambda} \ln\left[\frac{1}{2}\right]$$

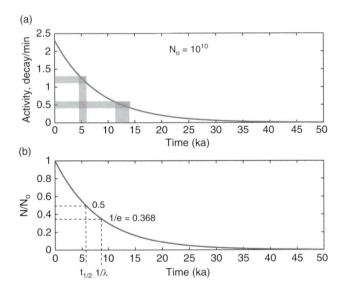

(a)

(b)

Figure 6.8 (a) Rate of decay of ^{14}C concentration in a sample (measured by its "activity"), and (b) concentration of ^{14}C relative to its initial concentration, N_o. Note that the half-life corresponds to $N/N_o = 0.5$, while the mean life ($1/\lambda$) corresponds to $N/N_o = 1/e = 0.368$. By 30–40 ka, the concentration and hence the activity of the ^{14}C system is so low that the signal becomes difficult to measure relative to the noise (gray bar). Gray horizontal bars representing measurement uncertainty correspond to increasing uncertainty in sample age as the age increases.

The first step is simply algebra, and the second step requires taking the natural logarithm of both sides and rearranging. The half-life is therefore related to the inverse of the decay constant (as the decay constant describing the likelihood of decay at any moment goes up, the half-life goes down). As $\ln(0.5) = -0.693$, this relationship simplifies to $t_{1/2} = 0.693(1/\lambda)$. Just to be complete about this discussion, note that the units of λ must be the inverse of time, or $1/T$. Its inverse is then a timescale, which is called the "mean life" of the radioactive system. This is the average lifetime for a ^{14}C atom, given the probability of its decay at any moment, λ. It may also be interpreted as the time it takes for the concentration of ^{14}C to fall to $(1/e)$ of its original value. We now see that the half-life is 69% of the mean life. Returning to Figure 6.8, we can graph both the half-life and the mean life.

Now consider an application. We have found a piece of wood in a landslide deposit (or a river terrace, etc.), and wish to use the ^{14}C method to date the wood and hence establish the age of the landslide. How do we measure this? There are two methods, one old and less expensive, the other new and more expensive. In the first method we count the number of decays of ^{14}C atoms per unit mass of carbon in what amounts to a large Geiger counter. In essence, we are using the equation for the decay rate, $dN/dt = -\lambda N$, to estimate N, the concentration of parent atoms. Here the sample is placed in a sealed container surrounded by scintillation counters that detect and count decays. The higher the decay rate, or "activity," the faster we get an answer. You can also see that the larger the sample is, the more atoms will decay within it in a given time. However, if the sample is very old, and now has very low N, the decay rate will be similarly low and we will have to wait a long time to detect enough decays for us to have much confidence in the answer. In addition, the lower the decay rate, the more likely the signal will be contaminated by counting of events that are not from the sample itself but from other energetic particles such as cosmic rays. Such laboratories are typically sited below ground in order to minimize counting of stray events. The facility at the Quaternary Research Center at the University of Washington was also lined with lead bricks constructed of pre-WWII materials. These issues place an effective limit on the ages that can be documented using this method (see Figure 6.8) of about 40–50 ka. Sample ages that are beyond this age are said to have yielded an infinite age, which merely means that the reported age cannot be distinguished from twice or ten times this age, or for that matter from an infinite age.

An alternative method has revolutionized ^{14}C dating within the last couple of decades. In this method one counts not only those atoms that decay within some reasonable time, but the ratio of ^{14}C to ^{12}C in the whole sample. Because this uses all of the C atoms in the sample, and not that tiny fraction of them that decay, it requires a much smaller sample. We can now date individual foraminifera, for example, or individual seeds, or a small thread from the Shroud of Turin. The measurement uses a mass spectrometer, but a fancy one. The method is dubbed the accelerator mass spectrometric, or AMS, method because it requires an accelerator to generate high speeds in the isotopes so that their small mass differences can be measured.

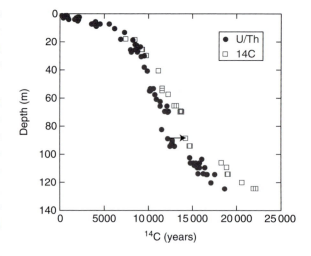

Figure 6.9 Discrepancy between ^{14}C and U/Th dates on a set of coral samples from Barbados. The U/Th ages are consistently older, the discrepancy being greatest for older samples (after Fairbanks, 1989, Figure 2, reproduced with permission of Nature Publishing Group).

Non-steady production

As with any dating method, there are several problems with which the user of the method must be aware. The radiocarbon method is not perfect, even within its 40–50 ka range of application. This is primarily because the production rate of ^{14}C in the atmosphere is not steady. We have known this for some time simply by comparing the date derived using ^{14}C with ring-counts on trees. Ideally, these methods should yield the same results. That they do not documents the variability of the production rate through time. But why does it vary? There are two culprits. While the cosmic ray flux to our position in the solar system should be steady through time, the magnetic fields of both the Earth and Sun vary with time. Both influence the deflection of the incoming cosmic ray beam. The strength of the Sun's magnetic field is reflected in the number of sunspots on its surface. This varies with an 11-year period, although there are longer timescale variations in which this 11-year cycle is embedded. The Earth's magnetic field varies both in its strength and orientation. This directly affects the steering of charged particles (protons, hydrogen nuclei, which constitute most of the cosmic rays impacting the Earth's upper atmosphere): times of low magnetic field strength result in less effective shielding and higher production of ^{14}C.

Given the importance of the ^{14}C method in geology and archeology, significant effort has been expended to calibrate the ^{14}C method against others. We have seen that when a tree ring chronology is available, this has been fruitful in demonstrating the unsteadiness of the production. But what about that part of the timescale that is beyond available tree ring chronologies? The longest such chronology is roughly 10 ka. In a now-famous paper in *Nature* in 1989, Fairbanks compared the ^{14}C clock with the U/Th clock on corals. Happily, the C in corals can be used for ^{14}C, while U substitutes for Ca in coral skeletons. The two methods could be used on the same samples. Fairbanks showed that the ^{14}C clock was off from the U/Th clock by at least 2 to 20 ka (Figure 6.9). Given that the U/Th clock is a purely atomic one, and is immune from the production rate variability to which ^{14}C is subjected, the U/Th results are considered the more reliable. The radiocarbon clock has subsequently been calibrated to that derived from the U/Th results, or more accurately, the production rate history for cosmogenic radionuclides, of which ^{14}C is a member, has been back-calculated from these and other results.

The reservoir effect

Recall that a living organism will incorporate ^{14}C from the atmosphere or the ocean from which it draws its carbon. If the residence time of ^{14}C in that fluid is long, some of the ^{14}C will have decayed by the time the organism incorporates it in its shell or plant tissue. The concentration of ^{14}C will therefore be lower than expected, causing a shift in the calculated age. This is called the reservoir age. While the mean residence time for C in the atmosphere is short, it is a few hundred years in the ocean. This shift is therefore applied to the calculation of the age of ocean-dwelling organisms, and can be several hundred years. It has also been shown by careful comparison of the ^{14}C ages of marine organisms in varves (annual layers) that can be counted from some specially sited oceanic sediments that this reservoir effect is not steady over time. The shift shifts. This reflects variation in the water balance of the ocean, which is seriously tweaked in times of large rates of change in ice sheets.

This effect may be quantified if we know independently the age of a deposit. This is the case in the wonderfully varved deposits of the Cariaco basin in South America. Varves have now been counted back to roughly 50 ka (Hughen *et al.*, 1998, 2000, 2004). The ^{14}C ages of the sediments can then be compared with the ages derived from independently dated records such as the Greenland ice core from GISP2. If the production of ^{14}C in the atmosphere has been steady, and the reservoir effect has not changed through time, these chronologies should line up on a 1:1 line. As seen in Figure 6.10, they do not, and the departures can be used to deduce how the reservoir effect has evolved through the last glacial cycle.

Cosmogenic radionuclides

These rare isotopes are born of interaction of cosmic rays with atmospheric and Earth surface materials. ^{10}Be, ^{26}Al, ^{36}Cl, ^{14}C are radioactive isotopes, and therefore decay with a characteristic timescale (see Table 6.1). It is this timescale that determines the temporal range of application of the nuclide. We have just discussed ^{14}C, which is produced by interaction of cosmic rays with N in the atmosphere. Two other commonly used cosmically produced nuclides, ^{3}He and ^{21}Ne, are stable, meaning they do not decay

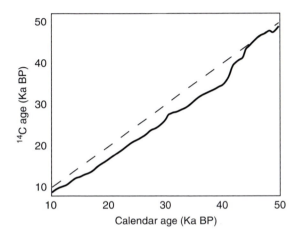

Figure 6.10 Reduction of ^{14}C age below calendar age for interval including the LGM. Equivalent ages would lie on 1:1 line (dashed). Departure reflects the effective reservoir age of the oceans, which varies greatly through the LGM (redrawn from Hughen *et al.*, 2004, Figure 2, with permission from the American Association for the Advancement of Science).

Box 6.2 Use as a tracer for documentation of fossil fuel burning

An interesting twist on this reservoir effect is that associated with the burning of fossil fuel. As all fossil fuel (coal, oil, natural gas) is extracted from deposits that are very old (tens of millions of years) relative to the half-life of ^{14}C (thousands of years), the CO_2 emitted by fossil fuel-burning power plants will all be "^{14}C dead." The ^{14}C concentration of the CO_2 will be zero. This gas is then mixed efficiently with the rest of the atmosphere as the wind blows the emissions from power plants downwind. The air downwind will therefore display a lower than expected ^{14}C concentration, reported as $\Delta^{14}CO_2$. That the degree of lowering is directly proportional to the emission rate means that the ^{14}C concentration of this air may be used to deduce the emission rate from power plants (e.g., Figure 6.11). This method has been proposed as a quantitative means of monitoring emission rates, replacing present methods that involve estimation based upon economic metrics such as gross national product.

Table 6.1 **Radionuclides and their half-lives**

Nuclide	$t_{1/2}$, half-life (yr)	λ, decay constant (1/yr)	τ, mean life (years)
^{10}Be	1.387×10^6	5.0×10^{-7}	2.00×10^6
^{14}C	5.73×10^3	1.21×10^{-4}	8.27×10^3
^{26}Al	7.05×10^5	1.42×10^{-7}	1.02×10^6
^{36}Cl	3.01×10^5	2.30×10^{-6}	4.34×10^5
^{41}Ca	1.04×10^5	6.67×10^{-6}	1.50×10^5
^{129}I	1.56×10^7	4.44×10^{-8}	2.25×10^7

Source: Half-lives after Prime lab web site, with updated ^{10}Be half-life after Chmeleff *et al.*, 2009; Korschinek *et al.*, 2009.

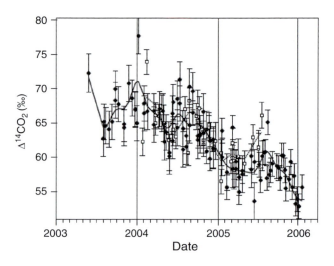

Figure 6.11 Time series and best-fitting filtered curve of $\Delta^{14}CO_2$ from Niwot Ridge, CO (closed diamonds; solid line) and upper tropospheric samples from New England (open boxes; dashed line) over a three-year period. Downward trend is attributable to increasing contributions of ^{14}C-dead CO_2 to the atmosphere from power plants (from Turnbull *et al.*, 2007, Figure 6, with permission from the American Geophysical Union).

time over which most of the Earth's landscape has evolved significantly, over which many of its geomorphic features have been produced (see Bierman and Nichols, 2004).

The cosmic rays responsible for the nuclear reactions are energetic particles originally accelerated to great speeds in supernova explosions elsewhere in the cosmos; hence their appellation "cosmic." The particles responsible are originally protons, or the nuclei of hydrogen atoms, which are stripped of charge upon encountering the Earth's magnetic field to become neutrons. Neutrons are far less reactive than protons, so can then penetrate more deeply into the atmosphere. As these particles descend into the atmosphere they encounter ever-increasing densities of gas, increasing the likelihood and frequency of nuclear interactions. The production rate at first increases with depth into the atmosphere, as there are more atoms with which to react. It then declines, as significant numbers of reactions sop up energy in the incoming particle rain. An example of the cascade of interactions and particles produced from a single incoming particle is shown in Figure 6.12. Such a diagram inspires wonder at the complexity of the process. It is the most energetic of these that are most capable of producing the nuclear reactions in atmospheric and near-surface materials.

Because these interactions are cumulative, the production rate of nuclides depends upon the depth into the material, and on the density of the material. In effect, what matters is the probability of the incoming particle impacting an atom, which therefore depends upon the number of atoms per unit volume, and the depth into the material. The parameter characterizing this interaction likelihood is Λ, the "mean free path." This may be translated into a characteristic length scale by dividing by the density of the material. For example, if $\Lambda = 160$ gm/cm^2, then a length scale over which the production rate declines significantly is Λ/ρ. We call this length scale z_* (although in the literature it is common to see Λ referred to as a length). Given that the difference in density between rock or soil and the atmosphere is about 2000-fold, the characteristic length scales will differ accordingly. For most of the nuclides of concern, the length scale z_* within rock is a large fraction of 1 m, while that in the atmosphere is about 1.5 km. That the production rate falls off with a 1 m scale in the Earth means

through time once produced. While we will focus on ^{10}Be and ^{26}Al, much of the discussion is relevant to all of the nuclides. That these nuclides are produced by interaction with common minerals on the Earth's surface (e.g., quartz, SiO$_2$), and that the half-lives of the nuclides are of the order of one million years, means that they are readily available, ubiquitous, and can be used to establish timing on the Earth's surface over most of the Plio-Pleistocene. This is the

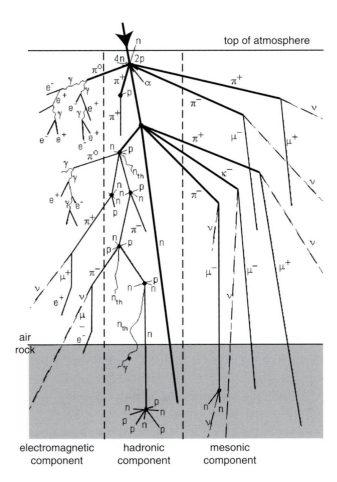

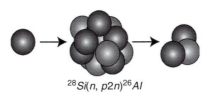

$^{28}Si(n, p2n)^{26}Al$

Figure 6.13 Diagram of spallation reaction producing ^{26}Al from ^{28}Si (from D. Granger, website, Purdue).

Figure 6.12 Cascade of particle interactions generated by the entrance of a high-energy particle at the top of the atmosphere. Cosmogenic nuclides produced both in the atmosphere and in the top few meters of rock most commonly result from at least secondary particles.

particle has spent within this zone – exactly the zone in which most geomorphic processes operate. Similarly, that the production rate depends upon altitude with a length scale of 1.5 km means that by 4.5 km, say the mean altitude of the Tibetan Plateau, or of the altiplano in South America, the production rate will be e^3, or roughly 25-fold higher than at sea level. Much to the benefit of the field geomorphologist who has to carry out his/her samples from the field, this implies that samples from such places can be much smaller (see Problem 3 on page 157).

Production occurs by several mechanisms, each of which may be written as a nuclear reaction. The most important reaction is the splitting of an atom, or "spallation" in which the target nucleus shatters into several smaller shards, as depicted in Figure 6.13. In the reaction shown, ^{26}Al is produced from splitting of ^{28}Si. The other shards produced are typically smaller, e.g., an alpha (α) particle, or nucleus of ^{4}He. In general, the reactions are numerous and quite complex. Another process is neutron capture, which is important for example as one mechanism involved in the production of ^{36}Cl. Here it is slow neutrons, braked by the numerous interactions within the atmosphere and near-surface rock, that are now slow enough to be captured into a nucleus, transmuting ^{35}Cl to ^{36}Cl.

In situ production profiles within rock

In situ production means production in place, in this case within the rock or soil near the Earth's surface. The production rate profile for ^{10}Be and ^{26}Al can be characterized by the sum of three exponential profiles, each with its surface production rate, P_o, and with a characteristic decay length scale. These are shown in Figure 6.14 in both linear and log plots. Recall that exponentials appear as straight lines when shown on a log–linear plot. It is apparent that near the surface the spallation mechanism dominates, while below some trade-off depth muogenic production mechanisms dominate. This cross-over depth is several meters, by which time the production rate is down to only a few percent of its surface value. In only a few applications must we worry about muogenic production.

that by a depth of several meters the production rate will be negligible. Since cosmogenic radionuclide accumulation will occur only within this top few meters, the concentration of cosmogenic radionuclides will reflect the time a parcel of rock or a sand

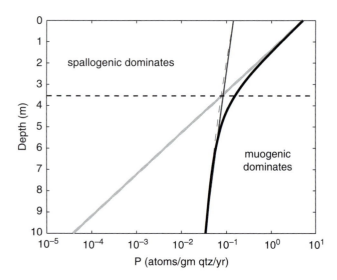

Figure 6.14 Profiles of production rate due to three main production mechanisms. Bold gray: spallation; black: total production. Dashed line is sum of production due to two muogenic processes, which are well approximated by the single exponential shown in the thin solid line. The cross-over depth (dashed horizontal line), given the assumed density, is roughly 3.6 m.

Variations in production rate in space and time

CRN production varies in both space and time. While the flux of cosmic rays arriving from the cosmos in the vicinity of the Earth is likely both isotropic and steady, the production of radionuclides in the Earth's atmosphere and at its surface is neither uniform nor steady. The primary particles involved are protons (the nuclei of hydrogen). Because they are charged, they are affected by magnetic fields. Production of cosmogenic nuclides is therefore modulated by the spatial pattern of the magnetic field of the Earth, by temporal variations in it, and by the solar wind which in turn is modulated by the Sun's magnetic field. Solar variations give rise to the well-documented 11-year cycle that is documented in atmospherically produced ^{10}Be measured in ice cores. The Earth's magnetic field steers charged particles toward the poles (witness the aurora borealis), and shields them from equatorial latitudes. This causes higher cosmogenic nuclide production at higher latitudes than at lower latitudes. The magnetic field of the Earth changes in both its effective dipole axis (magnetic pole location) and strength. Secular variation in the strength of the Earth's magnetic field results in high production rates at times of low field strength (and presumably a spike in production associated with a magnetic field reversal, during which the dipole strength lessens dramatically). It is largely this variation in the magnetic field that wreaks havoc with the CRN-based clock. We can use the tool with confidence only when we know the rate at which the clock has ticked in the past. In ^{14}C dating, the ^{14}C dates have been calibrated against tree-ring and varve chronologies that are independent and that extend almost the entire timescale over which the method is useful (say 40–50 ka). For other nuclides with million-year timescales (^{10}Be and ^{26}Al, for example), no such independent chronometer exists. Much effort in both European and US scientific communities has recently been expended to produce a reliable production rate history. This was done largely through measurement of CRN concentrations in surfaces of known age, from which the mean production rate over the age of the surface may be calculated. In the early use of the method, samples of glacially polished bedrock immediately up-valley of moraine-ponded lakes were measured (e.g., Nishiizumi et al., 1989). The independent age of the surface came from ^{14}C dating of basal peats in the lake sediments.

One other element of reality must be faced in CRN dating. Topography can shield or partially block the cosmic rays responsible for production, thereby lowering the production rate on a surface. We have already noted that production varies with elevation – the atmosphere serves to attenuate the cosmic ray flux. In calculating the topographic shielding at a site, we must assess the fraction of the incoming cosmic ray beam that is blocked by the local horizon. This is made somewhat complicated by the fact that the beam is not uniform with angle from the vertical (the zenith angle). The incoming beam of cosmic rays is presumed to be isotropic. But the travel path within the attenuating atmosphere is longer the higher the zenith angle. This results in a non-uniform contribution to the production of CRNs from different parts of the dome, lowest efficiency for near-horizontal angles, highest for vertical. It is this bell-shaped pattern that is truncated by the horizon. The user of the method therefore documents the angle to the horizon in many directions (commonly 8). For most low-relief sites the shielding factor is between 0.9 and 1.0, but can drop significantly in mountainous terrain.

Theoretical backdrop

It is the concentration of these rare nuclides in soil and rock that we use to deduce timing in the landscape. Given that nuclides are both produced and decay, we must take both processes into account in crafting an equation that governs the evolution of their concentration through time. Just as in many other systems described in this book, the word picture for the system is:

> rate of change of number of atoms in the box = production of atoms in the box – decay of atoms in the box

This is illustrated in Figure 6.15. If we cast this mathematically, we have

$$\frac{dC\,\rho dxdydz}{dt} = P\rho dxdydz - \lambda C\,\rho dxdydz \qquad (6.6)$$

where C is the number of atoms per unit mass of quartz, ρ the density of the rock, P the production rate of new nuclides per unit mass of quartz, λ the decay constant for the nuclide, and $dxdydz$ the volume of the box. This can be converted to an equation for the rate of change of nuclide concentration per unit mass of material (say atoms ^{10}Be per gram of quartz, something we measure) by dividing by the mass, $\rho dxdydz$, leaving

$$\frac{dC}{dt} = P - \lambda C \qquad (6.7)$$

We can anticipate the shape of the solution by inspecting this equation. Early in the history of accumulation, low concentrations of nuclides will result in only small contributions from the second term, and the rate of growth of the concentration should be steady, set by the production rate P. As concentration increases, however, the decay term will grow, forcing the rate of increase of concentration to decline. Ultimately, the concentration should become high enough that the production and decay terms balance, and a steady concentration should then be achieved. Setting the left-hand side of Equation 6.7 to zero, we find that this would occur when $C = P/\lambda$. Indeed, the solution to this equation, shown in Figure 6.16, reveals just this behavior:

$$C = \frac{P}{\lambda}\left[1 - e^{-\lambda t}\right] \qquad (6.8)$$

The early growth rate (slope on the plot) is set by the production rate, P. The asymptotic concentration is

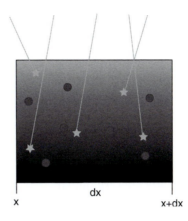

Figure 6.15 Production and decay of cosmogenic radionuclides. Cosmic rays interact with atoms in surficial materials to produce new nuclides (stars). Radionuclides decay (circles) with probability set by the decay constant. The rate of change of the concentration of radionuclides is therefore set by any mismatch between the birth and death rates.

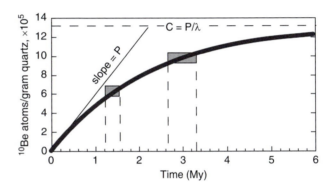

Figure 6.16 Approach of ^{10}Be concentration toward secular equilibrium. Initial slope is set by the production rate. The timescale for the approach is set by the decay constant, λ, the inverse of which, the "mean life," is 2.00 Ma. Uncertainty in measurement of [^{10}Be] (heights of gray boxes) leads to uncertainty in age (dashed lines), which increases with the age. At ages of several times the mean life, the concentration is no longer a good clock.

P/λ, at which time the system is said to be in "secular equilibrium." This plot also reveals the timescale over which the concentration of a radionuclide will be useful as a clock. Once the concentration gets close to P/λ, and the concentration is no longer changing significantly with time, the concentration can no longer be used to reveal time. The real limiting time for the method is dictated both by the characteristic time of decay (or the "mean life," $1/\lambda$) and by our ability to resolve concentrations in the method (see Figure 6.16, with error bars).

How are the measurements made: processing and AMS measurement

The measurement of the concentration of cosmogenic radionuclides from rocks is not easy. It includes mineral separation, purification of the CRN nuclides to generate a target, and measurement of the CRN nuclide concentrations in an accelerator mass spectrometer. The first step is the separation of quartz from all other minerals in the sample. This in itself is time-consuming, as the properties of quartz are very similar to those of feldspar. The most common method entails slow dissolution of the sand-sized grains in a dilute HF acid cocktail. As the dissolution of feldspar is slightly faster than that of quartz, many hours of stirred acid baths result in the sample "cleaning up" to pure quartz. After addition of a known quantity of very pure ^{9}Be or ^{27}Al spike (the common stable nuclides of these elements), the quartz is then dissolved and the Si fumed off to leave a scum on a vial that contains all the impurities in the quartz, including cosmogenically produced atoms of ^{10}Be and ^{26}Al, along with the spike. Isolation of these nuclides by ion chromatographic columns results in a sample target into which the CRN nuclides are tamped. Measurement of the CRN concentrations is done in an accelerator mass spectrometer facility, of which there are not many in the world. Most of these facilities are dedicated largely to ^{14}C measurements, mostly for medical analyses; the time windows in which ^{10}Be, ^{26}Al and other nuclides of importance are available to geomorphologists are short. The facility is very complex, as revealed in diagrammatic form in Figure 6.17. The heart of the procedure is a tandem accelerator in which the atoms sputtered from the sample target are accelerated to great speeds. The ratios of ^{10}Be/^{9}Be or ^{26}Al/^{27}Al are then measured by bending the resulting high-speed beam of ions around corners using magnets that are always the root of a mass spectrometer. What is remarkable is that we can routinely measure these ratios to levels of 10^{-15}, with errors of a few percent. It is hard to grasp this number, but here is one way. If Be atoms were sand grains, with a diameter of 0.1 mm (fine sand), and ^{10}Be were red while ^{9}Be were white, then measuring to levels of 10^{-15} corresponds to finding one red grain in a volume of 10^3 m^3, or roughly the volume of a typical one story house. Enough said! These are difficult measurements to make.

Over the last two decades geomorphologists have used the concentrations of cosmogenic radionuclides in surface materials to deduce the rates of many surface processes. We have dated abandoned bedrock surfaces, abandoned depositional surfaces, and cave deposits. We have obtained measurements of the rates of surface lowering (erosion) at points, and averaged over basins. We are also beginning to see the method employed to date stratigraphic sections going back into the Pliocene (>2 Ma). We will describe briefly each of these applications.

Dating bedrock surfaces

Once abandoned, a bedrock surface carved by some process, for example glacially or fluvially, will simply accumulate cosmogenic radionuclides at a rate dictated by the local surface production rate. As the concentration will follow the curve depicted in Figure 6.16, one may deduce an age for the

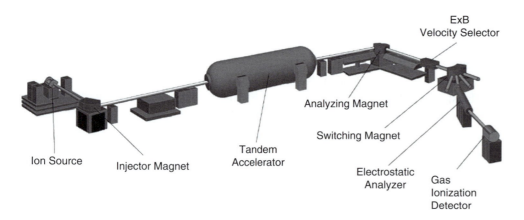

Figure 6.17 Diagram of AMS facility used to measure small concentrations of radionuclides (from PRIME lab website).

abandonment of the surface by solving Equation 6.8 for time. This is simply

$$t = -\tau \ln\left(1 - \frac{\lambda C}{P_{\mathrm{o}}}\right) \qquad (6.9)$$

where C is the measured radionuclide concentration and $\tau = 1/\lambda$. In the limiting case of young ages and hence low concentrations, the decay term in Equation 6.7 can be neglected and the age is simply the concentration divided by the surface production rate,

$$t = C/P_{\mathrm{o}} \qquad (6.10)$$

In only rare instances have very old surfaces been discovered. The reason is that all surfaces are experiencing *some* rate of surface lowering due to weathering; original surfaces that have experienced no erosion since their exposure by an erosion event in the past are therefore difficult to find. In general, one must use the existence of some kind of surface indicator, such as the patina of fluvial wear, or glacial polish, whose thicknesses are very small, to document the lack of surface lowering since abandonment of a surface. It is just such glacially polished surfaces that are used to document long-term mean surface production rates; Equation 6.8 can just as easily be solved for P_{o} if we know independently the surface age. In the case of glacial polish, such independent ages come from ^{14}C dating of peats in the base of moraine-ponded lakes, which were presumably roughly coeval with the formation of the last bit of polish on the bedrock just up-valley.

In Figure 6.18 we show an example of such a bedrock surface, sculpted by the channel of the Indus river, and then abandoned, and now more than 100 m above the river. Cosmogenic radionuclide dating of this and several other surfaces with similarly river-worn bedrock morphology has revealed a spatial pattern of long-term incision of the Indus river through its middle gorge as it tangles with the rapidly uplifting rock near Nanga Parbat (Burbank *et al.*, 1996).

Using a set of samples collected from glacially polished surfaces, Guido *et al.* (2007) documented the deglaciation history of a 90 km long valley draining the San Juan mountains in southwest Colorado. The measured ^{10}Be concentrations were first interpreted to be surface ages using Equation 6.9. All but one of the surface ages lined up in a monotonically declining trend with up-valley distance from the terminal moraine complex (Figure 6.19), which

Figure 6.18 Sampling bedrock strath more than 100 m above the Indus River, Pakistan, as it incises its Middle Gorge through the Himalayas. Exposure dating of this and other scraps of strath terraces revealed rates of incision as high as 1 cm/yr (photograph by R. S. Anderson).

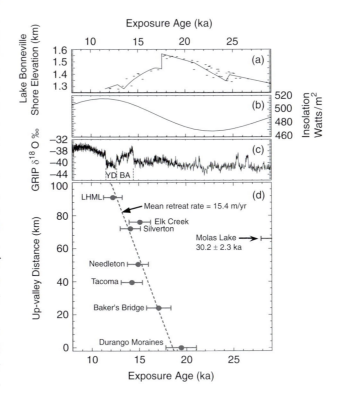

Figure 6.19 Ages of sites in Animas Valley, San Juan Mountains Colorado, based upon ^{10}Be concentration in LGM terrace, and in glacially polished bedrock (d), and various proxies for climate: (a) Lake Bonneville shoreline elevation; (b) insolation; (c) δ^{18}O from GRIP ice core, Greenland. Deglaciation takes at least 7 ka (after Guido *et al.*, 2007, Figure 3).

lends confidence to interpretation of the trend as a deglaciation history: the samples record the time since they were uncovered by retreat of glacial ice from the last glacial maximum. The one outlier, sampled from a resistant quartzite ridge, both illustrates the potential problem with the method and its potential utility as an erosion meter. Sampled from roughly half way up the valley, the outlier appeared to be twice as old (had twice the ^{10}Be concentration) as most of the samples, making it impossible to interpret as a time since deglaciation of that portion of the valley. It is instead interpreted to have experienced too little erosion during the last glacial cycle to remove the full inventory of ^{10}Be obtained during exposure to cosmic rays in the last interglacial. Given the decay of production with depth in rock, characterized by z_* ($= 0.7$ m), "full" resetting of the cosmogenic clock requires removal of at least $3z_*$ or $4z_*$ of rock, leaving 3% ($e^{-3} \sim 0.03$) to 1% ($e^{-4} \sim 0.01$) of the original inventory. Lack of full resetting therefore indicates less than say 2–3 m of erosion in the glacial cycle. If the glacial cycle at that point in the valley lasted 30 ka, this would translate into long-term glacial erosion rates of less than $3\,\text{m}/30\,\text{ka} = 0.1\,\text{mm/yr}$. It appears that glacial erosion rates are for the most part sufficient to reset the cosmogenic nuclide clock to zero. In only rare geologic settings is significant inheritance measured: either in very hard rock such as quartzite, or in massive granites such as some of the joint-free intrusives of Yosemite Valley.

Dating depositional surfaces

More common than bedrock surfaces are sediment-capped surfaces, some simply mantling bedrock surfaces with several meters of sediment (marine terraces, fluvial terraces, pediments), others filling the landscape more deeply (alluvial fans, fill terraces, moraines). These too, if sampled well away from edges of the surface that might be experiencing significant modification by either erosion or deposition, can be dated using cosmogenic radionuclides. There is, however, a problem. All of the sediment that accumulated to form the deposit came from elsewhere, and therefore spent some time within a few meters of the Earth's surface, and therefore accumulated cosmogenic radionuclides (e.g., Anderson *et al.*, 1996). This "inherited" component can be large compared to that obtained while sitting on or in the surface we wish to date.

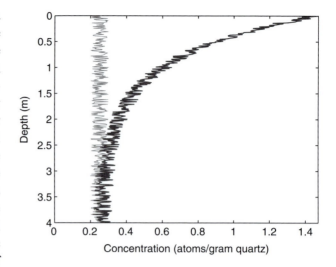

Figure 6.20 The expected shifted exponential of cosmogenic radionuclides in a deposit. The gray profile reflects the inheritance of radionuclides, averaged over many grains. The bold profile is the total concentration, which includes both inheritance and post-depositional grow-in of nuclides. Scatter at any depth largely reflects variation in the inheritance, as post-depositional production should be determined well by the depth into the deposit (in other words, it is deterministic).

In addition, the inheritance will inevitably vary from one grain to another. Each grain has its own history of exposure, and hence will arrive on the surface with its own inheritance. This gives rise to considerable scatter if one dates single cobbles on a surface. How do we see through this problem of inheritance? The solution is expensive. A method has evolved in which one measures the concentrations of several samples (hence the expense) in a vertical profile into the surface, each sample being an amalgamation of equal mass from many clasts. The amalgamation process effectively averages out the inheritance, the stochastic component in the concentration, to which has been added a deterministic component that varies systematically with depth due to the decline in production with depth. The expected profile is a shifted exponential, as seen in Figure 6.20, the shift being the mean inheritance of the deposit, and the exponential being the post-depositional accumulation of nuclides. Once the shift is constrained, which is best accomplished with one or more samples from several meters depth, the remaining exponential can be solved for the time since deposition.

This method has been used to date both marine terraces (see "Whole landscapes," Chapter 18, as it is

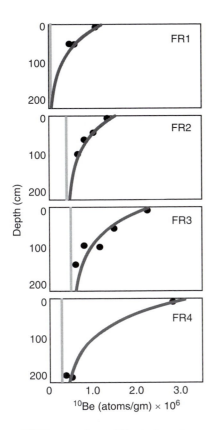

Figure 6.21 Profiles of ^{10}Be in four river terraces beside the modern Fremont River, Utah (from Repka *et al.*, 1997, *EPSL*, with permission from Elsevier).

employed on the Santa Cruz terraces), and fluvial terraces. In a fluvial example, the terraces that bound both the Wind River in Wyoming and the Fremont River in Utah show this shifted exponential well (Figure 6.21). As we will see below, the inheritance itself can be used to constrain rates of exhumation in the landscape contributing sediment to the river.

One must exercise caution in applying this method, as even flat landforms can fool us into thinking that they are static. For example, Hancock *et al.* (1999) showed that the very extensive, flat terraces of the Wind River in Wyoming, while flat and immune to hillslope erosion or deposition, have been buried significantly by loess. This windblown mantle of silt can apparently come and go. We find on many present terraces, in valleys subjected to high winds in the Pleistocene, that loess or silt caps are common in soils. These might only be a few tens of centimeters thick. In attempting to date such surfaces, where we had an independent date from a volcanic ash, the age we predicted from cosmogenic concentrations was

significantly too young – they were far lower than they should have been. The cosmogenic results therefore imply that the assumed production rates to which the samples were subjected were too low. This can best be explained by the surfaces having been buried by on average at least a meter of material, presumably loess, since their formation (Hancock *et al.*, 1999). As the loess is not there during the present interglacial, it must have been at least this thick during glacial episodes.

This problem of landform evolution is even worse on landforms that are not flat to begin with. For example, the method has been applied to date glacial moraines (see discussion in Briner *et al.*, 2005; and modeling of Putkonen and Hallet, 1994). These moraines begin as relatively sharp-crested landforms, mounds of material deposited around the fringe of a glacier, which are then abandoned as the glacier recedes. The landform subsequently evolves toward a more rounded form, through processes discussed in the Hillslopes chapter. The peaks decline, and the materials removed from them are deposited low on the flanks of the form. This means that a sample taken from the crest of the moraine was once in the subsurface; its production history will not have been steady, but will have grown through time to its present maximum value. The opposite is the case well down on the flank of the form. Researchers have either (1) ignored the problem altogether, (2) acknowledged the direction in which this process would push their age, and assert that the age deduced from the concentration is a lower limit, or (3) chosen the largest possible boulders on the moraine crest and hoped that these boulders reflect the original crest of the moraine.

While we have pointed out problems with the method and its application to date geomorphic surfaces of various types, we stress that this method is often the only quantitative absolute dating method available. The materials used, quartz, are almost ubiquitous.

Exhumation rates

We have already admitted that it is difficult to find surfaces that are not eroding by one or another mechanism, for example, by weathering and removal of grains. Rather than sulk about not having well-behaved geomorphic surfaces, we can make the best of the situation and use the cosmogenic radionuclide concentration to document this rate of lowering. Imagine the production rate history that would be experienced by a rock that is slowly and steadily being exhumed at a rate ε. We show in Figure 6.22 that as the surface approaches the rock parcel, the production rate will climb exponentially. Ignoring for the moment any decay of nuclides, the cosmogenic radionuclide concentration ought simply to climb at an ever-increasing rate. Upon being sampled at the surface, the concentration will be the integral of the production rate history. Cast mathematically, the concentration upon exposure will be

$$C = \int_0^\infty P_o e^{-\varepsilon\, t/z*} \mathrm{d}t = P_o z*/\varepsilon \qquad (6.11)$$

Here we have neglected decay. The concentration simply equals the product of the surface production rate, P_o, with the time it takes the rock parcel to pass through the last z_* to the surface, i.e., z_*/ε. If we include the decay of nuclides, the equation is slightly altered to

$$C = \frac{P_o}{(\varepsilon/z*) + \lambda} \qquad (6.12)$$

By inspection of this equation, one can see that in the limiting case of very low λ (long half-life), or very rapid erosion, high ε, this equation reduces to Equation 6.11. In other words, as long as decay is minor over the relevant time it takes to erode through the region in which production occurs, we are safe using Equation 6.11.

This method has been used to determine bedrock erosion rates at many sites around the world (e.g., see an early summary in Bierman, 1994). Examples of bedrock lowering rates in the alpine and desert areas of western North America reveal very slow rates of exhumation that are all only a few microns per year, or a few meters per millions of years (Figure 6.23; see also work on Australian desert landscapes by Bierman and Caffee, 2002).

Researchers have also placed constraints on the amount of erosion accomplished by the Last Glacial Maximum (LGM) ice sheet covering the eastern edge of Baffin Island (Briner *et al.*, 2006). As shown in Figure 6.24, they found that on the inter-fjord flats the ^{10}Be concentrations were much higher than could be accounted for by post-glacial accumulation of ^{10}Be. Instead, the LGM ice sheet must not have beveled

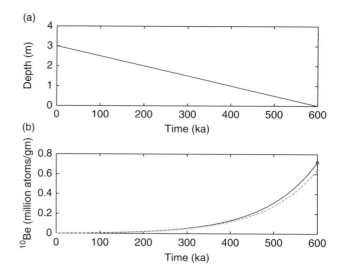

(a)

(b)

Figure 6.22 Numerical simulation of (a) depth history of a block of bedrock as it is exhumed at a rate of 5 microns/year over 600 ka, and (b) its history of accumulation of ^{10}Be atoms. The rate of accumulation accelerates as it comes closer to the surface, reflecting the rise in the production rate. Solid dot corresponds to the analytic solution ignoring decay (Equation 6.11). Dashed line: concentration history including decay of nuclides; open dot represents the analytic solution (Equation 6.12).

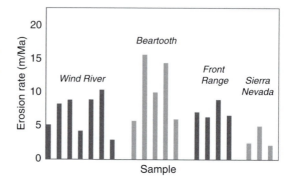

Figure 6.23 CRN-based erosion rates from bedrock in western US mountain ranges. Rates shown are based on ^{10}Be; ^{26}Al-based ages are comparable. Note that the average rate is ~7 m/Ma or 7 microns/year (after Small *et al.*, 1997, *EPSL*, with permission from Elsevier).

the surface sufficiently to reset the profile to near-zero. It had inherited ^{10}Be from prior interglacial exposure, meaning that LGM erosion was less than a few meters. In contrast, on the subaerial walls of the adjacent fjords, the ^{10}Be concentrations could be interpreted as reasonable deglaciation times, meaning that all inheritance from exposure during

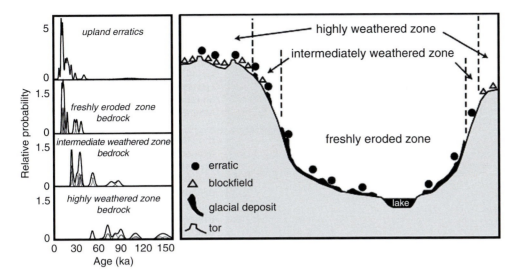

Figure 6.24 CRN dates on glacial erratics and exposure ages on bedrock in the fjorded edge of eastern Baffin Island. (a) Age distributions from sites in various landscape positions shown schematically in (b). Low erosion rates on highly weathered upland surfaces result in high effective ages there, contrasting with low (LGM) ages from the freshly eroded fjords. LGM ages of glacial erratics on the upland surfaces speak to gentle (non-erosive, and presumably cold-based) ice covering the inter-fjord highlands, while erosive (presumably temperate) ice gouges the fjords (after Briner *et al.*, 2006, Figures 2a and 5).

previous interglacial periods had been removed. Ice in the fjords is therefore more erosive than ice on the inter-fjord flats, meaning that the ice is sliding less rapidly on the inter-fjord flats than in the deep fjords. This may imply that ice was thinner and therefore polar (frozen to the bed) on the inter-fjord flats, and, being frozen to its bed, could accomplish no erosion.

Basin-averaged erosion rates

The method of documenting erosion rates at points in a landscape has now been extended to obtain average rates of exhumation within whole basins. Granger *et al.* (1996) showed that the measured average concentration of many sand grains can be used to deduce the average erosion rate of the contributing basin. This is a very powerful tool, as it achieves at once both a spatial and a temporal average rate. The spatial average comes from the fact that sand grains representing myriad sites within the basin can be sampled using just one sample from, say, the point bar on the exit stream. As you can hold a million sand grains in your hand, the average should be a very robust one! The temporal average is inherent in the calculation of the lowering rate from any particular bedrock site. The averaging time is the time it takes the parcel to traverse z_*. For example, if the erosion

rate ε is a common 10 m/Ma, and the length scale $z_* = 0.7$ m, then the timescale is $0.7/10$ Ma, or 70 thousand years. One must be aware of this timescale in interpreting a single rate, and in comparing this rate with others.

Granger *et al.* (1996) have successfully and elegantly tested this method in a small basin in Nevada in which a small alluvial fan built out onto and covered a shoreline of Lake Lahontan (Figure 6.25). As this shoreline was independently dated (using ^{14}C) to be 14000 years old, the volume of the fan could be divided by this time to derive a long-term (14 ka) average erosion rate in the contributing basin. Using this same sediment, the cosmogenic radionuclide concentration could be used to solve for the basin average erosion rate. The rates derived using the different methods were amazingly similar.

The method has since been used in many settings to map out the role of lithology, climate and tectonics in setting the basin-averaged lowering rates. Such cosmogenic radionuclide-based long-term average rates have been contrasted with stream sediment-gaging records to suggest that the gaging records miss large rare events (Kirchner *et al.*, 2001; Figure 6.26). Cosmogenic radionuclide-based basin erosion rates in Sri Lanka (von Blankenburg *et al.*, 2004) are surprisingly low, despite the fact that Sri Lanka has both

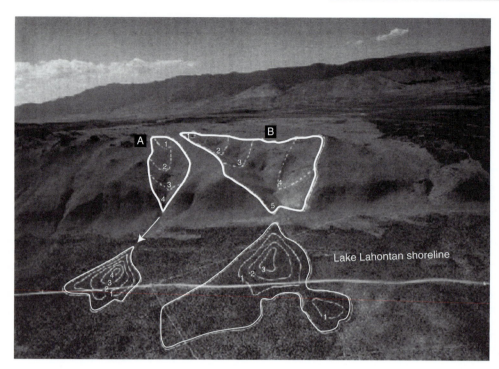

Lake Lahontan shoreline

Figure 6.25 Field site in which the basin-averaged erosion rate method employing [10]Be was tested on two adjacent catchments draining the edge of an escarpment on the Fort Sage Mountains. Fan volumes accumulated on a Lahontan Lake shoreline were documented to deduce basin-averaged erosion since 16 ka. Isopachs at 1 m intervals are shown in the deposits. [10]Be and [26]Al concentrations were measured in the same fan sediments, from which the basin-averaged erosion rate may be estimated. The estimates from the two methods correspond very well for both catchments: 5.8 cm/ka for catchment A and 3.0 cm/ka for catchment B (after Granger et al., 1996, Figure 1, *Journal of Geology*, with permission from the University of Chicago Press).

high relief and huge amounts of rainfall, both of which have been thought to drive high erosion rates. The method has also been employed to derive basin-averaged erosion rates in the past (paleo-erosion rates). In this twist on the method, researchers utilize the concentration of radionuclides attributable to inheritance from river terraces as the measure of basin-averaged erosion rates at the time of deposition of the terrace. As shown in Figure 6.27, if the fluvial system has many terraces, one can document the erosion rate history of the basin contributing sediment (e.g., Schaller *et al.*, 2004).

The key here is that documentation of these rates has not been available to geomorphologists until very recently. Surface lowering rates of the order of several microns per year are not measurable on PhD (or funding) timescales. Stream-gage-based sediment transport rates, from which erosion rates can be calculated, are spotty in space, and are at best 100 years in duration.

Burial ages

We can even date cave deposits using cosmogenic radionuclides. The method is based upon the fact that both [10]Be and [26]Al are produced in the same

materials (quartz), and that [26]Al decays roughly twice as fast as [10]Be ($\tau_{\mathrm{Be}10} = 2.00\,\mathrm{Ma}$; $\tau_{\mathrm{Al}26} = 1.02\,\mathrm{Ma}$). If quartz-rich sediment is washed into the cave by the river responsible for the dissolution of the rock, and is then sequestered far enough underground to prevent further production of radionuclides within the sediment, then the differential decay of the nuclides results in decay in the ratio of their concentrations. This ratio can then be used as a clock. Mathematically, the ratio may be expressed as

$$R = \frac{N_{\mathrm{oBe}}\mathrm{e}^{-t/\tau_{\mathrm{Be}}}}{N_{\mathrm{oAl}}\mathrm{e}^{-t/\tau_{\mathrm{Al}}}} = R_{\mathrm{o}}\mathrm{e}^{-t/\tau_R} \tag{6.13}$$

where

$$\tau_R = \frac{\tau_{\mathrm{Be}}\tau_{\mathrm{Al}}}{\tau_{\mathrm{Be}} - \tau_{\mathrm{Al}}} \tag{6.14}$$

Here R_{o} is the initial ratio (most often taken to be the production ratio, 6.75), and the mean life of the ratio, τ_R, is 2.08 Ma. In principal, therefore, the ratio can be used to date sediment as old as several of these timescales, or roughly 5 Ma. This is the essence of what has become known as the "burial age method" (Granger *et al.*, 1997; Granger and Muzikar, 2001). As with most methods, it has several advantages, but a few drawbacks. One principal advantage is that the

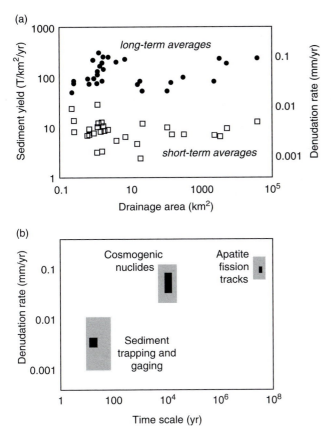

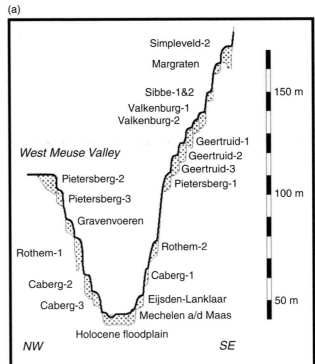

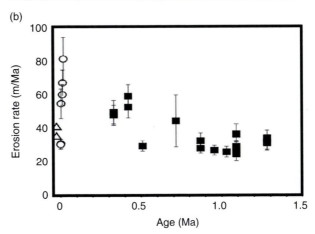

Figure 6.26 Comparison of sediment-gaging-based average sediment yields and cosmogenic ^{10}Be-based erosion rate estimates for catchments draining the Idaho batholith. (a) ^{10}Be-based methods (filled circles) suggest rates an order of magnitude higher than those derived from sediment gaging and trapping (closed squares), independent of drainage area. (b) Results from these methods are compared with several million-year average exhumation rates deduced from apatite fission tracks, which support the higher rates based on cosmogenic radionuclides (after Kirchner *et al.*, 2001, Figures 1 and 2).

Figure 6.27 The River Meuse displays a number of terraces up to more than 100 m above the modern river, (a) dating back to more than 1.3 Ma. ^{10}Be concentration profiles on the alluvial cover from these terraces can be used to deduce basin-averaged erosion rates in the contributing basin. Analysis of many such profiles yields a history of erosion rate in the River Meuse headwaters. (b) In general, these show roughly twofold acceleration of the erosion rate from early to middle Pleistocene, with the highest rates in late Pleistocene (redrawn from Schaller *et al.*, 2004, Figures 2 and 6, with permission from the University of Chicago Press).

method is immune to any temporal variations in the production rate, as it is simply decay that is the clock. The necessary depth of burial is a couple of tens of meters, deep enough to prevent production by both spallogenic and muogenic processes; most caves are at least this deep. The cave sediment must be quartz-rich, meaning that somewhere in the headwaters of the cave there must be a quartz-rich source of sediment. As it is only the inherited nuclides that are counted in this method, the concentrations of cosmogenic radionuclides start out small, and decay from there, halving every 2 Ma. Finally, we must know well the initial ratio of the nuclides, R_o. That we know this

to only roughly 10% (6.75 being the presently accepted value of the ratio of production rates, R_o), the ratio can only be known to that level. This places a lower limit on the utility of the method. Enough decay must

(a)

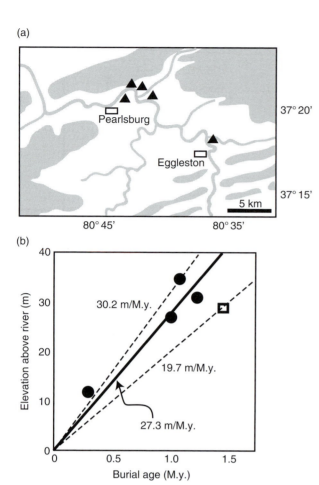

(b)

Figure 6.29 Phreatic passage in Snailshell Cave, Borneo. Such passages were at one time completely filled with water, leaving the walls sculpted from roof to floor (photograph by Greg Stock, with permission to reproduce).

Figure 6.28 (a) New River, Virginia (triangles = caves) as it passes through the valley and ridge province of the Appalachian mountains. Gray = ridges. (b) Incision history based upon the ratio of cosmogenic radionuclides (burial dating) of sediments in caves along the river. Bold line represents average incision rate based on all data (redrawn from Granger *et al.*, 1997, Figures 1 and 3).

have occurred to lower the ratio by more than 10% in order to have any confidence in the age. Practically, this means that caves less than 300 ka cannot be dated reliably at the moment. Nonetheless, that we can date these voids beneath the Earth's surface at all is remarkable, and opens up their use as a means of documenting the rates of incision of the streams responsible for them.

Several cave systems have been dated in this way. The original study was that of Granger *et al.* (1997), who demonstrated that the New River in Virginia has been incising at rates of 27 m/Ma averaged over the last million years (Figure 6.28). Subsequent work on

caves bounding other rivers in the Ohio-Mississippi drainage, including the extensive Mammoth cave system in Kentucky (Granger *et al.*, 2001), date back at least 2 Ma, and reveal a story of river incision that possibly reflects rearrangement of the Ohio drainage in glacial times.

Caves in the Sierra Nevada also tell a story of river incision history, in this case dating back to more than 3 Ma (Stock *et al.*, 2004, 2005). Most are surprised to hear that there are caves in the Sierras at all; in fact there are more than 300 mapped caves. While the bulk of the Sierras are granitic, the wall rocks of the batholiths were significantly metamorphosed, in places taking limestones up to marbles. It is in these marble septa in the walls of the river valleys draining the western slope of the Sierras that useful caves have been developed. The situation is ideal for burial dating because the granitic sediment in the headwaters is quartz-rich, and has been washed into several of these caves. Users of this method must be careful in selecting proper sedimentary deposits within the cave. Only those in passages that are demonstrably carved by the stream itself are useful if the goal is to determine river incision rates. Such passages are called phreatic passages, and have a distinctive cross-sectional shape, with evidence of full occupation by rapidly flowing water. The photograph in Figure 6.29 shows a particularly clear-cut example of such a passage from a cave in Borneo. The incision of Kings Canyon has been documented using several caves

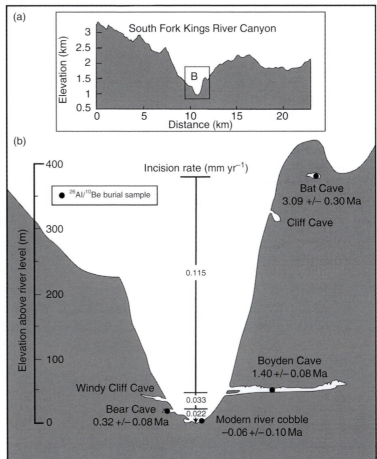

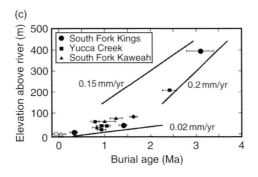

Figure 6.30 Use of cosmogenic radionuclide burial dating of sediments in Sierran caves to derive an incision history for the Kings River Canyon. (a) Cross section of Kings Canyon, indicating location of the caved inner gorge detailed in (b) and shown in photograph at left. Ages of caves shown span the last 400 m of incision. (c) Plot of cave ages vs. elevation emphasizes that the rate of incision was rapid from 3–1.5 Ma, and slowed thereafter (after Stock *et al.*, 2004 and 2005, *EPSL*, with permission from Elsevier).

at varying heights above the modern river. In effect, these caves act as strath terraces internal to the mountain. Using ages of four such caves, as shown in Figure 6.30, the incision is revealed to have been rapid from 3–1.5 Ma, and then slowed significantly in the Quaternary. This information greatly enriches our understanding of Sierran erosion, as prior to this work only the long-term rate of incision, between 9 Ma and the present, was known from dates on volcanic flows that came down ancient valleys.

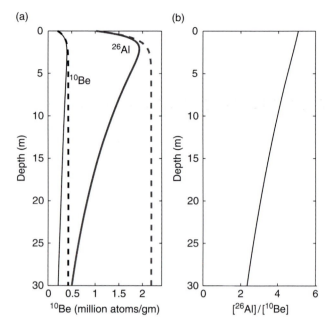

Figure 6.31 Profiles of ^{10}Be and ^{26}Al concentrations (a) and their ratio (b) in an accreting deposit. Initial concentrations (inheritance) are 1×10^5 and 4.8×10^5 atoms/gram quartz for ^{10}Be and ^{26}Al, respectively. Porosity of the deposit is taken to be 35%. The deposition rate is 20 m/Ma, and the total duration of deposition is 1.5 Ma. The production rate ratio is taken to be 4.8. Concentrations initially increase with depth while the sediment is in the zone of significant production, and then decline once decay overwhelms new production. Dashed curves ignore decay, and reveal the pattern due simply to accumulation in the production zone. The ratio monotonically declines with depth.

Use to date stratigraphy

Only very recently has this burial method been adapted to explore the depositional history of a sedimentary deposit. The twist here is that the burial to depths needed to halt all new production is not instantaneous, as in caves, but is progressive. Many deposits lack material that would allow dating by any other means. In particular, loess sequences, which in China go back to at least 3 Ma, are commonly dated using only magnetic stratigraphy (depths at which the samples are magnetically reversed and normal). Thick piles of gravel also lack datable material. As the sample is buried within the accumulating deposit, it continues to accumulate new nuclides until it falls below the production zone. Beyond this depth, the rate of decay exceeds the rate of production, and the concentration of each nuclide begins to decline. Only then does the ratio begin to decline as the rate of loss

of ^{26}Al is greater than that of ^{10}Be. The history of the concentrations of both nuclides, and of the ratio, might therefore look something like those shown in Figure 6.31.

This method can be pushed back to the beginning of the Plio-Pleistocene glaciations. Recently, Balco *et al.* (2005a,b,c) have used a modified version of the burial method to date buried tills from the great ice sheets that once covered the Midwest of North America. They target the paleosols developed in the tops of the tills, as these were stable surfaces for some potentially long period of time, and were subsequently buried quickly by a later till. The first feature allows the initial concentrations to be high. The second feature prevents any significant production of nuclides during the burial process – it is effectively instantaneously buried, just as sediment is instantaneously deeply buried in a cave. They report the age of the earliest glacial till to be around 2.4 Ma (Figure 6.32; Balco *et al.*, 2005a), which corroborates interpretations of onset of North American glaciation from the δ^{18}O record from the Gulf of Mexico. Later tills of 0.7 to 1.5 Ma are also reported at other sites (Balco *et al.*, 2005b).

Shallow geothermometry: establishing long-term rates of exhumation

We often wish to determine how rapidly a landscape has evolved over long timescales. In general, one would like to know how long it took a parcel of rock to reach the surface from some fixed depth. Ideally, one would like to measure the time since the rock crossed below a certain pressure. Knowing how to derive depth, z, from pressure ($P = \rho g z$), we could then determine the long-term exhumation rate (total depth/time). Unfortunately, information about *pressure* is not well preserved in a rock. Instead, a set of methods has been developed to enable us to document the time since a rock crossed below a particular *temperature*, as proxies for this are preserved within rocks. If we know how to deduce temperature from depth, then this serves the same purpose. We will see that the interpretation of information derived from these temperature proxies requires awareness of what controls the thermal field within the Earth, and how this evolves in time beneath an evolving landscape.

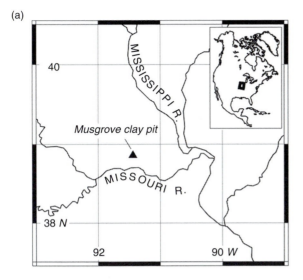

(a)

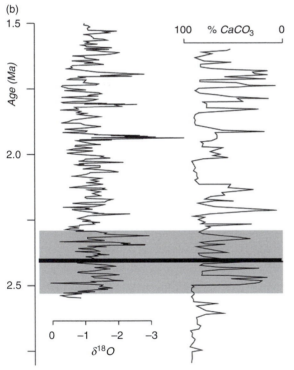

(b)

Figure 6.32 Dating the onset of North American glaciation from burial ages in early till sequence of the mid-west USA. (a) Map of sample site north of Missouri River, (b) $\delta^{18}O$ and $CaCO_3$ time series from deep sea cores, and cosmogenic date from sample (bar with shaded error band) (after Balco *et al.*, 2005a, Figure 1, with permission from the American Association for the Advancement of Science).

We have learned to make use of metamorphic minerals in rocks that emerge at the surface to deduce their pressure–temperature (P–T) paths through an orogen. Minerals form at specific P–T conditions.

One method therefore entails dating when the mineral formed, and hence when it crossed this phase boundary. In general, these reactions take place at high temperatures and tell us little about the lower temperature history of the rock as it then is exhumed. The revolution in the last few decades has involved the development of proxies for temperature histories of minerals subsequent to their formation at great pressures and temperatures. The temperature proxies come in two flavors: fission tracks and trace gases. Both are born of the decay of radioactive elements in mineral grains. Fission tracks are damage zones in a crystal lattice generated by the rare splitting of atomic nuclei (mostly U) into two pieces. The trace gasses are small shards from the fission itself, in particular ^4He nuclei (two neutrons and two protons) that are trapped within the crystal. In both cases the method is based upon the fact that the mineral grains can "forget" that these fission events ever happened if the grain is maintained above a certain temperature: the fission tracks heal, or anneal, and the small gas atoms diffuse out of the mineral. As the events are "remembered" at all lower temperatures, the proxies become a recording of how much time the grains have spent below these critical temperatures. The methods differ in the critical (or closure) temperatures (Table 6.2). Therefore, if used in concert with one another, they can document the times at which a sample crossed several temperatures en route to the Earth's surface, producing a temperature history (see example of this in Figure 6.33).

Fission tracks

Fission tracks (FT) are generated in any mineral in which elements capable of nuclear fission occur (^{238}U). These events are rare – about one for every two million alpha-decay events. But the heavy shards of the original ^{238}U nucleus (usually with mass numbers of about 90 and 140) do considerable damage to the crystal. The tracks are invisible, but can be revealed by lightly etching a polished section of the grain. Fission tracks began to be used as a chronometer in the 1960s by Naeser (1967) (see reviews in Gallagher *et al.*, 1998, and Tagami and O'Sullivan, 2005). The minerals commonly used in shallow thermochronometry are apatite and zircon. The damage zones, or fission tracks, are initially

Table 6.2 **Common thermochronometers and temperature ranges**

System	Mineral	Precision (%, 1 sigma)	Closure temperature (°C)	Activation energy (kJ/mol)
(U-Th)/Pb	Zircon	1–2	>900	550
	Titanite	1–2	550–650	330
	Monazite	1–2	~700	590
	Apatite	1–2	425–500	230
^{40}Ar/^{39}Ar	Hornblende	1	400–600	270
	Biotite	1	350–400	210
	Muscovite	1	300–350	180
	K-feldspar	1	150–350	170–210
Fission track	Titanite	6	380–420	440–480
	Zircon	6	230	210
	Apatite	8	90–120	190
(U-Th)/He	Titanite	3–4	160–220	190
	Zircon	3–4	160–220	170
	Apatite	3–4	55–80	140

(after Reiners *et al.*, 2005, Table 1; see this for full references to the methods)

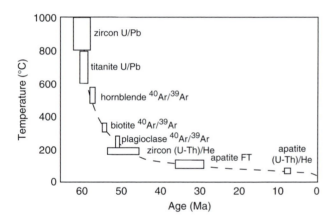

Figure 6.33 Temperature history of the Speel pluton, Alaska, derived from several thermochronometers. The data reveal rapid cooling after emplacement, followed by rapid cooling caused by exhumation after 10 Ma (after Reiners, 2005, Figure 3A and references therein, with permission from the American Geophysical Union).

measured as well, as a measure of how rapidly fission tracks should be produced in the mineral. The temperature below which the tracks heal at such slow rates that they do not change significantly over geologic time is called the closure temperature. The temperature range within which the healing rate is fast enough to alter track lengths but not fast enough to erase them over the time spent at that temperature is called the partial annealing zone, or PAZ. In Figure 6.34 we reproduce an example in which this partial annealing zone has been documented in samples taken from a borehole. Samples taken along vertical transects in mountainous topography have also been used to detect exhumation of such a PAZ by recent erosion.

Ar/Ar thermochronometry

^{40}K is a radioactive isotope, constituting roughly 0.012% of the naturally occurring potassium, and decays to ^{40}Ar with a half-life of 1.25 Ga. Ar is a noble gas capable of diffusing out of mineral grains either through the lattice of the grain (volume diffusion), or along grain boundaries (grain boundary diffusion). As in all diffusion problems, the rate of diffusion is strongly dependent upon the temperature of the medium. At high temperatures the daughter ^{40}Ar is free to escape the grain, and the system behaves as an open system. As the temperature drops,

about 15 microns long. Above a certain temperature, the tracks shorten in length by annealing at each tip, the rate of annealing being a strong function of the temperature. In modern FT analyses, both the number and the length distribution of tracks are measured. The mineral grain is mounted, beveled, etched lightly with acid to widen the tracks sufficiently to make them visible in a lab microscope, and the track number and lengths are documented. The concentration of U in the sample must be

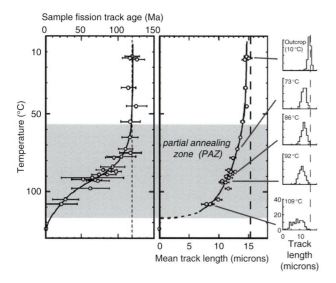

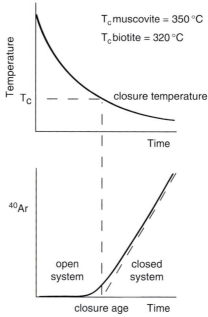

Figure 6.34 Fission track analyses of samples from a depth profile in the Otway Basin, southeastern Australia, shown against present temperatures. Stratigraphic age of the package is 120 Ma (thin dashed line). Track length distributions (right) shown for several of the samples, and mean track lengths for all samples (middle) show significant shortening of track lengths above temperatures of about 55 °C, defining the lower temperature boundary of the partial annealing zone (PAZ). Vertical dashed lines: mean original track lengths (~16 microns). For greater temperatures and hence depths, the age deduced from track numbers (left) will under-represent the true age of the rock (after Gallagher *et al.*, 1998, Figure 6, with permission from Annual Reviews).

Figure 6.35 ^{40}Ar/^{39}Ar system. The system is open to diffusive loss of ^{40}Ar at high temperatures, becoming progressively more closed and retentive of ^{40}Ar as temperature cools during exhumation.

however, and ^{40}Ar continues to be produced, more and more of it is retained in the grain, and the system moves toward being quantitatively closed. Below a closure temperature of roughly 350 °C, muscovite is effectively closed to such diffusion, and quantitatively retains the gas. The closure temperature for biotite is slightly lower, ~320 °C (Figure 6.35). Note that this is a relatively high closure temperature. In contrast, the ^{4}He atoms generated by alpha-decay in the U/Th system discussed below are much smaller and can diffuse much more readily in crystal lattices. This results in the low closure temperatures of the U/Th-He thermochronometer. (They are the He bicycles, as opposed to the Ar trucks, cruising down the narrow crystal lattice streets.)

The ^{40}Ar/^{39}Ar tool has been traditionally used to infer cooling ages of rocks. More recently analytical improvements have allowed its use on individual grains, promoting the popularity of new detrital methods. We illustrate in Figures 6.36 and 6.37 the use of rock samples in deriving an exhumation history from a site in Nepal. Here Wobus *et al.* (2008) analyzed biotite grains separated from rock samples collected in a vertical transect on a steep valley wall. As shown in the schematic in Figure 6.36, the slope on the age–elevation plot may be interpreted as the exhumation rate. At this site in Nepal, the exhumation rate appears to have increased dramatically at about 10 Ma.

With the development of the ^{40}Ar/^{39}Ar laser microprobe, we can now assess the quantity of Ar in individual grains. This has greatly refined the ability to date not only whole rocks but many individual grains within a single rock. And it has also opened the door to the use of this tool in detrital thermochronometry, in which the distribution of cooling ages of individual grains from a sedimentary deposit, or from a modern fluvial system, can be compiled. Armed with this information, we can now invert the cooling age distribution for the cooling history of the catchment comprising the headwaters from which the sample was derived. The target grains are K-bearing minerals of K-feldspar, muscovite and biotite. Ages from each of more than 100 grains are determined by laser fusion of the grain, from which a graph of the

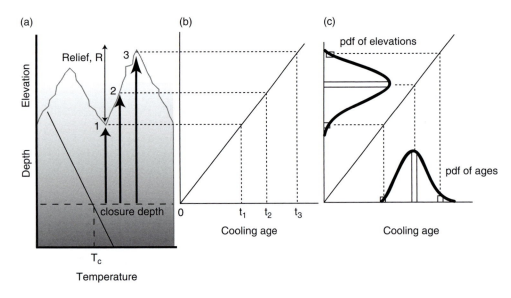

Figure 6.36 (a) Use of elevation profiles to infer exhumation rate in a steady-state landscape. (b) Expected age-elevation distribution in a uniformly eroding landscape such that the rate of erosion in the valley bottom is identical to that at the crest of the mountain. The long-term erosion rate can then be determined to be $R/\Delta t$, where Δt is the difference between ages at the mountain crest and valley trough. (c) The detrital method in which the pdf of ages from a sediment sample in a stream draining this landscape is interpreted in the light of the pdf of elevations in the catchment (its hypsometry). In a uniformly eroding landscape and steadily eroding landscape the pdf of elevations should map directly onto the pdf of cooling ages. Note that the interpretation of the cooling ages depends upon one's assumption about the geothermal gradient, which dictates the assumed cooling depth (after Brewer *et al.*, 2003, Figure 2, with permission from Blackwell Publishing).

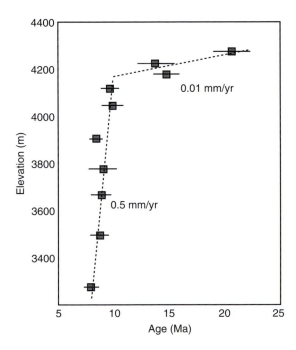

Figure 6.37 Elevation profile of biotite ^{40}Ar/^{39}Ar ages from Langtang, Nepal, showing 50-fold acceleration of exhumation rate at roughly 10 Ma. Bars depict full range of measured ages from ten laser-fusion analyses at each elevation (after Wobus *et al.*, 2008a, Figure 3, *EPSL*, with permission from Elsevier).

sort shown in Figure 6.36 is obtained. This is the sum of the probability density function of the individual grain ages (the synoptic pdf, or SPDF). It is this graphic that is then interpreted in the geological context of the sample.

Several uses of the ^{40}Ar/^{39}Ar method relevant to geomorphology have evolved over the last decade (summarized in Hodges *et al.*, 2005). These include (1) determination of the provenance of a sample, where the various bumps on the SPDF are attributed to rock from a portion of the catchment with a known age distribution; (2) timing of the exhumation in the catchment; (3) determining the lag time between exhumation and deposition, which in the case of small lag allows estimation of the mean exhumation rate in the source region; and (4) constraining modern erosion rates in a catchment (e.g., Brewer *et al.*, 2003). As noted by Stock and Montgomery (1996), in this method one must acknowledge the probability distribution of elevations from which the grains have been sampled within the catchment (the hypsometry, shown in Figure 6.36). And (5) documenting the location of strong gradients in exhumation, which in turn

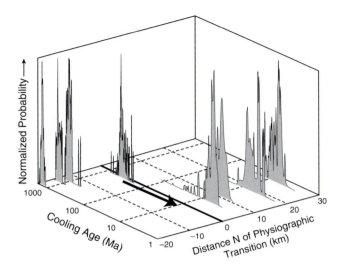

Figure 6.38 Distributions of $^{40}Ar/^{39}Ar$ muscovite ages from detrital samples from small catchments north and south of a major physiographic break in the Himalayan front, central Nepal. The 20–30-fold drop in detrital age across the physiographic break (arrow) implies significant activity on a surface-breaking fault, driving rapid exhumation of rocks to its north (after Hodges *et al.*, 2005, Figure 5; see also Wobus *et al.*, 2003).

can illuminate the presence of geologic structures that are otherwise difficult to observe (e.g., Hodges *et al.*, 2004; Wobus *et al.*, 2008a, as shown in Figure 6.38).

(U-Th)/He method

Several ^{4}He nuclei, or α particles, are emitted in the decay of U and Th to stable Pb. As in fission track analyses, both zircon and apatite are the primary minerals of choice, although some use titanite (sphene). The closure temperature of apatite to He loss is $68 \pm 5\,^{\circ}$C (Farley quoted in Reiners, 2002). This roughly $70\,^{\circ}$C closure temperature is much less than the $110\,^{\circ}$C closure temperature for fission tracks in apatite, making (U-Th)/He the lowest temperature thermochronometer in common use at present. This method can reveal the time it has taken a rock parcel to move through the last 2–3 km to the Earth's surface ($70\,^{\circ}$C/($25\,^{\circ}$C/km)$=3$ km), making it relevant to the evolution of mountainous topography whose relief is often of the same order. The method entails mass spectrometric measurement of the concentration of ^{4}He from a very small mass of the mineral, sometimes derived from laser ablation of the surface. The concentrations of the parent nuclides U and Th are then measured in ICP-MS.

The basis of a U-Th/He geochronometer is that the ^{4}He concentration in a mineral grain is a clock. The basic equation governing the system is a modification of the radioactive decay equation that tracks not the parent but the daughter population:

rate of change of daughter population = rate of gain by decay of parent(s) – loss by diffusion

In symbols, this becomes

$$\frac{dD}{dt} = -n\frac{dN}{dt} - \text{diffusion} \tag{6.15}$$

where n is the number of daughter atoms generated by the decay of a single parent atom. We can also proceed by recognizing that the number of daughter atoms will be the product of n with the number of parent atoms that have decayed to that time, i.e.,

$$D = n(N_o - N) = n(N_o - N_o e^{-\lambda t})$$
$$= nN_o(1 - e^{-\lambda t}) \tag{6.16}$$

by appeal to the radioactive decay equation for parent atoms. The problem is that in this system what we measure is not the number of original parent atoms, but the present number, N. Solving the radioactive decay equation for N_o and replacing N_o in Equation 6.14 yields

$$D = nN e^{\lambda t}(1 - e^{-\lambda t}) = nN(e^{\lambda t} - 1) \tag{6.17}$$

Knowing n, and measuring N and D, we can invert this equation for time, t. For the (U/Th)/He system, this becomes (see Reiners, 2005, Equation 1):

$$^4\text{He} = 8^{238}\text{U}[e^{\lambda_{238}t} - 1] + 7^{235}\text{U}[e^{\lambda_{235}t} - 1]$$
$$+ 6^{232}\text{Th}[e^{\lambda_{232}t} - 1] \tag{6.18}$$

The number of ^{4}He or alpha particle daughters produced from decay of these three parents varies from six to eight.

Use of such shallow, low-temperature geochronometers requires that we acknowledge another source of complexity in our interpretation of the concentrations. As we show in Figure 6.39, because the $70\,^{\circ}$C isotherm is only about 2–3 km deep, it is likely to be significantly warped within high mountainous relief. When interpreting the results to derive an exhumation rate for a particular sample, one must therefore pay particularly close attention to the thermal structure through which these rocks have passed en route to the surface. The isotherms are not only warped by the topography, they will evolve as the topography

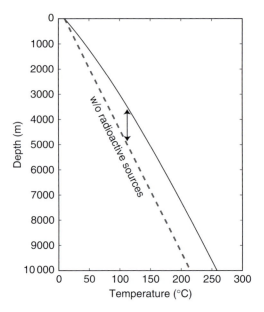

Figure 6.39 Modeled temperature profiles in the upper crust demonstrating the role of minerals containing short-lived radioactive elements in altering the thermal structure. Steady profiles show curvature reflecting the need to pass increasing amounts of heat as the surface is approached. Mantle heat flux taken to be 41 mW/m²; conductivity of rock = 2 W/m-K. Radioactive heat source is distributed exponentially, with surface source of 10^{-5} W/m³, and scale for decay of source with depth of 3 km. Arrow depicts the difference in depths at which a thermochronometer would pass through its closure temperature (here taken to be 120 °C).

evolves. A rock parcel exhumed in the valley bottom will have experienced significantly greater thermal gradients than a rock exhumed on the ridge crest. In modeling the thermal structure, it is also important to acknowledge the role of radiogenic elements in the shallow crust, as their concentration in the top few km of the rock column will increase the thermal gradient, which in turn alters the depth at which a particular isotherm is crossed. One can see this bend in the geotherm easily in Figure 6.39.

It is also possible for the temperatures of rocks at the surface itself to be raised above 70 °C by natural processes (see our discussion of hot processes in the Weathering chapter). If the surface of the rock is dark enough, direct exposure of the rock surface to the sun can raise the surface temperature to nearly 70 °C, inducing some diffusive loss of He. Forest or brush fires burn at much higher temperatures than this. Although the exposure to such temperatures is brief, and can cause damage to the rock surface (see

Weathering chapter), they can again cause loss of He. This fact has been cleverly used to determine the ages of natural coal-seam fires in Wyoming, where brush fires sweeping across the landscape have ignited outcropping Cretaceous coal seams. The fires burn downward along the dipping coal seam into the sub-surface, dying only when they can no longer obtain enough oxygen. The result is the baking of the surrounding shales to red clinker (Figure 6.40). As these rocks are the hardest in the landscape, red clinker hills dominate the subtle landscape around the edges of northern Wyoming's Powder River Basin. He dating of these beds has revealed how long ago these fires occurred, documenting the timing of late Cenozoic basin exhumation (Heffern et al., 2008).

Because the closure temperature is so low, it is also possible for the clock to be reset by surface processes involving heating, in particular wild fires (Mitchell and Reiners, 2003). While this represents a problem for documenting exhumation, the problem can be turned around and used to assess the intensity of past wildfires. The depth in the rock to which the resetting occurs reflects the temperature of the fire and its duration, as we discuss in Chapter 7 on weathering. The use of two minerals with different closure temperatures allows documentation of the penetration of the thermal wave (Mitchell and Reiners, 2003).

Another characteristic of these systems is that the ejected alpha particle travels a fair distance from the original atom in the fission event. This is indeed the root of the fission track method. But it poses problems in the (U-Th)/He method because some fraction of the fission events will result in ejection of the alpha particle from the mineral grain. This loss must be corrected for. While the calculation is somewhat involved, the principle of the correction is straightforward, as illustrated in Figure 6.41. One must assume a mineral shape and know the characteristic stopping distance, call it δ. Assuming that the ejection angle is random with respect to the grain boundary, statistically speaking those events that occur within half the stopping distance of the wall will result in ejection. Calling the stopping distance δ, the expected under-representation of ^{4}He in the grain is simply the fraction of the grain volume represented by a shell of thickness $\delta/2$. Consider an idealized case. Let the grain be spherical of diameter D. Then the expected loss is the shell volume divided by the grain volume:

$$F = \frac{(\delta/2)4\pi D^2/4}{\pi D^3/6} = 3\frac{\delta}{D} \qquad (6.19)$$

Calculations for other more realistic grain shapes result in similar expressions. They always go as δ/D. The effect is inversely proportional to the grain diameter, meaning the smaller the grain the larger the correction.

Yet another twist on this method has evolved in the last couple of years. Researchers are now beginning to exploit the concentration profiles within apatite crystals as a signal of thermal history to which the rock has been subjected (Shuster and Farley, 2003, 2005). The method is called the $^4\text{He}/^3\text{He}$ method because the profile is documented by much more precise measurement of the ratio of He isotopes. The ^3He is emplaced

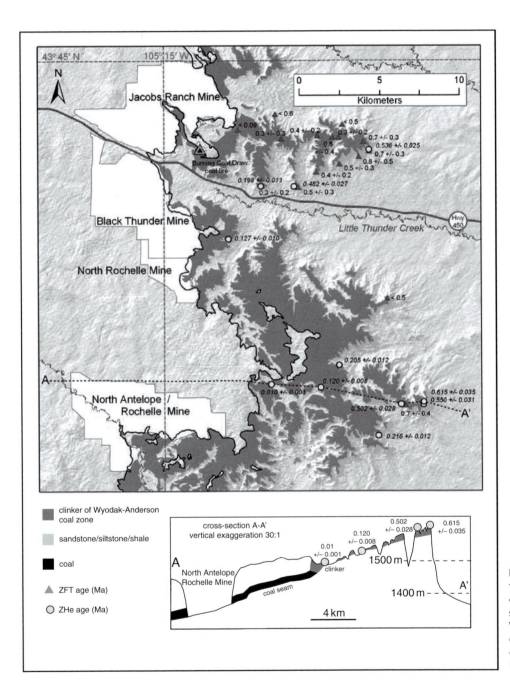

Figure 6.40 Clinker beds from the natural burning of the outcrop of the Wyodak coal seam, Powder River Basin, Wyoming. Ages shown are derived from FT and U/Th-He on clinker beds (after Heffern et al., 2008).

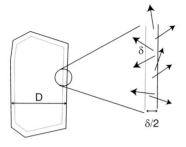

Figure 6.41 Alpha rejection. Fission of radioactive U and Th atoms results in rejection of some fraction of the ^{4}He atoms (alpha particles) from the mineral. Statistically, half of the events within an outer shell of width $\delta/2$ will produce alpha particles that are lost from the mineral.

uniformly in the crystal by intense radiation, usually in a medical irradiation laboratory. This method can theoretically allow us to deduce temperature paths down to about 30 °C. On a typical geotherm (25–30 °C/km) this temperature range is within about 1 km of the surface, meaning that the method should record the timing of the last 1 km of erosional history of the rock. Documentation of the concentration profile within apatite grains is done indirectly, by progressively increasing the temperatures to which the apatite grains are subjected; the first gases emitted are from nearest the grain edges, the later gases from the interiors of the grains. This method has recently

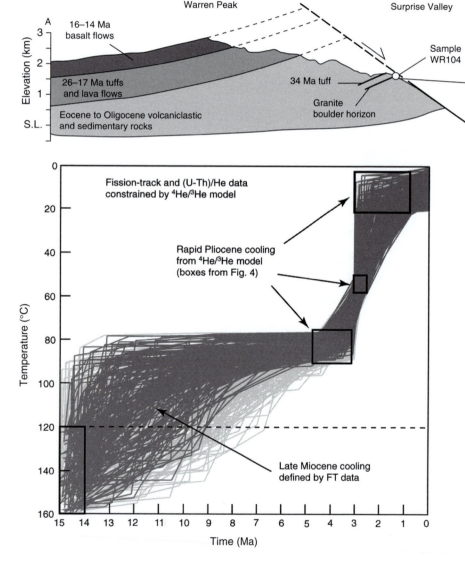

Figure 6.42 Cooling history derived from thermochronometric analysis of a single sample from normal-faulted Warren Peak bounding Surprise Valley, Nevada (top cross section). Geologic cross section suggests faulting-related exhumation and associated cooling must begin after 16–14 Ma basalt flows cap the landscape. Three methods (fission track, (U-Th)/He, and ^{4}He/^{3}He) are combined in a single model to deduce the cooling history. Two episodes of rapid cooling are separated by a period of slower cooling (after Colgan et al., 2008, Figures 1 and 5).

been used to deduce erosion of a fjord into the mountains of British Columbia, telling us for the first time that at least at this site the fjord was rapidly emplaced at around 1.8 Ma (Shuster *et al.*, 2005).

The two-stage exhumation history of a mountain range in the Basin and Range province of the western USA has also been assessed by using multiple thermochronometers, including the ^{4}He/^{3}He method (Figure 6.42; Colgan *et al.*, 2008). The long-term exhumation history of the eastern margin of Tibet has also been assessed by application of several thermochronometers shown in Figure 6.43, revealing rapid exhumation initiated in the Mio-Pliocene.

The history of exhumation can also be deduced by applying thermochronometry to populations of detrital grains within a stratigraphic section. An example is illustrated in Figure 6.44.

To derive depth histories and hence exhumation histories, the temperatures obtained from thermochronometric methods must be turned into depths. This requires either simplifying assumptions about the thermal structure of the Earth, or models of how the thermal structure will evolve over the course of the exhumation. For high-temperature methods involving closure temperatures of $> 200\,°C$, it is reasonable to assume that the thermal field at the time of

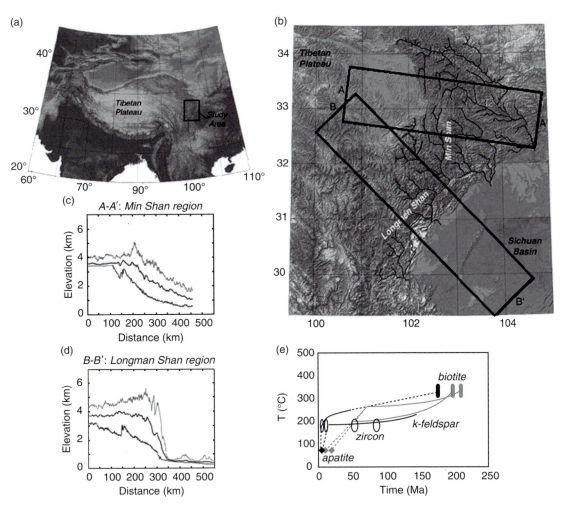

Figure 6.43 Topography of eastern margin of Tibetan Plateau shown in (a), detailed in (b). Swath profiles of topography roughly perpendicular to the margin of the plateau in two areas (c and d). (e) Thermal history of the region as deduced from several thermochronometric systems. Lines connect data from individual samples. The thermal data indicate slow cooling through the Mesozoic and early Cenozoic, with rapid cooling, implying rapid exhumation of the Tibetan margin, in the Mio-Pliocene (after Kirby *et al.*, 2002, Figures 1 and 9, with permission from the American Geophysical Union).

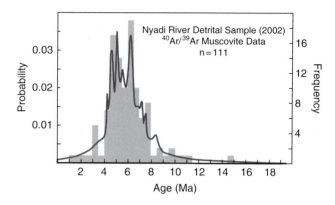

Figure 6.44 Example SPDF of detrital ^{40}Ar/^{39}Ar muscovite ages. Relatively young ages suggest rapid exhumation in the headwaters of the Nyadi River, Nepal (after Hodges *et al.*, 2005, Figure 1, with permission of the Mineralogical Society of America).

closure was simple (essentially one-dimensional), that the isotherms were horizontal and were uninfluenced by the details of the topography above. One must only assume a geothermal gradient at the time. If enough information exists in vertical transects of multiple thermochronometers, it is possible to constrain this paleo-gradient.

The picture is not so simple in lower closure temperature systems. Here the proximity of the Earth's surface warps the isotherms. In general, isotherms are compressed beneath valleys, and expanded beneath ridges as the movement of heat becomes two- to three-dimensional rather than one-dimensional. Because the topography influences the thermal structure to depths of perhaps twice the total relief in the landscape, it is also possible that evolution of the topography will influence the signal (see Safran, 2003). This is shown in Figure 6.45. In other words, one must acknowledge not only the spatial distribution of temperatures in the rock mass, but the evolution of the topography and co-evolution of the thermal field in the interpretation of these temperature data. Most recently, these complexities have led to the development of numerical codes in which both topography and thermal fields are modeled to predict the T-paths of rocks emerging at any location on the Earth's surface (e.g., Braun, 2005; Ehlers, 2005; Ehlers and Farley, 2003; see discussion in Reiners *et al.*, 2005).

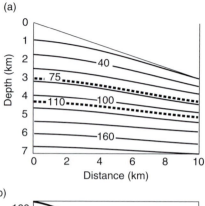

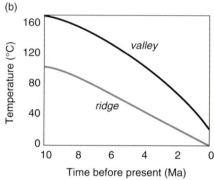

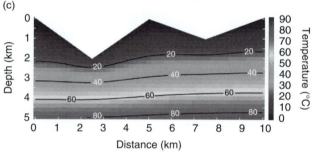

Figure 6.45 Thermal structure in a steadily eroding ridge-valley transect (a) and the resulting temperature history of rocks emerging on the ridge and in the valley (b). Isotherms corresponding to FT (110 °C) and (U-TH)/He thermochronometers (75 °C) are shown as bold lines. Isotherms are compressed beneath the valley floor and are expanded near the ridge crest. Closure temperatures are shown as bands in (b). A valley bottom sample would cross through the FT closure temperature at ~4.5 Ma, and through the (U-Th)/He closure temperature at ~2 Ma (after Safran, 2003, Figure 2, with permission from the American Geophysical Union). (c) Steady-state temperature field in a cross section through mountainous topography. Mantle heat flux is taken to be uniform in space, and no radioactive elements contribute to the heat flux. Isotherms, reported in 20° contours, are compressed beneath valleys, and are expanded beneath ridges. The effect of topography is minor below a depth equivalent to the relief of the valley (calculation courtesy of Dylan Ward).

Summary

A technological revolution within the last two decades has placed geomorphology on a much more rigorous timeline that extends millions of years, covering the timescale over which most landscapes have evolved. The materials that constitute the clocks now in use are commonly available in most rocks, from the quartz used in CRN dating to the accessory minerals employed in thermochronometry. We can now assess how rapidly mountains are being exhumed at timescales of millions of years to thousands of years. We can compare basin-averaged rates to rates at a point. This revolution in the documentation of timing has gone hand in hand with development of models of landscape evolution that can now be rigorously tested.

On the flip side, the careful interpretation of the cosmogenic and thermochronometric data requires increasingly sophisticated models in which we acknowledge the geomorphic setting of a sample. The evolution of the landscape governs both the production rate history and the thermal history of any particular sample. The models require yet again that we craft formal equations for the conservation of some quantity, in this case the numbers of elements, nuclides, or fission tracks.

Problems

1. You are asked to date a shoreline that is found well above the margin of a present day lake. This is important paleoclimatic information for the basin, and you need to know how long ago the lake was at this elevation. You find a tree stump and sample its outer rings to date using the radiocarbon method. You have this analyzed at Beta Analytic Labs in Florida, and they report back that only 21% of the original ^{14}C remains in the sample.

   (a) How old is it, given that the half-life of ^{14}C is 5730 ± 40 years? (This was determined by a set of researchers at Cambridge, and is known in the community as the Cambridge half-life.)

   (b) Given the range of estimates of the half-life, calculate the associated range in age estimates. (This will be a minimum estimate of the range for several reasons, in part because there will be additional error associated with error in the lab analysis.)

2. On the Huon Peninsula of Papua New Guinea, a set of coral marine terraces grace the coastline.

These are datable using U/Th methods, and have yielded good dates back to several hundred thousand years. One of these terraces 120 m above modern mean sea level is dated to be 80 ka.

   (a) What is the long-term rate of rock uplift on this coastline?

   (b) What assumptions have you made in performing this calculation?

3. You are asked to date a moraine in the Indian Peaks of the Front Range, Colorado, and choose to do so using ^{10}Be concentrations in moraine boulders. One of the key calculations you have to make in this business is how big a sample to collect.

   (a) Using the following constraints, estimate how much sample you will need to collect. Report the answer in kilograms. The AMS method requires that you provide them with two million atoms of ^{10}Be.

   • The moraines are located out in the valley well away from any valley walls that would mask the radiation from cosmic rays.

- We expect the moraines to be last glacial maximum (LGM) in age.
- The rock type is a granite with roughly 30% quartz.
- The elevation at the site is 8236 ft.
- The ^{10}Be production rate at sea level is 4.8 atoms per gram of quartz per year.
- The rate of increase of production rate with elevation is an exponential with a length scale of 1.5 km. (In other words, the equation for production rate is $P_o = P_{osealevel}{}^*e^{(z/z*)}$, where $z* = 1.5\,km$)

(b) If this mass is a cube, what is the length of a side of the cube, in meters? (In other words, just how big is this?)

4. You have now gone through the agony of processing the sample and have had it analyzed at the Purdue PRIME lab. They report back and provide a concentration from which you calculate that the concentration of ^{10}Be is 6.1×10^5 atoms/g of quartz. How long has the boulder been exposed to cosmic rays – in other words, how old is the moraine? Do the calculation in two ways:

(a) assuming that there is no decay of ^{10}Be over the age of the boulder;

(b) taking into account radioactive decay. (*Hint*: here you will have to use a slightly modified equation to calculate the age. See the text for guidance.)

5. *Dating a terrace*. Given the ^{10}Be concentrations in the table below, and assuming sediment bulk density $\rho = 2100\,kg/m^3$ and characteristic attenuation length [fast neutrons] $\Lambda = 1600\,kg/m^2$,

(a) determine the best fitting shifted exponential profile;

(b) calculate the age of the surface, assuming that the local production rate $P_o = 30$ atoms/(g yr) in quartz, and ignoring decay.

(c) Using the inheritance determined from the shift in the profile, calculate the basin-averaged erosion rate, assuming the same production rate.

Depth (m)	^{10}Be concentration (atoms/g)	^{26}Al concentration (atoms/g)
0	4.05×10^5	2.43×10^6
0.5	2.43×10^5	1.47×10^6
0.9	1.66×10^5	1.05×10^6
1.3	9.81×10^4	5.97×10^5
1.7	4.81×10^4	3.06×10^5

6. Using the Speel River pluton cooling history shown in Figure 6.33 (from Reiners, 2005), estimate at what depth the pluton is likely to have been emplaced. Assume a geothermal gradient of $30\,°C/km$. If the surface temperature is $0\,°C$, what is the minimum exhumation rate implied by the U/Th-He date at 8 Ma?

7. *Weathering rinds*. The average weathering rind thickness on basalt clasts on an 8 ka flow is 1.2 mm. If the rind thickness grows as the square root of time, what is the age of a moraine on which basalt clasts are measured to have average rind thicknesses of 2.2 mm?

8. *Ratio clock*. Quartz-rich sediment in a cave in the Sierras is measured to have a ^{26}Al/^{10}Be ratio of 2.8, and the ^{10}Be concentration is 2.2×10^4 atoms/g in quartz.

(a) What is the age of the deposit (and hence the age of the void in which they are found)?

(b) Calculate the inherited concentration upon deposition, and estimate the basin-average erosion rate in the catchment, having calculated that the average production rate in the catchment is 35 atoms/(g yr) in quartz.

9. You are in Pakistan and do not have access to the web. You are sampling a strath terrace above the Indus in its Middle Gorge near Nanga Parbat, and need to know how large a sample to collect. You cannot simply say "I don't know" and take a sample the size of your vehicle, because (i) there is limited space in the vehicle, and (ii) it costs money to ship the samples home. You are 200 m above the river at an elevation of 3700 m. The rock you are sampling appears to be a highly metamorphosed granite. List all the assumptions you make in the calculation.

10. Plot the expected profile of ^{10}Be in the sandy cover deposits on a marine terrace. The local production rate of ^{10}Be is 6 atoms/(g yr) in quartz. The age of the deposit is 130 ka. Let's assume that this terrace is young enough to ignore decay. Plot the profile for the following two cases:

    (a) assume that there is no bioturbation;

    (b) assume that bioturbation homogenizes perfectly the top 50 cm of the deposit.

11. Estimate how many sand grains you can hold in your hand. (This is how many presumably independent samples you have of erosion rates at spots in the landscape when you sample a handful of sand from a river draining a basin.)

12. You measure the ^{10}Be concentration of a sample from a granitic outcrop atop Mt Osborn in the Wind River Mountains of Wyoming to be 9.2×10^5 atoms/g in quartz. The elevation is 3500 m, and the sea level production rate of ^{10}Be is 6 atoms/(g yr) in quartz. What is the local erosion rate of the outcrop?

13. What is the effective "averaging time" implicit when you use ^{10}Be concentrations to deduce erosion rates in a catchment? Assume that the relevant length scale for decay of production rate with depth is 0.7 m. Calculate this for erosion rates of 10 and 100 microns/yr.

14. *Thought question.* List and discuss the several means by which cosmogenic radionuclides have been employed in the last two decades. Contrast the state of the art of establishing timing in the landscape in the absence of this tool.

15. *Thought question.* Compare the timescales over which we document average exhumation rates when using ^{10}Be, apatite fission tracks, and U/Th-He methods. Assume that the real exhumation rate is 10 m/Ma.

Further reading

Bierman, P. R., 2007, Cosmogenic glacial dating, 20 years and counting, *Geology* **35** (6): 575–576.
This is a quick review of the method as employed in glacial settings, looking back on progress over the first two decades of its use.

Cerling, T. E. and H. Craig, 1994, Geomorphology and in-situ cosmogenic isotopes, *Annual Review of Earth and Planetary Sciences* **22**: 273–317.
This is an early snapshot of the use of this method, with attention to both stable and radionuclides.

Gosse, J. C. and F. M. Phillips, 2001, Terrestrial in situ cosmogenic nuclides: theory and application, *Quaternary Science Reviews* **20**: 1475–1560.
This is a lengthy but useful review of use of nuclides in geomorphology as of the turn of the century.

Granger, D. E. and P. Muzikar, 2001, Dating sediment burial with cosmogenic nuclides: theory, techniques, and limitations, *Earth and Planetary Science Letters* **188** (1–2): 269–281.
A valuable resource for practitioners, this review article provides sufficient information to allow full understanding of the methods and pitfalls of the use of cosmogenic nuclides to date buried surfaces (e.g., caves, deep deposits).

Reiners, P. W. and T. A. Ehlers, eds., 2005, Thermochronology, *Reviews in Mineralogy and Geochemistry* **58**.
The editors have brought together articles that collectively review the field of thermochronology as it is used to assess the thermal history of the shallow crust. It is a very useful entrance point into the literature, which is fast-evolving with new methods.

Weathering

Nature does not hurry, yet everything is accomplished

Lao Tzu

In this chapter

In most landscapes on Earth the surface is mantled with a mobile layer of material derived from bedrock that is permeable to water and gases, supports growth of plants, and can readily be transported by geomorphic processes. The rock beneath the mobile debris layer is chemically altered and fractured to varying degrees. The processes that produce and alter these near-surface materials are collectively called weathering. Because the Earth's surface layer – its skin – is significant for sustaining human society, it was recently designated the Critical Zone (National Research Council, 2001). The Critical Zone is characterized by complex interactions between water, gases, minerals, biological agents, and physical processes. Weathered rock and soils are manifestations of processes in the Critical Zone.

A spheroidally weathered granite boulder in Buttermilk country, eastern Sierra Nevada, California, is delaminating in sheets several centimeters thick. Chips from the spallation of these sheets are scattered around the boulder (photograph by R. S. Anderson).

In this chapter we delve into the processes that break down bedrock as it comes in contact with the fluids and stresses found at the surface. We begin by considering a generalized profile of the near-surface region, which allows us to define some terms and outline some of the important drivers of weathering processes. We will consider processes that generate fractures in rock first, and then processes that chemically alter rocks. Part of our focus will be on ways to quantify the strength changes and the mass transfers associated with these processes. Against this process backdrop, we then examine the roles of weathering and erosion in the geologic carbon cycle, and finally examine the important geomorphic question of production of mobile debris from bedrock.

Weathering as part of erosion

The notion that weathering is an important component of erosion was articulated by G. K. Gilbert in the "Land sculpture" chapter of his *Report on the Geology of the Henry Mountains*, first published in 1877:

Stated in their natural order, the three general divisions of the process of erosion are (1) *weathering*, (2) *transportation*, and (3) *corrasion*. The rocks of the general surface of the land are disintegrated by *weathering*. The material thus loosened is *transported* by streams to the ocean. In transit it helps to *corrade* from the channels other material. (Gilbert, 1877, p. 100)

In this passage, Gilbert defines weathering as a process that disintegrates and *loosens* rock. This emphasis on the mobility of products of weathering is characteristic of how many geomorphologists consider weathering, because it is this mobilization that sets the stage for shaping landscapes.

Having identified weathering as the first step in erosion, Gilbert then observed that the presence of a mobile regolith implies that erosion is not limited by weathering rates:

Over nearly the whole of the earth's surface there is a soil, and wherever this exists we know that the conditions are more favorable to weathering than to transportation. Hence it is true in general that the conditions which limit transportation are those which limit the general degradation of the surface. (p. 105)

This condition, now called *transport-limited erosion* (Carson and Kirkby, 1972), is defined as the state where the rate of erosion depends only on the rate of processes that transport sediment. In transport-limited erosion regimes, erosion can be modeled without considering the rate of weathering, because the supply of mobile debris is equal to or exceeds the rate at which it can be transported. The geomorphologist need only focus on the capacity of transport processes to understand the rate and patterns of landscape lowering. The thickness of mobile debris is either unchanging through time (the case where erosion rate equals the weathering or loosening rate), or increasing through time (the case where the rate of weathering or loosening exceeds the erosion rate). As Gilbert notes, this will be the situation where a soil blankets the land surface.

The complement to transport-limited erosion is *weathering-limited erosion*, the regime where the rate of loosening of debris limits the rate of erosion. Weathering-limited erosion characterizes areas of bare bedrock or areas thinly mantled with soil. These are areas where wresting of material from bedrock occurs slowly in relation to the rate at which the weathering products are swept away. It is important to note that weathering in this context refers to the full suite of processes that produce mobile debris, and in fact the most important processes on bare rock are likely to involve mechanical wear and growth of fractures that lead to flaking or breaking of fragments.

The weathered profile

Rocks brought to the Earth's surface are placed in a reactive zone in which a variety of weathering processes occur. Although tectonic uplift moves rock upward relative to some datum, such as the geoid, it is erosion, the net removal of material from the surface, that brings rock toward the surface and into the reactor. As rock moves up toward the surface, it undergoes progressive chemical and mechanical alterations that produce fractured and weathered rock, and in some circumstances soils. A generalized weathered profile in Figure 7.1 shows the typical effects of weathering processes on rock, and introduces some of the nomenclature used to describe weathered rock. Rock that is weathered to any degree is considered *regolith*. The term regolith derives from *rhegos*, Greek for blanket, and *-lith* for rock (the same root as lithology and lithograph). Within regolith we can identify three layers that provide useful designations for geomorphologists. The deepest, and hence earliest, manifestations of weathering are often fractures in bedrock.

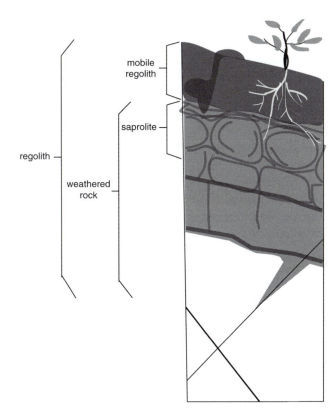

Figure 7.1 Schematic diagram of a weathered profile on a hillslope. Here the contact between the weathered and unweathered rock is shown as gradational, weathering progressing most rapidly down avenues of cracks and joints in the rock. Continued weathering of the isolated blocks can take the form of spheroidal weathering. The mobile regolith, the layer of mobile material on the surface, in which most of the organic content is concentrated, is affected by both plants and animals, which not only help to move the material, but produce the acids that promote further weathering reactions.

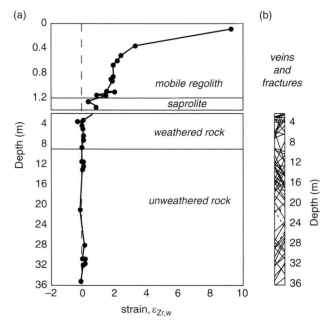

Figure 7.2 (a) Profile of strain in weathered profile of the headwater CB1 catchment, near Coos Bay, Oregon. Strain is the relative volume change from the parent rock, a greywacke sandstone, calculated based on Zr concentrations. Note the break in depth scale. (b) Orientation and density of fractures in CB-1 weathered profile, as observed in a bedrock core (after S. P. Anderson *et al.*, 2002, Figures 2 and 5).

Chemical alteration of the rock may appear to radiate away from the deepest fractures. *Weathered rock* is fractured and/or chemically weathered rock that has not been mobilized by hillslope or bioturbation processes. *Saprolite* differs from weathered rock in degree of alteration. It is material that is sufficiently broken down that it can readily be augered through or dug with a shovel, yet saprolite retains the original rock structure. The retention of rock structure is significant in two ways. First, it implies that the rock has not undergone strain. Because of this it is commonly assumed to have undergone isovolumetric weathering (no change in volume). Second, intact rock structure implies that saprolite has not been displaced or mobilized by weathering (this is another way of saying the material has not undergone strain). This fact is always worth bearing in mind when you encounter saprolite

in a road cut or soil pit. This material, which crumbles easily in your hand, has not been displaced by the forces exerted on it. Saprolite will convert to mobile regolith given sufficient provocation. Its presence is indicative of the vigor (or lack, rather) of hillslope and bioturbation processes. The final layer in our tour of the weathered profile is *mobile regolith*. Often this mobile layer of debris is called soil, but we will save that designation for mobile regolith organized into horizons by soil-forming processes. Mobile regolith has been detached from the weathered rock below it, and is in motion both vertically and laterally.

We have emphasized two physical characteristics in designating layers within the weathered profile: the strength and mobility of the material. The hillslope processes we discuss in subsequent chapters, particularly Chapter 10, respond to these characteristics. An example of the physical variations of material in the weathered profile is seen in measured strain for a 36 m-deep profile in the Oregon Coast Range in western North America shown in Figure 7.2. Strain, $\epsilon_{Zr,w}$, is calculated from the concentration of Zr, an element unaffected by chemical weathering processes, using a mass balance method we will develop later in this

Figure 7.3 Contrasting weathered profiles. The top image shows the upper 2.5 m of a weathered profile in granodiorite in the Colorado Front Range. A thin mobile regolith layer overlies highly fractured weathered rock. A saprolite layer, of thickness apparently controlled by fractures, is visible. High angle fractures dominate the alteration of the weathered rock layer. The lower image shows the top 15 m of a roughly 30 m-thick weathered profile developed on gneiss in southern India. The mobile regolith layer is about 2 m thick (note roots), while the saprolite with intact rock structure extends down many meters below the surface (upper photo by S. P. Anderson, lower photo by J. Riotte, with permission).

chapter. For now, note that the strain, the change in volume of a sample relative to its unweathered parent rock, is essentially zero over most of the profile and averages about 3 (implying 300% expansion relative to parent rock) in the mobile regolith. Fractures increase in density and change in orientation in the weathered rock portion of the profile. The deepest weathering features found in the profile are high-angle fractures, probably of tectonic origin. Moving up in the profile, these fractures begin to show oxidized stains on the fracture faces, and higher still, are surrounded by haloes of oxidized rock. Above this, the rock is pervasively oxidized but still intact. The density of fractures cutting the rock increases upward through this profile, and their orientation becomes near surface-parallel. In this example, no fractures are shown in the saprolite, which was only examined in augured holes. Auguring disrupts this type of structure.

Weathered profiles vary substantially in different environments, as illustrated in Figure 7.3. For instance, in mountain landscapes unweathered rock may be present at the surface, and fractures constitute the primary manifestation of weathering processes. In contrast, weathered profiles can be tens of meters thick in places where chemical weathering has worked for millions of years and erosion rates are low. These variations in thickness and degree of alteration of the weathered profile reflect differences in the rate at which material moves through the near-surface reactor and the rate at which weathering of the rock occurs.

The Critical Zone

A recent National Research Council report called attention to the Critical Zone (National Research Council, 2001), which they defined as the region from the top of the vegetation canopy to the base of groundwater. In this heterogeneous near-surface region the fluid envelopes of the Earth overlap with the crust. The interactions between rock, water, air, and living organisms control the availability of nutrients and the structure of the environment (Brantley *et al.*, 2007). The weathered profile, from the land surface down to fresh, unaltered bedrock, provides the physical structure in which Critical Zone processes occur, and is itself a product of these processes.

It can be useful to consider the Critical Zone as a giant reactor in which rocks are bathed in solutions produced from percolating rainwater and where surface-down perturbations such as temperature fluctuations or bioturbation are important (S. P. Anderson, *et al.*, 2007). Percolating water may be enriched with organic exudates and charged with dissolved gases, most notably CO_2. The hydrologic cycle renews these solutions, bringing new water into the top of the reactor with each rainstorm. Most minerals are out of equilibrium with these solutions, and hence chemical weathering reactions proceed. These reactions produce altered minerals (typically clays and oxides) and solutes that are carried away in solution. Biological agents often promote weathering. For example, microbes catalyze reactions with minerals that can be harnessed for their metabolic needs or that release essential nutrients. Root growth pries open cracks in rocks, and burrowing organisms effectively mix and stir up mobile regolith. The upper part of the reactor is subjected to temperature fluctuations that are more rapid and of greater magnitude than at greater depths. These drive thermal expansion or contraction of the minerals themselves or growth of ice lenses, producing stresses that break rocks apart. These mechanical processes open new avenues for water flow through rock, and hence for chemical reactions to occur.

The Critical Zone reactor is that part of the crust that is strongly affected by the surface, an influence that tends to decline with depth in the rock. Rock breaks down in this reactor in response to mechanical stresses and a reactive chemical environment. The balance in the tug of war between the weathering processes that break down rock and the transport processes that carry off the broken down products sets the structure of the weathered profile, in particular the depth of the mobile regolith layer at the top.

Denudation

We have seen now that weathering covers everything from processes that fracture rock to those that dissolve rock, and that weathering produces layers of altered rock and mobile regolith. Given these different processes and products, how can we measure the rate of weathering? Weathering rates are reported for all of the processes and outcomes we can comprehend – the rate of mineral transformation, the rate of conversion from immobile to mobile regolith, the solute flux from a watershed. It is very important when you see a weathering rate to ask and understand exactly what is being measured. Is it a chemical reaction rate normalized to the surface area of the mineral under consideration? Is it a surface lowering rate? Is it a chemical flux normalized to the area of a watershed? Each of these measures of weathering give us different information, and none are directly comparable.

Mass loss

One way to measure weathering is simply to monitor the material removed from a landscape in rivers.

To denude is to strip something of its covering; in geomorphology, denudation connotes a loss of mass from a landscape. Mass is lost from landscapes both as solids and as solutes. The *total specific denudation rate*, $\dot{E}_{tot}$ [M/L^2/t], of a landscape is defined as the rate of mass loss per unit land surface area. The total denudation includes both particulate and dissolved mass losses, and is most easily determined from the material loads in rivers. We write the total denudation rate as the sum: $\dot{E}_{tot} = \dot{E}_{diss} + \dot{E}_{part}$, where $\dot{E}_{diss}$ is the specific dissolved load [M/L^2/t], and $\dot{E}_{part}$ is the specific particulate load [M/L^2/t]. Note that the designation *specific* here indicates that we have divided by the basin or watershed area, so that differences in the rates are not due to differences in basin size. Chemical weathering, the dissolution or chemical alteration of minerals, releases solutes that constitute chemical denudation. Does the largely invisible dissolved load in water actually amount to a significant fraction of the total mass loss from a basin? It depends on the particular basin under consideration. The best way to answer this question is to look at the dissolved loads and sediment loads in rivers.

A survey of major rivers of the world in Figure 7.4 shows that solid particles dominate the mass transported out of terrestrial landscapes to the ocean. In the figure, rivers are ordered by increasing total specific denudation rate. There is a general trend of increasing dissolved load with increasing particulate sediment load, but the trend is far from perfect. A similar survey is presented in Table 7.1 (minor differences between Figure 7.4 and Table 7.1 give some sense of the error in our estimation of these riverine fluxes). The table is organized in order of decreasing river discharge, an approach that does not organize well the suspended or dissolved load variations observed in the rivers. Globally, rivers transport about four times more mass in particulate form (global total of 5277 Tg/yr) than in dissolved form (1383 Tg/yr). The numbers reported here include only suspended sediment in the particulate sediment flux. Since bedload (see Chapter 14) is a significant, but difficult to measure, component of the particulate sediment transported by rivers, the fraction of total denudation in dissolved form is probably less than indicated in Table 7.1.

The average total denudation rate for the major river basins (sum of suspended and dissolved loads, divided by the drainage area and by the density of rock, here taken to be 2650 kg/m^3) is 129 μm/yr,

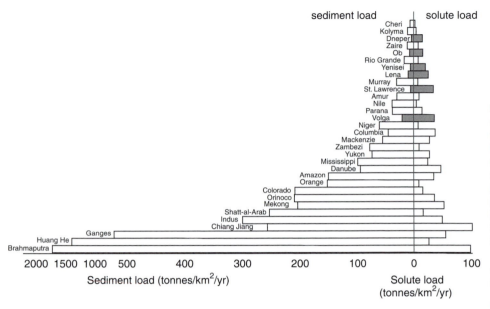

Figure 7.4 Riverine loads of sediments and solutes for the major rivers of the world. Highlighted in gray are those rivers draining landscapes that are dominated by periglacial processes. They are dominated by solute rather than sediment loads. Note change in scale at 500 tonnes/(km² yr) (after Summerfield, 1991, Figure 15.11, with permission from John Wiley & Sons; data primarily from Maybeck, 1976, and Milliman and Meade, 1983).

or about one-eighth of a millimeter per year. Denudation is not evenly apportioned between dissolved and suspended sediment however. Compare the summed global load of dissolved to that for suspended material (in Tg/yr). These total global riverine mass fluxes show that about four times as much mass is transported in solid than in dissolved form (5277 Tg suspended versus 1383 Tg dissolved load per year). The ratio of dissolved to suspended varies considerably from one river to another, however, ranging from 0.02 in the Huang He to 11.3 in the St. Lawrence. The average ratio of dissolved to suspended sediment flux for the rivers reported here is 1.72, a value much greater than expected from the ratio of the total global loads. In a handful of the rivers shown in Figure 7.4 dissolved loads exceed particulate sediment loads. Most of the rivers with dissolved/suspended ratios greater than 1 are found in the Arctic, where low river gradients and extensive tundra wetlands are probably responsible for the relatively high dissolved loads. If we omit these Arctic rivers, the dissolved/suspended radio is 0.42. A few sediment-laden rivers dominate the total mass flux to the oceans: the Huang He and Amazon rivers together account for nearly 50% of the global suspended sediment load. This difference between the ratio of the total loads, and the mean of the ratios, shows that proportions of sediment and solute delivery vary widely from one watershed to another. In many rivers, dissolved load constitutes an important component of the total denudation rate.

Processes that fracture rock

Many weathering processes produce fragments and reduce particle sizes, an action known as comminution. Fragmentation of rock influences its properties in the following important ways:

1. Smaller chunks are more easily transported. The size at which a rock chunk becomes movable depends on what transport mechanisms are available, but it is certainly the case that smaller chunks are more easily dislodged and transported.

2. Fractured materials are not as strong as the same material without fractures. It should be obvious that a collection of disaggregated blocks is weaker than the same mass in one intact block. Stress is not transmitted across voids; hence, fractures provide planes of weakness where material in shear can slip. Perhaps less obvious is that flawed material, in which microfractures do not form through-going cracks, is weakened. In addition, stress concentrations at crack edges can lead to further crack growth and disintegration.

Table 7.1 Major rivers, and their water, sediment and solute discharges (after Berner and Berner, 1996)

River	Location	Drainage area (10^6 km^2)	Water discharge (km^3/yr)	Specific discharge (m/yr)	Dissolved load (Tg/yr)	Specific dissolved flux (tonnes/(km^2 yr))	Suspended load (Tg/yr)	Specific suspended flux (tonnes/ km^2 yr))	dissolved/suspended	Total denudation rate (µm/yr)
Amazon	South America	6.15	6300	1.02	275	45	1200	195	0.23	90
Zaire	Africa	3.82	1250	0.32	41	11	43	11	0.95	8
Orinoco	South America	0.99	1100	1.11	32	32	150	152	0.21	69
Yangtze	Asia	1.94	900	0.46	247	127	478	246	0.53	141
Brahmaputra	Asia	0.58	603	1.04	61	105	540	931	0.11	391
Mississippi	North America	3.27	580	0.18	125	38	400	122	0.60	60
Yenisei	Asia	2.58	560	0.22	68	26	13	5	5.20	12
Lena	Asia	2.49	525	0.21	49	20	18	7	2.70	10
Mekong	Asia	0.79	470	0.59	57	72	160	203	0.36	104
Ganges	Asia	0.98	450	0.46	75	77	520	533	0.14	230
St. Lawrence	North America	1.03	447	0.43	45	44	4	4	11.30	18
Parana	South America	2.60	429	0.16	16	6	79	30	0.20	14
Irrawaddy	Asia	0.43	428	0.99	92	214	265	616	0.35	313
Mackenzie	North America	1.81	306	0.17	64	36	42	23	1.50	22
Columbia	North America	0.67	251	0.37	35	52	15	22	3.50	28
Indus	Asia	0.98	238	0.24	79	81	250	256	1.30	127
Huanghe (Yellow)	Asia	0.77	59	0.08	22	29	1100	1429	0.02	550
Sum		31.87	14 896		1383		5277			
Mean				0.48		60		282	1.72	129

3. Fractures provide avenues for water and air to move through the rock. The rate at which water moves through the voids found in most rocks is vanishingly small, but if the rock has fractures in it, or if the rock is completely disaggregated, then water can penetrate and move through the subsurface. The rate of water flow through the subsurface controls the hydrologic response of a watershed (storms produce less change in stream flow where deep permeable soils exist than where impervious surfaces such as unfractured rock or pavement predominate).

4. Fracturing and fragmentation produce surface area. Chemical weathering processes involve interaction of water and minerals at the surface of the mineral. Just as coarse rock salt dissolves more slowly than finely ground salt (such as the stuff we sprinkle on our French fries), big rock chunks weather more slowly than finely divided rock. Soils are typically made of very small particles indeed – sand, silt, and clay – and consequently the surface area of a small quantity of soil can be very large. In addition to being a surface for chemical processes to operate on, these surfaces provide habitat for bacteria, fungi and other small living things.

Consider, for example, the surface area of a cube of rock 10 cm on a side. The surface area of one face of the cube is 10 cm × 10 cm, or 100 cm². The cube has six faces, so the total surface area of the cube is 600 cm². Now dice the cube into eight equal sized chunks, each new cube having sides of 5 cm × 5 cm. The total surface area is (5 cm × 5 cm) × 6 faces × 8 cubes = 1200 cm². If we keep dividing down to coarse silt size particles (50 μm), we'll find that our original 10 cm × 10 cm × 10 cm chunk of rock is now broken into 8 billion (8×10^9) pieces, and has a surface area of 1 200 000 cm²! Crushing the rock has increased the chemically reactive area and the space for microbial habitat by about four orders of magnitude.

Here we explore the processes commonly considered physical processes, as their primary result is comminution and fracture generation. While there are many processes, and we do not pretend to come to grips with them all here, we will discuss a few in detail. The mechanical processes often have to do with the thermal state of the rock, either because the rates of chemical processes are modulated strongly by temperature, because there are changes in the state of water within the rock mass, or because differences in temperature can set up differential expansion of the rock minerals themselves.

Thermal stress and strain

Rocks, like most solids, expand when heated. Non-uniform expansion (or contraction) of rock or its mineral constituents produces stress; if the stress exceeds the strength of the material, it fails. Gradients in thermally induced strain (expansion or contraction) sufficient to produce fractures are generally confined within a few centimeters of the ground surface, because this is where temperature gradients are greatest. Two modes of rock failure are commonly attributed to thermal stress: granular disintegration, in which breaks occur along boundaries of grains at the rock surface, and spallation, in which plates break off the outer rock surface. Granular disintegration produces rough rock surfaces, while spallation tends to round boulders. In some situations, thermal effects extend below this surface layer. In Chapter 9, for instance, we will consider ice wedge polygons, which form when frozen ground cracks in rectilinear patterns with spacing of the order of 10 m during protracted cold episodes.

The propensity of a material to change volume with temperature is expressed with the volumetric thermal expansion coefficient, α, defined by $\alpha = \frac{1}{V}\left(\frac{\partial V}{\partial T}\right)_P$, where V is the volume, T is temperature, and the subscript P indicates constant pressure. An analogous linear thermal expansion coefficient can also be defined. Common rock-forming minerals have α values of the order of 1–$10 \times 10^{-6}\,°C^{-1}$. For the case of $\alpha = 10 \times 10^{-6}\,°C^{-1}$, a 100 °C increase in temperature will cause a rock mass 1 m long to increase in length by $(1\,m)(10 \times 10^{-6}\,°C^{-1})(100\,°C) = 10^{-3}$ m, or 1 mm. It experiences a 0.1% strain. The stress, σ, generated from this strain can be computed from

$$\sigma = \frac{\alpha E \Delta T}{1 - \upsilon} \tag{7.1}$$

where E is Young's modulus (the stiffness, or ratio of tensile stress to tensile strain in a material), ΔT is the change in temperature, and υ is Poisson's ratio (the dimensionless ratio of strain transverse to a load to the strain along or axial to a load). By itself, a 0.1% strain may not seem such a big deal. If the whole rock mass expands and contracts uniformly – breathing ever so gently and subtly – no serious damage would occur. But this is not necessarily the case, for several reasons. First, the thermal expansion coefficient can vary from one mineral to the next within a rock mass. Second, the temperatures within a rock are rarely uniform. Both of these effects lead to non-uniform stresses that can lead to crack growth.

Consider spatial variations in the thermal expansion coefficient. Most rock is composed of more than one mineral, each with its own coefficient of expansion. Furthermore, many minerals are anisotropic in their thermal expansion properties, which means that a single crystal expands non-uniformly when heated. The values of α in quartz range from $7.7 \times 10^{-6}\,°C^{-1}$ parallel to the c-axis of the crystal to $13.3 \times 10^{-6}\,°C^{-1}$ perpendicular to the c-axis (Hall *et al.*, 2008; Siegesmund *et al.*, 2008). Calcite crystals are even more anisotropic: they expand with heating parallel to the c-axis ($\alpha = 26 \times 10^{-6}\,K^{-1}$), but actually contract perpendicular to the c-axis ($\alpha = -6 \times 10^{-6}\,K^{-1}$) (Siegesmund *et al.*, 2008). Of the common rock-forming minerals, quartz has the highest coefficient of thermal expansion over typical Earth surface temperature ranges, and calcium feldspars have the lowest values of about $3–4 \times 10^{-6}\,K^{-1}$ (Lane, 1994). The mismatch in thermal expansion coefficients of adjacent mineral grains due to differences in mineralogy or in crystallographic orientation of anisotropic minerals in a rock can lead to stresses that generate cracks. The magnitude of the intercrystalline stresses generated increases with crystal size, which explains the propensity of coarse crystalline rocks to weather by surficial granular disintegration (Gómez-Heras *et al.*, 2006; Hall *et al.*, 2008).

Stresses due to thermal expansion also arise from non-uniform heating and cooling. Rock surfaces heat non-uniformly due to differences in their orientation relative to the sun and differences in albedo. Daily cycles of non-uniform heating and cooling caused by insolation, modulated by weather and seasons, are thought to drive granular disintegration, the detachment of individual mineral grains from surfaces (Gómez-Heras *et al.*, 2006). Rapid heating events, or thermal shock, produce steep temperature gradients both in terms of temperature changes with time (dT/dt) and with depth in the rock (dT/dz).

Figure 7.5 Granitic boulders at head of Tuttle Creek alluvial fan, near Lone Pine, eastern Sierras, showing spalling of surficial flakes that occurred during a brief grass fire. Within a few years the soot that allowed discrimination of the spall sites was washed off, making it very difficult to discern the fire impact. Lon Abbott for scale (photograph by R. S. Anderson).

In Figure 7.5, we see that the thermal shock from wildfires can lead to spalling of flakes off of rock surfaces. In both granular disintegration and thermal shock spalling, the important condition for failure is a temperature gradient that produces a gradient in thermal expansion and stresses in the rock.

Consider thermal shock during grass or forest fires. Obviously, while rocks don't burn, nearby vegetation can. The temperatures to which the rock surface can be heated can exceed several hundred degrees centigrade. Importantly, the pulse of heat is ephemeral, lasting only minutes to perhaps an hour before the fire runs out of local fuel. Very large vertical (or normal to the rock surface) gradients in the rock expansion set up very strong stresses that can operate on existing flaws in the rock to rive it. The spalls one sees near boulders on a recently burned alluvial fan, such as that shown in Figure 7.5, are of the order of 10 mm thick, not 100 mm thick, and stand out as light patches on a fire-scorched rock face (Shakesby and Doerr, 2006). Can we explain these observations with knowledge of the thermal perturbation to which the boulder surfaces have been subjected? As we have seen in our discussion of the thickening of oceanic lithosphere in Chapter 3, and will again in discussion of permafrost in Chapter 9, in diffusion problems, length scales, δ, can be estimated using the timescale and the thermal diffusivity: $\delta \sim \sqrt{\kappa t}$. Using a rock diffusivity of $1\,mm^2/s$, and a timescale (for a forest or grass fire) of 30 minutes (1800 s), the depth into the rock this thermal disturbance would have reached is $(1 \, mm^2/s \times 1800 \, s)^{1/2} = 42 \, mm$. This calculation estimates the depth to which a significant change in temperature would propagate in the time given. We will go through a slightly more formal analysis of this depth scale below. This quick analysis shows that the observed thickness of plates spalled from boulders is not unreasonable. Furthermore, the spalling will occur after the thermal wave has propagated to some depth, which is likely to be after the fire has scorched the rock surface. Crudely, one can see in the field that the larger, thicker plates spalled from rocks that were adjacent to larger vegetation – trees as opposed to bushes. This too makes sense, given that the tree would have burned for longer than the bush.

The rate of heating (or cooling) also affects thermal stress damage. Yatsu (1988) noted that gradients (dT/dt) of greater than $2 \, °C/min$ lead to irreversible failure. In the fire example discussed above, it is easy to see that a fire that heats surfaces by order $100 \, °C$ in 30 minutes will exceed this limit. Detailed observations have also shown that values of dT/dt that exceed Yatsu's limit can occur on rock surfaces in response to normal daily insolation and weather. A passing cloud, for instance, can lead to rapid cooling of a surface. In addition to these temporal gradients, spatially uneven heating of rock surfaces can occur due to orientation of surfaces to insolation (shadowing) and albedo of minerals (Gómez-Heras et al., 2006; Hall et al., 2008). In coarse, crystalline rocks, surface roughness can produce shadowing that significantly affects surface temperatures. Moreover, as shown in Figure 7.6 the difference in albedo of felsic and mafic minerals produces measurable differences in surface temperature. Indeed, to assess thermal stresses one should do a full energy balance that considers effects of shadowing, albedo, heat capacity of each mineral (the heat required to change the temperature of a given mass), the dissipation of heat through conduction to the rock interior, and the thermal mass of each mineral grain. Rather than going through the surface energy balance, which determines the surface temperature of the rock, we consider next how temperatures vary with depth and time in response to a given surface temperature variation. This provides a fairly simple tool to assess the depth of influence of thermal stress in response to heating or cooling events.

Our goal is to determine the depth to which a temperature fluctuation at the surface penetrates into

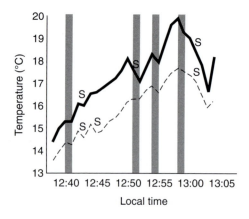

Figure 7.6 Fluctuations in surface temperatures of a granite building side over a 30-minute period, measured with an infrared thermometer. The two lines show temperatures in adjacent areas with a concentration of mafic minerals (solid line) and of felsic minerals (dotted line). Gray bars indicate periods of cooling due to wind, while S indicates cloud or tree shading events (after Gómez-Heras et al., 2006, Figure 6, with permission from Elsevier).

the subsurface, and hence the depths where thermal stress and strain might be significant. The depth of penetration of a thermal wave into a rock mass is controlled by two factors: the timescale of the thermal cycling at the surface and the thermal diffusivity of the rock. While the thermal diffusivity does not vary significantly – perhaps only by a factor of two or three – the timescales of the thermal fluctuations at the surface of a rock certainly range widely. A lightning strike lasts perhaps a second; a brush fire perhaps half an hour; clouds passing by the sun minutes to hours; and of course there are the astronomically set timescales of the day and year. In all cases, the means by which heat is moving about in the rock is principally by conduction, meaning that we can appeal to the diffusion equation introduced in Equation 3.7 in Chapter 3. What makes one problem differ from another is the nature of the boundary condition, in other words the history of the temperature at the surface (boundary) of the rock mass. Unfortunately, this equation is not simple to solve for most situations (i.e., most boundary conditions), at least not with paper and pencil (analytically, that is). There are, however, solutions for specific situations that, happily, are at least closely approximated by natural systems. For variations with a regular period, for example, one may often approximate the surface temperatures as varying sinusoidally. If on the other hand the temperature changes radically in a short period of time and is then held at this new temperature for a while, the problem is classified as an "instantaneous step change in temperature" at the top of a half-space. Once one knows how to classify the thermal problem, we can lean heavily on existing work on conduction of heat in solids, which has been extensively studied for 200 years (e.g., see Carslaw and Jaeger, 1967).

The solution to the diffusion equation in response to a sinusoidally varying surface temperature with mean of T_o, amplitude of the variation T_{amp}, and period P, is

$$T = T_o + T_{amp}e^{-z/z^*}\sin((2\pi t/P) - (z/z^*)) \qquad (7.2)$$

where z is the depth into the rock mass, and the natural length scale, $z^* = \sqrt{\kappa P/\pi}$. This equation describes the temperature, T, at every depth, z, at every time, t. We will discuss this equation again in connection with periglacial processes in Chapter 9, since permafrost is a thermally defined condition. For now, our interest is in defining the depth at which a temperature fluctuation at the surface (a lightning strike or fire or daily solar heating event) has damped so much that it is no longer significant. The exponential expression in front of the sine term sets the amplitude of the temperature fluctuation at each depth. Note that the sine term varies, as all sines do, between -1 and $+1$. To find the maximum temperature reached at any depth, we set the sine term to $+1$, and to find the minimum temperature at any depth, we set the sine term to -1. The magnitude of these temperature extremes declines exponentially with depth, such that the range of possible temperatures in response to a temperature fluctuation of T_{amp} at the surface is to be found within an envelope that narrows with depth, as shown in Figure 9.5. The length scale z_* sets how rapidly this envelope narrows with depth. At a depth of z_*, the amplitude is $1/e$ (about $1/3$) of the amplitude at the surface (T_{amp}). At a depth of several times z_*, usually taken as $3z_*$, the temperature does not vary significantly; the amplitude at that depth is e^{-3} of T_{amp}, or about 5% of the amplitude at the surface. Because the thermal stress we are discussing here is driven by either spatial or temporal changes in the temperature within a rock, these processes will shut off as the envelope narrows. It is this depth, of several times z_*, that we approximated with the length scale δ above.

Now inspect the sine term in the equation. It looks like a regular sinusoidal fluctuation with time except for the z/z_* term. This last term shows the lag of the

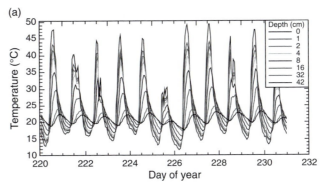

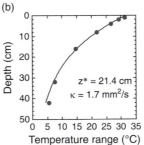

Figure 7.7 Temperature data from Vedauwoo site in the Laramie Mountains. (a) Temperature history from a ten-day period in mid-summer during which the temperatures were nearly sinusoidal owing to the lack of clouds. Records show damping of temperature swing amplitudes with depth (cm) and lagging of the peak. (b) Full daily temperature range decays exponentially with depth with a characteristic length scale z_* of 21 cm. Corresponding thermal diffusivity κ is 1.7 mm²/s (after R. S. Anderson, 1998, Figures 6 and 7, with permission from INSTAAR).

thermal wave with depth. It takes time for the thermal information to be handed down into the rock, meaning that rock at depth will experience a thermal cycle that is out of phase with that at the surface. Again, the length scale that dictates this lag is z_*. At a depth of z_* the thermal signal will be 1 radian out of phase (about 1/6th of a full cycle). By a depth of $z = 2\pi z_*$, the thermal signal will be 2π radians or a full period out of phase. At a depth of several times z_* the amplitude of temperature fluctuations will be very small (5% of the surface, as we discussed above), and highly lagged relative to the surface fluctuations. This lag can set thermal stress up in rock, as it is the gradients in temperature that set the stress gradients.

The lag and the decay in temperature amplitude with depth can be seen in Figure 7.7, which shows temperature profiles measured every 5 minutes in a granite bedrock surface in the Laramie Mountains,

Box 7.1 Quantification of mechanical weathering

The cumulative effects of thermal stress on rock disintegration and denudation are difficult to quantify. While thermal shock from individual fires can spall large fractions (50% or more) of rock surfaces by 2–70 mm (Shakesby and Doerr, 2006), the long-term denudation depends on the frequency of these events. Rounding of large boulders on alluvial fans may reflect the long-term effect of fire spalling, while vertical fractures, also attributed to fire spalling, tend to split smaller rocks on these same surfaces (<30 cm diameter). The effects of daily heating and cooling of rock surfaces are more difficult to estimate. The damage is likely to be incremental, microfractures that accumulate over time and progressively fatigue the rock. Temperature measurements with the spatial and temporal resolution such as that shown in Figure 7.6 support the notion that thermal stresses will damage rock. Further evidence comes from observations of rock damage. Figure 7.8 shows observations by MacFadden *et al.* (2005) of crack orientation in boulders on desert alluvial fans. Although there is considerable cracking that can be attributed to foliation, or existing planes of weakness, the cracks in boulders showed a strong tendency to be aligned nearly north–south. They hypothesized that the N–S cracks reflect stresses set up in the rock from repeated (daily) differential warming of the east and west sides of the boulders by insolation. To first order, the larger the rock, the stronger will be the gradients in heating and in stresses that can be established on a diurnal timescale.

A number of methods are used to characterize intact rock strength and hardness, properties that change as rocks weather. The most direct is the uniaxial compressive strength, a measure of the resistance of rock to crushing (Selby, 1980). This test is done in the lab with precisely cut core samples. Field measurements can be done with a Schmidt hammer, a percussive device that measures the in-situ elastic properties of rock (Goudie, 2006). The rebound height of a spring-loaded piston off the rock surface provides an index of surface hardness called the R value. Weak rocks, such as chalk, have R values of about 10, while strong rocks such as fine-grained igneous rocks have R values greater than 60. For a given rock type, changes in the R values scale with degree of weathering. Another measure of weathering is the velocity of seismic p-waves in rock. The p-wave speed can be measured on samples in the lab or with a field seismic survey. The p-wave speed declines as the density of pores and microfractures increases, and is also lower in secondary than primary minerals. Another approach is to measure changes in the porosity of rock. Bulk density generally declines as rock is progressively weathered, a change attributed to increasing porosity. The porosity, and indeed, the pore size distribution can be measured directly with mercury intrusion porosimetry, a technique that is sensitive to pore sizes ranging from nanometers to micrometers. These methods generally measure some physical characteristic of the rock that reflects increasing numbers of microscopic fractures or changes in mineralogy. This ambiguity makes it difficult to assign directly change to a property (density or R value, for instance) to a specific process.

Wyoming. Daily temperature fluctuations of more than 30 °C at the surface during the period of observation, decline to about 5 °C at 0.4 m depth. This information is used to calculate the z_*, and hence the thermal diffusivity, κ, of the rock, yielding values of 0.21 m and 1.7 mm/s, respectively. The temperature oscillations at the deepest 0.42 m thermister lag the surface by nearly half a day. Surface temperature fluctuations visible on some of the days, due to rare clouds during the period of observations, are not discernable in the deepest thermisters. Every timescale of thermal perturbation at the rock surface will influence thermal stress and strain to different depths. If we hold the diffusivity constant, at 1 mm²/s (a reasonable value for most rocks), the value of z_* for a 15 minute perturbation is 17 mm, for a one day perturbation is 0.17 m, and for an annual perturbation is 3.2 m. Thus we might expect thermal stresses due to passing clouds or wildfires to influence a region of a few tens of millimeters, daily cycles to influence a few decimeters, and annual cycles to reach depths of a few meters.

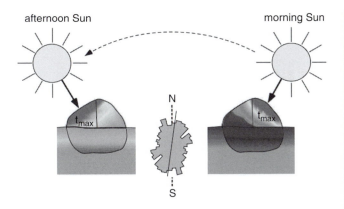

Figure 7.8 Daily cycle of solar radiative heating drives thermal cracking of rocks in the Mojave desert. Shading shows morning and afternoon isotherms in rock cross sections. Rose diagram of crack orientation displays strong propensity for N–S crack orientation; average crack orientation shown with thin line (after MacFadden *et al.*, 2005, Figure 3).

Frost cracking

We all know or at least suspect that freezing can be a significant cause of rock breakdown. The process is called frost cracking. Note the nearly complete disaggregation of the rock on a Little Ice Age moraine (only ~150 years old), shown in Figure 7.9. In recent years our knowledge of the frost-cracking process has been altered significantly by both theoretical and experimental work. This work shows that frost cracking is not the simple process one might suppose from having seen a pop bottle disintegrate in the freezer. Although water expands some 11% when it changes phase to ice, this is not the primary driver of frost cracking. The geomorphic systems we are dealing with are not closed, as a sealed bottle is, but are rather best treated as open systems. Rocks are permeable to some degree, and water migrates in them. In the case of water in a porous media at freezing temperatures, water actually migrates toward sites of ice growth. Thus, frost cracking differs from the thermally induced fracturing that we have just discussed, in that water transport as well as temperature and thermal expansion must be considered.

Water migration to the freezing front is a well-known phenomenon in soils (see discussion of frost heave in Chapter 9). After all, the ice lenses that grow in soils and are responsible for heave of the surface are visible with the naked eye. Walder and Hallet (1985) applied the physics of ice-lens growth during

Figure 7.9 Frost shattering of a shale block on a Little Ice Age moraine near Mt Robson, Canadian Rockies. 50 mm lens cap for scale (photograph by R. S. Anderson).

freezing of soils to that in permeable rock. They argued that, just as in frost-susceptible soils, water moves toward a freezing front and then freezes most readily in existing openings – cracks. Figure 7.10 depicts the pre-existing microcracks in rock (microcracks have one or two dimensions much smaller than the third, and lengths typically less than 100 μm; Kranz, 1983), and the migration of water toward cracks at subfreezing temperatures. The water migrates along grain boundaries in a thin film that remains unfrozen owing to slight ordering of the water molecules on mineral surfaces. Upon encountering a large pore or an ice lens, these surface forces decline and water can finally freeze. The phase change to ice both increases the size of the lens in the crack, promoting growth of the crack tip, and acts to generate a tension on the film of water within the unfrozen rock. This process operates most effectively at sub-(bulk) freezing temperatures around −3 to −10 °C. At higher temperatures water is thermodynamically stable in the

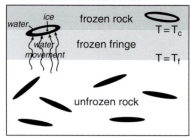

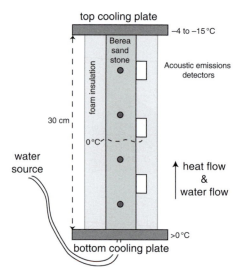

Figure 7.10 Schematic of the role of ice lens growth in pre-existing flaws (shown here as ellipses) in frost-cracking of the rock. Water migrates from the unfrozen rock through the frozen fringe in which the temperatures are below the bulk water freezing point, T_f, to accumulate in growing lenses (light-colored areas inside the cracks) (after Walder and Hallet, 1985, Figure 1).

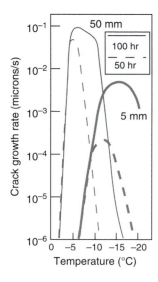

Figure 7.11 Model of crack growth rate due to frost cracking. For Westerly Granite with two initial crack lengths of 5 mm (bold lines) and 50 mm (thin lines), crack growth rate is modeled as a function of temperature, for two different durations of freezing. Crack growth rate is very strongly controlled by the temperature (note log scale), reaching maximum rates at temperatures well below 0 °C (after Walder and Hallet, 1985, Figure 2).

Figure 7.12 Experimental apparatus to explore frost cracking. Heat flows conductively from bottom to top aluminum plates through the well-characterized Berea sandstone bar. Foam insulation assures that the heat travels only in the vertical direction. Temperatures are measured with thermistors (dots), while acoustic emissions are monitored at three locations (boxes on right of bar). Water is available to be drawn into the unfrozen base of the rock bar from a tank. The 0 °C isotherm is maintained at a given location, and the experiment is allowed to run for several days (after Hallet et al., 1991, Figure 2, with permission from John Wiley & Sons).

thin films and microcracks, and hence will not freeze. At lower temperatures the effective viscosity of water in the films increases, slowing the transport of water to the growing ice lenses.

Walder and Hallet performed theoretical calculations to explore the dependence upon rock properties and upon the thermal scenario. The crack growth rates they calculated for two different initial crack sizes, shown in Figure 7.11, are at a maximum at −5 to −15 °C (the range owing to hydraulic characteristics of the system). They called the temperature range most conducive to frost cracking the frost cracking window. It took a set of clever experiments, however, to begin convincing the geomorphic community that the open system treatment was appropriate. Using a small column of permeable, homogeneous sandstone, with one end of the column held at subfreezing temperatures while the other end was kept above 0 °C and connected to a water source, they "listened" for cracking events using the apparatus shown in Figure 7.12. Three acoustic transducers (essentially little seismic sensors) recorded the time of cracking, and could also be used to determine the locations of cracking events. Their results are shown in Figure 7.13. These nicely show two salient features of frost-cracking in rock:

- Cracking continued as long as the temperatures were maintained in the frost cracking window (−3 to −10 °C), well below freezing, and the rock had access to water.
- Most cracking occurred where the rock was within a narrow range of subfreezing temperatures, the frost cracking window.

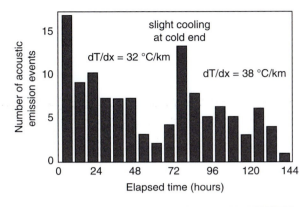

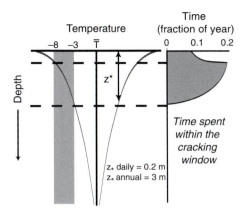

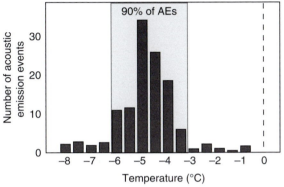

Figure 7.14 Temperature envelope funneling to depths with a characteristic decay scale of z_* that depends on the period of the surface temperature oscillation. The frost cracking window temperature range, here taken as −3 to −8 °C, is shaded. Right panel plots time spent within the frost cracking window, showing how this depends on depth. The shape of this integral depends on the particular choice of mean annual temperature (MAT) and amplitude of annual temperature swing (redrawn from R. S. Anderson, 1998, Figure 1, with permission of INSTAAR).

Figure 7.13 Results of a six-day sustained freezing experiment with Berea sandstone, using apparatus shown in Figure 7.12. (a) History of acoustic emissions (AEs) showing sustained cracking during experiment for two different vertical temperature gradients (dT/dx). (b) Distribution of temperatures at which cracking occurred, deduced from acoustic-emissions-located events (after Hallet *et al.*, 1991, Figure 5, with permission from John Wiley & Sons).

These aspects of frost cracking, that it continues at sustained low temperatures, and that it occurs in portions of the rock in the frost cracking window temperature range, cannot be explained by water freezing in sealed pores. In fact, the open system model predicts that frost cracking is most effective if rocks are slowly cooled and held at low temperatures (Walder and Hallet, 1986). These conditions produce low temperature gradients and provide time for water migration to sites of freezing in the interior of the rock. It is the annual frost cycle, rather than quick overnight frosts, that does the most damage to rocks.

Several recent experiments demonstrate the implications of the Walder and Hallet open system model of frost cracking. R. S. Anderson (1998) explored numerically the depth over which temperatures in the frost cracking window occur, and the time in an annual cycle spent within this optimal temperature range at each depth. His results, shown in Figure 7.14, illustrate the trade off between extremes in temperature fluctuations (which dominate the surface), and the slow temperature oscillations at depth. For the particular conditions (MAT and T_{amp}) shown in Figure 7.14, it is depths well below the surface that spend the most time in the frost cracking window. Murton *et al.* (2006) applied these ideas to explain shattered outcrops in southern England, a region impacted by cold periglacial climates during the Pleistocene. They conducted freezing experiments on blocks of chalk in environmental chambers for two conditions, one in which freezing proceeded down from the top of the block, and the second in which freezing proceeded down from the top and up from the base of the block (bi-directional freezing). The latter case simulated the active layer over permafrost. Although both blocks cracked significantly, as shown in Figure 7.15, the block exposed to bidirectional freezing displays greater frost-cracking, as shown in the heave of the surface. The lower temperature gradients and slower cooling in this scenario provide more water flow to the ice lenses at the lower freezing front. In fact, much of the crack growth occurs during the thawing of the active layer, which they argue allows draining of now free water to feed the ice lensing process in the deeper still-frozen material. The authors

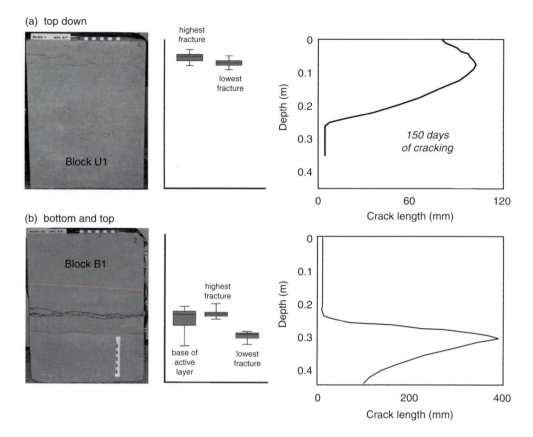

(a) top down

highest fracture

lowest fracture

Block U1

150 days of cracking

Depth (m)

Crack length (mm)

(b) bottom and top

Block B1

highest fracture

base of active layer

lowest fracture

Depth (m)

Crack length (mm)

Figure 7.15 Frost cracking experiments on blocks of chalk 0.45 m tall. Photographs show vertical sections of blocks at conclusion of experiments in (a) unidirectional freezing, and (b) bidirectional freezing conditions. Ice lenses in cracks are visible at depths of less than 100 mm in the unidirectional freezing case, although the freezing front reached > 350 mm in the experiments. Ice lenses are visible at 250–300 mm depth in the bidirectional freezing cases, about the depth of the base of the active layer. Box-and-whisker plots show documented positions of cracks. Graphs on right show models of crack growth for the two cases using the Walder and Hallet open system theory. The modeled distributions of cracks agree well with the measured crack locations (after Murton *et al.*, 2006, Figures 3 and 4, with permission from the American Association for the Advancement of Science).

argue that this process should promote efficient brecciation of rock at the base of the active layer. If frost cracking can shatter flat lying bedrock well below the surface, then it can also be expected to be important along cliff faces. Hales and Roering (2005) mapped active, unvegetated scree slopes in New Zealand's Southern Alps. Their data show that much of the scree (rockfall) production occurs in an elevation band that corresponds to a range of temperatures that dip into the frost-cracking window, as shown in Figure 7.16.

Other fracturing processes

We have focused on thermal stress and frost cracking, two processes that operate widely on Earth's surface, and that demonstrate the principles of stressing rock to the point of breaking (fracture). There are other means to exert stress at the small scale, which we will simply mention without going into great detail. These involve salt and plants. In each process, the growth rate of the cracks will depend upon the stresses that can be achieved, and by the strength of the material. Salt cracking is important in arid environments. This is analogous to frost cracking, in the sense that salt crystals grow in microcracks, exerting stress that leads to growth of the cracks. Roots growing in fractures can exert stresses that widen pre-existing cracks over time. The stresses are limited by the turgor pressure of the root and the lifetime of the tree. One purpose of roots is to anchor vegetation against the stresses exerted by wind or hillslope creep. The structure of root systems in trees growing on hillslopes

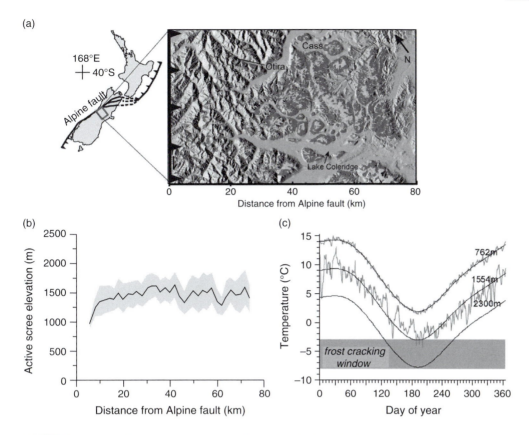

Figure 7.16 Scree production in the Southern Alps of New Zealand. (a) Map of study region showing regions of unvegetated scree production in medium gray. (b) Distribution of elevations of high scree production. (c) Envelope of temperatures for several elevations, showing dip into the frost cracking window in specific elevation band (after Hales and Roering, 2005, Figures 1–3).

reflects this role, as shown in Figure 7.17. The depth to which wind stress impacts soil and rock ought to be governed by the architecture of the root network. Indeed, when one listens seismically to the vibration of the ground near a tree, the ground is significantly shaken when the wind blows. Seismologists know this well enough to avoid siting their seismometers anywhere near trees if possible.

The deeper history of fractures

Our discussion has focused on processes that either operate very close to the surface (thermal stress) or require existing microfractures (frost cracking, salt cracking, root growth). These processes work to disintegrate rock surfaces and shatter rock with pre-existing flaws. Some of the most impressive landscapes, however, are those that seem largely untouched by these decay processes, such as the big walls and domes found in Yosemite National Park in California and shown in Figure 7.18. Theoretically, Terzhagi (1962) suggests that a rock wall should be able to stand to a height given by its uniaxial compressive strength, $h = \frac{\text{uniaxial compressive strength (N/m}^2)}{\text{unit weight of rock (N/m}^3)}$. For strong rocks, this limit is greater than 1500 m (Selby, 1980), yet these heights are not found. The great wall of El Capitan in Yosemite Valley, for instance, and those that bound the fjords of eastern Baffin Island (e.g., Figure 8.54) present just over 1000 m of vertical challenge to climbers. We first address what controls the rock mass strength, focusing on those processes that insert flaws into rock, and then turn to a specific process that generates surface parallel joints in flaw-free rock.

The stress of denudation

Let us consider for a moment how a rock makes its way to the Earth's surface. In any scenario, the

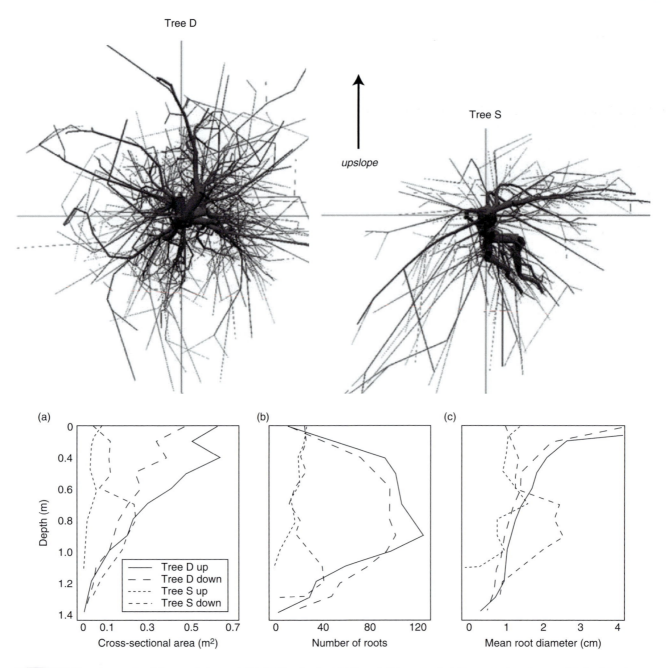

Figure 7.17 Plan view maps of tree root size and distribution for two white oak (*Quercus alba*) trees growing on a moderate slope in a stand of 60–100-year-old trees in the Whitehall Research Forest near Athens, Georgia, USA. The root systems were progressively excavated using high-velocity air-lances, mapped and digitized. The larger dominant tree (tree D) grew in soil > 2 m deep, while the suppressed tree (tree S) grew over a boulder 0.45 m below the surface. Root distribution statistics for the upslope (up) and downslope (down) sides of the trees shown in plots: (a) cross-sectional tree root area, (b) number of roots, and (c) mean root diameter as functions of depth below the soil surface (based on Danjon *et al.*, 2008, Figures 2 and 3, with permission of Oxford University Press).

overlying rock must be removed. In general this entails the surface processes we have been talking about in this book. One exception is tectonic denudation, in which overlying rock mass is removed by large-scale low-angle normal faulting, as occurs in southern Tibet. In general, however, the geomorphic processes that together allow denudation are rather gentle. The rock does not feel the influence of the

Figure 7.18 Half Dome, Yosemite National Park, California, showing results of sheet jointing, and glacial erosion of the U-shaped valley to steepen its north face. The vertical face on Half Dome stands some 600 m. View from Glacier Point (photograph by Greg Stock).

Earth's surface until it is within a few meters of the surface, at which point it will begin to experience thermal oscillations. In a volcanic arc, for example, the batholith that lies beneath the volcanic arc will slowly come to the surface as the overlying volcanic pile is removed by erosion. The rock never experiences much in the way of strain subsequent to its solidification from a magma. In contrast, consider a parcel of rock in an active compressional orogen. Its trajectory to the surface differs from that of the batholith. Compressional tectonics results in thrust faults in the brittle crust that serve to overlap the rock masses. This requires vertical motion, which in turn comes about through bends in faults that allow horizontal motion to translate into vertical motion. Motion of a rock mass through a listric fault can result in cracking of the rock as it is put through the compression of the interior of the bend. We call this the tectonic "rock crusher" (Figure 7.19), and argued in Molnar *et al.* (2007) that tectonic stresses break rock deep in the crust, such that rock in such settings arrives at the surface pre-fractured. That the rock does indeed experience stresses sufficient to crack it is reflected in the seismic activity in such settings. Seismicity is not constrained to the thrust fault itself, but is generated from a large rock volume nearby. The size distribution of seismic events suggests that faults (brittle failures) as small as a few meters are activated.

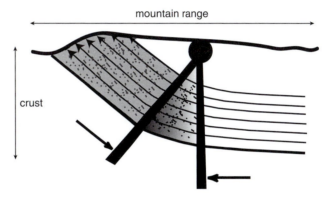

Figure 7.19 One of the roles of tectonics is the generation of cracks within rock that is ultimately delivered to the surface of the earth. In the example shown here, cracks are formed within the strain field associated with a bend on a fault, which serves as a tectonic, crustal-scale jaw crusher. Crack lengths and densities increase with time spent in the bend. The resulting fracture field is then translated to the surface, where the fractures influence the surface processes and their efficiencies (after Molnar *et al.*, 2007, Figure 2, with permission from the American Geophysical Union).

It is therefore not surprising that the crystalline rock now exposed at the surface in thrust-fault bounded ranges such as the Laramide ranges of western North American is so riddled with cracks, while that at the surface in the batholithic core of the Sierras is so flaw-free. This dramatic difference in the character of the rock manifests itself in many ways.

Figure 7.20 Comparison of two domes of similar scale in massive rock. (a) North dome in Yosemite National Park, formed in Half Dome granodiorite, the westernmost unit of the Tuolumne Intrusive Series, and (b) dome in Archaean rocks of the Periyar drainage in Kerala state, western Ghats, India (photographs by R. S. Anderson, from Molnar *et al.*, 2007, Figure 9).

It will certainly control the hydrology; fractures limit the height of rock walls. We see in our discussion of glacial erosion in Chapter 8 that it governs the processes available for a glacier to damage the landscape. In the massive rock of the Sierras, for example, in places the glacial erosion is limited to the less efficient process of abrasion.

The origin of sheeting joints

We address briefly the exceptions to the rule in these massive rocks left unscathed by dodging the rock crusher. In Figure 7.18 one can see the classic surface-parallel sheeting joints that typify the domes of the Yosemite landscape. In Figure 7.20 we compare photographs of a Yosemite dome with another granitic dome in the Periyar valley of the Western Ghats of India. They are remarkably similar in shape and scale. Both sites sport massive, flaw-free rock in which high compressive stresses can be sustained. As we will see, these stresses are required to generate sheet-jointing, which in turn governs the evolution of domes. Sheeting joints (also called exfoliation joints) are defined to be opening mode fractures that form nearly parallel to the topography. By opening mode we mean that they do not involve shear along

the fracture, but rather open normal to the surface of the fracture. Sheeting joints can develop to depths of the order of 100 m. They are always associated with massive rock, be it granite, gneiss, sandstone, or marble. They commonly increase in spacing with depth, and the curvature of the deeper fractures tends to decline. In other words, they flatten out a bit. All of these observations cry out for some reasonable explanation.

Martel (2006) has recently shown that the requirement for generating surface-parallel sheeting joints is that a tensile stress must arise that is normal to a convex traction-free surface, and that this stress exceeds the tensile strength of the rock. Consider the force balance sketched in Figure 7.21(a) on a section of rock beneath a landscape with negative curvature. The normal stresses acting on a potential fracture plane parallel to the surface include the radial component of the weight of the rock, and the net force imposed by the compressive stress directed vertically away from the center of curvature. For equilibrium to exist, there must be tensile tractions at the base of the element. When these exceed the tensile strength of the rock, the rock fractures along the surface-parallel plane. In simplest terms, the weight of the rock serves to clamp the potential fracture shut while the

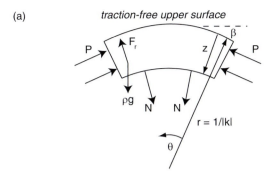

(a) traction-free upper surface

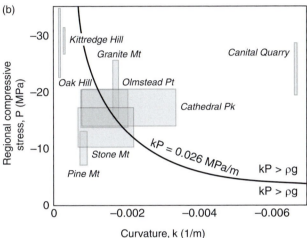

(b)

Figure 7.21 (a) Force balance diagram for a segment of a rock dome, showing compressive regional stresses, P, and resulting tensile surface-normal stresses, N. When the tensile stresses exceed the tensile strength of the rock, sheet fractures should form parallel to the topography. (b) Regional stresses required to generate tensile failure in a rock dome of a given curvature (solid line, assuming tensile strength = 0.026 MPa/m), and empirical combinations of curvature and regional stress for labeled sites (after Martel, 2006, Figures 2 and 4, with permission from the American Geophysical Union).

compressive stress acting parallel to the surface serves to pop it open.

Martel summarizes this situation in a single equation reflecting the equilibrium stress state of the rock. At equilibrium, at the surface, $z = 0$, and the variation in stress, N, with depth is given by

$$\frac{\partial N}{\partial z} = kP - \rho g \cos\beta \tag{7.3}$$

where k is the curvature of the surface, P is the regional pressure, ρ is the density, g is the acceleration due to gravity, and β is the surface slope. The sign convention is such that in compression $P < 0$, and on

convex surfaces, the curvature $k < 0$. Where sheeting joints occur, we can assume, therefore, that $kP > 0$. Consider the apex of the convex surface, where $\beta = 0$. $N = 0$ at the surface because it is stress-free. If $\partial N/\partial z$ is positive, then a tensile stress ($N > 0$) must develop in the subsurface. This happens wherever $kP > \rho g$. We display the line at which $kP = \rho g = 2600 \times 9.8$ Pa/m $= 0.026$ MPa/m in Figure 7.21(b). The tensile strength of the rock will dictate how deeply one must go beneath the surface for the tensile stress to exceed the tensile strength. Given that the tensile strength of massive rock is no more than a few MPa, the stresses will exceed the tensile strength of the rock at shallow depths.

Can we explain the observed increase in spacing and decline in curvature with depth? Martel proposes the following explanation. Once a fracture develops, N is reset to zero on this new fracture surface (no tension can exist across a crack). The tensile stresses then again increase with depth, but this time they increase a little less rapidly because the curvature on the next fracture plane (k) is slightly lower. One must therefore go more deeply into the rock to find the same threshold tensile stress. The increase in spacing is therefore linked physically to the decline in curvature with depth.

Two requirements for sheet-jointed landscapes are therefore: massive rock, and significant compressive stress parallel to the Earth's surface. Such sites are indeed rare, the hardest requirement to fulfill being the strength of the rock. Most rock at the Earth's surface is too chopped up by other fractures to retain sufficient strength.

While the sheeting joints that ornament the domes of Yosemite, the Periyar valley, and other massive rocky terrains give aesthetic character to the landscape, they will also play important roles as guides to subsurface water. The sheet joints ought to govern strongly the hydrology of such settings.

Fractures and rock strength

The observation that fractured rock is weaker than intact rock is trivial, but how does one quantify this? A number of schemes for classifying rock strength have been developed, many for civil engineering applications (e.g., Bieniawski, 1973; Barton et al., 1974). We present in Table 7.2 the rock mass

Table 7.2 **Selby's rock mass strength classification**

Parameter	Very strong	Strong	Moderate	Weak	Very weak
Intact rock strength (N-type Schmidt hammer R value)	100–60	60–50	50–40	40–35	35–10
Rating:	*20*	*18*	*14*	*10*	*5*
Weathering state	Unweathered	Slightly weathered	Moderately weathered	Highly weathered	Completely weathered
Rating:	*10*	*9*	*7*	*5*	*3*
Joint spacing	>3 m	3–1 m	1–0.3 m	300–50 mm	<50 mm
Rating:	*30*	*28*	*21*	*15*	*8*
Joint orientations	Very favorable; steep dips into slope, cross joints interlock	Favorable; moderate dips into slope	Fair; horizontal dips, or nearly vertical (hard rocks only)	Unfavorable; moderate dips out of slope	Very unfavorable; Steep dips out of slope
Rating:	*20*	*18*	*14*	*9*	*5*
Width of joints	<0.1 mm	0.1–1 mm	1–5 mm	5–20 mm	>20 mm
Rating:	*7*	*6*	*5*	*4*	*2*
Continuity of joints	None continuous	Few continuous	Continuous; no infill	Continuous; thin infill	Continuous; thick infill
Rating:	*7*	*6*	*5*	*4*	*1*
Outflow of groundwater	None	Trace	Slight (<25 L/(min 10 m^2))	moderate (25–125 L/(min 10 m^2))	great (>125 L/(min 10 m^2))
Rating:	*6*	*5*	*4*	*3*	*1*
Total rating:	100–91	90–71	70–51	50–26	<26
Classification:	Very strong	Strong	Moderate	Weak	Very weak

Source: Selby (1980)

strength classification system that Selby (1980) developed for landscape evolution problems. Like the engineering schemes, Selby's rock mass strength classification accounts for the measured strength of intact rock (i.e., unfractured blocks), state of weathering, water movement, and the number, orientation, size, continuity, and infilling of joints. Each parameter is determined from simple measurements or characterizations in the field. The summed parameters yield the rock mass strength classification, ranging from very strong to very weak. This empirically derived classification scheme places greatest weight on the intact rock strength, joint spacing, and joint orientations, all of which are measurable parameters for bedrock outcrops.

Parameters such as groundwater flow and joint continuity, while recognized as playing some role, have little effect in Selby's classification. Selby's scheme can be applied to any rock mass having sufficient exposure to make the observations needed, such as cliffs or layers of differing composition within a cliff band. A recent application of rock mass strength criteria to alpine talus (Moore *et al.*, 2009) found that joint orientation was the single most important control on cliff recession, while topographic attributes such as elevation, aspect, and cliff slope angle and length were not correlated with cliff erosion. Using rock mass strength classification for regolith mantled rock masses, however, presents serious observation challenges.

Chemical alteration of rock

Rocks at the Earth's surface are exposed to air, water, dissolved exudates from plants, and a community of organisms ranging from bacteria to Giant Sequoias, from earthworms to grizzly bears. In this environment, minerals can be altered profoundly, producing familiar red-brown colors of soils, caverns in limestone, and white caliche under desert rocks. The chemical transformation of rocks and minerals at the Earth's surface is called chemical weathering. These transformations are of fundamental interest to geomorphology for several reasons:

- Dissolution of minerals is a form of erosion, in some environments constituting the primary erosive process. As we discussed earlier, and as shown in Figure 7.4, the dissolved load in rivers must be considered in any study of the mass gains and losses from a landscape. Importantly, and different from many other transport processes, removals and additions of mass in solution may occur in the subsurface. Chemical weathering processes occur on mineral surfaces. However, mineral surfaces are connected through pore space and fractures so that chemical processes occur throughout a volume of rock, rather than just at the land surface.
- Chemical weathering alters the physical properties of rocks. Weathered rocks often have lower strength than unweathered rocks because of greater porosity and formation of weak clay minerals. It is also possible for material strength to increase due to chemical alterations, particularly where solutes released from one area are precipitated in another. In soils, precipitation of silica (e.g., opal), gypsum, or carbonate can cement horizons to form duripan, petrogypsic, or petrocalcic horizons, respectively. These cemented horizons impede water and root penetration. Thus, weathering changes the resistance of material to physical erosion processes.
- Weathering reactions form one stroke of the great geochemical cycles of the Earth, in which elements are transferred between the reservoirs of the atmosphere, the oceans, and the crust. Chemical weathering of silicate minerals is part of the geologic carbon cycle, which controls

atmospheric CO_2, an important part of the climate system.
- There are two different ways to look at weathering: from the perspective of solid phase properties (mineralogy, porosity) in regolith, and from the perspective of dissolved materials released into solution. The solid phase represents a record of chemical processes integrated over time. The dissolved materials in soil water and streams reflect weathering and exchange processes during the transit of water to the point it is sampled. Solutes provide a snapshot of present chemical processes, while soil and rock chemistry provides a history of past chemical processes. Both perspectives are useful. We will begin by discussing aqueous geochemistry, as this provides insight into the chemical processes taking place.

Chemical equilibrium

Rocks produced by magmatic or metamorphic processes in the Earth's interior are generally out of equilibrium with conditions at the Earth's surface, particularly with the fluids moving through the Critical Zone. Although thermodynamic equilibrium is rarely achieved in natural systems, describing the equilibrium state is useful. Equilibrium may in fact be a reasonable approximation of a natural system (of rock and water), and it is the condition toward which a system will trend. Thermodynamics does not define the rate (or kinetics) of change in the system. However, the rates of processes are greater for systems far from equilibrium, and slower near equilibrium; hence we gain some insight into rates of chemical processes.

At equilibrium, the system attains its minimum energy, expressed as the thermodynamic quantity Gibbs free energy (G). The Gibbs free energy is a measure of the work a system can do at constant pressure (P) and temperature (T). It is related to the enthalpy (heat content), H, and entropy, S, of the system by

$$G = H - TS \tag{7.4}$$

where T is the temperature in kelvin, G and H have units of kJ/mol, and S has units of kJ/(mol K). At constant T and pressure, P, changes in Gibbs free energy are given by

$$\Delta G = \Delta H - T\Delta S \tag{7.5}$$

For processes occurring at equilibrium, $\Delta G = 0$. Spontaneous processes release Gibbs free energy, and so have negative ΔG values. It is worth remembering at this point that chemical reactions do not stop at equilibrium. At equilibrium, the rate of the forward reaction is equal to the rate of the back reaction, and so the system does not change in composition over time.

We define chemical equilibrium for a system that we specify. For instance, we might be interested in calcite in water, a system containing components $CaCO_3$ and H_2O. Of course, any real system is contained within a vessel (perhaps a beaker if a lab, perhaps a pore in a soil or rock). We idealize systems to the simplest form possible, hence ignoring the possible role of vessel walls, or perhaps the presence of a gas phase, or other components in the system. Each chemical component contributes to the Gibbs free energy of the system, an idea expressed by the chemical potential (μ). The chemical potential is defined as the change in Gibbs free energy for a change in the number of particles of a component species in a system:

$$\mu_i = \left(\frac{\partial G}{\partial n_i}\right)_{T,P} \tag{7.6}$$

where n_i is the number of moles of component i being added to (or removed from) the system. The T and P subscripts indicate that temperature and pressure are held constant. From this expression, we can see that the Gibbs free energy of a component can be written as $G = n_i \mu_i$. The chemical potential for a component can also be defined from the activity (equivalent in value to concentration for most solutions) for liquid or solid phases:

$$\mu_i = \mu_i^\circ + RT \ln a_i \tag{7.7}$$

where μ_i° is a constant, the chemical potential of pure component i in its standard state, and R is the gas constant, which has a value of 8.3143 J/(mol K). The $^\circ$ superscript always indicates a standard state reference condition; the standard state is an arbitrary but defined temperature and pressure condition, commonly 25 °C and 1 bar. Other standard states can be used, and this will change the value of the constant, μ_i°.

Consider the following reaction:

$$aA + bB \rightleftharpoons cC + dD \tag{7.8}$$

in which a moles of A and b moles of B react to form c moles of C and d moles of D. The change in the

Gibbs free energy for the reaction, ΔG_R, is the difference between the Gibbs free energy of the products and the reactants:

$$\Delta G_R = G_{\text{products}} - G_{\text{reactants}} \tag{7.9}$$

which, using the definition $G = n_i \mu_i$, can be written in terms of chemical potentials:

$$\Delta G_R = c\mu_C + d\mu_D - a\mu_A - b\mu_B \tag{7.10}$$

Using the definition of chemical potential in Equation 7.7, we find:

$$\Delta G_R = c\mu_C^\circ + cRT \ln a_C + d\mu_D^\circ + dRT \ln a_D - a\mu_A^\circ \\ - aRT \ln a_A - b\mu_B^\circ - bRT \ln a_B \tag{7.11}$$

which can be simplified to

$$\Delta G_R = c\mu_C^\circ + d\mu_D^\circ - a\mu_A^\circ - b\mu_B^\circ + RT \ln\left(\frac{a_C^c \cdot a_D^d}{a_A^a \cdot a_B^b}\right) \tag{7.12}$$

or the even simpler form:

$$\Delta G_R = \Delta G_R^\circ + RT \ln\left(\frac{a_C^c \cdot a_D^d}{a_A^a \cdot a_B^b}\right) \text{(7.6)} \tag{7.13}$$

where ΔG° is the standard free energy of the reaction. This is the change in free energy when a moles of A and b moles of B produce c moles of C and d moles of D, all in standard state. At equilibrium, the change in free energy is zero (recall Equation 7.5), which means that $\Delta G = 0$. We can make this substitution and rearrange Equation 7.13 to yield:

$$\frac{a_C^c \cdot a_D^d}{a_A^a \cdot a_B^b} = \exp\left(\frac{-\Delta G_R^\circ}{RT}\right) \tag{7.14}$$
$$= K_{\text{eq}}$$

Notice that we removed the natural logarithm (ln) by taking the exponential of both sides of the expression. You perhaps recognize the left-hand side of Equation 7.14 as a definition of the equilibrium constant, K_{eq}. From Equation 7.14 we now have two expressions for the equilibrium constant, one defining it in terms of activities (or concentrations) of products and reactants for a reaction at equilibrium and stoichiometric coefficients for the reaction, and the other defining it in terms of the standard free energy of reaction, ΔG°, which can be found tabulated for many compounds of interest (e.g., Drever, 1997). The activity of pure solids and of water is 1, and so equilibrium constants in dissolution reactions can be related to solute activities (equal to concentration in most dilute natural waters).

Solubility and saturation

The equilibrium constant for dissolution of a mineral in water is called the solubility product, K_{sp}. If we consider, for example, the dissolution of gypsum:

$$CaSO_4 \cdot 2H_2O \rightleftharpoons Ca^{2+} + SO_4^{2-} + 2H_2O \quad (7.15)$$

the equilibrium constant and solubility product are:

$$K_{eq} = K_{sp} = a_{Ca^{2+}} a_{SO_4^{2-}} \quad (7.16)$$

Note that the activity of water and of the solid phase (gypsum) are both taken to be 1, so the equilibrium constant simplifies to the product of the activities of the ions produced. The equilibrium constant can be calculated from tabulated Gibbs free energies (using Equation 7.14), or from tables of equilibrium constants (e.g., Drever, 1997). We can use solubility products to determine how soluble a particular mineral is, and whether a solution is undersaturated or oversaturated with respect to that mineral.

The solubility, S, of a solid is the number of moles of that solid that can be dissolved in a liter of water. For the example of gypsum, $S = (K_{sp\text{-}gypsum})^{1/2}$, since $K_{sp\text{-}gypsum} = [Ca^{2+}][SO_4^{2-}]$, and each mole of gypsum releases 1 mole of Ca^{2+} and 1 mole of SO_4^{2-}. Note that we have used square brackets, [ ], which indicate solute concentrations in mol/L, in place of activities, since in dilute solutions activities are approximately equal to molar concentrations. At 25 °C, $K_{sp\text{-}gypsum} = 10^{-4.58}$ and therefore $S_{gypsum} = 5.1$ mmol/L. A solution in equilibrium with gypsum will have $[Ca^{2+}] = [SO_4^{2-}] = 5.1$ mmol/L.

Bench River, a glacier outlet stream in Alaska, had $[Ca^{2+}] = 0.27$ mmol/L and $[SO_4^{2-}] = 0.13$ mmol/L (S. P. Anderson et al., 2000). We can see by inspection that Bench River is undersaturated in gypsum, but we might want to know how far from equilibrium the system is. We use the saturation index, SI, for this purpose. Consider again the reaction shown in Equation 7.8, in which A and B react to form C and D. If the reaction is not at equilibrium, then the left-hand side of the expression:

$$\frac{a_C^c \cdot a_D^d}{a_A^a \cdot a_B^b} = Q \quad (7.17)$$

will not be equal to the equilibrium constant, and it is called the activity product, Q (or the ion activity product, if all the species are ions). At equilibrium,

$Q/K_{eq} = 1$. This ratio provides a measure of the direction the reaction will proceed and departure from equilibrium. For instance, if $Q/K_{eq} > 1$, the reaction will tend to the left, producing more A and B, and consuming C and D. Conversely, if $Q/K_{eq} < 1$, the reaction will tend to the right, producing more C and D, and consuming A and B. The saturation index is defined as

$$SI = \log \frac{Q}{K_{sp}} \quad (7.18)$$

where we have used K_{sp} to emphasize that SI applies to dissolution reactions. Because the saturation index is defined as the log of the ratio of the activity product and solubility product, we see that SI = 0 at equilibrium, SI > 0 for supersaturated systems, and SI < 0 for undersaturated systems.

Let's return to our Bench River example. With respect to gypsum:

$$SI = \log \left[\frac{a_{Ca^{2+}} a_{SO_4^{2-}}}{K_{sp\text{-}gypsum}} \right] \quad (7.19)$$

If we again assume that concentrations can be used for activities (an assumption that the activity coefficient = 1), then for the Bench River,

$$SI = \log \frac{(0.27 \times 10^{-3})(0.13 \times 10^{-3})}{10^{-4.58}} = -2.9$$

This strongly negative value for SI indicates undersaturation with respect to gypsum in Bench River. Any gypsum present in the channel will dissolve.

These equilibrium considerations do not indicate whether a dissolution reaction will take place at an appreciable rate. Gypsum dissolves rapidly, so in the case of the gypsum in Bench River, both thermodynamics (saturation state) and kinetics (rapid reaction rate) favor its dissolution. It would therefore be surprising to find gypsum in the suspended sediments in Bench River. However, the case is different for quartz. Bench River carries a dissolved silica concentration of 0.022 mmol/L, which indicates it is undersaturated with respect to quartz ($K_{sp\text{-}quartz} = 10^{-3.98}$ at 25 °C), and for this system, $SI_{quartz} = -0.67$. The rate of dissolution of quartz in water is very low at near-neutral pH, so even though it is thermodynamically favored, quartz dissolution is negligible in the system. Therefore, quartz will easily survive transport in the suspended load of Bench River. We will return to the subject of kinetics (rates of reactions) shortly.

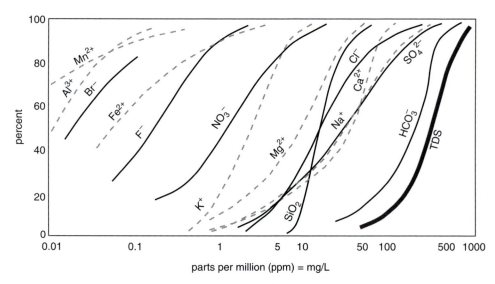

Figure 7.22 Cumulative probability curves for concentration of chemical species in natural terrestrial river waters. Dashed lines: cations; solid lines: anions (after Stumm and Morgan, 1996, Figure 15.1, with permission of John Wiley & Sons, Inc.).

Thus far, we have considered systems consisting of one mineral dissolving in water, a situation that is rarely encountered in natural systems. The solubility of a particular mineral can be affected by ions in solution from other sources. The presence of a common ion, that is one of the dissolution products for a mineral from another source, will reduce the solubility of that mineral. For a solid that dissolves to release a cation and an anion of equal charge ($NaCl$ or $CaSO_4$, for example), the solubility in a solution containing a concentration, C, of one of the dissolution products can be determined as follows. The solubility product for the reaction $MN = M^+ + N^-$ is given by $K_{sp} = [M^+][N^-]$, for cation M and anion N, here shown with $+1$ and -1 charges, respectively. Consider a solution containing a concentration C of the cation M^+ (for example), balanced by some other anion. The equilibrium concentration of the anion N^- would be $[N^-] = S$, the solubility of MN. At equilibrium, the concentration of the cation M^+ is $[M^+] = C + S$. Putting these equilibrium concentrations into the solubility product expression yields $K_{sp} = (C + S)S$. For high concentrations C, the term $(C + S)$ can be approximated as C, and the solubility is $S = K_{sp}/C$ (approximately, for $C \gg S$). For very low concentrations C, $S = K_{sp}^{1/2}$, as in the simple system. In systems of minerals, the more readily soluble minerals (or kinetically favored reaction products) can impact the solubility of other minerals.

In silicate mineral weathering systems, the number of phases and solutes to consider makes assessing solubility more complex. For instance, aluminum, which is relatively insoluble, even at low concentrations affects the solubility of various secondary phases. To analyze equilibrium in systems of multiple minerals and solutes, it quickly becomes advantageous to use software with mineral thermodynamic data and routines for solving simultaneous equations. A widely used program is PHREEQC (Parkhurst and Appelo, 1999).

Rivers, continental crust, and common chemical weathering reactions

River waters from around the world contain a fairly consistent set of solutes in a relatively restricted range of concentrations, as illustrated in the probability diagram in Figure 7.22. By far the dominant solute in rivers by mass is bicarbonate ion (HCO_3^-). Rounding out, the top seven solutes are SO_4^{2-}, Ca^{2+}, Na^+, Cl^-, SiO_2, and Mg^{2+}. (There is nothing magic about seven solutes, just a sizeable drop – note the log concentration scale – to the next pair of solutes, K^+ and NO_3^-.) When river solute concentrations are compared with the abundance of elements in the continental crust shown in Table 7.3, there are some notable discrepancies. Cl, S and C are all present in less than 0.1 wt % abundances in continental crust, yet figure importantly in river

Table 7.3 **Elemental and mineral composition of the continental crust**

Element	Upper continental crust (weight %)	Mineral or mineral group	Lithosphere abundance (weight %)
Si	30.8	Feldspars	41
Al	8.0	Quartz	12
Fe	3.5	Pyroxenes	11
Ca	3.0	Amphiboles	5
Na	2.9	Micas	5
K	2.8	Clay	4.6
Mg	1.3	minerals	
		Olivine	3
		Calcite/	2
		dolomite	
		Magnetite	1.5

Adapted from Taylor and McLennan (1995) and Taylor and Eggleton (2001)

solute loads. Al is the third most abundant element in the crust (after Si and O, the latter not shown in Table 7.3), yet is a very minor component in riverine dissolved loads.

The discrepancies arise because river solute loads reflect the abundance and solubility of the minerals in the Critical Zone, as well as the rate at which the minerals weather. Looking at the mineral abundances in Table 7.3, we see that feldspars rule the crust, but highly weatherable minerals with low abundance, such as calcite, are also significant contributors to river solute loads. Biological processes contribute to the solutes in rivers as well. This is most easily seen in NO_3^-. Nitrogen is found in few minerals, but vigorously cycles between the atmosphere and biosphere and therefore appears prominently in river solute loads. Here, we focus on the primarily inorganic reactions involving rock minerals. Given the relatively short list of minerals and mineral groups that make up the bulk of continental crust, we can learn much about chemical breakdown of rock by considering only a few mineral weathering reactions.

Congruent dissolution

Relatively few minerals dissolve congruently, the simple process of solution of a solid mineral into

dissolved products. Quartz, halite, and gypsum all dissolve congruently. For example:

$$SiO_{2(quartz)} + H_2O \rightleftharpoons H_4SiO_{4(aq)} \qquad (7.20)$$

shows the simple dissolution of quartz into silicic acid, the dissolved form of silica at pH values less than about 9. From the equilibrium constant (the solubility product) for this reaction, we can determine the solubility of quartz, the concentration of a solute in equilibrium with the solid phase. Given the relationship $K_{eq} = K_{sp} = a_{H_4SiO_4}$ (recall that water and pure minerals have an activity of 1), and the value of K_{eq} at $25\,^\circ C$ of $10^{-3.8}$, the concentration of H_4SiO_4 in water in equilibrium with quartz is $10^{-3.98}$ mols/kg. The concentration units used here are *molal*, the number of mols per kilogram of solution. For the dilute solutions typical of groundwater and rivers, activities, which the equilibrium constant is defined by, are very close to molal concentrations. We can further assume that solutions are dilute enough that molality (mols of solute per kg of solution) is equivalent to molarity (number of mols of solute per liter of solution). Thus, the solubility of quartz in water at $25\,^\circ C$ is $10^{-3.98}$ mols/kg or $10^{-3.98}$ mols/L. Solute concentrations are often expressed in parts per million (ppm), equivalent to mg/L. The conversion from molarity to ppm $SiO_{2(aq)}$ can be done as follows:

$$\left(\frac{10^{-3.98}\ \text{mols}\ H_4SiO_4}{L}\right)\left(\frac{1\ \text{mol}\ SiO_2}{\text{mol}\ H_4SiO_4}\right)$$
$$\left(\frac{60\ g}{\text{mol}\ SiO_2}\right)\left(\frac{10^3\ \text{mg}}{g}\right) = 6\ \text{ppm}\ SiO_{2(aq)} \qquad (7.21)$$

This concentration is at the very low end of the range of river dissolved silica concentrations shown in Figure 7.22. In general, groundwater and river water is supersaturated with respect to quartz. The dissolved silica comes from other mineral weathering reactions or from amorphous silica, which has a higher equilibrium constant.

Calcium carbonate dissolves congruently as follows:

$$CaCO_3 + H_2CO_3 \rightleftharpoons Ca^{2+} + 2HCO_3^- \qquad (7.22)$$

In this reaction the aqueous reactant, carbonic acid (H_2CO_3), strongly affects the solubility of calcium carbonate. Any process that increases the supply of H_2CO_3 will drive dissolution of calcium carbonate, as will processes that remove HCO_3^- (bicarbonate).

Carbonic acid is a ubiquitous weak acid formed by $CO_{2(g)}$ dissolved in water:

$$H_2O + CO_2 \rightleftharpoons H_2CO_3 \tag{7.23}$$

Rainwater is slightly acidic (pH ~5.5) due to equilibrium with atmospheric CO_2, but soil waters can be more acidic owing to a partial pressure of CO_2 that is 10–100 times greater than atmospheric values, which drives Equation 7.23 to the right. Thus the solubility of calcium carbonate is greater where organic rich soils are present. The equilibrium constant for Equation 7.23 decreases with increasing temperature because the solubility of CO_2 gas is greater at lower temperatures. Thus, calcium carbonate is more soluble in colder environments. At 25 °C and atmospheric CO_2 partial pressure ($P_{CO_2} = 10^{-3.5}$, or 0.035%), the Ca^{2+} concentration in equilibrium with calcium carbonate is about 19 ppm, and the corresponding HCO_3^- concentration is about 57 ppm. If the CO_2 partial pressure is increased by a factor of 100 to $10^{-1.5}$, a value that has been observed in soils, the equilibrium Ca^{2+} concentration increases to about 87 ppm. Carbonate equilibria are discussed in greater detail in Drever (1997) and elsewhere.

Incongruent dissolution

Feldspars, the most abundant minerals in the crust, and many other aluminosilicate minerals, dissolve incongruently, releasing solutes and forming one or more new minerals. The solutes released do not match the stoichiometry of the weathering mineral (the primary mineral) because of the formation of a new solid phase – commonly a clay mineral or an oxide – during the reaction. A general aluminosilicate mineral reaction looks like:

$$\text{aluminosilicate} + H_2O + H_2CO_3 \rightleftharpoons \text{clay mineral}$$
$$+ \textit{cations} + HCO_3^- + H_4SiO_4 \tag{7.24}$$

where carbonic acid and water react to produce a new clay mineral and release cations, dissolved silica, and bicarbonate. The minerals produced by incongruent dissolution reactions are called secondary minerals. Although written here with carbonic acid, other acids, such as organic acids from decaying organic matter, sulfuric acid, or nitric acid, can be the source of protons in weathering reactions. If these acids are involved, then anions other than bicarbonate will

result. However, much of the abundant bicarbonate (HCO_3^-) in river water (Figure 7.22) derives from carbonic acid driven weathering of aluminosilicates. Figure 7.23 shows the weathering sources of solutes in major rivers of the world, ranked by increasing dominance of silicate weathering reactions, most of which are incongruent. The "atmospheric" contribution in Figure 7.23 is the bicarbonate in river water due to reaction of carbonic acid (from the atmosphere) with silicate or carbonate minerals. Calculations of the sort shown in Figure 7.23 show that carbonic acid is the most important acid source for silicate weathering.

A number of possible weathering reactions for feldspar minerals are shown in Table 7.4. In the table, the reactions are written without specifying the acid source; these can be written with carbonic acid if one remembers that $H_2CO_3 \rightleftharpoons H^+ + HCO_3^-$. The proportions of cations and dissolved silica released depend on the clay mineral formed, as well as the composition of the feldspar. The reactions in Table 7.4 are written for pure end-members of the feldspar mineral solid solutions, but most feldspars are solid solutions of either the plagioclase (Ca- and Na-rich) feldspar or the alkali (Na- and K-rich) feldspar. Given the uncertainty in primary mineral and clay mineral composition, deciphering weathering reactions from solute loads alone can be a tricky business. Nonetheless, the general pattern of cation and silica release (in most cases, but note the pure anorthite weathering reaction in Table 7.4), anion production (generally bicarbonate), and formation of cation-depleted, aluminous clays holds.

Incongruent weathering may proceed through a series of steps, in which cations and silica are progressively lost and the secondary mineral evolves to more aluminum-rich and cation-poor forms. For instance, feldspars may first weather to smectite, which has a higher Si:Al ratio and contains cations, and then progress to kaolinite upon further loss of silica and cations. Under extreme leaching, all silica can be removed and the pure aluminum oxide gibbsite is produced. A generalized summary of these steps for major silicate mineral groups is illustrated in Figure 7.24.

Oxidation

Oxidation and reduction reactions involve transfer of electrons between atoms. One atom gives up some of its electrons, becoming oxidized in the process, while another atom takes the electrons, becoming reduced in

Table 7.4   **Examples of some feldspar weathering equations for pure end-members**

$KAlSi_3O_8 + H^+ + \frac{9}{2}H_2O = \frac{1}{2}Al_2Si_2O_5(OH)_4 + 2H_4SiO_4 + K^+$
orthoclase   kaolinite

$KAlSi_3O_8 + \frac{2}{3}H^+ + 4H_2O = \frac{1}{3}KAl_3Si_3O_{10}(OH)_2 + 2H_4SiO_4 + \frac{2}{3}K^+$
orthoclase   illite

$NaAlSi_3O_8 + H^+ + \frac{9}{2}H_2O = \frac{1}{2}Al_2Si_2O_5(OH)_4 + 2H_4SiO_4 + Na^+$
albite   kaolinite

$NaAlSi_3O_8 + \frac{3}{4}H^+ + \frac{7}{2}H_2O = \frac{3}{8}Na_{0.66}Al_{2.66}Si_{3.33}O_{10}(OH)_2 + \frac{7}{4}H_4SiO_4 + \frac{3}{4}Na^+$
albite   smectite

$CaAl_2Si_2O_8 + 2H^+ + H_2O = Al_2Si_2O_5(OH)_4 + Ca^{2+}$
anorthite   kaolinite

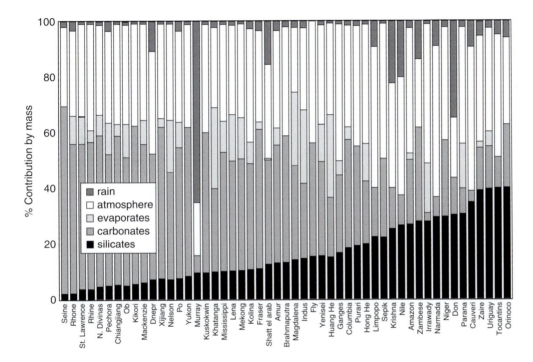

Figure 7.23 Contributions of various sources to solutes in world rivers, ranked left to right according to increasing role of silicate weathering. The bars show the fraction of total solute concentration (% of total dissolved solids in mg/L) from each source. Contribution from the atmosphere is bicarbonate ions of atmospheric origin from silicate and carbonate weathering. Contribution from rain is principally Na and Cl from sea salt (after Gaillardet *et al.*, 1999, Figure 5, with permission from Elsevier).

the process. The most important redox reactions in weathering involve oxidation of ferrous (Fe^{2+}) to ferric (Fe^{3+}) iron and oxidation of sulfides (S^{2-}) to sulfate (SO_4^{2-}). In Figure 7.22 we saw that sulfate was the second most common anion in river water, after bicarbonate. Sulfur occurs nearly equally as sulfates (e.g., gypsum and anhydrite) and sulfides (e.g., pyrite) in the crust (Holland, 1978). We have already discussed the simple congruent dissolution of gypsum and anhydrite.

The most common sulfide, pyrite, weathers through oxidation to sulfate. The full process is viewed as composed of steps, the first of which is dissolution of pyrite and oxidation of the sulfur to sulfate:

$$FeS_2 + \tfrac{7}{2}O_2 + H_2O \rightarrow Fe^{2+} + 2SO_4^{2-} + 2H^+ \qquad (7.25)$$

The iron is then oxidized:

$$Fe^{2+} + \tfrac{1}{4}O_2 + H^+ \rightarrow Fe^{3+} + \tfrac{1}{2}H_2O \qquad (7.26)$$

Box 7.2 Saltiness of the ocean

The solutes derived from chemical weathering of minerals near the Earth's surface are carried by rivers to the oceans, where they add their chemical load to that of the oceans. Can we use river solute loads to understand the saltiness of the sea? The ocean is well mixed and has a salinity of 35 parts per thousand, or 0.35%, by mass. Roughly 90% of this salinity is attributable to the high concentrations of Na and Cl ions. It was long ago thought that one could determine the age of the ocean, and hence constrain the age of the Earth, by dividing the mass of salt in the ocean by the rate at which salt was being added to the ocean from the world's rivers. According to Drever (1988, p. 268), this calculation was first proposed in 1715 by the astronomer Edmund Halley, and was actually first performed by Joly (1898), in the midst of the great debate about the age of the Earth. This is a classic reservoir age calculation. You take the volume of the reservoir, V, and divide by the rate of input, Q, to determine how long the reservoir would take to reach that size: $T = V/Q$.

Let us try it, using a few major ions (tabulated in Drever, 1988, Table 12–2). To be specific, we modify the simple equation above for our problem: $T = \frac{\rho_{sw} V_o C_{sw}}{\rho_w Q_r C_r}$, where ρ_{sw} and ρ_r are the densities of seawater and river waters, respectively, V_o is the volume of the ocean, and C_{sw} and C_r are the concentrations by mass of some chemical species in the seawater and river, respectively. The product of the density of seawater and the volume of the ocean suggests that the mass of the world's ocean is roughly 1.35×10^{21} kg. The concentration of Cl is 1.93×10^{-2} kg/kg of seawater. The product of these numbers is the mass of Cl in the ocean: 2.61×10^{19} kg, which is the numerator in our equation. Now we need the inputs, the denominator. The mean concentration of Cl in river water is 5.75×10^{-6} kg/kg. The sum of river water inputs to the ocean is 3.74×10^{16} kg/year. The product of these is the mass input rate of Cl to the ocean: 2.15×10^{11} kg/yr. Performing the division yields 1.21×10^8 years, or 121 million years. Interestingly, when we try this for the next highest concentration species in the ocean, Na, we calculate an age of 75 million years. For other species, the calculated ages are even shorter, ranging down to only 20 000 years for Si. These numbers raise two questions: (1) why are all of the ages so much smaller than the 4.5 billion year age of the Earth, even accounting for the time it might have taken to form the first oceans on Earth, and (2) why do they differ? It is now clear that this will not work as a means of dating the Earth! The reason is that we have neglected to account for any losses of these ions from the ocean. And the reason that they differ so greatly is that the mechanisms of removal differ among the chemical species. For example, the efficient removal of Si by diatoms reduces the concentration of Si in seawater to less than 1 ppm. As this is in the numerator of the timescale equation, the timescale for Si residence time is correspondingly short. Cl and Na are much less easily removed from the ocean, but nonetheless are. Thus, the saltiness of the sea, and the details of abundance of different solutes, reflects both their inputs from rivers as well as the rates of processes that remove them (to aerosols, zeolites, fecal pellets, saline pore waters, and so on).

Commonly, the solubility product of an iron hydroxide phase is exceeded, and an oxidized iron hydroxide precipitates:

$$Fe^{3+} + 3H_2O \rightarrow Fe(OH)_3 + 3H^+ \tag{7.27}$$

The sum of these reactions shows a net production of iron hydroxide, sulfate and hydrogen ion:

$$FeS_2 + \tfrac{15}{4}O_2 + \tfrac{7}{2}H_2O \rightarrow Fe(OH)_3 + 2SO_4^{2-} + 4H^+ \tag{7.28}$$

The acid produced by the oxidation of pyrite can attack other minerals, including calcium carbonate and primary silicate minerals. Thus, the presence of pyrite can catalyze chemical breakdown of the surrounding rock when it is exposed to oxidizing surface waters. Reactions of this sort are responsible for the generation of acid mine drainage from tailings of sulfide bearing ores.

Oxidation of iron within the structure of silicate minerals can be the first step in weathering. For instance,

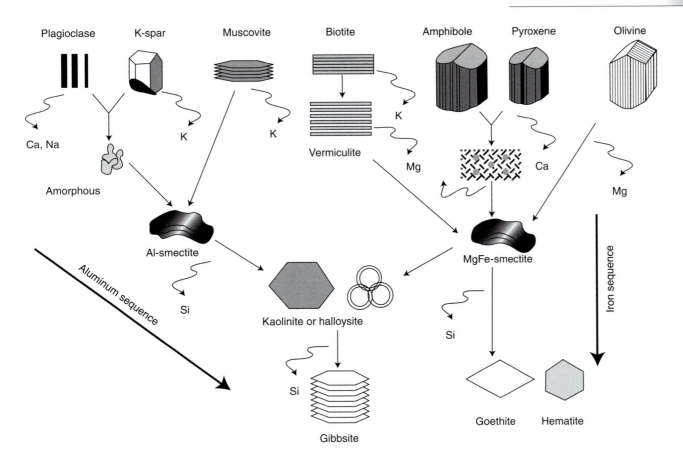

Figure 7.24 Typical weathering steps for major silicate mineral groups. Primary minerals (across top) loose cations to solution (squiggly lines) to form secondary mineral products. Continued weathering leads to more cation-poor, Al-rich products such as kaolinite or gibbsite, or iron oxides goethite and hematite (after Taylor and Eggleton, 2001, Figure 8.12, with permission of John Wiley & Sons).

oxidation of Fe^{2+} to Fe^{3+} in the tetrahedral layer (the siloxane sheets) of biotite leads to ejection of interlayer K^+ to maintain charge balance in the crystal (Blum and Erel, 1997; Buss *et al.*, 2008). The weathered product is the clay mineral vermiculite. Under oxidizing conditions, structural iron in pyroxenes and amphiboles is oxidized to Fe^{3+} in an outer surface layer, but this layer does not strongly affect the subsequent weathering rate of the minerals (Schott and Berner, 1984).

Ferrous iron (Fe^{2+}) is more soluble than the oxidized ferric iron (Fe^{3+}). In oxidizing environments iron released by H_2CO_3 or H_2SO_4 reaction with mafic silicates is therefore commonly immobilized into iron oxides such as goethite, hematite, maghemite and ferrihydrite (Birkeland, 1999). These minerals are responsible for the red-yellow-brown hues of soils. In reducing environments, however, reduced ferrous

iron is removed in solution. This loss of iron oxides is responsible for the green-blue-gray colors of redoximorphic features found in reducing environments, such as water-saturated soils.

Chemical kinetics

The rates at which minerals release solutes and dissolve or transform play important roles in landscape evolution, since these chemical processes affect rock strength, soil formation, global geochemical cycles and ecosystem productivity. The important mineral groups in the crust (Table 7.3) break down slowly in the reactor of the Critical Zone. However, there is a wide range of reaction rates among silicate minerals, as illustrated by the lifetimes of 1 mm-radius crystals shown in Table 7.5. The table is based on laboratory

Table 7.5   **Dissolution rates of silicate minerals and crystal lifetimes**
Crystal lifetimes calculated for 1 mm crystals at 25 °C, pH = 5, and far from equilibrium conditions (table from Lasaga et al., 1994).

Mineral	Mineral family	Formula	Log (rate) (mol/(m^2 s))	Molar volume (cm^3/mol)	Lifetime (yr)
Quartz	Tectosilicate	SiO_2	−13.39	22.688	34 000 000
Kaolinite	Phyllosilicate – clay mineral group	$Al_2Si_2O_5(OH)_4$	−13.28	99.52	6 000 000
Muscovite	Phyllosilicate – mica group	$KAl_2(AlSi_3O_{10})(OH)_2$	−13.07	140.71	2 600 000
Epidote	Sorosilicate	$(Al,Fe)_3O(SiO_4)(Si_2O_7)(OH)$	−12.61	139.2	923 000
Microcline	Tectosilicate – K feldspar	$KAlSi_3O_8$	−12.50	108.741	921 000
Albite	Tectosilicate – plagioclase feldspar	$NaAl Si_3O_8$	−12.26	100.07	575 000
Sanidine	Tectosilicate – K feldspar	$KAlSi_3O_8$	−12.00	109.008	291 000
Gibbsite	Oxide – analogous to mica structure	$Al(OH)_3$	−11.45	31.956	276 000
Enstatite	Single chain inosilicate – pyroxene group	$MgSiO_3$	−10.00	31.276	10 100
Diopside	Single chain inosilicate – pyroxene group	$CaMgSi_2O_6$	−10.15	66.09	6 800
Forsterite	Nesosilicate – olivine group	Mg_2SiO_4	−9.5	43.79	2 300
Nepheline	Tectosilicate – feldspathoid	$(Na,K)AlSiO_4$	−8.55	55.16	211
Anorthite	Tectosilicate – plagioclase feldspar	$CaAl_2Si_2O_8$	−8.55	100.79	112
Wollastonite	Single chain inosilicate – pyroxenoid group	$CaSiO_3$	−8.00	39.93	79

measured dissolution rates of individual minerals (Lasaga, 1984; Lasaga et al., 1994). It effectively quantifies the field-based classification of mineral persistence in weathering systems known as the Goldich Stability Series (Goldich, 1938). The Goldich series itself mirrors the order in which minerals crystallize from a melt, the Bowen Reaction Series. Minerals that crystallize from melts at high temperatures (e.g., olivines, Ca-plagioclase) tend to be less resistant to weathering, while minerals that crystallize at low temperatures (e.g., quartz) tend to be more resistant to weathering. The rates or lifetimes of crystals shown in Table 7.5 vary by a staggering five orders of magnitude. The two secondary minerals in the table (kaolinite and gibbsite) have dissolution rates that fall within the range of primary silicates. Understanding the mechanisms that explain these differences in dissolution rates is beyond this text. Instead, we will explore how different weathering environments affect these intrinsic dissolution rates.

Measuring reaction rates and developing understanding of the mechanisms underlying these rates is the realm of chemical kinetics. For a reaction $A + B \rightarrow Z$, the rate might be proportional to the concentration of A, or to the concentration of B, or to the square of the concentration of one of these species, or some other dependence. In general, $r = k[A]^\alpha[B]^\beta[Z]^\sigma$,

where r is the reaction rate, k is the rate constant, and the exponents α, β, and σ are constants. Note that the rate is expressed as a function of the concentrations of A, B, and/or Z (hence the square brackets), rather than the solute activities found in thermodynamic equilibrium constants. This is because the concentration determines the rate of molecular collisions, which partially controls the rate of reaction (Brantley and Conrad, 2008). Weathering reactions are heterogeneous, meaning they take place between two or more phases, commonly a solid and a solution. The reactions take place on or near the mineral surface, and so the mineral surface area (or density of reactive sites on the mineral surface) per unit volume of fluid becomes an important characteristic of the system in rate laws.

The exponents in the rate law α, β, and σ are the partial orders of reaction, and their sum $(\alpha + \beta + \sigma)$ is the overall order of the reaction. A first-order rate equation is one that depends linearly on only one concentration (A or B, for instance). A reaction rate that is proportional to the concentrations of both A and B would be second order overall. Alternatively, a second-order reaction might be proportional to the square of the concentration of one of the reactants. An example of a first-order rate equation is

Box 7.3 Calculating a crystal lifetime

It would seem a pretty straightforward thing to calculate the rate of disappearance of a crystal if the dissolution rate is known. Mineral dissolution rates, k, are reported in $mol_{mineral}/(m^2\ s)$, the number of moles (formula units) of the mineral per unit surface area of the mineral per unit time. Thus, the number of moles of mineral i dissolved, n_i, per unit time for a spherical crystal of surface area A and radius R is given by

$$\frac{dn_i}{dt} = Ak_i = 4\pi R^2 k_i \qquad (7.29)$$

Dissolution will diminish the size of the crystal, so we need to relate the moles of mineral loss to volume loss. Using the molar volume, $\bar{V}_i$, the volume occupied per mole of the mineral, we calculate the rate of volume loss as

$$\frac{dV}{dt} = -4\pi R^2 k_i \bar{V}_i \qquad (7.30)$$

From the volume of a sphere, $V = \frac{4}{3}\pi R^3$, we can derive the rate at which the volume diminishes:

$$\frac{dV}{dt} = \frac{dV}{dR}\frac{dR}{dt} = 4\pi R^2 \frac{dR}{dt} \qquad (7.31)$$

Combining Equations 7.30 and 7.31, we find:

$$\frac{dR}{dt} = -k_i \bar{V}_i \qquad (7.32)$$

from which it follows, for conditions in which the dissolution rate k_i is constant, that $R = R^0 - k_i\bar{V}_i t$. This expression gives the radius of a crystal at time t from an initial radius R^0, and can be rearranged to solve for the time at which $R = 0$. This is how the crystal lifetimes were calculated in Table 7.5. While there are several assumptions embedded (spherical crystals, uniform conditions of temperature, and far from equilibrium conditions), the resulting crystal lifetimes are useful in comprehending the difference between $k_{quartz} = 10^{-13.39}\ mol/(m^2 s)$ and $k_{anorthite} = 10^{-8.55}\ mol/(m^2 s)$.

radioactive decay. In the decay equation (see Equation 6.4 in Chapter 6), the amount of radioactive element lost per unit time (the reaction rate) is proportional to the amount of the radioactive element present.

Weathering reactions such as those shown in Table 7.4 are actually the result of multiple elementary reactions that describe the smallest chemical steps in a complex mechanism, usually involving two or at most three chemical species. The slowest of these elementary reactions is the rate-controlling step. Kinetics generally focuses on these rate-controlling steps rather than the overall reaction rate. For most silicate minerals, the rate law can be written for the release of a solute i from the mineral θ as follows (Lasaga, 1984):

$$r = \frac{dc_i}{dt}\bigg|_{diss} = \frac{A_\theta}{V}\beta_{i\theta}k_\theta \qquad (7.33)$$

where dc_i/dt is the change in concentration of solute i due to the dissolution of mineral θ, A_θ is the surface area of mineral θ, V is the volume of fluid in contact with the mineral, $\beta_{i\theta}$ is the stoichiometric content of i in mineral θ, and k_θ is the rate constant in $mol/(m^2 s)$. Here, the mineral surface area stands in for the concentration of a reactant in the general rate law. Because reactions take place on mineral surfaces, the rates of reactions are sensitive to the nature of these surfaces – crystal defects and surface coatings are among the properties that can strongly affect reaction rates. Natural mineral surfaces are not smooth as a rule, and so geochemists measure the surface area using a gas adsorption method, known as the BET method (named for the authors of the original paper, Brunauer, Emmett and Teller, 1938). The BET method detects the nooks and crannies in etch pits and other roughness elements on mineral surfaces. As a result, BET surface area is generally much greater than the surface area one would calculate for spheres (like the 1 mm crystals considered in the box) or cubes of a known size or size distribution. While the BET surface area is easily measured, it is not a property of the system that can be easily predicted. One approach is to estimate the surface roughness, λ, which is defined as $\lambda = A_{\mathrm{BET}}/A_{\mathrm{geom}}$, the ratio of the BET surface area to the geometric surface area. The volume term in Equation 7.33 is another parameter difficult to estimate in natural systems. The most straightforward approach is to use porosity in place of volume.

Two general mechanisms control the rate of dissolution reactions (and actually of precipitation reactions as well). The first is control by a reaction occurring on the mineral surface itself. The rate-limiting elementary reaction may be the detachment of an ion from the mineral surface, or the formation of an activated complex – a transient, high-energy intermediate on the mineral surface. These are examples of reaction control or interface controls on the dissolution rate. Under this type of control, the solute concentrations in the fluid at the mineral surface are indistinguishable from that in the bulk fluid (the pore fluid some distance away from a mineral surface). Alternatively, transport of ions or molecules away from the mineral surface control the rate of reaction. Under transport control, solute concentrations in fluid adjacent to the mineral surface are higher than in the bulk fluid. Diffusion of dissolved products away from the mineral surface is the rate-limiting step in this situation. A single mineral may display both control mechanisms under different conditions. For instance, dissolution of calcite is thought to be transport controlled at low pH, and interface controlled at higher (near neutral) pH (Brantley, 2008). Most silicate minerals show surface reaction control except at very low pH, where the rate of surface reaction is high enough that the reaction becomes limited instead by diffusion of reaction products away from the surface. This situation implies that mineral surface effects are the most important in controlling mineral weathering rates.

Formulations based on Equation 7.33 have been used to determine weathering rate constants (values of k_{θ}) from measured solute fluxes in field and laboratory settings, or conversely to estimate the solute flux from watersheds. Velbel (1985) evaluated mineral weathering rates in the Coweeta watershed (North Carolina) from measured stream chemistry and a detailed estimation of mineral surface area. Based on petrographic observations, he determined a characteristic grain size and shape for minerals in the regolith. The total surface area was determined from modal abundances and the mean thickness of saprolite in the catchment. Velbel found that his field-based rate constants were about an order of magnitude lower than laboratory-based measurements, although uncertainties in field surface area are certainly of this order. Swoboda-Colberg and Drever (1993) and Drever and Clow (1995) conducted much smaller field studies, using irrigated soil profiles and an irrigated "nano-catchment" respectively to measure weathering rate constants. In these essentially one-dimensional, shallow (maximum 50 cm depth) plots, the mineral surface area could be determined with much greater precision than Velbel could in his watershed. Even so, field-based measurements yielded rates much slower than rates determined in a laboratory flow-through reactor with samples from their plots. The conclusion from these studies was that the mineral surface areas measured with BET overestimates the surface area that is in contact with moving fluids in the soil (Drever and Clow, 1995). In other words, either not all pores are water filled, or flow in the very smallest pores is highly restricted, and therefore the mineral surface area that is in communication with soil water is less than the total mineral surface area. Laboratory reactors used to measure mineral weathering rate constants are designed to keep mineral grains suspended in rapidly flowing water precisely to eliminate uncertainty about solute transport from mineral surfaces. In a glacial outlet stream, where the assumption of rapid water flow through a water saturated system (the glacier bed) seemed justified, S. P. Anderson (2005) found a good correspondence between measured solute flux and the solute flux calculated from laboratory rates and mineral surface area estimated from grain size, sediment concentration, and mineral surface roughness.

Temperature dependence

The Arrhenius equation, named for the famous Swedish chemist, Svante Arrhenius, who first put this forth in 1889, captures the temperature dependence of the reaction rate:

$$k_{\theta} = A \exp\left(\frac{-E_{\mathrm{a}}}{RT}\right) \tag{7.34}$$

in which A is the pre-exponential factor, E_a is the activation energy, R is the gas constant, and T is the temperature in degrees kelvin. Values of the activation energy are determined from measurements of k at different temperatures, which are plotted in Arrhenius plots of $\ln k$ vs. $1/T$. Most silicate minerals display values of E_a between 40 and 80 kJ/mol, with a mean around 60 kJ/mol (Lasaga, 1984). The activation energy can be thought of as the energy required for an intermediate step or energy barrier in a reaction between products and reactants. The magnitude of E_a is related to the mechanism of the rate-controlling process. Activation energies for surface reaction controlled mechanisms, in which bonds must be formed or broken, tend to be higher than for transport-controlled mechanisms. One way to identify reactions for which transport is the rate-limiting step is through low activation energies of around 20 kJ/mol. While the pre-exponential factor and activation energy are generally observed to be independent of temperature (Brantley and Conrad, 2008), the rate-limiting step for dissolution reactions is not. At higher temperatures, the rate-limiting step is more likely to be transport control, and therefore a lower activation energy applies, while

---

Box 7.4 Temperature and reaction rates in a soil profile

Given the strong and nonlinear dependence of dissolution rates on temperature seen in Figure 7.25, and knowing that temperatures vary below the ground surface (Figure 7.7), how do the effective rate constants vary with depth in soil? We know that temperatures vary on many timescales, but most importantly on daily and annual cycles. The exponential increase in reaction rate with temperature (the Arrhenius relationship) disproportionately weights periods of high temperature, so that the time-averaged rate constant will be greater than that predicted from the mean temperature. Combining the expression for the depth dependence of temperature (Equation 7.2) with the Arrhenius expression (Equation 7.34) yields the expected instantaneous reaction rate. We must then average over the full period of the oscillation of the temperature, here the annual cycle, P. To accomplish this we turn again to the mean value theorem, obtaining an average reaction rate, $\bar{k}$, over the period P:

$$\bar{k} = \frac{1}{P} \int_0^P A \exp\left(\frac{-E_a}{R(\bar{T} + \Delta T \sin(2\pi t/P))}\right) dt \tag{7.35}$$

Although the temperature oscillates sinusoidally about its mean, $\bar{T}$, the time-averaged reaction rate increases with the amplitude of the oscillation, ΔT, where $\Delta T = T_{amp} e^{-z/z^*}$ (see Lasaga et al., 1994). This phenomenon is germane to our problem of weathering in a soil because the amplitude of temperature swings varies strongly with depth. Because the reaction rates increase nonlinearly with temperature (Figure 7.25), the extra reaction progress accomplished during excursions above the mean is not fully compensated by lower rates when the temperature is below the mean. This is illustrated in Figure 7.26 for the case in which both annual and daily cycles of temperature are taken into account. In the numerical integration of Equation 7.35 illustrated in the figure, we assumed that the temperature varies sinusoidally with a prescribed surface amplitude for both the annual and daily cycles (in the case illustrated, both are set to $\Delta T = 20\,°C$). This analysis parallels that performed by those concerned with the dating of surfaces and stratigraphic sections using amino acid racemization (e.g., McCoy, 1987), as this too is a chemical reaction modulated by temperature.

In summary, one role of the mean annual temperature, and hence the climate of the site, is to set the mean reaction constant at several meters depth; this is properly captured by Equation 7.34 using $T = \bar{T}$. However, as seen in Figure 7.26, the effect of the increase in amplitude of the temperature swing to a maximum at the surface is to enhance the weathering rate near the surface relative to that at depth. Both the daily and the annual cycles play a role, operating with depth scales commensurate with the diffusion length scale for the period of oscillation.

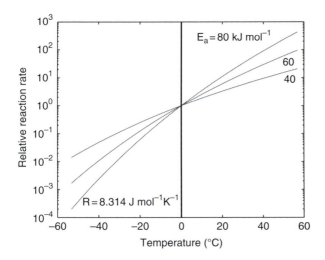

Figure 7.25 Reaction rate relative to that at $0\,°C$ ($k_T/k_{0\,°C}$) for three typical values for the activation energy over the temperature range relevant to Earth surface processes.

the same dissolution may be surface reaction controlled at lower temperatures and therefore display a higher activation energy. The primary message of Equation 7.34 is that the higher the temperature, or the lower the activation energy, the faster the reaction progresses. Clearly, as temperature rises, the reaction rate will rise as shown in Figure 7.25. This is one means by which climate comes into the weathering problem. Note that for a rise from 0 to $25\,°C$ the reaction rate increases by about a factor of nearly ten for the typical activation energy used in Figure 7.25.

pH dependence

Most silicate minerals dissolve at greater rates at both low and high pH, with a region of relatively low dissolution rates in the neutral pH range. The typical flattened-V shape pattern of dissolution rates as a function of pH is shown in Figure 7.27. At low pH, rate constants increase by a factor of about $10^{0.5}$ (about 3) for each unit decline in pH. In the middle range of pH, from about pH 4 to about pH 8, rate constants are independent of pH. Above about pH 8, rate constants again increase, by a factor of $10^{0.3}$ (about 2) for each unit increase in pH. The slopes of the pH-rate constant relationship and the pH range of pH independence differ from one mineral to another, but most follow the same pattern. The pH dependence is expressed as the sum of rates in each of the pH regions:

$$r = k_H a_{H^+}^n + k_N + k_{OH} a_{OH}^m \qquad (7.36)$$

The value of the exponent n is usually -0.5 to -1, while the exponent m that controls the alkaline part of the curve is 0.3 to 0.5. The relationship can be expressed in an additive fashion because the activity of H^+ in the alkaline region and of OH^- in the acid region are small enough to render these terms insignificant.

The pH dependence of dissolution rates is evidence of the proton-promoted dissolution mechanism. The idea is that protons (H^+) or hydroxyls (OH^-) adsorb to cations on mineral surfaces. This weakens the bond between the cation and the crystal lattice, and accelerates release of the cation into solution. The rate is proportional to the concentration of H^+ or OH^-, but nonlinearly (hence, n and m are generally <1).

Biological controls

Plants and microbes affect weathering rates in numerous ways (Taylor *et al.*, 2009). Here we focus specifically on the effect of biology on dissolution kinetics. Two direct effects on kinetics stand out: elevated P_{CO2} due to respiration and the role of organic exudates. Typical soil atmospheres have P_{CO_2} elevated to 10–100 times atmospheric values. The pH for a system with water and CO_2 present is described by several equilibria:

$$CO_2 + H_2O \rightleftharpoons H_2CO_3 \qquad K_{CO_2} = \frac{a_{H_2CO_3}}{P_{CO_2}} \qquad (7.37)$$

$$H_2CO_3 \rightleftharpoons H^+ + HCO_3^- \qquad K_1 = \frac{a_{H^+} a_{HCO_3^-}}{a_{H_2CO_3}} \qquad (7.38)$$

The equilibrium constant expressions in Equations 7.37 and 7.38 can be rearranged:

$$a_{H^+} a_{HCO_3^-} = K_1 K_{CO_2} P_{CO_2} \qquad (7.39)$$

Charge balance for the solution is given by $[H^+] = [HCO_3^-] + 2[CO_3^{2-}] + [OH^-]$, where the square brackets indicate concentrations and the coefficient of 2 reflects the charge per mole for CO_3^{2-}. In the acidic pH range (pH < 7), the carbonate ion (CO_3^{2-}) and hydroxyl (OH^-) are found in insignificant concentrations, and so the change balance simplifies to $[H^+] = [HCO_3^-]$. Assuming that $[H^+]^2$ is the same as H^{+2} (i.e., concentration is interchangeable with activity), then

$$a_{H^+} = \sqrt{K_1 K_{CO_2} P_{CO_2}} \qquad (7.40)$$

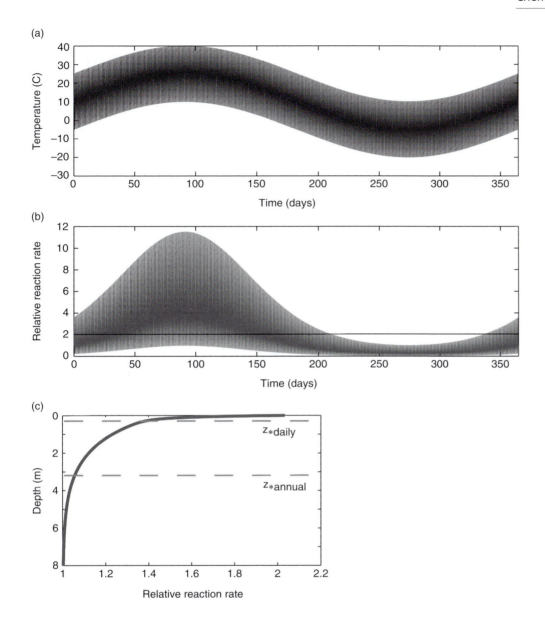

Figure 7.26 (a) Time series of temperature with daily cycles riding atop annual cycle, each with 15 °C amplitude. (b) Time series of instantaneous reaction rate, using $E_a = 60$ kJ/mol, normalized by reaction rate at the mean annual temperature. Asymmetry arises from the nonlinear Arrhenius dependence of reaction rate on temperature. Black bar represents mean of the instantaneous reaction rates, showing a roughly twofold enhancement of the rate over that expected at the mean temperature. (c) Depth profile of mean-annual reaction rate, normalized with the rate at depths where temperature variations vanish. Temperature history at the surface is that shown in (a). The rates become enhanced in the near-surface region where both daily and annual temperature swings are significant.

Given values at 25 °C of $K_1 = 10^{-6.35}$, $K_{CO_2} = 10^{-1.47}$, and the definition $pH = -\log a_{H^+}$, we can now calculate the pH of solutions for various CO_2 concentrations. Under normal atmospheric conditions, $P_{CO_2} = 10^{-3.5}$ and pH = 5.6. For a soil atmosphere of $P_{CO_2} = 10^{-2.5}$, the soil solution drops to pH = 5.1, and at $P_{CO_2} = 10^{-1.5}$, the soil solution pH = 4.7. The significance of this drop in pH on dissolution rate depends on the minerals present. For feldspars, hydrogen-promoted dissolution becomes significant at pH values below this range (about 4.5). Although the elevated CO_2 partial pressure in soil atmospheres does lower solution pH, it does not necessarily bring it into the realm where lower pH accelerates dissolution rates.

Organic exudates from plants and microbes present another mechanism by which biological systems

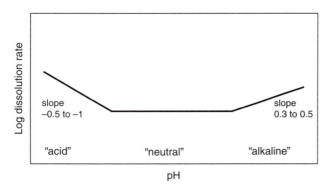

Figure 7.27 Schematic diagram of the pH dependence in dissolution rate constants. Rate constants increase at both high pH and low pH, and are insensitive to pH changes in neutral pH ranges. The locations of the transition points between each regime, and the slope of the rate constant dependence vary between minerals (based on Drever, 1997, Figure 11.13, with permission from Prentice Hall and Pearson Higher Education).

directly affect kinetics (Berner *et al.*, 2003). Dissolved organic molecules can bind to cations or metals in solution or at the surface of minerals. Any molecule or functional group on an organic compound that can form these usually reversible associations with a metal or cation is called a ligand, and the ligand–cation or ligand–metal association is called a coordination complex. As in proton-promoted dissolution, the formation of a coordination complex on the mineral surface weakens the bond of the metal or cation to the crystal lattice (Brantley, 2008). This ligand-promoted dissolution has been treated as

$$r = k_l a_{\text{ligand}}^p \tag{7.41}$$

and is thought to be additive to the effects of proton-promoted dissolution in Equation 7.36. Drever (1997) discusses controversy over the significance of ligand-promoted dissolution at the concentrations of organic ligands found in natural systems. Organic compounds may influence dissolution rates through more indirect mechanisms than the ligand-promoted dissolution model. These indirect effects include the binding of cations at mineral surfaces, which allows protons otherwise blocked by charge balance to attack the surface. Alternatively, the formation of organic ligand–Al complexes in solution reduces the free Al concentration in solution, affecting saturation state and hence dissolution rates, which alters solubility (Brantley, 2008).

Chemical affinity

At chemical equilibrium, the forward chemical reaction rate is equal to the backward chemical reaction rate, so that the net result is no change in concentration of products and reactants. Our discussion of dissolution rates has thus far been for conditions that are far from equilibrium. Thus the back reaction rate could be neglected and only the forward reaction considered. As a system approaches equilibrium, however, this assumption no longer holds, and the net reaction rate declines as equilibrium is approached. Chemical affinity is the departure from equilibrium; it strongly affects reaction rates near equilibrium. The theory to describe the effects of chemical affinity on reaction rate is developed for simple, elementary dissolution reactions, although it is often applied to complex silicate dissolution reactions. The rate law is:

$$r = k_+ \left(1 - \exp\left(\frac{\Delta G_R}{n_1 RT}\right) \right) \tag{7.42}$$

where k_+ is the forward reaction rate constant, n_1 is a stoichiometric number conventionally set to 1, R is the gas constant, T is temperature in kelvin, and ΔG_R is the chemical affinity. The exponential term in Equation 7.42 is effectively the rate of the back reaction; the net rate is the difference between the forward and the back reaction rate. The chemical affinity is defined by $\Delta G_R = RT \ln\left(Q/K_{\text{eq}}\right)$, where Q is the activity product and K_{eq} is the equilibrium constant. In effect, chemical affinity (ΔG_R) is simply another measure of the saturation, and is closely related to the saturation index, SI, defined in Equation 7.18. For $n_1 = 1$, and given the definition of ΔG_R, Equation 7.42 rearranges to $r = k_+(1 - Q/K_{\text{eq}})$. This expression can be recast in terms of SI to $r = k_+\left(1 - 10^{\text{SI}}\right)$. When the system is far from equilibrium, SI is large and negative (for undersaturated conditions), and the net reaction rate is essentially equal to the forward reaction rate. As saturation and equilibrium is approached, SI increases to 0, and the net reaction rate declines. This behavior is shown in Figure 7.28 as a ratio of the net reaction rate to the far-from-equilibrium net rate plotted against *SI*.

The important question is how far from equilibrium does this chemical affinity or saturation state effect become important. Drever (1997) and Drever and Clow (1995) suggest that chemical affinity effects should be taken into account when the net reaction

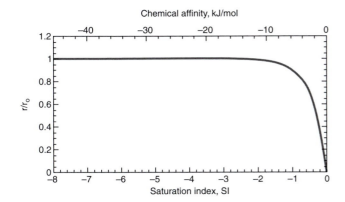

Figure 7.28 Relative reaction rate (ratio of reaction rate to the reaction rate at far-from-equilibrium conditions) as a function of the saturation index (SI) or the chemical affinity of the reaction. The chemical affinity axis here is for $T = 25°C$ (based on Drever (1997), Figure 11.14, with permission from Prentice Hall and Pearson Higher Education).

rate is 97% of the far-from-equilibrium rate (i.e., the back reaction rate is 3% of the forward reaction rate). This is an arbitrarily chosen point, although from Figure 7.28, the change in the reaction rate with SI steepens noticeably around this value. This transition corresponds to $SI = -1.52$. Most silicate minerals in natural weathering systems will be far from saturation ($SI \ll -1.52$), so the theory suggests that chemical affinity effects are unlikely to be important. An exception is potassium feldspar, which has low solubility, and so approaches saturation in more dilute solutions. White *et al.* (2001) show that K-feldspar in a weathered granite profile at Panola Mountain, Georgia remains unaltered in the lower part of the profile, while plagioclase feldspars around it are visibly decayed. They reason that percolating solutions reach saturation with respect to K-feldspar at shallower depths than for plagioclase. This effectively reduces the dissolution rate for K-feldspar, while plagioclase feldspar dissolution continues at the far-from-equilibrium rate. At present, however, while chemical affinity controls are recognized as important, there is very little work to demonstrate this in natural systems (White, 2008).

Mineral surface age

Our knowledge of kinetic rate constants is based on empirical observations. Considerable effort has gone into compiling rate constants from laboratory measurements, where conditions of temperature, mineralogy, pH and saturation state can be known and controlled. Fewer studies have focused on determining kinetic rate constants from field observations. A long-standing issue has been a discrepancy between field and lab-based measurements. The laboratory rates tend to be higher than field rates by an order of magnitude or more. We have already discussed the notion that the mineral surface area in contact with moving soil solutions may be lower than in the reactors used in laboratories and that this is responsible in part for the rate discrepancy. Another hypothesis is that mineral surfaces alter as they age in ways that affect the dissolution rate. This idea was explored by White and Brantley (2003), using a large data base of rate constants determined from laboratory and field observations, as well as data from a long-running (six years) laboratory dissolution experiment. When dissolution rates were plotted as a function of time, with time being the duration of the experiment (for laboratory rates) or the estimated age of the weathering environment (for field rates), a power-law relationship emerged. The data for plagioclase, K-feldspar, hornblende and biotite, fit a general rate law:

$$r = 3.1 \times 10^{-14} t^{-0.61} \tag{7.43}$$

for r in mol/(m^{2}s) and t in years. The negative exponent indicates a decline in the dissolution rate over time, amounting to a tenfold decrease over 50 ka, and steeper declines in dissolution rate for older weathering environments. The fitting of a single rate law to different minerals goes against the observations shown in Table 7.5 and embedded in the Goldich stability series. Given the orders of magnitude differences in dissolution rates over time, the distinctions between minerals become less clear.

How does aging affect mineral surfaces? If minerals are orderly crystal structures, it would seem that peeling off atoms from the surface over time would simply reveal the next layer, each unchanging, a sort of endless series of onion skins. (As an aside, Stumm and Morgan (1996) estimate that some 6–60 monomolecular layers are dissolved from mineral surfaces each year for typical dissolution rate constants.) However, mineral surfaces are not perfect arrays of atoms. Steps, pits, and crystal defects such as dislocations give important topography to the mineral surface. These features strongly impact mineral dissolution (and its inverse, precipitation), as the atoms at steps or dislocations are more easily removed from the crystal lattice. Mineral surfaces tend to roughen with age, due to pitting and etching of the surface, and hence the BET surface area will increase over time. In long-term experiments, solute losses from laboratory reactors must be normalized to increasing surface areas, and the result is a decline in the dissolution rate in moles per square meter of mineral per second. In natural systems, these rougher surfaces may also lead to stagnation of fluid in micropores; chemical affinity effects will reduce the dissolution rate in these pores. At the same time, aged mineral surfaces lose reactive sites and/or become occluded by precipitates or formation of leached layers. These effects will decrease the dissolution rate normalized to the mineral surface area.

The decline in dissolution rates over time documented by White and Brantley (2003) can be explained in part with these mineral surface effects. However, some of the decline at long times (their plots cover time from 10^0 to 10^6 years) probably stem from external controls, such as solutes released from more soluble minerals leading to saturation for less soluble minerals. The fact that one rate law fits the time dependence for many silicate minerals, however, suggests a general underlying effect of surface age on silicate dissolution rates.

Long-term carbon cycle

Chemical weathering of minerals plays a central role in global cycling of elements. Perhaps the most important is the global carbon cycle (Figure 7.29), which regulates climate through the greenhouse activity of CO_2 (Ruddiman, 2008). Exchange of carbon between the biosphere and the atmosphere regulates the CO_2 content of the atmosphere over short timescales. The familiar plot of CO_2 concentrations measured daily at Mauna Loa over the last 60 years shows seasonal variations due to these atmosphere–biosphere exchanges superimposed on the general rising trend due to burning of fossil fuels (Figure 1.7). Over longer timescales, however, carbon is transferred between the atmosphere and lithosphere through processes of chemical weathering, precipitation of carbonate minerals, and volcanic and metamorphic outgassing. The geologic carbon cycle modulates the

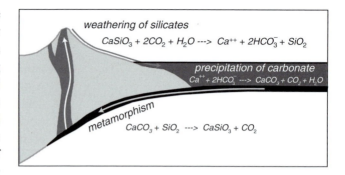

Figure 7.29 Long-term carbon cycle showing major components of the rock cycle, and sites in which reactions take place to move carbon between rock and atmospheric reservoirs.

CO_2 in the atmosphere, and hence Earth's temperature over timescales of millions of years.

Removal of CO_2 from the atmosphere occurs through a combination of weathering of calcium silicate minerals and precipitation of carbonates. We discuss calcium here, but magnesium silicates and magnesium carbonates work in the same way. A generic silicate weathering reaction looks like:

$$CaSiO_3 + 2H_2CO_3 = Ca^{2+} + 2HCO_3^- + SiO_2 + H_2O \quad (7.44)$$

Remember that the carbonic acid (H_2CO_3) in the reactants is formed by dissolution of CO_2 in water. Carbonic is a weak acid, meaning that it does not dissociate completely in water, but does provide a steady source of H^+ to drive weathering reactions. We write the reaction with the mineral formula for wollastonite, chosen not because it is common or significant, but for simplicity. It stands in for calcium (or magnesium) silicates. The importance of carbonic acid-silicate mineral weathering globally can be seen in the partitioning of river solute sources in Figure 7.23. The black bars in the diagram show the proportion of solutes derived from silicate weathering, which despite the low solubilities of silicate minerals exceeds the proportion from carbonate minerals in many rivers.

The bicarbonate (HCO_3^-) and Ca^{2+} ions produced by weathering will precipitate as calcium carbonate:

$$Ca^{2+} + 2HCO_3^- = CaCO_3 + H_2CO_3 \quad (7.45)$$

The carbonate precipitation reaction occurs most readily in the ocean, with the help of marine organisms. Carbonate precipitates also form in soils in arid environments. Water percolating through the soil column gains HCO_3^- and Ca^{2+} ions from weathering, then evapotranspiration of the water concentrates solutions and carbonate precipitation occurs. Combined, the silicate weathering and carbonate precipitation reactions transfer CO_2 from the atmosphere into minerals in soils or carbonate rocks on the seafloor:

$$CO_2 + CaSiO_3 = CaCO_3 + SiO_2 \qquad (7.46)$$

The silica in Equation 7.46 represents silica released by weathering, which may precipitate as clay in the weathering environment, or precipitate into opal or diatoms (organisms with silica cell walls called frustules) elsewhere.

The atmosphere contains about 760 Gt (gigatonnes) of C (a gigatonne is 10^9 tonnes, or 10^{12} kg). The rate of CO_2 consumption by silicate weathering globally is 0.03 Gt C/yr. How long would it take weathering to remove all the CO_2 from the atmosphere? We can calculate this by dividing the size of the reservoir (760 Gt) by the rate at which it is leaking (0.03 Gt/yr). The result is: $760/0.03 \approx 20\,000$ years (in round numbers). Although 20 000 years is a long time for humankind, it is an eyeblink in geologic time. Samples of the atmosphere preserved in bubbles in ice cores from Antarctica reveal CO_2 levels in the atmosphere over the last 650 000 years cycling between values of $\sim$180 and 280 ppmv. The concentration of CO_2 does not plummet because CO_2 is returned to the atmosphere by volcanic outgassing and metamorphism. These return processes can be represented with a general metamorphic reaction:

$$CaCO_3 + SiO_2 = CO_2 + CaSiO_3 \qquad (7.47)$$

Here the limestone and clay (or opal or diatom or silicate mineral) bearing rocks metamorphose into calcium silicates, releasing CO_2 that escapes through volcanic gases, or through seepage in hot springs. You can see that the reactions in Equations 7.46 and 7.47 together form a closed loop in which carbon moves from atmosphere (CO_2) to lithosphere (carbonate rocks and minerals) and back again.

The metamorphic and volcanic processes that return CO_2 to the atmosphere are driven by heat loss from the interior of the planet, while weathering processes respond to variations in climate (Walker et al., 1981). The interplay of tectonics (driving CO_2 emissions to the atmosphere) and climate-driven weathering (driving CO_2 removal from the atmosphere) over geologic time has been explored in models, such as the BLAG model (Berner et al., 1983), and GEOCARB (Berner, 2004). In these models, chemical weathering processes increase in global effectiveness as CO_2 increases (and climate warms) because of the Arrhenius effect on dissolution rates and increased vigor of the hydrologic cycle in warmer climates. Increases in CO_2 are driven by increasing rates of sea floor spreading, which controls both outgassing at mid-ocean ridges and volcanic outgassing and metamorphism at subduction zones. Thus the sea-floor spreading rates drive CO_2 inputs, and weathering rates respond to CO_2 in a way that stabilizes climate. An alternative view on weathering–climate feedbacks proposed by Raymo and her colleagues in a series of papers (Raymo et al., 1988; Raymo and Ruddiman, 1992) suggests that tectonics influences the carbon cycle (and hence climate) through links between chemical weathering and physical erosion processes.

The tectonically driven climate change hypothesis (Ruddiman, 1997) states that uplift brings about an increase in erosion rates, and that chemical weathering rates will increase in response. The increase in chemical weathering rates lowers atmospheric CO_2 levels, which then cools the climate. This hypothesis was developed to explain the cooling during the Cenozoic that led to the onset of northern hemisphere glaciation. Tectonics is still a key driver, yet this model focuses on its influence on the weathering part of the carbon cycle, rather than on the metamorphism/volcanism part of the carbon cycle. The tectonically driven climate change hypothesis provoked extensive discussion of evidence for uplift, connections between uplift and erosion, and how physical and chemical weathering interact (Ruddiman et al., 1997).

Several points about this debate are relevant to geomorphology. The uplift–weathering connection can be broken into two key questions: does total denudation increase in regions of high uplift rate, and do chemical weathering fluxes increase with total denudation rate? The uplift–erosion question is taken up in Chapter 3. But do chemical weathering losses increase if erosion rates are higher? This question is about the weathering flux from a reservoir (the continents, or a watershed, or a vertical weathering profile), not just the rate of solute release from a mineral surface we discussed in the section on chemical kinetics.

A number of workers have argued that erosion rate and chemical weathering fluxes do increase together.

If we think of the Critical Zone at the Earth's surface as a reactor (Figure 7.1), into which rock is fed from the bottom, and through which mobile regolith moves laterally producing erosion, then we can estimate the residence time for material in this reactor. In the one-dimensional case, the rate of uplift (or of denudation) sets the residence time, t_r, for a given, steady-state thickness for the reactor: $t_r = Z/\dot{e}$, where Z is the thickness of the reactor (the hydrologically active weathered profile), and $\dot{e}$ is the total denudation rate (for steady state, assumed equal to the uplift rate). Studies based on analysis of the extent of weathering in soil and rock have found that greater weathering fluxes are associated with higher total denudation rates ($\dot{e}$) and short residence time (t_r) of rock particles in the Critical Zone (Stallard, 1995; S. P. Anderson et al., 2002; Riebe et al., 2001a, 2004; Waldbauer and Chamberlain, 2005). The correlation between erosion rate and weathering fluxes based on river data is less clear. In a large database of solute fluxes from granitic watersheds, White and Blum (1995) found no correlation between the fluxes and proxies for physical erosion rates (extent of recent glaciation, relief). In another analysis of solute fluxes and sediment fluxes from small watersheds, West et al. (2005) found that erosion rates controlled chemical weathering fluxes at low erosion rates, but not at high erosion rates. They emphasize the tradeoffs in control mechanisms of mineral supply (surface area) and climate factors (temperature and precipitation). In glacial catchments, where erosion rates are extremely high, but temperatures are low, S. P. Anderson (2007) found no enhancement of chemical weathering fluxes.

Work on controls on the kinetics of weathering reactions suggests several mechanisms for higher erosion rates to be correlated with higher chemical weathering fluxes. First, aging of mineral surfaces appears to reduce intrinsic mineral dissolution rate constants (White and Brantley, 2003). Thus, systems in which particles have a short residence time in the reactor of the Critical Zone (due to high erosion rates) will maintain higher dissolution rate constants on average. Also more weatherable minerals are more likely to be found throughout the Critical Zone where particle residence times are short, which means the mean dissolution rate of the ensemble of minerals in the reactor will be higher. To the extent that higher erosion rates are associated with a thinner Critical Zone, soil water is likely to be less chemically evolved,

and therefore a reduction in dissolution rates due to chemical affinity effects is less likely.

S. P. Anderson (2007) outlined a strategy to understand the impacts of glacial erosion on the carbon cycle and chemical weathering that could be extended to any erosional system. She noted that chemical weathering fluxes are greatest where high surface area material is kept bathed in chemically aggressive solutions, but this is not necessarily in the same location where erosion occurs. In other words, sediment (and its reactive surface area) is the link between erosion and weathering. Searching for enhancement of weathering at sites of active erosion may not reveal the full connections between these processes.

We have focused on inorganic carbon transfers in the geologic carbon cycle. Organic carbon also cycles through the rock reservoir and impacts atmospheric composition. The fixing of CO_2 by photosynthesis,

$$CO_2 + H_2O \rightleftharpoons CH_2O + O_2 \qquad (7.48)$$

produces organic matter and oxygen gas. Some organic matter is buried and incorporated into sedimentary rocks, primarily in a recalcitrant form known as kerogen, but also as coal, oil and gas. Erosion exposes buried organic matter. Oxidation, which reverses the photosynthesis reaction, then returns CO_2 to the atmosphere. Changes in global erosion rates or in rates of burial of organic matter will impact the carbon cycle over geologic time.

Effects of chemical alteration of rock

Our discussion of chemical processes has focused on aqueous geochemistry and the interaction of rock and water. Over time, these processes transform rock to saprolite and soil. When looking at weathered rock, we see the accumulated effects of weathering processes over time.

Assessing mass losses (or gains) in regolith

How does one measure the mass transfers due to weathering? This commonly amounts to asking what has been removed from rock, although in some instances, mass additions are also important. If one assumes that chemical weathering removes mass from rock, but does not change its volume, then it is possible to use bulk density to quantify total mass losses.

Mass balance in a soil
(after George Brimhall, Jr.)

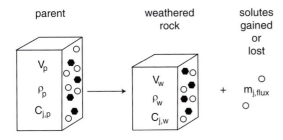

parent   weathered   solutes
          rock   gained
                       or
                       lost

Figure 7.30 Schematic of mass balance for an element j (open circles) in a weathering system relative to an immobile element i (filled hexagons). While the mass of the immobile element in a volume of parent material does not change as the volume weathers (or inflates by eolian deposition), the mass of the mobile element will decrease (or increase) in concentration by the amount $m_{j,\text{flux}}$ (after Brimhall and Dietrich, 1987, Figure 1, with permission from Elsevier).

The concept of isovolumetric weathering does apply to weathered rock (recall that saprolite is defined as isovolumetrically weathered rock) (see Figure 7.1), but it becomes less viable in mobile regolith. The mixing and transport processes that disrupt mobile regolith can affect its bulk density.

Where volume changes and mass transfers (gains or losses) accompany weathering, another approach must be used. Brimhall and Dietrich (1987) developed a method that makes use of elements or minerals that are relatively inert in weathering systems. The method considers changes to a volume of unweathered parent material, accounting for mass transfers in and out of that volume by normalizing to the mass of an immobile element (i.e., one that is inert, or unaffected by weathering). We start with a statement of mass balance for any chemical element j as diagrammed in Figure 7.30:

$$\tfrac{1}{100}\left(V_p\rho_p C_{j,p}\right) = \tfrac{1}{100}\left(V_w\rho_w C_{j,w}\right) - m_{j,\text{flux}} \qquad (7.49)$$

where V is volume, ρ is bulk density, C is concentration in weight percent, and $m_{j,\text{flux}}$ is mass transfer of j in or out of the volume. The subscripts indicate whether the quantity is for the parent material (p) or weathered material (w). The left-hand side of Equation 7.49 gives the mass of element j in a volume of parent material, while the first term on the right-hand side of the equation is the mass of element j in the weathered volume (i.e., the volume that the original

parent material occupies after weathering). The $m_{j,\text{flux}}$ term is what we want to know: the mass fluxes into or out of the volume during weathering. Mass additions will be reflected in positive values of $m_{j,\text{flux}}$, while mass losses give negative values of $m_{j,\text{flux}}$.

For an element that is immobile in weathering, and does not have an external source (such as dust), the $m_{j,\text{flux}}$ term is zero. In this case, Equation 7.49 simplifies to (using now subscript i to specify that the element is immobile):

$$\tfrac{1}{100}\left(V_p\rho_p C_{i,p}\right) = \tfrac{1}{100}\left(V_w\rho_w C_{i,w}\right) \qquad (7.50)$$

The mass of immobile element i in a volume of weathered material V_w is equal to the mass of i in the volume of parent material V_p that weathered to produce V_w. One question that can be addressed with this simplified mass balance is how much the volume of the material has changed due to weathering processes. Recall that volumetric strain is defined as the change in volume relative to the original volume. Rearrangement of Equation 7.50 yields the volumetric strain from weathering:

$$\varepsilon_{i,w} = \frac{V_w}{V_p} - 1 = \frac{\rho_p C_{i,p}}{\rho_w C_{i,w}} - 1 \qquad (7.51)$$

This expression shows that volumetric strain associated with weathering can be quantified from measurable quantities – the bulk density and immobile element concentration of weathered and parent material. This is how the strain was determined for the weathered profile shown in Figure 7.2.

Mass gains or losses of an element j can be computed as a percent change in the mass of element j relative to its mass in the parent material. This is expressed in a dimensionless mass transfer coefficient, τ_j:

$$\tau_j \equiv \frac{m_{j,\text{flux}}}{V_p\rho_p C_{j,p}} \times 100 \qquad (7.52)$$

Rearranging Equation 7.49, and using the strain definition in Equation 7.51, yields an expression for τ_j:

$$\tau_j = \frac{C_{j,w} C_{i,p}}{C_{j,p} C_{i,w}} - 1 \qquad (7.53)$$

Values of τ_j can be negative or positive: mass additions yield positive values, while mass losses are reflected in negative values of the mass transfer coefficient. For instance, $\tau_j = -0.5$ indicates that 50% of the mass of j in the parent material has been lost during weathering. A value of $\tau_j = 0.5$ indicates, on

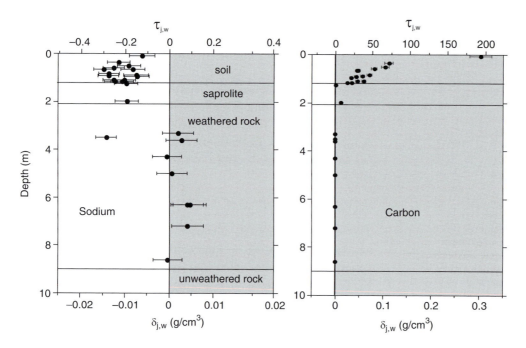

Figure 7.31 Profiles showing mass transfers for elements Na and C in the top 10 m of the weathered profile at the Coos Bay site, Oregon Coast Range. Na shows mass losses, primarily in the soil, while C shows strong gains in the soil. The carbon additions reflect the importance of organic matter additions in these forest soils. Note the difference in mass transfer coefficient scales for the two plots. Parent material here is greywacke sandstone, which shows considerable heterogeneity and explains some of the scatter in calculated τ_j and δ_j values (after S. P. Anderson et al., 2002, Figure 6).

the other hand, an addition of j of 50% relative to the parent material. If there are no gains or losses of element j, then $\tau_j = 0$.

Alternatively, the mass transfers can be computed as mass gains or losses per unit volume of the parent material. Unlike the dimensionless τ_j, the volumetric mass transfer coefficient, δ_j, has units of mass per unit volume:

$$\delta_j \equiv \frac{m_{j,\text{flux}}}{V_p} = \frac{\rho_p}{100}\left(C_{j,w}\frac{C_{i,p}}{C_{i,w}} - C_{j,p}\right) \tag{7.54}$$

The two mass transfer coefficients are simply related to each other:

$$\delta_j = \left(\frac{C_{j,p}\rho_p}{100}\right)\tau_j \tag{7.55}$$

Examples of τ_j and δ_j for two elements in the weathered profile at the Coos Bay site in the Oregon Coast Range are shown in Figure 7.31. Negative values of τ_{Na} in the soil show that sodium is lost in the soil by up to 30% relative to the parent material (characterized from three samples at >30 m depth in the profile). In contrast, τ_C in the soil reaches more

than 150, indicating 15 000% enrichment in carbon relative to the parent material. The carbon data shows the additions of organic matter in the soil.

Both the mass transfer coefficient, τ_j, and the volumetric mass transfer coefficient, δ_j, have utility. Values of τ_j can be determined from the chemical concentration of elements in weathered and parent rock alone. Although the volumetric mass transfer coefficient, δ_j, requires knowledge of parent rock bulk density in addition to the elemental concentrations, it can be used to compute absolute mass gains or losses in a soil profile. The total mass gains or losses of an element j in a weathered profile, $M_{j,\text{flux}}$, can be determined by summing (integrating) δ_j over the profile:

$$M_{j,\text{flux}} = \int_0^H \delta_j dz \tag{7.56}$$

where z is depth, and H is the total thickness of the profile. The mass transfers can also be summed for all elements at a single depth:

$$m_{\text{total,flux}} = \sum_{j=1}^n \delta_j \tag{7.57}$$

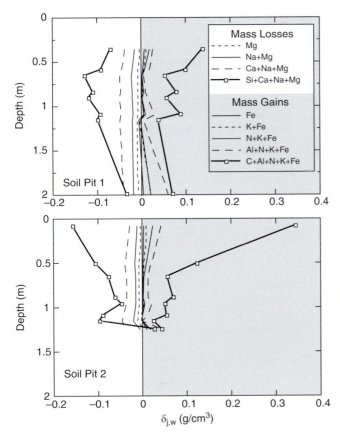

Figure 7.32 Summary of absolute mass gains and mass losses (δ_j) in two soil pits in the CB2 catchment. Si and cations Ca, Na, and Mg are depleted throughout the soil depth; C, N, and Al show mass additions in the soil (shaded side of plot). Each line represents the mass gains or mass losses for the sum of the elements shown in the key; the δ_j for any individual element is the distance between two lines. The mass losses can be summarized as Si > Ca > Na > Mg; mass gains are dominated by C. The deepest sample in each profile is saprolite collected from the bottom of the soil pits (after S. P. Anderson *et al.*, 2002, Figure 7).

where n is the number of elements considered. A calculation of this sort is shown in Figure 7.32 for two soil profiles at the Coos Bay site.

The mass transfer coefficients from Equations (7.53) and (7.54) provide powerful tools for quantifying changes in weathering systems. There are several issues that must be addressed when applying this approach, however. First, the computations are based on a homogeneous and identifiable parent material. Although this seems an obvious statement, there are many settings where it can be tricky to identify the parent material or where the parent material is not homogeneous. In sedimentary rock such as the sandstone shown in

Figure 7.31, small variations in depositional environment can influence the composition of the parent rock. Layered or foliated rocks are other examples where heterogeneity in parent material can limit the utility of the elemental mass balance approach.

A second issue is identifying an immobile element or mineral in the system. Zirconium is a common choice, since this element is found almost entirely in the mineral zircon, which is highly resistant to weathering. Titanium is also used. In this case, titanium is immobilized not because it resides in a recalcitrant mineral, but because it is retained in secondary phases. Niobium has also been used as an immobile element. In choosing any of these trace elements as the immobile element for mass balance calculations, an assumption is that the element is uniformly distributed in the parent material. This can be problematic in sedimentary rocks, where heavy minerals can be segregated, or in strongly foliated rocks. Another approach is to use quartz, a major component of many silicate rocks, and an effectively unweatherable mineral (see Table 7.5). The advantage of choosing an abundant mineral as the immobile component is that small variations in its abundance in the parent material will have little impact on calculations.

Chemical alteration of rock strength

Chemical alteration of rocks affects rock strength through formation of porosity, changing density, growth of clays (usually weaker than primary minerals), growth of cements (often stronger than original material), and growth of cracks. Some of these effects are considered here. The strength of intact rock can be assessed with the uniaxial compressive strength measured in a laboratory apparatus, or estimated from field Schmidt hammer tests. The intact rock strength (one of the parameters in the rock mass strength classification, see Table 7.2) varies with rock type, but also depends on the degree of weathering or the rock. In Table 7.6, values of uniaxial compressive strength are classified by field descriptions of mechanical behavior and typical rock types. A specific example from weathered profiles of Oporto granite in Portugal (Begonha and Sequeira Braga, 2002) is shown for comparison. The fresh granite samples have uniaxial compressive strengths of 130–157 MPa (very strong), and have very low free porosity as expected in unweathered granite. Weathered granite samples with

Table 7.6   **Field estimates of uniaxial compressive strength (based on Hoek and Brown, 1997), and an example of Oporto granite (Begonha and Sequeira Braga, 2002)**

Strength descriptor	Uniaxial compressive strength (MPa)	Field description	Examples	Oporto granite example		
				Degree of weathering	Uniaxial compressive strength (MPa)	Free porosity (%)
Extremely strong	>250	Can only be chipped with rock hammer	Fresh basalt, chert, diabase, gneiss, granite, quartzite			
Very strong	100–250	Many blows of rock hammer to fracture	Amphibolite, sandstone, basalt, gabbro, gneiss, granodiorite, limestone, marble, rhyolite, tuff	Fresh	130–157	0.5–1.00
Strong	50–100	More than one blow of rock hammer to fracture	Limestone, marble, phyllite, sandstone, schist, shale	Slightly weathered Ave. 2.4% secondary minerals	97–132	1.5–1.9
Medium strong	25–50	Fractures with one blow of rock hammer	Claystone, coal, schist, shale, siltstone			
Weak	5–25	Indentation made by firm blow with rock hammer	Chalk, rocksalt, potash	Moderate–highly weathered Ave. 8.7% secondary minerals	20–29	7.8–10.8
Very weak	1–5	Crumbles under firm blow with rock hammer	Highly weathered or altered rock			
Extremely weak	0.25–1	Indented by thumbnail	Stiff fault gouge			

only a few percent secondary minerals show reductions in uniaxial compressive strength. As the secondary mineral fraction approaches 10%, rock free porosity increases by an order of magnitude, and uniaxial compressive strength declines by a factor of five relative to the fresh samples.

Chemical alterations can lead to microcrack growth, meaning that chemical alteration yields mechanical alteration. The chemical–mechanical process linkage can take the form of chemical alteration reducing the strength of rock, or of chemical alteration exerting stress on rock. In both cases, chemical alteration drives crack formation. In a study of weathered basalts, Moon and Jayawardane (2004) noted significant declines in rock strength associated with the loss of alkaline earth cations (e.g., Ca and Mg) from the rock. In the profiles they analyzed, strength loss occurred in rock without measurable secondary mineral formation. They reasoned that cations (Ca^{2+}, Mg^{2+}, and Fe^{2+}) lost during incipient weathering were replaced with smaller H^+ or Al^{3+}. The smaller ions distort the crystal lattice and weaken it, rendering the rock susceptible to growth of microcracks in response to the surrounding stress field. They noted progressive development of horizontal cracks in the weathering profile, the orientation expected for cracks formed in response to the maximum tensile stress, which is aligned in the vertical direction.

A different chemical alteration–mechanical weakening connection is associated with incongruent reactions in which the secondary phase produced has a larger molar volume than the primary mineral. Weathering of biotite to hydrobiotite or vermiculite is the most common weathering reaction of this type (Wahrhaftig, 1965; Eggler *et al.*, 1969; Isherwood and Street, 1976). Biotite weathering begins with diffusion of oxygen into rock to oxidize Fe^{2+} in the biotite lattice to Fe^{3+}, and ejection of interlayer K^+ to maintain charge balance (Buss *et al.*, 2008). This oxidation-leaching reaction increases the basal layer spacing (the spacing of layers of tetrahedral-octohedral-tetrahedral sheets) from 10 to 10.5 Å, and the expansion exerts stress on the surrounding rock. Fletcher *et al.* (2006) modeled the elastic strain accumulation due to the oxidation and expansion of biotite in granitic rock, and found they could explain the spacing of macroscopic fractures that typify spheroidally weathered rock.

In these examples, modest, nearly undetectable chemical transformations reduce the strength (replacement of large cations with smaller ones), or exert stress (volumetric expansion) on rock. Microcracks form, which allow greater access of water and oxygen to the interior of the rock, and other weathering reactions ensue, which lead to the highly or completely weathered states in which compressive rock strength is a fraction of that of the fresh rock.

The conversion of bedrock to mobile regolith

In geomorphology the most important boundary in the weathered profile presented in Figure 7.1 is that between mobile regolith and the underlying weathered rock. In order to transport material down a slope, material must be available that is small enough to be transported by the available surface processes (Anderson and Humphrey, 1989). In many landscape evolution models, weathering amounts to the production of mobile regolith from underlying rock, a source term for material to the moving blanket of mobile regolith (e.g., Ahnert, 1970). A leading edge of geomorphology is to develop models (both conceptual and physical) of how this mobilization front progresses down into the Critical Zone reactor. The processes involved in the generation of regolith from bedrock include chemical, biological and mechanical phenomena, with many linkages between them. For instance, cracks produced through the growth of small

ice lenses within a rock can damage the minerals such that subsequent wetting of the rock leads to chemical corrosion at the crack tips. The key to the production of mobile-regolith production is not chemical transformation or mechanical breakdown, but the transport of material. As such, the boundary identifies the depth to which transport processes operate given the bedrock, climate, and vegetation.

Mobile-regolith production functions

In work dating back to that of G. K. Gilbert in his work on the Henry Mountains, geomorphologists have proposed that mobile regolith production be dependent upon the thickness of mobile debris in some way. For a brief history of thought on depth-dependent mobile regolith production see Humphreys and Wilkinson (2007). Gilbert hypothesized in particular that at least in semi-arid environments regolith is produced most efficiently when there is some finite thickness of regolith already mantling the bedrock interface that is being attacked. He reasoned that while water runs rapidly off a bare bedrock surface, any existing regolith would hold water against the bedrock, allowing more efficient chemical attack of the bedrock. For very thick mobile regolith, however, there may never be sufficient water in the regolith to perform this task (for instance, it may simply get wicked back to the surface to be evaporated), and the rate of conversion would decline. The concepts elaborated by Gilbert were formalized by Carson and Kirkby (1972) in a graph showing soil (mobile regolith) production as a function of soil depth (Figure 7.33). The function has a maximum production rate at some depth, and is therefore known as the "humped" function.

The humped model, while conceptually appealing, is difficult to test. In a test of the performance of different functions in the landscape of the Oregon Coast Range. Dietrich *et al.* (1995) proposed an alternative function, in which mobile regolith production declines exponentially with increasing soil depth. The exponential model better produced the distribution of soil depth and outcrop area than the humped model. Direct measurement of regolith production rates using cosmogenic radionuclides is fueling the debate. Heimsath *et al.* (1997, 1999, 2000) made use of steady-state landscapes (i.e., mobile regolith production rates equal total denudation rates, so that soil thickness is steady). They sampled bedrock from the base of the mobile regolith, and evaluated the rate at which

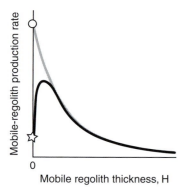

Figure 7.33 Mobile-regolith production functions presently debated in the literature. In both models, the production rate of mobile regolith falls off rapidly at large regolith thickness. They differ largely in whether the bare bedrock ($H = 0$) rate is low (star: the "humped" regolith production function) or is high (circle: the exponential model).

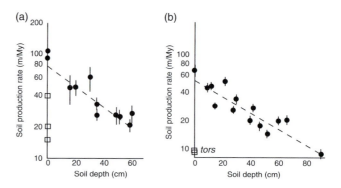

Figure 7.34 Production of mobile regolith deduced from concentrations of ^{10}Be from samples at the mobile regolith–bedrock interface (a) from sites in California (after Heimsath et al., 1997, Figure 3, with permission of the Nature Publishing Group), and (b) from Australia (after Heimsath et al., 2000, Figure 3). Boxes denote bare bedrock exposures, in some cases of rock differing from the dominant bedrock of the catchment. Note the semi-log axes in both plots; a straight line implies an exponential dependence of mobile regolith production rate on mobile regolith thickness.

it was eroding. The steady-state assumption allows them to use the same equation that is used for steady erosion, but simply reduce the production rate at the bedrock surface by that associated with the thickness of the regolith, i.e., $P_{\text{bedrock}} = P_{\text{o}}e^{-H/z}{}_*$, where z_* is the characteristic length scale that is appropriate for regolith densities, and H is the regolith thickness. Their plot of regolith production rate as a function of regolith thickness is shown as Figure 7.34. Ignoring the samples on bare bedrock, this shows an exponential decline with no significant "hump" in it. The rate declines by a factor of $1/e$ (roughly $1/3$) by a regolith thickness of about 50 cm, although this length scale varies from climate to climate. Small et al. (1997) performed a similar study in an alpine landscape high in the Wind River Mountains, Wyoming, but in a landscape in which the regolith thickness was essentially uniform (a Gilbert convex hilltop as described in Chapter 10). Here they found that the rates of lowering for bare bedrock, found in

tors dotting the surface, are about a factor of two lower than those for the regolith-buried bedrock surface (7 vs. 15 microns per year). At least in this landscape, then, the long-term average regolith production rate is indeed lower on bare bedrock than beneath a finite thickness of regolith. This would imply a humped regolith production function in which the regolith production on bare bedrock is finite but low, small regolith thicknesses allow rapid rock breakdown at the bedrock interface, and regolith thicknesses greater than this result in reduction of regolith production. The details, then, remain to be worked out. The largest unknown at present is just how the weathering rates vary in very thin to nonexistent soils, and under what conditions if any the maximum weathering rate is found on bare bedrock.

Summary

In the near-surface region of Earth, rock of the lithosphere interacts with atmospheric gases, water in the hydrosphere, and life in the biosphere. Rock is out of chemical equilibrium with the surface environments, and is subject to a variety of stresses as it approaches the surface. Together these various assaults work to disintegrate bedrock, generating the weathered profile that forms the physical structure of the Critical Zone. The depth of the weathered profile and its internal structure (weathered rock, saprolite, mobile regolith) varies across landscapes.

The total denudation rate is the mass loss per unit area per unit time, and comprises both solute and solid phase losses from the system being considered. Data from rivers around the world show wide ranges in both the total denudation rates and the ratios of dissolved to suspended load. The dissolved to suspended sediment ratio reflects differences in the efficiency of chemical dissolution and removal of material by sediment transport.

Fracturing is often the deepest manifestation of weathering in the Critical Zone. Rocks fracture when stresses exerted exceed the strength of the rock. In the near surface, temperature fluctuations generate stress within rocks because of the differences in thermal expansion coefficients of minerals. This process is most effective within a few centimeters of the rock surface, where the magnitude and frequency of temperature fluctuations are greatest. Frost cracking is another effective mechanism to break rock. Here stresses are generated by growth of ice lenses in existing microcracks in rock. These ice lenses grow fastest (and frost cracking is most likely) at temperatures slightly below freezing. Water remains unfrozen in thin films on mineral and ice surfaces at sub-freezing temperatures. Flow through these films feeds growing ice lenses in the rock as long as the temperatures are appropriate and water is available. Frost cracking can operate at meter scale depths in rock, especially porous rocks. Other processes ranging from root growth to pounding of waves on seacliffs can fracture rock as well; in every case some agent must stress some part of the rock. If the strain rate is too high or the strain too great, the rock will fracture.

Joints are regularly spaced and oriented fractures in rock. Joints may form due to tectonic (long-range) stresses, such as those associated with motion of rock through bends in faults. Rocks that reach the Critical Zone unfractured by tectonic processes may reach the surface with sufficient strength to break dominantly in surface-parallel sheeting joints.

Minerals in rocks dissolve in waters percolating through the Critical Zone because they are out of equilibrium with surface environments. The equilibrium state for a system can be assessed from the standard free energy for the reaction, $\Delta G°$, or from the activities of reactants and products (and the activities of solutes are generally equal to the solute concentration in mol/L). While the equilibrium thermodynamics tell us what the stable state of a system is, they do not provide information on the rate of reactions. The rate of dissolution varies by orders of magnitudes from one mineral type to another, with quartz being among the slowest, and evaporates being among the fastest to dissolve. Dissolution rates are affected by temperature, pH, organic ligands, solution saturation state (chemical affinity), and mineral surface age. Rates of mineral dissolution measured in laboratory reactors tend to be much faster than those measured in the field. The discrepancy can be attributed in part to mineral surface aging, but may also result from details of water movement through the soil. In particular, incomplete filling of pores with water is likely to greatly reduce the mineral surface area in contact with rapidly moving water. Leached layers and coatings that occlude mineral surfaces may also reduce the field dissolution rates.

Chemical weathering of silicate minerals is driven by protons (H^+), primarily derived from CO_2 dissolved in water. The atmospheric CO_2 consumed by weathering of calcium silicate minerals is precipitated in carbonate minerals, effecting a transfer of carbon from the atmosphere to the lithosphere. Hence, weathering of calcium (and magnesium) silicates coupled with carbonate mineral precipitation is part of the carbon cycle. Imbalances in the global CO_2 consumption by weathering and precipitation and the emission of CO_2 from volcanoes and metamorphic processes alter the CO_2 content of the atmosphere over geologic time. In considering the global carbon cycle, the total CO_2 consumption due to weathering will be affected by the kinetic parameters that influence the rate of mineral dissolution at the mineral–water interface, but more importantly, perhaps, by the exposure of fresh mineral surfaces to weathering processes. The processes that fracture rock and transport regolith thus play a role in controlling the carbon cycle.

The extent of chemical alteration can be quantified using an elemental mass balance in systems for which a homogeneous and well-characterized parent material can be identified. The mass balance technique relies on there being some element or mineral that is unaffected by chemical alteration to which other mass losses or gains can be normalized. Chemically altered rock is transformed in physical properties. Changes in uniaxial compressive strength can appear with very minor changes in chemical composition. The strength of a rock mass reflects both the strength of the intact rock mass, as most easily quantified with uniaxial compressive strength, and the number, extent, and orientation of fractures in the rock. A number of schemes to quantify these weathered rock characteristics exist. We discussed the rock mass strength classification of Selby.

For many geomorphologists, weathering can be summarized as the process (or perhaps, in generous moments, the suite of processes) that detaches fragments from intact bedrock, releasing material to the mobile blanket of debris found over much of the Earth's surface. The production of mobile debris is important, since it introduces material into the mobile and chaotic layer that can creep or slide or otherwise move downslope. However, as the change from fixed to free is neither a purely chemical nor a purely physical transformation, the process (or suite of processes) is difficult to study. Current models suggest that the production rate of mobile regolith is controlled by the thickness of the mobile layer, either as an exponentially declining rate with increasing depth, or as a rate that rises to a maximum at an intermediate depth, before falling exponentially with depth. There is a desperate need for models that explicitly incorporate the physical, chemical, and biological processes that we have cataloged in this chapter that can then be employed to understand these depth dependences.

Think of the Earth's surface as a Critical Zone in which water, air, rock, and living systems react and interact with each other. Weathering processes – the mechanisms that break and breakdown rock and minerals – set the scene for the behavior of the Critical Zone. Water and gases (including dissolved gases) move in pore spaces and along fractures generated by weathering. Soil creep and landsliding are possible in mobile debris. Of course, these interactions all entail significant feedbacks – weathering takes place where water can reach, and erosion is easier where rocks are weakened. But the scene is set, once weathering – or, in Gilbert's words, the disintegration of rock – has prepared the Critical Zone for erosion.

Problems

1. *Erosion rate of a catchment.* You are asked to assess the erosion rate of a catchment. The granitic rocks are essentially uniform within the catchment, which has a map area of $32 \, \text{km}^2$. The sediment transport rate, averaged over 30 years of gaging records, is $1.9 \times 10^6 \, \text{kg/yr}$.

   Calculate the average erosion rate of the catchment over this timescale, reporting it in both microns/year and mm/year. Please be sure to describe all the assumptions you are making in this analysis.

2. Convert the axis of Figure 7.4 on river loads around the world to one of lowering rates. What is the mean lowering rate of the Nile River basin? The Mississippi?

3. Calculate the expected vertical profile of reaction progress in the case in which the variation is solely due to temperature (the Arrhenius equation). Assume the temperatures to be driven by sinusoidal variation at the surface with decay with depth being determined by the thermal diffusivity, $\kappa = 1 \, \text{mm}^2/\text{s}$. Use the constants provided in the text for the universal gas constant and the activation energy of

   plagioclase-clay. How much higher will this chemical reaction rate be at 25 °C than at 15 °C?

4. Perform a calculation of the timescale analogous to the one shown in Box 7.2 about the saltiness of the sea, this time for Ca ions. Their mean concentration in river water is 1.34×10^{-5} kg/kg in seawater, and 4.12×10^{-4} kg/kg in seawater. Can you think of a major removal mechanism for this ion in the ocean?

5. Perform the same calculation to estimate your age. Taking your weight as a starting point, estimate the rate at which you add weight on a daily basis by summing the weight of your meals. Divide the two to determine how many days old you are. You are likely to severely underestimate your age. What is wrong with this calculation?

6. Calculate and plot the expected "frost cracking index" for the following conditions. $T_{\text{bar}} = +2\,°\text{C}$, $\Delta T = 14\,°\text{C}$, $\kappa = 1 \, \text{mm}^2/\text{s}$. Assume that frost cracking occurs in the window $-8\,°\text{C} < T < -3\,°\text{C}$, and ignore the role of latent heat (equivalently, assume that the surface materials have essentially no porosity). Discuss how this

would change if we were to add in the effect of latent heat.

7. Convert a thermal diffusivity of $\kappa = 1\,mm^2/s$ into m^2/yr.

8. If fire spall dominates the weathering and erosion of boulder surfaces on Sierran glacial moraines, and fire spalls are typically 1 cm thick, calculate the effect of fire spall on the CRN-based age of a boulder. Assume that the true age of the surface is an LGM one, roughly 18 ka, and that the recurrence interval of fire spall is (i) 1000 years, and (ii) 200 years. First, think about the problem and discuss what direction this effect would move the age (is it younger or older than the true age?).

9. For the same fire spall of 1 cm thickness, estimate the duration of the heating of the boulder surface. This is an instantaneous heating problem that is perfectly analogous to the instantaneous cooling problem we employed to assess the thickness of the lithosphere. Assume the thermal diffusivity of the rock is $1\,mm^2/s$.

10. Halite dissolves congruently according to: $NaCl \rightleftharpoons Na^+ + Cl^-$. Given the equilibrium constant at $25\,°C$, $K_{eq} = 10^{1.58}$, what is the solubility of halite in water at $25\,°C$? (*Hint*: you will need to calculate separately the concentration of Na^+ and Cl^- in equilibrium with halite, and sum the result to get the total solubility.) Given a global mean average runoff of $0.3\,m/yr$, and assuming that this runoff is saturated with respect to halite, what is the average denudation rate of halite? How does this compare with a global average total denudation rate of $0.13\,kg/(m^2\,yr)$? Under what conditions would you expect to see halite outcrops on the surface?

11. The activation energy (E_a) for augite (a pyroxene) is $79\,kJ/mol$, and the activation energy for anorthite (a plagioclase) is $35\,kJ/mol$. For each of these minerals, calculate the ratio of the dissolution rate constant at $T = 25\,°C$ ($k_{T=25}$) to the rate constant at $T = 15\,°C$ ($k_{T=15}$). The gas constant R has a value of $8.314\,472\,JK^{-1}mol^{-1}$. (*Note*: you will need to convert the temperatures to degrees kelvin.)

Further reading

Birkeland, P., 1999, *Soils and Geomorphology*, 3rd edition, New York: Oxford University Press, 448 pp.
A good reference on soils, which we have not covered in this textbook. The first few chapters review the mineralogy of rocks and soils, and the chemical and physical processes that are most important to soil formation. Later chapters discuss the factors of soil formation, and are studded with field examples.

Drever, J. I., 1997, *The Geochemistry of Natural Waters: Surface and Groundwater Environments*, 3rd edition, Upper Saddle River, NJ: Prentice Hall, 436 pp.
This textbook is an accessible introduction to the geochemistry relevant to the study of chemical weathering processes. The text includes applications of the concepts to real world studies from the literature, which provides a nice dose of reality along with the more clear-cut theory.

Glaciers and glacial geology

Fire and Ice
Some say the world will end in fire
Some say in ice.
From what I've tasted of desire
I hold with those who favor fire.
But if I had to perish twice,
I think I know enough of hate
To say that for destruction ice
Is also great
And would suffice.

Robert Frost

In this chapter

We address the processes that modify landscapes once occupied by glaciers. These large mobile chunks of ice are very effective agents of change in the landscape, sculpting distinctive landforms, and generating prodigious amounts of sediment. Our task is broken into two parts, which form the major divisions of the chapter: (1) understanding the physics of how glaciers work, the discipline of glaciology, which is prerequisite for (2) understanding how glaciers erode the landscape, the discipline of glacial geology.

Kennicott Glacier, Wrangell–St. Elias National Park, Alaska. Mt. Blackburn graces the left skyline. Little Ice Age moraines bound the sides of the glacier roughly 16 km from its present terminus near the town of McCarthy (photograph by R. S. Anderson).

Although glaciers are interesting in their own right, and lend to alpine environments an element of beauty of their own, they are also important geomorphic actors. Occupation of alpine valleys by glaciers leads to the generation of such classic glacial signatures as U-shaped valleys, steps and overdeepenings now occupied by lakes in the long valley profiles, and hanging valleys that now spout waterfalls. Even coastlines have been greatly affected by glacial processes. Major fjords, some of them extending to water depths of over 1 km, punctuate the coastlines of western North America, New Zealand and Norway. We will discuss these features and the glacial erosion processes that lead to their formation.

It is the variation in the extent of continental scale ice sheets in the northern hemisphere that has driven the 120–150 m fluctuations in sea level over the last three million years. On tectonically rising coastlines, this has resulted in the generation of marine terraces, each carved at a sea level highstand corresponding to an interglacial period.

Ice sheets and glaciers also contain high-resolution records of climate change. Extraction and analysis of cores of ice reveal detailed layering, chemistry, and air bubble contents that are our best terrestrial paleoclimate archive for comparison with the deep sea records obtained through ocean drilling programs.

The interest in glaciers is not limited to Earth, either. As we learn more about other planets in the solar system, attention has begun to focus on the potential that Martian ice caps could also contain paleoclimate information. And further out into the solar system are bodies whose surfaces are mostly water ice. Just how these surfaces deform upon impacts of bolides, and the potential for unfrozen water at depth, are topics of considerable interest in the planetary sciences community.

Finally, glaciers are worth understanding in their own right, as they are pathways for hikers, suppliers of water to downstream communities, and sources of catastrophic floods and ice avalanches. The surfaces of glaciers are littered with cracks and holes, some of which are dangerous. But these hazards can either be avoided or lessened if we approach them with some knowledge of their origin. How deep are crevasses? How are crevasses typically oriented, and why? What is a moulin? What is a medial moraine, what is a lateral moraine? Are they mostly ice or mostly rock?

How does the water discharge from a glacial outlet stream vary through a day, and through a year?

Glaciology: what are glaciers and how do they work?

A glacier is a natural accumulation of ice that is in motion due to its own weight and the slope of its surface. The ice is derived from snow, which slowly loses porosity to approach a density of pure ice. This evolution is shown in Figure 8.1. Consider a small alpine valley. In general, it snows more at high altitudes than it does at low altitudes. And it melts more at low altitudes than it does at high. If there is some place in the valley where it snows more in the winter than it melts in the summer, there will be a net accumulation of snow there. The down-valley limit of this accumulation is the snowline, the first place where you would encounter snow on a climb of the valley in the late fall. If this happened year after year, a wedge of snow would accumulate each year, compressing the previous years' accumulations. The snow slowly compacts to produce firn in a process akin to the metamorphic reactions in a mono-mineralic rock near its pressure melting point. Once thick enough,

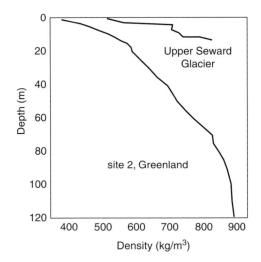

Figure 8.1 Density profiles in two very different glaciers, the upper Seward Glacier in coastal Alaska being very wet, the Greenland site being very dry. The metamorphism of snow is much more rapid in the wetter case; firn achieves full ice densities by 20 m on the upper Seward and takes 100 m in Greenland. Ice with no pore space has a density of 917 kg/m³ (after Paterson, 1994, Figure 2.2, reproduced with permission from Elsevier).

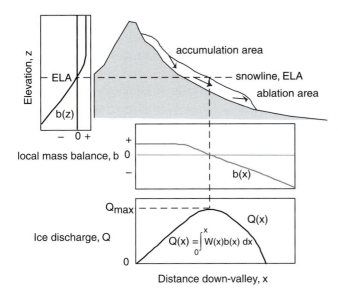

Figure 8.2 Schematic diagrams of a glacier (white) in mountainous topography (gray) showing accumulation and ablation areas on either side of the equilibrium line. Mapped into the vertical, z (left-hand diagram), the net mass balance profile, $b(z)$, is negative at elevations below the ELA and positive above it. We also show the net balance mapped onto the valley, parallel axis, x (follow dashed line downward), generating the net balance profile $b(x)$. At steady state the ice discharge of the glacier must reflect the integral of this net balance profile (bottom diagram). The maximum discharge should occur at roughly the down-valley position of the ELA. Where the discharge goes again to zero determines the terminus position.

this growing wedge of snow-ice can begin to deform under its own weight, and to move downhill. At this point, we would call the object a glacier. It is only by the motion of the ice that ice can be found further down the valley than the annual snowline.

A glacier can be broken into two parts, as summarized in Figure 8.2: the accumulation area, where there is net accumulation of ice over the course of a year, and the ablation area, where there is net loss of ice. The two are separated by the equilibrium line, at which a balance (or equilibrium) exists between accumulation and ablation. This corresponds to a long-term average of the snowline position. The equilibrium line altitude, or the ELA, is a very important attribute of a glacier. In a given climate, it is remarkably consistent among close-by valleys, and at least crudely approximates the elevation at which the mean annual temperature is $0\,°C$.

The ice of the world is contained primarily in the great ice sheets. Antarctica represents roughly 70 m of sea level equivalent of water, Greenland roughly 7 m,

and all the small glaciers and ice caps of the world roughly 2 m. We note, however, that the small ice bodies of the globe are contributing disproportionately to the present sea level rise.

Types of glaciers: a bestiary of ice

First of all, note that sea ice is fundamentally different from glacier ice. Sea ice is frozen seawater; it is not born of snow. It is usually a few meters thick at the best, with pressure ridges and their associated much deeper keels being a few tens of meters thick. Icebreakers can plow through sea ice. They cannot plow through icebergs, which are calved from the fronts of tidewater glaciers, and can be more than a hundred meters thick. It is icebergs that pose a threat to shipping.

Glaciers can be classified in several ways, using size, the thermal regime, the location in the landscape, and even the steadiness of a glacier's speed. Some of these classifications overlap, as we will see. We will start with the thermal distinctions, as they play perhaps the most important role in determining the degree to which a glacier can modify the landscape.

The temperatures of *polar glaciers* are well below the freezing point of water throughout except, in some cases, at the bed. They are found at both very high latitudes and very high altitudes, reflecting the very cold mean annual temperatures there. As is seen in Figure 8.3, to first order, a thermal profile in these glaciers would look like one in rock, increasing with depth in a geothermal profile that differs from one in rock only in that the conductivity and density of ice is different from that of rock.

In contrast to these glaciers, *temperate glaciers* are those in which the mean annual temperature is very close to the pressure-melting point of ice, all the way to the bed. The distinction is clearly seen in Figure 8.3. They derive their name from their location in temperate climates whose mean annual temperatures are closer to $0\,°C$ than at much higher elevations or latitudes. The importance of the thermal regime lies in the fact that being close to the melting point at the base allows the ice to slide along the bed in a process called regelation, which we will discuss later in the chapter. It is this process of sliding that allows temperate glaciers to erode their beds through both abrasion and quarrying. Polar glaciers are gentle on the landscape, perhaps even protecting it from subaerial mechanical weathering

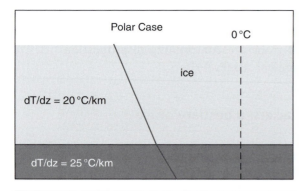

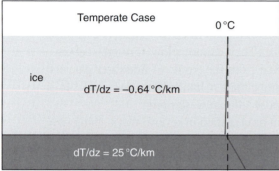

Figure 8.3 Temperature profiles in polar (top) and temperate (bottom) glacier cases. Slight kink in profile in the polar case reflects the different thermal conductivities of rock and ice. Roughly isothermal profile in the temperate case is allowed by the downward advection of heat by melt water. Temperature is kept very near the pressure-melting point throughout, meaning it declines slightly (see phase diagram of water, Figure 1.2).

processes that would otherwise attack it. How does a temperate glacier remain close to the melting point throughout, being almost isothermal? Recall that associated with the phase change of water is a huge amount of energy. In a temperate glacier, significant water melts at the surface, and is translated to depth within the firn, and even deeper in the glacier along three-grain intersections. Glacier ice is after all a porous substance. If this water encounters any site that is below the freezing point, it will freeze, yielding its energy, which in turn warms up the surrounding ice. So heat is efficiently moved from the surface to depth by moving water – it is advected. This is a much more efficient process of heat transport than is conduction, and can maintain the entire body of a glacier at very near the freezing point.

A straightforward distinction can be made in terms of size. Valley glaciers occupy single valleys. Ice caps cover the tops of peaks and drain down several valleys on the sides of the peak. Ice sheets can exist

in the absence of any pre-existing topography, and can be larger by orders of magnitude than ice caps. The greatest contemporary examples are the ice sheets of east and west Antarctica, and of Greenland, with diameters of thousands of kilometers, and thicknesses of kilometers. Their even larger relatives in the last glacial maximum (LGM), the combined Laurentide and Cordillera, and the Fennoscandian Ice Sheets, covered half of North America and half of Europe, respectively.

Tidal glaciers are those that dip their toes in the sea, and lose some fraction of their mass through the calving of icebergs (as opposed to loss solely by melting). These are the glaciers of concern to shipping, be it the shipping plying the waters of the Alaskan coastline, or the ocean liners plying the waters off Greenland.

Most glaciers obey what we mean when we use the adjective "glacial." Glacial speeds might be a few meters to a few kilometers per year, and will be the same the next year and the next. We speak of glacial speeds as being slow and steady. The exceptions to this are surging glaciers and their cousins embedded in ice sheet margins, the ice streams, which appear to be in semi-perpetual surge. These ill-behaved glaciers (meaning they don't fit our expectations) are the subjects of intense modern study. They may hold the key to understanding the rapid fluctuations of climate in the late Pleistocene, which in turn are important to understand as they lay the context for the modern climate system that humans are modifying significantly.

All of these we will visit in turn, but first let us lay out the basics of how glaciers work.

Mass balance

The glaciological community, traditionally an intimate mix of mountain climbers and geophysicists, has a long and proud tradition of being quite formal in its approach to the health of glaciers and their mechanics. Once again, the problem comes down to a balance, this time of mass of ice. One may find in the bible of the glaciologists, Paterson's *The Physics of Glaciers*, now in its third edition (Paterson, 1994), at least one chapter on mass balance alone (see Further reading in this chapter for more suggestions of other excellent textbooks). One may formalize the illustration of the mass balance shown in Figure 8.2 with the following equation:

$$\frac{\partial H}{\partial t} = b(z) - \frac{1}{W(z)} \frac{\partial Q}{\partial x} \tag{8.1}$$

where H is ice thickness, W is the glacier width, and Q the ice discharge per unit width [= L^2/T]. Mass can be lost or gained through all edges of the block we have depicted (the top, the base, and the up- and down-ice sides). Here b represents the "local mass balance" on the glacier surface, the mass lost or gained over an annual cycle. It is usually expressed as meters of water equivalent per year. This quantity is positive where there is a net gain of ice mass over an annual cycle, and negative where there is a net loss. The elevation at which the mass balance crosses zero defines the "equilibrium line altitude," or the ELA, of the glacier. Because it is an altitude, it is a horizontal line in Figure 8.2. The mass balance reflects all of the meteorological forcing of the glacier, both the snow added over the course of the year, and the losses dealt by the combined effects of ablation (melt) and sublimation. Where the annual mass balance is positive, it has snowed more than it melts in a year, and vice versa. To first order, because it snows more at higher altitudes, and melts more at lower altitudes, the mass balance always has a positive gradient with elevation. Examples of mass balance profiles from a variety of glaciers in differing climates are shown in Figure 8.4. Note the positive mass balance gradient in each case, which is especially well marked in the ablation or wastage zones. One can easily pick out the ELA for each glacier. The ELA varies greatly, being lowest in high latitudes (where it is cold and the ablation is low) and nearest coastlines (where the winter accumulation is high due to the proximity of oceanic water sources). A classic illustration of the latitudinal dependence is drawn from work of Skinner *et al.*, reproduced in Figure 8.5. The modern ELAs, as deduced from snowlines and mass balance surveys, are everywhere much higher than the ELAs reconstructed (more on how to do this later; see Figure 8.19) from the last glacial maximum (LGM) at roughly 18 ka. In places the rise in ELA is up to 1 km!

One may measure the health of a glacier by the total mass balance, reflecting whether in a given year there has been a net loss or gain of ice from the entire

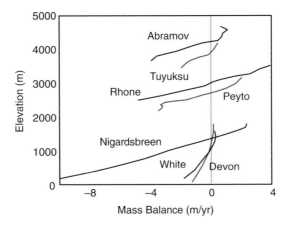

Figure 8.4 Specific mass balance profiles from several glaciers around the world, showing the variability of the shape of the profiles. Mass balance gradients (slopes on this plot) are quite similar, especially in ablation zones (where the local balance $b < 0$) except for those in the Canadian Arctic (Devon and White ice caps) (adapted from Oerlemans and Fortuin, 1992, Figure 1, with permission of the American Association for the Advancement of Science).

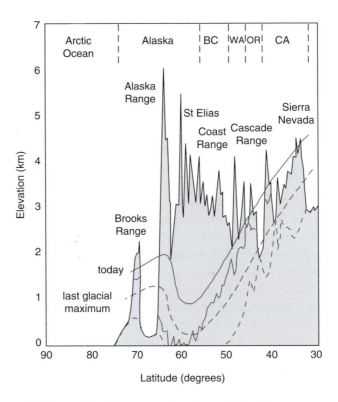

Figure 8.5 Profiles of topography (gray), equilibrium line elevation (ELA, top) and glacial extent (bottom) (solid, present day; dashed, last glacial maximum (LGM)) along the spine of Western North America from California to the Arctic Ocean. Note the many-hundred meter lowering of the ELA in the LGM, and the corresponding greater extent of the glacial coverage of the topography (after Skinner *et al.*, 1999, with permission from John Wiley & Sons, Inc.).

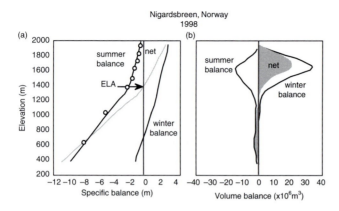

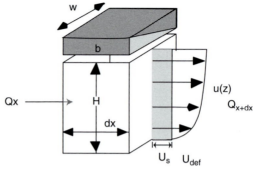

Figure 8.7 (a) Mass balance for a section of glacier of width W, down-glacier length dx, and height H. Inputs or outputs through the top of the box dictate the local mass balance, b. Down-glacier discharge of ice into the left-hand side of the box, Q_x, and out the right-hand side, Q_{x+dx}, includes contributions from basal sliding (shading) and internal ice deformation (after MacGregor *et al.*, 2000, Figure 2).

Figure 8.6 Mass balance profiles for the year 1998 on the Nigardsbreen, a coastal Norwegian glacier. (a) Specific balance in meters of water equivalent. Winter balance from snow probe surveys, summer balance from stake network (circles). Net balance is shown in gray; net balance is zero at 1350 m, which is the ELA. (b) The volume balance derived by the product of the specific balance with the altitudinal distribution or hypsometry of the glacier. That the glacier has so much more area at high elevations is reflected in the high contribution of accumulation to the net balance of the glacier (gray fill). In 1998, the net balance is highly positive; there is more gray area to the right of the 0 balance line than to the left, so that the integral of the gray fill is > 0. In this year the positive total balance represents a net increase of roughly 1 m water equivalent over the entire glacier (after data in Kjøllmoen, 1999).

glacier. This is simply the spatial integral of the product of the local mass balance with the hypsometry (area vs. elevation) of the valley:

$$B = \int_0^{z_{max}} b(z)W(z)\mathrm{d}z \qquad (8.2)$$

This exercise is carried out annually on numerous glaciers worldwide. See Figure 8.6 for an example from the Nigardsbreen, Norway. The Norwegians are interested in the health of their glaciers because they control fresh water supplies, but also because a significant portion of their electrical power comes from subglacially tapped hydropower sources.

It is a common misconception that a considerable amount of melting takes place at the base of a glacier, because after all the Earth is hot. Note the scales on the mass balance profiles. In places, many meters can be lost by melting associated with solar radiation. Recall that the heat flux through the Earth's crust is about $41\,\mathrm{mW/m^2}$ (defined as one heat flow unit, HFU), a trivial flux when contrasted with the high heat fluxes powered in one or another way by the sun (about $1000\,\mathrm{W/m^2}$). The upward

heat flux from the Earth is sufficient to melt about 5 cm of ice per year. As far as the mass balance of a glacier is concerned, then, there is little melt at the base.

If nothing else were happening but the local mass gain or loss from the ice surface, a new lens of snow would accumulate, which would be tapered off by melt to a tip at the ELA (or snowline) each year. Each successive wedge would thicken the entire wedge of snow above the snowline, and would increase the slope everywhere. But something else *must* happen, because we find glaciers poking their snouts well below the ELA, below the snowline. How does this happen? Ice is in motion. This is an essential ingredient in the definition of a glacier. Otherwise we are dealing with a snowfield. The Q terms in the mass balance expression reflect the fact that ice can move downhill, powered by its own weight. Ice has two technologies for moving, one by basal sliding, in which the entire glacier moves at a rate dictated by the slip at the bed, the other by internal deformation, like any other fluid (see Figure 8.7). We will return to a more detailed treatment of these processes in a bit. Know for now that the ice discharge per unit width of glacier, Q, is the product of the mean velocity of the ice column, $\bar{U}$, and the thickness of the glacier, H.

Given only this knowledge, we can construct a model of a glacier in steady state, one in which none of the variables of concern in the mass balance expression are changing with time. Setting the

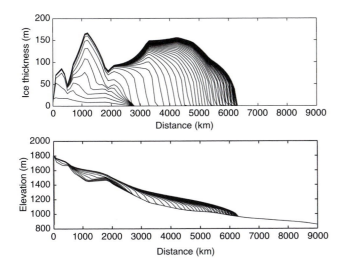

Figure 8.8 Model of glacier evolution on bedrock profile from Bench Glacier valley, Alaska, shown in evenly spaced time steps out to 600 years. Climate is assumed to be steady, with a prescribed mass balance profile. Top: profiles of ice thickness through time. Bottom: glacier draped on bedrock profile. The glacier reaches approximately steady state at ~500 years. Measured maximum ice thickness of 180 m is well reproduced by the final model glacier, implying that the mass balance profile $b(x)$ is well chosen.

simulations shown represent 600 years, and the glacier comes into roughly steady state within 400 years. Similar modeling exercises have been used recently to explore the sensitivity of alpine glaciers to climate changes in the past and in the future (e.g., Oerlemans 1994, 2001, 2005).

This exercise also yields another interesting result. In steady state, we find that within the accumulation area, the ice discharge must be increasing down-valley in order to accommodate the new snow (ultimately ice) arriving on its top. Conversely, the ice discharge must be decreasing with down-valley distance in the ablation region. This has several important glaciological and glacial geological consequences. First, the vertical component of the trajectories of the ice parcels must be downward in the accumulation zone and upward in the ablation zone, as shown in all elementary figures of glaciers, including Figure 8.2. As a corollary, debris embedded in the ice is taken toward the bed in the accumulation zone and away from it in the ablation zone. Glaciers tend to have concave up-valley contours above the ELA, and convex contours below (hence you can approximately locate the ELA on a map of a glacier simply by finding the contour that most directly crosses the glacier without bending either up- or down-valley). Debris therefore moves away from the valley walls in the accumulation zone and toward them in the ablation zone. This is reflected in the fact that lateral moraines begin at roughly the ELA. This observation is useful if one is trying to reconstruct past positions of glaciers in a valley, or more particularly to locate the past position of the ELA. As the ELA is often taken as a proxy for the 0° isotherm, it is a strong measure of climate, and hence a strong target for paleoclimate studies.

This straightforward exercise should serve as a motivation for understanding the mechanics of ice motion. These mechanics are at the core of all such simulations. It is what separates one type of glacier from another. And whether a glacier can slide on its bed or not dictates whether it can erode the bed or not – and hence whether the glacier can be an effective means of modifying the landscape.

left-hand side of Equation 8.1 to zero, we see that there must be a balance between the local mass balance of ice dictated by the meteorological forcing (the climate) and the local gradient in the ice discharge:

$$Q(x) = \int_0^x b(x)W(x)\mathrm{d}x \qquad (8.3)$$

Here we have taken x to be 0 at the up-valley end of the glacier. If we ignore for the moment the width function $W(x)$, reflecting the geometry (or really the hypsometry) of the valley, the discharge will follow the integral of the mass balance. For small x, high up in the valley, since the local mass balance is positive there the ice discharge must increase with distance down-valley; conversely, it must decrease with distance below that associated with the ELA, as the mass balance is negative there. The ice discharge must therefore go through a maximum at the ELA. To illustrate this, we show in Figure 8.8 a simulation of the evolution of a small alpine glacier in its valley, starting with no ice and evolving to steady state. We impose a mass balance profile, and hold it steady from the start of the model run. The

Ice deformation

Like any other fluid on a slope, ice deforms under its own weight. It does so at very slow rates, which are

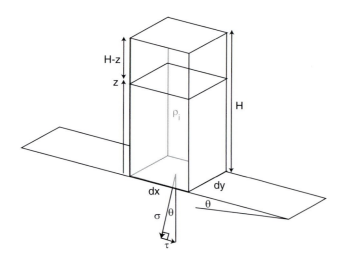

Figure 8.9 Definition of normal and shear stresses imposed by a column of material (here ice) resting on a sloping plane.

gradient of the velocity in the vertical direction. We will then have to integrate this expression to obtain the velocity.

In a simple fluid, Newton demonstrated that there is a linear relation between the shear stress acting on a parcel of the fluid and the shear strain rate of that fluid. These are therefore called "linear" or Newtonian fluids. Although ice is more complicated, we will walk through the derivation using a linear fluid first, and then take the parallel path through the expressions relevant to ice. The problem requires several steps:

1. Development of an expression for the pattern of shear stress within the material.
2. Development of an expression to describe the rheology of the material.
3. Combination of these to obtain an expression relating the rate of strain to the position within the material.
4. Integration of the strain rate to obtain the velocity profile.

dictated by the high viscosity of the ice. As the viscosity is temperature dependent, increasing greatly as the temperature declines, the colder the glacier is the slower it deforms. Although the real picture is considerably more complicated than that we will describe here (see Hooke, 2005, and Paterson, 1994, for recent detailed treatments), the essence of the physics is as follows. Consider a slab of ice resting on a plane inclined at an angle to the horizontal, as sketched in Figure 8.9. We wish to write a force balance for this chunk of ice. It is acted upon by body forces (fields like gravitational fields and magnetic fields). As ice is not magnetic, the relevant body force is simply that due to gravity. The weight of any element of ice is *mg* where *m* is the mass of the slab, or its density times its volume (d*x*d*y*d*z*). One may decompose the weight vector into one acting parallel to the bed and one acting normal to the bed. As shown in the figure, the normal stress, σ, (recall that a stress is a force divided by the area of the surface, d*x*d*y*) acting on the slab on its top side is $\rho g(H - z)\cos\theta$, and at its base $\rho g H \cos\theta$.

Now that we have an expression for the stresses within the slab, we introduce its material behavior, or rheology, the relationship describing the reaction of the material to the stresses acting upon it. To anticipate, our goal is to derive an expression for the velocity of the ice as a function of height above the bedrock–ice interface, or the bed of the glacier. The rheology will relate the stresses to the spatial

The pattern of stress

At any level within a column of material resting on a slope, the shear stress is the component of the weight of the overlying material that acts parallel to the bed, divided by the cross-sectional area of the column, while the normal stress is that acting normal to the surface. These are illustrated in Figure 8.9. The weight is of course the mass times the acceleration, here that due to gravity, and the mass is the density times the volume. If we take the density to be uniform with depth in the column, this yields the expression for the shear stress as a function of height above the bed, z:

$$\tau = \rho_i g (H - z)\sin(\theta) \qquad (8.4)$$

Here the quantity $H - z$ represents the height of the overlying column of material, which is exerting the stress on the underlying material. Note that we have not yet identified the nature of the material – i.e., we have not yet specified how the material responds to this stress. This is a general expression for the vertical profile of shear stress within a brick, a column of rock on a slope, or in a fluid such as water, lava, or ice on a slope. As long as the material density is uniform, the stress increases linearly with depth into the material, as plotted in Figure 8.10, reaching a maximum at the bed. Importantly, the shear stress is said to "vanish"

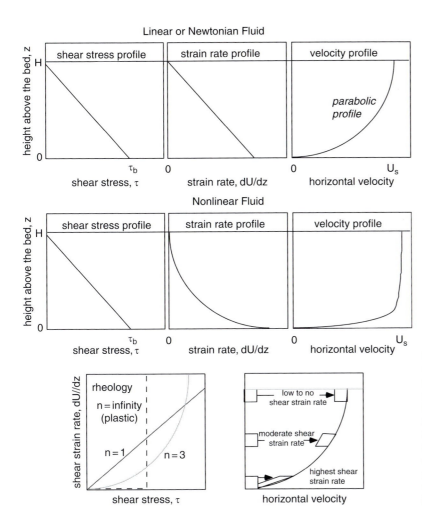

Figure 8.10 Diagrams to aid in the derivation of the velocity profiles in linear fluids (top row) and in nonlinear fluids (second row). In each row the shear stress profile is the same, linearly increasing from 0 at the top of the fluid to the basal shear stress τ_b at the base. The middle box shows the shear deformation rate profile, $dU/dz(z)$, and the third box shows its integral, the velocity profile, $U(z)$. The bottom box shows graphically the shear strain associated with three different levels in the fluid. Shear strain can be measured by the change in angles in a box with originally orthogonal sides.

at the surface; nothing magic here, it is simply zero where $z = H$. The shear stress exerted by the overlying column of air is negligible (until we begin worrying about entrainment of small sand and dust particles in Chapter 14).

The rheology

Now we must address the response of the material to this applied shear stress. This is called the rheology of the material. In the case of a solid or elastic rheology, there is a finite and specific strain of the material that results from an applied stress. Consider a rubber band. You apply a stress to it, a force per unit area of the rubber. The band stretches a certain amount. The strain of the rubber, ε, is defined as the change in length divided by the original length:

$$\varepsilon = \frac{\Delta L}{L_o} \tag{8.5}$$

where L_o is the original length. This is called the linear strain, the strain of the material along a line. This is associated with the changes of length in the direction of the applied force, here a normal force. Note that strain is dimensionless. In the case of elastic solids, this strain is both finite and reversible: when the force is taken away, the material returns to its original shape. The relationship between the stress and the resulting strain is captured by this simplified rheological statement for an elastic solid:

$$\varepsilon = \frac{1}{E}\sigma \tag{8.6}$$

where σ is the applied stress, and E is Young's modulus. The higher the Young's modulus, the more stress it takes to accomplish a given strain. For completeness, we must recognize that in an elastic material the strain in one dimension is connected to the strain in another direction; the rubber band thins as you stretch it. The material constant that relates strain in one dimension to strain in orthogonal directions is Poisson's ratio, ν.

There is another type of strain, called a shear strain, which results from a shear stress. Rather than changes in length, this is captured as changes in angle. Consider a material on which we have scribed a right angle. Shear strain of the material results in a change to this angle. Again, it is dimensionless (radians). A given stress results in a given strain, here a shear stress and a shear strain:

$$\varepsilon_{xz} = \frac{1 + \nu}{E} \tau_{xz} \tag{8.7}$$

For further discussion of strains and stresses in elastic materials, see for example Turcotte and Schubert (2002).

Now let us consider a fluid rather than a solid. They differ fundamentally because an applied stress can result in an infinite strain of the material. Imagine a plate on which you pour some molasses or treacle. Tip the plate; in so doing you exert a shear stress on the molasses. The molasses keeps moving as long as you keep the plate tilted. There is no specific strain associated with an applied stress, and we cannot therefore use an elastic rheology to describe the behavior. However, one could instead relate a specific *rate* of strain to the applied stress. In the case of a shear stress, the resulting shear strain rate is equivalent to the velocity gradient in the direction of shear (see Figure 8.10). In other words, for the case at hand of a fluid on a tipped plate (or bedrock valley floor), the shear stress results in a strain rate that is captured in the gradient of the horizontal velocity with respect to the vertical, dU/dz:

$$\dot{\varepsilon}_{xz} = \frac{dU}{dz} = \frac{1}{\mu} \tau_{xz} \tag{8.8}$$

The parameter μ that dictates the scale of the response is called the viscosity of the fluid. This is called variously a Newtonian viscous rheology, or a linear viscous rheology. The shear strain rate is related linearly to the shear stress.

Combining this mathematical representation of a linear viscous fluid with that for the shear stress as a function of depth (the stress profile), we obtain an expression for the shear strain rate at all levels within the fluid:

$$\frac{dU}{dz} = \frac{\rho_i g \sin(\theta)}{\mu}(H - z) \tag{8.9}$$

This is shown in Figure 8.10. Note that the profile of shear strain rate mimics the shear stress profile in that it is zero at the surface of the fluid, and linearly increases to a maximum at the bed. What does this mean for the velocity profile? At the surface, there can be no shear strain rate. Equivalently, the velocity profile must have no slope to it at the surface. The gradient (slope) of the velocity profile then increases linearly toward the bed.

We obtain the velocity itself by integrating the strain rate with respect to z. This is a definite integral, from 0 to some level z in the fluid:

$$U(z) = \frac{\rho_i g \sin(\theta)}{\mu}\left(Hz - \frac{z^2}{2}\right) \tag{8.10}$$

While the contribution due to internal deformation (flow) is indeed zero at the bed, to this must be added any slip along the bed. In most fluids, the "no slip condition" is applied, as molecules within the fluid interact with stationary ones in the bed to bring the velocity smoothly to zero at the bed. Ice, we will see below, is different. First, it can change phase at the bed, and second, it can slide as a block against the bed. Ignoring this for the moment, the resulting profile of velocity is shown in Figure 8.10, where you can see that the features we expected are displayed: there is no gradient at the top of the flow, and the highest gradient in the velocity is found at the bed.

Three other quantities are easily extracted from this analysis: the surface velocity, which we would like to have because we can measure it, the average velocity, and the integral of the velocity, which is equivalent to the ice discharge per unit width of the glacier. The surface velocity is simply $U(H)$, which is

$$U_s = \frac{\rho g H^2 \sin \theta}{2\mu} \tag{8.11}$$

The average velocity can be obtained formally by application of the mean value theorem to the problem (see Appendix B):

$$\bar{y} = \frac{1}{b - a}\int_a^b y(x)dx \tag{8.12}$$

In the case at hand, the variable is the velocity, and the limits are 0 and H:

$$\bar{U} = \frac{1}{H - 0}\int_0^H U(z)dz \tag{8.13}$$

We find that the average velocity is

$$\bar{U} = \frac{\rho g H^2 \sin \theta}{3\mu} = \frac{2}{3} U_s \quad (8.14)$$

or two-thirds of the surface velocity. Note on the graph of velocity vs. depth in Figures 8.10 and 8.11(a) where this mean velocity would be encountered in the profile. It is closer to the bed than to the surface, and is in fact at about six-tenths of the way to the bed from the surface.

The integral of the velocity, or the discharge per unit width of flow, is the product of the mean velocity and the flow depth, and is therefore

$$Q = \bar{U}H = \frac{\rho g H^3 \sin \theta}{3\mu} \quad (8.15)$$

Note the strong (cubic) dependence on the depth of the flow.

Ice wrinkles 1: Glen's flow law

While the equations derived above illustrate the approach one takes to flow problems in general, they are not appropriate for ice. Ice differs from many fluids in that the relationship between the shear stress and the strain rate (the rheology) is not linear. Instead, the rheology is roughly cubic, as shown in Figure 8.10. Ice is therefore said to have a nonlinear rheology. A more general rheological relation can be written

$$\frac{dU}{dz} = A\tau^n = \left[A\tau^{n-1}\right]\tau \quad (8.16)$$

where n is an exponent that one needs to determine experimentally, and the constant A is called the "flow-law parameter." We have already dealt with the linear ($n = 1$) case. Glen's experiments (Glen, 1952) revealed that n is approximately 3 for ice. The term in brackets represents the inverse of an effective viscosity: $1/\left[A\tau^{n-1}\right]$. This expression implies that, as the shear stress increases, the effective viscosity declines, and radically so for all $n > 1$. The consequence is that ice near the bed (under high shear stress) behaves as if it is much less stiff than ice near the surface. We can now again combine this equation for the relationship between the shear strain rate and the shear stress (the rheology) with that for shear stress profile to obtain the profile of shear strain rate:

$$\frac{dU}{dz} = A[\rho_i g \sin(\theta)]^3 (H - z)^3 \quad (8.17)$$

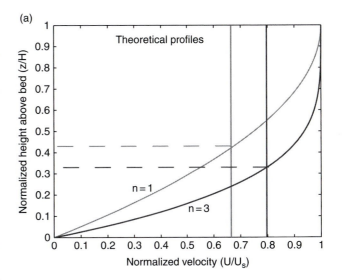

(a)

Figure 8.11(a) Theoretical flow profiles for $n = 1$ (thin line) and $n = 3$ (bold line) fluids, normalized against maximum height above bed and maximum flow speed. The mean speed is shown as the vertical lines, and the position above the bed at which this mean speed would be measured is signified by the dashed horizontal lines. As the nonlinearity of the rheology increases, the mean speed approaches the surface speed, and the depth at which it would be measured is found nearer the bed.

This may then be integrated to yield the velocity profile:

$$U(z) = A[\rho_i g \sin(\theta)]^3 \left[H^3 z - \frac{3z^2 H^2}{2} + z^3 H - \frac{z^4}{4}\right] \quad (8.18)$$

The resulting velocity profile is significantly different from that for the linear rheology case. In Figures 8.10 we show it crudely, and in Figure 8.11(a) more formally. In particular, much more of the change in speed of the ice with distance from the bed is accomplished very near the bed. The flow looks much more "plug-like."

Again, we can obtain the surface speed by evaluation of the velocity at $z = H$:

$$U(H) = U_s = A[\rho_i g \sin(\theta)]^3 \left[\frac{H^4}{4}\right] \quad (8.19)$$

The specific discharge of ice is obtained by integrating the profile from 0 to H:

$$Q = A[\rho_i g \sin(\theta)]^3 \left[\frac{H^5}{5}\right] \quad (8.20)$$

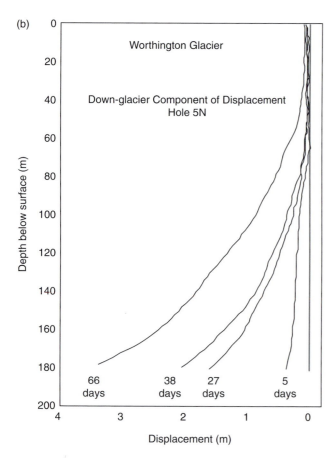

(b)

Worthington Glacier

Down-glacier Component of Displacement
Hole 5N

Depth below surface (m)

66
days

38
days

27
days

5
days

Displacement (m)

Figure 8.11(b) Four measured deformation profiles of an initially straight vertical borehole drilled almost to the bed of the Worthington Glacier, Alaska. Ice depth at this location is roughly 190 m. Measurements made with borehole inclinometer; only the down-glacier component of deformation is shown. Profiles are shown relative to the surface position (after J. T. Harper, pers. comm. 1996; see Harper et al., 1998, Figure 2, with permission from the American Association for the Advancement of Science).

and the average speed is obtained by using the mean value theorem, or by recognizing that the average speed is simply Q/H:

$$\bar{U} = A[\rho_i g \sin(\theta)]^3 \left[\frac{H^4}{5} \right] \qquad (8.21)$$

This is shown as the dashed line in Figure 8.11(a). Note that this average speed is related to the surface speed through

$$\bar{U} = \frac{4}{5} U_s = \frac{n+1}{n+2} U_s \qquad (8.22)$$

As the nonlinearity of the flow law, expressed by n, increases, the mean speed approaches the surface

speed. The flow profiles for $n = 1$ and $n = 3$ cases are compared in Figure 8.11(a), in which we also show the mean speeds for both cases. That the velocities are normalized to the maximum speed (that at the surface, U_s) makes it straightforward to see how the mean speed relates to the maximum in both cases. As n increases, the maximum speed becomes a better proxy for the mean speed, and the mean speed should occur closer to the bed.

It is also important to realize that the nonlinearity of the flow law results in a very sensitive dependence of the ice discharge on both ice thickness and ice surface slope. The ice discharge, $Q = \bar{U}H$, varies as the fifth power of ice thickness and the third power of the ice surface slope. A doubling of the ice discharge on a given slope can be accomplished by ice that is only $2^{(1/5)}$ or about 15% thicker!

Note that, in the formulations above, we have assumed that the flow-law parameter, A, is uniform with depth. While this is a good approximation in temperate glaciers, in which the temperatures are close to the pressure melting point throughout, the assumption breaks down badly in polar glaciers. Both experiments and theory show that the flow-law parameter is sensitive to temperature:

$$A = f(T_k) = A_o e^{-\frac{E_a}{RT_k}} \qquad (8.23)$$

where A_o is a reference flow-law parameter, E_a is the activation energy, R is the universal gas constant, and T_k is the absolute temperature. Recommended values for A_o are as follows: at $0\,°C$: $2.1 \times 10^{-16}\,\mathrm{yr}^{-1}\,\mathrm{Pa}^{-3}$, at $-5\,°C$: $7.5 \times 10^{-17}\,\mathrm{yr}^{-1}\,\mathrm{Pa}^{-3}$, at $-10\,°C$: $1.5 \times 10^{-17}\,\mathrm{yr}^{-1}\,\mathrm{Pa}^{-3}$.

As the temperature decreases, the argument of the exponential factor becomes more negative and A declines. Since the effective viscosity varies as $1/A$, the viscosity therefore increases. (For discussion see Turcotte and Schubert, 2002, chapter 7.) Using the values for activation energy for ice (61×10^3 J/mole) and the universal gas constant (8.31 J/mole-K), the flow-law parameter and hence the effective viscosity (both shown in Figure 8.13) are expected to vary over four orders of magnitude in the temperature range relevant to Earth's glaciers and ice sheets. Let's think about the implications of this for the shape of the velocity profile. As temperature decreases with height above the bed, z, the flow-law parameter will decrease rapidly. The ice effectively

(a)

Athabasca Glacier cross-sectional view

Raymond (1971) observations

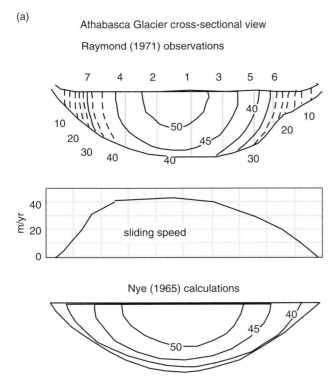

Nye (1965) calculations

(b)

Athabasca Glacier speed profile
Horizontal Velocity (m/yr)

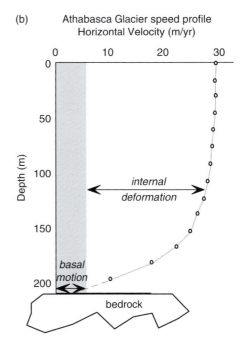

Figure 8.12(a) Top: distribution of down-valley ice speeds in cross section of the Athabasca Glacier, Canada, as interpolated from measurements in seven boreholes (labeled gray lines), from Raymond (1971). Center: cross-valley distribution of sliding speed as deduced by the intersection of the velocity contours with the bed in the top panel. Note strong broad peak in the sliding speed in the center of the glacier. Bottom: flow field as predicted by Nye (1965) theory for flow in a parabolic channel, to which a uniform sliding speed has been added (after Paterson, 1994, Figure 11.11, reproduced with permission from Elsevier).

Figure 8.12(b) Velocity profile of the Athabasca Glacier, Canada, derived from inclinometry of a borehole and measurement of surface displacement of the borehole top. Projection to the base yields estimate of the contribution from motion of the ice relative to the rock, or basal motion (gray box) (data from Savage and Paterson (1963)).

stiffens with height above the bed. This reduces the rate of strain, or the velocity gradient, and the flow profile should look even more plug-like than in the uniform temperature case. This extreme sensitivity of the rheology to temperature requires that the modeling of ice sheets incorporates the evolution of the temperature field. Such models are said to require thermo-mechanical coupling.

While experiments on small blocks of ice inspired the exploration of the nonlinear rheology of ice, it is field measurements of entire glacial profiles that have been used to test the theory. This information comes largely from the deformation of boreholes in glaciers. Boreholes are initially drilled straight downward using hot tips, and later steam drills. Using an inclinometer, the dip of the hole is measured at each

of many depths, from which the profile may be constructed. The location of the top of the hole may also be tracked using either GPS or optical surveying, so that we know its speed. The deformation speeds may then be deduced as a function of depth, to be contrasted with theory, as shown in Figure 8.11(b) and in Figure 8.12(b) (see Harper *et al.*, 1998, 2001). Note that the difference in motion between the bottom and top of the hole may now be calculated. Knowing the speed of the top and the difference in the speed of the top and bottom of the hole, one can subtract the two to determine any motion of the base of the hole that is unaccounted for. This leftover motion we call "basal motion." This consists of motion of the ice relative to the rock, and can be either direct sliding of the ice over rock, or deformation of an intervening layer of water-saturated sediments (basal till).

Ice wrinkles 2: sliding/regelation

While the physics of internal deformation are interesting and can accomplish the translation of large masses

(a)

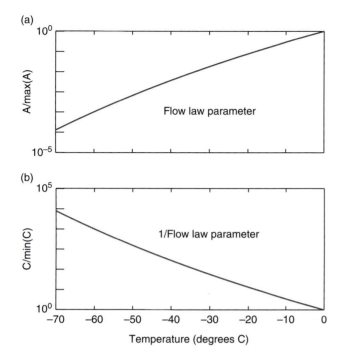

(b)

Figure 8.13 Dependence of flow-law parameter on temperature. (a) Flow law parameter, *A*, and (b) inverse of flow law parameter, which scales the effective viscosity. Vertical axes are normalized to their values at the pressure melting point. Noting the logarithmic vertical axis, the effective viscosity will rise by four orders of magnitude over the 70 °C range depicted.

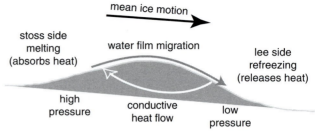

Figure 8.14 Schematic diagram of the regelation process by which temperate glaciers move around small bumps on the glacier bed. Ice melts on the high-pressure (stoss) side of the bump, moves around it as a thin water film, and refreezes in the low-pressure shadow on the lee side. The heat released by refreezing is conducted back through the bump to be used in the melting process. This is therefore an excellent case of coupling between thermal and fluid mechanics problems.

of ice down valleys, some large segment of the glacier population has yet another process to allow transport of ice down valleys. If the ice near the bed of the glacier is near the melting point, the ice can slide across the bed. As it is only by this mechanism that the bed of the glacier can be modified by the motion of ice above it, it is sliding that is the focus of glacial geologic studies.

Sliding of the ice is permitted by a special property of water: high pressure promotes melting. A corollary to this is that the high pressure phase of water is ice, its solid phase. This makes ice very different from, say, quartz or olivine, whose melted liquid state is lighter than their solid, and for which higher pressure therefore promotes the change of phase from the solid to the liquid. This can be seen in the phase diagram for the water system we first introduced in Figure 1.2. The negative slope (of −0.0074 °C/bar, or 7.4×10^{-8} °C/Pa) on the P–T plot, separating the water and ice phases, is what differentiates water from most other substances.

Consider the conditions at the bed of a temperate glacier, which by definition it is at its freezing point

throughout (meaning at all points the temperature lies along the phase boundary, declining at 0.0074 °C/11 m of depth). The glacier rests on a sloping valley floor that is not perfectly smooth, but has bumps and swales in it depicted in Figure 8.14. We probably would all agree that the ice is not accelerating (changing its velocity) very much. If it is doing so, it is doing so very slowly, meaning that the accelerations are very slight. This means, through Newton's law $F = ma$, that the forces on the ice are essentially balanced. There is also heat arriving at the base of the glacier from beneath the glacier, at a rate sufficient to melt about 5 cm of ice per year. This is not much melt, but given that the ice is already at the pressure melting point, there ought to be a thin layer of water present at the bed. You might think this would make it pretty slippery. For a slab of ice dx long in the down-valley direction, the force promoting down-valley motion is the down-valley component of the weight of the ice, or d$x \rho g$ Hsin(θ). What is resisting this, especially if there is a thin layer of water there, which is very weak in shear? Given that the shear resistance is therefore nearly zero, the forces resisting the down-valley motion of the ice are those associated with pressure variations associated with the bumps in the bed. In order to prevent acceleration of the ice, there must be a net component of the normal stress that is directly up-glacier. There must therefore be higher pressures on the up-valley sides of bumps than on the down-valley sides of bumps. Of course in the direction normal to the mean bed, the pressure must be that exerted by the normal component of the ice weight, or $dx \rho g\, H\cos(\theta)$.

But if the pressure fluctuates about some mean, say that of the pressure melting point of ice, then increasing the pressure a little bit on the up-valley side of a bump will promote melting of the ice, and decreasing it a little bit on the back of the bump will promote freezing of water. Given the phase diagram of the H_2O system, this must happen. In fact, the water film so generated on the up-valley side of the bumps is forced into motion for the very same reason: liquid water responds to pressure gradients by flowing from high pressures toward low pressures. Putting the two patterns together, we find that a parcel of basal ice performs something of a magic act to get past bumps in the bed. The ice melts on the up-valley sides of bumps, flows around the bump as a thin water film, and refreezes on the low-pressure down-valley sides of the bump. This process is called regelation, which is French for "refreezing."

Regelation is most effective in moving ice past small-scale bumps in the bed. Melting of water consumes energy, and refreezing of water releases energy, the same amount per unit volume of ice. That's nice – as no net energy must be added to the system. The problem is that the site where energy is needed to melt ice is different from where it is released upon refreezing. They are separated by the length of the bump. This heat energy must be transported through the bump or through the ice above it by conduction as shown in Figure 8.14. Heat conduction is dictated by the temperature gradient: the temperature difference between the two sites, divided by the distance between them. Therefore, the closer the sites, or the smaller the wavelength the bump, the more efficient the process.

It turns out that very large bumps can be circumvented by another process that makes them easy to get around as well. The situation is diagrammed in Figure 8.15. For long wavelength bumps, only a small-scale perturbation of the flow field in the ice itself is required to move past the bump, meaning that the ice does not have to regelate to get by the bump. This leaves intermediate sized bumps, with wavelengths of around 0.5 to 1 m, as the hardest bumps for the basal ice to move past. These have been called the "controlling wavelengths" for the basal sliding process. We will see that these details of the sliding process are strongly reflected in the patterns of erosion at the bed of a temperate glacier.

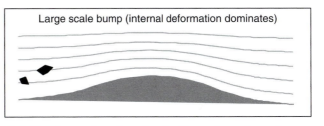
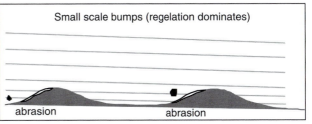

Figure 8.15 Trajectories of clasts embedded in basal ice as it encounters big (top) and little (bottom) bumps in the bed. Ice can deform sufficiently to accommodate the larger bumps, allowing clasts in the ice to ride over the bumps. In the small-bump case, the ice trajectories intersect the bed, reflecting the regelation mechanism. Clasts in the ice will be brought forcefully into contact with the bed, and cause abrasion of the front (stoss) sides of these bumps, leading to their elimination.

Direct evidence for the existence of this thin film of subglacial water, and the operation of the regelation mechanism, comes from several sources. One of the more striking is to be found on limestone bedrock, where the susceptibility of calcite to solution allows the subglacial water system to be read in great detail (Hallet, 1976). One sees on the upslope sides of small bumps little dissolution pits, and on the down-glacier sides precipitates. In fact, the precipitates take the form of small stalactites that grow almost horizontally, anchored to the downslope sides of the bumps. These interesting forms are easily visible in the photographs of Figure 8.16. This pattern of solution and re-precipitation is argued to represent the solution of calcite by the very pure water film, and its expulsion from solution as the water refreezes on the downslope side of the bump. The film is thought to be only microns thick. The ice produced by the regelation process is distinct in at least two senses from that produced originally from snow. It has a strong isotopic signature associated with fractionation that occurs upon both melting and refreezing. And it is largely bubble-free. Typical glacial ice is bubbly from air originally trapped in the ice as it metamorphoses from firn to ice. The bubbles give rise to the white

Figure 8.16 Details of the recently deglaciated bed of Blackfoot Glacier, Montana. Top: view down-glacier. Bumps in the bed localized by argillitic partings in the limestone bedrock show dissolutional roughening on the stoss (up-valley) sides, and re-precipitation of calcite (white) in the lee (see Hallet, 1976). Bottom: detail of the subglacially precipitated calcite, glacier flow top to bottom (photographs by R. S. Anderson).

color of the ice. In the regeneration process, the air in the bubbles is allowed to escape upon melting on the stoss sides of bumps, and is not incorporated into the regeneration ice in the lee of the bumps. Thus basal ice can take on a beautifully complex blue and white streaked look, clear blue in the regeneration ice and white due to bubbles in the original ice.

Concurrent measurement of glacier sliding and of local basal water pressure has led to the hypothesis that sliding is promoted by high water pressures. This is captured in the expression

$$U_{\text{slide}} = c \frac{\tau_{\text{b}}^{p}}{\{P_{\text{i}} - P_{\text{w}}\}^{q}} \tag{8.24}$$

where c is a constant that serves to scale the sliding speed and whose dimension depends upon p and q, and p and q determine the sensitivity of the sliding to τ_{b}, and to $P_i - P_{\text{w}}$, respectively. The expression in the denominator, the difference between the normal stress exerted by the ice overburden ($\rho_i gH$) and the local water pressure, P_{w}, is called the effective stress.

As the water pressure approaches that of the ice overburden, the effective stress goes to zero and sliding ought to become very rapid (infinite, if we take it to the limit of zero effective pressure). This state we also call the flotation condition: the full pressure of the column of ice overhead is being supported by the water pressure. Given the density difference between water and ice, this would correspond to a water table in the glacier at a height of ρ_i/ρ_{w}, or roughly nine-tenths of the ice thickness. Note as well that within this expression it is the water pressure at the bed that can change rapidly, while both the normal and shear stresses associated with the ice column cannot. This suggests that a way to document the sliding or basal motion of a glacier is by measuring the temporal variations in the speed of a monument on the surface of the glacier.

The details of the relation between water pressure and sliding rate are the target of modern glacial research. Water pressures are very difficult to measure in the field, as they require drilling holes in the glacier

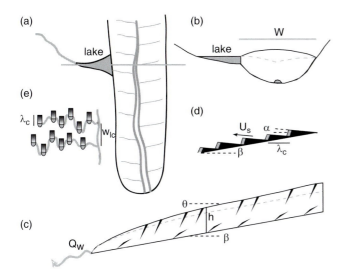

Figure 8.17 Sketch of the hydrological system in a glacier. (a) Map view showing subglacial tunnel system and tributaries to it that, in turn, connect sets of cavities in the lee of bumps (shown in plan view in E and in cross section in D). (b) Cross-valley profile through the glacier at the location of a side-glacier lake ponded by the ice, showing subglacial tunnel and a hypothetical water table (dashed). (c) Long-valley cross section of the glacier showing crevasses and the water table, with water discharging in the exit stream (after Kessler and Anderson, 2004, Figure 1, with permission from the American Geophysical Union).

that connect to the water system at the bed. When this is done, the pressures vary considerably both in time and in space. We do not know at present how best to average these pressures, nor the length scale over which such a measurement must be made in order to be relevant to sliding. The bottom line, however, is that the glacial hydrologic system evolves anew every year (see review in Fountain and Walder, 1998). This system is complex, and consists of several interacting elements shown schematically in Figure 8.17. Water is generated at the glacier surface by melt. This percolates into the snow and/or runs off on the ice surface to find a conduit that takes it into the subsurface. This often consists of a moulin (a vertical hollow shaft) or the base of a crevasse. At the bed, the hydrologic system consists of three elements: a thin film of water at the ice–bedrock interface we have already talked about, a set of cavities in the lee of bumps in the bed, and conduits or tunnels. These two larger-scale elements close down significantly in the winter when the melt water input is turned off: most of the water drains out of the system, leaving cavities and tunnels as voids that collapse by viscous closure of the ice.

These structures, the pipes and little distributed reservoirs of the subglacial system, must therefore be born anew each melt season. This is what makes the glacial hydrologic system so interesting, and it is intimately related to the seasonal cycle of sliding that is now well documented. As the melt season begins, water that makes its way toward the bed does not have an efficient set of conduits through which to drain. It therefore backs up in the glacier, raising the water table and therefore pressurizing the subglacial system as the column of water piles up. This allows sliding to begin, which in turn opens up cavities in the lees of bumps. Nearer the terminus, a tunnel system begins to grow, forced open by high rates of melt as water flows through the tunnel under a high water pressure gradient toward the terminus. As water flows through this nascent system, it widens due to frictional dissipation of heat, out-competing the tendency of the ice to move toward the conduit. The lower pressure conduit is therefore inserted into the glacier from the terminus up-glacier. As it reaches a particular location, it serves as a low-pressure boundary condition for the adjacent cavities, and can serve to bleed the water out of cavities, which in turns lowers the water pressure in the local glacier. As the conduit system elongates, it therefore bleeds off the pressures that were sustaining the sliding of the glacier, and terminates the sliding event. In the meanwhile, the conduit system grows until it can accommodate the rate of water input from the glacier surface. This basic explanation of "spring sliding events" has been modeled as well, and shows the up-glacier evolution of the system (see Kessler and Anderson, 2004).

Documentation of such dynamics requires high-frequency measurements of glacier position. This was accomplished first with computer-controlled laser distance ranging systems that automatically ranged to targets on the ice, for example at Storglacieren, Sweden. Recent work on small to medium alpine glaciers has begun to utilize GPS measurements to document the detailed surface motion history of a glacier through a melt season. One example from the Bench Glacier in Alaska (e.g., R. Anderson et al., 2004c; MacGregor et al., 2005) is shown in Figure 8.18. While the uplift of glaciers during these speedup events has led to models of enhanced sliding over up-glacier tilted blocks in the bed for some time (e.g., Iken and Truffer (1997)), these new measurements in concert with records of stream discharge

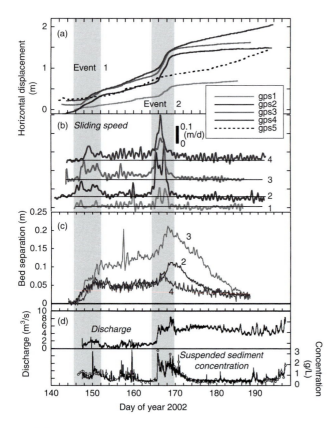

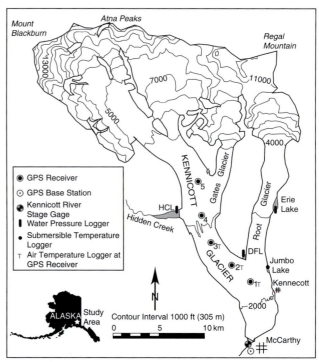

Figure 8.18 The record of surface motion of five GPS monuments on the surface of 100–180 m-thick Bench Glacier, Alaska (a), their speeds (b), the uplift of the surface not attributable to surface-parallel motion (c), and both the discharge and the suspended sediment concentration of the exit stream (d). Acceleration associated with increases in sliding occurred in two events separated by two weeks of roughly steady sliding. The termination of the second event coincides with a major increase in stream discharge, interpreted to reflect completion of the subglacial conduit that bleeds high pressures from the glacier bed. Uplift of the surface reflects block sliding up stoss slopes of bumps in the bed, and collapse of the resulting cavities after termination of sliding (after R. S. Anderson *et al.*, 2004c, Figure 7, with permission from the American Geophysical Union).

Figure 8.19 Map of Kennicott Glacier, Alaska, with instrumentation deployed in summer 2006. Outburst floods from Hidden Creek Lake (HCL) serve as a probe of the relationship between the hydrologic system of the glacier and its sliding. History of displacements of GPS monuments on the glacier surface allow the separation of steady flow and non-steady basal motion. Pressure gages at HCL and at Donoho Falls lake (DFL), and river gaging at the exit river in McCarthy provide constraints on how the hydrologic system behaves (after Bartholomaus *et al.*, 2007, Figure 1, with permission from Nature Publishing Group).

have generated a coherent model of alpine glacier sliding mechanics. The picture of basal motion that has emerged from study of the Kennicott glacier in Alaska, shown in the map of Figure 8.19, is one in which the glacier slides whenever the water inputs to the glacier exceed the capacity of the plumbing system of the glacier to pass that water (Bartholomaus *et al.*, 2007). This is shown in the melt-season record in Figure 8.20, and is broken out into several timescales in Figure 8.21. The water therefore accumulates in the glacier, which must result in pressurization of the basal water system.

The Kennicott Glacier serves as a particularly good natural experiment in that a side-glacier lake called Hidden Creek Lake visible in Figure 8.19 outbursts each year, and has done so for at least the last century. This slug of water passes through a subglacial tunnel to the terminus, generating a flood that for decades washed out the railroad bridge across which copper ore from the Kennicott mines was taken to market. Passage of this water through the tunnel greatly perturbs the subglacial water system, promoting a sliding event documented in both Figures 8.20 and 8.21 that exceeds background speeds by a factor of six. The detailed trajectory of the GPS monument on the ice surface also holds clues for what must be happening at the base of the glacier. In both the Bench Glacier and the Kennicott Glaciers, rapid sliding coincides with departure of the trajectory of

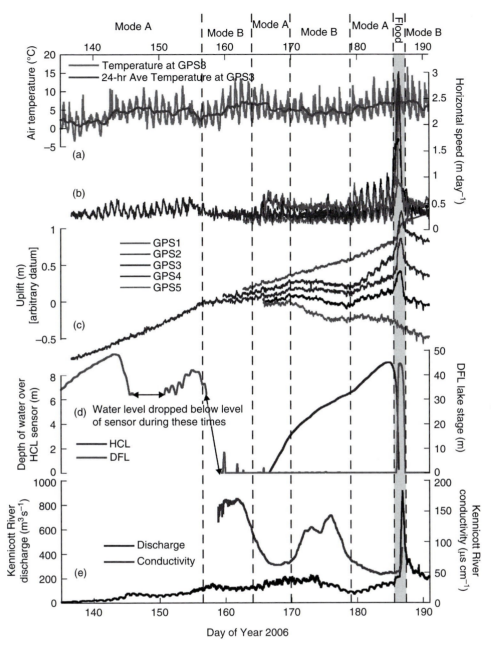

Figure 8.20 Melt season 2006 record of glacier motion and associated meteorological and hydrological histories. Vertical dashed lines identify two distinct glacier modes, A and B. Period of HCL outburst flood is lightly shaded. (a) Half-hour and 24-hr averaged air temperatures at GPS3. (b) 4-hr averaged horizontal ice speeds at each GPS receiver. (c) Uplift (vertical motion minus the surface-parallel trajectory) at each GPS receiver. (d) Lake level record at Hidden Creek Lake (HCL) and Donoho Falls Lake (DFL). DFL stage is relative to the lake basin floor, while HCL record captures the uppermost 9 m (of ~100 m) of lake filling and draining. (e) Kennicott River discharge and electrical conductivity (after Bartholomaus et al., 2007, Figure 2, with permission from Nature Publishing Group).

the ice surface from bed-parallel: during rapid sliding the ice appears to rise above this trajectory, while during slow-down of the ice after such events the ice surface appears to subside, eventually returning to bed-parallel motion. This should not be viewed as the insertion of a slab of water at the bed, but instead as sliding of the glacier up stoss sides of bumps in the bed. In the aftermath of the rapid sliding, the drop in water pressure that promoted the slow-down also

allows the cavities in the lees of the bumps to collapse, which in turn causes the subsidence of the ice surface. Interestingly, by contrasting Figures 8.18 and 8.20, the collapse appears to be roughly eightfold faster (about 1 day) beneath the thick (400 m) ice of the Kennicott than beneath the thin (180 m) ice of the Bench Glacier (about 1 week). We argue that this difference is what one would expect from the difference in the stresses causing collapse: the nonlinear flow law of ice should

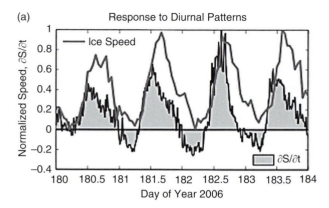

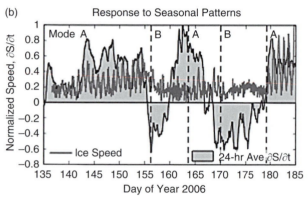

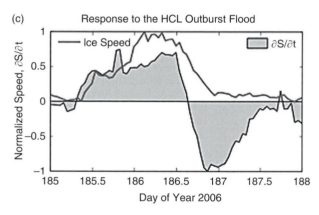

Figure 8.21 Relations between sliding and water budget of the Kennicott Glacier on three timescales: (a) diurnal, (b) seasonal, (c) the outburst flood from Hidden Creek Lake. In all cases, the motion of the glacier associated with basal motion (normalized by its maximum value in the period depicted) occurs whenever more water is being supplied to the glacier than its plumbing system can accommodate. This occurs when the rate of change of storage within the glacier, dS/dt, is positive (also depicted as normalized within the specified period) (after Bartholomaus *et al.*, 2007, Figure 3, with permission from Nature Publishing Group).

allow collapse of voids to be 2^n faster for stresses that are twofold higher. If $n = 3$, as in Glen's flow law, this should result in an eightfold difference in the rate of collapse.

Basal motion by till deformation

In some and perhaps many glaciers, the glacial ice is separated from the bedrock floor of the valley by a layer of till. In these cases, the basal motion is accomplished not by sliding of the glacier over bedrock bumps, but by deformation of water-saturated till. Till is glacially produced sediment with a wide distribution of grain sizes. The rate of deformation of this granular mixture is dictated by whether the water in the till is frozen or not, and by the pressures of the water if it is liquid. In Black Rapids Glacier in Alaska, which occupies a strike valley localized by the Denali Fault, concurrent observations of ice motion, and of till deformation at the bed using measurements in a borehole, reveal seasonal variations in deformation rate. These too are no doubt associated with the water pressures at the base of the glacier.

Boreholes through the great ice streams that drain the Antarctic Ice Sheet toward the Ross Sea reveal that they ride on layers of till. These streams have been shown to turn on and off at century timescales in an ice stream cycle. When the ice is thick and slow the thermal structure through the ice is such that the temperatures in the till can reach the freezing temperature, allowing it to deform. Given the high stresses at those times, the till deforms rapidly which in turn drains ice from the path of the stream, thinning it. The thinning brings the very cold surface temperatures closer to the bed, allowing heat to be lost more rapidly, which in turn eventually freezes the water in the till. Once the speed slows, the ice stream re-loads, thickening over the next centuries until the cycle is repeated.

Applications of glaciology

Glacier simulations

Simulations of alpine glaciers are now being performed in two plan view dimensions (e.g., Kessler *et al.*, 2006, on Kings Canyon; Plummer and Phillips, 2003, on Bishop Creek glaciers, California). This allows modeling of glaciers on real-world topographies represented by DEMs, on which one may explore the proper characterization of the climate required to generate glaciers that extend to LGM positions documented by end-moraines, the effects of

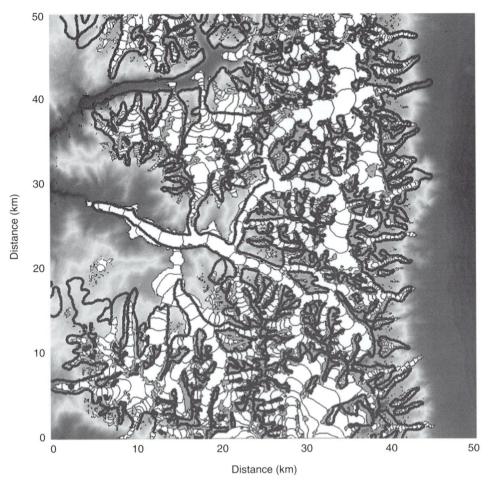

Figure 8.22 Steady-state glacier extents for experiment 4 from Kessler *et al.* (2006), in which the mass balance was determined using an orographic model and a positive degree day melt model. Bold line indicates mapped LGM glacial extents extracted from Moore (2000) for this 50 × 50 km area of the southern Sierras centered on Kings Canyon (after Kessler *et al.*, 2006, Figure 13, with permission from the American Geophysical Union).

radiation input to the glacier surface, and so on. One stringent test of such a coupled model of climate and glacier response is in the southern Sierras, where the tilted mountain block results in a strong difference in the precipitation across the range. The mapped moraines reveal a strong asymmetry in glacier length, with much longer glaciers on the western windward side of the range than on the steep, faulted eastern range front. The two-dimensional glacier simulation shown in Figure 8.22 can reproduce this LGM moraine pattern only if a strong gradient in precipitation is employed. The resulting difference in the estimated ELA, plotted in Figure 8.23, is about 150 m lower on the west (windward) side of the range than on the east.

Paleo-climate estimates from glacial valleys

Now that we know something about how glaciers work, let us use this knowledge to address questions raised by past glaciers. One of the goals of glacial geology is to estimate the past footprints of glaciers, and from them determine the past climate needed to produce glaciers of that size. It is often assumed that the mean annual 0 °C isotherm roughly coincides with the ELA. Above it, temperatures are too cold to melt all the snow that arrives, and vice versa below it. If one assumes that the temperature structure of the atmosphere obeys a lapse rate of say 6.5 °C/km or 0.0065 °C/m, then the depression of the ELA in meters can be translated into a cooling of the climate in °C. But how do we find the paleo-ELA? We use two methods illustrated in Figure 8.24. First, we note that the up-valley ends of lateral moraines correspond roughly with the ELA. If these can be located in a particular valley, then one can estimate the paleo-ELA from the up-valley elevation of their ends. We will discuss these moraines further at the end of the chapter. The second method utilizes an empirical relationship of the map view of

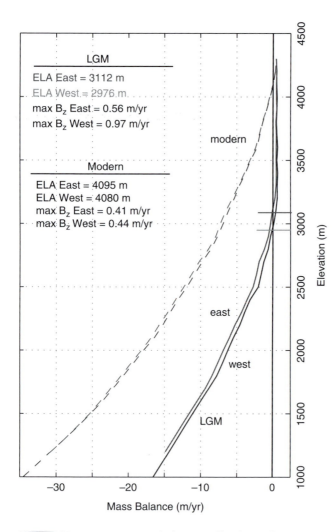

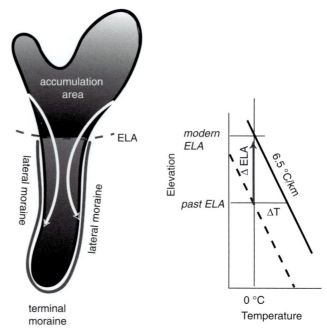

Figure 8.24 Methods for estimating paleo-ELAs are based upon features of glaciers illustrated. Up-valley ends of lateral moraines coincide with the ELA on a glacier because the trajectories of blocks falling onto the ice are as shown in white arrows: they move toward the glacier center in the accumulation area, and outward to the edge in the ablation zone, responding to the local slopes of the glacier. Typical AARs, ratios of accumulation area to total area of a glacier, are 0.65. If the paleo-area of a past glacier can be measured using terminal moraines, the ELA can be estimated using this AAR. Translation of the change in ELA into an estimate of change in mean annual temperature is based upon application of an assumed lapse rate shown in the figure at the right.

Figure 8.23 Average net mass balance profiles for modern and LGM (–7 °C) climates (dashed and solid lines, respectively). Profiles are derived from an orographic precipitation model and a positive degree day melt model. Mass balance profiles on the eastern and western flanks are shown. Note 140 m offset of ELAs at LGM time required to match the moraines (after Kessler et al., 2006, Figure 12, with permission from the American Geophysical Union).

modern glaciers: the accumulation area is roughly 65% of the total area of the glacier. The accumulation area ratio (AAR = accumulation area/total area) is therefore 0.65. If this holds true for glaciers in the past, and one can map the footprint of the past glacier using its terminal moraine, then one can determine at what elevation a contour would enclose an accumulation area of 65% of this. This method is more commonly used than is that based upon lateral moraines because terminal moraines are more likely to remain visible than are the upper ends of lateral moraines. We have already seen in Figure 8.2 how this has been used to estimate

ELA lowering during the LGM. In many ranges such estimates suggest a lowering of many hundred meters, from which climatic cooling is estimated to have been 6–10 °C. While this is valuable information, we note several weaknesses in the method: AARs vary from 0.6 to 0.8 on modern glaciers, past lapse rates are not necessarily equivalent to those of today, and glacial health is dictated not solely by temperatures, but by rates and patterns of snowfall that can vary with climate as well.

Ice sheet profiles

Ice sheets are always steepest at their outer margins, and decline in surface slope toward their centers. Why is this? The argument goes as follows. An interesting manifestation of the nonlinearity of ice rheology captured in Glen's flow law is that one may treat ice to

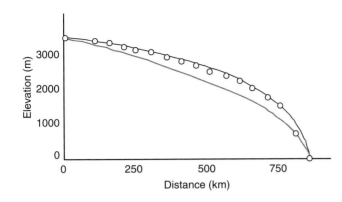

Figure 8.25 Profile of the Antarctic Ice Sheet from Mirny (circles), along with theoretical profiles: parabola (gray line) and curve that incorporates a uniform accumulation rate (black line) (after Paterson, 1994, Figure 11.4, reproduced with permission from Elsevier).

first order as a plastic substance. A plastic material is one in which there is no deformation up to some "yield" stress, beyond which there is an infinite ability to flow. It has no strength beyond this limit. On a rheological plot, this looks like the broken line in Figure 8.10. As you can see from this plot, a non-linear substance with a power $\gg 1$ approaches the behavior of the plastic substance. If a glacier were plastic, there must be no place within it that falls above this failure envelope.

Now consider the ice sheet profile sketched in Figure 8.25. At some point on the bed, the shear stress is $\rho g H \sin(\theta)$. If the substance is plastic, and this shear stress is above the yield strength, then infinite strain rates would occur and the ice at the bed would rapidly move, thinning the ice there. This thinning would continue until the shear stress diminished below the yield strength. The argument is therefore that the ice sheet maintains a shape at which the shear stress at the base of the sheet is exactly the yield strength for ice. All we need to know is what this value is, and the handy number to carry in your head is: 1 bar, or, in SI units, 10^5 Pa. Conceptually, then, given that the shear stress involves the product of surface slope and ice thickness, near the edge of the ice sheet the slope must be high in order to reach the yield stress, and as the ice thickens toward the center the surface slope must decline. We can cast this a little more mathematically. Let's consider an ice sheet on a flat continental surface. All we need is to set the shear stress to be a constant:

$$\rho_i g H \sin \theta = C \tag{8.25}$$

Given that the surface slope can be expressed as dH/dx, where H is the thickness of the ice, we can separate the variable H, yielding the ordinary differential equation

$$H dH = \frac{C}{\rho_i g} dx \tag{8.26}$$

This can be solved by integrating both sides:

$$H^2 = \frac{2C}{\rho_i g} x$$

or

$$H = \sqrt{\frac{2C}{\rho_i g} x} \tag{8.27}$$

The shape of the profile should be a very specific one, obeying a square root dependence on distance from the terminus. There are several things we can do with this relationship. If we have an ice sheet to measure, given that we know the density of ice, the acceleration due to gravity, and can measure both H and x, we can deduce the best-fitting value of C, the yield strength of ice. We show in Figure 8.25 the shape of the present day ice sheet in Antarctica. The simple model we have just described is shown as one of the lines on the plot. While the fit is not great, only slight modification of the theory is required to accommodate the data very well.

Another use of this simple analysis is reconstruction of past ice sheet thickness profiles. For this reconstruction problem, all we know is the outline of the ice sheet, derived from say the map view pattern of terminal moraines. Using the yield strength derived from present day ice sheets in the above exercise, we can reconstruct how thick the ice must have been as a function of distance from the margin.

Note that we have made many simplifying assumptions in the above analysis, including the characterization of the rheology as a simple plastic. More detailed reconstructions of ice sheets take into account better representations of the nonlinear (Glen's flow law) rheology, and must also handle the thermal problem that dictates where an ice sheet is frozen to its bed (behaving as a Type I polar glacier), and where it is Type II polar or temperate, and therefore can slide.

Surging glaciers and the stability of ice sheets

We have focused so far on glaciers that will likely look the same next year as they do this year. Exceptions to this rule are surging glaciers, whose surface speeds can increase by orders of magnitude during a surge phase. Surges of valley glaciers may last one or two years, and be separated by decades to centuries. Surging glaciers generate beautifully looped medial moraines, making the glacier tongue look more like a marbled cake than a glacier. A famous example from the Susitna Glacier in Alaska is shown in Figure 8.26. A surge also leaves the glacier with a chaotic pattern of crevasses that should be avoided at all costs as a climbing route. (You would be hard-pressed to climb the glacier photographed in Figure 8.29 in its state at that time.) And on a larger scale, ice streams in Antarctica bear some resemblance to surging glaciers surrounded by non-surging ice. Understanding surge behavior, therefore, has been a major focus within the glacial community.

Our understanding of surges comes largely from the intense and long-lasting study of the 25 km-long Variegated Glacier near Yakutat, in coastal southeast Alaska, mapped in Figure 8.27. This glacier was apparently in surge when photographed in 1905, again in the 1920s, then in the 1940s, and was well documented by a pair of air photos in 1964 and 1965. The apparently approximately 20-year intervals between surges suggested that the next surge might happen in the mid-1980s, and a study was initiated in the late 1970s with the intent of characterizing a glacier as it approached its surge, and then to document well the behavior during the surge. Several teams from at least four institutions poked and prodded and photographed and surveyed the glacier as it approached its surge, which culminated in a two-phase period of rapid motion in 1982 and 1983.

The surge itself can be characterized as a wave of rapid motion that translates through the glacier. In front of the wave, the glacier might be moving at 0.1–0.3 m/day, while in the middle of the high-velocity region it might be moving at 60 m/day, as shown in Figure 8.28. This wave of high velocity translates at yet higher velocities (several times 60 m/day). The crevasse patterns result from the pattern of velocities. At the leading edge of the wave, the ice is in extreme compression, to which it responds by thickening, by thrust faulting, and by folding, in a dramatic analog

Figure 8.26 Air photograph by Austin Post of Susitna Glacier, Alaska, taken on September 3, 1970. Medial moraines are highly contorted due to repeated surging of a tributary glacier in the headwaters (Image susitna1970090301, held at National Snow and Ice Data Center/World Data Center for Glaciology, Boulder).

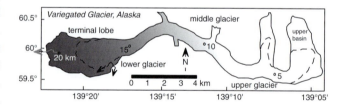

Figure 8.27 Map of the Variegated Glacier, Alaska, showing the up-glacier and down-glacier limits of the area involved in the 1982–1983 surge (dashed lines), a few of the stake locations spaced at 1 km intervals, and the locations of the major outlet streams near the terminus (after Kamb *et al.*, 1985, with permission from the American Association for the Advancement of Science).

to the structural geologic evolution one might expect in convergent tectonics. It thickens so much more in the center of the ice than at the edges that it actually fails in tension, generating longitudinal crevasses. On the trailing limb of the high-velocity wave, velocities again drop to lower values, and the ice is put into tension in the longitudinal direction, which in turn generates transverse crevasses. The two sets of crevasses chop up the glacier surface into an amazing chaos of ice pillars that are the signature of the

(a)

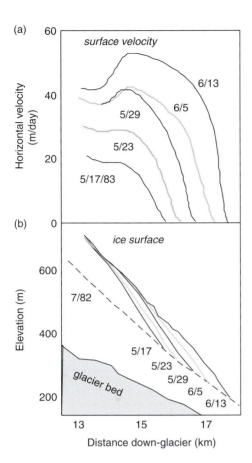

(b)

Figure 8.28 (a) Ice surface velocity profile, $u(x)$, and (b) ice surface topography, $z(x)$, in a 3 km reach of the lower part of the Variegated Glacier, during the 1983 surge (after Kamb *et al.*, 1985, Figure 4, with permission from the American Association for the Advancement of Science).

Figure 8.29 Aerial view looking up-valley of Variegated Glacier near the termination of its 1982–1983 surge. Both longitudinal crevasses, formed early in the surge, and transverse crevasses, formed late, serve to dissect the glacier surface into square pillars of ice, some of them many tens of meters tall (photograph by R. S. Anderson).

passage of a surge that is obvious in Figure 8.29. Interestingly, the abrupt halt of the surge on July 4, 1983 was coincident with a huge flood of very turbid water reported in Figure 8.30 (see Humphrey *et al.*, 1986).

The data taken together lead to this simplified schematic of a surge. In the aftermath of a surge, the glacier is left with a lower than average surface slope throughout, and has been significantly thinned. It therefore moves slowly, as both slope and thickness dictate lower internal deformation and sliding speeds. The net balance in the following years then rebuilds the glacier accumulation area, and thins the ablation area, as there is smaller than average transfer of ice from accumulation to ablation areas. So the glacier thickens and steepens, as shown in Figure 8.31. Accordingly, you can see in Figure 8.32 that each

year its maximum sliding speed rises. Finally, the sliding speed increases sufficiently to disrupt the subglacial fluvial network, and the water that would normally find its way out the glacial conduit system is trapped beneath the glacier. High water pressures result, with attendant positive feedback on the sliding speed. This runaway process is the surge. The surge terminates when, for whatever reason, the water finds a way out, usually resulting in a catastrophic flood.

Tidewater glaciers

When a glacier extends its terminus down to the ocean, we call it a tidewater glacier. Some of the

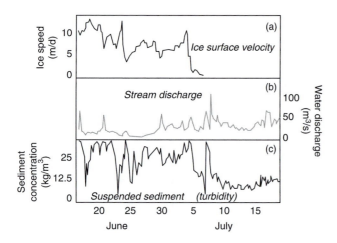

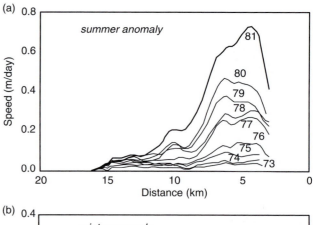

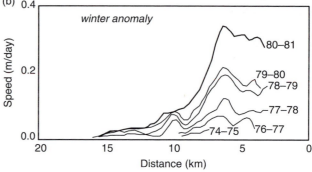

Figure 8.30 (a) Ice surface velocity at the 9.5 km point on Variegated Glacier, Alaska (see Kamb *et al*., 1985), (b) water discharge, and (c) sediment concentration from turbidity measurements in the weeks surrounding the abrupt termination of the 1983 surge (after Humphrey and Raymond, 1994, Figure 6, with permission of the authors and the International Glaciological Society).

Figure 8.32 Summer (a) and winter (b) velocity anomalies on Variegated Glacier centerline in the decade preceding the 1982–1983 surge. Note different scale for the summer vs. winter anomalies, attesting to the enhancement of sliding in the summer melt season. The anomaly grows by at least an order of magnitude over the decade (after Raymond and Harrison, 1988, Figure 5, with permission of the authors and the International Glaciological Society).

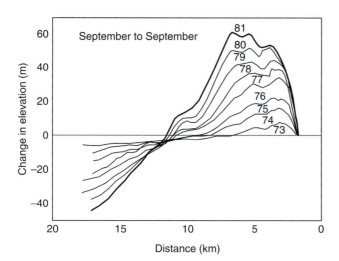

Figure 8.31 Evolution of the elevation anomaly in the decade leading up to the 1982–1983 surge of Variegated Glacier, Alaska. The glacier thickened more than 60 m in the accumulation area, while thinning by more than 50 m in the ablation area (after Raymond and Harrison, 1988, Figure 4, with permission of the authors and the International Glaciological Society).

largest and fastest glaciers are tidewater glaciers. While the snout of a tidewater glacier may be in the water, the glacier is still grounded, still in contact with a solid substrate. The ice tongue is not floating. In Antarctica now and not long ago in the Arctic, ice shelves also exist, which are indeed floating beyond a point called the grounding line. Tidewater glaciers differ from typical alpine glaciers in several ways, all due to their interaction with the ocean. Most importantly, they can lose mass through a mechanism other than melting – they can lose mass by calving of icebergs, which can be seen cluttering the terminus area in Figure 8.33. This can be very efficient. On Alaska's Pacific coastline the termini of these tidewater glaciers are the targets of tourist ships, as the terminal cliff is the scene of dramatic iceberg calving events. But this is not all. Because the ice extends into a water body, the lowest the water table within the glacier can get is sea level. We have already seen the importance of the state of the hydrologic system within the glacier on its sliding rate. In particular, as we have seen, our simplest working model for sliding speed involves the effective pressure at the base of the glacier in the denominator of the expression (Equation 8.24). As

Figure 8.33 Aerial photograph of tidewater glaciers calving into a fjord near NyAlesund, Svalbard archipelago. Note prominent plume of sediment-laden water exiting from beneath the uppermost portion of the foreground glacier. The medial moraines separate ice emanating from various tributary valleys in the headwaters. Ice cliff at the sea is roughly 100 m tall (photograph by Suzanne Anderson).

the flotation condition is approached, or as water pressure increases, sliding speeds ought to increase dramatically. As the glacier extends into deepening water, the effective pressure at the bed must dramatically plummet. If the sliding speeds obey our simple rule, then the sliding speed ought to increase dramatically as well. This pattern of sliding is one of the factors that leads to the extensive fracturing of tidewater glacier snouts. In effect, the glacier ice therefore arrives at the terminus already prepared for calving. It is already riddled with fractures.

Calving

Given its importance in the operation of these massive glaciers, we know surprisingly little about the mechanics of calving. This is a complicated process involving the propagation of fractures within the ice, either from the bed upward or the surface downward. Interactions with the body of water include melt-notching of the terminal cliff by seawater, buoyancy of the seawater, and tidal swings in sea level. Where calving rates have been measured, they have been shown to be related closely to water depth; the greater the water depth the greater the rate of calving. In Figure 8.34 we show data collected by Brown *et al.* (1982), which reveal a nearly linear relationship of calving rate to water depth, D:

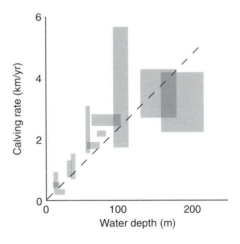

Figure 8.34 Dependence of calving rate on water depth (after Brown *et al.*, 1982).

$$U_{\text{calve}} = aD^p \tag{8.28}$$

From the plot one may deduce that $p = 0.92$, which is sufficiently close to 1 to allow simplification to $U_{\text{calve}} = 0.02D$.

Tidewater glacier cycle

The implications of this dependence of calving on water depth were recognized in the 1960s by Meier and Post, who proposed what has been called the tidewater glacier cycle. If a tidewater glacier is to extend across deepening water, it must effectively bring along some protection against this calving loss. The proposed mechanism involves a morainal shield. These are big glaciers, and they deliver large amounts of sediment to the sea, presumably both as till at the base of the ice and as sediment delivered to the terminus in the subglacial drainage system. One can see well the muddy plume of water marking the position of one such exit stream in the photo in Figure 8.33. As well, one can imagine sediments at the bed of the fjord being distorted or pushed by the advancing front of ice. All of these sediments get smeared along the glacier terminus by marine processes (Powell and Molnia, 1989). The resulting shoal acts to guard the glacier terminus against the more efficient calving that would otherwise attack the ice front. This allows the glacier tongue to propagate into deeper water by bringing with it the protective shoal. The rate of advance is slow, and is presumably dictated by both the rate of delivery of ice to the front and the rate at which the shoal can be built into deeper water. Notice the vulnerability as the ice extends, however. If for some reason the snout retreats off its shoal, the terminus will experience increased water depth, and the calving rate should increase. Once started, this should result in a retreat that will last until the glacier is once again in shallowing water. Meier and Post suggested that tidewater glaciers undergo periodic cycles of slow advance and drastic retreat in this sort of bathymetric setting. These cycles can have little to do with the mass balance of a glacier. This is an instance in which the retreat of a major tidewater glacier could be completely decoupled from climate. One tidewater glacier may be advancing slowly while another in an adjacent fjord may be in dramatic retreat. The behavior is self-organized.

In coastal Alaska, these tidewater glaciers have retreated up their fjords, exposing brand new landscapes within the last century. When John Muir visited what is now Glacier Bay National Park in the late 1800s, the bay was fully occupied by a large glacier. Now one must paddle a kayak tens of kilometers up the fjord to find the remnants of this glacier, each tributary of which has pulled its toe out of the water.

The most recent example of this retreat is presently occurring on Columbia Glacier, just north of Valdez.

The retreat has been dramatic, as captured in repeat photographs summarized in Figure 8.35. This glacier was of special interest due to its proximity to the terminus of the Alaska pipeline. As icebergs calved from the Columbia could potentially enter the shipping lanes traversed by oil tankers, the behavior of the calving front of the Columbia has been closely watched. Indeed, the retreat predicted in the 1980s is now in full swing. Luckily, the terminal moraine serves as a very effective barrier to the transport of icebergs. They must melt, grind, and smash each other into smaller bits before they can escape the moraine sill, and are therefore small enough not to pose a significant risk before they are cast into the open ocean. We expect the retreat to last into next decade, opening a new fjord: Columbia Bay. The story of the retreat, and of the scientists who have studied this glacier for now a century, serves as an introduction to glaciology as a science. It is lovingly told and beautifully illustrated by Tad Pfeffer in his *The Opening of a New Landscape: Columbia Glacier at Mid-Retreat* (2007).

Contribution to sea level change

Sea level is now rising at a rate of about 3 mm/yr. The rise reflects two phenomena: (1) new water added to the oceans, much of which is coming from glaciers, and (2) warming and expansion of the oceans (see Nerem *et al.* (2006)). The prospect for the coming century is not good, as both of these rates will increase. As is summarized in Figure 8.36, much of the contribution of new water comes from glaciers and ice caps, despite their being dwarfed in total ice volume and area by the big ice sheets of Greenland and Antarctica.

That tidewater glaciers extend to the sea implies that they have large accumulation areas in high elevations, that the accumulation rates are high due to proximity to the storms coming off the ocean, or both. The low elevations of the ablation zones also lead to very high melt rates, rates of many meters per year. In a recent study of glacier change in Alaska (Arendt *et al.*, 2002), comparison of the glacier centerline profiles collected using a laser altimeter mounted on a small plane with those derived from 1950s USGS maps showed that the glaciers in Alaska alone could account for roughly half of the sea level rise attributable to glaciers worldwide.

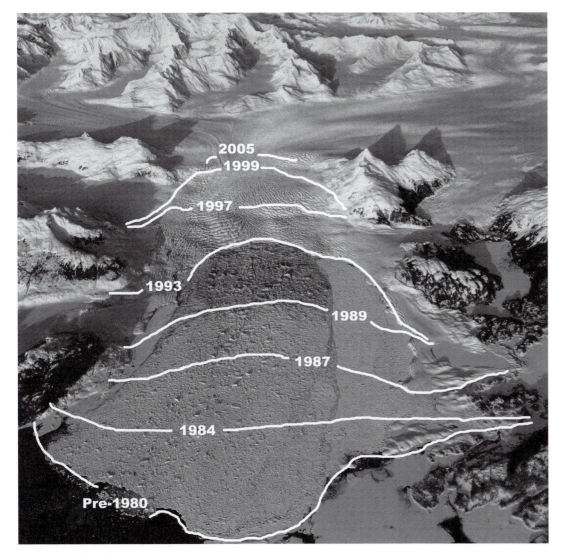

Figure 8.35 Columbia Glacier in retreat, view looking roughly east toward headwaters in the Chugach Range, Alaska. The retreat between 1988 and 2005 is 13 km, averaging roughly 680 m/yr (image courtesy of R. M. Krimmel and W. T. Pfeffer, INSTAAR/cu).

The great ice sheets: Antarctica and Greenland

Most of the mass of ice in the present cryosphere is contained in the two great ice sheets that cover Greenland and Antarctica. It is from these ice sheets that the longest terrestrial records of climate change have been extracted from ice cores now reaching back almost a million years in Antarctica. Antarctica has had major ice cover since roughly 30 Ma. The Antarctic Ice Sheet presently holds the equivalent of 70 m of sea level, while that in Greenland represents a potential sea level rise of 7 m.

These are cold places, with mean annual surface temperatures that fall as low as −50 °C. In Antarctica, there is very little melt of ice on the surface. The ice is polar, meaning that for the majority of the volume of each ice sheet the temperature is well below the freezing point. As we have discussed above, this means that the transport of ice is accomplished solely by internal deformation. Basal sliding is limited to a few special places; it is on these places that we focus because it is here that ice can be pulled rapidly from the continents to contribute to sea level rise in the short term – on human timescales.

As you can see from the image in Figure 8.37, Antarctica is subdivided into two ice masses, the west

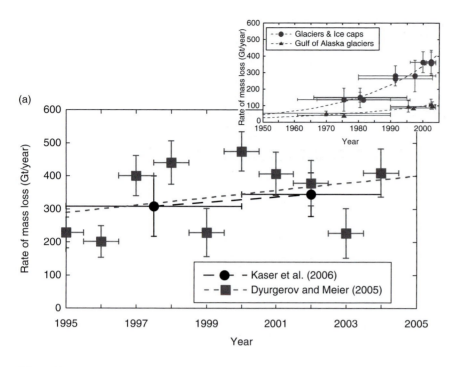

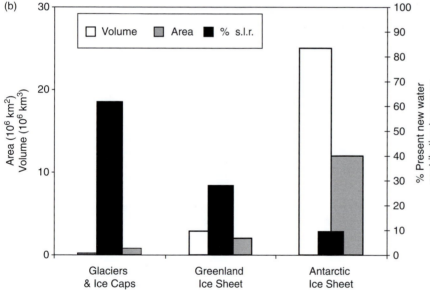

Figure 8.36 (a) Rate of loss of ice mass from small glaciers worldwide over the last decade (1995–2005), and since 1950 (inset). Also shown in inset is the loss attributable to glaciers draining into the Gulf of Alaska. (b) Comparison of present area (gray) and volume (white) of glaciers and ice sheets. The relative contributions to sea level rise (s.l.r., black) reveal that while the area and volume of ice recently stored in glaciers and ice caps is small, they contribute 60% of the new water to the world's oceans. Projections over the next century suggest that this will continue (after Meier *et al.*, 2007, Figures 1 and 2, with permission of the American Association for the Advancement of Science).

and east ice sheets. The east is wider, thicker, and contains the majority of the ice. The West Antarctic Ice Sheet (WAIS), however, is the target of considerable research because it is viewed to be more vulnerable to rapid ice loss. This comes from the fact that it is grounded well below sea level, information that we have gleaned by flying airborne radar surveys that can detect both internal layering in the ice and its bed. It is also warmer, with the Antarctic peninsula poking well north. It is from this peninsula that Ernest

Shackleton landed his party after abandoning their ship, *Endurance*, and from which he launched his journey across the raging southern ocean, ultimately to find help in South Georgia (read Alfred Lansing's *Endurance*, or Roland Huntsford's *Shackleton*).

Antarctica is bounded by a set of thick ice shelves also shown in Figure 8.37 that serve as intermediaries between the continent and the ocean. The boundary between the ice sheet and an adjacent ice shelf occurs where the ice lifts off the bed to become floating.

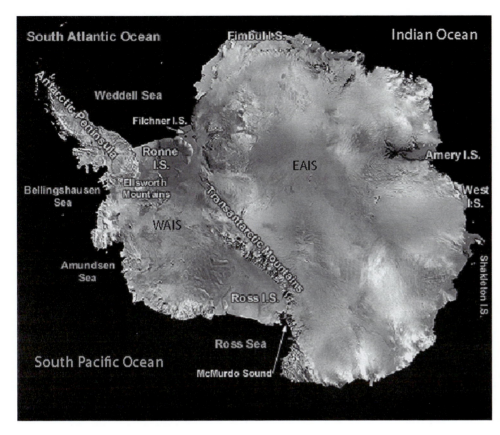

Figure 8.37 Map of Antarctica showing main topographic features, from AVHRR Mosaic color composite. EAIS and WAIS are the East and West Antarctic Ice Sheets, separated by the Transantarctic Mountains, a rift-flank uplift. I.S. = Ice Shelf (modified from USGS http:// TerraWeb.wr.usgs.gov/TRS/ projects/Antarctica/AVHRR.html).

Subglacially, this is an important interface, as it is here that sediment delivered from the continent is dumped in the ocean, generating grounding line fans. Subaerially, the location of the grounding line can be sensed using inSAR, as the ice shelf responds to tidal forcing while the grounded ice does not. Interest in these bounding ice shelves has increased recently as more than one of these shelves has experienced disintegration. The demise of a large portion of the Larsen B ice shelf has effectively unpinned the grounded glaciers that feed the ice shelf, leading to their acceleration (Hulbe *et al.*, 2008).

With very few exceptions, there is no melting of ice in Antarctica. In such cold places the ice budget must be closed by a different mechanism. Ice leaves the continent either by calving of icebergs, or by blowing of snow in the sustained high winds that drain the chilled air from the polar plateau (see discussion of katabatic winds in Chapter 5).

Greenland's ice sheet shown in Figure 8.38 also covers the majority of the landmass, and in its center is more than 2 miles (about 3 km) thick. Its center is also a smooth, windblown plateau. Those small portions of the landmass that are visible reveal highly dissected plateau-like edges, deeply indented by steep-walled fjords. In contrast to Antarctica, the outer edges of the Greenland Ice Sheet experience melting. While some of this snow and ice meltwater makes its way to the bed and out of the ice sheet, some refreezes in the subsurface and cannot be counted as part of the negative mass balance.

Most of the ice is lost from these ice sheets through a relatively few rapidly moving "ice streams". These can be tens of kilometers across, and move at rates of a large fraction of kilometers per year. These speeds are best documented from satellite using the same inSAR methods employed to assess tectonic deformation. A recent summary image of ice stream speeds is presented in Figure 8.39. The best known of the ice streams drain toward the Ross Ice Shelf from the West Antarctic Ice Sheet. First named A through E, some have now begun to take on the names of researchers who have long studied them. And the ice streams are odd. It has now been shown that they move very rapidly for a while, and then shut down to

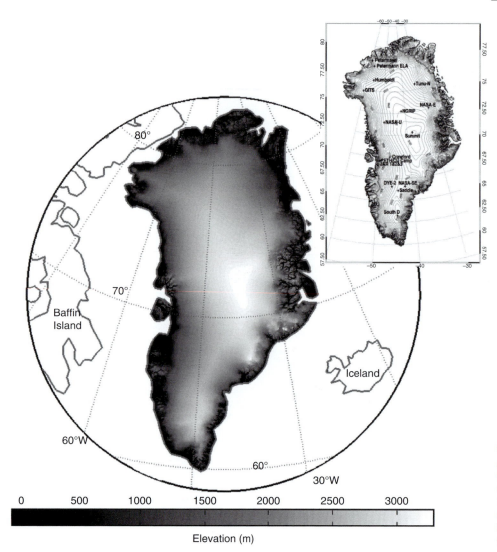

Figure 8.38 Satellite radar produced map of Greenland (from A. Bamber, courtesy Liam Colgan, after Fahnestock *et al.*, 1997, *Digital SAR Mosaic and Elevation Map of the Greenland Ice Sheet*. Boulder, CO: National Snow and Ice Data Center).

become dormant for decades to centuries before another burst of activity. Drilling to the base of ice stream B has shown that the base of the ice stream rides on a layer of till that is implicated in the rapid motion (e.g., Tulaczyk *et al.*, 2000a and 2000b; Bougamont *et al.*, 2003). Briefly, we think this works as follows. When the temperature at the base of the ice stream is near the melting point, the water-saturated till shears rapidly and the ice stream above it goes along for the ride. But this slowly drains the ice from upstream and the ice stream thins to the point where the cold surface temperatures can conduct to the base to refreeze the till. Now that the ice spigot has been turned off, ice can build back up the thickness that was lost. This slow thickening in turn decreases the temperature gradient through the ice, reducing the rate at which heat is lost to the atmosphere, and

the base begins to warm. The cycle repeats when the temperature rises sufficiently to melt the water in the till. At present, ice stream C is turned off, and the huge crevasse field that delineates the boundary of the stream when it is in motion is now buried in tens of meters of new snow.

In Greenland, ice leaves the ice sheet through major fjords that presently compete with Alaska's tidewater Columbia Glacier for the fastest ice on Earth. The recent increases in the speeds of these outlet glaciers, largely documented using inSAR methods and shown in Figure 8.40 (e.g., Rignot and Kanagaratnam, 2006; Howat *et al.*, 2008; Nick *et al.*, 2009; Joughin *et al.*, 2008), have drained a significant volume of ice from the outer fringe of the ice sheet. Just as in tidewater glaciers in Alaska, the acceleration has resulted in a thinning of the ice, which in turn results in further

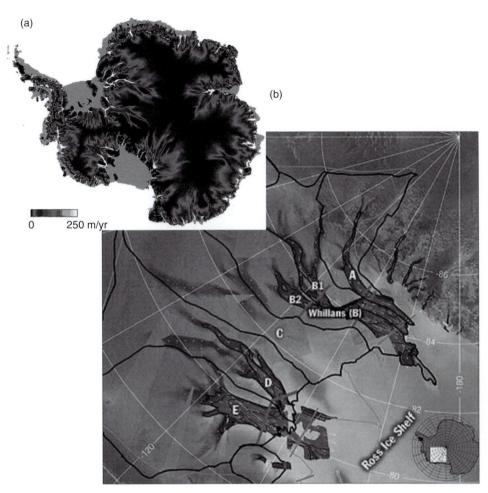

(a)

(b)

0   250 m/yr

Figure 8.39 Antarctic ice speeds documented using inSAR methods, (a) covering all of the continent, and (b) focused on the ice streams draining the West Antarctic Ice Sheet toward the Ross Ice Shelf (after Ian Joughin, image courtesy RADARSAT Antarctic Mapping Project).

acceleration. The termini of several of the largest outlet glaciers have receded many kilometers.

Glacial geology: erosional forms and processes

Now that we have some understanding of how glaciers work, let us turn our focus to how glaciers modify their physical environment. Glaciers both erode landscapes, generating characteristic glacial signatures (see N. Iverson, 1995, 2002), and they deposit these erosional products in distinctive landforms. We will describe the origins of these features, starting with erosional forms, and starting at the small scale. Again, the goal is not to be encyclopedic, but to address the physics and chemistry of the processes that gave rise to each of the forms.

Erosional processes

Any visit to a recently deglaciated landscape will reveal clues that it was once beneath a glacier. The bedrock is often smooth, sometimes so smooth that it glints in the sun as if it has been polished. The bedrock is also commonly scratched. The small bumps have been removed, leaving a landscape dominated by larger bumps and knobs that have characteristic asymmetry. And at the large scale we see U-shaped valley cross sections, hanging valleys, fjords, and strings of lakes dotting the landscape. At an even larger scale, fjords bite through the margins of most continents that have been subjected to repeated occupation by ice sheets. Each of these features cries out for an explanation, beyond "this valley was once occupied by a big moving chunk of ice."

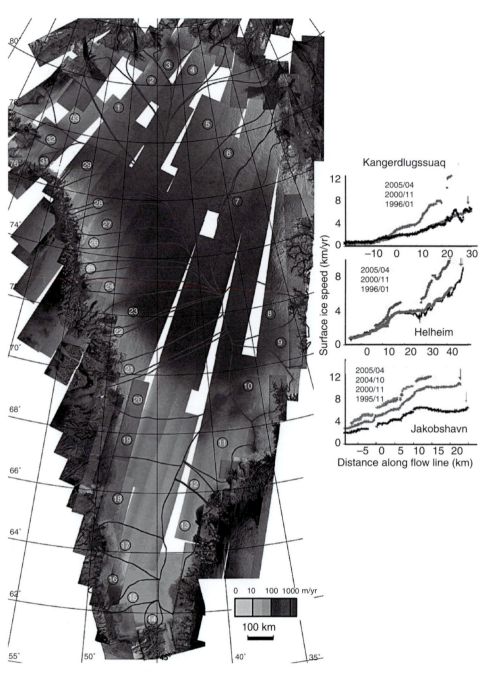

Figure 8.40 Surface ice speed mosaic of the Greenland Ice Sheet assembled from year 2000 Radarsat-1 radar data, color coded on a logarithmic scale from 1 m/yr (light) to 3 km/yr (dark), overlaid on a map of radar brightness from ERS-1/Radarsat-1/Envisat. Drainage boundaries are shown. Right: Ice speed profiles down flow lines, showing speed-up over the decade 1995–2005 (after Rignot and Kanagaratnam, 2006, Figures 1 and 2, with permission of the American Association for the Advancement of Science).

Abrasion

The scratches, called striations, are the products of individual rocks that were embedded in the sole of the glacial ice as it slid across the bedrock. The erosion resulting from the sum of all these scratches is the process called subglacial abrasion. It is abrasion that smoothes the bed. It is the short wavelength bumps in the bed that are gone, making it look smoother. This should trigger recollection of the regelation process by which glaciers move past small bumps. Consider again that bump, and think about a rock embedded in the ice approaching the bump. The bottom line is that while the ice can get around the bump with its regelation magic act, the rock cannot. Instead, it is forcefully pushed against the bump, as we illustrated in Figure 8.14, the force being great enough to cause the rock to indent, scratch or striate the stoss side of the bump. As the

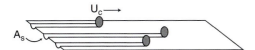

Figure 8.41 Sketch of glacial erosion by abrasion. Multiple striations of the bedrock by rocks embedded in the sole of the glacier, moving at the sliding speed U_{slide}, lead to lowering or erosion of the bed. The rate depends upon the number of striators per unit area of bed, the cross-sectional area of the eroded striation, A_s, and the rate of lengthening of the striation, U_c. The cross-sectional area of the groove is proportional to the drag force imposed by the ice as it regelates past the rock clast.

rock is still being dragged along by the ice, this indentation stretches out to gouge a micro-channel that is the striation.

The theory of glacial abrasion requires that we handle the effects of each such striator, and that we have some knowledge of the number of them in the basal ice. The theory was worked out by Bernard Hallet (1979), and is summarized in Figure 8.41. The abrasion rate, e, goes as the number of indentors per unit area of bed, C, the cross-sectional area of the indentation that they induce in the bedrock, A_s, and the rate at which the striators are dragged across the rock, U_c:

$$e = CA_sU_c \qquad (8.29)$$

We must determine what controls the cross-sectional area of the indentation. Experimental work shows that the depth of the indentation increases as the force imparted by the indentor increases. Hallet argued that one should consider the clast doing the work as being embedded in a fluid (ice) that is moving toward the bedrock at a rate comparable to the sliding rate, U_{slide}. If ice is disappearing at the stoss side of the bump, and the rock is not, then ice has to move past, around the rock. In this view, the force involved is that associated with the viscous drag of the ice as it moves around the rock to disappear at the stoss side of the bump. From the theory of fluid flow past obstacles, such as that employed in thinking about settling of clasts in a fluid, and recognizing that the flow is likely to be very low Reynolds number (viscous as opposed to turbulent) flow, the drag force can be written

$$F_d = \frac{1}{2}A\rho_f C_d U_{rel}^2 \qquad (8.30)$$

where C_d is the drag coefficient, A the cross-sectional area of the clast, and U_{rel} the relative velocity of the fluid and the rock. For low Reynolds numbers, Re $[= UD\rho/\mu]$, the drag coefficient is $24/Re$, or $24\mu/D\rho U_{rel}$. This leaves the expression for the drag force as

$$F_d = 3\pi D\mu U_{rel} \qquad (8.31)$$

But the relative velocity of the ice relative to the clast is some fraction of the sliding speed, U_{slide}. Combining these expressions reveals that the abrasion rate should go as the square of the sliding speed:

$$e = \gamma U_{slide}^2 \qquad (8.32)$$

We note for completeness the other view of the origin of the contact force. Why is the relevant force not the weight of the overlying column of ice? This overlying column of ice has to be pretty heavy, given that ice is about one-third as dense as rock. The problem with this view is that the ice is not a solid, but must be considered a fluid. The rocks at the base of the glacier are engulfed in the fluid, so they not only do not support the weight of a column of ice overhead, but instead are buoyed up by the surrounding fluid. It is the buoyant weight of the rock that counts. In this view, then, the depth of the scratches that could be imparted to the bed are greater when you drag a rock across the bed *in air* than they would be under ice. Try it. The scratches are very minor compared to the striations that must have occurred subglacially.

We note that recent work (Iverson *et al.*, 2003) has suggested that the rock–rock friction associated with these striators provides a significant brake on the sliding of the glacier. This would serve as a negative feedback in the glacial erosion system, as the more clasts there are the more the glacier will be slowed, and the less rapid the erosion will be.

Given a pre-glacial irregular bed consisting of bumps of all wavelengths, this abrasion process will most effectively remove bumps of small wavelength, leaving the bed smoother at these scales. At some scale, however, larger bumps persist. No glacial valley is perfectly smooth. When we look closely at these larger bumps, they often have an asymmetry, with a smooth up-valley stoss slope that shows signs of abrasion, and an abrupt, steep down-valley or lee side that looks like it has been quarried. These forms are called roche moutonee, apparently named for the similarity of their shapes with that of the rear of a sheep. Their

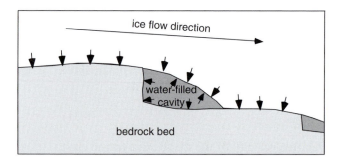

Figure 8.42 Schematic diagram of a water-filled cavity beneath a temperate glacier. Arrows represent normal stresses.

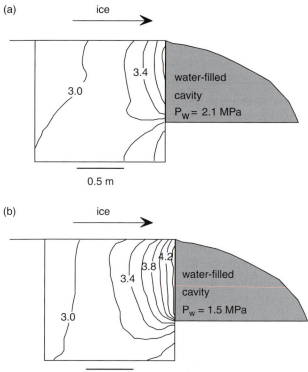

Figure 8.43 Contours of stresses (in MPa) within bedrock corner at ledge edge. Ice sliding from left to right. No vertical exaggeration. (a) With water pressure steady at 2.1 MPa, and (b) with water pressure reduced to 1.5 MPa. Note strong stress concentrations at ledge face (after N. Iverson, 1991, Figure 6, with permission of the author and the International Glaciological Society).

origin requires introduction of the other major subglacial erosional process: quarrying.

Quarrying

Larger bumps in the bed can generate the subglacial cavities on their lee sides illustrated in Figure 8.42, as the ice flowing over the top of the bump separates from the bed in the lee. All but those subglacial cavities at the very edges of the glacier will be filled with water. The quarrying process involves the generation of cracks in the bedrock at the edge of the cavity, and results from imbalance of forces on the top and sides of the rock bump. The theory of subglacial quarrying has been addressed by Neal Iverson (1991; see also 1995 and 2002) and by Bernard Hallet (1996), quite independently. Cracks grow at a rate dictated by the stresses at their tips. It is therefore the magnitude and the orientation of the stresses within the bedrock bump that are crucial to the problem.

Consider the down-glacier edge of the bump, which is often quite abrupt due to quarrying and can be idealized as a square corner. The ice presses downward against the bedrock that comprises the top of the bump. The water pressure in the cavity, which acts equally in all directions, presses against the lee side of the bump. If there is a large difference between the water pressure and that exerted by the overlying ice overburden, then there will be a large stress difference, and crack growth will be promoted. Iverson generated a model of the magnitude and orientation of the stresses, and hence the expected orientation of cracks under various pressure difference scenarios. Figure 8.43 summarizes the stress fields associated

with two different water pressures. One can see that the orientation of cracking to be expected is that parallel to the lee face of the cavity. The growth rate would be proportional to the pressure difference. An end-member situation is one in which the water in the cavity is abruptly removed, reducing its pressure to atmospheric. For the period of time required for the cavity to fill with ice as it relaxes into the low-pressure void, the entire weight of the ice overburden would be borne by the top of the bump, very large pressure gradients would exist, and cracks would grow that are nearly vertical. It is therefore the temporal fluctuations in water pressure at the bed that matter: it is transience in water pressure associated with pressure drops that generates the greatest stress gradients in the rock mass.

In a clever physical experiment installed beneath a Norwegian glacier, Cohen and coworkers (2006) have

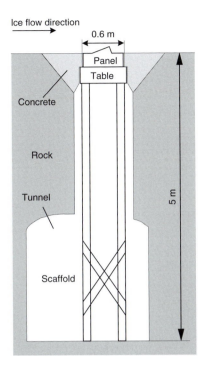

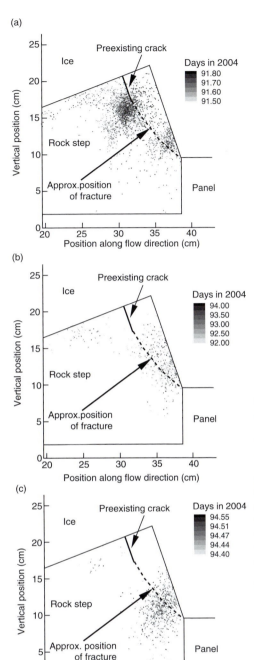

Figure 8.44 Experiment beneath Engabreen, Norway, designed to explore the dependence of transient water pressures on subglacial quarrying. Left: experimental setup showing how the artificial triangular bump was inserted at the bed of the glacier through a vertical shaft from a tunnel. Right: locations of acoustic emissions (AEs), reflecting crack growth, after three pump tests during the experiment. Early AEs (a) cluster near the tip of a pre-cut initial crack, but progressively migrate downward toward the base of the bump (b and c). The bump ultimately failed along a plane (dashed) roughly following the trend of the AEs (after Cohen *et al.*, 2006, Figures 1 and 8, with permission of the American Geophysical Union).

demonstrated that the basic ideas captured in this quarrying model are appropriate. They took advantage of the fact that in several places in Norway hydropower generation taps subglacial water, which in turn leads to considerable funding for subglacial research. Researchers can access the bed of Engabreen (*breen* is Norwegian for *glacier*), through a tunnel that ends beneath 210 m of ice. They installed an artificial ramp in the bed into which a cut had been made to simulate an existing crack in the stoss side of a bed bump. This is shown in Figure 8.44. Ports in the lee of the ramp allowed them to pump in water to

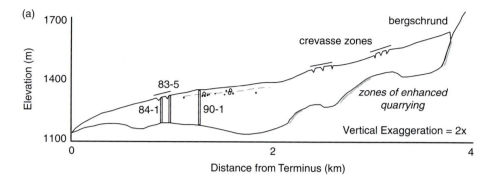

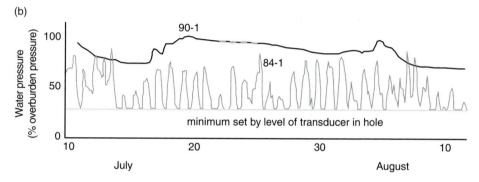

Figure 8.45 Topographic profiles of ice surface and bed (a), and water pressure records (b) from Storglacieren, Sweden. Profile shows several overdeepenings of the bed and major crevasse zones in regions of extension. Borehole locations in major overdeepening and at crest of bedrock bump are shown, along with spot measurements of the water pressure in the middle of the overdeepening. These measurements are all close to the level expected for flotation of the ice: 90% of the ice thickness, shown in gray line. Pressure records are very different for two sites, that in the middle of the overdeepening showing little variation around 90–100% of flotation, that at the crests of the bump showing major diurnal fluctuations between 70–90% flotation and that pressure associated with the depth of the transducer (gray line). Similar pressure fluctuations are inferred to promote enhanced quarrying of the bed at sites shown in (a) (redrawn from Hooke, 1991, Figures 2 and 3).

melt away a cavity in the lee. When the water pump was turned off, the pressure in the lee of the bump decayed rapidly as cavity collapse could not keep pace with the rate of pressure drop. Iverson's theory would lead us to expect that at these times large gradients in stress should occur that are conducive to the propagation of existing cracks normal to the stoss face of the bump. The experiment allowed them to check this notion: they instrumented the bump with eight acoustic emission (AE) transducers to listen to any cracking that occurred, and to triangulate the location of the cracking within the bump. Remarkably, after several pumping episodes that resulted in growth and decay of the cavity in the lee, and the associated cycling of pressures to which the bump was subjected, they could see the enhancement in the number of acoustic emissions, and the migration of the AEs downward in the bump. Inspection of the bump at the end of the experiment revealed that the edge of the bump had been quarried; the crack had indeed propagated as the AEs had indicated.

This experimental confirmation of the basic quarrying mechanism leads us to ask why pressures in a subglacial cavity system should vary at all? The origin of the water in the cavity system is melt of the glacier surface. At some level, variations in the rate of production of melt, which occur on both daily and seasonal timescales, are the most likely culprits. Boreholes drilled to the base of glaciers have recorded huge swings in the subglacial water pressure on quite short timescales. In the example shown in Figure 8.45, from Storglacieren in Sweden, the water pressure fluctuations are much greater over the crest of a bedrock rise than within a slight overdeepening. This may lead to large spatial variations in the rate of quarrying of the glacier bed.

As discussed in the section on glacial sliding, the detailed records of surface velocity during sliding events on alpine glaciers all suggest that cavities grow during rapid sliding induced by high water pressures, and that they subsequently collapse as the sliding event wanes. In order for the pressure differences in the edges of subglacial bumps to rise above the threshold for driving cracks, the water pressure in the cavity must drop faster than the ice roof can collapse to maintain contact with the water in the cavity. The intimate connection between water pressure, sliding, and quarrying should be apparent. So too should its complexity.

How do we know these cavities exist at all? There are a couple ways. The most direct evidence is from air-filled cavities (caves) at the edges of some glaciers. Here subglacial spelunking allows us to walk around in the cavity system and to observe first-hand the operation of the plucking mechanism. By their nature, such air-filled cavities are overlain by only minor thicknesses of ice. Nonetheless, direct observation of the edges of the ledges from one year to the next have shown that significant modification of the ledge takes place in a single year's sliding (Anderson *et al.*, 1982). A time lapse movie taken of the ice roof as it slides by such a ledge edge beneath the Grinnell Glacier, Montana, reveals that numerous small chips and blocks are torn off the ledge over a single winter. They become embedded in the basal ice (the roof of the cavity), and are therefore available to act as abraders when the ice roof reattaches to the bed at the downstream boundary of the cavity.

In limestone bedrock, the nature of the subglacial water system may be carefully mapped out after deglaciation of the valley. Numerous studies, notably Hallet and Anderson (1981), produced maps of these surfaces, making use of the solutional forms of the bedrock when it is in intimate contact with liquid water, and the clear evidence of abrasional erosion elsewhere. These maps commonly show sets of cavities, occupying significant fractions of the bed, linked to one another through relatively narrow passages. As we have seen above, this "linked cavity" system is now a dominant feature of many conceptual models of the subglacial hydrologic network.

Large-scale erosional forms

This brings us to the large-scale erosional forms imposed by glaciers on the landscape. We all learn in introductory geology classes that one can identify

Figure 8.46 U-shaped valley in Glacier National Park, Montana. View from Angel Wing looking east. Valley long profile is punctuated by numerous water-filled depressions separated by bedrock sills (photograph by R. S. Anderson).

a valley that has been shaped by glaciers when it has a U-shaped cross section, and when it is dotted with alpine lakes. Hanging valleys whose floors are perched well above the floor of a main valley are also attributed to glaciers. But just how do these features become imprinted on the landscape? This is in fact a topic of considerable recent and ongoing research.

The U-shaped valley

First let us deal with the cross section of a valley, an example of which is shown in the photograph in Figure 8.46. How does a classic V-shaped cross section indicative of fluvial occupation of a valley get replaced by a U-shape? And how long does it take?

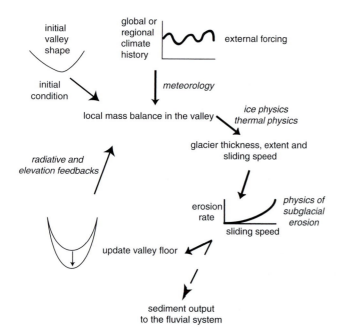

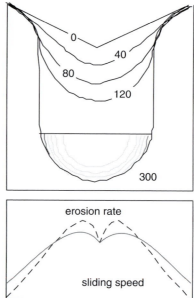

Figure 8.47 Structure of a glacial landscape evolution model. Relevant processes are in italics. Model is initiated by a specified valley geometry (the initial condition), and is driven by a specific climate history (external forcing, which here serves as a boundary condition on the glacier surface). Local valley geometry modifies the global or regional climate to generate the local mass balance profile on the glacier. Ice physics determines motion of the ice, which is thermally dependent in that both the viscosity of the ice and the ability to slide are dictated by the temperature of the ice. The sliding pattern dictates the pattern of subglacial erosion, which then both modifies the bed and generates sediment that is passed to the river downstream. The modified valley profile can then modify the local mass balance by both radiative feedbacks (hiding in a deeper valley) and elevational feedbacks that dictate both local temperatures and rates of snow accumulation. The arrows form a loop that is repeated, interacting with the external world through climate forcing and delivery of sediment from the system.

Figure 8.48 Numerical simulation of cross-valley profile evolution during steady occupation of the valley by a glacier. Initial fluvial V-shaped profile evolves to U-shaped profile characteristic of glacial valleys in roughly 100 ka, given the sliding and erosion rules used. Bottom graph shows initial distribution of sliding speed and corresponding erosion rate. Low erosion rates in valley center allow faster rates along the walls to catch up. Final erosion rate is roughly uniform, causing simple downwearing of the U-shaped form (redrawn from Harbor, 1992, Figure 5).

Jon Harbor has addressed this problem using a two-dimensional glacier model (Harbor, 1992, 1995). We summarize the model strategy in Figure 8.47. As in most models of landscape evolution, an initial landscape form is assumed (the "initial condition"), the glacier flow through that landscape is assessed, the pattern of erosion is calculated, the landscape shape is updated, and the flow is then recalculated. In the first suite of model runs, the glacier is allowed to achieve a steady discharge through the cross section, after which the ice discharge remains the same. In other words, climate change is ignored. The pattern of sliding, and hence of erosion, in the valley bottom is non-uniform, leading to the evolution of the shape. In particular, the

sliding is reduced in the bottom of the triangle, and the sliding rate reaches a maximum in the middle of the valley walls. This results from the dependence of the sliding rate on the water pressure field assumed in the glacier, which Harbor assumed to be steady at 80% of flotation. As seen in Figure 8.48, this leads to widening of the valley walls with time. The form of the valley evolves toward one in which the form no longer changes (a steady form), and that subsequently propagates vertically as a U-shape form. In more elaborate model runs, Harbor explored the role of glacial cycles on the evolution of the valley cross section. He found that sufficient shape change occurred over roughly 100 ka that the imprint of glacial occupation of alpine valleys could be imparted in as little as one major glacial cycle.

Cirques, steps, and overdeepenings: the long valley profile

The shape of a glacial valley in the other dimension is just as characteristic. In contrast to fluvial bedrock

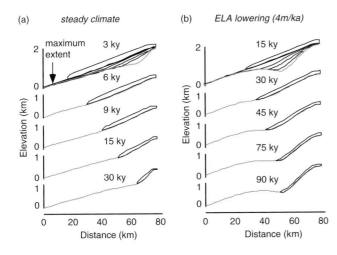

(a) *steady climate*   (b) *ELA lowering (4m/ka)*

Figure 8.49 Numerical models of long valley profile evolution of a glaciated valley in the face of (a) steady climate, over 30 ka, and (b) climate in which the ELA lowers at 4 m per thousand years, for 90 ka. Initial profile is a linear (uniform slope) valley. Top diagram shows all time slices in the respective simulation as light lines (redrawn from Oerlemans, 1984, Figure 4, reproduced with permission of *Zeitschrift für Eletscherkunde und Glaciologie*).

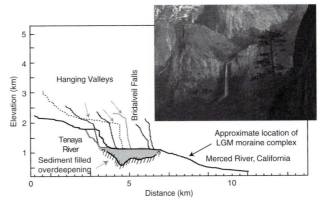

Figure 8.50 Longitudinal profile of Yosemite Valley the Merced and Tenaya Rivers, and several of their tributaries, showing steps, hanging valleys and overdeepenings (after MacGregor *et al.*, 2000, Fig. 1), and photograph of Bridal Veil Falls, which occupies a classic hanging valley tributary to the main valley of the Merced River (photograph by Greg Stock).

valleys, which display smooth concave up profiles, glacial valley profiles display steps and flats. Many of these steps occur at tributary junctions. In addition, the upper portions of alpine glacial valleys display knobby less well-organized topography rather than the expected smooth U-shape. Finally, many tributary valleys are said to hang above the trunk valleys upon deglaciation. The valley system disobeys what has come to be called "Playfair's law," which states that the trunk and tributary streams join "at grade" (see discussion of this in Chapter 13 on bedrock rivers).

Oerlemans (1984) was the first to attempt to model these features numerically. By embedding an erosional rule in a one-dimensional glacial model, he simulated the evolution of valley long profiles. His chief results in alpine glacial settings are shown in Figure 8.49. The model glaciers were driven by a simple climate in which the ELA was allowed to vary sinusoidally through time, which in turn moved a mass balance profile up and down. Interestingly, the glacial valley deepened so dramatically over several glacial cycles that the glaciers declined in size through time. While Oerlemans countered this by asserting a net cooling of the climate, the effect may be real. Glaciers likely do deepen their valleys through time. In so doing they both hide more effectively from direct insolation, reducing the melt rate, and decline in elevation,

which should both reduce accumulation and increase melt rate.

More recently, MacGregor *et al.* (2000) targeted the stepped nature of the long valley profile. The bedrock profile from Yosemite Valley reproduced in Figure 8.50 serves as a target for such models. Again a one-dimensional glacial long valley model was employed in which a net mass balance pattern was imposed. A simple balance profile pivoted about the ELA, and climate change was modeled by an imposed history of ELA scaled to the $\delta^{18}O$ deep sea record. Glacier dynamics included both internal deformation, and a rule for sliding. As in Harbor's models, the algorithm for sliding included an imposed water table, and assessed the sliding rate based upon the effective stress. The rule for the erosion rate was varied between model runs, but in all cases involved the sliding rate. The basic patterns produced were found to be insensitive to the rule chosen. As you can see in Figure 8.51, models of the single trunk glacier valley inevitably resulted in flattening of the valley floor down-valley of the long-term mean ELA and steepening of the valley profile up-valley of this. Only when more complicated valleys were modeled did a discrete step in the valley floor appear. In model runs including a single tributary valley, the erosion of the tributary valley floor was outpaced by that of the adjacent trunk stream, leading to disconnection of the

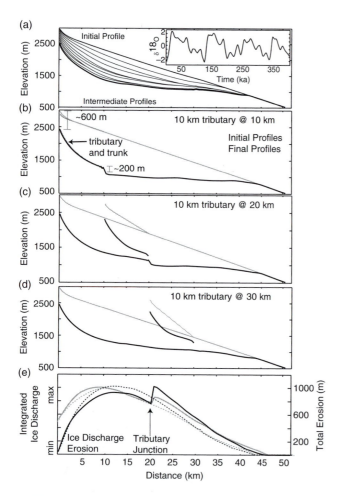

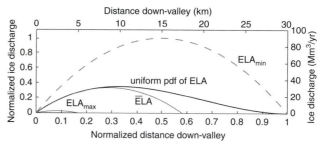

Figure 8.52 Analytic solution for ice discharge patterns resulting from steady-state glaciers driven by a simple mass balance function, in a linear valley profile, and a uniform distribution of ELAs. Maximum and minimum glaciers corresponding to minimum and maximum ELAs, respectively, are shown along with the glacier discharge expected from the average ELA, and a long-term average resulting from a uniform distribution of ELAs between maximum and minimum. The symmetry is broken when a distribution of ELAs is permitted. Maximum discharge occurs down-valley from that corresponding to the mean climate, at roughly one-third of the glacial limit (the maximum terminus position), and smoothly tapers to zero discharge there. Distance and ice discharge are normalized using glacier length and ice discharge associated with the lowest ELA (from Anderson *et al.*, 2006a, Figure 7, with permission of the American Geophysical Union).

Figure 8.51 Modeled evolution of the long profile of a valley with one tributary subjected to repeated glaciation. Steps develop in the valley profile, and the tributary valley is hung. The height of the step and the height of the hang depend upon the position of the tributary, and the long-term ice discharge (after MacGregor *et al.*, 2000, Figure 3).

two profiles, and a step appeared in the main valley. This led to the generation of model hanging valleys, and associated steps in trunk valleys depicted in Figure 8.51. A chief result of the calculations was that the long-term pattern of erosion mimicked the pattern of long-term discharge of ice. Tributary valleys see less ice discharge than trunk streams that drain larger areas. Trunk streams immediately down-valley of a tributary must accommodate the long-term discharge of ice from the tributary, and therefore see more ice.

That the time steps in these complex models had to be very small in order to maintain numerical stability meant that it was difficult to explore a wide range of valley shapes. Anderson *et al.* (2006) build upon the

observation that ice discharge was a faithful proxy of erosion rate in MacGregor's models. They employ analytic models of ice discharge patterns based upon the assumption that the glacier is at all times close to steady state. This implies that the discharge must at all times reflect the integral of the mass balance up-valley of a point, as we deduced from Equation 8.3. These simplifying assumptions allow efficient exploration of the roles of climate variability and valley hypsometry (distribution of valley width with elevation). They show in Figure 8.52 that in the simplest case of a linear mass balance profile and a uniform width glacier, a parabolic divot should be taken out of the valley floor, with the maximum of erosion located where the ELA intersects the valley profile. As in MacGregor's models, the valley flattens down-valley of the ELA and steepens up-valley of it. Incorporation of climate variability can be accomplished in a couple of ways. One could simply assert a history of the ELA, and numerically chop out parabolas of differing magnitudes and lengths. More efficient still, one can assert a probability distribution function (or probability density function, pdf, see Appendix B) that captures the spread of ELAs. The pdf of the $\delta^{18}O$ curve, reported in Figure 8.53, looks approximately Gaussian over the last 3 Ma since global scale glaciations began in the last glacial cycle.

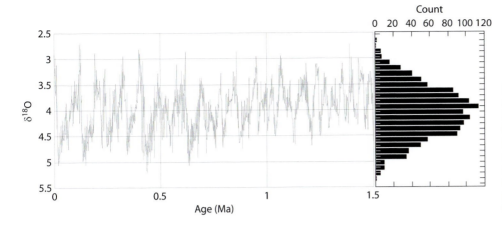

Figure 8.53 Marine isotopic record for last 1.5 Ma (after Zachos *et al.*, 2001), and a histogram of it. The distribution is roughly symmetric, and appears to be normally distributed (after Anderson *et al.*, 2006a, Figure 10, with permission of the American Geophysical Union).

Using this pdf, they show that the erosion pattern should feather into the fluvial profile down-valley, and that again the maximum of erosion occurs where the long-term ELA crosses the valley profile. This simple expectation can break down when more realistic plan view shapes of valleys are acknowledged. Most glacial valley systems are much wider in the headwaters than they are in their lower elevations, reflecting the inheritance of the dendritic fluvial network. The ice discharge pattern predicted from Equation 3 must therefore include both the pattern of mass balance and of valley geometry (width). But recall that the principal driver of erosion is the sliding rate. Ice discharge per unit width of the valley should be a better proxy for the sliding rate. This means that where the valley narrows, more ice is being shoved through per unit width than in a wider portion of the valley. When this element of reality is incorporated in the analytic models, the pattern of erosion begins to include a bench in the upper valley, below which the erosion rate increases dramatically. This reflects funneling of the ice into a narrower throat as the tributary valleys coalesce in the headwaters.

Fjords

The ornamentation of coastlines and shelves by fjords is a characteristic of continents that have harbored large ice sheets in the past. An example from eastern Baffin Island is shown in the photograph of Figure 8.54. The coastlines of Greenland, Fennoscandia, New Zealand, Patagonia, and Antarctica are

Figure 8.54 Kilometer-tall walls of Precambrian granite tower above Sam Ford Fjord, eastern coast of Baffin Island. Sea ice covers an 800 m-deep fjord, which reaches its greatest depth where it crosses the island-bounding mountain range (photograph by R. S. Anderson).

decorated with deep fjords that indent the coastline by many tens of kilometers. Fjords reach depths of more than 1 km below present sea level. For example, Sognefjord in Sweden is more than 1.2 km deep. This greatly exceeds the lowstands of sea level during glacials (order 100–150 m), and therefore requires submarine erosion of rock. Rivers cannot accomplish such erosion, as their power dissipates upon entering the ocean, which in turn generates the sedimentary deltas that characterize their entrance. Fjords are instead gouged into the continental margins by glaciers. They bite deeply into the landscape, reaching more than 100 km inland from the coast, in places punching through coastal mountain ranges (for example, in Baffin (Kessler *et al.*, 2008), in the Transantarctic Mountains (Stern *et al.*, 2005), and in British Columbia). The first two cases involve mountain ranges that correspond to the uplifted flanks of rift margins. Here recall the isostatic rebound calculation of rebound associated with these fjords, reproduced in Figure 4.35. In such situations, the greatest depths of the fjords occur as they pass through the crest of the range. You can see this by inspecting the profiles in Figure 8.55. At their upstream heads, fjords gradually lose definition and feather into the landscape in "onset zones." At their downstream edges, fjords extend onto the continental shelf as fjord troughs and often terminate in a bedrock sill. The deep water in the base of a fjord is therefore somewhat isolated from the ocean, and houses a distinct ecology. In detailed mapping and cosmogenic radionuclide analysis of fjorded landscapes on Baffin Island, Briner *et al.* (2006) have shown that significant erosion in the last glacial occupation of the landscape was limited to the fjords themselves. They interpret the lack of erosion in inter-fjord regions to the polar thermal condition of the bed there. Ice was presumably thin, cold, and slow on the inter-fjord flats, while it was thick, temperate, and fast in the fjords.

Fjords serve as the major conduits through which ice from present-day ice sheets enters the ocean. In Greenland, for example, 40% of the ice leaving the continent leaves through the three main fjord-occupying outlet glaciers of Jacobshaven, Kangerdlugssuag and Helheim (Bell, 2008). Combining this observation with the fact that fjords owe their origin to erosion by glacial ice leads us to ask (1) how the early icesheets of the last glacial cycle, say 2–3 Ma, operated in the absence of such efficient conduits for ice from the interior, and (2) how rapidly the present fjords were eroded into the continental edges?

While the latter question awaits analysis through thermochronology, the former can be addressed through modeling. Employing a two-dimensional

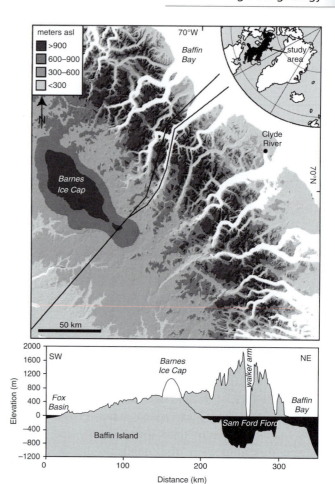

Figure 8.55 Baffin Island fjorded topography. Top: the fjorded continental edge of northeastern Baffin Island showing the inland plateau, on which the Barnes Ice Cap sits, and the fringing mountain range. Inset: location of Baffin Island in the western Arctic. Bottom: topographic cross section of Baffin Island, along Sam Ford Fjord and neighboring topography, following transect lines shown in map (after Kessler *et al.*, 2008, Figure 1, with permission from Nature Publishing Group).

glacier model, Kessler *et al.* (2008) have simulated how fjords might have evolved through time in a simplified landscape that includes a circular continent with a bounding mountain range meant to mimic a rift flank range. The model setup is shown in Figure 8.56. They demonstrate that once the ice sheet is thick enough to overtop the mountain range, ice from the interior is steered through any mountain pass. As erosion rate is tightly correlated with ice discharge, this leads to preferential erosion of the passes relative to the surrounding range. This dimples the contours of the ice sheet well upstream from the mountain pass, which steers yet more ice toward

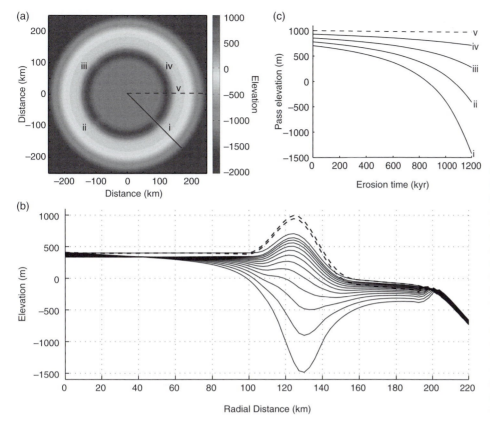

Figure 8.56 Evolution of model topography. (a) Initial synthetic bed topography. Fringing mountain range is dissected by four valleys with varying depths. (b) Evolution of a fjord; thin solid lines indicate bed profiles at 100 ka time steps for the deepest valley at transect "i" in (a); dashed bold lines indicate the initial and final topographic profile at the range crest at transect "v" in A. (c) Elevation through time at the radial position of the initial crest for each of the four passes (i, ii, iii, iv) and the crest (v) (after Kessler *et al.*, 2008, Figure 2, with permission from Nature Publishing Group).

the pass. The deeper mountain passes out-compete their neighbors, leading to a few dominant fjords that efficiently tap ice from the interior. Just as those in Baffin Island, the simulated fjords are deepest as they cross the range crest. This reflects the fact that ice discharge per unit width is greatest there: as seen in Figure 8.57, upstream, the ice is converging; downstream, the ice diverges. Because the two-dimensional model employed in these simulations did not have a thermal component, meaning that it could not simulate the variation in thermal conditions at the bed, the authors argued that the ice steering–erosion feedback was sufficient to allow deep fjords to form. The thermal feedbacks should serve to magnify this pattern.

The exploration of the effect of presence or absence of fjords and associated outlet glaciers on the behavior of ice sheets has only begun. In preliminary experiments, Kessler *et al.* (2008) have shown that the presence of fjords indeed influences the behavior of the ice sheet, for example by altering its response time to variations in climate. This could potentially influence the degree to which ice sheets are controlled by the various periods of orbitally controlled insolation.

Depositional forms

Glacial deposits come in many shapes and forms, including isolated glacial erratics (an example of which is shown in Figure 8.58), and are the subject of numerous books. The reader is pointed to Menzies (1996), to Bennett and Glasser (1996), and to Benn and Evans (1998) for more comprehensive summaries of this portion of the glacial system. We treat here only the basics. Many glacial deposits consist of poorly sorted sediments that run the gamut in terms of grain size; they were once called boulder clays and are now called simply tills. The first-order classification comes from distinguishing those materials that have seen the bed of the glacier, from those that have not.

Moraines

Many moraines, be they lateral (on the edges), terminal (at the terminus), or medial (down the middle), are derived from supra-glacial debris, debris that has never seen the bed of the glacier. While this might be

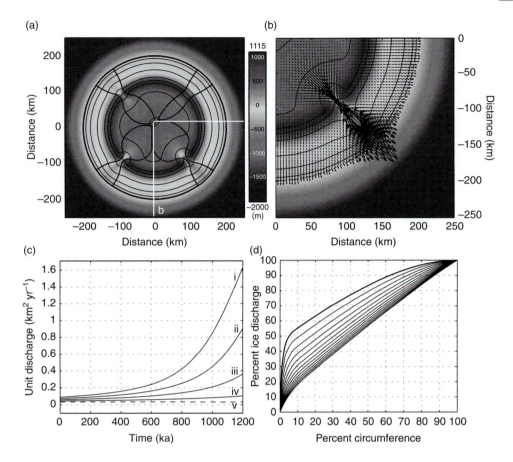

Figure 8.57 Modeled ice surface, bed elevation, and ice velocity pattern after 1.2 Ma. (a) Bed elevation in grayscale. Thin black lines denote ice surface contours. Bold lines demark interior regions that drain through each of the four valleys. (b) Ice velocity vectors in the quadrant of the deepest valley (outlined in (a)). (c) Maximum specific ice discharge versus time along the five transects in Figure 8.56(a). (d) Percentage ice discharge along the crest (125 km from center) versus percent crest length through which that discharge passes; curves show time intervals of 100 ka. The discharge of ice from the landmass becomes increasingly dominated by that exiting the fjords (after Kessler *et al.*, 2008, Figure 3, with permission from Nature Publishing Group).

Figure 8.58 Glacial erratic in glacial valley floor east of Mt Whitney in the Sierras. Isolated block of granite lies on granitic bedrock, but was transported into place in the last glacial maximum glacier (photograph by R. S. Anderson).

obvious in the case of the lateral moraines, which are accumulations of debris found along the margins of a glacier, and are often delivered to those sites by hillslope processes, it is less clear for the other two classes of moraines. Lateral moraines consist largely of angular blocks riven from the valley walls by periglacial processes, and delivered by individual rockfall events, landslides, or avalanches that incorporate rocky debris as they slide against the wall. Rarely are there blocks that show striated facets indicating time spent at the glacier bed. A point of caution: it is worth noting that some lateral moraines near present day margins of glaciers are ice-cored. These are not the simple triangular piles of rock that they look to be, but are instead a rather thin carapace of angular rock covering glacial ice.

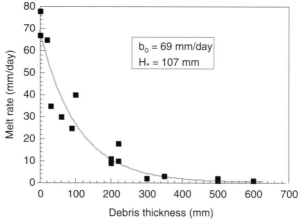

Figure 8.59 Barnard Glacier, in Alaska's Wrangell–St. Elias National Park. This glacier's many tributaries make it famous for its medial moraines. Inset: Austin Post air photograph held at World Climate Data Center. Main image: lower flight allows view of the main medial moraine in the glacier centerline, showing how thin the debris cover is – one clast thick in most places. It also reveals how each moraine is a ridge with debris-rich ice emerging at its center (photograph by R. S. Anderson).

Figure 8.60 Dependence of ice ablation rate on debris thickness. Note scale length for exponential decay of ablation rate is roughly 11 cm (after Anderson, 2000, Figure 5, with permission from International Glaciological Society; data from S. Lundstrom, 1992).

Terminal moraines are accumulations of debris at the snout or terminus of a glacier. The larger terminal moraines are generated when a glacier remains in the same position for some time, continuing to deliver debris to the margin on an annual basis. Upon retreat, this pile of debris is left behind. Smaller piles may be generated each year (annual moraines) as a signature of the retreat history.

Medial moraines are the dark stripes we see gracing the interiors of many longer glaciers. Some of the more famous are in Alaska. There is no less striking evidence that the flow of ice is laminar, and that most glaciers are very steady in their motion, than that these debris stripes are so parallel to the margins of the glacier. In fact, it is the rare exceptions to the straightness of these medial moraines that indicate that something has "gone wrong" up one or another tributary; the looped moraine signature of a surging glacier is telltale, as seen in Figure 8.26. While there are exceptions, by far the majority of the medial moraines result from the welding together of lateral moraines at interior tributary junctions such as those on Barnard Glacier in the photos of Figure 8.59. The debris is swept into the seam between the two ice streams. It is angular debris, and has never seen the bed. The ice on either side of the seam has very little debris content, while that in the seam has at least some. In the ablation zone, below the ELA, this debris emerges at the surface as the ice melts. Because the ice can melt (and therefore vanish), while the debris cannot, the debris accumulates on the surface of the glacier. At this point another element comes into play, however. The presence of any significant debris on the surface of the ice leads to reduction of the melt rate of the ice. As shown in Figure 8.60, only 10 cm of debris reduces the melt rate by a factor of e (2.72). So the adjacent debris-free glacier surface melts faster, leaving the debris covered ice higher. This produces a slope away from the debris cover, which in turn promotes the downslope motion of

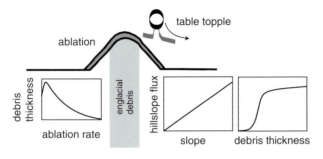

Figure 8.61 Schematic of medial moraine problem. Debris from a debris-rich septum of ice (gray) emerges at the surface at a rate dictated by the ablation rate. Debris travels down the slope by the "table topple" mechanism at a rate that depends upon slope. The total debris discharge depends on both local slope and debris concentration. The moraine widens beyond the margins of the debris-rich septum through time. The moraine crest is convex, while the side slopes beyond the debris-rich ice septum are straight.

the debris sketched in Figure 8.61. The result is a stripe of debris that widens down-glacier as more debris emerges on the surface, and is continually being spread laterally. Walking on practically any portion of a medial moraine will reveal that the hillslope processes are indeed very efficient. The debris is everywhere very thin, usually only one clast thick (see this in crevasses that cut the medial moraines in the low-level air photo of Figure 8.59). Medial moraines typically display a rounded crest and relatively straight limbs. Indeed, this problem has been treated as a hillslope evolution problem (Anderson, 2000), resulting in parabolic profiles. We will see in Chapter 10 that this is the expected steady-state profile of a hilltop. Finally, we note that while these debris stripes add to the visual drama of a glacier, the debris is so thin that medial moraines are rarely preserved in the geomorphic record after deglaciation.

An interesting feedback occurs in complicated glacier networks. As each tributary junction spawns a new medial moraine, and as every moraine widens down-glacier, it is common for these medial moraines to merge, eventually covering the entire glacial cross section with debris. This will inevitably lead to the reduction in the mean ablation rate, and will promote the longer extent of a glacier. This appears to be the case, for example, in the Karakoram Range, where highly debris covered snouts of glaciers extend well down the valleys. Similarly, in the easily eroded volcanic rocks of the Wrangell Range in Alaska, debris-covered glacial snouts abound, as seen in Figure 8.62, as do rock glaciers.

Figure 8.62 Debris-covered snout of a glacier draining the high Wrangell Mountains of Alaska. The easily eroded volcanics that dominate the geology of this range generate debris in sufficient quantity to create wide medial moraines that merge in the terminal reaches of the glacier. This in turn prevents ablation, and promotes longer glaciers than the climate would otherwise suggest. These debris-mantled terminal reaches are very difficult to travel across. They are dotted by numerous small lakes, the steep walls of which are often the only bare ice in the area (photo by R. S. Anderson).

Eskers

There is one distinctive depositional feature attributable to the subglacial fluvial system: eskers. These are sinuous ridges with triangular cross sections that consist largely of gravels. They can be up to several tens of meters in relief. An esker represents the operation of a subglacial tunnel through which significant volumes of surface meltwater passed, entraining and transporting the gravel as it did so. As such, their present shape and location in the landscape provides evidence of the hydrologic and glaciological conditions during which they formed. The most extensive systems of eskers were formed beneath the continental scale ice sheets.

One of the most striking features of eskers is that they do not obey the contours of the landscape on which they currently rest. This could be explained in two ways. The first is that these conduits in which the gravels were initially deposited were englacial rather than subglacial conduits, and were simply laid down over the landscape as the glacier retreated. The second is that they are indeed subglacial, and that something inherent in the subglacial hydrologic system allows water to flow uphill. Sounds outrageous, but this is the most likely. Arguments against the englacial explanation include the fact that the gravel deposits have intact stratigraphy that is not deformed by such a superposition on uneven ground. Most importantly, however, the apparent uphill flow of subglacial water, the details of where this uphill flow occurs, and the style of deposition in the various divides, can all be explained. We are very used to thinking about water responding only to topographic gradients, as it does when running in open channel flow on the surface of the Earth. But eskers represent flow in closed conduits – pipes – beneath the glacier. Just as water in a hose can be made to flow upward by water pressure gradients, so too is the flow in the conduits dictated by horizontal gradients in pressure, running down-gradient. The pressure field is governed not only by the bed topography, but also by the ice thickness profile, which involves both the ice surface and the bed profiles. The tilted nature of the equipotentials is plotted in the schematic cross section of a glacier in Figure 8.63. The equation dictating the potential field is that for the total head, which is the sum of the elevation head, and the pressure:

$$\phi = \rho_w g z + \rho_i g (H - (z - z_b)) \tag{8.33}$$

where z is the elevation, z_b the elevation of the bed, and H the thickness of the ice. Water within a glacier or an ice sheet flows perpendicular to lines of equipotential within the glacier. These lines, or planes of equipotential, can be calculated by setting the gradient, $d\phi/dx$ to zero:

$$\frac{d\phi}{dx} = \rho_w g \frac{dz}{dx} + \rho_i g \frac{d(H - (z - z_b))}{dx} = 0 \tag{8.34}$$

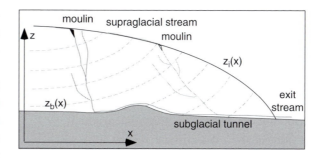

Figure 8.63 Schematic of the water flow paths within a glacier profile. Melt water enters the glacier through moulins and crevasses. Englacial water flows normal to equipotential lines (dashed) until it encounters the bed. Flow at the bed occurs in tunnels in the ice and/or the bed; flow speed is dictated by pressure gradients. These are more strongly determined by the ice surface gradient than by the bed gradient (after Shreve, 1985b, Figure 4).

Solving this for the dip of the equipotential lines, dz/dx, we find that

$$\frac{dz}{dx} = -\left(\frac{\rho_i}{\rho_w - \rho_i}\right) \frac{d(z_b + H)}{dx} = -\left(\frac{\rho_i}{\rho_w - \rho_i}\right) \frac{dz_s}{dx} \tag{8.35}$$

As we can see from Equation 8.35, and from Figure 8.63, the lines of equipotential essentially mirror the topographic contours of the ice surface, but dip 11 times more steeply than does the surface. The flow field of water within the glacier is therefore canted at a high angle to the surface, pointing down-glacier. When the englacial channels intersect the bed, of course, the flow is trapped at the ice–rock interface. It must still obey the potential gradient. The equation for the head gradient (the derivative of Equation 8.33) has two terms, one related to the ice surface slope, the other to the bed slope. We now seek the expression for the potential gradient along the ice–rock interface, i.e., when $z = z_b$:

$$\frac{d\phi}{dx}\Big|_{bed} = (\rho_w - \rho_i)g \frac{dz_b}{dx} + \rho_i g \frac{dz_s}{dx} \tag{8.36}$$

The term associated with the bed slope, dz_b/dx, is one-eleventh as important as that associated with the slope of the surface of the ice, dz_s/dx. The flow will go in the down-glacier direction, i.e., in a positive x-direction, as long as this gradient is negative (flow is driven down potential gradients). This requires that

$$\frac{dz_b}{dx} < -11 \frac{dz_s}{dx} \tag{8.37}$$

For example, if the surface of the ice slopes at −0.01, or 10 m/km downglacier, then water will continue to flow down-glacier as long as the bed slope is less than 0.1. This is a steep *upward* slope. When the ice sheet is removed to reveal the esker deposit, it looks as if the water in the subglacial tunnel had flowed uphill, while in fact it had flowed down the potential gradient. The bottom line is that the surface topography of the glacier is 11-fold more important in dictating the direction of flow than is the topography of the landscape over which it flowed.

One of the classic examples of esker systems is the Katahdin esker system, in Maine. This system extends 150 km from central Maine to the calving margin at roughly the present coast. In a pair of papers, Shreve (1985a,b) uses the physics of water flow in a subglacial conduit to explain a wide variety of the features of these eskers. As one follows any particular esker down-ice, the esker disobeys present day contours in a particular way, and it changes shape in a characteristic pattern. Shreve first explained these patterns, and then made use of them to reconstruct the shape of the ice sheet covering this portion of Maine around 14–16 ka. As shown schematically in Figure 8.64, the shape change is such that on the flat the esker tends to be sharp-crested, and triangular in cross section. As it climbs toward a divide, it first becomes multiple-crested, and then broad-crested, losing its triangular cross section altogether. Over the top of the divide, there may be a gap in the esker before it resumes its course, again sharp-crested. Shreve focused on what is happening to the water in the ice-walled tunnel. Heat is generated as the water flows, due to frictional interaction with the walls. This energy can be used either to warm up the water or to melt the walls or both. Note also that as the water flows into regions of thinner ice, it is moving into warmer ice (recalling the thermal profile within a Type II polar glacier). The faster the ice is thinning, the more of this energy being released must be consumed in warming the water to keep pace with the warming of the walls. If this pace of warming is not sufficient, the water will be below the pressure melting point of ice, and will freeze onto the wall. Shreve deduced that the condition for freezing is when the bed gradient is positive, and is more than 1.7 times the magnitude of the ice surface gradient. It is at this point that the ice is thinning too rapidly for

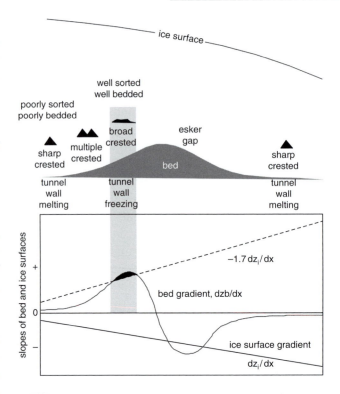

Figure 8.64 Esker styles. Types of eskers and their locations in the topography beneath an ice sheet. Eskers evolve from sharp-crested on flat topography to multi-crested and broad-crested as the water in the subglacial tunnel is forced over a pass in the basal topography (after Shreve, 1985b, Figure 6, with permission from Elsevier).

the water to warm up fast enough. Rapidly melting walls allow ice with all its basal debris to move toward the ice conduit; in addition, the shape of the conduit is tall, the melting rates being highest where the water is deepest. Freezing walls pose the opposite case. With no new ice moving toward the conduit, the roof is low-slung, the freezing rates being highest there. New sediments are not supplied, and sediment already in the pipe is simply reworked, generating better sorted deposits.

Shreve also reconstructed the ice surface morphology. Wherever he found well-defined transitions from sharp-to-broad crested, he could pin down the ice surface slope. In addition, he made use of the divergence between the modern topographic gradient and the path of the esker over it to deduce the ice surface slope. As reproduced in Figure 8.65, at seven places along the 140 km Katahdin esker system, surface gradients constrained the shape of

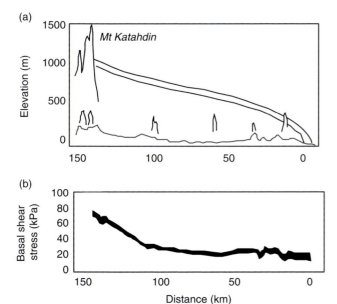

(a)

(b)

Figure 8.65 (a) Profile of the ice sheet along flow line over Maine, through Mt. Katahdin, at roughly 11 ka, as deduced from esker topography (bottom continuous line) as it crosses various passes (isolated bumps). Two profiles shown bracket maximum and minimum estimates. (b) Basal shear stress profile for same section of the ice sheet. Stress is everywhere below 1 bar (100 ka) and over much of the profile is only 20–30 kPa (after Shreve, 1985b, Figure 3, with permission from Elsevier).

the ice sheet, allowing recreation of the ice sheet profile at the time of operation of the esker system. The thickness profile was then integrated, constrained to have the proper surface gradient at each location. In other words, knowing dz_i/dx at several locations, $z(x)$ was obtained by integration. Note that no assumptions were made about the rheology of the ice, its thermal profile, whether or not it was sliding, or the nature of the underlying material. This independently documented ice sheet profile could then be used to assess the pattern of the basal shear stress of the ice sheet. Recall that we used the assumption that the basal shear stress was uniform, and of the order of 1 bar (10^5 Pa), in our simple reconstructions earlier. Shreve found that the shear stress varied from 0.2–0.3 × 10^5 Pa (relatively low compared to estimates from modern ice sheets of about 1 × 10^5 Pa), and viewed this as supporting the claims of others that the history of the demise of the local portion of the Laurentide ice sheet from its last glacial maximum had been first to deflate (lower in

slope), and then to retreat from its margins. This implies that the eskers were formed when the ice sheet was in its deflated state.

Erosion rates

Just how efficient are glaciers at eroding the landscape? What means do we have of constraining erosion rates beneath hundreds of meters of ice? Our principal probe is the sediment output from glacial streams. By its nature, this yields only a spatially (and perhaps temporally) averaged measurement of the erosion rate. It is impossible to tell unambiguously where the sediment has come from beneath the glacier. Even so, this number is useful in that it allows comparison with fluvially dominated systems, and it allows comparison amongst glacial systems from which one might ferret out what variables are the dominant ones. If we can measure well the sediment discharge, Q_{sed}, then the spatially averaged rate of erosion of the glacier whose basal area is $A_{glacier}$ is

$$\bar{e} = \frac{Q_{sed}}{A_{glacier}} \tag{8.38}$$

It is not at all trivial to measure sediment load in a river, however. As discussed in the fluvial chapter (Chapter 12), one must measure both suspended sediment and bedload transport rates. Using this method, one can obtain a near real-time record of the sediment output from the glacier. There is no doubt that this records the sediment output from the glacier. We are not assured, however, that this is equivalent to the rate at which sediment is being produced by erosion at the glacier bed. The problem is that sediment can be stored temporarily at the glacier bed, to be exported from the subglacial system later.

An intermediate timescale (one to several years) estimate of erosion rates can be obtained in special circumstances. In Norway, some of the hydropower used in the nation derives from water tapped subglacially. Sediment loads in the subglacial water, which could damage the turbine blades, is diverted into reservoirs to trap the sediment. The traps are emptied periodically, and the number of truck loads can be tallied to estimate the volume of sediment involved since the last cleaning.

Another technique yields yet longer-term rates of glacial erosion. This relies upon the volume of sediment in a depositional basin that was transported to the site from a glacial basin over a given period of time. For example, the basin may be a lake (e.g., Loso *et al.*, 2004), the interior of a fjord (e.g., Powell and Molnia, 1989), or the continental shelf of Alaska. Drilling the deposit, or seismically imaging it, can constrain the geometry and hence the volume of the deposit. And time lines can be drawn within it by obtaining an age from one or another marker horizon within the deposit.

The results of many studies utilizing these methods have been summarized by Hallet *et al.* (1996). Their summary plot is reproduced in Figure 8.66. The rates vary greatly, but can range up to several millimeters per year. They also vary in a systematic way geographically. The glacial basins in Alaska are very productive of sediment, while those in Norway are one to two orders of magnitude less so. This presumably reflects both the erodibility of the substrate and the delivery of snow to the region. The Alaskan coast is dominated by young rocks of an accretionary prism, while the Norwegian coast is a Precambrian shield. And the Alaskan bight enjoys huge snowfalls relative to Norway. A steady-state Alaskan glacier

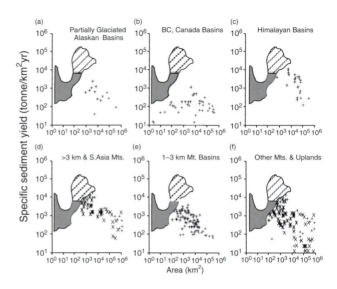

Figure 8.66 Sediment yield from glacial basins around the world (shaded pattern), and comparison with fluvially dominated basins of specified size or location (marked with X's) (after Hallet *et al.*, 1996, Figure 3, with permission from Elsevier).

must therefore transport much more ice down-valley than its Norwegian counterpart. To the degree to which this is accomplished by enhanced sliding, the erosion rate ought to be enhanced.

Box 8.1 Yosemite Valley

We are now in a position to ponder the origin and evolution of the classic glacial landscape of Yosemite Valley and, more broadly, Yosemite National Park. This granitic landscape is dominated by smooth rock walls with large patches of glacial polish that makes the white granitic surfaces yet brighter. These surfaces reflect the light that dazzled John Muir, and that so attracted photographers Ansel Adams and later Galen Rowell. The deep glacial valleys are interrupted here and there by smooth almost hemispherical domes looking like state capitols that peel away in huge graceful onion-like sheets. Several of these, including the graceful Half Dome, are shown in Figure 8.67. The rock of the Park is all granite. It formed 180–80 million years ago from intrusions in a collisional arc setting. These plutons are the roots of the arc volcanoes that once capped the range. The rock has come to the surface simply through erosion and is now well above sea level due to isostatic adjustment to this erosion. The volcanic rocks once on the surface have been eroded from the surface and deposited to the west in what was the forearc, and is now the Great Valley. No tectonic forces pushed the rock up against gravity, cracking it as it was moved along curved faults, as has happened for example in the Laramide ranges of the American West. The rock moved toward the surface gently, and this gentle unroofing leads to the intact, massive character of the rock. This character in turn translates into all of the major features that together are the signature of Yosemite. The lack of cracks means the walls are strong and can maintain near verticality for hundreds of meters. Hence El Capitan. The precious few cracks challenge climbers on far-flung routes up the vertical walls. That the rock is so intact also promotes the formation of the domes. As discussed in Chapter 10, sheet jointing requires massive rock so that compressional stresses do not simply move myriad chunks of rock on pre-existing cracks, but instead generate surface-normal stresses on curved surfaces. But our interest here is in the glacial features. The smooth, glinting polished landscape, dotted with scattered glacial erratics as exemplified by the scene in Figure 8.68, is yet another corollary to the massive rock quality. The lack of cracks robs the glaciers that repeatedly slid across this landscape of the most efficient of the erosional mechanisms: quarrying was limited, meaning that abrasion was the only method remaining for the glaciers to employ. Hence the polish. The few erratics scattered about on these surfaces speak to the low debris concentration in the ice, also a manifestation of the difficulty of plucking blocks from the bed. These effects are pronounced in the Cathedral Peak Pluton, which is the most massive, least fractured of the plutons in the Park. The limited erosion of the portion of the Yosemite landscape underlain by the Cathedral Peak pluton is now documented by ^{10}Be concentrations that are far too high to record the time of deglaciation (see Guido *et al.*, 2007; Dühnforth *et al.*, 2008; Ward *et al.*, 2009). Where the spacing between fractures is small, quarrying dominates, sufficient erosion occurs to reset the ^{10}Be clock, and the measured ^{10}Be concentration is that one would expect from the time of deglaciation (i.e., $C = P_{o}T$, see Chapter 6). Where the spacing between fractures is large, insufficient erosion occurs during the last glacial cycle, and the measured ^{10}Be reflects both the accumulation since deglaciation and some remainder of the inventory from the last interglacial exposure (called inheritance). Perhaps the low erodibility of the Cathedral Pluton is why Tuolumne Meadows sits so high, with the deep glacial troughs of the Tuolumne and Merced Valleys plunging off it to the northwest and to the southwest, respectively.

Figure 8.67 Upper Yosemite Valley seen from Glacier Point looking east. Half Dome graces the right skyline. All rock in view is granitic. The Ahwiyah Point rock fall below Half Dome is visible as a light streak. It occurred on the morning of March 28, 2009 and is estimated to have involved about 40 000 cubic meters of rock, making it the largest in Yosemite in 22 years (pers. comm., Greg Stock, NPS) (photograph by R. S. Anderson).

Summary

Glaciers modify alpine landscapes significantly, leaving their marks on the landscape in the form of U-shaped valleys, cirques, hanging valleys, stepped longitudinal profiles, and fjords in the erosional portion of the landscape, and characteristic deposits in moraines and eskers. We have reviewed the physics of how these features originate. Sedimentary records from glacial terrain imply that glaciers are efficient at eroding the headwaters. Their actions therefore impact the fluvial landscape downstream by strongly modulating the sediment supply to the river system, charging it with slugs of sediment when glaciers are large, and starving it when they are small. We will see in Chapter 12 that this generates stacks of fluvial terraces.

Figure 8.68 The essence of the upper Yosemite landscape includes broad bare bedrock surfaces sporting glacial polish and scattered erratics. Note the massive (fracture-free) nature of the bedrock in Tuolumne Meadows (photograph by Miriam Dühnforth).

In order to understand the erosion by glaciers, we had to understand how glaciers work. Glaciers are charged with the task of transporting snow (metamorphosed into denser pure ice) from the accumulation zone above the ELA into the ablation zone below it, where it is lost to melt or to calving. At steady state, the pattern of ice discharge required to accomplish this task increases to a maximum at the ELA, and declines to zero at the terminus. Glaciers transport ice by two mechanisms: internal deformation of ice, and basal motion that we loosely call sliding. Internal deformation is governed by the nonlinearity of the rheology of ice. This makes the ice discharge very sensitive to ice thickness, meaning that only small changes in ice thickness are required for a glacier to accommodate climate change. Sliding remains a major focus of modern glaciological research both because it governs the erosion by glaciers and because it controls the rate of delivery of ice to the margins of ice sheets. We do know that sliding is strongly governed by the state of the glacial hydrologic system. It appears that whenever the inputs of water to the complex hydrologic system of tunnels and cavities beneath a glacier exceed its ability to pass that water onward to the terminus, sliding is promoted. In addition, basal motion can be accomplished by shearing of underlying till. The rapid motion of the ice streams of Antarctica appear to be governed by the thermal state of this till.

Sliding of the glacier against its bed is the only means by which a glacier can damage the underlying rock. Both abrasion and quarrying rates increase with sliding speed. Quarrying focuses on the down-valley edges of bumps and ledges in the bed. Abrasion smoothes the stoss or upstream sides of these bumps, and serves to remove small-scale bumps, leading to smooth surfaces.

Larger scale forms such as glacial cirques, U-shaped valleys, hanging valleys and fjords must be understood in the light of repeated occupation of the landscape by glaciers that have come and gone through the last several million years. Spatial variation in the long-term ice discharge, some fraction of it accomplished by sliding, is responsible for at least some portion of this pattern of erosion. That alpine valley profiles evolve to flat-floored valleys with steepened headwalls is an inevitable consequence of the pattern of ice discharge, which to some degree mimics the pattern of erosion. Hanging valleys result from the large discrepancy of long-term ice discharge through the tributary vs. the trunk valley. So also is the punctuation of continental edges by glacial fjords. Topographic steering of the ice through any preexisting drainage allows fjords to bite deeply into the landscape. The details of the erosion pattern, however, will also be governed by spatial variations in the susceptibility of rock to erosion. The pattern of glacial erosion in the last glacial cycle in the Yosemite landscape, for example, suggests that fracture spacing strongly controls the erodibility of rock.

Problems

1. The table below shows mass balance data collected in 2006–2007 at Hardangerjøkulen, the sixth largest glacier in Norway. Hardangerjøkulen is a small ice cap that drains in several directions through outlet glaciers. The data we have are for the outlet called Rembesdalsskåka.

Your primary task is to determine whether this was a positive or negative mass balance year for this glacier (the outlet Rembesdalsskåka, for which we have data).

Steps:

(i) Compute the specific net balance for each altitude band.

(ii) Plot the specific winter, summer and net balances as a function of altitude.

(iii) Compute the volume for winter, summer and net balance in each elevation band.

(vi) Plot the volume balances (winter, summer, net) as a function of altitude.

(v) Compute the total net balance for the glacier both as a specific balance (units of depth of water equivalent) and as a volume balance (volume of water equivalent).

In addition to a table showing the results of these calculations, and the plots requested, answer the following questions:

(a) What is the equilibrium line altitude for this glacier in 2006–2007?

(b) Was this a positive or negative mass balance year for this glacier?

Mass balance Hardangerjøkulen 2006–2007

Altitude (masl)	Area (km²)	Specific winter balance (m w.e.)	Specific summer balance (m w.e.)
1850–1865	0.09	2.70	−1.45
1800–1850	3.93	3.45	−1.50
1750–1800	4.03	3.60	−1.60
1700–1750	3.46	3.50	−1.75
1650–1700	1.94	3.20	−1.90
1600–1650	0.75	2.70	−2.10
1550–1600	0.59	2.35	−2.30
1500–1550	0.57	1.95	−2.50
1450–1500	0.29	1.73	−2.70
1400–1450	0.19	1.56	−2.95
1350–1400	0.10	1.39	−3.20
1300–1350	0.10	1.22	−3.45
1250–1300	0.27	1.05	−3.70
1200–1250	0.36	0.80	−3.95
1150–1200	0.28	0.55	−4.25
1100–1150	0.11	0.30	−4.55
1020–1100	0.05	−0.05	−4.90

2. *Glacial water balance.* We are interested in knowing when water is accumulating within a glacier, when it is steady, and when it is losing water. Kennicott Glacier is 40 km long and 4 km wide. The average (water-equivalent) melt rate at the glacier surface over the melt season is 5 cm/day, as measured using ablation stakes. Water discharge is also measured at the terminus of the glacier.

Estimate what the discharge of water should be, in m³/s, when the water outputs from the glacier equal the water inputs.

3. *Glacier thickness.* We wish to estimate the local thickness of a glacier given only a topographic map of the glacier. Assume that the basal shear stress $\tau_b = 0.8$ bars $= 8 \times 10^4$ Pa, and that the walls are far enough away (the glacier is wide enough) that they play no role in supporting the ice. The contours on the map show that the ice slope is 10 m/km or 0.01 at this location. Given the density of ice $\rho = 917$ kg/m³ and $g = 9.8$ m/s², estimate the ice thickness H at this location.

4. *Glacial terminus position.* Given a simple glacial valley with a slope of 0.01, such that the elevation of the bed $z = 3500 - 0.01x$, a uniform valley width, and a mass balance profile $b(z) = 0.01*(z - \text{ELA})$, where the equilibrium line elevation $\text{ELA} = 3000$ m, calculate the expected terminus position for a steady glacier. The maximum elevation of the valley, at $x = 0$, is 3500 m. Calculate and plot the down-valley pattern of ice discharge.

5. *Sea level rise.* Given a recent estimate of the rate of loss of glacial ice to the ocean of 400 Gtonne/year, what rate of sea level rise would this produce?

6. Calculate the expected thickness of ice that can be melted from the base of a glacier due to geothermal heat flux. Assume that the glacier is temperate to its base (why is this important?), and that the geothermal heat flux is 60 mW/m^2.

7. From the plot of GPS-derived horizontal speeds on the Bench Glacier in Figure 8.18(a), estimate (a) the background speed of the glacier that is presumably attributable to internal deformation of the ice, and (b) scale the maximum sliding speeds during the two speed-up events at the GPS 3 site.

8. Estimate the flow-law parameter A from the ice speed profile of the Worthington Glacier in Figure 8.11(b). Assume the local slope of the glacier at the borehole location to be 60 m/km, and that the ice obey's Glen's flow law ($n = 3$).

9. How thick would ice have to be to remain grounded in the middle of Sam Ford Fjord on the eastern edge of Baffin Island, where present water depths are 800 m. How thick would it have to be if sea level were 120 m lower (i.e., during a glacial lowstand)? List what assumptions you are making in such a calculation. Note that it is required that the ice be grounded for some fraction of the time in order to erode the bedrock floor of the fjord.

10. *Thought question.* Given the climate oscillations that have characterized the Quaternary, as summarized in the marine isotope record in Figure 8.53, discuss how the alpine landscape ought to respond. The answer should include addressing how glaciers might respond, and how sediment released from the glaciers might drive landscape change in the river system downstream.

11. *Thought question.* Ponder the design of a means of measuring the flow speed within a subglacial conduit through which water enters from the glacier surface through a moulin and exits the glacier in an exit stream. For safety reasons, you are not allowed to enter the conduit yourself, meaning you will need to find some remote means of measuring it.

12. *Thought question.* Design a field experiment to measure the sliding history of a glacier over the course of a full annual cycle. Talk about the instrumentation that will be required, and where and how you will deploy it. In other words, produce a field plan.

Further reading

Benn, D. I. and D. J. A. Evans, 1998, *Glaciers and Glaciation*, London: Arnold, 734 pp.
This broad compilation of the literature provides the student with a snapshot of our knowledge of the field as of the late 1990s.

Hooke, R. and B. Le, 2005, *Principles of Glacier Mechanics*, 2nd edition, Cambridge: Cambridge University Press.

This book provides an alternative entrance point for the student of glaciology wishing to learn about the mechanics of glaciers. It is both rigorous and accessible, leaning heavily upon the author's broad experience in the field.

Menzies, J. (ed.), 1996, *Modern Glacial Environments: Processes, Dynamics and Sediments, Vol. 1:*

Glacial Environments, Oxford: GBR, Butterworth-Heinemann Ltd.
An edited volume with chapters by leaders in the field, this book crosses the discipline from glaciers to their influence on the landscape.

Paterson, W. S. B., 1994, *The Physics of Glaciers*, 3rd edition. Pergamon, Elsevier Science, 480 pp.
Long the bible of glaciology, and now in its third edition, this textbook provides the basic reference in the field.

Pfeffer, W. T., 2007, *The Opening of a New Landscape: Columbia Glacier at Mid-Retreat*, AGU Special Publications Series, **59**, 116 pp.
This slim book provides, quite literally, a snapshot of the landscape emerging from beneath this fast-retreating tidewater glacier in Alaska. Because so many early glaciologists contributed to its study, and the issues involved in such glaciers touch upon many aspects of glaciers, the text also provides something of a history of glaciology. The photography is stunning.

Sharp, R. P., 1991, *Living Ice: Understanding Glaciers and Glaciation*, Cambridge: Cambridge University Press, 228 pp.
As with most things Bob Sharp touched, this little book is charmed. It provides an avenue for the general reader to access how glaciers work, and how the landscape has evolved as they have come and gone in alpine landscapes.

Periglacial processes and forms

In the bleak midwinter, Frosty wind made moan,
Earth stood hard as iron, Water like a stone . . .

Christina Rossetti

In this chapter

The actions of frost and freezing temperatures produce unique landscapes shaped by frost heave and by soils that alternate between solid and semi-liquid state. As much of the land surface is either currently subject to freezing conditions, or has been shaped by freezing conditions in the past, the study of periglacial processes is relevant to much of the Earth's surface. At present, mean annual temperatures are below 0 °C for 35% of the terrestrial land surface. During past glaciations, periglacial processes operated over even greater areas of the earth. The word "periglacial" was first coined by a Polish geomorphologist (Lozinski, 1909) to describe Pleistocene-age features found in present-day Romania beyond the margin of the former Fennoscandian ice sheet. Today, however, the term periglacial describes non-glacial freeze–thaw processes in any setting. Hence, periglacial processes and landforms are not restricted to glacier or ice sheet margins. In periglacial environments, the phase change between water and ice can drive processes not found elsewhere. Periglacial processes are active today in alpine environments, in Siberia, in Tibet, and in vast tracts of Alaska and Canada.

The Arctic coastal plain in Alaska is patterned by ice wedge polygons in many areas, and dotted with innumerable lakes each summer. In winter, this landscape will resemble the imagery of Rossetti's mid-nineteenth century carol (photograph by R. S. Anderson).

Freezing and thawing brings profound change to regolith materials, with important geomorphic consequences. The near surface undergoes annual cycles of freezing and thawing in which the water content, soil strength and bulk density of the soil change dramatically. These changeable properties can conspire to produce spectacular patterned ground and dramatic solifluction lobes. In very cold climates, below this seasonally thawed "active layer" there may be perennially frozen ground called permafrost, and in places the permafrost can be quite thick. The presence of permafrost and seasonally frozen ground influence hydrology, impeding flow at times, providing a water source at others. Both the active layer and permafrost present engineering challenges: rates of creep in the summer may be abnormally high on low slopes, rocks and posts may be heaved out of the ground, thawed soils may slump. Global warming does not make these issues irrelevant; indeed, engineering problems may increase as the depth and duration of seasonal thaw increase, and ecosystem responses to warming affect carbon cycling and treeline. Ironically, one of the best indicators of global warming may actually be found in responses of periglacial landscapes.

Because it is the sub-freezing thermal state that defines periglacial landscapes, and the physical property changes associated with freezing and thawing that make periglacial processes unique, we begin this chapter with a detailed look at the thermal structure of the landscape. We will then progress to discuss the processes and landforms unique to periglacial environments. We conclude with a call to action for new research in these settings that are especially subject to rapid change in the face of global warming.

Definition and distribution of permafrost

The surface of the ground must freeze in order to trigger periglacial processes and their geomorphic consequences. The longer it remains frozen, and the deeper the freeze, the more extensive and the more intense these processes become. In temperate latitudes, surface temperatures can dip below zero, freezing parts of the subsurface. This can happen even if the mean annual temperature remains above 0 °C. Frost-penetration depths in such landscapes can be as much as several meters depending on local conditions. Something different happens when the mean annual temperature is less than 0 °C. While the surface

of the ground will freeze and thaw annually, below some depth it will remain frozen. *Permafrost*, or perennially frozen ground, is defined as any material (rock or soil) that remains below 0 °C for two years or more. Note that permafrost is defined by its temperature, and not by presence of ice. There are regions with dry permafrost, in which rock or sediments remain below 0 °C, but little ice is present. Periglacial regions are those subjected to cold, non-glacial processes. This includes regions with permafrost, but also the much broader regions where frost processes operate but permafrost has not formed.

At present, permafrost underlies 20–25% of Earth's land surface area, with most of this found in the northern hemisphere (French, 2007). We display a map of the northern hemisphere permafrost in Figure 9.1. The presence or absence of permafrost is controlled by the history of surface energy balance, the accounting of energy inputs and losses. Heat loss from Earth's interior and solar energy inputs at the surface are the energy sources at play. The tug-of-war between these energy sources leads to patchy development of permafrost in many areas. Where the surface temperature is very low, permafrost is *continuous*, meaning it underlies >90% of the land surface. The *discontinuous* permafrost zone is where 50–90% of the land surface is underlain by permafrost, while permafrost is considered *sporadic* where it underlies 10–50% of the land surface and *isolated* where it is found under less than 10% of the land surface. Subsea permafrost is found on the continental shelves around the Arctic Ocean basin, much of it probably relict permafrost from sea level lowstands during Quaternary glaciations. The area of discontinuous and sporadic permafrost greatly exceeds the area of continuous permafrost in the northern hemisphere. The present area of permafrost regions (of all types) is more than 1.5 times the area presently covered by glaciers in the form of ice sheets, ice caps, and small glaciers. The portions of the Earth's surface directly impacted by ice, permafrost, and periglacial processes were more extensive during the Quaternary, when glacial climates were the norm.

Thermal structure

The temperature structure, or architecture, of a permafrost region is shown in Figure 9.2. The top layer, called the active layer, seasonally thaws. We can think of the

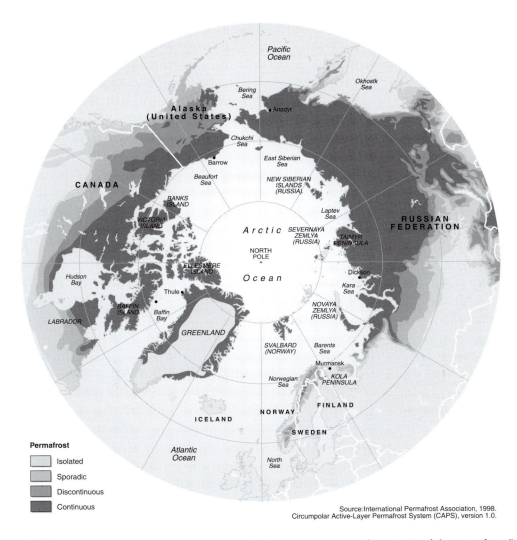

Figure 9.1 Permafrost distribution in the Arctic. Shading depicts degree of continuity of the permafrost (in UNEP/GRID-Arendal Maps and Graphics Library. Retrieved 14:29, May 20, 2009 from http://maps.grida.no/go/graphic/permafrost-distribution-in-the-arctic).

active layer as a thermal boundary layer in which temperatures shift over diurnal and seasonal cycles, reflecting the atmospheric and solar delivery of heat. Formally, the active layer is the depth at which the annual maximum temperature reaches $0\,°C$. Below this boundary layer, mean annual temperatures increase monotonically, following the geothermal gradient. At some depth, the temperature rises above zero again. This depth, where temperature re-crosses $0\,°C$, defines the base of permafrost. In Figure 9.3 we show a cross section through northern Canada in which both the active layer and the base of the permafrost are depicted. The active layer thins and the base of the permafrost deepens as one travels north into regions in which the mean annual surface temperature declines.

Let us be more formal about this heat flow problem. We introduced the concepts of conservation of

heat and of conduction of heat in solids in our treatment of the growth of the lithosphere away from spreading ridges in Chapter 3. Here the problem is slightly different in that we must deal with much shorter timescales of years to centuries as opposed to millions of years, and we must face the fact that the phase change of water absorbs heat upon melting and releases heat upon freezing.

Base of the permafrost

Consider first what sets the depth of the base of permafrost. We wish to determine what about the rock, and what about the climate sets the depth to the base of permafrost. Recall that the geotherm, the profile of temperature into the Earth, is set by the mean annual temperature, which pins the top of the

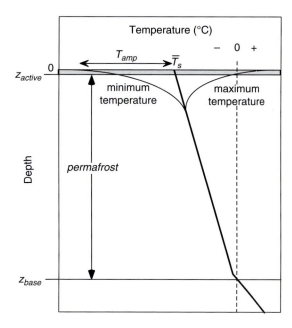

Figure 9.2 Schematic temperature profile through permafrost, showing the major features including the active layer (shaded), and the base of the permafrost (z_{base}). The mean annual surface temperature, $\overline{T}_s$, and the geothermal gradient, Q/k, set the mean annual temperature profile, shown in bold. The change in slope just above z_{base} reflects the difference in thermal conductivity (k) between ice-bearing and wet material. At depth, the pressure melting point is below 0 °C, hence in thick permafrost this kink may be located above z_{base}. Temperatures fluctuate within limits shown by the curves labeled "maximum temperature" and "minimum temperature." The amplitude of these fluctuations, T_{amp}, decreases with depth. Permafrost is the region where the temperature is below 0 °C for two or more years; this definition is met only below the active layer.

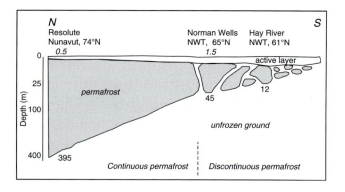

Figure 9.3 Cross section through permafrost in northern Canada, showing thinning of the permafrost and thickening of the active layer from north to south. Depths (in meters) to base of the permafrost are noted at three locations, while thicknesses of active layer are shown in italics. Note the nonlinear depth scale. Discontinuous permafrost is depicted as beginning about 66°N, and degrades to scattered patches south of 61°N (after R. J. E. Brown, 1970, Figure 4, reprinted with permission of University of Toronto Press, Inc.).

profile, and the geothermal gradient (the slope of the profile, dT/dz), which is set by the local geothermal heat flow, Q, and the thermal conductivity of the local rock, k. This is the steady-state profile, the one we would expect if the mean annual temperature at the surface, $\overline{T}_s$, had remained constant for a very long time. In other words, it comes from taking the left-hand side of the thermal diffusion equation (Equation 3.7), dT/dt, to be zero and then integrating twice. The equation for the mean thermal structure is therefore

$$\overline{T} = \overline{T}_s - \frac{Q}{k}z \qquad (9.1)$$

At first glance it may appear that Equation 9.1 shows temperatures decreasing with depth; note, however, that the heat flux Q is negative in the negative z-direction (toward the surface). We can solve

Equation 9.1 for the depth of the base of the permafrost by inverting the equation to one for z, and setting the temperature to be $\overline{T} = 0$ °C. The solution is

$$z_{base} = \overline{T}_s \frac{k}{Q} \qquad (9.2)$$

Since it is unlikely that you will have memorized typical values for heat flow or thermal conductivity, recall at least that typical geothermal gradients are of the order of 25–30 °C/km. Recalling Fourier's law (Equation 3.2), the base of the permafrost is $z_{base} = -\overline{T}_s/(dT/dz)$. For example, if the geothermal gradient is 25 °C/km, and the mean annual temperature is −30 °C, the base of the permafrost is 30/25 = 1.2 km. Everything between depths of about 1 or 2 m (the base of the active layer) and 1200 m is frozen stiff – "permanently." Note that this material does not have to contain water; the temperature simply has to be below 0 °C, the temperature at which bulk water freezes.

We plot in Figure 9.4 the temperature profiles for several mean annual surface temperatures, and the resulting depth to the base of the permafrost. We see that the thickness of permafrost is most strongly controlled by the mean annual surface temperature, $\overline{T}_s$. The geothermal gradient also controls permafrost depth, but since this parameter does not show as wide a range of variation as surface temperatures, this is a secondary effect. The latitudinal variation in

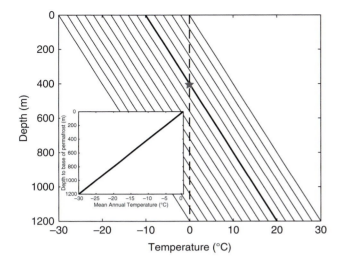

Figure 9.4 Mean annual temperature profiles in permafrost terrain for a range of mean annual surface temperatures, $\overline{T}_s$, from –30 to 0 °C. The temperature gradient, dT/dz, is held fixed at a typical 25 °C/km. Heavy line shows profile for $\overline{T}_s = -10$ °C, with depth of permafrost (*) at 400 m. Depth of permafrost, z_{base}, is linearly dependent upon the mean annual surface temperature (inset).

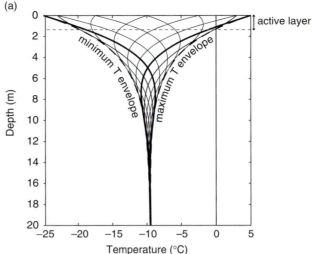

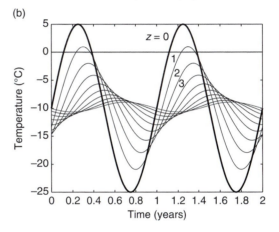

Figure 9.5 Temperature variations in the near-surface region for a mean annual temperature ($\overline{T}_s$) of –10 °C and amplitude at the surface (T_{amp}) of 15 °C. (a) Temperature profiles at intervals throughout a year. Heavy dashed lines show the envelopes of the maximum and minimum temperatures reached as a function of depth. Note that these extremes are reached at different times at each depth, and hence these envelopes do not show a snapshot of the temperature at a given time. Heavy lines show instantaneous temperature profiles for the times of maximum and minimum surface temperature. (b) Temperature over a two-year period at 1 m depth intervals from 0–10 m. Note the decay in amplitude and the temporal lag with increasing depth.

permafrost thickness shown in Figure 9.3 shows the effects of surface temperature variation.

Active layer depth

Now we ask what sets the depth of the active layer? This question is of importance to engineers designing structures for periglacial regions and to understand biogeochemical activity in periglacial settings. Because of the high organic matter content in many tundra soils and underlying permafrost, increases in active layer depth are a possible source of CO_2 emissions to the atmosphere. As the active layer is underlain by essentially impermeable permafrost, this layer dictates the hydrology of such landscapes. Hillslope processes in periglacial environments are strongly controlled by depth of thawing and water content. All of these problems – engineering, biology, hydrology, mass wasting – revolve around seasonal development of the active layer.

Temperatures change significantly in the "thermal boundary layer" of the ground, driven by annual oscillations in surface temperatures. Figure 9.5 displays in detail the temperature fluctuations in an idealized active layer and upper permafrost. We use two simplifications in this idealization: first, we ignore the latent heat associated with the freezing and thawing, and

second, we treat the surface temperature fluctuations as a sinusoid. These simplifications reduce the thermal problem to one of pure conduction driven by an easily characterized forcing. This approach is the same as our treatment of temperature fluctuations driving frost-weathering processes in Chapter 6. In fact, temperatures vary in the subsurface in the manner similar to that depicted in Figure 9.5 anywhere that has a seasonal climate signal. The temperature at depth varies as a damped and lagged version of the sinusoidally varying surface temperature:

$$T = \overline{T}_s - \frac{Q}{k}z + \left\{ T_{\mathrm{amp}} \exp\left(-\frac{z}{z_*}\right) \sin\left(\frac{2\pi t}{P} - \frac{z}{z_*}\right) \right\} \quad (9.3)$$

This is the most important equation describing the thermal behavior of periglacial landscapes. The first two terms on the right-hand side correspond to the mean annual thermal structure (just like the geotherm we used to determine the depth of the base of permafrost). The last term describes the oscillation of temperatures about that mean. There are two characteristics captured in this term: the decline in the magnitude of temperature oscillations with depth and the lag in the timing of those variations with depth. The annual temperature swings of magnitude T_{amp} at the surface fall off exponentially with depth, and at depth z_*, the amplitude is $1/e$ or about $1/3$ of that at the surface. The depth scale, $z_* = \sqrt{\kappa P / \pi}$, is set by the thermal diffusivity, κ, of the regolith, and the period of the oscillation, P. The depth scale for annual oscillations, determined from typical values of κ of about 1 mm^2/s, and, by noting that the period P of 1 year is about $\pi \times 10^7$ s, is about $\sqrt{10^7}$ mm, or $z_* \approx 3$ m. In an area with $T_{\mathrm{amp}} = 15\,°$C at the surface, temperatures will have an annual amplitude of about $5\,°$C at about 3 m depth (see Figure 9.5(a)).

The timing of temperature maxima and minima also varies with depth. Examine Figure 9.5(b) to see that at a depth of about 10 m, the annual temperature maximum occurs at about the time of the annual temperature minimum at the surface. The lag of the oscillation in temperature with depth is described by the sine part of the last term in Equation 9.3.

We can now address the question of the depth of the active layer. Since the base of the active layer is where the annual temperature maximum reaches $0\,°$C, we can answer this question by solving Equation 9.3 for this depth. We can simplify this equation by noting that the sine term simply bounces between -1 and 1. We will therefore find the maximum temperature at any depth by setting this sine term to 1 (or, conversely, find the minimum by setting it to -1). We can also eliminate the need to include the geothermal gradient by evaluating its importance within the region of concern – the top few meters. Given that the gradient is about $25\,°$C/km, or $0.025\,°$C/m, over a few meters the temperature change associated with geothermal heat flow is less

than $0.1\,°$C. This is trivial compared to the temperature oscillations at the surface, and we can neglect them (see this minor effect in the base of the profile in Figure 9.5(a)). The equation can now be inverted to solve for z_{active}:

$$z_{\mathrm{active}} = -z_* \ln\left[\frac{-\overline{T}_s}{T_{\mathrm{amp}}}\right] \quad (9.4)$$

Graphically, we have simply solved for the intersection of the right limb of the envelope within which temperatures range (the curve labeled "maximum T envelope" on Figure 9.5(a)) with $0\,°$C. Typical values for the amplitude of the thermal swing, T_{amp}, are about $15\,°$C. Obviously, if $T_{\mathrm{amp}} < \overline{T}_s$, there is never any thaw and the active layer depth goes to zero. A more typical case, with $T_s = -10\,°$C and $T_{\mathrm{amp}} = 15\,°$C, $z_{\mathrm{active}} = 0.4z_*$, or about 1.2 m, given the 3 m value of z_* we calculated above. As can be seen in Figure 9.3, this in the range of values for active layer depths. The dependence on the mean annual temperature is as expected from the theoretical Equation 9.4: the layer thickness declines as the mean annual temperature declines, as we depict in Figure 9.6.

Another question we might ask of the thermal state of permafrost is when within the year the maximum thaw depth occurs. To address this, we return to the sine term in Equation 9.3, which prescribes how temperatures vary around the mean (in this case sinusoidally; for other types of oscillations, a different expression is needed). The lag in temperature oscillations at a particular depth z is given by z/z_*. For example, at $z = 0$, there is no lag and the sinusoidal temperature variations are given by $\sin(2\pi t/P)$, which describes the temporal variation of the surface temperature. At the depth where $z/z_* = 2\pi$, the thermal state lags a full cycle (2π) behind that at the surface. (Compare the argument of the sine term at $z/z_* = 0$ and at $z/z_* = 2\pi$ to convince yourself that the second case is a full sinusoidal cycle behind the first case.) To determine when maximum thaw depth occurs, we assess the lag for the depth of maximum thaw, z_{active}. By dividing Equation 9.4 by z_*, we see that

$$\mathrm{lag}(z_{\mathrm{active}}) = \frac{z_{\mathrm{active}}}{z_*} = -\ln\left[\frac{-\overline{T}_s}{T_{\mathrm{amp}}}\right] \quad (9.5)$$

For the case we examined above, in which $T_s = -10\,°$C and $T_{\mathrm{amp}} = 15\,°$C, we find the lag $= 0.4$ radians, or $0.4/(2\pi)$ year. This is 0.063 yr, or 23 days. The maximum thaw depth should occur almost a month after the maximum surface temperature, or mid to late August

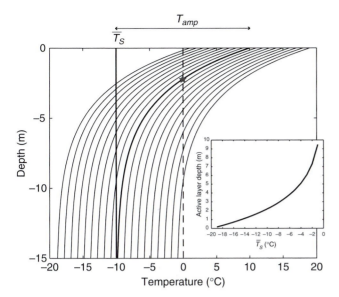

Figure 9.6 Curves showing the maximum temperature reached as a function of depth, for a sinusoidally varying surface temperature with an amplitude, T_{amp}, of 20 °C. Each curve represents a different mean annual temperature, the value of which is given by the lower asymptote. Heavy curve shows the maximum temperature envelope for the mean annual surface temperature, $\overline{T}_s$, of –10 °C (heavy vertical line). The depth at which the curve crosses 0 °C defines the base of the active layer (* for $\overline{T}_s = 10$ °C; relationship summarized for all $\overline{T}_s$ in inset). The active layer vanishes for a mean annual temperature of –20 °C, and is infinite for a mean annual temperature of 0 °C. These curves are *not* temperature profiles; the time of the annual maximum temperature varies with depth, as shown in Figure 9.5(b).

in the northern hemisphere. This lag can be seen in Figure 9.5(a), paying particular attention to the two profiles with the warmest surface temperatures. The mid-summer thermal profile, that with the warmest surface temperature (about 5 °C), intersects 0 °C at a depth of about 1 m, while the next month's profile, while having a lower surface temperature of about 2.5 °C, has warmed up at shallow depths to become tangent to the maximum temperature envelope at a depth of about 1.2 m.

Latent heat

We have made the simplifying assumption that we can ignore heat associated with the phase change of water. While we have captured the essence of the problem, showing the down profile changes in the magnitude of temperature oscillations, and the lag of those oscillations with depth, melting and

freezing ice in soils causes some departures from the patterns we have seen. Thermal problems in a system in which a phase change occurs have been called Stefan problems in honor of the Austrian physicist Josef Stefan (1835–1893), who in 1891 published a physical analysis of the formation of ice on polar seas. (This is the same Stefan who determined empirically that radiative heat flux varies as the fourth power of the absolute temperature, a relationship used in Chapter 5 in our discussion of the radiation balance of the Earth as a whole.)

Freezing or melting water in soil introduces either a heat source (in the case of freezing) or a heat sink (in the case of melting) into the thermal problem we have been considering. Water has a very high latent heat of fusion, so the energy implications of a phase change are significant. The heat capacity of water, the energy required to heat or cool 1 g of water by 1 °C, is 4.2 J, while the latent heat released upon freezing or absorbed during melting of that same 1 g of water is 334 J. If we think about the energy deposited on the ground surface as a finite resource (the Sun delivers only so much energy to the surface), then it is easy to see that the absorption of heat to melt ice will diminish the depth to which a thaw front will penetrate. Thus, our calculation of the active layer depth without a latent heat source term overestimated the active layer depth by an amount that depends on the water content. Similarly, the penetration of a freezing front into the subsurface is slowed by the need to conduct the latent heat released out to the atmosphere.

We will make use of the final result of a full development of the Stefan problem (which can be found in texts such as Turcotte and Schubert, 1982). Consider a half space initially at uniform temperature below the melting temperature (i.e., it is frozen). If this half space is subjected to a step change to a surface temperature above 0 °C, the temperature evolution looks much like the square root of time solution for a boundary layer like the lithosphere (Equation 3.8), but with different values for the constants. The depth of thaw, z_{thaw}, is

$$z_{thaw} = \left(\frac{2k \, \Delta T_s \, t}{\rho_{uf} L} \right)^{1/2} \tag{9.6}$$

where k is thermal conductivity of the unfrozen soil, t is time, ΔT_s is the step change in temperature at the surface, ρ_{uf} is the bulk density of the unfrozen soil,

(a)

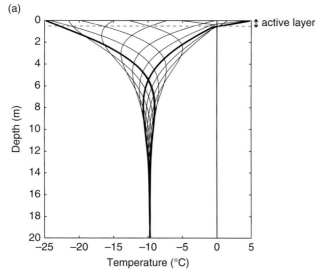

(b)

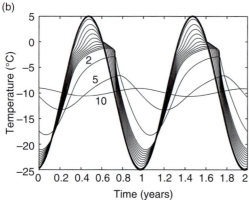

Figure 9.7 Temperature variations in the near-surface region for the surface temperature conditions identical to Figure 9.5, but now with latent heat effects included. Saturated regolith with 30% porosity extends to 1 m, underlain by bedrock with 3% porosity. (a) Temperature profiles at intervals through a year. The maximum temperature envelope departs from the exponential funnel predicted in the pure diffusion case (Figure 9.5(a)) and, as anticipated, the active layer is thinner, here about 0.5 m. (b) Two-year temperature histories at various depths show the important slowing of freezing front advance, known as the zero-curtain. The deepest temperature history is 10 m in depth, but depth intervals differ from those in Figure 9.5(b) to emphasize the role of latent heat effects (depths: 0–1 m at 0.1 m intervals; 2, 5, and 10 m).

and L is the latent heat of the soil (in J/kg). For example, for pure ice, with latent heat of fusion $L = 3.34 \times 10^5$ J/kg, and half a month (15 days) of surface temperature change at 5 °C, the thaw front would be roughly 0.15 m below the surface. For soils, the latent heat and bulk density terms need to be scaled to the water content (on mass basis). Nonetheless, this crude calculation yields the proper scale for the evolving depth of the frost table.

When we explicitly incorporate treatment of the phase change, the calculated temperatures in the subsurface display departures from the pure diffusion/ conduction case in expected ways, as shown in Figure 9.7. First, the need to extract the latent heat of fusion stalls the penetration of the thaw front, resulting in an active layer thickness that is much shallower than we have shown analytically in Figure 9.6. Second the subsurface temperatures remain very near the freezing temperature for significant periods of time as the phase change occurs. This is known as the "zero-curtain," and occurs during both freeze-up and thaw.

The active layer is much more than an academic concept. The permafrost table at the base of the active layer serves as a perfect aquiclude. Any water that penetrates this boundary, for example through a contraction crack, will freeze. The underlying permafrost might as well be concrete. As such this interface prevents penetration of roots, and defines the base of the biologically active Arctic sod or peat layer. In many places, the landscape is soggy, as the surface water has nowhere to go. When walking across the soggy tundra, the concrete-like base of the thawed layer is a welcome bottom to the muck, which otherwise feels bottomless.

Departures from the steady-state geotherm

Thick permafrost presents a relatively simple thermal system at depths below the active layer and below the influence of annual surface temperature fluctuations. The absence of liquid water means that heat flux is entirely due to conduction, described by the heat flow equation, $Q = -k \, (dT/dz)$. Careful measurements of the thermal structure of the upper few hundred meters of permafrost in northern Alaska and Canada, made possible in part by the many abandoned exploratory wells on the North Slope, have shown a persistent and ubiquitous departure from the expected linear temperature profiles, as revealed in Figure 9.8(a). In all cases, the top of the profile appears to be warmer than one would expect by projection of the geothermal gradient up from the base of the permafrost. Changes in the geothermal gradient, dT/dz, in a profile might reflect differences in thermal conductivity, k, with depth in the subsurface. However, measurements of the *in situ* thermal conductivity of the rock have eliminated non-uniformity in the thermal conductivity as an explanation for these departures. The warm tops of these profiles

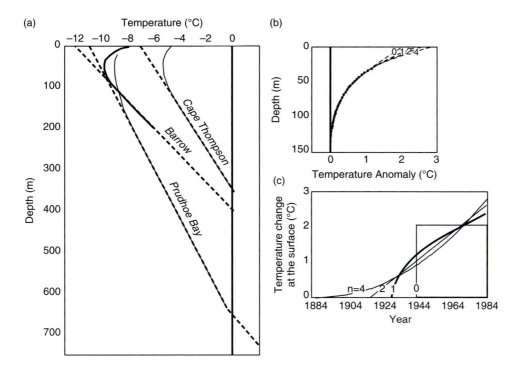

Figure 9.8 (a) Temperature profiles from three sites on the northern coast of Alaska. Cape Thompson and Prudhoe Bay have similar geothermal heat flow, Q, but significantly different mean annual surface temperatures, $\overline{T}_s$. Barrow and Prudhoe have similar $\overline{T}_s$ but differ in Q. All profiles show significant departures from the expected linear profiles (dashed) within the top 100–200 m (after Gold and Lachenbruch, 1973; Lachenbruch and Marshall, 1986). (b) Modeled temperature profiles following a change in $\overline{T}_s$ plotted as the difference from the steady-state geotherm, show that the temperature anomaly is greatest at the surface, and for the cases tested extends to depths of about 100 m. Four different cases of $\overline{T}_s$ history are modeled. (c) The four surface temperature histories used to produce the temperature anomaly profiles in (b). These include a step change 40 years ago ($n = 0$), a linear ramp ($n = 2$), and two more complex histories ($n = 2$ and $n = 4$), all constrained to yield a total addition of heat at the surface of about 200 MJ/m². Surface temperature changes of the order of 2–3 °C over 50–100 years are implicated by these models. Calculations are carried out assuming a thermal diffusivity of 1 mm²/s (after Lachenbruch and Marshall, 1986, Figures 8 and 12, with permission of the American Association for the Advancement of Science).

have instead been attributed to the long-term warming of the regional climate in this arctic environment. Lachenbruch and Marshall (1986) tested the influence of a variety of scenarios of surface temperature change – ranging from a simple step increase in temperature to a slow ramp up in temperature – on deep temperature profiles in permafrost. Their results, plotted as temperature anomaly profiles (Figure 9.8(b)), show that the depth and magnitude of the perturbation of the profile is well explained by rising $\overline{T}_s$. While they cannot resolve the details of warming histories (all histories in Figure 9.8c yield equally good fits), the general magnitude and timing of warming is discernable. As Lachenbruch and Marshall point out, these temperature profiles must be explained with changing surface temperatures. Departures from the steady-state isotherm cannot be attributed to changes in geothermal heat flux or

thermal conductivity. The only remaining parameter in Equation 9.1 is $\overline{T}_s$. The mathematical structure of this question is perfectly analogous to that of the depth of the lithosphere problem. The perturbation deepens with the square root of time dependence (sound familiar?). The 100 m-deep perturbation that is commonly observed in thick permafrost temperature profiles corresponds to about a century of warmer surface temperatures. Interestingly, this is the timescale over which humans have significantly altered the greenhouse gas content of the atmosphere. Unfortunately, it also corresponds to the end of the Little Ice Age, meaning that there has been room for debate about the role of greenhouse warming vs. the role of natural variations in climatic variables (such as solar radiation). In any case, the importance of these records of local climate change is that they are direct measurements of temperature – they are not

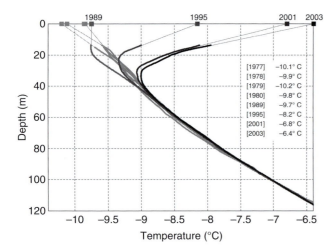

Figure 9.9 Permafrost temperature profiles were measured in borehole at East Teshekpuk Lake, Alaska (70° 35′ N, 153° 30′ W), over the last three decades, as part of the Global Terrestrial Network-Permafrost (GTN-P) deep borehole array. Repeated temperature profile measurements from 1977–2003 are plotted. Surface temperatures deduced from temperature profiles are shown as boxes at depth = 0, with the year indicated above the box. Rapid warming of the surface since 1990 is evident. Wind-driven coastal erosion encroached on the site, which was capped in 2008 (graphs provided by Gary Clow, USGS).

seen through the filter of a proxy such as vegetation change or a change in the isotopic content of snow, or water, or shell. Because we know well the physics of thermal conduction, and that in permafrost the only means of moving heat around is through conduction (no free water moves in this system), this thermal archive is an excellent one. In addition, it is in the right place on the globe. Global circulation models of the response of the Earth to changes in greenhouse gas content demonstrate that the strongest signals are to be expected at high latitudes.

To close this line of reasoning, we present recent borehole profiles obtained by Gary Clow of the US Geological Survey. Repeat measurement of borehole temperature profiles in the north slope of Alaska, and elsewhere around the Arctic is being done in the Global Terrestrial Network-Permafrost (GTN-P). Profiles from one network borehole at East Teshekpuk Lake, about 160 km southeast of Barrow, Alaska, over the last three decades are shown in Figure 9.9. While the profiles showed little change through the late 1970s and 1980s, they reveal large increases in temperature within the last 15 years, most dramatically since the mid-1990s. Extrapolation to a mean annual ground surface temperature shows

warming of ~3.8 °C at East Teshekpuk Lake from the early 1980s to 2005 (Figure 9.9); warming of ~3 °C is obtained in an array of boreholes across the Arctic Coastal Plain over this time period (Clow and Urban, 2002). In the course of our discussion of periglacial landscapes, we must bear in mind the potential impacts this warming might have on the types of processes available to the geomorphic system, and the rates at which they operate.

Geomorphology of periglacial regions

Ice is responsible for the interesting processes found in periglacial regions. It can be found in a variety of forms in the ground. *Pore ice* is water frozen in the interstitial space between particles. In fine-grained sediments, discrete macroscopic (mm to cm) layers or lenses of ice may grow, with an orientation orthogonal to the heat flow direction (commonly parallel to the ground surface). This is called *segregation ice* to highlight the differentiation into frozen sediment and layers of nearly pure ice. The ice content of sediments containing segregation ice can exceed the porosity, and hence can be rather high. The last form of ice is *massive ground ice*. As the name suggests, massive ground ice consists of very large (meter-scale) bodies of nearly pure ice. The ice masses can be oriented as horizontal tabular bodies, or as vertically oriented ice wedges. In some areas, massive ground ice comprises 20% or more of the volume of the subsurface. Our discussions will focus on segregation ice and on massive ground ice in the form of ice wedges.

Segregation ice and frost heave

In wet, fine-grained material, freezing generates discrete lenses of ice, forming segregation ice. Water actually migrates to the site of freezing to feed these ice lenses. If the freezing front is advancing downward from the surface (the top of the active layer), water will move up against gravity to contribute to the growing ice lenses. Segregation ice can also grow at the lower freezing front found at the top of permafrost or base of the active layer. Growth of ice lenses redistributes moisture, locally increasing the water content of the soil. As shown in Figure 9.10, is not uncommon to see ice contents in excess of the pore space in the unfrozen material, with the excess found

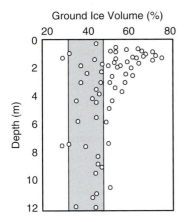

Figure 9.10 Ice content as a function of depth. Shaded box represents approximate saturation in soils with 30–45% pore space. That the near-surface materials contain such excess ice is evidence of the efficiency of the segregation ice process (redrawn from Pollard and French, 1980, with permission of the NRC Research Press).

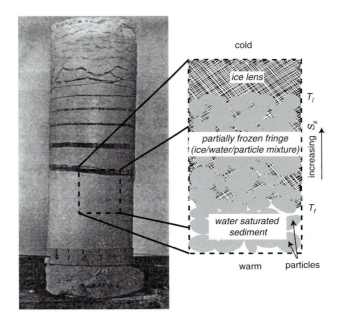

Figure 9.11 Photograph of frozen silt showing ice lenses (dark horizontal layers) from an experiment described by Taber (1930), reprinted with permission of the University of Chicago Press. Schematic of enlarged segment of the soil depicts transition from unfrozen water-saturated sediment through partially frozen fringe into pure ice of the basal ice lens. T_f corresponds to freezing temperature of water, T_l to temperature at the base of the ice lens, and S_s is the ice saturation, the volume fraction of the pore space occupied by ice (from Rempel, 2007, Figure 1, reprinted with permission of the American Geophysical Union).

in layers of nearly pure ice. Growth of these ice lenses deforms the soil and lifts the ground surface, a phenomenon called *frost heave*. Upon thaw, the elevated water content reduces the strength of the soil, in extreme cases to a soupy mess.

Stephen Taber did some of the most illuminating work on segregation ice and frost heave through a series of careful experiments in the early twentieth century using cutting edge laboratory apparatus – an electric refrigerator. An important observation is that frost heave was not solely due to the volumetric expansion of ice upon freezing. The final state of one of Taber's experiments is reproduced in Figure 9.11. A cylinder of clay was exposed to freezing temperatures at its the top surface, while a reservoir supplied water at the base. Macroscopic ice lenses are readily visible as dark layers. The surface of the clay cylinder heaved by an amount equal to the sum of the thickness of all the visible ice lenses. In many cases, frost heaving strains exceed by a factor of three or four the strain expected due to the volumetric expansion of water upon freezing. These observations showed that frost heave requires migration of water to the freezing front to grow discrete lenses of ice. Taber drove this point home by producing heave in a soil saturated with benzene, a fluid that contracts upon freezing. Applying confining pressures on the experimental apparatus slowed, but did not prevent growth of ice lenses and heave of the surface. Observations also reveal that

frost heave is greater in silty material, rather than sand or clay (Williams and Smith, 1989; p. 196).

Taber's experiments and subsequent work have shown that ice lenses grow at temperatures below the bulk water freezing temperature (the temperature at which water freezes in a large open vessel). The ice lenses are separated from unfrozen soil by the frozen fringe, a region in which liquid films and pore ice fill pore spaces. Water can move against gravity in porous media due to capillarity, the result of the tendency of water to wet mineral surfaces and the strength of the water–air interface (surface tension). However, recent work shows that growth of ice lenses arises from other phenomenon. Unfrozen water exists at sub-zero temperatures in porous media for two reasons. As the temperature falls below 0 °C, ice grows first in the largest pores, a manifestation of the effect of curvature of an ice surface on freezing point. Highly curved surfaces have greater surface energy, and hence require lower temperatures to

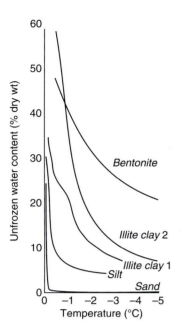

Figure 9.12 Unfrozen water content at sub-0 °C temperatures in several fine-textured materials (reproduced from Williams and Smith, 1989, Figure 1.4, with permission from Cambridge University Press).

form. This is called the Gibbs–Thomson effect, and is responsible for phenomena such as the growth of large crystals (with lower surface energy) at the expense of smaller crystals (with higher surface energy) in systems ranging from snow packs to crystallizing magma. Because the freezing point of water in small pores is lower than in large pores, ice forms over a range of temperatures in a porous media. The second reason unfrozen water can be maintained in soils below 0 °C is that any solid near its bulk freezing temperature is cloaked with a "premelted" layer, hundreds of molecules thick, in a quasi-liquid state (Rempel *et al.*, 2004). This layer of mobile molecules arises due to long-range intermolecular forces, such as van der Waals forces (recall that these are weak forces between molecules due to the interactions of dipoles, both induced and permanent), which disrupt the crystal structure. The premelted layer is found on ice surfaces at both ice–atmosphere and ice–mineral interfaces, and the layer increases in thickness as the bulk melting temperature is approached. These effects together can result in significant unfrozen water content in soils below the bulk melting point. The temperature range over which ice forms is greater in finer textured materials (Figure 9.12). In clay, for instance, unfrozen water contents of several tens of percent can be found at

temperatures as low as −4 °C, while sand has very little unfrozen water at sub-zero temperatures.

Water flows through the frozen fringe (Figure 9.11) from warmer to colder areas, where it can lower its free energy by changing phase to ice. Normal hydrodynamics govern water flow in the network of thin water films that feed water to a growing ice lens. Water pressure within the films is low due to a repulsive force between mineral and ice surfaces, arising from the same long-range intermolecular forces that maintain the premelted layer at the ice surface (Rempel *et al.*, 2004). The water flux down the pressure gradient is mediated by permeability of the soil and viscosity of the water. Thicker films promote higher fluxes, and higher viscosity reduces fluxes. As both the film thickness and the viscosity of water decline rapidly with lower temperatures, the rate of water transmission to the ice lens declines at lower temperatures. Interplay between water flow rate toward an ice lens and downward advance of the 0 °C isotherm (the freezing rate) controls the growth of ice lenses. When the freezing rate is greater than the rate of ice lens growth, the ice lens is abandoned as a preferred site of freezing, and a new ice lens begins to form at a slightly warmer temperature. If the freezing rate is lower than the rate of ice lens growth, a single ice lens can grow indefinitely.

Frost heave of the surface is favored in materials and under conditions that promote growth of ice lenses. Materials in which frost heaving is significant are considered to be *frost susceptible*. The presence of silt-sized material greatly enhances the frost susceptibility of a soil. Engineers label materials with >3% silt content as frost susceptible. The mobility of unfrozen water in silt compensates for the lower unfrozen water contents compared to clayey material. Environmental conditions conducive to frost heave include the presence of water and low rates of freezing. Slower freezing simply allows more time for water migration to growing ice lenses. In many cases, water supply places strong controls on ice lens growth. For a flat-lying area, with no upslope water source, the water content in the active layer at the beginning of winter places a limit on frost heave. As the freezing front advances down through the active layer, the supply of water declines, even as the rate of cooling slows at greater depths. Frost heave can be observed in summer in permafrost terrain due to freezing at the top of the permafrost, with the water released from thawing in the active layer.

Box 9.1 Art Lachenbruch and the Trans Alaska pipeline

One of the engineering marvels of arctic North America is the Trans Alaska Pipeline, maintained by the Alyeska Pipeline Service Company. This 1.2 m-diameter pipe traverses 1280 km of complex terrain from Alaska's North Slope oil fields to the ice-free port of Valdez, crossing landscapes underlain by continuous and discontinuous permafrost. Among many challenges, engineers had to accommodate the weak soils of recently thawed permafrost. Somewhat amazingly, the first pipeline designs called for the burial of a heated pipeline across this landscape. Dr. Arthur Lachenbruch, a USGS Arctic scientist who had studied the thermal state of permafrost in Alaska for a number of years, learned of this design in the men's restroom one day, in a conversation with one of the Alyeska engineers. Art called up the hearing commissioner to suggest that this might be a significant issue, and was asked to submit a letter to this effect. Art stayed up for most of the week between then and the next hearing, and performed the calculations that eventually became the core of a USGS Circular 632, "Some estimates of the thermal effects of a heated pipeline in permafrost" (Lachenbruch, 1970). His argument was that within a few years a heated pipeline would inevitably thaw the permafrost surrounding it. High ice contents found in many areas due to segregation ice and massive ground ice meant that the thawed material would be too weak to support a pipe. The likelihood of material failure and pipe rupture was especially high where the pipeline was running up or down a slope. Obviously, the consequences of a rupture were enormous both to the pipeline company and to the environment. The hearing commissioner suspended development of the pipeline (on which contracts for initial construction had already been let!), and the engineers were told to go back to the drawing board. Any proposed solution had to pass through Lachenbruch. The delay was a couple of years, but the redesigned pipeline has been operating since 1977 without serious mishap, pumping on average nearly 650 000 barrels per day. Over half of the pipeline is above-ground over regions of ice-rich permafrost, and as shown in the photograph in Figure 9.13 is supported by H-shaped vertical support members (VSM) that are anchored deeply in the ground. In particularly sensitive permafrost terrain, each VSM serves as a passive refrigerator, pumping heat from the ground when it is warmer than the air, and dumping it from heat fins into the local atmosphere. The active layer is de-activated, and is maintained in its strongest, frozen state. The maintenance of the pipeline includes overflights in helicopters that image each of the refrigeration units using infrared photography.

(See the Lachenbruch story on the website of USGS retirees: http://menlocampus.wr.usgs.gov/50years/accomplishments/pipeline.html)

The amount of heave varies spatially over small distances because of differences in water content, soil texture, and rate of freezing as controlled by surface energy balance. Undulations in the ground surface due to differential frost heave have hampered road construction, railways, and other human constructs. Differential frost heave disrupts and distorts soil profiles, an effect known as cryoturbation.

Upfreezing of stones

A consequence of frost heave is the upward transport of stones to the ground surface, a process called *upfreezing* or frost-jacking. The phenomenon is well known to farmers in cold regions, such as northern Europe and New England, and influenced some of Robert Frost's poetry (e.g., Stone Boats). Particle size sorting found in periglacial environments is attributed to the long-term effects of upfreezing on initially unsorted mixed-grain size sediments. For decades, the mechanism of upfreezing was debated in the geomorphology community. Some argued that clasts embedded in mixed grain size regolith were expelled toward the surface by the formation of ice lenses beneath them. This became known as the frost-push hypothesis. Others claimed that clasts were pulled upward with frost heaving soil within which the clast was embedded (see Figure 9.14). An experiment

Figure 9.13 The Trans Alaska Pipeline System transports oil across Alaska from the oilfields on the North Slope to the ice-free port of Valdez. For over half of its >1200 km length, the pipeline is above ground on vertical supports that are passively cooled to protect ice-rich permafrost. Aluminum cooling fins project above the pipe. The pipeline is placed in a zig-zag configuration, and can slide laterally between pairs of vertical supports to allow for thermal expansion and earthquake motion.

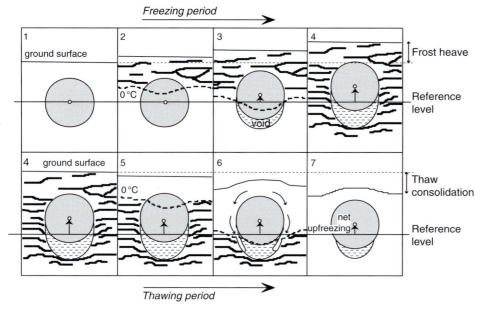

Figure 9.14 Schematic diagram of the frost-pull hypothesis after Beskow (1930). Downward freezing is depicted in the top set of panels, while downward propagating thawing is shown in the bottom set. Motion of clast and ground surface is shown relative to a reference level, which initially passes through the clast mid-point. Heavy lines are ice lenses. In panel 2, the clast has just adhered to freezing soil; in panels 3 and 4, the clast moves up with the frost heaving soil, leaving a void in its wake. Partial infilling of the void upon thaw, shown in panel 6, prevents the clast from returning to its starting position.

designed to test these hypotheses demonstrated the frost-pull mechanism (S. P. Anderson, 1988a) through observations of temperatures within and displacement of a clast and surrounding soil in a laboratory. Rapid cooling of the rock, the basis of the frost-push hypothesis, and predicted due to the higher thermal diffusivity of rock than soil, was not observed. Instead, through seven freeze–thaw cycles, the $90 \times 40 \times 40$ mm clast began moving upward with the frost heaving soil when the $0\,°C$ isotherm has progressed about one-third of the way through the clast. The inference is that upward motion is allowed only when the adfreeze bond between the frozen soil

and the clast has become strong enough to support the weight of the clast. Upward motion of the clast produces a void beneath the clast into which unfrozen silt slumps, as in the illustration from Beskow reproduced in Figure 9.14. It is this slumped material that prevents the clast from reoccupying its previous position upon thaw, and results in net upward motion of the clast relative to the surrounding soil.

The rate of upward migration of stones due to upfreezing can be cast as the product of the length of the stone (measured in the heat flow direction), the strain of the surrounding soil due to frost heave, and the frequency of freeze–thaw cycles. The process occurs in mixed grain size material in which frost-susceptible soil is studded with large objects or stones. Larger stones will tend to move toward the surface faster. Objects with tall aspect ratios such as pilings and fence posts are subject to severe frost jacking if no precautions are taken to limit the frost heave of surrounding soil. Objects subject to upfreezing must have a dimension in the heat flow direction that is larger than typical ice lenses, so that frost-heaving strain in the surrounding material will pull it upward. If the object is too small, it will move with soil particles, without a void opening beneath from differential motion.

Patterned ground

A variety of forms of patterned ground are found in periglacial environments, which have been extensively cataloged and discussed by Washburn (1979). As the name implies, patterned ground consists of geometric or repeated patterns on the ground surface. These patterns are manifested in sorting of surface materials (in sorted patterned ground) or not (non-sorted patterned ground), variations in vegetation, or micro-topography, or in some cases by a network of cracks. While discussing each form is beyond the scope of this text, we spend a little time on sorted circles, one of the more spectacular features of periglacial landscapes. Sorted circles are perhaps best exemplified in coastal sites in Svalbard (Figure 9.15). These circles are not giant bird nests, but are instead some of the best examples of self-organization in geomorphology. In the case shown, they have evolved from what was once beach deposits on a bedrock platform that has since been raised well above sea level by postglacial rebound. The slightly mounded centers of the circles are silt-rich, which contrast markedly with the clean pebbles and cobbles found in raised borders, and in the intervening non-patterned area. Sorted circles typically have diameters from one to a few meters.

As in any naturally occurring pattern, one must ask what processes allow the form to arise seemingly out of nowhere, and what sets the length scale of the pattern. Why choose 2 m and not 20 cm or 20 m? These features have been well studied over the last two decades. Major field investigations of the Svalbard circles by Hallet and coworkers (S. P. Anderson, 1988b; Hallet and Prestrud, 1986; Hallet et al., 1988; Hallet, 1990; Hallet and Waddington, 1992) documented the internal structure of the forms, the

Figure 9.15 Sorted circles developed in raised beaches of Kvadehuksletta, Svalbard. Fine-grained circle centers are bounded by coarse borders that are raised relative to the mean topography. Circle diameters 1–2 m (photograph by R. S. Anderson).

meteorology of the sites, the seasonal evolution of the thermal field, the heaving and subsidence of the fine centers, and the horizontal displacements of both the fine and coarse domains. The distinct sorting and sharp microtopography of the sorted circles in Figure 9.15 must be maintained by an active process or suite of processes; otherwise the actions of running foxes, needle ice, wind and gravity will tend to smooth the features away. Radial surface motion has been documented, revealing a pattern depicted in Figure 9.16 that suggests, when averaged over a year or more, something like convection. But it is a halting sort of motion best explained by the seasonal heaving of the active layer. In Svalbard, the active layer is about 0.8 m thick. In the winter the fine center heaves significantly upon freezing, as the frost-susceptible silt develops numerous ice lenses. Heave is greatest in the circle center, where the fine-grained layer is thickest, which in turn steepens the outward slope. The borders do not heave significantly as they consist primarily, although not exclusively, of coarse particles. In the summer, the thaw progresses downward from the surface. As it does so, the soil becomes weak upon melting of the ice lenses, as the underlying frozen silt acts as an aquitard. The thawed soil then creeps radially down the slightly steepened slope, as can be deduced from the pattern of displacement shown in Figure 9.16(a). In fact, excavation of trenches across these features has shown that a thin organic carpet on the surface is carried down below the inner edge of the coarse borders in a pattern

(a)

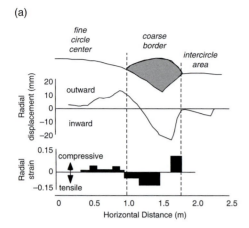

(b)

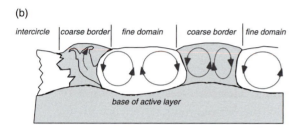

Figure 9.16 (a) Measured net displacements, and calculated strains in the summer 1984 thaw season at Kvadehuksletta, Svalbard. Inward displacements are toward circle center; outward are toward circle exterior. Strains calculated from gradients in the radial displacement field. (b) Patterns of inferred motions within the fine centers and coarse borders that are most compatible with the measured displacements (after Hallet and Prestrud, 1986, Figures 5 and 7, with permission from Elsevier).

eerily similar to subduction of lithosphere. The coarse borders in turn perform a rolling motion, rolling inward as their base is sheared. But this pattern of motion fails to explain the extreme sorting that makes these features so dramatic. The sorting comes from the upward motion of the coarse stones as they are ejected from the fine matrix by the process of upfreezing described above. Over the course of many turnovers, the pattern cleans itself up, leaving the coarse stones on the borders with no way to get back into the middle.

This suite of linked periglacial processes and the resulting array of sorted circles has been modeled by Kessler *et al.* (2001), as seen in Figure 9.17. The pattern emerges spontaneously from an initially randomly mixed soil. While the circles are small when first formed, they grow to a stable size that effectively fills the calculation space. The spacing appears to be controlled by the depth of the system, here controlled by the depth of the active layer; the final wavelength of the

features is roughly three to four times the active layer thickness. As a strong test of the viability of the model, many other oddities of these patterns are captured. For example, (1) initially oblong features are seen first to neck in the middle and then to split or bifurcate, and (2) the pattern elongates when evolving on a slope.

Ice wedge polygons

Extensive networks of ice wedge polygons are another striking form of patterned ground in the Arctic. An example is shown in Figure 9.18 from Alaska's North Slope. Tapering vertical wedges of nearly pure, foliated ice, up to several meters wide in cross section at the top (Figure 9.19), extend laterally and scribe polygonal networks across large areas. Visible ground ice contents of 10–20% are common in regions patterned by ice wedge polygons, given the width and spacing of the ice wedges. The diameters of the polygons bounded by ice wedges can be tens of m, and the ice wedges most commonly intersect at right angles, but can be hexagonal. In cross section, sediment adjacent to an ice wedge is bowed up, and the wedge is covered with a layer of vegetation and soil. These observations form the tests against which any theory of the phenomenon must be gauged. The questions are: what sets the length scale of the wedges, what dictates the preponderance of crack intersections to be orthogonal, and what processes conspire to form the cross sections of these forms? The prevailing theory, sketched in Figure 9.20, is that ice wedges grow by repeated thermal contraction cracking of frozen ground in winter, followed by ice growth in the cracks from meltwater in summer.

The theory of thermal contraction cracking of permafrost during deep winter cooling was put forth most completely by Lachenbruch (1962) in another seminal paper on periglacial phenomena. As the permafrost cools well below 0 °C (so that no or very little water remains unfrozen), the frozen ground, like any solid, contracts. If the cooling layer is laterally extensive, and therefore constrained, thermal contraction produces a horizontal tensile stress. The stresses are initially horizontally uniform, but are dependent upon depth because the temperature field varies with depth (see Figure 9.5). When the stresses exceed the tensile strength of the frozen ground, a crack will form that is perpendicular to the greatest extensile stress, which means it will be a near-vertical crack, but could be oriented in any horizontal direction

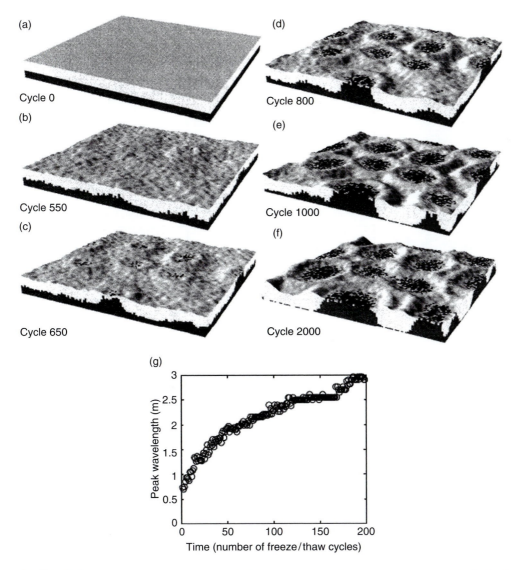

Figure 9.17 Simulation of emergence of sorted circles from an initially layered stratigraphy, (a) (black = fine-grained material; white = stones), through 2000 cycles (b–f) of freeze–thaw to the final state. Peak wavelength in the pattern evolves rapidly at first and more slowly through time (g) (from Kessler *et al.*, 2001, Figures 4 and 5, with permission from the American Geophysical Union).

(Figure 9.21). These cracking events can be quite dramatic, and have been described as sounding like a rifle shot. The crack will propagate downward as the thermal stresses continue to build up with further cooling of the surface. These cracks at least partially fill with hoar frost in winter, or water during thawing of the active layer, producing a thin ice-filled fracture. Repeated cycles of thermal contraction cracking, localized at the site of the previous crack, lead to growth of an ice wedge over time.

If cooling of frozen ground is slow, the stresses that arise from thermal contraction lead to creep deformation of the frozen ground. Non-recoverable (inelastic) creep and deformation relieve the tensile stress, and

preclude cracking formation. For this reason, thermal contraction cracks are associated with rapidly falling temperatures during cold mid-winter temperatures. Typically, ice wedge cracking occurs when air temperatures are $< -15\,°C$ and fall by $1.8\,°C/d$ for four or more days, although snow and other factors confound simple correlations (Mackay, 1993).

Let us scale the expected width of a crack if it takes up all the extension in a zone roughly $100\,m$ in length. Recalling our discussion of the thermal effects of cooling the lithosphere (Chapter 3), the thermal contraction of materials can be scaled using the thermal coefficient of expansion, α, which is about 10^{-5} per $°C$ for frozen soil. If, once frozen, the surface

Figure 9.18 Network of ice wedge polygons near the Beaufort Sea coast. Average length of the polygons is approximately 25 m. Ice wedges underlie the moats bounding the polygons. Raised rims around the polygons generate small differences in wetness that are reflected in the vegetation (photograph by R. S. Anderson).

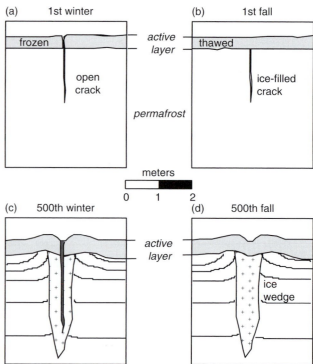

Figure 9.20 Schematic of ice wedge growth, showing winter and summer cross sections in both the first and much later (e.g., 500th) years. Open contraction crack is filled each year with water upon thaw at the surface, which refreezes as the water flows from the active layer into the permafrost. Incremental growth allowed by re-cracking at the same site for many years can lead to wedges of 1–2 m width. Stratigraphy in the soil is warped near the wedge, most abruptly at the surface, and is reflected in surface morphology of parallel ridges adjacent to the crack (after Lachenbruch, 1962, Figure 1).

Figure 9.19 Ice wedge exposed on Beaufort Sea coastal bluff. Vertical foliation in the ice wedge reflects incremental growth of the wedge through time. Warping of the ice-rich silt to make room for the wedge is evident on the right margin. The wedge is capped by the vegetation mat of the active layer, here 35 cm thick (photograph by R. S. Anderson).

temperature is lowered another 20 °C, say, then any 1 m strip of ground will contract by $20 \times 10^{-5} \times 1$ m, or 0.2 mm. Over 30 m, the width of the crack might become 6 mm. Clearly, this is nowhere near the full width of observed ice wedges. The discrepancy can be explained as follows. During the thaw season, as the surface material thaws first, water can dribble down the walls of the cracks to depths at which the surrounding soil is still frozen. Water that freezes in the crack at depths below the active layer will not melt. It is the re-cracking at the same sites year after year that allows the ice wedges to grow to their observed widths of up to several meters.

The other manifestation of ice wedge growth is that the surrounding frozen soils are forced to deform as they expand against the new little wedge of ice upon warming up in the summer. The deformation field is such that the deformation increases with proximity to the surface; the surface itself must warp upward to accommodate the growing ice mass at depth, as shown in the photograph in Figure 9.19. It is this double ridge bounding the crack that is the most easily identified surface manifestation of the cracks in the subsurface to produce images of the sort shown

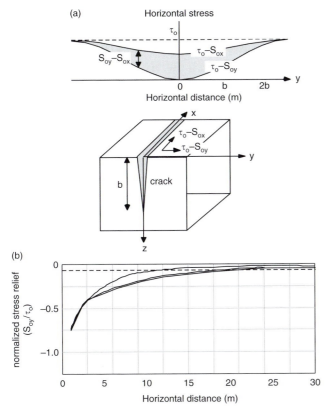

Figure 9.21 (a) Reduction in the extensile stress, τ_o, in crack-parallel (x) and crack-normal (y) horizontal components of the stress associated with the presence of a crack shown schematically in (b). As the crack-normal extensile stresses ($\tau_0 - S_{oy}$) are relieved more than those parallel to the crack ($\tau_0 - S_{ox}$), subsequent cracks will turn normal to the existing crack. (c) Length scale of the stress relief is scaled by the depth of the existing crack, such that cracking is expected to be prevented within some zone (after Lachenbruch, 1962, Figures 11 and 12).

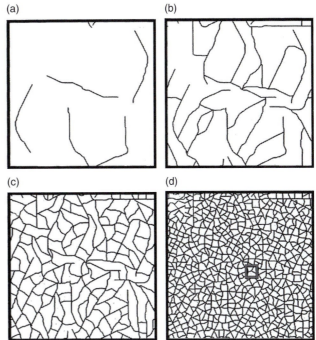

Figure 9.22 Model of evolution of ice wedge polygon landscape by thermal contraction cracking. Evolution from initial set of seed cracks (a) emplaced in a medium in which the stresses are uniformly tensile, through to (d), the final stable network of cracks in which the stresses everywhere are below the threshold for initiation of new cracks. The box in (d) highlights the occasional 120° intersections, occurring where new cracks approach to bisect bends in prior cracks. Most other intersections are orthogonal. The final mean distance across the polygons is 22 m, which is similar to that found in natural frost-wedge polygonal networks (after Plug and Werner, 2001, Figure 4, with permission from the American Geophysical Union).

in Figure 9.18. The underlying stratigraphy can even be exhumed on edge as horizontal layers become vertical against the ice wedge. This raised rim beside the wedges is often the highest topography in the landscape, and serves as a natural dam to pond water above the polygon centers. In addition, in the presently warming Arctic, small ponds are forming above the ice wedges; these ponds form a network that mirrors the ice wedge geometry beneath (see the chapter frontispiece).

But what sets the network geometry and the scale of the polygons? The key here is that the existence of a crack perturbs the state of stress near the crack in a predictable way (Figure 9.21). In particular, a crack relieves stresses that have led to its formation, i.e., tension normal to its trace. This leaves only those extensile stresses oriented parallel to the crack. Imagine now that a later crack nucleated elsewhere propagates toward the existing crack. As it enters the zone of influence of the existing crack, the new crack will turn to become normal to the crack, and will ultimately intersect the crack at right angles as shown in the simulations of Figure 9.22. This pattern is seen time and time again in contracting media (in soils, such as dessication cracks in vertisols, but also cracks in varnished and painted surfaces, mudcracks, etc.). A large portion of the cracks meet orthogonally, giving rise to what Lachenbruch called random orthogonal networks: cracks nucleate in random directions, but intersect one another at right angles.

The scale of the network is governed as well by the influence of existing cracks. The stress relief around an existing crack means that it will be difficult to nucleate another crack within this zone. The length scale of the zone of influence is set by the fracture properties of the material, and by the depth of the crack into the subsurface. This is important because it tells us that the crack spacing within a network can be used as a remote probe of the depth of cracking. Any time that surface information can be used as a probe of properties at depth is a great boon, and makes basic geomorphic observations useful as proxies for processes in the other dimension. Air photos of Arctic tundra become important databases.

Recent work on the evolution of ice wedge polygon landscapes has entailed detailed modeling of the network evolution (Plug and Werner, 2001, 2002; Figure 9.22). The network is allowed to develop from a set of initially random seed cracks, with rules for the interactions of new cracks with existing cracks. The key is that each crack modifies the stress field around it, and the direction of propagation of cracks twists to become orthogonal to the maximum tensile stress. Plug and Werner (2001, 2002) documented the statistics of natural frost wedge polygonal networks, and the model crack systems were tested against these statistics. Importantly, the mean spacing across the polygons was roughly 20 m, akin to that measured in natural settings.

Interestingly, the relics of these ice wedge networks can be seen in many northern latitude sites at the fringe of the Pleistocene ice sheets. In Denmark, for example, paleo-networks can be seen from the air at certain times of the growing season. Here, when the old ice wedges finally melted out after the LGM, the wedges were filled in with blowing sand. The crops grown on the well-draining sand mature at different rates than those over the glacial till and outwash that dominate the Danish landscape. In the late summer, these then show up as yellow networks in a sea of green. Similar relic patterns are seen in Wyoming.

Solifluction lobes

Soil mantled hillslopes in periglacial environments are often festooned with lobate or terraced features that are reminiscent of gobs of frosting running down a cake. These are solifluction lobes, landforms produced by slow creep processes associated with frost action. In map view, the lobes have sinuous fronts. They stand of the order of 0.1 to 1 m above the surrounding slope, and are fronted by rocks or rolls of tundra vegetation. In places, an advancing solifluction lobe front has overridden vegetation (Benedict, 1970). The remarkable aspect, dubbed the "mystery" of solifluction by Williams and Smith (1989), is that solifluction can occur in shallow sheets on low-angle slopes.

Washburn studied solifluction lobes from 1956–1961 as part of a compehensive analysis of periglacial processes at the Mesters Vig region in Greenland (Washburn, 1965, 1967). Observations from one solifluction lobe he surveyed, situated on a relatively steep slope (15–23°) are shown in Figure 9.23. The lobe has a fairly tall, steep front standing up to 3 m above the general surface. Note however, the more gentle slope of top surface of the lobe (10–15°) relative to the general slope of around 20° on which it sits. The lobe consists of clay, silt and gravel with larger clasts of glacial erratics and local bedrock, and had higher moisture content than surrounding talus slope materials, perhaps explaining the more dense vegetation on the lobe. Solifluction lobes often form in hillslope concavities, where water flowlines would converge. In other cases, lobes make their own collection topography: Benedict (1970) describes relatively low-gradient lobes (he called them turf banked or stone banked terraces) with ponded water on their low-gradient top surfaces. Washburn's two-year motion survey (1957–1959) at Mesters Vig shows maximum surface velocities in the center of the lobe of approximately 10 cm/yr, and a mean velocity for his 13 targets of 7.6 ± 3.1 cm/yr. A resurvey of three targets still in place in 1998 yielded remarkably similar rates (mean 7.6 ± 4.1 cm/yr) over the 40-year period 1957–1998 (Carver et al., 2002).

Surface velocity on solifluction lobes in Arctic, sub-Arctic and alpine settings range from 0.02–12 cm/yr (French, 2007; p. 226). The velocity information compiled by French shows a general increase in rate with slope, but with considerable scatter. This scatter presumably reflects differences in slope materials, vegetation, and moisture content. We discuss more fully the physics of the solifluction in the chapter on hillsopes (Chapter 10).

Pingos

A flight over the northern slope of Alaska, or over northern Canada and the islands of the high Arctic,

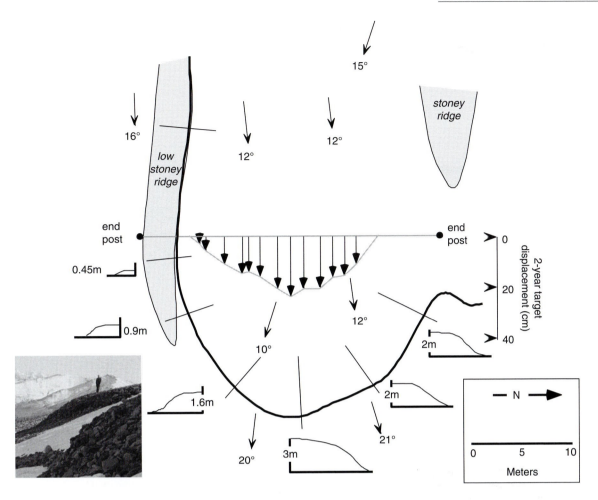

Figure 9.23 Map of solifluction lobe on Hesteskoen, Mesters Vig, Northeast Greenland (Washburn's site 17). Small posts embedded to 10 cm depths on August 5, 1957 were re-measured in July 12, 1959, the displacements denoted by arrows (scale on right). Isolated arrows show dip of the surface, and slope angles. Topographic profiles (without vertical exaggeration) are shown for selected points along the lobe front. Inset photograph shows the lobe front in 1998 (diagram after Washburn, 1967, p. 88, Figure 38; photo from Carver *et al.*, 2002).

would reveal a very flat landscape dominated by low-lying tundra. This is interrupted by the occasional river course, and by the minor topographic fringes of ice wedge polygons already discussed. But the largest positive topographic features in many places are conical mounds called pingos, from the Inuit word for a small hill. As sketched in Figure 9.24, pingos are cored by massive ice, and in fact, owe their topography to the blister of ice lying below the surface. These are no small feature: pingos may have diameters of 50–150 m, and heights of 1–10 meters. They are often circular in plan view, although some are significantly eccentric. Their crests are commonly breached by cracks, coalescing in a depression termed the pingo crater. There are smaller perennial ice-cored

features called frost mounds, but the term pingo is reserved for these larger ice-cored hills.

Pingos can only form in permafrost environments, so their presence is a signature of permafrost in the subsurface. About 25% of the pingos in the world are found in the Tuktoyaktuk Peninsula in northern Canada, where they generally form on the beds of drained lakes. These are closed-system pingos, in which the volume of water in the talik from which the pingo grew limits the size of the pingo. Open system pingos form where groundwater feeds a growing pingo, typically at the base of a hillslope. Pingos are dynamic forms that grow, persist for some time, and then collapse. The events that lead to pingo growth are well documented for closed system pingos.

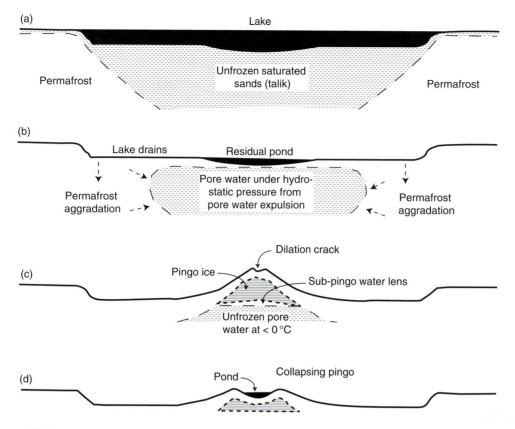

Figure 9.24 Diagrams illustrating the growth and decay of a closed system pingo in sandy lake sediments. (a) Lake basin with underlying talik. (b) Lake drainage leads to aggradation of permafrost on the lakebed, and expulsion of pore water and rejected solutes into talik. (c) Pingo ice gradually grows down from the surface, fed with water from talik. If water supply by expulsion from aggrading lakebed permafrost is greater than pingo ice growth, a pressurized sub-pingo water lens forms. (d) Pingo collapse occurs when pingo ice begins to thaw, or when mass wasting or fracture processes breach the pingo cover (after Mackay, 1998, Figure 4, with permission from the University of Montreal Press).

A typical growth sequence for a closed system pingo in the Tuktoyaktuk region begins with rapid drainage of a lake, which is a not uncommon occurrence in this low-gradient Arctic coastal plain. If the lake was deep enough, it is underlain by a talik, a body of unfrozen sediments (Figure 9.24). Aggradation of permafrost into the talik proceeds, and because the sediments in this region tend to be sandy, excess pore water is expelled into the talik rather than forming segregation ice. Solutes rejected from the aggrading permafrost lead to freezing point depression of a few tenths of a degree. At some point, the talik is fully encased by ice-bonded permafrost. Continued expulsion of pore water pressurizes the unfrozen water and a lens of pressurized water develops. The water pressure in the lens is sufficient to hydraulically lift the overlying frozen sediment cap; indeed the potentiometric surface can be several meters above the ground surface. Pingo ice, called injection or intrusive ice (although it is the water rather than the ice that intrudes), grows down into this water lens. Pingos grow rapidly at initiation, and more slowly through time, reflecting the square root of time dependence of the increase in thickness of the ice lens and of permafrost aggradation. The growth of a pingo ceases when the talik is fully frozen, thus depleting the water source. Decay ensues when either the thermal regime induces thaw, or when mechanical processes breach the sediment cap.

The sequence of pingo growth and decay outlined above was documented over many decades and described in detail by Canadian geomorphologist Ross Mackay. He noted an analogy to the formation of laccoliths, intrusions of magma into layered sedimentary rocks that bow up the overburden (Mackay, 1987). In other words, pingos can be thought of as hydrolaccoliths. The theoretical framework for the laccolith problem was first established

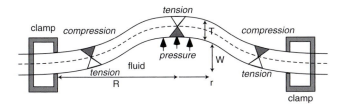

Figure 9.25 Schematic diagram of a pingo treated as a hydrolaccolith. The fluid pressure exerts an upward force that flexes the elastic plate of thickness T into a broad circular dome of radius R. The deflection of the plate is denoted $W(r)$. The theoretical treatment assumes that the plate is clamped at the edges, forcing the gradient of the deflection to be zero at the edges. The local curvature of the plate results in stresses within it that are tensile on the outsides of bends and compressive on the insides.

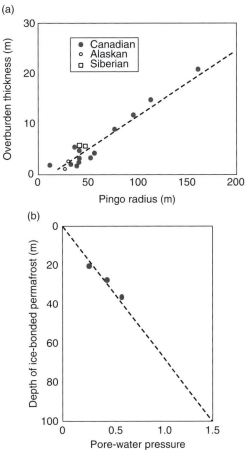

Figure 9.26 Observed relationships between thickness of overburden (or ice-bonded permafrost) and (a) pingo size or (b) pore water pressure (after MacKay, 1987, Figures 4 and 8, reproduced with permission from NRC Research Press, © 2008 NRC Canada).

by G. K. Gilbert (big surprise!) in his investigations of the Henry Mountains in southern Utah, and has been revisited in detail and with considerably more sophistication by Pollard and Johnson (1973). In the pingo case, the overburden is the frozen sediment warped upward by the upward-directed force imposed by the water lens and associated water pressure depicted in Figure 9.25. A key assumption is that the overburden may be treated as an elastic beam or membrane. Given that the solution for the flexural deflection of an elastic membrane under a radially symmetric distributed load exists, one can calculate the pattern of deflection, i.e., its dependence upon distance from the water lens. The maximum deflection, W_{max}, of the elastic plate of thickness T should go as the fourth power of the radius, R, of the pingo (Mackay 1987):

$$W_{max} = \frac{3}{16} \frac{q R^4 (1 - \nu^2)}{E T^3} \qquad (9.7)$$

where ν is Poisson's ratio (for these materials it is roughly 0.3–0.4), E is Young's elastic modulus, and q is the driving pressure beneath the plate. The effective pressure is taken to be the water pressure, P_w, less the overburden pressure, $\rho_b g T$. Hence the expression becomes

$$W_{max} = \frac{3}{16} \frac{(P_w - \rho_b g T) R^4 (1 - \nu^2)}{E T^3} \qquad (9.8)$$

Although ice-rich soil behaves not as a perfect elastic solid, but can creep (flow, like ice), this elastic pingo theory can reproduce many of the salient features of these pingos. For instance, as plotted in Figure 9.26,

pingo radius scales with thickness of the overburden, T. The pingo should grow in amplitude during the summer, as the elastic modulus of the lid declines with increasing temperature. As hinted in the sketch in Figure 9.25, the frozen sediment cover commonly fails in a set of normal tension cracks in the center of the pingo, where the tension in the lid ought to be greatest. As well, springs occasionally erupt at the margins of the pingo, where tension in the base of the lid is greatest (corresponding to greatest curvature).

Thaw lakes

A remarkable feature of Arctic coastal plain landscapes is the many thousands of lakes (Figure 9.27). In most other geologic settings lakes are accidents

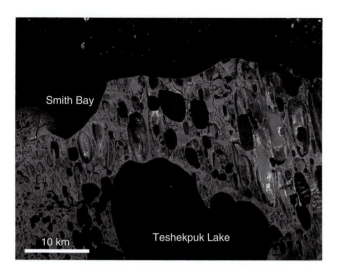

Figure 9.27 Landsat 7 image of Arctic coastal plain between Barrow and Prudhoe Bay, Alaska. The landscape is dominated by present shallow thaw lakes, and basins that have been occupied by lakes in the recent past.

$$T = T_{\mathrm{g}} + \frac{\mathrm{d}T}{\mathrm{d}z}z + (T_{\mathrm{w}} - T_{\mathrm{g}})\left(1 - \frac{z}{\sqrt{z^2 + R^2}}\right) \qquad (9.9)$$

where $\mathrm{d}T/\mathrm{d}z$ is the geothermal gradient, T_{g} is the surface temperature of undisturbed ground, and T_{w} is the temperature at the bottom of the thaw lake. Lateral temperature gradients are significant for small lakes. Use of the steady-state temperature approximation is only correct long after the thermal disturbance associated with a newly formed thaw lake. Therefore, numerical models for thaw depth are used that incorporate both horizontal heat flow and effects of latent heat. Thaw consolidation, assumed to occur instantly (although it certainly would be delayed in nature), is determined from the thaw depth and excess ice content (the volume of water in excess of the porosity of the sediment). West and Plug (2008) use such a model shown in Figure 9.28 to explore the effect of ice content on lake basin evolution.

As expected, the numerical model reproduced in Figure 9.28 shows that both the thickness of the talik and the depth of the lake increase roughly as the square root of time. The lake depth depends strongly as well on the ice content of the substrate over which the lake is forming: the higher the ice content, the greater the subsidence when the ice is melted. West and Plug (2008) note that this may explain the contrast between the deeper (10–20 m) lakes in their Seward Peninsula site, and the shallower (1–3 m) lakes on the Yukon coastal plain. Many lakes in the Arctic appear to have deeper centers, surrounded by shallow shelves. Lake ice persists in the lake centers longer than on the edges. Some have proposed that this reflects recent widening of the lakes. The discrete break in slope at the shelf edge should subsequently smooth through time due to the diffusion of the gradient in temperature there, as again modeled by Plug and West (2009), and as shown in Figure 9.29.

These systems are yet more interesting because there appears to be a lake cycle in which a thaw lake evolves and persists for some time and then drains. The Arctic slope in many circum-Arctic settings is almost entirely covered with thaw lakes or ghosts of prior lakes that have since drained. These are easily seen in Figure 9.27. The more recently drained lakes may be identified, as ice wedge polygons evident elsewhere are not well developed. Finally, in some settings the lakes appear to be preferentially oriented; there is a distinct grain to the environment.

(see Chapter 17). Here lakes cover 20–40% of the land surface. These lakes grow, coalesce, and migrate across the land surface, and can drain rapidly due to changes in drainage routes. Some 50–75% of the land surface can be classified as lakes or has been covered by lakes in the past (Frohn *et al.*, 2005). The lakes are an inevitable consequence of the thawing of the uppermost portion of the permafrost (for example see Hopkins, 1949; Burn, 2002), and may therefore be thought of as a class of thermokarst feature (see below). Melting of ice-rich permafrost leads to subsidence of the surface, as the excess pore water is expelled. Any process that promotes thawing of the upper layer of permafrost will therefore form a depression. The surrounding permafrost forms an impermeable bowl, so water accumulates into a lake. Once deep enough (over about 2 m), the lake will not freeze completely in the winter. Holding the base of the lake at 0 °C over winter will promote thawing of the underlying material, forming a thaw bulb or talik. The lake should therefore deepen, through the continued (but slowing) thaw and consolidation of underlying sediments.

The evolution of a thaw lake and associated talik has been modeled by several workers (for example, Ling and Zhang, 2003). The steady-state temperature profile under a lake is similar to Equation 9.1, but with a term added to account for horizontal temperature gradients. For a circular lake of radius R, the temperature at any depth z beneath the lake center is

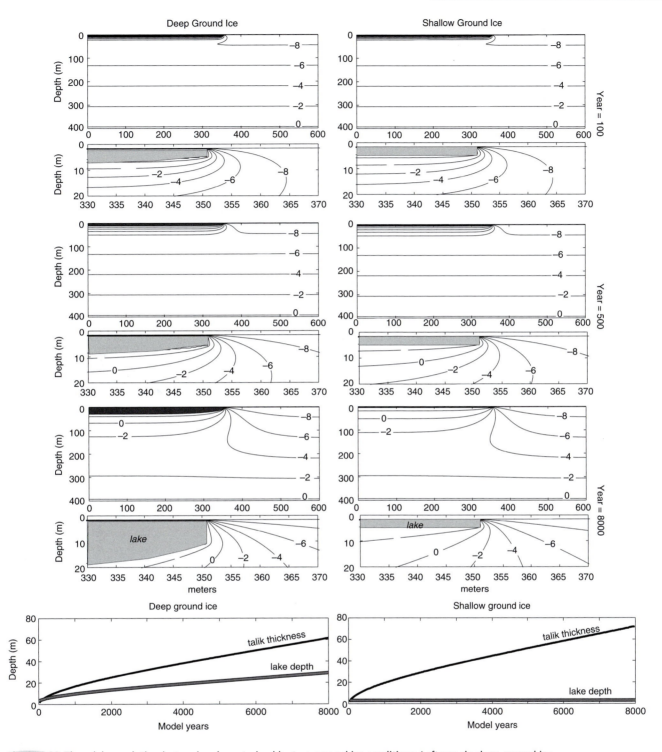

Figure 9.28 Thaw lake evolution in terrains characterized by two ground ice conditions. Left panels: deep ground ice (300 m of 30% excess ice permafrost); right panels: shallow ground ice (30 m of 30% excess ground ice permafrost). Simulations of the ground temperatures at initial, 500 year, and 8000 year model times. At each time step, the top panel shows temperature contours and lake bathymetry over the full 400 m model space, while the bottom panel shows detail in the top 20 m. Thaw of the ground proceeds similarly in both cases (time series shown in plots at bottom), but the lake deepens considerably due to greater settling in the deep ground ice case (after West and Plug, 2008, Figures 1 and 2, with permission of the American Geophysical Union).

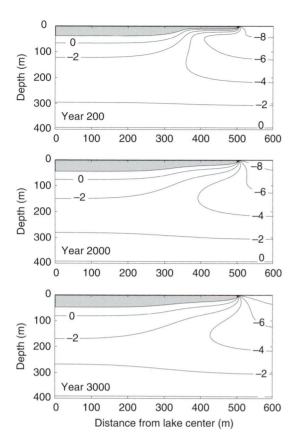

Figure 9.29 Evolution of the temperature field and lake depth subsequent to the widening of the lake from 350 m to 500 m. Slow deepening of the transient thermal response results in a ramp from the initial flat bottom to the new lake margin (after Plug and West, 2009, Figure 3, with permission of the American Geophysical Union).

The most prevalent explanation for this orientation is that they are oriented with their long axes aligned with the prevailing winds (Black and Barksdale, 1949; Hinkel *et al.*, 2005), although there are other views (see for example Pelletier, 2005). The full theory for this phenomenon, and for the full lake cycle, is a topic of present research, and must include an understanding of the mechanics of how shorelines of lakes evolve in a permafrost environment.

The present rapidly changing Arctic

The high latitudes are regions of amplified global warming. The Arctic is particularly sensitive because of the albedo feedback associated with changes in sea ice over the Arctic basin. A small increase in melt reduces the albedo of the surface (the albedo of sea ice ranges up to 0.8, while sea water albedo is less than 0.1), which increases the absorption of radiation, warming the ground and atmosphere. As a result, in recent decades, the Arctic is one of the most rapidly warming areas on Earth (Serreze *et al.*, 2000). Interest in effects of change in the Arctic has therefore been high, as evidenced by the recent publication of the Arctic Climate Impact Assessment (ACIA, 2005). The impacts of warming range from deepening the active layer, rising ground surface temperatures, decreasing snow depths, greater density of shrubs in tundra ecosystems, to changes in carbon flux in tundra soils from net sink to net source to the atmosphere (Hinzman *et al.*, 2005). Geomorphic processes are influenced by changes in climate conditions, and play important roles in the environmental and ecological changes in the Arctic.

Thermokarst

One of the more dramatic and visible impacts of climate warming in the Arctic is thermokarst, the collapse, erosion and slope instability associated with thawing permafrost (Nelson *et al.*, 2001). Thermokarst produces slumps and sinks, topography reminiscent of the deranged drainages and sinkholes of true karst. While karst results from dissolution of soluble rock, thermokarst features are a consequence of deepening of the active layer and thawing of the top of the permafrost. Most of the features considered thermokarst can be attributed to changes in shear strength and subsidence of recently thawed permafrost. Thermal erosion by running water on ice-rich permafrost also produces features considered as thermokarst.

The deepening of the active layer that sets off thermokarst can arise from rising air temperatures, but may also be triggered by changes in surface conditions that affect the surface energy balance. For instance, snow depths greater than about 0.4 m provide insulation of the ground from low winter air temperatures and will reduce winter cooling of the active layer. Vegetation impacts surface energy balance as well, through albedo, moisture–latent heat effects, and thermal conductivity. Loss of vegetation due to construction or fires generally leads to deepening the active layer.

Figure 9.30 Detachment thermokarst failure on a 4° slope near Feniak Lake, Alaska. Two people near the head of the failure are circled (photograph courtesy of M. Gooseff, from Gooseff *et al.*, 2009, with permission of the American Geophysical Union).

Thawed ice-rich permafrost has very low strength due to high pore pressures, which contributes to slope instability. The role of excess ice in the flow of soil is well known as a component of solifluction, and will be discussed in more detail in Chapter 10. However, where thawing involves the top layers of ice-rich permafrost, much more dramatic failures can occur. Figure 9.30 depicts a recent massive thermokarst detachment failure on a low gradient (4°) hillslope likely caused by elevated pore pressures at the top of the permafrost table. Thaw slumping of embankments near streams and shorelines of lakes involves the landward march of headscarps that feed soupy saturated sediments downhill. Scarps can retreat at meters per year (Lantz and Kokelj, 2008).

More commonly, thermokarst features are associated with subsidence and consolidation of thawed materials. Sites that are most susceptible to thermokarst degradation are those in which the ice content of the near-surface materials is very high. Remember that the ice content of these materials can be well above saturation; they hold much more water than there are pores in the original sediment. The amount of thaw consolidation or subsidence expected depends on the increase in active layer depth and the ice content of the thawed permafrost. Because settling goes on simultaneously with permafrost thawing, resolving the degree of settling can be difficult. Pullman *et al.* (2007) define the thaw strain, δ, the change in height of thawed permafrost due to expulsion of excess water and compaction, from measurable bulk densities: $\delta = 1 - \rho_F/\rho_T$, where ρ_F and ρ_T are the dry bulk densities of frozen and thawed substrate, respectively. Thaw settlement, S, can then be calculated from the measured active layer thickness, H, and the thaw strain, as shown in Figure 9.31:

$$S = \frac{(H_2 - H_1)\delta}{1 - \delta} \tag{9.10}$$

The depth of thaw settlement is seen to increase as the ice content of the permafrost subject to thaw increases (lower ρ_F), and the change in active layer thickness increases. Equation 9.10 does not address the time required for thaw settling to occur. We can expect clayey materials to expel excess water slowly, and remain waterlogged over extended periods.

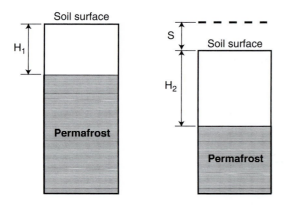

Figure 9.31 Thaw settlement, *S*, associated with deepening of the active layer over ice-rich permafrost from an initial thickness, H_1, to a thickness H_2 (after Pullman *et al.*, 2007, Figure 2, with permission from INSTAAR).

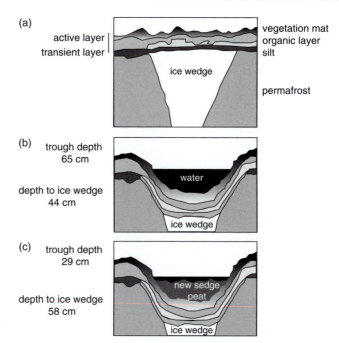

Figure 9.32 One style of thermokarst degradation of permafrost terrain. (a) Initial ice wedge overlain by silt and vegetative mat active layer with thin transient layer. (b) State of advanced degradation: the transient layer thaws upon warming, generating a trough that deepens through time as the top of the ice wedge melts. Water then fills the trough. (c) State of advanced stabilization: growth of new sedge within the trough provides new insulation, which stalls further melting of the ice wedge (after Jorgenson *et al.*, 2006, Figure 2, with permission of the American Geophysical Union).

Where permafrost thawing is uneven (for instance due to differences in ice content of the subsurface or snow depth variations), depressions will develop that may be poorly connected and hence poorly drained. Jorgenson and Osterkamp (2005) detail 16 modes of permafrost degradation, the majority of which produce thermokarst pits, depressions, or basins that may or may not be water-filled. An example of uneven thaw subsidence and thermokarst evolution is found in degradation of ice wedge polygons that dominate the Arctic coastal plain (Jorgenson *et al.*, 2006). If the thin active layer capping the tops of the ice wedges thickens, the top of the ice wedge melts, deepening a trough above it. Standing water collects to create sets of thin linear lakes that mirror the underlying pattern of ice wedges, as sketched in Figure 9.32, and the low albedo water absorbs yet more heat. In the Alaskan sites studied by the Jorgenson team, these new small lakes occupied almost 4% of the land area in 2001, up from 0.5% in the 1940s. But they also argued that there exists a negative feedback in the system that ought to limit the depth to which these lakes can grow, much short of melting the entire wedge. Ponding of water creates conditions for rapid growth of sedges, which in turn fill the ponds with peat. The new peat layer will retard or reverse degradation through its insulating role. A surprise in this Jorgenson study was that permafrost degradation was so widespread in a region thought to be protected by very low ground temperatures. The albedo feedback from incipient ponding seems to drive the degradation.

Coastal erosion

The coastline of the Arctic Ocean is rimmed with permafrost and underlain by subsea permafrost. In the recent warming climate, the coast is undergoing accelerated erosion. Consider the role of sea ice in coastal evolution. In the winter, sea ice is frozen to the seabed near shore (and as such is called shorefast ice). In such a setting, waves cannot impact the coastline, and cannot cause damage to the coast. Instead, the only geomorphic signature of sea ice is low ridges of gravel pushed about by the sea ice as it moves ashore, called push-ridges. But sea ice has been thinning over the last few decades. We know this by comparing now de-classified reports from submarine-based observations of the base of the sea ice, with present measurements. Sea ice has thinned by tens of percent in the last half-century. As it does so, it also decreases in its areal coverage of the Arctic Ocean.

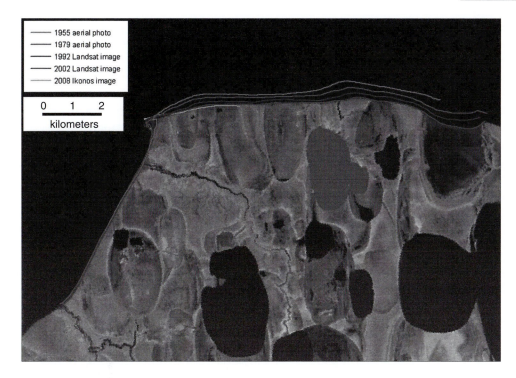

Figure 9.33 Evolution of coastline at Drew Point, Alaska, as deduced from analysis of repeat satellite imagery. Coastal erosion is roughly uniform except in the recently drained lake basin at the right of the image. Several hundred meters of retreat has occurred over the last 50 years (image courtesy of Nora Matell).

A corollary to this is that the sea ice detaches from the shoreline earlier in the spring and re-attaches later in the fall than it has historically. This exposes the shoreline to wave attack from which it used to be immune. So the external forcing is growing through time. In addition, the resistance of the land to attack is declining. It appears that, as the permafrost warms, the mechanical properties of the frozen sands and silts of the coastline alter toward being more susceptible to erosion. The result is extreme rates of retreat of the shoreline, as plotted on the Landsat image in Figure 9.33. In the summer 2004, the coastline retreated approximately 100 m in at least one site in northern Alaska, a rate that is unheard of, and that brings into harm's way human structures we had thought would be immune for centuries.

As discussed more fully in Chapter 16, the principal process is the melting of the ice-rich sediments, which generates a notch that promotes toppling of segments of coastal bluff. Although there is subaerial melt of the bluff edge (at cm/day rates) the dominant thermal process is convective heat transfer from the ocean to the ice-rich bluffs. The feedbacks in the thermal system on the northern Alaskan coastline include the following: as the sea ice is gone for longer from a particular site, the sea surface temperatures in the ice-free ocean probably increase; in addition, the retreat of the sea ice edge results in a longer fetch over which winds can blow, resulting in higher waves that are more efficient at transferring the oceanic heat to the shoreline. If the sediment freed by the melting of the toppled blocks is silt and clay, as it is on sections of the Beaufort Coast, the silt wafts away onto the shelf and can play no role in tripping the waves as a beach, thereby stalling erosion. The absence of such a beach implies that there is no brake on the system.

Permafrost and carbon

The interactions of ecosystems and permafrost degradation can be quite profound. We have seen that ice wedge degradation leads to ponding and a shift from tussock to water-tolerant sedge vegetation (Figure 9.32). Invasion of shrubs into tundra influences snow depth, which can trigger thermokarst processes (Sturm *et al.*, 2001). The depth of the active layer limits rooting depth, and influences the drainage of the soil. Arctic soils contain perhaps half the global soil carbon; the amount is poorly known because of heterogeneity and uncertainty about the depth profiles of the organic matter content of frozen ground (Schuur *et al.*, 2008). Although primary productivity (net C fixation into organic matter each year) in Arctic ecosystems is low, cold temperatures limit the rates of degradation, so carbon stores tend to increase over time. Permafrost thaw exposes peat or organic-rich soil to microbial decay, which returns soil carbon back to the atmosphere as CO_2 or CH_4, depending on whether the soils are aerobic (drained) or anaerobic (waterlogged). Observations have shown tundra soils transforming from carbon sinks to carbon sources in recent years. Increased plant growth in longer, warmer growing seasons will increase C uptake, but this is unlikely to offset carbon losses by decay (Schuur *et al.*, 2008). Progress in understanding the carbon dynamics of permafrost regions is closely tied to understanding the thermal behavior of the soil and thermokarst development.

Summary

Permafrost underlies nearly one quarter of Earth's present landmass, and was significantly more extensive during the glacial cycles of the Plio-Pleistocene. The geomorphology of these landscapes is dictated by transfers of heat and water within the materials of the subsurface. A formal understanding of these transfers again requires that we craft balances of heat and water.

The depth of permafrost, which can be hundreds of meters thick, is dictated by the mean annual temperature, the thermal conductivity of near-surface materials, and the geothermal heat flux. The thickness of the active layer, on which all ecological systems of periglacial landscapes depends, and on which all human infrastructure is placed, is dictated by the annual cycle of ground surface temperature, and by the surface energy balance. Here water content is crucial, as this stalls the penetration of the cold wave by requiring latent heat of fusion to be conducted away.

When a porous material freezes, more interesting things happen than simple freezing of water in pores. Water migrates in thin films at temperatures well below the bulk freezing point in fine-grained (especially silty) substrates. This results in growth of segregation ice, or discrete ice lenses in frost-susceptible soils. Some soil lenses are just a few millimeters thick. But massive ground ice many meters across can result over long periods of time. This phenomenon of excess or segregation ice in turn catalyzes most periglacial geomorphic processes.

Frost heave is a direct result. The slope-normal heave and the vertical subsidence upon thaw result in downhill frost creep. Upon thaw, the excess ice in near-surface materials melts, and the impermeable nature of the underlying frozen material prevents this water from escape. The resulting very weak soil can flow downhill on slopes of only a few degrees. Frost heave also results in jacking of larger clasts upward through the soil over repeated thermal cycles. A combination of frost heave and sorting of larger clasts can explain the remarkably well-organized sorted circles that dot flat Arctic landscapes.

Larger scale features of periglacial landscapes include ice wedge polygons, pingos, and thaw lakes. These landforms are characteristic of permafrost conditions.

Current warming of air temperatures at high latitudes is leading to rapid changes in these landscapes. These include the warming of temperatures at depth in the permafrost, which constitute a robust means of extracting paleo-temperature histories of the ground surface. Such profiles speak strongly of warming that began roughly 1850–1900, and of accelerated warming in the last couple of decades. Degradation of the permafrost is occurring through a set of thermokarst processes, including pervasive thaw slumping on slopes, and the growth of thermokarst lakes above degrading ice wedges. Finally, the reduction in sea ice cover of the Arctic Ocean is allowing warmer ocean water to melt ice-rich coastal bluffs at rates of many tens of meters per year.

Problems

1. If the mean annual temperature is $-12\,°C$, the geothermal heat flux $40\,mW/m^2$, and the thermal conductivity of the near-surface Earth materials $2.5\,W/m\text{-}K$, what is the depth to the base of the permafrost?

2. Ignoring the role of latent heat, calculate the thickness of the active layer at the same site as in problem 1. Assume that the amplitude (half the full range) of the temperature swing at the surface is $15\,°C$.

3. Does the approximation $1\ \text{year} = \pi \times 10^7$ seconds underestimate or overestimate the number of seconds in a year? What is the percentage error in using this approximation?

4. Write the equation for the envelope of annual *maximum* temperatures as a function of depth. How do you solve this expression for the active layer depth? Calculate the active layer depth for a region with a mean annual surface temperature of $-3\,°C$ and T_{amp} of $10\,°C$, assuming that the thermal diffusivity, κ, is $1\,mm^2/s$.

5. Calculate the depth at which *daily* temperature fluctuations are negligible (defined as a few percent of surface), assuming that the thermal diffusivity, κ, is $1\,mm^2/s$.

6. Estimate the maximum depth to which frost penetrates the ground at a location where the temperature ranges between $-15\,°C$ and $25\,°C$. Assume the temperature history at the surface is sinusoidal, and that the water content of the soil is low enough that we can ignore the latent heat associated with the phase change of water (in other words, your calculation works best in low porosity rock). Assume that the thermal diffusivity, κ, is $0.8\,mm^2/s$.

   At this depth, how far out of phase (by what fraction of a year) will the temperature be relative to the surface? In other words, the maximum frost penetration is likely to occur *after* the minimum in surface temperature; we are asking how *much* later this will occur.

7. Consider a peat (organic-rich soil) with $\kappa = 0.12$ mm^2/s and a mineral soil with $\kappa = 0.98\,mm^2/s$.
   (a) Calculate z_* for the annual temperature wave for each of these soils.
   (b) Which soil will have the greatest decline in temperature fluctuations with depth (tightest annual temperature maximum envelope)?
   (c) Prove your prediction by calculating the ratio of the amplitude of the annual temperature fluctuations at $1\,m$ depth to the amplitude at the surface for both soils. If $T_{amp} = 10\,°C$ at the surface, what is the amplitude of fluctuations at $1\,m$ depth for the two soils?
   (d) Calculate the lag, in days, of the annual temperature wave at $1\,m$ depth for each soil.

8. *Thanksgiving turkey problem* (with apologies to vegetarian readers). If it takes 2 hours to bake a turkey with a radius of 20 cm, how long would it take to bake one that has a radius of 25 cm? Present your formula and your reasoning, not just the answer.

9. Given the graph below of the *envelope* that encloses all temperatures measured over the course of many days in the summer, using thermistors at many depths into a borehole in a rock surface, answer the following questions:
   (a) Estimate the length scale, z_*, *just by inspection of the graph*. This length scale arises in setting the rate of decay of the amplitude of the temperature swings with depth. Show this depth on the plot.

(*Note*: you can do this with no knowledge of the formula for z_*.)

(b) Estimate the thermal diffusivity, κ, of the rock. This time, show your work.

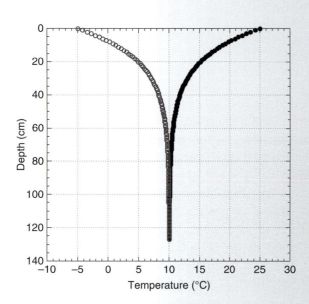

infer the temperature history of the surface in many places on the Earth's surface.

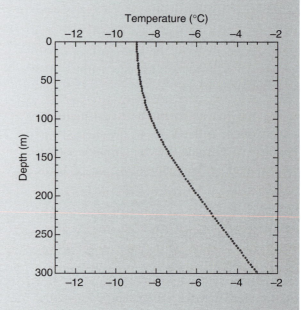

10. You are in South Africa, and are employed by a mining company to mine gold in the famous deep vein systems of what was once Rhodesia. The mine is 13 000 feet deep and you are working at this deepest level. The temperature gradient there is a typical 25 °C/km, the rock is typical with a thermal diffusivity of 1 mm²/s, the mean annual temperature at the surface is 20 °C, and the *total* annual temperature swing at the surface is 30 °C.

What is the *mean annual temperature* at the level in the mine in which you are working? (Show a diagram of the problem as you are setting it up.)

11. Is the surface temperature of the Earth warming? In the plot below we show a typical temperature profile from the North Slope of Alaska. This kind of temperature profile has been used to

12. Frost wedges in the Arctic can be several meters wide, but they are built incrementally. The annual cracks that fill in with water in the thaw, which subsequently freezes to grow the wedge, are much narrower. Consider a site in which the current frost wedges are 2.5 m wide, the minimum annual temperature is −30 °C, and the mean frost wedge polygon diameter (hence the spacing between ice wedges) is 20 m. If the coefficient of thermal expansion is $1 \times 10^{-5}/°C$, and the temperature falls rapidly from −15 to −30 °C, how wide will the resulting contraction crack be? How old is the frost wedge?

13. How does one contraction crack affect the growth of subsequent cracks?

(a) Draw an example of the cracking pattern to be expected in a material experiencing uniform contraction, starting with a single initial crack shown in the diagram. Choose cracks that originate at random within the space, and propagate in both directions. Assume that a crack alters the stress field over a distance of about 1 cm.

(b) How do you know when to stop this drawing? (*Hint*: is the pattern a fractal?)

(c) Find an example of such a surface. (For example, the glaze on a coffee mug.)

14. *Thought question.* Discuss how the world would be different if snow were black instead of white.

15. *Thought question.* List and briefly discuss several potential impacts of climate change (e.g., warming, changes in precipitation, etc.) on periglacial landscapes.

Further reading

French, H. M., 2007, *The Periglacial Environment*, 3rd edition, Chichester: Wiley.
This up-to-date comprehensive reference covers the history and distribution of permafrost, types and distribution of ground ice, periglacial processes, and applied periglacial geomorphology.

Washburn, A. L., 1979, *Geocryology – A Survey of Periglacial Processes and Environments*, London: Edward Arnold.
This book was the bible of the periglacial geomorphological community through the latter part of the twentieth century.

It is encyclopedic in its coverage, providing access to international study of periglacial phenomena through the late 1970s.

Williams, P. J. and M. W. Smith, 1989, *The Frozen Earth – Fundamentals of Geocryology*, Cambridge: Cambridge University Press.
An excellent reference on the physics of frozen ground, including thermal profiles, surface energy balance, solifluction, the mechanics of frozen ground, and frost heave.

CHAPTER 10

Hillslopes

The Hills erect their Purple Heads
The Rivers lean to see
Yet Man has not of all the Throng
A Curiosity.

Emily Dickinson, *The Hills erect their Purple Heads*

In this chapter

Every place in the terrestrial world that is not a channel or a beach is a hillslope of one or another stripe. The hillslope forms, the processes that shape them, and the rates at which these processes act vary radically in both space and time. How can we organize our thoughts about these fundamental elements in the geomorphic system? What is in common among hillslopes? And how can we impose some conceptual order on how they vary from one climate to another?

(a) Jack Schmitt and LRV on the Apollo 17 mission, at the edge of Shorty Crater. The Moon's surface is all hillslopes (image #AS17–137–21010, from http://www.apolloarchive.com/apollo_gallery.html).
(b) Pinnacled limestone landscape of Tsingy de Beuaraha National Park, Madagascar. This is as far from a diffusive landscape as one can get (copyright 2009 Frans Lanting, www.lanting.com, photograph 006446–01).

Consider the Moon for a moment. As shown in frontispiece figure(a), the entire surface of the Moon is composed of hillslopes. There are no channels. That lunar hills look as familiar as they do, with convex hilltops and concave footslopes, indicates that despite the lack of an atmosphere and of precipitation, and of vegetation or any other biological processes, the form of the landscape is not significantly different. In contrast, consider a terrestrial landscape like that in the limestone terrain called the Tsingy in Madagascar, shown in frontispiece figure(b). Here the landscape looks very unfamiliar. Tsingy means pointed hills, referring to the 100 m-tall fins in the photograph. The rarity of this kind of landscape tells us the processes dominating here are special, out of the ordinary. Indeed, this entire landscape is formed by the dissolution of soluble rock. We hope that by the end of this chapter you will be aware of the processes acting in more common terrestrial landscapes, which can serve as a means of highlighting the differences between these and other end-member landscapes, either rare terrestrial forms, or forms in other planetary surfaces.

One way to classify hillslopes is by whether they are soil-mantled or are bare bedrock. The shapes of these two types of slopes will be quite different: soil-mantled slopes tend to be smooth and rounded, while rocky slopes are generally steeper and more jagged. The processes dominating in each case differ too. We must address both the production of mobile material (which we attempt to do in Chapter 7), and the movement of this material. We will begin by examining soil-mantled slopes, where mass balance principles can be applied. In later sections we will address the force balance issues that are important in the evolution of rock slopes.

As always, we will begin by enforcing the conservation of some quantity, this time the mass of regolith. Regolith is generated by the weathering of bedrock (Chapter 7), and can move from one place to another by a variety of processes. In general, therefore, we must know something about the processes involved in the transportation of regolith. We must decipher to what extent the process depends upon the local slope of the land surface, whether it depends upon the position on the slope – the distance from the hillcrest. We must also address what happens at the foot of a slope: whether the material moved downhill accumulates at the foot, or is taken away by the conveyor belt of the local stream channel.

Hillslopes, in other words, do not exist in isolation, but are linked to their bounding channels. To put it mathematically, channels serve as the bottom boundary conditions for the hillslopes.

One may break many hillslope processes into those that result from some single "continuous" process, for instance the slow viscous creep of material down a slope; or the integrated or summed effects of discrete small-scale events – even slow creep is often the result of a sum of smaller motion events. For instance, both the higgledy-piggledy motion of individual sand grains blasted from a soil surface by the impacts of raindrops, and the helter-skelter motion of material driven by the digging of animals, can result in the net downslope motion of material. In such instances, we must know something about both the geomorphic result of an individual "event," and the intensity or areal density of such events: how many events occur in a unit area of surface per unit of time. Some parts of the problem may be deterministic, meaning that the outcome of an event is knowable once we know the initial conditions (e.g., the trajectory of a grain is calculable once we specify the angle and speed at which it is launched), while others are likely stochastic, meaning they can be treated only in a statistical way (e.g., the size distribution of raindrops). We will see that the resulting downslope transport often results from the asymmetry of such events; it is slightly more likely that rainsplashed grains will be ejected downhill than uphill (and those ejected downhill travel longer paths), and it is more likely that gophers place their diggings downhill of their hole than uphill. In all cases, we wish to parse the problem into pieces that represent the several factors involved: the meteorology, the susceptibility of the landscape to the particular process, and the geomorphic "reaction" or response to the meteorological process.

In general these hillslope processes are not constant in time, but respond significantly to the variable forcing of the weather. This introduces yet another "stochastic" aspect to hillslopes that must be understood and quantified before the long-term behavior of hillslopes can be understood. We must know what questions to ask of the meteorology, what statistical measures of the weather must be documented. And given that the timescale over which hillslopes evolve is typically very long relative to the time over which climate remains steady, a reasonable (although difficult) goal of this research would be to quantify how specific processes depend upon different climates.

While most hillslope processes are dominated by the forces imposed by the action of gravity, we will see that in several instances biology plays an important role in the operation of hillslopes. For example, rodents and insects burrow into hillslopes and move material about. Tree roots act in several ways: they pry apart rock, they dilate soil, and they provide cohesion to the soil, strengthening it against the tendency to fail in landslides, and so on. When the trees fall down, the soil bound up in these same roots can be transported in the direction of the fall. We must learn to ask the proper questions of the biological system, just as we must learn to ask the proper questions of the meteorological system.

Convexity of hilltops

Most hilltops are convex upward. The slope increases monotonically with distance from the divide. They are rounded, not pointed; they look like hills on the Moon, not the fins in Tsingy, Madagascar. G. K. Gilbert was the first to treat this convexity of hilltops in a formal way. In a wonderful paper in which he describes in words the process of hillslope diffusion, he forces us to think in an orderly way about the controls on rates of soil creep (Gilbert, 1909). He described in words the conservation of mass at several points along what he called a "mature" hillslope shown schematically in Figure 10.1. We use the term "steady state" today to describe Gilbert's mature hillslope, meaning a slope that is not changing in form through time – the entire hilltop is migrating downward steadily, but the shape of the landscape is unchanging. Gilbert noted if the shape of the hillslope and depth of regolith is unchanging, the quantity of regolith passing each point must increase downslope. Because slopes commonly have a uniform thickness of regolith, the average velocity of creep must increase downslope. He then reasoned that because gravity provides a greater downslope impelling force on a steeper slope, that the slope at any point is adjusted to provide just the increased transportation of regolith required by the mass balance. Gilbert shows that creep rates depend directly on hillslope gradient, and that this understanding can explain an important part of the landscape, the hillcrest.

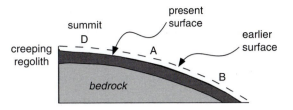

Figure 10.1 Schematic cross section of a steady-state convex hilltop. Summit at D and two equally spaced points A and B are shown. The hill surface is shown at two times – early (dashed) and present. The entire hilltop has been lowered by the same amount. Bedrock weathers to mobile regolith at the bedrock–regolith interface. Regolith produced between the summit, D, and the point A must be transported past A by creep, while all regolith produced between D and B must be transported past B. If the transport process is slope-dependent, the slope at B must be twice that at A. This is the essence of convex hilltops (after Gilbert, 1909, Figure 1).

Gilbert's analysis pertains to a precise set of conditions. He considers only places where creep occurs, and specifically does not consider places being eroded by running water. You know from experience that erosion by running water, exemplified by gullies and river channels, produces concave-upward profiles. Gilbert's discussion of only steady-state forms was primarily for convenience – we will see how to relax this assumption shortly. He also assumed that the regolith thickness remains uniform with distance downslope. This assumption, supported by field observation, is the one that allows him to make the inference that creep rate depends on slope.

Although Gilbert called upon creep of the regolith, one of the goals of this chapter is to show that his result is more universal than this. Note what happens if we assume that the downslope flux of regolith is linearly dependent upon the local slope of the hill. Since the amount of regolith that must be passed increases linearly with distance from the hillcrest, the slope must also increase linearly with distance. This is fundamentally why hilltops are convex. In fact, steady-state hilltops are not only convex, which is a loose term, but they are parabolic (the slope does not just increase, it increases *linearly* with distance from the divide) whenever the process moving regolith from place to place depends linearly upon the local slope. Figure 10.2 shows an example of a parabolic hillcrest, the shape made visible by the shadow of a straight-edge. While the one depicted is developed in the clay badlands of Utah, similar parabolic forms are found developed in

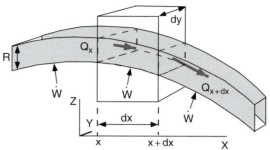

Figure 10.3 Conservation of mass on a hillslope element of size dx by dy, capturing the essence of Gilbert's (1909) argument for convex hilltops. Spatially uniform weathering supplies new regolith to each hillslope element. For a steady hillslope form, in which a uniform regolith thickness, R, is maintained, there must be a spatial gradient in the lateral transport of regolith, Q_x, increasing linearly with distance, x, from the hillcrest.

Figure 10.2 Hilltop profile in Blue Hills badlands in central Utah revealed by shadow cast by straight-edge. Bedrock is Tropic Shale of the Cretaceous interior seaway. Note the convex crest, with very high curvature (photograph by R. S. Anderson; Jon Stock and Melissa Swartz for scale).

Precambrian granites and other rock types – although the curvature is significantly less. We discuss these examples further later in the chapter.

Mass balance

Let us recast the problem in its most general form, a little more formally. Making no assumptions about process or that the system is in steady state, we can still conserve mass. Consider the mass of regolith within a box on the hillslope, sketched in Figure 10.3. The word statement is

rate of change of regolith mass [M/T] = rate of regolith inputs [M/T]− rate of regolith outputs [M/T]

Treating all sides of the box (except the top, through which, in the absence of very strong winds, we may safely assume no regolith is lost or gained), this becomes

$$\frac{\partial(\rho_b R dx dy)}{\partial t} = \rho_r \dot{W} dx dy + Q_x(x) dy$$
$$- Q_x(x + dx) dy + Q_y(y) dx$$
$$- Q_y(y + dy) dx \qquad (10.1)$$

where R is the regolith thickness, $\dot{W}$ is the weathering rate [L/T], or the rate of production of regolith, ρ_r is the density of rock, and ρ_b is the bulk density of the

regolith. If we align the x-direction of our box with the local slope, it is safe to ignore mass fluxes across the sides of the box (Q_y) and Equation 10.1 then becomes

$$\frac{\partial(\rho_b R dx dy)}{\partial t} = \rho_r \dot{W} dx dy + Q_x(x) dy$$
$$- Q_x(x + dx) dy \qquad (10.2)$$

Noting that we may hold the length of the sides of the box to be unchanging with time, we may remove them from the derivative on the left-hand side and divide through by dxdy. If we assume that the bulk density, ρ_b, also does not change, we may pull it out of the derivative as well, and the left-hand side then becomes simply the rate of change of regolith thickness with time. On the right-hand side the dy vanishes entirely. The equation can then be arranged to become

$$\frac{\partial R}{\partial t} = \frac{1}{\rho_b} \left[\rho_r \dot{W} - \frac{Q_x(x + dx) - Q_x(x)}{dx} \right] \qquad (10.3)$$

From our introduction to calculus, we recognize the term in brackets on the right-hand side to be the spatial derivative of the regolith discharge, $\partial Q_x/\partial x$:

$$\frac{\partial R}{\partial t} = \frac{\rho_r}{\rho_b} \dot{W} - \frac{1}{\rho_b} \frac{\partial Q_x}{\partial x} \qquad (10.4)$$

We have, in Equation 10.4, a general statement of conservation of mass on a hillslope in one dimension. The steady-state case that Gilbert postulated is captured by taking the time derivative to be zero, producing

$$\rho_r \dot{W} = \frac{dQ_x}{dx} \tag{10.5}$$

In words, Equation 10.5 requires that the local production rate of regolith by weathering of the bedrock beneath (the left-hand side) must be exactly balanced by the spatial change in the discharge of regolith with distance down slope (the right-hand side). Otherwise, regolith will pile up in the box, or be slowly drained from the box.

One may simply integrate this equation to arrive at a prediction of the spatial pattern of regolith discharge on a slope:

$$Q_x = \rho_r \dot{W} x \tag{10.6}$$

Regolith discharge simply increases linearly with distance from the divide ($x = 0$), at a rate governed by the regolith production rate.

The more general statement of the conservation of mass on a hillslope is the non-steady-state (or "transient") case, Equation 10.4, or its two-dimensional analog. In order to solve this general case, or in order to understand the evolution of a particular hillslope, we need to know about the processes that give rise to the two terms on the right-hand side of Equation 10.4. First, what establishes the rate of regolith production (see Chapter 7)? And second, what process or suite of processes is involved in moving regolith downslope? In order to determine how this process might depend upon slope, or upon position on the hillslope, or upon climate, we need to explore the physics (and chemistry ... and biology) of the specific hillslope processes involved.

Diffusive processes

Geomorphologists often talk about diffusive landscapes, or diffusive processes. Diffusion, whether it be of ink in a glass of water, smoke in air, or heat in a solid, tends to smooth out spikes in the concentration of some property. Landscapes dominated by diffusive erosion processes have rounded edges, because diffusion attacks sharp corners. Before turning to specific physical processes, let us consider briefly the general diffusive case. The goal is to develop the diffusion equation for a hillslope. The diffusion equation comes about by combining two concepts: conservation of mass, which we have just developed, and a rule for how flux (in our case, of regolith) responds to the controlling variables. For many processes, we might expect the flux rule to be something like:

$$Q = -kx^m \frac{\partial z^n}{\partial x} \tag{10.7}$$

where m and n are constants setting the relative importance of distance from the divide, x, and of slope, $\partial z/\partial x$, respectively, and k sets the efficiency (scale) of the process. In the diffusive case, $m = 0$ and $n = 1$, in other words the regolith discharge depends solely, and linearly, on the local slope, $\partial z/\partial x$. In this case, Equation 10.4 becomes

$$\frac{\partial R}{\partial t} = \frac{1}{\rho_b} \left[\rho_r \dot{W} - \frac{\partial(-k(\partial z/\partial x))}{\partial x} \right] \tag{10.8}$$

If we make the simplifying assumption that the hillslope efficiency constant, k, is spatially uniform, i.e., does not depend on x, then we can remove the k from the derivative to obtain the diffusion equation:

$$\frac{\partial R}{\partial t} = \left(\frac{\rho_r}{\rho_b} \right) \dot{W} + \kappa \frac{\partial^2 z}{\partial x^2} \tag{10.9}$$

where $\kappa = k/\rho_b$, and is called the landscape diffusivity. The ratio in front of the weathering rate reflects the inflation or bulking of the material as it is transformed from rock to soil. As in any other diffusion equation, the diffusivity must have units of L^2/T, as this does.

Equation 10.9 is a diffusion equation for a hillslope. This or an analogous diffusion equation comes up frequently in physical systems. Diffusion equations always describe the variations of some quantity in time (here, that quantity is regolith thickness, but it could also be concentration of a solute or temperature). We have already seen its cousin as a diffusion equation for temperature when heat is transferred conductively. The connection is that the transport of the quantity of concern (here regolith, or heat) is proportional to a spatial gradient in a related quantity (topography, or temperature). When we combine this concept with the need to conserve regolith (or heat), we find that the variations of that quantity always depend upon the second spatial derivative of something (here, that something is surface elevation). The second derivative describes the curvature of the surface, or the rate of change in slope. You can see from Equation 10.9 that regions with high curvatures will undergo the greatest changes in regolith thickness. This is the reason that diffusion tends to smooth topography: in regions of high curvature (dimples or bumps on the hillslope) the regolith will either be thickened or thinned fastest, leading to smooth slopes. We arrived at the diffusion equation when we chose a transport rule (Equation 10.7) in which the flux of sediment depended only on slope, and not on distance downslope. Had we chosen a transport rule with an x dependence ($m \neq 0$ in Equation 10.7), we would not have arrived at a diffusion equation. Transport of sediment by flowing water is an example of a process that does not lead to diffusive behavior, as we shall see. See the random walk box (Box 10.1) below to see how one can also arrive at a diffusion equation by appeal to summation over many potential particle paths.

The diffusion equation is a partial differential equation. In general, these can be solved analytically only under very simple conditions. Because the diffusion equation is so common, solutions for many particular conditions are available. One useful compendium is Carslaw and Jaeger's *Conduction of Heat in Solids* (1967), which has much broader application than its title implies. We can easily find the solution for the steady case, in which we set the left-hand side to zero. This transforms the equation into an ordinary differential equation, one in which the dependent variable, z, depends upon only one variable, x. The curvature is then a constant, set by the regolith production rate and the diffusivity, κ:

$$\frac{d^2 z}{dx^2} = -\frac{\rho_r \dot{W}}{\kappa \rho_b} \tag{10.10}$$

Note that the curvature is negative, meaning that the topography is convex up. Equation 10.10 can be integrated once to obtain the slope as a function of position:

$$\frac{dz}{dx} = -\left[\frac{\rho_r \dot{W}}{\kappa \rho_b}\right] x + c_1 \tag{10.11}$$

The constant of integration, c_1, is 0 because at $x = 0$ (the hillcrest) the slope is zero. A second integration yields the topography, $z(x)$:

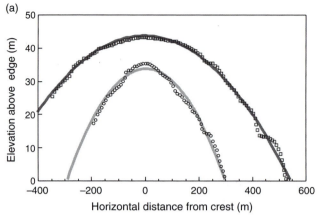

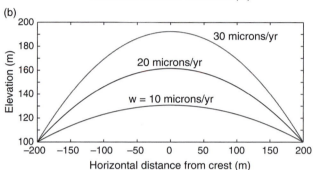

Figure 10.4 (a) Profiles of summit of Goat Flat in the Wind River mountains. Dots are data from hand-leveling of the topography. Curves are best-fitting parabolas through the data. Other than anomalous knicks into the hillslopes, usually associated with bedrock ledges or nivation hollows, the parabolas fit well. The curvature in this landscape is essentially uniform (after Anderson, 2002, Figure 3, with permission from Elsevier). (b) Analytic solutions for profiles of steady-state hilltops for three prescribed regolith production rates. As the rate of production increases, the curvature must increase to accommodate the required increase in downslope regolith transport.

$$z = -\left[\frac{\rho_r \dot{W}}{2\kappa \rho_b}\right] x^2 + c_2 \tag{10.12}$$

where c_2 is another constant of integration. Note that c_2 is an elevation. We may choose it by specifying the elevation at some location. In this case, shown in Figure 10.4, the easiest location is the hillcrest, where $x = 0$. Then $c_2 = z_{max}$. Then Equation 10.12 describes an inverted parabola, a simple geometrical shape with the characteristic that the slope is zero at the crest and increases linearly with distance from the crest. As G. K. Gilbert argued in 1909, this hillslope shape must arise if regolith is being produced everywhere at the same rate, in other words, if $\dot{W}$ is not a function of x (is uniform in x), and if the regolith discharge rule is simply

Box 10.1 From random walk to diffusion

Here we illustrate how one can move from a description of a microscopic process to a differential equation that describes the evolution of a system at a macroscopic level. The following derivation is a shorthand summary of Einstein's theory of Brownian motion (Einstein, 1905) (pers. comm. Greg Tucker).

Assume we have an ensemble of particles moving randomly and independently from one another, and for now simply consider motion left and right, in the x-direction. Let $p(\delta x, \tau)$ denote the probability of a particle (which could be a water molecule, a mosquito, a sediment grain, etc.) making a jump of distance δx in a time interval τ. Let $C(x, t)$ denote the concentration of particles at point x at time τ. (C can be read as the number per unit length in this one-dimensional treatment.) Our goal is to write an equation for $C(x, t)$ that captures the macroscopic outcome of zillions of moving particles without having to follow the dynamics of any particular particle.

If the concentration, C, of particles is a function of location and time, then the concentration at position x at some short time interval, τ, in the future is equal to the concentration at time t at a distance $-\delta x$ away times the probability of moving exactly δx during τ, integrated over all possible hop distances. Mathematically, this may be summarized as

$$C(x, t + \tau) = \int_{-\infty}^{\infty} p(\delta x, \tau) C(x - \delta x, t) \mathrm{d}(\delta x) \tag{10.13}$$

This equation would be far easier to handle if we could pull $C(x - \delta x, t)$ out of the integral. It turns out we can do this by performing a Taylor expansion around $C(x, t)$ (see Appendix B):

$$C(x - \delta x, t) \approx C(x, t) - \delta x \frac{\partial C(x, t)}{\partial x} + \frac{\delta x^2}{2} \frac{\partial^2 C(x, t)}{\partial x^2} - \frac{\delta x^3}{3!} \frac{\partial^3 C(x, t)}{\partial x^3} + \cdots \tag{10.14}$$

Substituting this expression for $C(x - \delta x, t)$ into the integral in Equation 10.13 yields

$$C(x, t + \tau) \approx \int_{-\infty}^{\infty} p(\delta x, \tau) \left(C(x, t) - \delta x \frac{\partial C(x, t)}{\partial x} + \frac{\delta x^2}{2} \frac{\partial^2 C(x, t)}{\partial x^2} - \frac{\delta x^3}{3!} \frac{\partial^3 C(x, t)}{\partial x^3} + \cdots \right) \mathrm{d}(\delta x) \tag{10.15}$$

This looks discouraging, but since $C(x, t)$ is not a function of δx, we can remove it from the integral and rearrange to get

$$C(x, t + \tau) \approx C(x, t) \int_{-\infty}^{\infty} p(\delta x, \tau) \mathrm{d}(\delta x) - \frac{\partial C(x, t)}{\partial x} \int_{-\infty}^{\infty} p(\delta x, \tau) \delta x \mathrm{d}(\delta x)$$
$$+ \frac{\partial^2 C(x, t)}{\partial x^2} \int_{-\infty}^{\infty} p(\delta x, \tau) \frac{\delta x^2}{2} \mathrm{d}(\delta x) - \frac{\partial^3 C(x, t)}{\partial x^3} \int_{-\infty}^{\infty} p(\delta x, \tau) \frac{\delta x^3}{3!} \mathrm{d}(\delta x) + \cdots \tag{10.16}$$

While again this looks more complex than where we began, we are poised to simplify it by appealing to some definitions of statistical moments (see Appendix B). First, note that it is a property of a probability density function that the integral over all possible probabilities must equal 1. The first term therefore becomes simply 1:

$$\int_{-\infty}^{\infty} p(\delta x, \tau) \mathrm{d}(\delta x) = 1 \tag{10.17}$$

allowing us to rewrite Equation 10.15 as

$$C(x, t + \tau) = C(x, t) - \frac{\partial C(x, t)}{\partial x} \int_{-\infty}^{\infty} p(\delta x, \tau) \delta x \mathrm{d}(\delta x)$$
$$+ \frac{1}{2} \frac{\partial^2 C(x, t)}{\partial x^2} \int_{-\infty}^{\infty} p(\delta x, \tau) \delta x^2 \mathrm{d}(\delta x)$$
$$- \frac{1}{3!} \frac{\partial^3 C(x, t)}{\partial x^3} \int_{-\infty}^{\infty} p(\delta x, \tau) \delta x^3 \mathrm{d}(\delta x) + \cdots \tag{10.18}$$

Box 10.1 (*cont.*)

The integral in the second term is the definition of the mean value of δx, or the first moment of the distribution $p(\delta x)$. Likewise, the integral in the second term looks almost like the variance. In fact, it is the second *raw moment*, which is equivalent to the variance but is calculated around zero rather than around the mean. In general, the nth raw moment of a probability distribution $P(x)$ is defined as

$$\mu'_n = \langle x^n \rangle = \int_{-\infty}^{\infty} P(x)x^n \mathrm{d}x \tag{10.19}$$

where the angle brackets denote an average. From this definition, you can see that each of the integral expressions in our Taylor series (Equation 10.17) corresponds to a raw moment of the distribution of hop-length distribution $p(\delta x, \tau)$. Equation 10.18 may now be rewritten:

$$C(x, t+\tau) - C(x, t) = -\frac{\partial C(x, t)}{\partial x}\langle \delta x \rangle + \frac{\partial^2 C(x, t)}{\partial x^2}\frac{\langle \delta x^2 \rangle}{2} - \frac{\partial^3 C(x, t)}{\partial x^3}\frac{\langle \delta x^3 \rangle}{3!} + \cdots \tag{10.20}$$

If we divide all the terms by τ, the left-hand side of the equation becomes the definition of a derivative in the limit that τ goes to zero:

$$\lim_{\tau \to 0}\left(\frac{C(x, t+\tau)-C(x, t)}{\tau}\right) = \frac{\partial C(x, t)}{\partial t} = \lim_{\tau \to 0}\left(-\frac{\partial C(x, t)}{\partial x}\frac{\langle \delta x \rangle}{\tau} + \frac{\partial^2 C(x, t)}{\partial x^2}\frac{\langle \delta x^2 \rangle}{2\tau} - \frac{\partial^3 C(x, t)}{\partial x^3}\frac{\langle \delta x^3 \rangle}{3!\tau} + K\right) \tag{10.21}$$

If we neglect the higher-order terms that contain cubes, etc., of the small distances δx, which is something we can do because as we shrink δx these terms shrink yet faster, and define the constants

$$v \equiv \lim_{\tau \to 0}\frac{\langle \delta x \rangle}{\tau} \tag{10.22}$$

and

$$D \equiv \lim_{\tau \to 0}\frac{\langle \delta x^2 \rangle}{2\tau} \tag{10.23}$$

we are left with a partial differential equation for the evolution of the concentration of particles:

$$\frac{\partial C(x, t)}{\partial t} = -v\frac{\partial C(x, t)}{\partial x} + D\frac{\partial^2 C(x, t)}{\partial x^2} \tag{10.24}$$

The first term on the right-hand side represents advection (or "drift"), while the second represents diffusion. This advection–diffusion equation appears, for example, in groundwater contaminant transport problems. From this derivation, you can see that the advection coefficient, v, represents the mean speed of a particle (that is, the average hop length divided by hop duration). Also note that if the average hop length is zero (i.e., there is an equal probability of going left or right), then the advection term disappears and we are left with a standard diffusion equation that we saw first in Chapter 3 in dealing with heat conduction problems. We also see that the diffusivity coefficient, D, represents the mean-squared particle displacement divided by hop duration (Equation 10.23).

In Figure 10.5 we illustrate this phenomenon by simulating a random walk of 10 000 particles. Each particle is allowed to take a random step drawn from a normal distribution with a standard deviation of 0.2 units, and a mean step length of 0.01 units. We report the new histogram of the positions of the particles after every 1000 steps. Through time, you can see the mean drift of the population to the right, reflecting the first term in the equation, and a spreading of the population reflecting the diffusive behavior of the population captured in the second term.

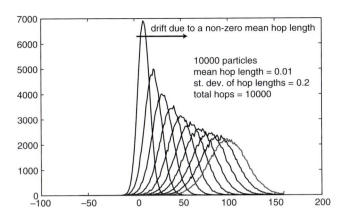

Figure 10.5 Random walk. Resulting migration and spread of a set of 10 000 particles whose behavior is governed by random hops. All particles originate at $x = 0$ (not shown). Histograms of positions shown at 1000 hop intervals. The mean position drifts to the right at speed = 0.01/timestep. The spread increases as the square root of time, while the magnitude of the peak in concentrations decays as the square root of time.

dependent on local slope. This is the essence of a diffusive landscape: hilltops are convex. Any sharp (high curvature) corners in the landscape are rapidly smoothed. One can see in Figure 10.4(a) that indeed some hilltops are not only convex, but are parabolic.

We now have a conceptual framework in which to explore both hillslope transport processes and hillslope forms. Convex slopes are likely to result from diffusive processes, in other words, regolith transport that is simply proportional to local slope. Other hillslope shapes, such as convexo-concave profiles, are likely to result from transport processes that depend upon something other than local slope, such as distance from the hillcrest. Let us now consider a few specific transport processes. Our goal in part will be to learn enough about the transport processes to allow us to construct models for Q that acknowledge its dependence on both climate and slope material properties.

Hillslope processes

One may classify hillslope processes in many different ways. In our treatment, we break these processes down into grain-by-grain processes, and mass (or continuous) processes. One could also classify them as deterministic and stochastic, or in many other ways. We will start with the process of rainsplash,

an example of a grain-by-grain process. Although rainsplash is not a dominant process in many landscapes, the physics that drive it are clear, and are amenable to experimentation. Attention to the rainsplash problem will allow us to develop experience with parsing a process into its component pieces, which can be applied to other more common cases.

Rainsplash

Bare soil surfaces are modified by the direct impact of raindrops, a process called *rainsplash*. Individual raindrops blast grains from the surface. On sloped surfaces, ejected grains travel longer trajectories downhill than uphill, resulting in a net downhill motion of material. Since the process may be cast as one in which the flux of material is proportional to the slope, this is a candidate for diffusive behavior. Note as well that it is a process that involves the summation over many individual particle motions, providing a linkage to the development of the diffusion equation through the random walk model described in the box above. Below we attempt to place the rainsplash process in a more formal setting, in order to assess the dependence of this process on the climate and the type of hillslope material. As hinted above, we will parse the problem into three parts:

- the number of ejected grains per impact and their velocities as a function of the impacting drop size;
- the resulting trajectories of the grains once ejected;
- the probability distribution of drop sizes and the total number of drops per unit area per unit time in a storm.

The first two are physics problems, while the third is a meteorological problem. In fact, these three pieces yield the transport associated with a particular storm. The long-term transport rate, averaged over even a year, and certainly over decades to millennia, requires averaging over a spectrum of storms. We must, however, start at the scale of a raindrop.

Studies have shown that the number and velocity of grains dislodged from soil surfaces by individual raindrops are roughly proportional to the kinetic energy of the impacting drop. While the response of the granular surface to raindrops is complex, having to do with the changes in the surface through the course of a storm, and to the specific grain size distribution of the surface, and so on, we wish here to focus on the fundamental physics. The details of the surface response are treated extensively in the literature, much of it in the soil science literature.

The problem of the delivery of kinetic energy can be broken into the need to know the number of raindrops per unit area of the surface per unit time (the raindrop flux), and the kinetic energy of each drop at impact. From the expression for kinetic energy,

$$E_k = \frac{1}{2}Mv^2 \tag{10.25}$$

we see that the latter calculation reduces to knowing the mass, M, and the impact speed, v, of a drop. Calculating the mass of a raindrop is easy. Despite the illustration on the Morton salt containers, raindrops are approximately spherical (actually the large ones pancake a bit, presenting a broader cross section to the air), and their mass is therefore $M = \rho_w \pi D^3/6$ for a drop of diameter D and water density ρ_w. We see already that the kinetic energy varies at least as the cube of the raindrop diameter.

What about their impact speeds? Like any object falling under the influence of gravity, a raindrop accelerates until the drag force opposing its motion is equal to the force pulling it downward, the weight of the particle. At this point, the vertical speed no longer increases, and the drop is said to be at its fall velocity (also called its terminal velocity or settling velocity). This balance is an example of Newton's law of motion, which states that the acceleration of an object is equal to the sum of the forces acting on it and is inversely proportional to the mass of the object. We have all memorized this as $F = ma$. By the time a raindrop hits the ground, it has achieved its fall velocity, which means that the acceleration in this familiar equation $a = F/m$ is zero. This condition occurs when $F = F_{drag} + F_w = 0$, where F_{drag} is the drag force acting on the drop, and F_w is the weight of the drop. We note that this weight should be the buoyant weight, in general, but that in this case the weight of the displaced fluid (air) is trivial and may be neglected. Since F_w and F_{drag} act in opposite directions, they have opposing signs, and the fall velocity is reached when their magnitudes are equal.

The force balance on a raindrop is illustrated in Figure 10.6. The vertical coordinate is z, taken to be positive upward, and, as is common in fluid mechanics, we call the vertical velocity w. The only

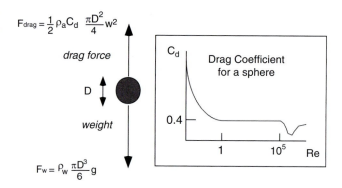

Figure 10.6 Forces on a grain falling in a still fluid. At the settling speed, the drag force and weight are exactly balanced. The drag coefficient, C_d, is a function of the Reynolds number, Re, showing an inverse relationship for low Re and near-constant values around 0.4 for high Re. This is interrupted by the "drag crisis" near Re = 10^5, in which the drag coefficient diminishes twofold.

acceleration in this case is in the vertical, meaning that we can write the acceleration of the drop to be the rate of change of the vertical speed, or dw/dt. The weight acts downward, in the negative z-direction, and may be written as $F_w = -Mg$. The opposing force acting to retard the fall, the drag force, is a little more complicated, as it depends strongly upon the speed of the object relative to the fluid. One can write a general formula for the drag force on a sphere,

$$F_{drag} = \frac{1}{2}\rho_a C_d \frac{\pi D^2}{4} w^2 \tag{10.26}$$

where ρ_a is the density of the fluid through which the particle is moving (air in this case), C_d is a non-dimensional parameter called the *drag coefficient*, and $\pi D^2/4$ is the cross-sectional area of a sphere presented to the flow. The drag coefficient is dependent upon the speed of the object relative to the fluid (and in general on the shape of the object – here we will assume a sphere). The value of the drag coefficient depends upon whether the flow around the object (our raindrop) is laminar or turbulent. In fluid mechanics, one relies upon a quantity called the Reynolds number, Re, to determine whether the flow field is laminar (low Re) or turbulent (high Re):

$$Re = \frac{\rho_w w D}{\mu} \tag{10.27}$$

where μ is the dynamic viscosity of the fluid. This is only one of the many non-dimensional numbers one encounters in fluid mechanics problems. In this case, the number represents the relative importance of inertial forces and viscous forces in the problem. The dependence of the drag coefficient upon the Reynolds number is shown in Figure 10.6. Ignoring the little dip out there at very high Re (a feature dubbed the drag crisis, at Re of 10^5 or so), there are essentially two asymptotic expressions, one at very high Re, the other at low Re. At high Re, the drag coefficient becomes a constant, at 0.4. And at low Re, the drag coefficient is inversely dependent upon Re, following the relationship $C_d = 24/\text{Re}$.

The general expression for the force balance on the particle is given by the sum of the buoyant weight of the particle (its weight less that of the fluid it displaces) and the drag force:

$$M\frac{dw}{dt} = -(\rho_w - \rho_a)\frac{\pi D^3}{6}g + \frac{1}{2}\rho_a C_d \frac{\pi D^2}{4}w^2 \quad (10.28)$$

To determine the settling velocity, which by definition occurs when the weight is exactly balanced by the drag force, we simply set the left-hand side of Equation 10.14 to zero, insert the relevant drag coefficient, and solve the equation. Breaking the relation C_d (Re) into two domains, reflecting low and high Re, results in two expressions for settling velocity, one for low and the other for high Reynolds numbers:

$$\text{for low Re: } w = \frac{gD^2\rho_w}{18\mu} \quad (10.29)$$

$$\text{for high Re: } w = \sqrt{\frac{gD\rho_w}{0.3\rho_a}} \quad (10.30)$$

Equation 10.29 demonstrates that the fluid viscosity, μ, which relates the rate of shear strain in a material to the local shear stress on that material, plays a role in governing the settling speeds at low Re.

If you are doing such a calculation, you should check to make sure you have chosen the correct formulation of the drag coefficient, by assessing the Re associated with the calculated settling velocity. If you used the low Re formulation, and find Re > 1 for the calculated settling velocity, or vice versa, you have used the wrong formulation. In Figure 10.7 we show the fall velocities for objects of several densities in both water and air, calculated using a fuller representation of the C_d(Re) function that spans the transition region between low and high Reynolds numbers (see Morsi and Alexander, 1972).

While the above discussion is general, let us return to the problem of raindrops. We are now in a position to assess the dependence of the kinetic

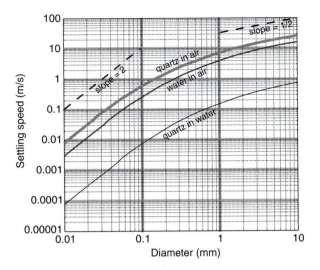

Figure 10.7 Settling velocity as a function of particle size for three cases: water droplets in air, quartz grains in air, and quartz grains in water, plotted in linear–linear space. Although the settling velocity increases monotonically with particle size, note the steeper increase (as D^2: slope of 2 on the log–log plot) for small particle sizes, and less dramatic increase (as $D^{1/2}$: slope of 1/2 on the log–log plot) for large particle sizes.

energy of a raindrop on drop size. Drops much larger than a fraction of a millimeter fall with high enough velocities to be in the high Reynolds number regime, and therefore obey Equation 10.30. Their velocities increase as the square root of the drop diameter. We noted earlier that the mass of a raindrop increases with the cube of the drop diameter. When we assess the dependence of kinetic energy on drop diameter in Equation 10.25, we see that the kinetic energy goes up as the fourth power of the drop diameter:

$$E_k = \frac{1}{2}Mv^2 \sim D^3\left[D^{1/2}\right]^2 \sim D^4 \quad (10.31)$$

where three powers come from the raindrop mass and one from the square of the velocity. This result has strong implications for the types of storms that are important in driving rainsplash erosion: those capable of producing large drops ought to be many times more effective in transporting sediment. Empirically, high-intensity rainstorms have much larger drops (e.g., Laws and Parsons, 1943; Kneale, 1982), reflecting the fact that high-intensity storms are typically convective storms with high columns within which the raindrops are allowed to grow to very large sizes (see Figure 10.8). The rainsplash effects of the

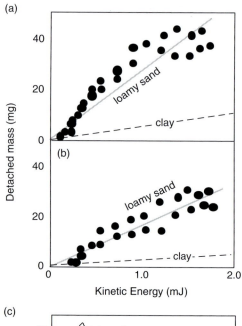

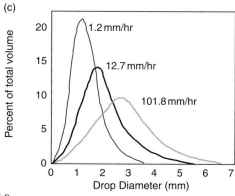

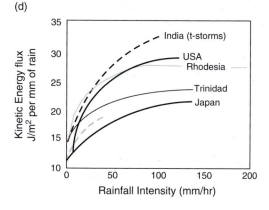

Figure 10.8 Raindrop sizes and their effects. (a and b) The mass ejected from a uniform surface for both a loamy sand and a clay surface as a function of kinetic energy of the raindrop size (after Sharma and Gupta, 1989). Data shown as filled circles are approximated as a linear function of kinetic energy. The difference between (a) and (b) is that the soil moisture in (b) is lower, meaning that grains are held more tightly by surface tension of the water in the pores. (c) Distribution of raindrop sizes for three different rainfall intensities (after Laws and Parsons, 1943). The dominant drop size increases by 1 mm for every order of magnitude increase in the rainfall intensity. (d) Distribution of kinetic energy flux as a

occasional summer cloudburst should therefore far outweigh the effects of many winter drizzles.

The flux of kinetic energy to the ground is the sum of the kinetic energy of all the drops. We can get at this impossible-sounding number if we know the distribution of drop sizes in a rainstorm, which is something that is measurable. The probability density of raindrop size, $p(D)$, describes the fraction of all the drops in a storm within each size category, dD. Since we can now calculate the kinetic energy of any drop size, the total kinetic energy delivered to the surface per unit time per unit area of the landscape is obtained by summing over the drop sizes. This is done formally with the integral

$$E_k = \sum_{i=1}^{N} E_{ki} = N \int_0^{\infty} E_k(D)p(D)\,dD \qquad (10.32)$$

where N is the total number of drops landing on a unit surface area in a unit of time (say an hour), or the "raindrop flux," and $p(D)$ is the probability density of raindrop size distribution (e.g., Figure 10.8(c)). The integral yields the mean kinetic energy of the raindrops. Note that if we replace the kinetic energy of the drop with the volume of the drop, the equation becomes one for the rainfall intensity (volume per unit area per unit time, which is a length per time, say mm/hr).

One could also think of this quantity as "rain power." Power has units of energy per time; for example, 1 watt = 1 joule/second. Rain power would be defined as $P = N\overline{E}_k$. We will see a similar quantity again when we explore the problem of stream incision into rock, in which case it is termed "stream power." In both cases we will be asking how efficiently this power is converted into geomorphic work.

But we are not yet done with the rainsplash problem. We have defined the meteorological forcing of the system, but have not yet assessed the response of the landscape. We must address the efficiency of the meteorological engine. We need to translate the ejection of grains from the surface to a transport distance, and we need to determine the circumstances in which

Caption for Figure 10.8 (cont.)
function of rainfall intensity for several localities. Note the different curves for India, with and without thunderstorms (after Selby, 1982, Figure 5.22, with permission of Oxford University Press). For a recent review, see van Dijk *et al.*, 2002, Figures 3–5.

Box 10.2 The parabolic grain trajectory

Assuming that the drag on the grain is negligible in this problem, the horizontal speed of the grain is unchanging between its launch and its landing. Its horizontal speed is $u = V_o\cos(\theta)$, where V_o is the initial speed of the grain. The vertical speed, on the other hand, changes due to gravitational acceleration. Again ignoring drag, the vertical acceleration is simply $-g$. The evolution of its vertical speed is therefore $w = w_o - gt$, where $w_o = V_o\sin(\theta)$. We may integrate the horizontal and vertical speeds to obtain horizontal and vertical positions, x and z, through time. These "parametric equations" for its position are: $x = ut = V_o\cos(\theta)t$, and $z = V_o\sin(\theta)t - (1/2)gt^2$. One may plot these positions through time to see that they are parabolas. But we may use a trick to discover the dependence of z on x, i.e., to solve for $z(x)$. Note that we may solve the x-position equation for time: $t = x/V_o\cos(\theta)$, and replace the t in the vertical equation with this formula. The resulting equation for $z(x)$ is $z = x\tan(\theta) - \frac{g}{2V_o^2\cos^2(\theta)}x^2$, which is indeed a parabola. Note that we recover the expected trajectory length by solving for where $z = 0$. This occurs at $x = 0$ and at $x = (2V_o^2/g)\cos(\theta)\sin(\theta)$. The advantage of casting the problem in this form is that we may now solve for where this parabola intersects a surface of prescribed slope. This treatment of ballistic trajectories (motion under gravity alone) and many more twists on the theme are elegantly presented in Neville DeMestre's book, *The Mathematics of Projectiles in Sport* (1990).

this results in net downslope transport. What is the fate of the ejected grains? If we know the speeds and the angles at which particles are ejected from the surface, and the slope of the surface, we can calculate the distance the particles will travel before re-encountering the local surface, as shown in Figure 10.9. Ignoring the drag of the air, and any wind that might accompany the storm (we are aiming at a first-order understanding here), the trajectories of ejected grains are parabolas (see Box 10.2). On a flat surface, they would travel a distance

$$L = \frac{2V_o^2}{g}\cos\theta\sin\theta \qquad (10.33)$$

where V_o is the initial speed of the ejection from the bed, and θ is the launch angle with respect to the horizontal. As is sketched in Figure 10.9, on a tilted surface, the distance traveled is shorter on the uphill side and longer on the downhill side of the ejection site. It is the average of the uphill and downhill distances that scales the net rainsplash transport. To first order, the mismatch in up- and downslope distances is proportional to the local slope, in other words, $L_{net} = aS$ where S is the local slope and a is a constant. We can determine crudely how big the discrepancy is between the hop length on a flat surface and the hop length on a canted surface by solving for the intersection of the grain trajectory (a parabola) with a

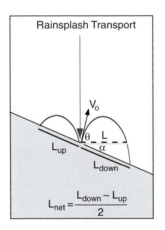

Rainsplash Transport

$$L_{net} = \frac{L_{down} - L_{up}}{2}$$

Figure 10.9 Schematic of effects of a raindrop on ejection of sediment. The length of the downslope trajectory, L_{down}, is greater than that upslope, L_{up}, leading to net downslope transport, L_{net}, associated with each drop impact.

sloping surface (a straight line). The discrepancy, or $dL = L_{down} - L$, is roughly

$$dL = \frac{2V_o^2\cos^2\theta}{g}\tan\alpha \qquad (10.34)$$

where α is the slope of the surface. The sum of the discrepancy from upslope truncation and downslope extension of the trajectories is roughly twice this. Noting that $\tan(\alpha)$ is approximately the negative of the local slope, dz/dx, we obtain the following expression for the net transport length:

$$L_{net} = -\frac{4V_o^2 \cos^2 \theta}{g}\frac{dz}{dx} \qquad (10.35)$$

Note that for a flat surface, $dz/dx = 0$, Equation 10.35 predicts a net transport of zero, as expected. For a given launch angle, the net distance increases essentially linearly with hillslope angle, and depends strongly upon the launch speed.

We have now determined that the transport rate ought to increase as the slope increases. This hints that the process may be diffusive. In a diffusion problem, however, we need to know how to weld together information about the mean transport rate of an ejected grain with information about the process intensity. In this case the intensity is set by the number of drops landing on a surface per unit area per unit time, the raindrop flux. The net transport rate of mass per unit width of the hillslope, also called the specific discharge with units of [M/LT], is

$$Q = m_p n L_{net} N \qquad (10.36)$$

where m_p is the mass of an individual splashed particle, n is the mean number of ejected grains per raindrop impact, L_{net} [L] is the net transport distance length, and N is the raindrop flux $[1/L^2T]$. We now have a transport rule for a given rainstorm:

$$Q_x = -m_p n N \frac{4V_o^2 \cos^2 \theta}{g}\frac{dz}{dx} \qquad (10.37)$$

Note that this boils down to a flux that is linearly related to the local slope, the transport constant in front of the slope being related to the meteorologically important variables in the problem. We therefore expect diffusion from rainsplash! To summarize, in models of landscape evolution a hillslope process is often invoked in which the flux of regolith is simply cast as $Q = -k(dz/dx)$.

We may break Equation 10.37 down into

$$Q = P\beta L_{net} \qquad (10.38)$$

where P is rain power and captures the meteorological forcing, β is the mass ejection efficiency (e.g., Sharma's data) which describes the susceptibility of the landscape to rainsplash transport, and L_{net} reflects the asymmetry of the trajectories. It is this latter effect that dictates the slope dependence of the transport

process (i.e., sets the roughly linear increase in transport with slope angle).

Recent experimental work on raindrop impacts has illuminated the details of the process. Using very high-speed video, Furbish *et al.* (2007) recorded the impact of raindrops of various sizes on sandy substrates at various inclinations. Several examples are shown in the photographs in Figure 10.10. They found that the asymmetry of the distribution of splashed grains was largely due to the asymmetry of the ejecta, rather than by the asymmetric truncation of parabolic trajectories described above. In other words, upon impact with an inclined surface, the redirection of the momentum of the water in the raindrop entrains more grains at higher speeds in the downslope direction than upslope direction. The net effect is similar, in that a mean downslope transport of grains occurs, but the physics behind the asymmetry of the resulting splash, leading to a nonzero L_{net}, differs.

The bottom line is therefore that one should expect a downslope drift of particles due to the summation of the effects of many raindrop impacts. The realization that this process is akin to a random walk has allowed the treatment of rainsplash in the statistical mechanical framework we have abbreviated in Box 10.1. A commonly observed phenomenon in the desert has motivated Furbish *et al.* (2009) to take the statistical mechanical approach to the next level. Most bushes in the desert sit on small mounds of sediment that are remarkably similar in their footprint to the canopy of the bush. These researchers suggest that the bush serves as something of an umbrella, reducing the number and intensity of raindrops beneath the canopy; in other words, as illustrated in Figure 10.11, the bushes generate non-uniform raindrop activity. The random walks of sediment driven by raindrop impacts will lead to a net accumulation of grains beneath the bush; while many raindrops in a ring outside the bush canopy drive sediment inward, the lower number of raindrop impacts beneath the canopy fail to drive sediment outward at the same rate. The net flux is therefore established simply by the gradient in the raindrop activity. But as the mound grows in height, the outward slope of the topography begins to take effect, creating an outward drift of particles. The mound reaches a steady configuration when the inward flux of particles due to the activity gradient in raindrops is balanced by the outward drift of particles due to the slope of the mound.

(a)

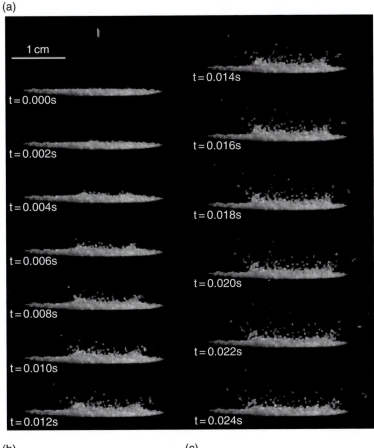

(b)

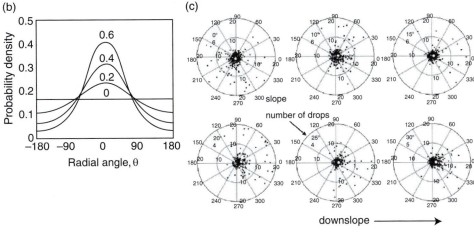

Figure 10.10 Rainsplash experiments. Raindrops of known size impact clean sand of selected diameter, and the pattern of ejected grains is documented. (a) Sequence of photographs over 1/40th of a second showing one impact, and the grain–grain collisions that dominate the ejection pattern. (b) Definition sketch for the concentration factor, dictating the asymmetry of the ejection pattern, with downslope corresponding to $\theta = 0$. (c) Representative impact ejection patterns on increasingly sloped targets, ranging from 0–30°. Note increasing asymmetry to the ejection pattern (increasing concentration factor) with slope (after Furbish *et al.*, 2007, Figures 3, 6, and 11, with permission from the American Geophysical Union).

They further point out that this is a means by which the bush can command, or harvest, any nutrients in the surrounding soil, self-generating a resource island. Clever bush! Of course, upon death of the bush and decay of the canopy, the slope effects take over, and the form diffuses due to the dominance of downslope transport, as again shown in Figure 10.11.

One might wonder how to go about measuring rainsplash transport. The drop size distribution is clearly important to document when attempting to assess the role of rainsplash sediment transport. This is done using a remarkably simple contraption: a pan filled with flour. When exposed to the rain, the drops ball up and become coated in a monolayer of flour dust. This is then

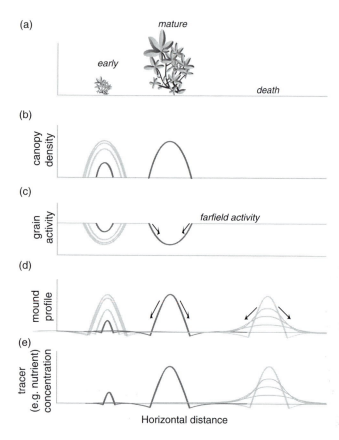

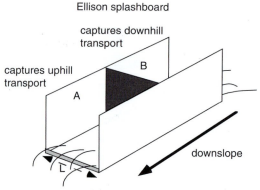

Figure 10.12 An Ellison splashboard, shown here without its lid, allows documentation of net downslope rainsplash transport by differencing that mass of grains caught in trough A, representing uphill-splashed grains, from that in trough B, representing downhill-splashed grains. Division by the width of the trough, L, and the time interval, T, over which the measurement occurred results in a net transport rate: $Q = (M_A - M_B)/LT$. The lid prevents grains from being splashed or sluiced out of the trough.

Figure 10.11 Rainsplash induced growth of mounds beneath growing plant canopies (see Furbish et al., 2009). (a) Canopy evolution from early to mature stages. (b) Canopy density at early (left) and mature (right) stage. Temporal patterns are shaded on left of (b)–(d), showing asymptotic growth to mature stage. (c) Reduction in grain activity below far-field activity due to canopy growth. Arrows depict inward transport of grains due to gradients in grain activity. (d) Evolution of mound growth beneath the plant. Arrows depict slope-driven outward "drift" of particles. At steady state, outward slope-induced drift equals inward activity-gradient transport. After death of the plant, drift, uncompensated by activity gradient related transport results in topographic diffusion of the form. (e) Concentration of near-surface soil tracers, showing net loss of tracers from the moat around the plant, and enhancement of concentrations beneath the plant. Plants harvest nearby nutrients, becoming resource islands.

dried, and the resulting mixture is sifted. The size distribution of pellets can then be converted to the size distribution of raindrops knowing that the volume of the pellet corresponds to the surface area of the drop.

Measurement of the rates of sediment transport driven by raindrops, or the rainsplash transport rate, is commonly accomplished using a device called the Ellison splash board, developed by Ellison in the 1940s. This device, shown in Figure 10.12, separates that material being splashed uphill from that being splashed downhill, using a small board to prevent cross-over between measurement cups. It is the difference between the mass collected on either side of the board that measures the net downhill transport of debris.

Creep

Creep is the slow downslope transport of soil or regolith by bulk motion under the influence of gravity. This motion occurs in a variety of settings. The motion is usually so slow that it is difficult to measure except on long timescales (months to years). Recall that if it can be shown that this motion is linearly related to the local slope, then the resulting steady hillslope form will be convex up and parabolic. What we need in order to solve the problem is the discharge of regolith through the upslope and downslope sides of the box. What is the physics of this slow motion? And what might the velocity profile of such motion look like?

Solifluction: frost creep and gelifluction

Slow downslope movement of soil is a significant process in periglacial areas, and is responsible for lobate steps or terraces on hillslopes called solifluction lobes. This slow downslope motion occurs in the short summers of Arctic and alpine areas, and results from

Box 10.3 Analogy to micro-meteorite bombardment

On other terrestrial planets on which the hydrologic cycle is either dormant or non-existent, there is obviously no raindrop transport. Nonetheless, the surfaces are barren of vegetation and are therefore subject to bombardment from other projectiles. These come in the form of meteorites. On planets without significant atmospheres, micro-meteorites are capable of reaching the surface of the planet before burning up by drag-heating in passage through the atmospheric blanket. Surfaces of the Moon, Mercury and Mars are therefore bombarded with the full size spectrum of interplanetary debris, summarized in Figure 10.13. As this debris is dominated by small diameter particles, the surfaces are subject to micro-meteoritic bombardment. This has many effects. First, a regolith cover consisting of a mixture of micro-meteoritic and primary crustal debris grows as these meteorites accumulate. Second, the regolith is continually being churned up by new impacts (a process cleverly termed "impact gardening"). Third, it will move material down slopes, for the same reasons that rainsplash accomplishes net downslope transport. Old surfaces that have not been subject to rejuvenation by a very large impact or by a volcanic event are therefore smoothed at the small scale by these impacts. One need only look at the beautiful medium format Apollo photographs of the Moon (e.g., the frontispiece for this chapter) to be impressed with the diffused nature of the landscapes we visited (see Light, 1999). The image of a typical view of the surface of Mars in Figure 10.14 reveals the competition between the roughening of the surface by impact of large debris, and the smoothing of the surface by the much more frequent small particles.

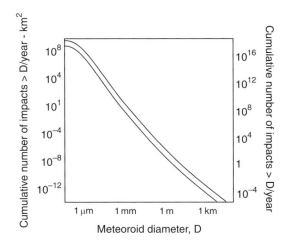

Figure 10.13 Present day impact rate at the top of Earth's atmosphere. The curve represents the cumulative number of impactors above that diameter to be expected in a given square km. The right axis is calculated by converting the impact rate to that over the entire surface area of the Earth (redrawn from Hartmann, 1999, Figure 6–4, with permission from Brooks-Cole).

the fact that these frozen soils attain excess water as they freeze through the growth of ice lenses. This same water super-saturates the near-surface soil upon thawing from the surface downward. The saturated soil is weak when subjected to the shear stresses associated with the downslope component of the weight of the material, allowing it to flow. Linc Washburn, in his extensive work in the Canadian Arctic and in Greenland, has documented carefully the velocities of solifluction lobes (see Figure 9.23). These can attain speeds of centimeters/day on even very shallow slopes. The velocity is clearly controlled by the thickness of the thawed layer, by the water content of the material, and by the slope. Let us consider a simple model of how the water content of the soil might affect the velocity profile. The latest soil to thaw is that immediately above the thaw front, at the base of the thawed layer. As soil thawed earlier will have had a chance to drain, and therefore to consolidate, it is likely that the soil is weakest immediately above the thaw front as well.

The most easily measurable quantity that might be used to test the relevance of this simple treatment is the surface velocity. Markers extending into the sub-surface, however, will experience tilting due to the finite velocity difference between the base of the marker and the surface. This solifluction case is interesting in that the velocity profile ought to vary substantially through the season as the thaw proceeds. No motion will occur in the winter. The net motion over the course of a year ought to depend sensitively upon the maximum depth of thaw, and upon the duration of thaw, both of which will vary from year to year,

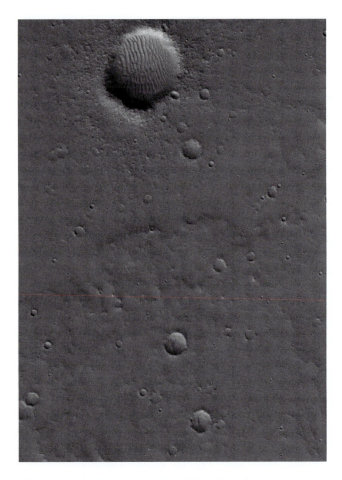

Figure 10.14 Mars Global Surveyor (MGS) Mars Orbiter Camera (MOC) image showing a cratered surface in the Tempe Terra region. A wide range of crater sizes is clear. What is less clear is the dominance of small impact craters. These serve to blur the older craters through time. Large crater in the upper left has sand dunes on its floor. Location near: 31.0°N, 84.1°W. Image width: ∼3 km (∼1.9 mi). Illumination from: lower left. Season: Northern Winter (image taken by and available from NASA/JPL/Malin Space Science Systems. MGS MOC Release No. MOC2–1349, January 21, 2006, S1100389).

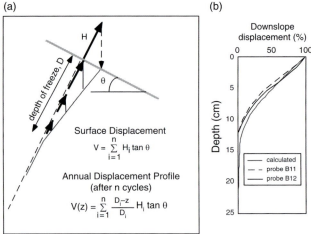

Figure 10.15 Frost heave displacement profiles. While the displacement, V, from a single frost heave cycle of magnitude H is roughly linear with depth, the sum over n cycles results in a displacement profile that more smoothly asymptotes to zero with depth. In (b) two measured profiles are contrasted with the calculated profile based upon the summed linear profiles shown in (a) (after Matsuoka and Moriwaka, 1992, Figure 10, with permission from INSTAAR).

and even from place to place within a basin. If one desires a long-term soil transport rate on a particular slope, one must integrate the expected velocity profile over both depth (to obtain the transport rate at any instant), and over the thaw season. Finally, as not all seasons yield the same maximum thaw depth or duration, one should also integrate over a distribution of seasonal thaw cycles.

In regions prone to periodic freezing and thawing, regolith moves downslope by another means. We have already touched upon this in our discussion of periglacial processes in Chapter 9. Rather than flowing as a very viscous fluid, the soils periodically expand in a

process called frost heave, and subsequently collapse upon thaw. While the expansion is usually normal to the slope, the collapse is more or less vertical. A single full cycle of motion therefore represents a ratcheting of material downslope (see Figure 10.15). The maximum downslope movement at the surface per heave cycle is set by the height of the heave and the local slope, while the total annual downslope movement requires summing over all heave (freeze–thaw) cycles. Matsuoka and Moriwaki (1992) have documented such movement over more than five years on hillslopes in the Sor Rondane Mountains of Antarctica using a clever device that allows real-time monitoring of the heave. Most previous studies used a set of dowel segments, or some other marker that is placed vertically in the soil, to be excavated at some later date, usually years after installation; the resulting pits are known as Young pits. This method yields information only about the total displacement profile, and yields no information about when within the period of the experiment the motion actually occurred. Previous workers had reported complex profiles, some of which were concave downslope, some concave upslope, some straight (see examples in Figure 10.16). Matsuoka and Moriwaki installed instead a flexible strip with a string of strain gages attached to it that

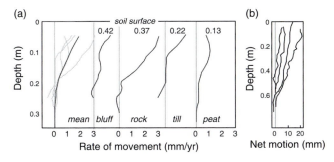

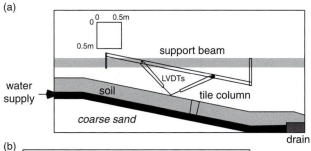

Figure 10.16 Displacement profiles from (a) Young pits as reported by Kirkby (1967, Figure 12, with permission to reproduce from the University of Chicago Press) and (b) segmented dowels as reported by Lehre (1987, Figure 4). Note slightly different vertical scales. Although the profiles vary in their detailed shape on various slopes (small numbers) and in different materials, the mean profile (left graph in (a)) demonstrates that the displacement declines monotonically to zero at depths of several tens of centimeters.

record the local curvature of the strip. By integrating the measurements of curvature, they can calculate the instantaneous displacement profiles of the strip. Simultaneous documentation of the temperature profiles, and hence the freezing depths, allowed them to determine the degree to which the strain events can be related to the freeze–thaw events. The agreement with simple theory (the summation of a set of linear displacement profiles) of heave displacement is striking. Their results support the notion that while the displacement profile from a single event may be linear, the addition of many such profiles, from events whose freeze depths vary, yields a complex profile that looks more like an exponential decay of displacement with depth. In their case, the distribution of freeze depths was such that the total displacement profile was convex up, as shown in Figure 10.15. This is another example of a process with both a deterministic element and a stochastic element. The deterministic aspect is the heave profile associated with a single event, while the stochastic one is the depth of a freeze–thaw event, which depends upon the vicissitudes of the weather.

Two recent sets of laboratory experiments shed more light on the process of downslope movement in the face of periodic freezing and thawing. Harris and Davies (2000) use a slab of soil 30 cm thick with a test section 3 m long on a 12° slope. The apparatus and their results are shown in Figure 10.17. They combine the dowel-profile technique (here using stacks of unglazed tiles) with a pair of transducers that allow real-time recording of both slope-normal

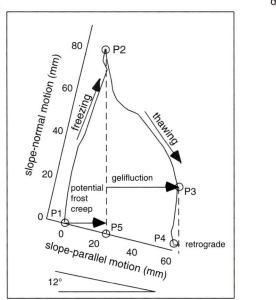

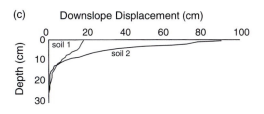

Figure 10.17 Gelifluction experiments. (a) Experimental set-up, with frost-susceptible soil on 12° slope, linear motion transducers attached to rigid beam above soil, and column of unglazed tiles embedded in soil. (b) Example of soil surface motion from LVDT records during a freeze–thaw cycle, showing contributions of frost creep and gelifluction (flow) to the downslope displacement of the soil. (c) Displacement profile after full four-cycle experiment, for two soil types, as deduced from tile column (after Harris and Davies, 2000, Figures 1, 3, and 6B, with permission from INSTAAR).

and slope-parallel motions of the surface to deduce the contributions of frost creep and of internal deformation to this downslope movement. The internal deformation component is called gelifluction. A typical freeze–thaw displacement record is shown in Figure 10.17. Heaving of the soil during freezing results in only slope-normal motion, reflecting growth of ice lenses within the soil column. The maximum frost creep would be achieved if the soils were to experience purely vertical motion during thaw. What is seen in the experiments is that in the early and middle of the thaw phase the soil moves both downward and parallel to the surface. In other words, it flows. Late in the thaw, the motion returns to slope-normal, undoing a little bit of the downslope motion (this has been termed "retrograde motion" by Washburn). The physics involved during the thaw process entails the thaw consolidation of the soil, and the sensitivity of soil rheology to the moisture content and associated pore pressures. In the middle of the thaw event, the soil has maximum moisture content, derived from melting of the ice lenses that grew during the freeze phase. The slightly elevated soil water pressures at and immediately behind the thaw front eventually dissipate late in the thaw phase. This effectively increases the soil viscosity, its resistance to shear under the downslope component of its own weight, slowing the rate of slope-parallel motion. The exhumed columns of tiles show the net downslope displacement profiles in the face of combined frost creep and gelifluction during freeze–thaw cycles: maximum displacement occurs at the surface, a minor convexity of the profile shows in the upper few centimeters, followed by a roughly exponential decline toward zero displacement at depths of 20–30 cm. The several-fold difference in the scales of the displacements of the two soils reflects differences in their frost-susceptibility: soil 1 is a gravelly silty sand, while soil 2 is a fine sandy silt.

In a second set of experiments, done in a $2 \times 1\,\text{m}$ box in which frost-susceptible soil was packed 0.3 m deep, Font *et al.* (2006) watched the displacements within the soil over 41 freeze–thaw cycles. The device and the final profiles are reproduced in Figure 10.18. They keep the room at a mean temperature below 0 °C, which assures that a permafrost table develops, atop which an active layer freezes and thaws in each cycle. As water cannot drain downward through the permanently frozen substrate, the soil is more likely to remain saturated during thaw, promoting surface parallel motion.

We can cast more formally now the expected dependence of regolith discharge on slope and on the important features of the local climate. Given the theoretical displacement profiles shown to occur

in a single freeze–thaw cycle with a given freeze depth, we may integrate this to obtain the total regolith discharge in such an event, as indeed Font *et al.* did with their tile profiles. In general, we must integrate over the distribution of freezing events of varying depths. The displacement profile as a function of depth into the regolith, z, may be written

$$V_i = H_i \tan \theta \left[\frac{\zeta_i - z}{\zeta_i} \right] \text{ for } z < \zeta_i \tag{10.39}$$

where the subscript i represents the ith freezing event, ζ is the frost-penetration depth, H the height of the maximum heave, and θ is the local slope. See Figure 10.19 for the definition sketch. The heave amplitude is set by the strain of the soil upon growth of ice lenses discussed in the periglacial chapter. In other words, we may say that $H = \beta \zeta$, where β is the soil strain upon freezing. Note also that the local slope may also be written $\partial z / \partial x$. We must now integrate this displacement profile with respect to depth in order to quantify the regolith discharge per event:

$$q_i = \int_0^{\zeta_i} \beta \zeta_i \rho_\text{b} \frac{\partial z}{\partial x} \left[\frac{\zeta_i - z}{\zeta_i} \right] \text{d}z = \frac{\rho_\text{b} \beta}{2} \frac{\partial z}{\partial x} \zeta_i^2 \tag{10.40}$$

This is written as a mass discharge, and therefore includes the bulk density, ρ_b, converting from volume to mass. Note that the discharge varies as the square of the frost penetration depth. This is because both sides of the triangle representing the displacement profile have ζ in them. The integral is simply the area of this triangle, which is half of the product of its depth, ζ, with its width at the top. As the displacement at the ground surface is set by $\beta \zeta_i (\partial z / \partial x)$, the dependence is on the square of ζ. This implies that deep frost events are more efficient at transporting regolith than shallow ones. In order to complete the analysis we must acknowledge that there exists a frequency of freezing events, or a number of events per year, and a range of frost penetration depths, both of which are set by the local climate.

In Figure 10.19 we show a plausible probability density function of frost depths, the most likely being shallow, and the likelihood falling off exponentially with depth (see Appendix B for this and other possible pdfs):

$$p(\zeta) = \frac{1}{\zeta_*} \text{e}^{-\zeta/\zeta_*} \tag{10.41}$$

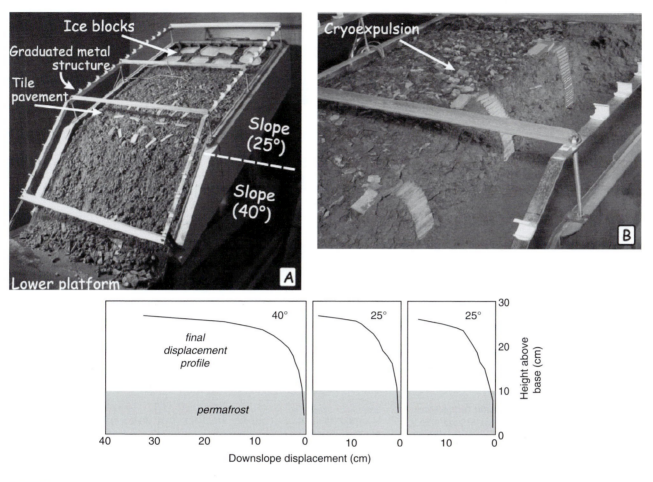

Figure 10.18 Experiments in periglacial activity on a sloping 2 × 1 m slab of soil. Top left: experimental set-up showing segments with 25 and 40° slopes. Top right: after 41 freeze–thaw cycles, exhumation of the tile columns tracking final soil displacement profiles. Bottom: tile displacement profiles shown after final cycle, as viewed through glass wall, showing greater displacement per cycle on the more steeply sloped site (leftmost) (after Font et al., 2006, Figures 3 and 6, with permission from John Wiley & Sons).

where ζ_* is the mean frost penetration depth. The total annual average regolith discharge is then

$$Q_i = f \int_0^{\infty} q_i p(\zeta) \mathrm{d}\zeta$$

$$= f \frac{\rho_b \beta}{2} \frac{\partial z}{\partial x} \int_0^{\infty} \zeta^2 \frac{1}{\zeta_*} e^{-\zeta/\zeta_*} \mathrm{d}\zeta \qquad (10.42)$$

$$= f \frac{\rho_b \beta}{2} \zeta_*^2 \frac{\partial z}{\partial x}$$

where f is the number of frost events per year. The flux of regolith is linearly dependent upon slope. Importantly, Equation 10.42 suggests how the flux ought to be governed by the climatic and by the material properties of the regolith. Note that the flux

increases as the square of the mean frost penetration depth, ζ_*. This treatment is considerably more informative than the simpler but process-blind formula $Q = -kS$.

Biogenic process examples

We have focused thus far on physical hillslope processes. In cases in which the landscape is entirely clothed in vegetation, surface processes such as rainsplash and even sheetwash are ineffective in moving particles from place to place. As can be seen in many such landscapes, however, the role of biological agents is clear. These roles can range from obvious to subtle, and can involve both animals and plants. Animal bioturbation includes the construction of mounds

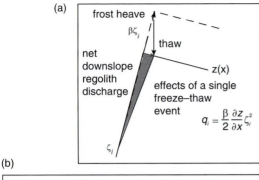

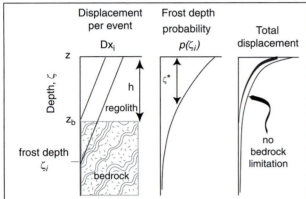

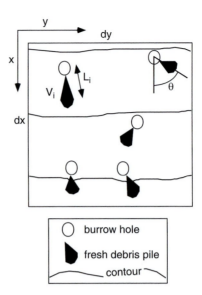

Figure 10.19 Frost heave transport process. (a) Displacement profile and its integral, regolith discharge, associated with a single frost-heave cycle. (b) Effects of both presence of bedrock (left) and probability distribution of freezing depths (middle) on the long-term average profile of regolith transport (right). Note that the presence of bedrock limits the displacement profiles for all events in which the freezing depth is greater than the depth to the bedrock (after Anderson, 2002, Figure 8, with permission from Elsevier).

Figure 10.20 Schematic of a burrow complex on a hillside. The set of holes and their debris piles are characterized by an areal density (N, number per unit area), by the volume of the deposit, and by a vector connecting the center of the burrow and the center of the deposit (magnitude L and angle θ). Finally, we must also know the timescale over which these deposits were generated, T.

above ground, burrows below ground, lateral ploughing of the surface, particle ingestion and evacuation during foraging activities, food caching and prey excavation, wallowing and trampling, and the infilling of abandoned burrow structures (e.g., see review in Meysman *et al*., 2006, and references therein). We can cast many of these processes into the same framework we have used to address the rainsplash and other processes, in that we can identify particular events, and must know something about the spatial intensity of these events.

Rodents

Consider, for example, the case of fossorial (meaning burrowing) rodents. Burrowing creatures excavate tunnel systems within the soil, bringing some fraction of the soil mass to the surface. This mass is deposited nearby the exit hole. It does not take much field inspection to realize that on a significant slope, these creatures avoid placing the debris on the upslope sides of their burrows. (In vegetated landscapes, this is the only debris that rainsplash and runoff can re-distribute, and it gets redistributed downhill ... .) The problem of determining how to characterize this biological process as a geomorphic transport rule requires a characterization of both the individual soil-moving events, shown in Figure 10.20, and the number of such events per unit area per unit time (i.e., what we might call their "intensity"). In other words, we have a problem that is similar in structure to the rainsplash problem. Following our equation for flux in the rainsplash case (Equation 10.37), we have

$$Q_x = \frac{1}{T\mathrm{d}x\mathrm{d}y} \sum_{i=1}^{N} \rho_\mathrm{b} V_i L_i \cos\theta_i \qquad (10.43)$$

where T is the period over which these events are summed, $\mathrm{d}x$ and $\mathrm{d}y$ are the lengths of the area in which N mounds were deposited, and ρ_b is the bulk density of the mound debris. This can be simplified to

$$Q_x = n\rho_\mathrm{b} \frac{\overline{V}}{T} \overline{L} \int_0^{2\pi} \cos\theta \, p(\theta)\mathrm{d}\theta \qquad (10.44)$$

where we have defined the angle θ to be the deviation from directly downslope (see Figure 10.20). Here n is the number of mounds per unit area of surface (e.g., $5/m^2$), and the overbars represent averages of volume and lengths. In this latter formulation, we have cast it in a way that is useful for field documentation. One must document the probability distribution of angles, $p(\theta)$, and have collected the volume and time together to indicate that one needs to document the volume removed from underground over a particular time interval. Notice that in each formulation the units of Q are M/LT. To reiterate, these odd units correspond to a particular mass passing across a unit length of slope in a unit time. If we want to know how much regolith passes a slope L wide, we must multiply this "lane discharge rate" by L.

What are the long-term effects of this sort of process and can we cast this as a simpler equation that will allow us to address the long-term shape-change of the landscape? All the constants in front of the integral relate to the *magnitude* of the process, the rate at which material is being distributed on the surface. On the other hand, all of the directionality of the process is embedded in the integral. If there is an even distribution of angles in each direction, i.e., if $p(\theta)$ is a constant, then the integral will be zero (cosine has as much above 0 and below it), and there will be no net transport in the x-direction. Characterization of the propensity for critters to place their diggings downhill reduces to defining the function $p(\theta)$. This is perfectly analogous to the rainsplash problem, and what Furbish *et al.* (2007) termed the "concentration factor." It seems that the probability $p(\theta)$ ought to depend upon the magnitude of the local slope. Say, for example, that

$$p(\theta) = \frac{1}{A} e^{-\left(\frac{\theta}{\sigma}\right)^2} \tag{10.45}$$

where the angle σ scales the sharpness of the propensity of an animal to place material downhill rather than uphill, and A is a constant that assures that the integral of $p(\theta) = 1.0$ (all probability densities must have this property). The value of the integral in Equation 10.44 is a function of the angular scale, σ, which represents the sharpness of this angular discrimination. As σ increases, the probability distribution flattens out, the value of the integral declines, and the *net downslope* motion vanishes. We should expect that the discrimination depends upon the species.

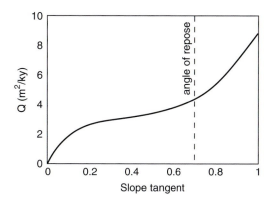

Figure 10.21 Theoretical dependence of regolith transport rate by gophers on slope (after Gabet *et al.*, 2003, Figure 8, reprinted, with permission, from the *Annual Review of Earth and Planetary Sciences*, Vol. 31, © 2003 by Annual Reviews, www.annualreviews.org).

The problem becomes one in animal behavior, and requires field determinations of the value of σ.

One can easily map the distribution of mounds, and their volumes. An excellent example of this is shown in Black and Montgomery (1991), in which they map gopher mounds in a small swale in the Marin headlands of northern California. This sort of work allows us to see the potential importance of critters in working the landscape.

Recent work on hillslopes in southern coastal California has added greatly to the data set on the biogeomorphological effects of rodents. Working in a landscape dominated by pocket gopher (*Thomomys bottae*) activity, Gabet (2000) provides us with quantitative information about the masses of the piles, the lengths from the exit portals, and the angles of the slopes on which the displacements occur. He proposes a nonlinear relation, illustrated in Figure 10.21, between the volumetric discharge of soil and the local slope in which discharge greatly increases with increasing slope.

Other complexities that must be addressed in this interesting arena of biogeomorphology include the need to assess to what degree creatures factor in other features of the environment as they choose where to live and to dig. For example, they likely avoid places where the regolith is less than some critical thickness (probably scaling with their own height). They may avoid slopes with a certain aspect, which would translate into an aspect asymmetry in their contributions to hillslope evolution. And so on.

Few have attempted to address this problem quantitatively. An example of a valiant attempt to incorporate both biological and physical processes in hillslope evolution is that of Jyotsna and Haff (1997). In the desert of eastern California, many hillslopes appear to be dotted with mounds associated with rodent holes. Physical processes of rainsplash and potentially slow regolith creep of an eolian mantle serve to redistribute this material once it is on the surface. As we have seen, each of these processes may be cast as a slope-dependent transport rule with coefficients reflecting the relative efficacy (efficiency) of the process. The roughness of the landscape will reflect the relative importance of the rodent-transport (mounding) process and the small-scale smoothing attributable to rainsplash and other processes.

Bioturbation has also been shown to be an important player at hazardous waste sites. At sites where contamination of soil is pervasive, and remediation activities focus on placing a non-eroding cap on the remaining contaminated soil, animal burrowing activities can lead to the much more rapid cycling of this material back up to the surface than anticipated (Smallwood *et al.*, 1998). Once brought to the surface, wind can re-suspend the finer material and transport it off site.

Tree-throw

A theory for tree-throw can be composed in the same manner. Here the event is a tree falling over. The tree brings with it, embedded in its root structure, a mass of soil and at times rock, called a root wad. This wad is displaced from the hole from which it came by some fraction of a root wad diameter. This soil and rock mass is slowly released from the roots as the roots decay, generating a small mound to one side of the hollow illustrated in Figure 10.22. On relatively flat ground, this produces a characteristic mound and hollow micro-topography that can persist for centuries. The forests of the eastern United States are dotted with these features (see, for example, Schaetzel and Follmer, 1990; also Denny and Goodlett, 1956). Importantly, again we need directional information to assess whether tree-throw can be implicated in the net downslope movement of regolith. Schaetzel and Follmer suggest that the direction of tree fall has more to do with wind direction than with local slope. A tree-throw event locally roughens the landscape.

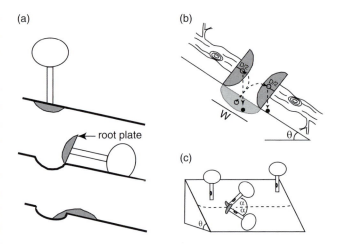

Figure 10.22 Diagram of the tree-throw process, in which a root plate is tipped out of its original site, and ultimately decays, accomplishing net downslope transport. (a) Basic geometry, (b) details of the displacement of the center of mass of the root plate, and (c) definitions of angles used to calculate net downslope transport (after Gabet *et al.*, 2003, Figures 3–5, reprinted, with permission, from the *Annual Review of Earth and Planetary Sciences*, vol. 31, © 2003 by Annual Reviews, www.annualreviews.org).

Its degradation through decay of the root wad and subsequent small-scale surface processes serves to smooth the landscape. It is the competition between these roughening events and smoothing processes that gives the topography a particular roughness at the scale of meters in such forested landscapes.

We note that these are not the only processes that give rise to diffusion, and that importantly there are other processes, mostly associated with water running over the surface and within the shallow surface, that result in non-diffusive behavior. Without these processes, there would be no channels, and the world would indeed be much smoother than it is. We will cover these processes in later chapters, after briefly addressing the landslide problem.

Pacing hillslopes

The rate of motion of regolith is in general very slow and difficult to document in many settings. We have turned to the use of cosmogenic radionuclides to establish long-term rates. Recall that if (and only if) a landscape is in steady state one may predict the spatial pattern of regolith discharge with distance from the divide. It must linearly increase with

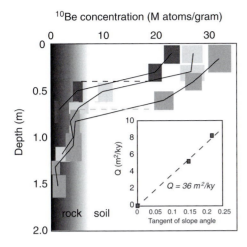

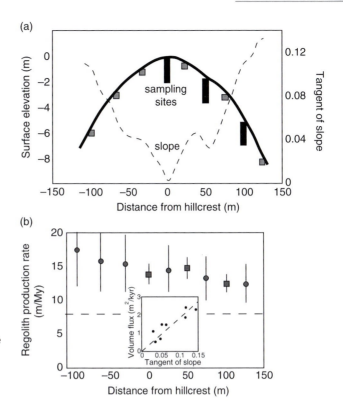

Figure 10.23 Profiles of [10]Be in three test pits from the Black Diamond Mines Preserve Regional Park, California. Dashed lines depict base of soil in each profile. Boxes represent sample depth interval and 1σ [10]Be values from two duplicate samples. Inset: relationship between slope and sediment discharge suggested by the [10]Be data. The approximately linear relationship between slope and regolith discharge support Gilbert's view of convex hilltops (after McKean *et al.*, 1993, Figures 2 and 3).

Figure 10.24 "Cosmo Hill" on Osborn Mountain, Wind River Range, Wyoming. (a) Topographic profile showing topography, slope, and locations of soil pits. Deep pits go to the bedrock interface at roughly 1 m. (b) Regolith production rates deduced from cosmogenic radionuclide concentrations in bedrock at base of pits; dashed line represents mean bedrock lowering rate on bare bedrock in this and other nearby ranges. Inset: volume flux of regolith deduced from cosmogenic radionuclide results, plotted against local slope. Linear fits indicate $Q = -kS$ flux law (after Small *et al.*, 1999, Figures 5, 8, and 10, with permission from Elsevier).

distance, at a rate governed by the rate of regolith production (Equation 10.6). As the regolith discharge is the product of regolith thickness with mean down-slope speed, the mean speed may be back-calculated using the measured regolith thickness: $U = Q/h$. And if the regolith is uniform in thickness, as it should be in steady state, then the mean speed must increase linearly from the divide.

Besides the instrumentation of hillslope profiles described above, there are two ways to estimate long-term speeds. We can use cosmogenic nuclides in two ways. McKean *et al.* (1993) used the inventories of atmospherically produced (so-called garden variety) [10]Be in clay-rich soils of the coast ranges of California as a clock. These are nuclides produced within the atmosphere and swept out in precipitation, and subsequently tightly attached to clay particles in the soil. The integral of the profile, or the full inventory, should be proportional to time spent on the hillslope. They collected vertical profiles of [10]Be at various distances from the divide. As reproduced in Figure 10.23, you can see that the inventory indeed increased with distance downslope at a rate that suggested the regolith discharge increased with slope. Their relationship suggested $\kappa = 36\,\text{m}^2/\text{ka}$, or $0.036\,\text{m}^2/\text{yr}$. Inspection of their plot in Figure 10.23 allows a quick calculation

of the mean speed. At a point on the landscape with slope = 0.15, the [10]Be-implied soil discharge is 5.2 m²/kyr. Dividing by the measured soil thickness of about 0.6 m, we can estimate the mean speed to be about 9 m/kyr or 9 mm/yr.

Using instead *in situ*-produced [10]Be, Small *et al.* (1999) documented the evolution of profiles with distance from the divide in an alpine landscape in Wyoming. Their profiles showed that [10]Be decreased slightly with depth into the regolith, likely reflecting persistent turbation over much of the profile. The inventory of [10]Be in the regolith suggested that regolith production rates, shown in Figure 10.24(b), were roughly 15 microns/yr, and that the topographic diffusivity was $\kappa = 18\,\text{m}^2/\text{ka}$. The implied volumetric discharge of regolith at a distance of 125 m from the hillcrest is 2200 m²/yr. The measured regolith thickness of 0.9 m allows us to estimate that the mean speed of the regolith at these distances, Q/h, is only about 0.0024 m/yr or 2.4 mm/yr. These would indeed be difficult to measure in real time.

Note that we can also predict timescales using a relatively simple calculation. One commonly calculates the residence time of water in a steady reservoir by dividing the volume of the water in the reservoir (L^3) by the inputs of water (L^3/T). Here we have a regolith reservoir. As long as the hillslope is in steady state, the mean residence time on a hillslope at a position x can be estimated from dividing the regolith mass per unit width, M, by the rate at which mass is being released from the underlying rock, I:

$$T_r = \frac{M}{I} = \frac{\rho_s h x}{\rho_r x \dot{W}} = \frac{h}{\dot{W}} \frac{\rho_s}{\rho_r} \tag{10.46}$$

(see discussion in Mudd and Furbish, 2006). The second ratio is simply the inverse of the bulking factor. On the alpine Osborn site studied by Small *et al.*, the implied mean residence time is roughly 40 ka.

It is somewhat counterintuitive that the mean residence time of the soil remains constant over the entire slope; Equation 10.46 is independent of x. How can that be? It seems that regolith ought to age with distance from the divide. But we must remember both that fresh, zero-age regolith is being supplied uniformly to the base of the regolith, and that regolith speeds up with distance downslope. Recall that mean regolith speed must increase with distance from the divide: $U = \dot{W}x/h$. These factors of distance traveled and speed of travel balance out. We can see this by redoing the calculation above based upon not the inputs but on the outputs from the regolith reservoir. At steady state, the mean residence time may equivalently be calculated by dividing the mass per unit width in the box by the rate at which it is being lost at any chosen distance from the divide, x:

$$T_r = \frac{M}{Q} = \frac{\rho_s h x}{\rho_r U h} = \frac{\rho_s}{\rho_r} \frac{x}{U} \tag{10.47}$$

In order for this expression to be equivalent to that in Equation 10.46, and in particular for the timescale not to depend on x, the mean velocity U must vary with x. And indeed it does, linearly so. The timescales described here have strong implications for the chemical evolution of regolith. The chemical weathering in soil will depend upon not only the intrinsic climatic variables of precipitation and temperature and plant-related acidity, but upon the time spent in the chemical reactor that is the regolith. To the extent that the mean residence time may be taken to reflect the degree of chemical weathering in a soil, we ought to expect that any index of weathering should be uniform on steady-state hillslopes. We can also see from Equation 10.46 that high rates of regolith production ought to translate into lower residence times. Recalling from our discussion of chemical weathering in Chapter 7 that fresh mineral surfaces weather much more rapidly than those that have been exposed for a long period of time (White and Brantley, 1995), rates of chemical weathering should decline rapidly with mean residence time. This is likely the explanation of the common observation (e.g., Riebe *et al.*, 2001a–c) that chemical weathering rates are strongly governed by (or slaved to) physical weathering (regolith production) rates.

Landslides

We have dealt a fair bit with hillslope processes that lead to the rounding of hilltops. Diffusive processes, those that result in regolith discharge that varies solely and linearly with the local slope, are indeed common. Rainsplash, frost heave, solifluction, all of these accomplish the rounding of the landscape. But on yet steeper slopes, further from the hillcrest, it is equally common to see more or less straight slopes on which the primary means of passing material from one place to another is not by slow creep of material, but by catastrophic landslide events. Landslide scars dot the hillslopes, showing as either bare rock or non-vegetated regolith patches amongst the trees and bushes. We will see that landslides play an important role in setting the locations of channel heads, the tips of the channel network, in some landscapes. Clearly, they can also pose some of the most lethal natural hazards a landscape can offer up.

The landslide problem is a classic case in geomorphology, involving a system in which there is a distinct threshold. Below this threshold, essentially nothing happens. Above it, a lot happens. In addition, landsliding is an inherently stochastic process. One cannot simply look at the landscape and expect to be able to deduce what it will look like ten years later. This is one of the things that makes geomorphology both interesting and difficult. We have only recently begun to recognize the complexity (richness) of this type of problem. Let us be explicit in what we mean by stochastic, as the term comes up time and time again in geomorphic circles. It can be explained best in distinction to its antonym, deterministic. If a system is deterministic, then with sufficient knowledge about the state of the system at one time, we can predict with certainty the state of the system at some future instant. This is not the case in stochastic systems. There is some randomizing element in the problem. We can describe its future state in only a probabilistic way. The trajectory of a cannonball is essentially deterministic (which is why there were look-up charts for gunners), while that of a dust mote is stochastic.

But back to landslides . . . There are several parts to the problem. We must first assess the requirements for generation of a failure – in other words, we must come up with an equation that spells out the conditions that define this failure threshold. Even the simple analysis we will present should reveal the salient, or important, parts of the problem: the dependence upon the slope, the materials, the degree of saturation, whether there are trees on the slope or not, and so on. Then we must worry about what happens in the aftermath of the failure. What sort of mixture of rock and water is going to race down the hill? Does it behave like an intact block, or does it pulverize into fist-sized blocks that travel as a granular flow, or does it transform into an intimate mixture of debris and water that we call a debris flow? Each of these event types has its own mechanics. We are frankly only beginning to be able to deal with the complexity of this post-failure physics.

The force balance at failure

Consider first the diagram of a segment of a planar hillslope shown in Figure 10.25. We wish to develop an expression that reflects the forces acting on this block of material at the time of failure. The plane at depth could be the interface between the regolith and the bedrock, or it could be a joint plane or plane of weakness within the bedrock. For now, you might want to think of it as a brick on a sloping plank of wood. Presumably right at the time of failure the forces are exactly balanced; it is not yet accelerating. The forces acting in the problem are all generated by

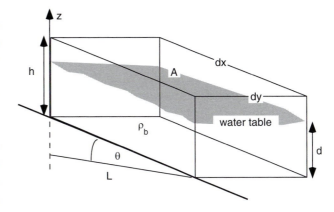

Definition Sketch for Planar Landslide

Figure 10.25 Segment of a planar hillslope showing geometry of a slab of regolith of area $A = dxdy$ and thickness h above a potential failure plane with slope θ relative to the horizontal. Also shown (bold line) is the water table with a height d above the failure plane.

the weight of the material. The weight is $F = mg$, where the mass of the box $m = \rho_b h dxdy$, ρ_b being the bulk density of the material. Of course, the weight acts vertically. The component of this weight acting to try to shear the block from the underlying material, what we will call the driving force, is therefore $F \sin\theta$. Given that the block of soil is on a slope, the volume that comes into calculation of the mass involves a parallelogram of width L and height h. Noting that the cross-sectional area of the parallelogram is Lh, and that $L = dx*\cos\theta$, the driving force becomes:

$$F_d = \rho_b(h dy dx \cos\theta)g \sin\theta \qquad (10.48)$$

The resisting forces are essentially frictional, arising from the micro-roughness of the surfaces in contact. The greater the interlocking of grains, the higher is the force necessary to break or fail the contacts. The frictional resistance is the product of the force normal to the contact times the friction coefficient that is a measure of the interlocking of roughness. The normal force is $F \cos\theta$, while the friction coefficient has been given the odd designation of "$\tan\phi$" in the soils engineering literature.

$$F_r = [\rho_b(h dy dx \cos\theta)g \cos\theta] \tan\phi \qquad (10.49)$$

Because we are dealing with arbitrary blocks of soil rather than discrete blocks, like the brick on the plank, it is more appropriate to cast the balance in terms of stresses rather than forces. A stress is simply

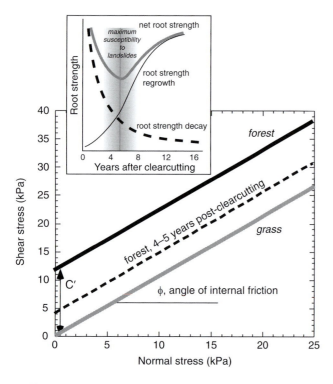

Figure 10.26 Coulomb rheology plot of shear stress vs. normal stress at failure for three soils, a grass soil with little to no cohesion, C', a forest soil with significant cohesion, and the same forest soil 4–5 years after clearcutting. This last case demonstrates the significant reduction of root-related cohesion as the fine roots decay. Inset: evolution of root strength after clearcutting, showing that the cross-over between root decay and root regrowth governs the timing of greatest susceptibility to landsliding (after Sidle *et al.*, 2006, Figure 2, with permission from Elsevier).

a force per unit area, which we can obtain by dividing through our expressions for force by the expression for the area of the base of the block, or $\mathrm{d}x\,\mathrm{d}y$. Note here that we use the along-slope distance $\mathrm{d}x$, rather than the horizontal length L. The balance of stresses *at failure* then becomes:

$$
\text{driving stresses} = \text{resisting stresses}
$$
$$
\rho_{\mathrm{b}}gh\sin\theta\cos\theta = \rho_{\mathrm{b}}gh\cos\theta\cos\theta\tan(\phi)
$$
$$
\text{or}
$$
$$
\rho_{\mathrm{b}}gh\sin\theta = \rho_{\mathrm{b}}gh\cos\theta\tan(\phi) \qquad (10.50)
$$

You can see from the plot in Figure 10.26 that the origin of the name $\tan\phi$ is simply that it is the slope on the plot of shear stress vs. normal stresses upon failure. So far, what we have described is the "dry, cohesion-less" case: the brick on a dry wood plank, or a dry pile of sand. The pile will fail when the angle anywhere is great enough that the left-hand side of

the equation, the driving stresses, is greater than the right-hand side, or the resisting stresses. Note that in this case (and this case only) the density, the thickness and the acceleration due to gravity cancel out from both sides of the equation. Recalling that $\sin/\cos = \tan$, the equation becomes simply

$$
\tan\theta = \tan\phi \qquad (10.51)
$$

at failure. This equation has a couple of important ramifications. First, one can use the angle of repose (or the angle at which the material fails as it is tipped to greater and greater angles) as a measure of the coefficient of friction. Because this failure takes place internal to the material, this angle $\tan\phi$ is also called the "internal angle of friction" or more simply the "friction angle." Second, note what will happen to the angle of repose as one goes from Earth to Mars, say – nothing! The angle of repose of dry material should be independent of the acceleration due to gravity. The lee slopes of sand dunes on Mars should be exactly the same as those on Earth, i.e., about 31°.

Unfortunately, this simple equation does not capture the most general case. We have to introduce three elements of reality: water, cohesion in the form of mineral cohesion, and cohesion in the form of roots. Let us start by adding water to the problem. Water enters in several ways. First, by filling the pores, it changes the bulk density of the material. The expression for the bulk density of a saturated porous medium is $\rho_{\mathrm{b}} = \rho_{\mathrm{g}}(1-\eta) + \rho_{\mathrm{w}}\eta$, where w denotes water, b bulk, and g grains, and η is the porosity of the medium. In general, we wish to know the mean density of the column of material over the potential failure plane. This is

$$
\overline{\rho}_{\mathrm{b}} = \frac{\rho_{\mathrm{sat}}^{d} + \rho_{\mathrm{dry}}(h-d)}{h} \qquad (10.52)
$$

where here d denotes the water table height above the failure plane. You can check this by noting that when $d = h$, i.e., the water table is all the way up to the surface and the entire soil mass is saturated, the mean density becomes the saturated density. Similarly, when it is all dry, $d = 0$, and the density is that of the dry material. Now if this were the only role that water played, i.e., making the material heavier, the effects would come in equally on both sides of the stress balance equation. They would cancel out. Merely changing the density of the brick does not change the slope on which it begins the slide. But that is not

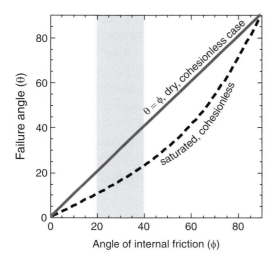

Figure 10.27 Failure angle as a function of the angle of internal friction, ϕ. For dry conditions, the relationship is 1:1, meaning that the angle of internal friction can be read from the angle of repose. Such experiments show that typical values of the angle of internal friction are between 20 and 40° (shaded region). In wet (fully saturated) conditions, the failure angle is reduced significantly, by roughly a factor of two in the region of typical angles of internal friction.

the only effect water has on the system. In a crude way, we can think of the water as producing a water pressure (again, a force per unit area) at the potential failure plane, that supports some portion of the normal stress of the material. This will promote failure, as the normal stress comes into the frictional resistance. Formally, we define an "effective stress" as the normal stress less the water pressure:

$$\sigma' = \overline{\rho}_b g h - \rho_w g d \qquad (10.53)$$

Once this effect is incorporated, the symmetry of the force balance is broken. No longer can the density simply be canceled out. The graph of the failure angle, θ, as a function of the coefficient of internal friction now has to include the degree of saturation (d/h) of the material. We show in Figure 10.27 the dry and fully saturated ($d = h$) cases. Over a reasonable range of angles of internal friction, the slope at which failure occurs is roughly half of that at which a dry slope will fail. Typical values of the angle of internal friction are 20–40° (for pure sand we have already shown it to be 31°), meaning that slopes above about half of this angle should not be expected to be stable if they are ever saturated.

This is not what we observe in nature. In the coast ranges of Oregon, for instance, where it rains several meters per year, sometimes at very high rates, slopes are often 30–40°. Either even these rains are insufficient to saturate the material, or there is some other factor that comes into play on the resistance side of the equation. There is, and it is called cohesion, denoted C in the balance equation. It comes in two flavors, mineral and biological. The origin of mineral cohesion is electrostatic attraction between platy particles. This augments the geometrical interlock mechanism for generating frictional resistance, as it represents bonds that must be broken in order to allow motion. The origin of biological cohesion is roots. Any plant sends out myriad roots into the subsurface, each one of which has a tensile strength that must be exceeded before it fails. Obviously, the larger ones provide more resistance – taking more stress either to sever them or to pull them out of the soil – but there are many more of the smaller ones than large ones. The resistance to failure on any plane in the subsurface therefore depends upon the plant type and its age. While this problem has not been worked out in any very formal way, we do have evidence from the field that is pertinent to deciphering the relative roles of different sizes of roots. It has been documented that freshly logged landscapes in the Pacific Northwest of the USA will generate the largest number of landslides about 4–5 years after the clearcutting. This is clearly visible by the offset of the pre- and post-cutting failure curves in Figure 10.26. This is about the time it takes for roots of the 1-mm diameter scale to rot, losing their binding strength.

To recap, the final equation for the stress balance at failure for a planar landslide, humbly called the Master equation, is

$$\overline{\rho}_b g h \sin\theta \cos\theta = [\overline{\rho}_b g h \cos\theta \cos\theta - \rho_w g d] \\ \times \tan(\phi) + C' \qquad (10.54)$$

where C' denotes the sum of the mineral and root contributions to cohesion at the failure interface. In geological engineering circles, it is common to take the ratio of the resisting stresses to driving stresses, which we would always want to be greater than 1. This ratio is called the "factor of safety," F_s. To be complete, the factor of safety is therefore:

$$F_s = \frac{[\overline{\rho}_b g h \cos\theta \cos\theta - \rho_w g d]\tan(\phi) + C'}{\overline{\rho}_b g h \sin\theta \cos\theta} \qquad (10.55)$$

When this quantity is well above 1.0, the slope should be stable even in the worst-case scenario ($d = h$), while if it drops below 1.0, the driving stress exceeds the resistance and the slope is likely to fail.

Plants do two more things that we ought not to ignore. First, all geologists know that plants suck. In particular, they suck water out of the soil and up into their stems as a necessary ingredient in the photosynthesis of new plant matter. This net transport of water out of the soil, in what is called the evapotranspiration process, reduces the likelihood of soil saturation, thereby reducing the likelihood of failure. In terms of the Master equation (Equation 10.54), this lowers the water table, d. Anyone who has cut down a tree or hauled firewood a long way knows that trees are also heavy. By felling a tree, and removing it from the hillslope altogether, the weight of the tree and hence both the contributed normal and shear stresses it represented, vanish from the stress balance. It turns out that the net effect of removing the weight of the trees is one of stabilization on relatively shallow slopes, and destabilization of slopes steeper than the internal angle of friction. Can you prove this to yourself?

We conclude with one last note on the failure mechanics. The above analysis is quite simplistic in many ways. First, while we assessed the role of water by addressing the position of the water table, in reality the relevant quantity is the pore water pressure (e.g., Iverson, 1997). In addition, we have treated the problem as one with planar hillslopes. This is never the case. Hillslopes are always crenulated in the cross-slope direction, giving rise to noses or ridges separated by troughs or hollows. The subsurface water we have been talking about is actually in motion (it is after all on a slope; even though it is in the subsurface, this causes flow). It will travel down the fall line, converging in the troughs and diverging on the noses, increasing the likelihood of saturation, and of the generation of high pore pressures, in the hollows while reducing these quantities on the intervening noses. As you might suspect, landscapes such as the coast ranges of Oregon show the largest number of failures in the troughs. In fact, this has huge implications for the whole landscape. These failures gouge out the troughs yet further, maintaining them as points of convergence for groundwater in the landscape. The failures race down the troughs, mix thoroughly, and generate very mobile debris flows, which in turn accomplish further scouring of the upper portions of the drainage network. Studies of this landscape have revealed a cycle of filling of the hollows by slow creep of material, punctuated by the periodic sluicing of the hollows by landslides.

It should be noted that there is a distinct spatial pattern to the generation of landslides in the aftermath of clearcutting in forests. It is commonly observed that most landslides occur in hollows immediately downslope of a site at which drainage from a logging road is routed.

A primer on the behavior of saturated granular materials

In recent experimental and theoretical work, researchers have carefully documented the behavior of saturated granular materials (e.g. Iverson 1997; Iverson *et al.*, 1997; 2000). While many models for the rheology of granular materials under these conditions have been proposed, recent work suggests that much of the behavior during deformation may be understood using the classical Coulomb description. As has been advocated for decades, this is best summarized in the Coulomb–Terzhagi equation,

$$\tau = \sigma' \tan \phi \tag{10.56}$$

where τ and σ' are the local shear and the effective normal stresses acting on a plane and ϕ is a friction angle characterizing the material. This can also be written as

$$\tau = (\sigma - P_w) \tan \phi \tag{10.57}$$

where P_w is the water pressure. We have already seen this in our analysis of landslide failures. Note in particular the role played by water pressure. The shear stress necessary to cause motion is reduced as the water pressure increases. In order to understand the behavior of the granular material we therefore must have a clear picture of the water pressure field.

The feedback between water pressure and displacement is well illustrated in a recent set of lab experiments. In the ring shear device shown in Figure 10.28(a), designed to allow large strains in the material, a saturated sandy loam was placed in the annulus along with several miniature pressure transducers and a bead column that would allow tracking of both local water pressure fluctuations and total displacement. The normal force and the shear force were carefully monitored,

(a)

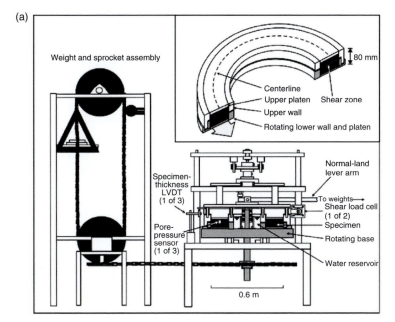

(b)

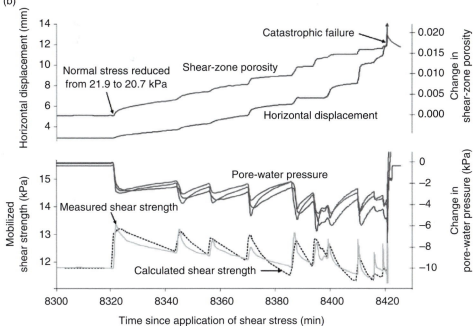

Figure 10.28 Ring shear device for exploring the roles of pore pressure on the shear straining of granular materials. (a) Ring shear device showing in top right the cross section of the annulus in which the granular material is placed. (b) Histories of horizontal displacement (top) and measured pore pressure (bottom) (after Moore and Iverson, 2002, Figures 1 and 2).

and the displacement of the top platen with respect to the bottom was measured using a set of linear motion transducers (devices that report micron-scale displacements). The resulting histories of displacement and water pressure are shown in Figure 10.28(b). Water pressure builds up until some threshold is crossed, at which time displacement begins to occur. The water pressure immediately drops, stalling the motion event. This series of events repeats itself quasi-periodically.

Interestingly, the experiment ends with an event of a different sort: the displacement is unchecked and the water pressure remains high. This is a catastrophic failure from which the system cannot recover. This and other similar experiments have been interpreted to reflect a strong feedback system that is inevitable in granular materials under shear. When a granular material is forced to shear, grains in one layer must rise to move out of their pockets. This dilation enlarges pores into which water can be pulled. This reduces the high water pressure that was promoting the displacement in the first place, and hence stalls the displacement event. This feedback has been dubbed "dilatant strengthening." The shearing occurs, then, in short displacement events; it is not continuous, as it would be in a viscous material. Only when these events are averaged over times large compared to their duration does the motion appear to be smooth.

These experiments point to the importance of two material properties: the hydraulic conductivity and the degree of compaction. It should be obvious that the efficiency with which water is drawn to the site of dilation should influence the duration of the slip event. Also, the more dilatant the material, meaning the more pore space is generated by motion of one grain layer over another, the stronger the feedback. More compact materials, in which grains must move a significant fraction of their diameter to escape the pockets with others grains, will be more dilatant. This is illustrated on two scales in recent experiments. First, it is this effect that is invoked to explain the final failure of the ring shear experiments. The material between the platens has slowly become more porous, more dilated, less compact through the course of the experiment. Finally, when water pressure is high enough to promote an event, the material has no more pore space to generate upon further shear, and the feedback is lost. The system continues to move with no dilatant strengthening brake. Second, in large-scale experiments in the USGS debris flow flume in Oregon (shown in the photograph in Figure 10.29; see Iverson et al., 1992), catastrophic failure of saturated soils was shown to be most likely for the least compacted soils, which were those placed most gently into the apparatus. This is illustrated in Figure 10.30. Experiments in which the soils were tamped upon emplacement in the chamber displayed instead the sets of displacement events that result from the dilatant-strengthening feedback.

This material behavior is pertinent to several geomorphic and geological systems. The relevance to landsliding and to subsequent deformation of the landslide mass is obvious. The shear of fault gouge, a saturated granular material caught between two rock walls that are moving with respect to one another, should sound a lot like the ring shear experiments with a linear rather than annular symmetry. Similarly, till at the base of large glaciers is caught in

Figure 10.29 USGS Debris flow flume in the foothills of the Cascades, Oregon. Flume is 95 m long, 2 m wide, and can contain flows as deep as 1.2 m. It slopes at 31°. Up to 20 m³ of debris can be loaded behind a headgate at the top of the flume. Flows travel the length of the flume and form deposits on the flat foreground. Experiments can therefore include the study of flow initiation, transport, and deposition (after Iverson et al., 1992, USGS Open-File Report 92–483).

an analogous situation. The walls enclosing the natural experiment are instead ice above and rock below, and again it is very likely to be saturated.

What oversteepens the slopes?

One might reasonably ask what processes are involved in the generation of a slope that is too steep. It must have become too steep via some process that somehow increased its slope. The mechanisms are several, although there are a couple primary culprits. One mechanism is tectonic or volcanic tilting.

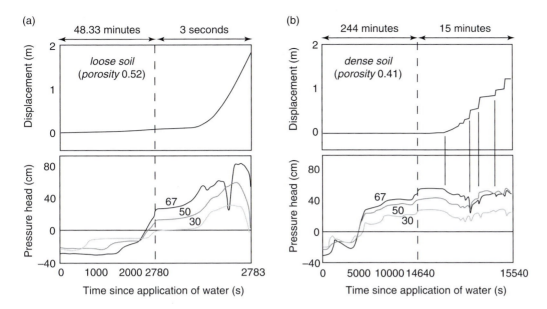

Figure 10.30 Experimental investigation of the role of soil compaction on landslide initiation. Experimental apparatus is the USGS debris flow laboratory in Oregon. The same sandy soil is used at densities that differ by only 10%. Top plots show the displacement history over two timescales. Bottom plots show corresponding pressure histories at three levels in the soil (30, 50, and 67 cm). (a) The loosely packed low-porosity soil fails catastrophically over 3 seconds, associated with pore pressure increases. (b) The dense soil deforms only slowly in incremental steps, taking 15 minutes to accomplish a comparable displacement. Each displacement event is clearly associated with a drop in pore pressure, shown by vertical lines connecting events in the two plots (redrawn from R. Iverson *et al.*, 2000, parts of Figures 3 and 4, reprinted with permission from the American Association for the Advancement of Science).

Tectonic tilting can be large in areal scale, but is typically small in amplitude. Tectonic tilt is the spatial derivative of the tectonic uplift pattern. Given that the scales of the uplift welts are many kilometers to tens of kilometers, and the amplitude of the surface deflections is of the order of meters, the tilts are typically less than 1 m/km. If we need to increase slopes by several degrees, this will not do. Volcanic mechanisms are occasionally to blame, as for instance in the failure of the edifice of Mt. St. Helens that triggered the May 18, 1980 lateral blast. Tiltmeters deployed on the flank of the volcano by the USGS at that time documented changes in tilt of several degrees as the magma neared the surface. And indeed the ensuing debris avalanche is one of the ten largest yet documented (e.g., the catalog of Costa and Schuster, 1988).

Yet by far the principal means of steepening a slope is not through tilting of any kind, but rather by removing mass from the toe of the slope. This can easily be accomplished by either a river or a glacier (or, for that matter, humans). Removal of the toe locally oversteepens the slope, which is therefore where the majority of landslides nucleate. One can simulate this mechanism very well with a simple apparatus, depicted in Figure 10.31. The lowering of the sliding wall is meant to mimic the lowering of the channel at the base of the hillslope. One can see in the profiles of Figure 10.32 several characteristics of real hillslopes in both the geometry of the slope at any snapshot in time, and in the dynamics of the process. The majority of the failures involve small numbers of grains, but the occasional slope-clearing landslide resets the entire slope.

The aftermath

Now that our piece of hillslope has failed, what happens to the mass of material in the failure? The result depends upon several factors, including the height of the drop once it fails, the degree of saturation of the rock and soil mass, and so on. An extensive nomenclature has sprung up around the classification of mass movements, the chief parameters being whether the mass is wet or dry, and whether it travels fast or slow. For now let us just consider the fast ones. When these also happen to be wet, they have a high likelihood of transforming into

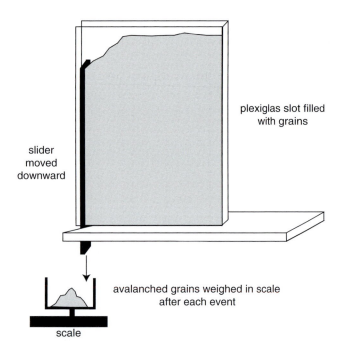

Figure 10.31 Sketch of grain-avalanche apparatus. Plexiglas-walled slot filled with grains (e.g., dry beans) is made to avalanche through downward motion of a slider on its margin. The weight of each resulting avalanche is recorded, producing a time series of landslides and a distribution of landslide volumes.

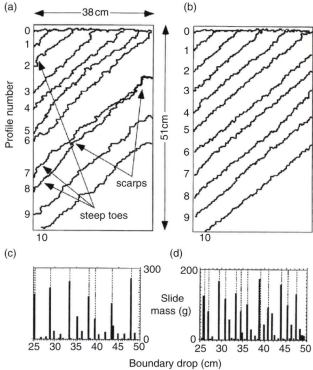

Figure 10.32 Hillslopes of beans. As shown in profiles of (a) and (b), beanslides occur sporadically as the boundary on the left-hand side of the box is dropped smoothly and slowly. The recorded slide masses shown in (c) and (d) result in a probability distribution of slides in which 70% of the slide mass occurs in 10% of the events. Large slope-clearing events reset the slope angle. Intermediate size events clear the lower slopes, steepening the toes of the slope, giving the appearance of an oversteepened "inner gorge" (after Densmore et al., 1997, Figure 1, with permission from the American Association for the Advancement of Science).

debris flows, which can travel huge distances away from the source.

First, consider a few examples. It is commonly found that the landslide mass ends up well away from its source. There are numerous examples of a landslide traveling all the way down a valley wall and well up onto the opposite side of the valley (wiping out everything in the path). In other words, they appear to travel with very little friction. These "long runout" slides have been characterized by the ratio of the length of the runout to the vertical height of the drop: $R = L/H$. These ratios can be very high. To develop some sense of what these numbers mean, we will use some simple scaling to assess the expected runouts, and the expected velocities of the emplacement of these landslide masses.

A landslide has fallen from its niche in a hillside shown in Figure 10.33, and has come to a screeching halt at an altitude of z' on the other side of the valley. We can use the values of z and z' to constrain the velocity history of the slide mass, or at the very least to constrain the speeds at which the landslide mass crossed the valley floor. If all the potential energy of

the slide mass (relative to that it would have at the valley bottom) were to be converted to kinetic energy, it would be traveling at a speed calculated from

$$mgz = \frac{1}{2}mv^2$$

or

$$v = \sqrt{2gz}$$

(10.58)

as it crossed the valley floor. On the other hand, we know that it had to have been going at least $v = \sqrt{2gz'}$ in order to have had enough kinetic energy to reconvert back to its potential energy at the site of deposition. Therefore, we know that the real velocity must have been $\sqrt{2gz} > v > \sqrt{2gz'}$, as it wiped out the village. In the case of the 1717 Troilet slide in the Italian Alps, which wiped out a town and many livestock, we find that the velocity must have been between 320 and 125 km/hr. A similar calculation for the prehistoric Saidmarreh slide in the Zagros Mountains of Iran results in a lower limit of 240 km/hr.

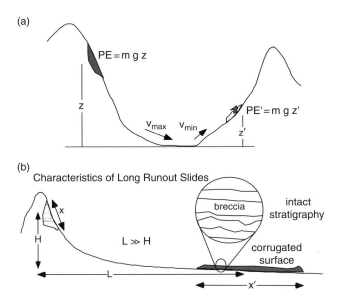

Figure 10.33 (a) Landslide source on left valley wall and deposit on right valley wall showing means of estimated maximum and minimum speeds at which the landslide crossed the valley floor. (b) Portrait of a long runout landslide, one in which $L > H$, including the failure zone, the runout path, and the brecciated deposit.

These slides can travel at remarkable speeds, and apparently experience very low frictional losses of energy.

How can this be? This has been a topic of hot debate for half a century. It has stimulated much detailed work on these major landslides in the field, and has led to the construction of complex numerical models. As sketched in Figure 10.33, these slide masses are characterized as openwork breccias in which the original stratigraphy is surprisingly intact. Despite the fact that the slide mass may be ten times thinner than its original thickness, and ten times longer than its original length, the strata in the top of the failed mass remains on top of the deposit. It has not been stirred up like a fluid, particles from the bottom mixed with particles from the top. These are strong constraints on any working theory of such events.

The hypotheses to explain long runout slides have been both varied and interesting. One of the most intriguing was that proposed by Ron Shreve, who in 1960 published his PhD research on the huge prehistoric Blackhawk slide (Shreve, 1968a,b). An aerial photo of this magnificent slide is shown in Figure 10.34. This 4 km-wide rockslide came down off the east side of the San Gabriel Mountains and flowed many kilometers out into the Mojave Desert. In order to explain the long runout of the Blackhawk slide, he called upon its traveling upon a cushion of compressed air. In careful

Figure 10.34 Blackhawk landslide. Roughly 17 000 years old, this slide was emplaced in only a couple minutes from its failure site in the San Gabriel Mountains (photograph by Kerry Sieh, with permission to reproduce).

mapping of the slide, he noted that it must have passed over a lip in the topography. Upon landing, the mass trapped the air, which formed a hydroplane-like low-viscosity fluid that could rapidly shear, transporting the mass until the air managed to escape (Shreve, 1968a).

Later arguments have tended to favor another mechanism involving simply the interactions of grains within the slide mass. One of the chief problems with the air cushion hypothesis is that slides with similar geometries have been seen in images of the Moon, which of course lacks an atmosphere. The most likely model at present is one in which the particles of the finely comminuted (meaning broken up) mass all bounce off one another in a hectic manner, one on one, those near the base of the slide bouncing force-fully off the bed over which the slide is moving. This "granular flow" behaves in some ways as if the grains are like atoms in a gas, the transfer of momentum and energy from place to place within the flow occurring through the binary collisions of the grains. Supercomputer simulations of a flow with up to 10^7 grains (Campbell *et al.*, 1995) have reproduced many of the salient features of the real world slides, in particular their long runout and the lack of disruption of the stratigraphy. This is shown in Figure 10.35 both in different times during the simulation, and at different sites in the final deposit.

Initiation of landslides is more likely when the ground is accelerated during an earthquake. The huge and numerous landslides that were triggered by the 2002 strike-slip Denali earthquake, as exemplified by Figure 10.36, are particularly well delineated by having traveled across glaciers occupying the valley floors. In tectonically active landscapes such as New Zealand, landslides are clearly a dominant geo-morphic agent. Careful mapping of almost 5000 land-slides across the Southern Alps that have occurred within the last 60 years has revealed an inverse power-law relationship between frequency of landslides and their size (here their mapped area). As can be seen in Figure 10.37, landslides that are 10 times larger in area occur roughly 20 times less frequently.

Debris flows

Another possible outcome of a landslide is the trans-formation of that mass into a debris flow. As land-slides are often saturated upon failure, the intimate

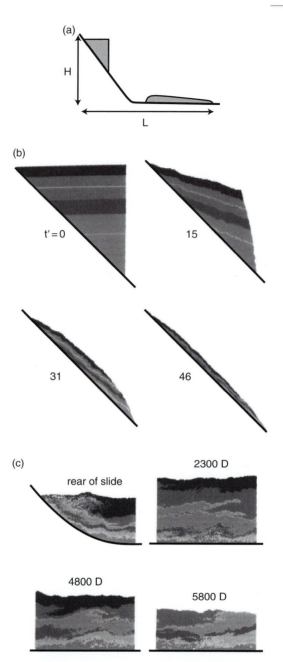

Figure 10.35 Results of numerical simulations of long-runout landslides using one million colored particles with initial horizontally layered stratigraphy. (a) Geometry for the simulations, with steep slope leading to a flat surface through a rounded corner. (b) Frames from several times in the simulation as the triangular stack of one million particles is released. Time displayed in non-dimensional form $t' = t\sqrt{g/D}$. (c) Final deposit shown at four positions (denoted by numbers of particles of width D). Note that the stratigraphy displayed by differently colored particles remains intact through the event, and in the final deposits shown in the bottom plots (from Campbell *et al.*, 1995, Figure 1 and Plates 1 and 3, with permission from the American Geophysical Union).

Figure 10.36 Long runout landslides triggered by the M7.9 Denali earthquake of November 2002. Several landslides flowed across the Black Rapids Glacier (photograph taken by Peter Haeussler, USGS Alaska Volcano Observatory, on November 4, 2002).

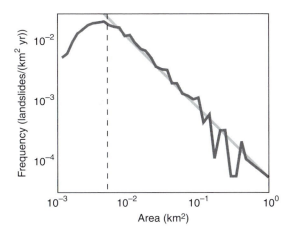

Figure 10.37 Probability distribution of landslides in New Zealand's Southern Alps. Landslides with areas below the dashed line are either not detected, given the mapping method, or are in fact absent. Straight line fit to landslides above this size indicates a power-law area distribution (after Hovius *et al.*, 1997, Figure 2).

mixing of water and rock particles during the initial failure can promote this transformation into a fluid, well-mixed mass. Many images of these debris flows are well captured in a USGS open file report (Costa and Williams, 1984), in which we see how varied the

behavior of these flows can be. Their speeds vary from less than 1 m/s to more than 10 m/s. Their high densities (roughly 2000 kg/m^3) allow transportation of large blocks and debris. These are a major means by which volcanoes extend their hazard well away from the volcanic cones themselves, as debris flows can travel tens of kilometers on low slopes. In Japan, the engineers charged with the duty of designing structures that will allow common debris flows to pass through major towns on the flanks of the volcanoes are highly revered. They are called "sabo engineers," and they have constructed concrete trapezoidal channels tens of meters across that serve to pass the flows safely by. The possibility that large-scale failures of these volcanic edifices can occur has only recently been established. A classic example of such a flow is the Osceola flow on which a portion of the city of Tacoma is built, in the shadow of Mt. Rainier. These pose a considerable hazard to surrounding civilization. (An excellent reference for this is the USGS film titled *Perilous Beauty, the Hidden Dangers of Mt Rainier*.)

But it is not only volcanic terrain that debris flows impact. They are common in arid regions, including the Los Angeles basin, and the eastern Sierras, for example. The alluvial fans of the eastern Sierras owe

Box 10.4 Estimation of flow speeds and debris flow discharge from paleo-flood evidence

There are two steps to the analysis. We wish to calculate mean velocities first, from which discharges can be calculated. We will use a technique that does not require knowledge of the material behavior in order to calculate the velocities. We employ the "super-elevation" of the flow as it rounds a corner to calculate the mean velocity of the flow. This may be measured from a survey of mudlines left on the banks of the channel. Once again, we call upon a balance of accelerations acting on the fluid, this time as it comes around a bend. In Figure 10.38, we show that the accelerations are: (1) centripetal acceleration, which may be written $<u>^2/R$, where $<u>$ is the mean velocity of the fluid in the channel cross section, and R is the radius of curvature of the bend; and (2) the component of gravity acting inward down the surface slope, written $g \sin(\beta)$, where g is the acceleration due to gravity (9.81 m/s^2), and β is the cross-channel slope of the debris flow surface as it rounds the bend. Setting these expressions equal to each other and solving for the mean velocity yields $<u> = \sqrt{Rg \sin \beta}$. Note that all we need to measure in the field is the radius of curvature of the channel, and the surface slope of the slop. Now that we have constrained the mean velocity of the debris flow, it is simple to calculate the discharge through this section of the channel: the discharge is the mean velocity times the cross-sectional area (which gives a volume per unit time): $Q = <u>A$, where A is the cross-sectional area of the channel. To get the mass discharge (mass per unit time) we need to multiply the volume discharge by the bulk density of the fluid, ρ_b, which is given by $\rho_b = \rho_g(1 - v) + s\rho_w v$, where ρ_g is the grain density, ρ_w is the density of water, v is the porosity of the matrix, usually roughly 0.35, and s is the degree of saturation. This is 1.0 if all pores are filled (i.e., saturated), and 0 if dry. The transformation of a landslide to a debris flow suggests a high degree of saturation, i.e., $s = 1$.

their shape to repeated deposition of debris flows with highly variable mobility (e.g., Whipple and Dunne, 1992). Those debris flows of low mobility tend to come to a halt in the channel, and serve to plug the channel so that subsequent flows are diverted. The channels and channel-plugging snouts on two adjacent debris fans are easily visible in the airborne laser swath mapping image of Figure 10.39.

Debris flows and long runout landslides belong to a class of flows that we could call granular mass flows. These include the pyroclastic flows that pose such hazards associated with volcanoes. All are concentrated mixtures of dense grains in a less dense liquid or gas driven downslope by gravity. In a recent summary, Iverson and Vallance (2001) emphasize that these flows are neither steady nor uniform, and that the state of the material mixture varies greatly both temporally and spatially from the time of initiation through its transit downslope to the final site of deposition. It is therefore not appropriate to assign such a flow a rheology in the sense we have described for a fluid, say, with a viscosity or a yield stress. These mixtures can exist in at least two states,

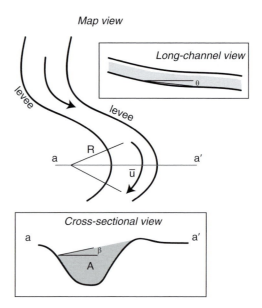

Figure 10.38 Sketch of a debris flow channel showing how one may deduce the mean velocity, u, in the channel from deposits and scour marks left in the aftermath of a debris flow. The radius of curvature of the channel bend, R, is documented from map view geometry. The super-elevation angle is documented from depositional and scour marks in the channel cross section, a–a'.

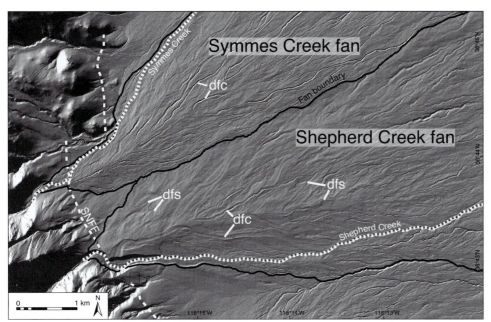

Figure 10.39 Shaded-relief map of Shepherd and Symmes Creek fans, bounding the eastern Sierras. Topography is derived from airborne laser swath mapping (ALSM) topographic data with resolution of 1 m/pixel. The fan boundaries are outlined by solid black lines; the active channels are shown by dotted white lines. SNFF, Sierra Nevada frontal fault. Examples of former debris flow channels and snouts are labeled as *dfc* and *dfs*, respectively. It is the plugging of the channel by a low-mobility debris flow that diverts subsequent flows, promoting the formation of a fan-like form (after Duhnforth *et al.*, 2007, Figure 2, with permission from the American Geophysical Union).

ranging from essentially a rigid solid to a fluid. The variables that control the state of the mixture are somewhat analogous to those that control the state of other materials we have dealt with (recall the phase plot of H_2O, which can change from liquid to solid (ice) along a phase boundary defined by a line in P–T space). Here the pressure involved is the fluid pressure, and more specifically the non-equilibrium pressure, that over or below that fluid pressure that would exist in hydrostatic equilibrium at some depth within the material. The temperature is more difficult to define, but represents the degree of agitation of the mixture; it involves the speeds of the particles just as true temperature involves the distribution of speeds of the molecules in a fluid. That the fluid pressures and the degree of agitation in the flow vary in both space and time as it travels from its source to its site of deposition dictates that even its state will vary in a complex way.

To a good approximation, the mixture behaves as a Coulomb material, in which the shear stresses are related to the normal stress as in Equation 10.56. And as in the landslide initiation problem, fluid pressures play a key role. We may write the equation for the normal stresses as

$$\sigma = (\rho_s - \rho_f) v_s g h \cos \theta - P_w \qquad (10.59)$$

where ρ_s and ρ_f are densities of the solid and fluid phases. Here the volumetric grain concentration, v_s, shows up explicitly, and again we see the role of fluid pressures in P_w. It turns out that the grain concentration varies only slightly in most flows, meaning that the dominant determinant of the normal stresses (and hence the shear stresses) is the fluid pressure. The essence of the problem of describing the flow, or of predicting the flow behavior, therefore lies in the spatial and temporal evolution of the fluid pressures within the material. Fluid pressures can diffuse within a mixture, moving from sites of high pressure toward sites of low pressure.

The problem gets even more complex (and interesting) when we acknowledge that the diffusivity that governs the redistribution of these pressures may involve the grainsize distribution, and that grains can become segregated within these mixtures while in motion. In particular, coarse grains come to the top (just as the hazelnuts come to the top in an agitated jar), and because the flow speed is fastest at the top, these coarse grains move to the fronts and sides of flows, where they exert high flow resistance. This can lead to an instability in which the flow fronts break up into fingers that characterize the leading edges of pyroclastic flows (e.g., Pouliquen *et al.*, 1997).

Hillslope models

We are now in a position to link models of hillslope transport with those of regolith production to generate animations of landscapes. We show in Figure 10.40 an example of modeled hillslopes in the face of both regolith production and regolith transport. They are meant to mimic the parabolic hilltops depicted in Figure 10.4 in the Wind River Mountains, although many alpine summits display similar smooth convex profiles. The regolith production rule used is that in which rates are slow on bare bedrock, increase to a maximum, and then decline with further regolith thickness. Initial conditions (in this case flat) are rapidly "forgotten" in the landscape as the form evolves toward a steady parabolic form with uniform regolith thickness. The steady curvature of the hilltop is dependent upon the efficiency of the transport process, all else being equal. The transport process being

mimicked is that of frost heave. By reducing the mean frost penetration depth by a factor of two, the transport rate on a given slope declines, and the curvature needed to accommodate the regolith being produced increases.

In a much more complex situation, Densmore et al. (1998) attempt to simulate the evolution of a landscape driven by normal faulting. There the channel incision rates through the rock of the rising footwall block steepen the adjacent slopes so that they become bedrock landslide-dominated. The triangular facets that signify a normal faulted landscape are evident (e.g., frontispiece to Chapter 4). Koons (1989) addressed the rapidly evolving landscape of New Zealand, and was among the first to quantitatively address the role of orographic precipitation.

In another tectonically active landscape, this time dominated by thrust and strike-slip faulting, the uplifted staircase of the Santa Cruz marine terraces decay topographically as they age. While these terraces saw

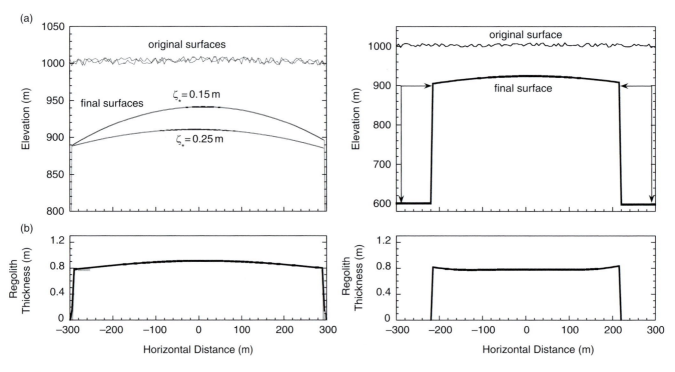

Figure 10.40 Simulations of high surface topographic and regolith cover evolution, and their dependence on boundary conditions and hillslope transport efficiency. Top panels: original and final profiles of topography. Bottom panels: final profiles of regolith thickness. Initial condition: flat surface with roughness, and regolith thickness of 5 m. Left simulation: boundary conditions involve only glacial down-wearing of adjacent valleys. Right simulation: boundary conditions include both glacial down-wearing of adjacent valleys, and lateral migration of glacial valley wall. Final profiles are nearly parabolic, and have uniform but thin (~0.8 m) regolith. Some simulations show bare bedrock at the crest of the hill, others do not, depending upon the particulars of the initial conditions (after Anderson, 2002, Figures 15 and 20, with permission from Elsevier).

early work based on the diffusion equation, later research addresses not only the transport of terrace sands but the generation of regolith from the underlying sandstones. Only then can the asymmetry of the decaying cliffs be understood (see discussion in Chapter 18, Figure 18.6).

We have discussed diffusive processes that tend to round hillcrests. As we will see in Chapter 11 in our discussion of water in the landscape, the overland flow of water down a hillslope is not expected to produce convex hilltops. However, as discussed in Chapter 11, Dunne (1990) has shown that the wide convex tops that bound long hillslopes in Kenya cannot be attributed to the motion of grains by rainsplash. Rainsplash is simply not efficient enough to accomplish this in the allotted time. Instead, his models, reproduced in Figure 11.18, demonstrate that the overland flow and associated sheetwash process can act as a diffusive process. Surface grains are indeed entrained in the overland flow, but the typical storm doesn't last long enough to transport them all the way down the slope. So they are dropped, and far from their initial positions, as if they had taken long leaps instead of the small hops associated with rainsplash. The result is a more efficient diffusive process, resulting in broad convex hilltops.

Summary

We formally address the question of hillslope form by casting the problem as one of conservation of mass, here of regolith. This approach is one we take at the start of many geomorphology problems, whether it be the conservation of mass of ice on a glacier, the mass of sediment on a channel bed, the volume of water in a channel segment or a lake, and so on. In this case the exercise serves to focus the need for knowledge of what sets both the rate of regolith production (Chapter 7), and the rate of regolith transport down a slope.

Many hillslopes have convex tops, which can be attributed to the dominance of diffusive transport processes at least near the hillcrests. The diffusion equation describing the evolution of convex hilltops and of the smoothing of microtopography arises when this expression for conservation of regolith mass is combined with an expression for regolith transport in which the transport rate is linearly related to the local hillslope angle.

We have explored several specific processes that can be characterized by a linear dependence on local slope, including rainsplash, the burrowing of fossorial rodents, tree-throw, frost heave, and solifluction. In each of these cases, we must describe the geomorphic result of individual events, and then sum over these events. In most cases the events themselves could be described as deterministic (once the initial conditions are known, the outcome is certain), while the summation takes place over a distribution of events that has a stochastic character and must therefore be described statistically.

The random nature of particle motion in the regolith when subjected to several of these mechanical processes has inspired some to embrace the statistical mechanical treatment of the problem. At present this work is on the leading edge, treating either the vertical motion (bioturbation) of sediment, or rainsplash. But the approach appears to be extendable to other processes, and may guide the collection of relevant field information.

Hillslopes are strongly slaved to the local conditions at their base, which are in general set by the fluvial or glacial systems. Rapid incision can trigger steepening of lower slopes and can lead to landslides. We have summarized the simplest calculation of hillslope stability in the "infinite slope" case, and have illustrated the role of slope, regolith thickness, and groundwater pressures. The latter guides an assessment of the climatic conditions under which failure is favored, and allows us to deduce the role of hillslope convergence or divergence in governing the likelihood of failure.

The post-failure fate of landslide material varies greatly with local conditions, but includes in long runout landslides some of the most remarkable events in geomorphology. These are likely governed by particle collisions in the granular material that become increasingly efficient as the grains are further crushed into very elastic cores of the original blocks. Debris flows are one end-member result of a landslide. They too are very mobile, and can transport debris from sources in mountainous terrain onto debris-flow fans that bound especially arid mountains.

Problems

1. On a hillslope with regolith 1.2 m thick, and a regolith production rate of 8 μm/yr, what would be the mean residence time of regolith in the regolith column?

   You then note that this slope is 100 m long and that the regolith is uniform in thickness. Calculate the residence time another way, this time by addressing the entire regolith "reservoir" on the hillslope, and the volumetric regolith discharge rate at the base of the slope. Are the calculated residence times equivalent?

2. Plot the expected steady-state pattern (as a function of distance from the crest) of discharge of regolith on a convex hillslope in a climate in which the production rate of regolith is 20 μm/yr. Report your answer in m^2/year.

   If the regolith was 0.8 m thick, what would be the spatial pattern of the mean downslope velocity of the regolith?

   How would these numbers change if 25% of the mass loss occurred in solution?

3. How much strain does a soil sustain from the growth of tree roots? Assume that the tree roots are 10 cm in diameter, are simple vertical taproots, and are spaced at 1.5 m spacing. (Obviously this is a very simple model of roots!)

   How much ought the soil to inflate vertically due merely to this taproot growth if the soil is 1 m thick? Do the experiment in a small beaker to illustrate this effect.

4. On a hillslope with a weathering rate of 20 μm/yr and an effective hillslope diffusivity of 0.02 m^2/yr, what would be the expected curvature of the hillslope? Plot the full topographic profile if the summit of the hillslope, at $x = 0$, is 1000 m. The channels at the edge of the hillslope are 200 m from the hillcrest.

5. *Steady regolith discharge.* On a 150 m long hillslope, regolith is being generated at a uniform rate of 40 microns/yr. Calculate and plot the distribution of downslope regolith discharge with distance from the crest.

6. *Slope stability.* Consider a long linear hillslope with a slope angle of 25°. It has soil that is 1 m thick, and a shear test on the soil reveals that its internal angle of friction is 31°, with a cohesion (including roots) of 5000 Pa and a porosity of 35%. Perform a stability analysis on this slope for the worst-case scenario of full saturation. If you find that it is indeed unstable under full saturation, assess the water table height that is required to cause failure.

7. *Runout of a large landslide.* Consider a landslide mass that fails from an elevation of 2500 m. The valley floor onto which the rockmass tumbles lies at 1700 m. The mass comes to a screeching halt on the opposite valley wall at an elevation of 1950 m. Calculate the maximum and minimum speed at which the mass is traveling as it crosses the valley floor. Given these estimates, estimate how much time a citizen of the valley floor would have to get out of the path of the slide if they were fortunate enough to have witnessed its initiation.

8. Consider the plot of landslide areas in Figure 10.38. Assess the slope of the straight line required to fit the data, and report the resulting probability density function for landslide area.

9. *Thought question.* Explain as simply as you can why a compacted granular material is less likely to fail catastrophically on a slope than is a loosely packed material.

10. *Thought question.* How might the odd spiky landscape illustrated in the photograph at the start of the chapter, from the Tsingy Park in Madagascar, form? It certainly violates our expectation of convex hilltops. What rule set might explain this cusp-like appearance?

Further reading

Butler, D., 1995, *Zoogeomorphology*, Cambridge: Cambridge University Press, 239 pp.
This is one of only a few summary treatments of the role of biology in geomorphic systems.

Carson, M. A. and M. J. Kirkby, 1972, *Hillslope Form and Process*, New York: Cambridge University Press, 475 pp.
A classic book on hillslopes by two of Britain's most prominent geomorphologists, this summary of the state of our understanding in the early 1970s is still a useful reference on the topic.

Selby, M. J., 1993, *Hillslope Materials and Processes*, 2nd edition, Oxford: Oxford University Press, 451 pp.
This encyclopedic treatment of the hillslope literature in the early 1990s remains a reference book. Much of this material was also included in Selby's geomorphology text The Earth's Changing Surface (1985).

CHAPTER 11

Water in the landscape

Nothing is softer or more flexible than water,
yet nothing can resist it.

Lao Tzu, *Tao Te Ching*

Once rainwater hits the ground surface or snow melts, it begins a journey that is transformative, both of landscapes and of the water itself. Water drives many geomorphic processes, from rainsplash erosion to landsliding, from karst formation to soil formation. Runoff from hillslopes and groundwater collects into rivers, which can carve landscapes and sweep away detritus. Even in arid landscapes water plays an important role, as the rare rainstorm can do more geomorphic work in a single event than occurs in the intervening years of dryness. The power of a desert rainstorm is amplified by the general lack of protective vegetation in these environments, but it also reflects the potential energy difference between flowing water and flowing air. Because water flows through landscapes, it is a great integrator. It is for this reason that most geomorphologists view the drainage basin, the total area that contributes runoff to a given cross section on a river, as the fundamental unit of landscapes. Transport of water, solutes, and sediment from an area is most easily measured if the area considered is a drainage basin. This method of parsing up landscapes is so common that areas lacking regularly branching rivers and well-defined divides (sensible drainage basins) are considered "deranged."

Downpour in Blue Hills badlands near Caineville, Utah. Raingage in the foreground records rainfall intensity, while PVC supports center sonic rangers over channel to record stage of resulting flashfloods (photograph by R. S. Anderson).

We begin this chapter with a consideration of the basic morphology of drainage basins. This leads to the acknowledgement that the locations of the tips of channels – the channel heads – matters significantly in setting the density of channels in a landscape. In order to explore what governs the channel heads in a landscape, we first address where water is stored in the subsurface, and how this water moves. We deal with infiltration into the soil, and briefly touch upon the lateral transport of water in both the unsaturated and saturated (groundwater) zones. Then we focus on the runoff mechanisms, most of which entail some portion of the flowpath in the subsurface. Here we ask what governs the character of the hydrograph in the adjacent stream – its peak and the lag between the rainfall and runoff peaks. Armed with this sense of how water moves over and within the landscape, we address the geomorphic consequences on hillslopes, focusing on processes that establish channel heads in the environment. We first address Horton's view of channel heads being determined by entrainment of sediment by overland flow (Horton, 1945). While we reserve the full derivation of the equations for open channel flow for the river chapter (Chapter 12), in this chapter we introduce the equations needed for the calculation of mean flow speeds relevant to overland flow. You will see that in reality it is subsurface flow mechanisms that localize the erosion that leads to channel tips, and that the preferred sites are in crenulations of the landscape – geometry matters. The architecture of the hydraulic conductivity field matters as well, and it reflects both the geologic and geomorphic history to which the parcel of rock has been subjected.

As in most chapters in this book, many of the problems we touch upon in this chapter are best approached using continuity as an organizing principle. We can cast the water balance in a whole catchment as a sum of inputs minus outputs. We can conserve water in a sheet of water running across the surface. We can conserve water in the groundwater system. In each case, setting up the required balance forces us to be organized about what we must then know about how something (in this case water) moves from one place to another. It is our representation of the physics of these transport processes that changes from unsaturated subsurface flow to saturated subsurface flow to overland flow.

Drainage basins

Large drainage basins contain a channel network that typically displays a tree-like or dendritic pattern. Little streams join to become big ones, big ones join to become bigger yet, and so on. We are all familiar with this pattern. In the early twentieth century, leading geomorphologists Arthur Strahler and Robert Horton developed a terminology to describe this network. Strahler (1952) defined "first-order" drainages as streams that do not have tributaries. As seen in Figure 11.1, these are found at the tips of the drainage network. Second-order streams are formed by joining two first-order streams, third-order where two second-order drainages join, etc. Several patterns emerge when we look closely at maps of drainage basins, ordering the streams in this fashion. One of

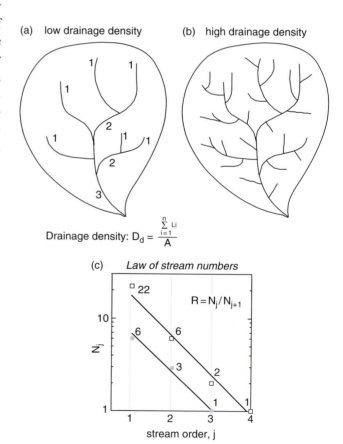

Drainage density: $D_d = \dfrac{\sum\limits_{i=1}^{n} L_i}{A}$

Figure 11.1 Ordering of drainage patterns in (a) low drainage density and (b) high drainage density basins. The drainage density is the total length of channels, *L*, per unit area of drainage basin, *A*. (c) Law of stream numbers revealed as a straight line on log plot of numbers of streams vs. stream order.

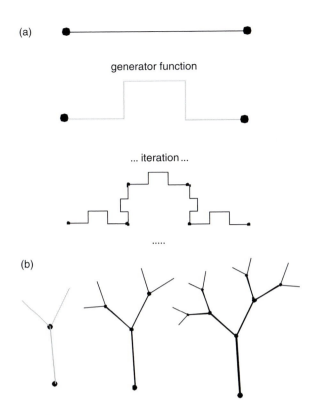

(a)

generator function

... iteration ...

.....

(b)

Figure 11.2 Fractal primer. Generation of a fractal object by iteratively applying a generator function: (a) top-hat, (b) tree.

the drainage divides and the channel heads in a large drainage basin is inversely proportional to the drainage density, D_d, defined as

$$D_d = \frac{\sum_{i=1}^{n} L_i}{A} \qquad (11.1)$$

where L_i is the length of tributary i and A is the drainage basin area. Note that this measure of the channel density has units of $1/L$. While the bifurcation ratio does not vary substantially between different landscapes, the drainage density does. Landscapes with high drainage density have on average shorter hillslopes than those of low drainage density, as can be seen in Figure 11.1.

What is important about the drainage density and therefore the location of the channel heads in a landscape? The rate at which water moves from one place to another in channels is much faster than it can move in unchannelized flows, either over the ground surface or in the subsurface. This is the key to the fundamental dichotomy in geomorphic systems: there are hillslopes and there are channels, and they are very different beasts. While differences in rates of sediment movement is one aspect of the slope–channel dichotomy, the significant difference for this discussion is the efficiency of transmitting water. Water is slowed on hillslopes by the thinness of the films and pores it must flow through, and by the tortuous routes it must take. The velocity of water moving toward a channel is much lower than the velocity water can attain in a channel.

Because hillslopes are so inefficient in passing water, the shorter the hillslopes in a landscape, the faster a rainfall event will be felt downstream. The peak of the hydrograph from a basin is strongly controlled by the lengths of the slopes over which water must flow before reaching channels, and hence on the drainage density of the landscape. High drainage density landscapes result in flashy discharges that respond quickly to rainfall inputs. Note, for example, the small flash floods depicted in Figure 11.3, in a very high drainage density shale badlands landscape. This focuses our attention, as geomorphologists, on what it is in a particular climatic and geologic setting that establishes the channel head locations. We will return to this quest after addressing where water can be stored in the landscape, and how it moves across and within the landscape.

the most commonly observed patterns is that the ratio, R_b, of the number of streams of order j, N_j, to the number of streams of the next largest order, N_{j+1}, is approximately constant. This results in straight lines on a plot of the log of N_j vs. j, also shown in Figure 11.1. R_b is called the bifurcation ratio, and is commonly about 3 or 4. As Robert E. Horton (1945) first recognized this relationship, it is known as Horton's law of stream numbers.

This regular branching has led some to describe river channel networks as a fractal. We sketch the basics of fractal generation in Figure 11.2 One of the chief characteristics of a fractal object is its self-similarity – it looks the same at any magnification. While the idea of fractals is beguiling, and is of some use for computer generation of realistic-looking landscapes, the fractal nature of the network ends at the channel heads. River networks do not go on forever. At the upstream end of each first order is a channel head. Above the channel head there is no discrete channel, only a slightly concave bowl (a zero order basin as it has been called). The average distance from

(a)

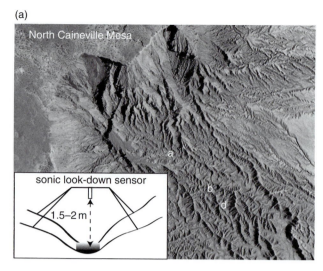

(b)

(c)

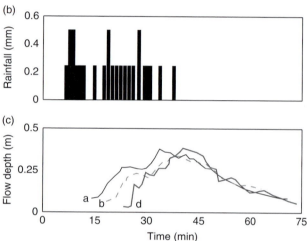

Figure 11.3 Flashflood in Blue Hills badlands draining southern edge of North Caineville Mesa, near Caineville, Utah. Top: image of badlands with locations of the sonic gages in the channel. Inset: sketch of the bridge support for sonic transducer. (b) Rainfall in this ~1 km² basin generated streamflow that passed three look-down sonic sensors at stations a, b, and d within 30 minutes of the initiation of rainfall. (c) Hydrographs reveal passage of the flood wave, the leading edge of which steepens with distance down channel. Different starting gage depths correspond to trigger thresholds for recording sonic stage values (after Dick *et al.*, 1997, Figures 2 and 4).

Water balance

One reason to use watersheds as a fundamental landscape element is that this choice simplifies the water balance. Water balance is a watershed accounting scheme that acknowledges conservation of water volume. Water inputs minus water outputs must equal changes in water storage: $I - O = \Delta S$. For a

drainage basin, inputs are generally limited to rain and snowmelt. Outputs include discharge of water in the stream, losses due to evapotranspiration, and export of groundwater. While stream discharge is often fairly easy to measure, the other two output terms are not. Evapotranspiration includes both direct evaporation from surfaces and transpiration, which amounts to evaporation mediated through plants. Groundwater flows occur through aquifers whose lateral dimensions and permeability are often poorly constrained. The fluxes of water in and out of a basin are balanced by changes in storage, in the form of soil moisture, (saturated) groundwater, snow, ice, and lakes. Thus we may write the water balance as

$$\int_T S dt = A \int_T R dt - \int_T Q_s dt - \int_T Q_{gw} dt - A \int_T ET dt \quad (11.2)$$

where the left-hand side is the storage, S, the first term on the right-hand side is the total rainfall over the period T, where R is the rainfall rate, the second term is the stream discharge, $Q_s(t)$ being the "hydrograph," the third term is discharge of groundwater, Q_{gw}, out of the basin, and the final term is the evapotranspiration, ET, loss.

The water balance can be simplified by considering only periods of time appropriate to the question at hand. For instance, over periods of a year or longer, the change in storage in soil or groundwater (or lakes or ice) is usually very small, and the left-hand side can be set to zero. Groundwater flows are often neglected because the magnitude is commonly much smaller than the other terms. With these simplifications and some rearrangement, the water balance can be written

$$\int_T R dt - \int_T ET dt = \frac{1}{A} \int_T Q_s dt \quad (11.3)$$

The rainfall minus evaporation term on the left is called the "effective precipitation," since it defines the water available to produce discharge in streams. This approach can provide an estimate of annual average evapotranspiration. Alternatively, if ET can be estimated, then Equation 11.3 can be used to estimate the mean annual discharge.

Let's return to the full water balance. The mass balance can be applied over short periods of time, such as a storm event. In this case, however, the change in storage term cannot readily be ignored. In fact, the rainfall term only contributes to the positive side of the balance for the duration of the rainstorm. After the rain stops, continued outflows (stream discharge, groundwater flow, evapotranspiration) are maintained only by withdrawals from the storage components of the water balance. If we restrict attention to a drainage basin with no lakes, snow, or ice, then the storage term only comprises changes in groundwater and soil water. A brief definition of terms is appropriate at this point. Although the study of groundwater covers any water below the land surface, the term groundwater generally refers to the saturated zone below the water table. In the saturated zone, the pores are completely filled, so that the porosity, ϕ, (the ratio of the volume of voids to the total volume) is equal to the volumetric water content, θ, (the ratio of the volume of water to the total volume). The change in storage in the saturated zone can be estimated from changes in height of the water table, given by $\int_T S_{gw} dt = A \int_T \phi (dz_{gw}/dt) dt$. Above the water table is the unsaturated or vadose zone, where pores are partially filled with water. The change in storage in the vadose zone is characterized by changes in soil moisture over the depth a wetting front reaches, L_f, (for a storm event), given by $\int_T S_v dt = A \int_T \Delta\theta (dL_f/dt) dt$, where $\Delta\theta$ is the mean change in soil moisture over the depth of wetting. Both of these terms are likely to be positive while rainwater or snowmelt infiltrates into the subsurface and percolates to the water table, and negative when the water table declines or as soil moisture declines.

Understanding in detail, then, the controls on timing of discharge in a river depends on understanding the storage reservoirs in a catchment and how water moves between them. While we will not completely unravel the problems in this arena, we will touch on some basic concepts. Rather than focusing on the full physics of subsurface flow, much headway can be made by looking at simplified components of the problem. We will explore infiltration, flowpaths, and flow rates along flow paths.

Soil moisture and its distribution with depth

We now describe more formally where water is to be found in the subsurface. Hillslopes consist of soils overlying rock. Both have a definable porosity. The porosity of a material is defined to be the fraction of a volumetric element of the material that consists of voids:

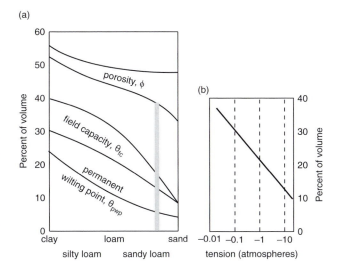

Figure 11.4 (a) Dependence of water content of soils on soil texture. Maximum water content is set by porosity. Field capacity is fully drained condition. Wilting point is the water content below which plants are incapable of producing sufficient suction to draw the water out of pores. (b) Tension of water held in pores in a sandy loam (gray bar in (a)). Suction required to extract water from the soil increases exponentially as the water content declines; at the permanent wilting point, this exceeds the suction that plants can produce (after Dunne and Leopold, 1978, Figures 6–7 and 6–9, with permission from W. H. Freeman).

$$\phi = \frac{V_{voids}}{V_{total}} \tag{11.4}$$

The voids can be occupied by some combination of fluids – in our applications usually air and water. Volumetric water content is defined as the fraction of the soil volume that is occupied by water:

$$\theta = \frac{V_{water}}{V_{total}} \tag{11.5}$$

The volumetric water content can therefore vary between zero and saturation, when all pores are occupied by water. At saturation, then, $\theta = \phi$. The porosity and the water content of a soil are strongly dependent on grain size, as shown in Figure 11.4. Coarse soils dominated by gravels have porosities of the order of 30–40%, while silts and clays can have porosities reaching 50%. We also show on this graph the field capacity of the soil and the permanent wilting point of the soil. The field capacity is defined to be that water content achieved when the soil is allowed to drain vertically for a long period of time. The water remaining in the soil is no longer moving, meaning that the force of gravity acting downward on

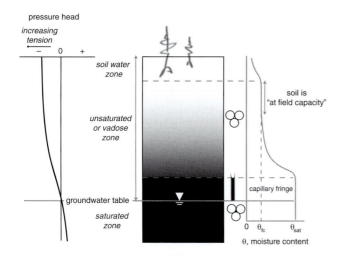

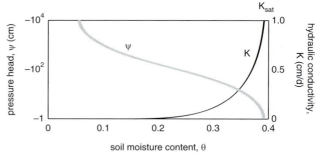

Figure 11.6 Characteristic curves for an unsaturated soil with porosity $= 0.4$. The tension (negative pressure) with which moisture is held in the soil increases with decreasing moisture content. The hydraulic conductivity of the soil increases dramatically with soil moisture content, reaching a maximum at saturation, the saturated hydraulic conductivity, K_{sat}.

Figure 11.5 Water in near-surface storage. The vadose zone encompasses all soil above the groundwater table. Some portion of the soil above the water table will be saturated as well in the capillary fringe, where it is held in tension within pores. The height of the fringe is dictated by soil texture, being greater in finer grained materials. The water content, θ, is drawn down by evapotranspiration of plants and evaporative losses from the soil surface in the soil water zone. Below this, the soil is held at "field capacity," the moisture content below which water is held tightly in pores against gravity (in part after Hornberger et al., 1998, Figure 8.1, with permission from The Johns Hopkins University Press).

it is balanced by the surface tension exerted by the mineral grains. The permanent wilting point is a yet lower moisture content. The water remaining in the pores of the soil under this condition is unavailable to plants (hence they will wilt if this is the only source of water for them). This water can neither be drained downward by gravity, nor be pulled upward by suction imposed by evapotranspiration of plants.

We are now armed to understand the profile of soil moisture content shown in Figure 11.5. At depths below the groundwater table, the soil is saturated. Immediately above the groundwater table, the soil is still saturated, but the pores are filled by capillary rise of water above the water table. Water there is in tension, and hence the zone is also called tension-saturated. Above this capillary fringe the moisture content declines. The pores are not completely filled with water, and the zone is unsaturated. This is also called the vadose zone (*vadose* is Latin for shallow). Nearest the surface, where roots can penetrate, the moisture content is raised by inputs of water from rain and snowmelt at the surface, and lowered by

losses to evapotranspiration. This entire profile is dynamic, fluctuating on storm-by-storm and seasonal timescales. The greatest variability is in the near-surface, but as we will see inflation and deflation of the groundwater table can play a significant role in the paths that water takes across and within the landscape.

Soil is a granular medium. In any given soil there exists a wide range of pore spaces to store and to pass water. The large spaces or macropores include cracks in soil, root holes, worm holes, and rodent burrows. The smaller pores are those between grains of the soil, whose sizes are dictated by the grainsize of the soil. In general, organic matter in the near-surface serves to bind particles into clumps or "peds." Water can rapidly infiltrate between peds, and is much more slowly wicked into them.

The mixture of pores of various sizes within a granular material manifests itself strongly in the behavior of water within the soil. The range of pore sizes governs both the forces with which the water is held in the soil, and the flux of water under an imposed force. This means that a given soil will possess a specific relationship between the water content of the soil and an imposed tension or suction, and another relationship between the water content and the hydraulic conductivity of the soil. These are called the characteristic curves of the soil, shown schematically in Figure 11.6. Water in smaller pores is more strongly held by surface tension than that in larger pores, for the same reason that water rises higher in a very thin capillary tube than in a wider tube. We have already seen that this is reflected in the field capacity of a soil. Imagine that a saturated block of soil is hooked up to a suction device. When we turn on the pump, the first water to be pulled out of the soil will be that from the largest pores. If this low suction is maintained, the flux of water from the block will eventually decline to zero. That water still held in the soil exists in pores small enough to exert a high enough tension to counteract the imposed suction. This gives us one point on the relationship between moisture content and suction. The pressure (which in unsaturated conditions is negative, or suction) is also termed the matric potential of the soil, referring to the interaction of the water with its solid matrix, the soil grains. At higher and higher suction, pores of smaller and smaller diameters yield their water. The resulting curve relating soil tension to water content is one of the characteristic curves; it is also called the water retention curve. Now consider how the flux of water in the soil will depend upon the moisture content when we impose a pressure gradient in, say, the horizontal direction. At saturation, all the pores are active, the pore water is not in tension anywhere, and the resistance to flow is simply that due to drag through the orifices in the soil (see box on hydraulic conductivity). Under conditions of water content below saturation, two effects reduce the flux of water. First, since that water that does exist resides in smaller pores, the flux through each pore constriction (or pipe) will be smaller because flux through pipes varies as the square of the pipe diameter (see Box 11.3). Second, the fraction of the pore space contributing to flow is smaller. Since the Darcy flux

is the product of the mean speed of the flow and the fraction of the cross section of the soil block contributing discharge, both effects reduce the flux and hence the hydraulic conductivity (the ratio of the flux to the imposed pressure gradient). The relationship between the water content and the hydraulic conductivity is the second characteristic curve. Armed with these relationships and a sense of where water can be found in the subsurface, let us now address how water moves in the subsurface.

Infiltration

The process by which water passes from the ground surface into the subsurface is called infiltration. Infiltration rates govern the recharge of soil moisture that will be available for plant growth, shown in Figure 11.4. Because water will run off the surface if the rainfall rate exceeds the infiltration capacity of the soil, infiltration also governs the timing and frequency of occurrence of overland flow. As overland flow is a much more efficient means of removing water from the landscape than is flow through the subsurface, infiltration capacity also dictates the flashiness of discharge in streams. Finally, overland flow is capable of entraining sediment, thereby setting the locations of channel heads in some landscapes.

As illustrated in Figure 11.7, within an individual storm event, infiltration rate declines with time, for several reasons. Long rainstorms can blast fine-grained sediment around on the surface, which finds

---

Box 11.1 Hydraulic conductivity, permeability, and soil moisture

Flow in porous media can be treated as a potential driven process, analogous to heat conduction, electrical currents, and other processes controlled by a potential gradient. In the case of groundwater, Darcy's law describes the flow rate as $Q_{gw} = -K(dh/dl)A$, where K is the hydraulic conductivity, dh/dl is the gradient in hydraulic head along a flowpath, and A is the cross-sectional area. The hydraulic conductivity in saturated materials depends on the porous medium, and the properties of the fluid. It can be shown that $K = k\rho g/\mu$, where k is the intrinsic permeability, ρ is the fluid density, g is the acceleration due to gravity, and μ is the dynamic viscosity. The permeability term contains information on the pore size and structure of the medium, and is the parameter of choice where fluids may vary, such as in petroleum geology. When water is the fluid of interest, hydraulic conductivity is the parameter most commonly used. Hydraulic conductivity, which has units of length per time [L/T], varies over many orders of magnitude in natural materials. For instance, gravel may have a hydraulic conductivity of up to 1 m/s, while unfractured shale may be as low as 10^{-13} m/s. See Table 11.1 for a list of hydraulic conductivities of some representative Earth materials.

Table 11.1 **Hydraulic conductivities of representative Earth materials**

K (cm/s)	10^2	10^1	1	10^{-1}	10^{-2}	10^{-3}	10^{-4}	10^{-5}	10^{-6}	10^{-7}	10^{-8}	10^{-9}	10^{-10}
Relative permeability	Pervious				Semi-pervious				Impervious				
Sediment	Well-sorted gravel	Well-sorted sand or sand and gravel			Very fine sand, silt, loess, loam				Unweathered clay				
Rocks	Highly fractured rocks				Oil reservoir rocks			Fresh sandstone		Fresh limestone, dolomite		Fresh granite	

Modified from Table 4.1 in Bear (1979)

Box 11.2 The Richards equation

Quantitative assessment of the redistribution of water in the vadose zone relies upon marriage of the continuity of moisture content with the physics of water movement in granular media. Consider a simple vertical column of soil. Continuity requires that the rate of change of volumetric moisture in a soil element reflects the difference between the downward volumetric flow rate (flux), q_z, through the top of the cell, at z, and the downward flux through the bottom of the cell, at $z + \mathrm{d}z$:

$$\frac{\partial\theta\mathrm{d}x\mathrm{d}y\mathrm{d}z}{\partial t} = q_z(z)\mathrm{d}x\mathrm{d}y - q_z(z + \mathrm{d}z)\mathrm{d}x\mathrm{d}y \qquad (11.6)$$

Dividing by $\mathrm{d}x\mathrm{d}y\mathrm{d}z$, and acknowledging the definition of the derivative, in the limit as $\mathrm{d}z \to 0$ we obtain

$$\frac{\partial\theta}{\partial t} = -\frac{\partial q_z}{\partial z} \qquad (11.7)$$

The downward Darcy flux, q_z, may be written as

$$q_z = -K(\theta)\frac{\partial(-z + \psi(\theta))}{\partial z} \qquad (11.8)$$

where K is the hydraulic conductivity. The first and second terms in the derivative represent the gravitational and the pressure head, respectively. We have added the minus sign to z to reflect that depth, z, increases downward while elevation decreases downward. Here we have acknowledged that the hydraulic conductivity, K, and the pressure head, ψ, are both functions of the moisture content, θ. Indeed, herein lies the complexity of the problem. Combining Equations 11.7 and 11.8 results in

$$\frac{\partial\theta}{\partial t} = \frac{\partial}{\partial z}\left[K(\theta)\frac{\partial(-z + \psi(\theta))}{\partial z}\right] \qquad (11.9)$$

which can be simplified to

$$\frac{\partial\theta}{\partial t} = -\frac{\partial K(\theta)}{\partial z} + \frac{\partial}{\partial z}\left[K(\theta)\frac{\partial\psi(\theta)}{\partial z}\right] \qquad (11.10)$$

This is the Richards equation, named for Lorenzo A. Richards who first published it in 1931. This is a partial differential equation, made strongly nonlinear by the dependence of hydraulic conductivity, K, and pressure head, ψ, on the moisture content, θ.

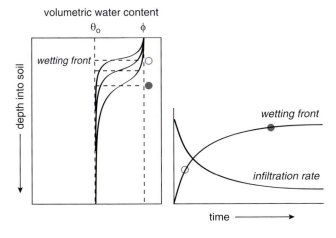

volumetric water content

θ_o ϕ

depth into soil

wetting front

wetting front

infiltration rate

time ——————⟶

Figure 11.7 History of moisture content shown in a vertical column of soil (left) and in time (right). The wetting front propagates rapidly downward at first but slows with time, mirroring the decline in infiltration rate. Here the initial moisture profile is assumed to be uniform with depth. Saturation is achieved in the near-surface soil.

its way into macropores and between peds to plug them. Clays in the soil can expand upon getting wet, which closes up cracks that will have grown in the wet season or between storms. And the soil can simply become saturated, in which case there is nowhere else to put the new water. The progress of a wetting front, defined as the midpoint of a strong moisture gradient, with time during a storm is shown in Figure 11.7, along with the history of infiltration at the surface.

Let us now explore the simplest case, that of a static vertical profile of moisture content such as that shown in Figure 11.5. We might expect this situation to occur well after a storm, or late in a dry season. If the water content is steady at all depths, and no water is being added to or lost from the top surface, the vertical fluxes must be zero everywhere. Mathematically, then, the two terms in the expression for the Darcy flux in Equation 11.8 must balance. In other words, the pressure head, ψ, must equal the elevation head, z:

$$\psi(\theta) = z \tag{11.11}$$

What we wish to know is the moisture profile, $\theta(z)$. Because the moisture content is a function of the tension (creating the characteristic curve), and tension linearly increases with elevation, the shape of the moisture profile must be the shape of the characteristic curve. In physical terms, the forces on the water in the soil must be balanced in order to prevent its moving.

The vertical force due to the weight of the water must therefore be balanced by the gradient in the pressure (tension). The water content profile takes on a gradient, which governs the gradient in the tension, such that this force balance is assured at all heights above the water table. This is why the profiles of water content, such as that shown schematically in Figure 11.5, mirror the sigmoidal look of a characteristic curve.

Infiltration is the movement of water into the soil, and hence is not represented by the static calculation we have just described. Infiltration has been modeled for decades with varying degrees of sophistication. At the high end, one would attempt to employ the full Richards equation (see Richards equation box) in all its nonlinearity. But this is computationally intensive, and is most often replaced by the simpler Green-and-Ampt or Green–Ampt approach (Green and Ampt, 1911) that captures the essence of a progressing wetting front. In both approaches one employs the continuity of moisture, assessing inputs and outputs of moisture from a stack of horizontal slabs of soil. As always the rate of change of the tracked quantity is simply the difference between the inputs and the outputs. Input to the topmost slab is dictated by the rainfall rate, or the saturated hydraulic conductivity, whichever is smallest. (If it rains faster than it infiltrates, the excess water runs off – see discussion of overland flow later.) The water content of the slab climbs until its hydraulic conductivity is high enough to pass water to the next slab. (Remember that K depends upon θ, and in particular increases to its maximum at saturation.)

Between storm events, the water content of near-surface soil changes due to both downward drainage of water toward the groundwater table, and upward withdrawal of water by both evaporation and plant transpiration. Gravitational drainage will lower the soil moisture content to a condition called its field capacity, θ_{fc}, where the water is now held strongly enough in the pores by surface tension in the pores to counter the force due to its own weight. Plants can reduce the moisture content yet further by exerting suction. Here the lower limit of moisture content is the permanent wilting point, θ_{pwp}. This depends upon the texture of the soil, as we have seen in Figure 11.4, and the ability of the particular vegetation to exert suction. The soil nearest the surface therefore undergoes the greatest variability of moisture content. The fate of water that falls on the landscape will depend intimately on the moisture state of the near-surface soils, and hence on the recent history of both rainfall and evapotranspiration.

When infiltration capacity is actually measured in the field, it is found to vary significantly from place to place, as well as with time. There are good reasons for this. The spatial dependence largely reflects the variation in soil type on a catena, or hillslope transect. The bases of slopes tend to be dominated by more clayey soils than those at the crests, and even the clay types vary, being less solute rich at the crests than at the bases. This tends to force the infiltration rate to decrease with distance downslope. The infiltration rate is also greatly influenced by vegetation. Roots disrupt the soil, providing avenues for water to penetrate into the subsurface more efficiently. Plants in arid regions often occur on mounds with anomalously high saturated conductivities (see our discussion of the potential origin of such mounds by rainsplash in the Hillslopes chapter). Dunne *et al.* (1995) have documented the threading of flow between vegetated mounds during overland flow, and have argued for the importance of swamping such mounds during high thickness overland flow events.

We have also touched upon the fact that infiltration capacity varies with time at a site. Fundamentally, this reflects the fact that soil moisture affects the rate at which water can be passed through a porous medium. In addition, large cracks can form in some soils as they dry, making them very permeable after a long period of dry conditions. As these cracks seal closed, through swelling of the clays, the surface becomes much less permeable. It is this influence of the existing moisture within a soil in setting its ability to infiltrate precipitation that makes the runoff from a landscape strongly dependent upon the "antecedent moisture" state of the landscape. The landscape has a memory. This memory varies strongly from one landscape to another, depending largely upon the clay content of the soil, and the rapidity with which the evapotranspiration and evaporation from the soil surface can withdraw water from the soil. In shale badlands, for instance, the memory of prior rain events only lasts about a day, while in coastal California a watershed may take a week to ten days of no rain to forget that it had ever been wet. This time constant is important in establishing the roles of sequences of storm events. It forces hydrologists and geomorphologists to characterize a climate not only

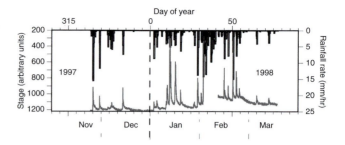

Figure 11.8 Hydrograph (bottom, left axis), and hyetograph (top, right axis) of Wilder Creek catchment during the winter of 1997–1998. This shows many typical features of the rainfall and hydrologic response in such systems. The rainfall is delivered in a few storms, all in the winter months. Early storms result in only small stream discharge, here depicted by the stage of this small creek (in units measured downward from a sonic look-down sensor). Later storms of the same intensity and duration, say in mid-January, result in much greater stage and hence discharge, demonstrating the role of antecedent moisture. The intense and enduring storms at the end of January 1998 knocked out the stream gage until it was repaired 2 weeks later.

with the distribution of storm sizes (durations and intensities), but with a range of time intervals between storms. Consider for example the hydrograph depicted in Figure 11.8. Storms early in the year were not very productive of runoff, while those later in the winter, and especially those late in a series of storms, were much more productive. The effective runoff coefficient (runoff/precipitation) rose from 7% early in the season to 70% later on.

Groundwater

The least visible portion of the water cycle involves water traveling through the subsurface. This transport and storage system is fed by infiltration from the surface (the inputs), and delivers water to streams and oceans (the outputs). As the speed of water through the pores of soil and rock is very small, and the volume of water involved is large, the time water spends in the subsurface can be quite long. The slow speeds serve to establish a long timescale filter to the inputs, allowing the outputs from the subsurface water system to be lagged significantly relative to the inputs. This lag is important in that the return of groundwater to the surface at the edges of our streams supports their baseflow long after the snowmelt or rainfall inputs have ended.

Geomorphologists are also interested in groundwater as a common trigger for landslides. The water table height, and the dynamic pressures associated with flow of water in the subsurface, can support a fraction of the weight of the overlying sediment or rock, reducing the effective normal stress, promoting its failure. Large landslides can occur well after water is delivered to the surface as either rain or snowmelt. The lag reflects the long timescale of water movement within the subsurface – the deeper the path, the longer the timescale.

Water from either rain or snowmelt must first move through the unsaturated zone near the surface before it reaches the saturated zone. The unsaturated zone is called the vadose zone, the saturated zone is the phreatic zone or groundwater, and the interface between the two is the groundwater table. Movement of water in the vadose zone is very complicated, as the efficiency with which water moves is highly dependent upon the water content of the soil – i.e., the degree to which the pores are filled. This is comparable to a heat problem in which the thermal conductivity depends upon the temperature. This was codified early in the 1900s first by Gardner (1919) and then by Richards (1931), in what is now known as the Richards equation (see review in Raats and van Genuchten, 2006). In this section we focus on flow within the saturated zone, fed by infiltration rates through the vadose zone that we will simply prescribe. We will see that this is complicated enough, and serves the present purpose of demonstrating the relevance of the conservation principle in setting up the problem.

As usual, we approach the problem by combining a statement for conservation with one for flux. Consider an element of the subsurface with a characteristic porosity and hydraulic permeability. As we are limiting the discussion to the groundwater system, which by definition is saturated, we may assume that all pores are filled with water. We seek a mathematical statement that captures this word statement:

> the rate of change of water in the element = the rate of inputs of water – the outputs of water + rate of local gain or loss of water

The last term reflects the possibility that water can be either captured or freed from its fluid form by mineral reactions. For example, in accretionary wedges at subduction zones, where water pressures are all-important in setting the material strength of the wedge, transformation of smectite to illite represents such a source

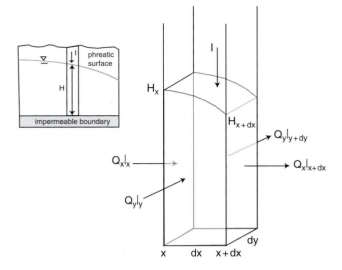

Figure 11.9 Schematic diagram for a water balance in a phreatic aquifer with source from above by infiltration, I, and lateral discharge of water through the sides of the element. Mismatches of inputs and outputs will drive change in the storage of water in the element, reflected in the height of the water table, H.

of water (e.g., Saffer and Bekins, 1998, 1999). We will ignore this complexity in the treatment below. So also, we will ignore the reality that the porosity and hence permeability of a rock or soil can be dynamic due to compaction and/or to cementation; we will assume that porosity and permeability are constant.

We seek an equation that describes the evolution of the groundwater table in both space and time. As sketched in Figure 11.9, we choose a control volume that is a vertical column extending down from the surface with horizontal dimensions dx and dy, within which the water table exists at a height h above some arbitrary datum.

We seek the simplest statements. The conservation statement then becomes

$$\frac{\partial (SHdxdy)}{\partial t} = q_x|_x - q_x|_{x+dx} + q_y|_y - q_y|_{y+dy} + Idxdy \quad (11.12)$$

where S is the storativity of the aquifer. Also called the effective porosity, this is the fraction of the sediment or rock that can be occupied by water that can freely come and go, the remainder being held firmly at mineral interfaces; it is smaller than the porosity. The left-hand side represents the rate of change of volume of mobile water within the groundwater in the control volume, where H is the top of the saturated zone.

Dividing both sides by the plan view area of the element, dxdy, and assuming that the storativity does not change in time, in the limit as dx and dy approach zero, we have

$$\frac{\partial H}{\partial t} = \frac{1}{S}\left[I - \frac{\partial q_x}{\partial x} - \frac{\partial q_y}{\partial y}\right] \tag{11.13}$$

This is a general evolution equation for the height of the groundwater table. It should look familiar, as it contains terms for a source, and for gradients in the fluxes. Without further analysis, we can already anticipate the behavior of the system in some simple situations. For example, consider the one-dimensional case, for groundwater flow only in the x-direction (all derivatives in y therefore vanish), when fed by a uniform input of water by infiltration from the vadose zone above. In other words, let I be uniform at I_o. Once the system has become steady (all time derivates vanish), the groundwater discharge in the x-direction must be

$$q_x = I_o x \tag{11.14}$$

Groundwater discharge should increase linearly with distance from the groundwater drainage divide. The reason for this behavior in the steady state directly mirrors other cases we have seen, for example the linear increase in either overland flow of water, or of regolith discharge on a hillslope. The discharge must increase downslope from the divide in order to transport the added increment of water or regolith.

To proceed further toward a prediction of the distribution of heights of the water table, we must assume some transport rule for the water. At this point in the development, we have not specified the physics of transport, but have simply exercised the principle of conservation. We need something analogous to Fourier's law to "close" the equation, to transform it from one in fluxes to one in water table heights. Here we appeal to Darcy's law, in which the transport in a given direction (here x) is given by

$$q_x = -K\frac{\partial h}{\partial x} \tag{11.15}$$

where h is the total head, and K is the hydraulic conductivity of the material. Let's inspect each of these components. The total head is defined as the head associated with elevation of the parcel of water (the elevation head) and that associated with the pressure (the pressure head). The hydraulic conductivity must depend not only upon the intrinsic

permeability of the material, k, but upon the viscosity of the fluid being pushed through it, μ:

$$K = \frac{k\rho g}{\mu} \tag{11.16}$$

In our problems this fluid will be water, and any variability in its viscosity will be due to the temperature dependence of viscosity. We discuss in Box 11.3 why intrinsic permeability is very strongly dependent upon the grain size of the material through which groundwater is passing.

In the case of a phreatic aquifer, as illustrated in Figure 11.9, the discharge Q through the side of the control volume is the product of this Darcy flux (q, with units of volume/area/time) with the height of the water table, H:

$$Q_x = -kH\frac{\partial h}{\partial x} \tag{11.17}$$

Here the combination KH is often called the transmissivity of the aquifer. Combining this with the equation for continuity of groundwater (Equation 11.13), we obtain

$$S\frac{\partial H}{\partial t} = \frac{\partial}{\partial x}\left(KH\frac{\partial h}{\partial x}\right) + \frac{\partial}{\partial y}\left(KH\frac{\partial h}{\partial y}\right) + I \tag{11.18}$$

This had better make sense in the simplest cases. For example, if the water table were flat and there were no flow, all terms on the right-hand side except the source term, I, disappear, and we are left with

$$\frac{\partial H}{\partial t} = \frac{I}{S} \tag{11.19}$$

The rate of water table rise depends linearly on the rate of infiltration and inversely as the storativity. For low storativity (effective porosity), the rate of water table rise is magnified, simply because there aren't as many pores into which to stick the water. This makes sense.

The Dupuit case

Now consider a steady case in which infiltration is spatially uniform, and flow is in one direction, call it x. If the lower boundary is in fact impermeable, and the length scale of the hillslope is much greater than the height of the water table, the flow in the majority of the aquifer will be roughly horizontal. Under these conditions, dubbed the Dupuit approximations after

Jules Dupuit (1804–1866, a French economist and civil engineer responsible for supervision of construction of the Paris sewer system), the equation above reduces to

$$0 = \frac{\mathrm{d}}{\mathrm{d}x}\left(KH\frac{\mathrm{d}h}{\mathrm{d}x}\right) + I_o \qquad (11.20)$$

This is at least now an ODE rather than a PDE. We have made progress. Further assuming that the hydraulic conductivity, K, is uniform, we may remove it from the derivative, yielding

$$\frac{\mathrm{d}}{\mathrm{d}x}\left(H\frac{\mathrm{d}h}{\mathrm{d}x}\right) = \frac{-I_o}{K} \qquad (11.21)$$

For the unconfined case represented here, the gradient in total head is equivalent to the gradient of the groundwater table height, meaning that we may replace $\mathrm{d}h/\mathrm{d}x$ with $\mathrm{d}H/\mathrm{d}x$. This looks messy, as it is nonlinear (the product of a variable and its derivative), but we may employ a trick. Recognizing that $\mathrm{d}(H^2)/\mathrm{d}x = 2H\mathrm{d}H/\mathrm{d}x$, this ordinary differential equation becomes

$$\frac{\mathrm{d}^2(H^2)}{\mathrm{d}x^2} = \frac{-2I_o}{K} \qquad (11.22)$$

which is a second-order ODE for H^2. We may now proceed to integrate both sides of the equation twice in order to solve for H^2, and then take the square root:

$$\frac{\mathrm{d}(H^2)}{\mathrm{d}x} = \frac{-2I_o}{K}x + c_1 \qquad (11.23)$$

where c_1 is a constant of integration. We will evaluate this after a second integration. This yields

$$H^2 = \frac{-I_o}{K}x^2 + c_1 x + c_2 \qquad (11.24)$$

We now impose boundary conditions, knowledge of the system at its boundaries. Here we will assert that the groundwater table intersects the top of the streams that bound the interfluve, here taken to be H_o at $x = 0$ and H_L at $x = L$. In general these could be at different elevations. The condition at $x = 0$ requires that $c_2 = H_o{}^2$. The condition at $x = L$ then results in

$$c_2 = \frac{(H_L{}^2 - H_o{}^2)}{L} + \frac{I_o}{K}L \qquad (11.25)$$

The final formula for the shape of the water table is therefore

$$H^2 = H_o{}^2 + \frac{I_o}{K}\left(Lx - x^2\right) + \left(\frac{x}{L}\right)\left(H_L{}^2 - H_o{}^2\right)$$

or $\qquad\qquad\qquad\qquad\qquad\qquad (11.26)$

$$H = \sqrt{H_o{}^2 + \frac{I_o}{K}\left(Lx - x^2\right) + \left(\frac{x}{L}\right)\left(H_L{}^2 - H_o{}^2\right)}$$

This is an equation for an ellipse, and is depicted in Figure 11.10 for several ratios of I_o/K. The height of the water table is scaled by the ratio I_o/K. The stronger the source term, I_o, the higher the arch of the water table, and the more efficient the flow of water within the groundwater system, scaled by K, the lower the water table arch.

We note that this form is perfectly analogous to the expected shape of the topography in the face of a source strength (weathering rate) and efficiency of regolith transport, as discussed in the hillslope section. In that system we expect parabolas; in this system we expect ellipses.

Obviously, pumping from a well into a water table will result in a transient lowering of the phreatic surface while the pump is on, and a rebound of the cone of depression after the pump is turned off. Transient cases of more geomorphic interest include the response of the groundwater table to oscillations in the stage of the adjacent river. We can anticipate that as the river stage rises, the water table may tip back toward the interfluve, while as it drops, the groundwater table gradient will increase toward the river. The former represents recharge of the groundwater table, the latter leakage from it.

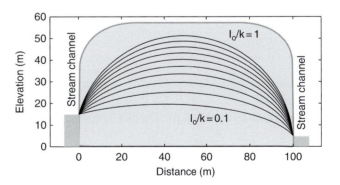

Figure 11.10 Calculated phreatic surfaces (groundwater tables) for ten ratios of I_o/K, in the subsurface of an interfluve between two bounding channels with differing stages, 15 m and 5 m. Each water table describes an ellipse. The greater the ratio of the infiltration rate to the hydraulic conductivity, the greater the arch in the groundwater table.

Box 11.3 Model of permeability as flow through small pipes

Models designed to provide a physical explanation for the constant of proportionality known as the hydraulic conductivity, K, assume that flow through a porous medium may be approximated as viscous, laminar flow through a set of very small diameter tubes or pipes. One may derive the flow speed distribution in a pipe of radius R, shown in Figure 11.11, to be

$$u = -\frac{1}{4\mu}\frac{dp}{dx}(R^2 - r^2) \tag{11.27}$$

where r is the radial distance from the centerline of the pipe, μ the viscosity of the fluid, and dp/dx the pressure gradient in the x-direction. The minus sign assures that the flow is driven down the pressure gradient. This may be integrated to yield the discharge of fluid in the pipe:

$$Q = -\frac{\pi R^4}{8\mu}\frac{dp}{dx} \tag{11.28}$$

The dependence on the radius of the pipe is very strong, and is sometimes called the fourth-power law for pipe discharge; doubling of pipe radius or diameter results in a 16-fold increase in discharge for the same pressure gradient. The mean speed through a single pipe is therefore Q/A, the cross-sectional area of the pipe ($A = \pi d^2/4$). This is

$$\bar{u} = -\frac{d^2}{32\mu}\frac{dp}{dx} \tag{11.29}$$

where now we have used the diameter, d, rather than the radius of the pipe. While this tells us how we ought to expect the flow to go in a porous medium if it consisted entirely of a set of pipes, it does not illuminate the permeability. This requires knowledge of how these pipes are distributed within the medium. We also need to account for the tortuosity of the real flowlines. Now consider a representative volume of a soil or rock through which groundwater is being forced. A first stab at this can be taken by assuming a simple cubic geometry in which the pipes of diameter d are spaced by a distance D, representing the grain size in a soil, or the block dimension in a rock mass. This is depicted in Figure 11.11. Note that in this simple geometry we could also define the porosity, the void space divided by the total volume. A single cube contains the equivalent of three tubes of length D. The porosity is then

$$\phi = \frac{3[(\pi d^2/4)D]}{D^3} = \frac{3\pi}{4}\frac{d^2}{D^2} \tag{11.30}$$

The flow through a single pipe accomplishes all of the flow contributed by a cross-sectional area D^2 of the porous medium. The mean velocity across the entire volume, consisting of pipes and non-conducting masses, is, without accounting for tortuosity,

$$\bar{u}_d = -\frac{d^2}{32\mu}\frac{\pi d^2}{D^2}\frac{dp}{dx} = -\frac{\pi d^4}{32\mu D^2}\frac{dp}{dx} \tag{11.31}$$

This is equivalent to the Darcy velocity, or the mean flow speed one would expect to measure through the porous medium.

Recalling that the equation for Darcy's law when considering horizontal flow can be written as

$$q = \bar{u}_d = -K\frac{dp}{dx} = -k\frac{\rho}{\mu}g\frac{dh}{dx} \tag{11.32}$$

we can see that the permeability of the medium is

$$k = \frac{\pi d^4}{32D^2} = C\phi d^2 \tag{11.33}$$

where C is a more general (dimensionless) constant reflecting the geometry of the connected pore space. This development allows us to see that the hydraulic conductivity, K, combines elements of the fluid (ρ/μ) and elements of the medium (ϕd^2) through which the fluid is passing. (After discussion in Turcotte and Schubert, 2002; Furbish, 1997; and Freeze and Cherry, 1979, following Hubbert, 1940.)

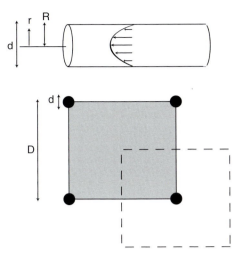

Figure 11.11 Top: definition sketch of flow in an individual pipe of radius *R*. Flow profile is parabolic between no-slip boundaries of the walls of the pipe. Bottom: schematic diagram of an idealized section of soil, characterized by flow tubes of effective diameter *d*, spaced by distance *D*. Flow through a cross section of dimension *D* (dashed) will be accomplished by one pipe of diameter *d*.

Groundwater rules of thumb

We conclude the discussion of groundwater by making some general observations that are relevant to geomorphic applications. The geomorphologist must be aware of the likely state of the geometry of the groundwater field beneath the topography. We have illustrated this for a very simple case in a one-dimensional landscape in order to develop some intuitive feel for how it will respond to rainfall and to soil properties.

The geometry of the groundwater mound, which in turn governs the distribution of groundwater egress from the landscape, is bound to the topography itself. In a broad fashion its shape should reflect some smoothed version of the topography. Importantly, the boundary conditions for the groundwater system are set by the location of the local river or sea. Therefore motion of these boundaries due to climate or erosion can influence the geometry of the groundwater system. We will see in discussion of amphitheater-headed alcoves that it is this geometrical effect that incites the feedback responsible for erosion of the alcoves in some situations.

In general, flow paths are nearly vertical near the surface, but turn at depth toward the bounding channel. One must know the groundwater flow field

in order to assess the evolution of water chemistry along the flowpaths, or the development of pore pressures within the subsurface that are in turn relevant to landsliding. It is the focusing of groundwater in crenulations in the topography that enhances the probability of landsliding in those locations. Conversely, the divergence of groundwater flow away from topographic noses lessens the likelihood of failure along ridges.

The architecture of porosity and permeability in the subsurface, the details of which we have so far ignored, in fact plays a crucial role in guiding flowpaths of subsurface water. In general, soils and regolith near the surface are more conductive than material at greater depths. For example, clay content in soils tends to increase with depth. This results in steering of the flow at conductivity contrasts, resulting in more horizontal flow than one would assume in a material of uniform conductivity. As an extreme example, consider a thin regolith over an impermeable rock (one that lacks even fracture permeability). Flow paths will be diverted to being parallel to the surface topography, flowing entirely in the regolith.

These elements of geological reality, reflected in geometrically complex and dynamic boundary conditions, and in the non-uniform architecture of the subsurface, must be acknowledged when pondering the impacts of subsurface flow on the geomorphic system. We now turn to some of the more obvious of these effects in considering the runoff of water from the landscape.

Runoff mechanisms

We have discussed how and where water resides in the subsurface of the landscape, how it finds its way into the subsurface through infiltration, and how it moves in the saturated zone, as groundwater. Here we address how water is transported to the channels, which then serve as efficient conduits to pass the flow toward the ocean, as discussed in Chapter 12. Our understanding of the mechanisms of runoff generation from a given landscape must be sufficient to explain at least the salient character of the hydrographs shown above, in particular the magnitude of the discharge peak and the time lag between rainfall

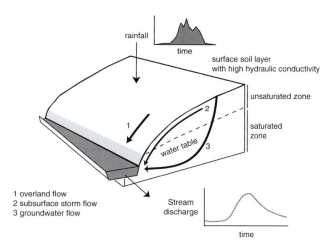

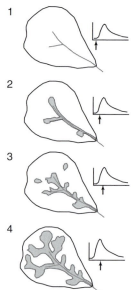

Figure 11.12 Mechanisms of delivery of precipitation to a stream channel from a hillslope (after Freeze, 1974, Figure 3, with permission from the American Geophysical Union).

Figure 11.13 Variable source area concept. The zone of saturation surrounding the channel grows through both a single event and through a season (shown at times 1–4). This leads to saturated overland flow (SOF) on this fraction of the catchment. As the zone of saturation is close to the channel, and SOF is rapid, the hydrographic response to a rainfall event is much faster, as shown in the hypothetical hydrographs to the right of each snapshot of the basin. The lag-to-peak from peak of rainfall (arrow) to peak discharge declines significantly as the catchment saturates (after Hewlett and Nutter, 1970, Figure 2, with permission from the American Society of Civil Engineers).

and stream discharge peaks. The chief elements of the runoff system are summarized in a classic paper by Alan Freeze (1974), the essence of which is reproduced in Figure 11.12. Rainfall intercepted by the landscape either runs off or infiltrates, which depends as we have seen upon the changing infiltration capacity of the soil. The process that most rapidly delivers water to the adjacent channel is overland flow, in which water in excess of the infiltration capacity runs over the land surface. It therefore encounters far less frictional resistance than water in the subsurface. The remainder of the rainfall infiltrates into the subsurface. Above the groundwater table, defined as the top of the saturated zone, water fails to saturate the pores of the soil; this is the vadose zone. The hydraulic conductivity of the vadose zone varies in both time and space. Water that infiltrates into any shallow conductive layer of soil can still travel relatively rapidly to the stream, in what we call subsurface storm flow. This corresponds to flow that may be perched on a less conductive subsurface layer. Flow rates through the groundwater table toward the stream are very slow, and dominate the baseflow of most hydrographs. Water that infiltrates through the vadose zone contributes to the groundwater table, raising its surface. If the groundwater table crops out at the base of a slope during a rainfall event, this portion of the landscape is saturated, and

will therefore generate overland flow. This "saturated overland flow" delivers water very rapidly to the channel, as it is both saturated and very close to the channel.

Dunne and Black (1970a and 1970b), working on forested, deglaciated hillslopes in Vermont, an environment that was decidedly not Hortonian, attempted to distinguish between the possible runoff mechanisms. They discovered the saturated overland flow mechanism (SOF), which allowed the rapid runoff of water over a portion of the landscape in which the slower saturated subsurface flow field outcropped. That the saturated portion of the landscape increases over the course of a storm has important ramifications for the understanding of temperature region hydrographs. It gave birth to the concept of the "variable source area" illustrated in Figure 11.13. Essentially, as a rainstorm progresses, the saturated portion of the landscape expands. As rainfall on this

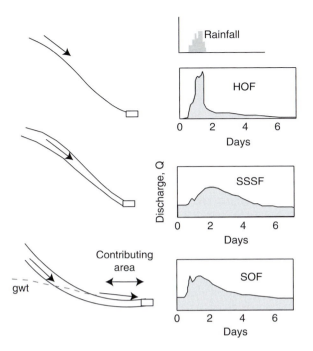

Figure 11.14 Schematic of hydrographic response to various runoff mechanisms. For a given rainfall event, runoff in a basin characterized by Horton overland flow (HOF) is quite flashy, showing small lag-to-peak, while those dominated by subsurface storm flow (SSSF) are more prolonged and less flashy. Saturated overland flow (SOF) can yield intermediate stormflow peaks as the rise of the groundwater table (gwt) results in an increasing fraction of the drainage area contributing overland flow to the channel.

saturated landscape has nowhere to go but across the land, it generates another class of overland flow. This saturated portion of the landscape might as well be a parking lot.

In Figure 11.14 we summarize the response of the landscape to rainfall in several situations, each controlled by a dominant runoff mechanism. The flashiest runoff occurs when overland flow dominates. Subsurface storm flow is the least flashy. And saturated overland flow can produce large peaks in runoff once the groundwater table has been inflated by prior rainfall or snowmelt infiltration.

Given the runoff mechanisms we have just listed, it is worth summarizing what controls each of the components of the water balance we have laid out. The rainfall intensity is largely if not solely determined by climate. The highest rainfall intensities in North America are found in the south-eastern United States. Of course, this portion of the continent is clothed in

forests, the soils of which are very conductive of rain. Runoff there is dominated by subsurface flow.

Infiltration capacity

We have discussed the process of infiltration above. Here we are interested in what governs the infiltration capacity, or the rate at which the soil will infiltrate water. We have already seen that this is highly dynamic, even within a storm. To summarize, the infiltration capacity of a soil is controlled by two of its characteristics – its texture and its structure – as well as the vegetation on the slope. The texture is governed largely by the grainsize of the soil mineral skeleton. What is most important to infiltration is not the total pore volume of a soil, but the sizes of the pores, and their interconnectedness. The wicking ability of a soil is allowed by capillary tension, which in turn is greatest for the smallest capillaries. The structure of a soil is dictated by the degree to which it aggregates into features that are larger than the mineral particles themselves – its "clumpiness." Aggregation in soils is promoted by organics, by cations that act as chemical bridges between clay particles, and by thin films of water. Aggregates promote the formation of large cracks in the ground surface, which in turn allow rapid egress of water into the subsurface. We have already discussed the water that does infiltrate into the ground.

Roles of vegetation

Vegetation plays a number of roles. First, as we have already addressed, it prevents direct bombardment of the soil by raindrops, and acts as little umbrellas. The vegetation itself, and its organic products upon decay, promote soil aggregation. Vegetation acts as a mechanical barrier to runoff on the one hand, and its disruption of the surface allows high conductivity conduits into the subsurface for rainwater.

Evapotranspiration

The rate at which water is returned to the atmosphere through the conduits of plants is determined by both climatic and vegetative properties. The direct evaporation is dictated by the temperature, the humidity of the atmosphere (read how hungry it is for water), and the wind speed (how rapidly air that has less water in it replaces air that has just taken on more water). In addition, the type of plant plays a role through its architecture (stomata size, etc.), and interestingly its color. The darker the leaf, the lower the albedo, and the more direct solar radiation it will take on. Finally, the rooting depth plays a role in determining the depth from which water can be wicked up to the surface through the plant water system.

Water storage in the soil

While water can be stored for long periods in the groundwater system, we are interested in the shorter-term storage afforded by the soil itself. This is what is important to the health and happiness of the local vegetation on the slope. There are several characteristic water contents that differentiate one soil from the next. Any soil has a maximum amount of water it can store – dictated by its pore space. At saturation, the pores are full. Another characteristic water content is the field capacity. This is the water content to which the soil would tend if allowed to drain freely under gravity alone. Field capacity varies widely, but largely reflects the clay content or the texture of the soil. The more clay in the soil, the larger the field capacity, i.e., the more water will be held against gravity by capillary forces. This is not a surprise. Unfortunately, nowhere near all of this water is actually available for consumption by local plants. The "available water content" of a soil is only some fraction of this field capacity. A plant can actually shrivel up and die even if the soil has water in it. This critical content is called the permanent wilting point of the soil, and can be quite high in clayey soils. Again, this reflects how strongly the mineral particles in clays actually hold onto the water. They don't let it drain by gravity, and they won't let plants suck it away either. It therefore turns out that there is an optimal soil texture (grainsize distribution) for growth of most plant types. This is a medium-textured soil, a loam (an even mix of sand, silt, and clay).

Overland flow generation

This is the simplest and the fastest mechanism for generating runoff. Runoff over the land surface is hindered only by friction against the bed, while that in the tortuous paths in the subsurface has to fight

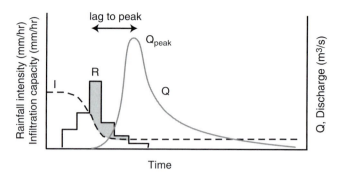

Figure 11.15 Stream response to rainfall input. Rainfall input varies through time to produce a given hyetograph. The infiltration capacity, I, declines through time as the landscape wets up, reaching a saturated hydraulic conductivity. Only the rainfall that exceeds the instantaneous infiltration capacity participates in the short-term hydrograph, or discharge record in the stream channel. The lag of the hydrograph from the hyetograph is called the lag to peak, and reflects both the drainage density of the landscape and the infiltration capacity, which in turn can depend highly on the antecedent moisture from past rainfall events.

friction and indeed water tension throughout its path. The flow speed differences are orders of magnitude. This immediately implies that anything done to a landscape that promotes the generation of overland flow will lead to larger and faster hydrograph peaks – what is called flashy response – for a given rainfall event. This is how precipitation makes its way off a pavement, or for that matter wherever there is a greater rate of rainfall, called the rainfall intensity, R, than the rate at which the surface can absorb the rain. The latter phenomenon is called infiltration, and the rate of infiltration is called the infiltration capacity, I. Overland flow is generated whenever $R > I$. The problem is that the infiltration capacity of a soil is very difficult to pin down. It is spatially variable because soil types vary spatially, and importantly it also changes in time. This immediately implies that one must know something of the history of a site in order to be able to quantify how it might respond to new rainfall. This plays the lead part in controlling the "role of antecedent moisture."

Consider, for instance, the hyetograph shown in Figure 11.15, and the plot of the history of infiltration capacity through the event. The integral of the excess rainfall will lead to overland flow. That the infiltration capacity decreases with time implies that the later portions of a rainstorm might be expected to contribute disproportionately to the generation of overland flow, while the earlier portion of the rainfall event will contribute more to the storage of water in the subsurface. Given that overland flow is by far the most rapid means of conveying water down hillslopes (in some environments), high rainfall intensity late in a storm can lead to the largest peaks in the hydrograph.

Overland flow of water and its geomorphic consequences

In arid regions, hillslopes commonly have smooth convex crests, below which they are broken into rills and gullies. These channels initiate at a distance down from the ridge crest that tends to be quite characteristic of a particular landscape. In badlands lacking vegetation, this distance can be as little as a few meters. In the late 1930s and early 1940s, a civil engineer named Robert Horton put forth a theory to explain the morphology of such arid region hillslopes (Horton, 1945). He paid particular attention to the lack of channels in a band within a fixed distance, X_c, from the ridge crests, which he called the belt of no erosion, and asked what sets this distance in a landscape. He based his theory on the notion that in these landscapes, where, when it does rain, it pours, overland flow should be the dominant runoff mechanism, and that the sparse vegetation allows this overland flow to transport sediment across the landscape. He argued his case very forcefully (in a 100-page paper in *GSA Bulletin*), so forcefully in fact that we now call this classic overland flow (when $R > I$) Horton overland flow (HOF). We follow his argument, in brief, below. He first develops an expression for the pattern of flow thickness that one might encounter in the midst of a long heavy rainstorm. This runoff he called sheetwash. He also knew that the main determinant of sediment entrainment was the shear stress at the base of the flow, τ_b, and that the sediment transport process was a thresholded one. This allowed him to assess the pattern of shear stress, and hence the pattern of sediment transport. We have already seen, in our discussions of sediment transport, that conservation of sediment, expressed as the erosion equation, implies that sites in which sediment transport rate is increasing in the downflow direction ought to be

sites of erosion, and vice versa. Recall that, in one dimension, this may be stated as

$$\frac{\partial z}{\partial t} = -\frac{1}{\rho_b} \frac{\partial Q}{\partial x} \tag{11.34}$$

where Q is the sediment transport rate in mass per unit contour length of slope per unit time. These arguments led Horton to propose a pattern of erosion and deposition on arid region hillslopes, and hence explain the forms of these hillslopes.

The movement of water across and within the landscape accomplishes most of the geomorphic work in molding the landscape. Now that we have set up the balance for regolith conservation, and have determined that it is the spatial pattern of regolith discharge that "counts" in thickening or thinning the regolith, one might imagine that the next step in addressing the evolution of landscapes would entail treating how sediment or regolith is transported by motion of water across it.

Under certain circumstances, water flows across the landscape as a sheet. In particular when rainfall rates are very much higher than infiltration rates, and the surface is not deeply crinkled into little gullies and rills, this process will occur. The sediment transport accomplished by overland flow is called sheetwash. Once again, we can formalize this problem by setting up a conservation of something in a control volume. Here the "something" is water atop the land surface.

Consider a simple case of a planar slope at an angle θ onto which rain is falling at a rate R (mm/hr, say) as set up in Figure 11.16. Rainwater is infiltrating at a rate I. The box is dx long along the slope, and again we ask you to imagine that the box in two dimensions has a length into the page of dy. In order to assess what the spatial pattern of sediment transport is, we must know the pattern of shear stress on the land, which in turn requires that we know the thickness of the overland flow, h, as a function of distance from the divide, x. The word picture that captures this sketch is:

> rate of change of volume of water in the box = rate of gain of water in the box − rate of loss of water from the box

Note that since the density of the material in transport does not evolve in any way, we can choose to

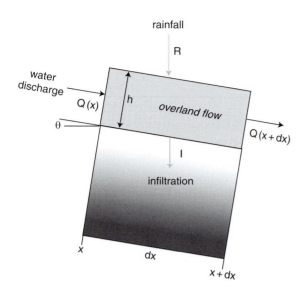

Figure 11.16 Water balance on a simple planar hillslope element given a rainfall rate R and an infiltration rate I. At steady state, the field of flow depths, $h(x)$, adjusts so that they are capable of passing downslope the excess rainfall $(R - I)$ in every element.

treat its volume, rather than its mass. The volume balance may be written

$$\frac{\partial(h dx dy)}{\partial t} = R dx dy - I dx dy + Q_x(x) dy$$
$$- Q_x(x + dx) dy \qquad (11.35)$$

where here the discharge is that of water on the slope. Dividing by the area of the box, $dx dy$, we obtain the evolution equation for the water thickness:

$$\frac{\partial h}{\partial t} = (R - I) - \frac{\partial Q_x}{\partial x} \qquad (11.36)$$

By now this should look pretty familiar. The rate of change is due to two processes: sources and sinks (here $R - I$), and the divergence of the flux or discharge of the medium. We see immediately that it is the excess rainfall $(R - I)$ that counts in setting the effective source of water on the hillslope. We now ask what must the discharge of water look like in steady state, after which we can ask what the pattern of thickness must be to achieve this.

At steady state, the left-hand side can be set to zero, leaving us with an ordinary differential equation for water discharge:

$$\frac{dQ_x}{dx} = (R - I) \qquad (11.37)$$

As the rainfall rate and the infiltration rate are assumed to be uniform in space and unchanging in time in the steady case, this may be simply integrated:

$$Q = (R - I)x \qquad (11.38)$$

As in the case of regolith on a hillslope, the steady discharge of water linearly increases with distance from the divide, at a rate dictated by the source strength. This is such a common finding in working these problems that it deserves a pause to think about its implications. The discharge *must* increase if continuity is to be obeyed. The system must figure out how to accomplish this. In general, the discharge (per unit contour length) is the product of the thickness of the material and its mean speed. The system then has two choices. Either the material (regolith, water) must thicken, or it must speed up, or some combination of the two.

In order to proceed, we need to relate the water discharge to the water thickness. In this simple case, the slope is uniform, and the only adjustable variable is flow thickness, h. The water discharge per unit contour width, Q, is equivalent to the product of the mean speed of the water and the flow depth: $Q = Uh$. We now need a formula for the mean speed. While there are several well-known formulas, we choose here the Darcy–Weisbach formula, in which

$$U = \frac{1}{f} \sqrt{ghS} \qquad (11.39)$$

where f is the Darcy–Weisbach friction factor and $S = \sin \theta$. The specific discharge of water is then

$$Q = Uh = \frac{1}{f}(gS)^{1/2}h^{3/2} \qquad (11.40)$$

Equating Equations 11.40 and 11.38, we can now solve for the dependence of the flow thickness on position, $h(x)$:

$$h = \left[\frac{f(R - I)}{(gS)^{1/2}}\right]^{2/3} x^{2/3} \qquad (11.41)$$

As shown in Figure 11.17 for several rainfall rates, the flow thickens with distance from the divide at a rate that is slightly less than linear, as can be seen by the 2/3 power of distance from the divide, x. The constant out in front sets the thickness scale. Again, it is worth inspecting the solution to see if we can make sense

(a)

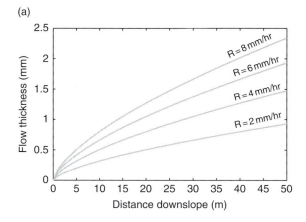

(b)

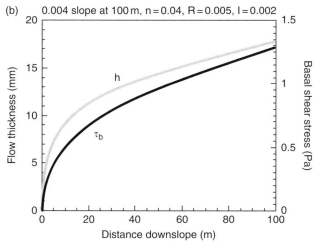

Figure 11.17 (a) Thickness of overland flow for four different rainfall intensities. Thickness increases as the two-thirds power of the distance downslope, and increases with rainfall intensity. All values of infiltration capacity, slope (uniform slope in all cases), and roughness factor are the same for all cases. (b) Expected steady-state flow thickness, h, and basal shear stress, τ_b, on a uniform slope of 0.004 for a prescribed steady rainfall rate of 5 mm/hr and infiltration rate of 2 mm/hr. Flow thickness and associated basal shear stress increase rapidly near the hillcrest, and more slowly thereafter.

distance downslope. For a uniform slope, S, the resulting pattern is reflected in the formula

$$\tau_b = \rho[gSf(R - I)x]^{2/3} \qquad (11.42)$$

This is also plotted in Figure 11.17. Although this equation is pretty ugly, it is well worth examining in some detail. Although it may be too specific, it has the essence of one runoff mechanism and its connection to the geomorphic consequences. Again, one can see the rapid initial increase in shear stress, and decline thereafter. Recalling that the sediment transport rate is thresholded, we can immediately locate the point on the hillslope above which there should be no sediment transport by sheetwash. This point Horton denoted X_c, delimiting a "belt of no erosion" near the hillcrest. Below this point, of course, there would be rapidly increasing sediment transport, as the transport rate nonlinearly increases with excess shear stress. We can see the dependence of the pattern on the excess rainfall, illustrating the role of both the rainfall intensity and the infiltration capacity.

Horton went on to argue that the sites of channel heads, the beginning of rills on the hillslopes, would have to be below this site of initiation of motion. He argued that the heads of rills in the landscape, at a distance denoted X_c from the hillcrests, were places where the sheetflow, sufficiently thick to incite sediment transport, converged, becoming yet thicker and yet more capable of transporting sediment. Conversely, on the intervening "noses," the divergent flow thinned, and lost its ability to transport sediment. He argued that this was a run-away process, leading to the incision of rills.

While this process of overland flow, and the sediment transport that it can accomplish, called sheetwash, is now thought to occur on only a small fraction of the Earth's surface, it is nonetheless instructive to have a solution for the end-member case against which to view all other processes. The analysis still pertains to parking lots, which cover a growing fraction of the Earth's surface.

There are many objections one might have to this simple story, and indeed it now has so many holes in it that it doesn't hold up very well as a viable theory. Only in completely unvegetated badland landscapes is this now viewed as the dominant mechanism. Horton's treatment is nonetheless a good starting point for discussion of the mechanisms operating in

of what is in the denominator and what is in the numerator. It indeed makes sense that the flow ought to be thicker at a given distance x if the excess rainfall is greater. If the slope is steeper, it also makes sense that the flow requires less thickness to accomplish the needed discharge, so that as S increases, h declines. The slope S should therefore be in the denominator, as it is.

Given this shape for the flow thickness, we can now think about the resulting sediment transport on the hillslope. With $h(x)$ and a prescribed $S(x)$, we can solve for the basal shear stress as a function of

other more common landscapes. Below we discuss a few of the objections, and observations that have forced significant modification of the story of arid region hillslope evolution.

The problem of drainage density

Horton focused our attention on the problem of what sets the drainage density in a landscape. In his terminology, the drainage density goes as $1/X_c$, the distance between the divide to the nearest channel head. Note that without any further discussion, we ought to be able to deduce the effect of the drainage density on the channel hydrograph. Since the slowest water paths in the landscape are on the hillslopes, their lengths will determine the timescales in the hydrological response of the basin to precipitation. The "finely divided" or "highly textured" landscapes of badlands (those having high drainage densities) ought to yield flashy discharge following closely on the heels of the storms, while sparsely channelized shallowly sloping vegetated landscapes ought to be sluggish in their response.

To recap, Horton claimed that the location of the channel head required at a minimum that there be sediment transport incited by overland flow, i.e., sheetwash. He then argued that once the flow was thick enough to cause sediment transport, that it would be unstable, and local areas would erode more rapidly than others, focusing the flow further there, and so on – a runaway process. This would be the eventual channel head, and would occur downhill from his X_c. While this notion of a cross-slope perturbation growing to form an incipient rill head has been formalized in a perturbation analysis published by Smith and Bretherton (1972), and yet again by Lowenhertz-Lawrence (1991, 1994), it has a few problems. First of all, as pointed out by Dunne, some interesting feedbacks kick in as the rill begins to form. The incipient nose between the proto-rills is now less protected by sheetflow, and begins to get pounded by raindrops. This diffuses the nose sideways, promoting the filling of the incipient rill. This competition will effectively delay the formation of rill heads until yet further downslope from Horton's predicted X_c. In addition, the simple Hortonian calculations ignored the role played by the sparse vegetation on such slopes. Any vegetation typically sits on small mounds from which the stalks protrude. It has been shown that these mounds are more conductive of water into the subsurface than is the unprotected inter-vegetative area, meaning that once the sheetflow is thick enough to access these conduits, it gets sucked into the subsurface. This slows the thickening of the overland flow sheet, and extends the distance to the channel head.

A third objection to the simple Hortonian model was raised again by Dunne (1991), working on very long Kenyan hillslopes. These are 1 km slopes, smoothly convex in shape, and sparsely vegetated. Dunne was concerned about the applicability of the Hortonian model in such landscapes for many reasons, not the least of which was that actual observations of Horton overland flow (HOF) were extremely rare. In the Hortonian model, the convex top of the slope is attributed to rainsplash – the only physical mechanism for moving sediment about in his "belt of no erosion." But how could rainsplash smooth something hundreds of meters in length, when individual rainsplash length scales were in millimeters? Could this be a reflection of rainsplash operating over millions of years? Working with a numerical model of rainfall and runoff, tied to a simple model of sediment transport by overland flow, Dunne explored the evolution of his numerical landscapes. He found that one could explain the broad convexity of the landscape with a stochastic set of rainfall events each of which generated sheetwash transport of sediment. The initial and final profiles are shown in Figure 11.18. Importantly, the rainfall events do not typically last long enough for the landscape to reach a steady hydrologic state. (This had been a simplification that allowed Horton to make considerable theoretical progress.) Instead, they effectively pick up sediment from the upper slopes, and begin to take it downslope, but are forced to put it back down again as the rain turns off, before they have accomplished their full task. This in effect allows sediment particles to take huge steps downhill in a single rain event, and summing over many such events yields an effectively diffusive process – like mega-raindrop impacts.

But the arguments about the location of channel heads are not limited to the arid landscapes. That the distance to the channel head is so important in setting not only the look but the hydrological character of a landscape has focused much attention on the problem of what sets the channel head location in the

(a)

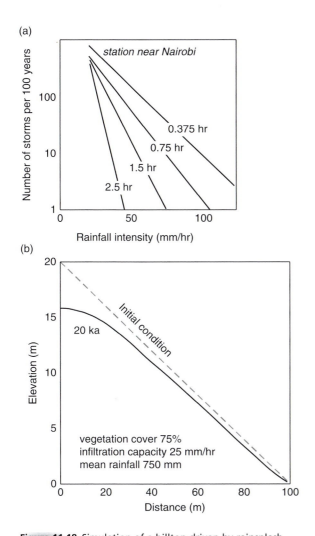

(b)

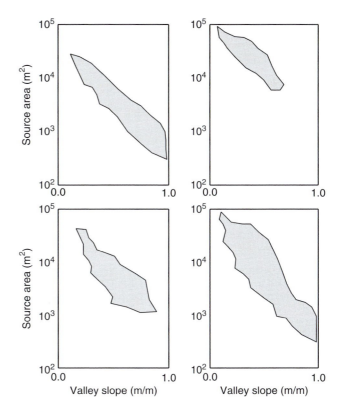

Figure 11.19 Inverse relationship between source area and valley slope at the site of the channel head, in several settings. This implies that the product of area and slope at the channel head is roughly a constant: steeper valleys will require less drainage area before a channel is initiated, and vice versa (after Montgomery and Dietrich, 1988, Figure 1, with permission from the Nature Publishing Group).

Figure 11.18 Simulation of a hilltop driven by rainsplash and overland flow. (a) Rainfall statistics near Nairobi, Kenya, and (b) simulated hilltop shown after 20 ka. Note the broad convexity of the hilltop resulting dominantly not from rainsplash transport, but from sheetwash from rainstorms that fail to reach steady state (after Dunne, 1991, Figure 6, with permission from the Japanese Geomorphological Union).

landscape. It has been shown repeatedly that there are other runoff mechanisms, and hence other means of setting the location of the channel head in landscapes that are fully clothed in vegetation. Working in several landscapes in California and Oregon, Montgomery and Dietrich (1988, 1992) documented that there is typically a well-defined relationship between the local slope and the contributing drainage area at the channel head: as the area increases, the slope necessary to generate a channel tip decreases, as is nicely shown in Figure 11.19. While this might in fact be predicted by the Hortonian mechanism as well,

these landscapes were subject instead to landslides and to the possibility of saturated overland flow at the channel head locations. In further work on this problem, these workers and their colleagues set out to define the landscape and climatic conditions that should cause channelization of the landscape. They found the role of convergence in the landscape to be extremely important, as indeed Horton admitted but did not formalize. Their argument goes as follows. Consider a water balance in the subsurface of an element of a hillslope shown in Figure 11.20, defined by the flow lines, the paths that water would take in the subsurface. The hillslope element has an area, A, called the contributing area because it is this area that contributes flow to a section of contour, c. Water infiltrates the ground through the top of this element, at a rate I_S, and can only escape through the plane beneath the contour, given that the edges of the

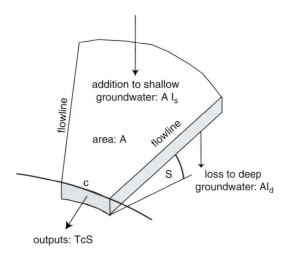

addition to shallow
groundwater: A I_s

flowline

area: A

flowline

c

S

loss to deep
groundwater: AI_d

outputs: TcS

Figure 11.20 Water balance on a hillslope element of area *A* and slope *S*, bounded by contours and flowlines (after O'Loughin, 1986, Figure 1, with permission of the American Geophysical Union).

element have been defined as flowlines. The water that leaks through the base of the element, at a rate I_d, and thereby contributes to the inflation of the deeper water table, may be included as a negative term in the balance. Application of Darcy's law, making the assumption that the groundwater flow across the downslope edge of the element is driven by a gradient set by the local surface slope of the land, dictates that the subsurface discharge from the cell is *TcS*, where *T* is the transmissivity of the soil, *S* is the local surface slope, and *c* is the contour length along the downslope edge of the cell. The balance then reads as

$$A(I_s - I_d) = TcS \tag{11.43}$$

If the water being contributed to the element is incapable of being passed through the base of the cell, it must exit over the surface as saturation overland flow. After slight rearrangement, the condition for saturation overland flow can be written as

$$\frac{A}{c} > \frac{TS}{(I_s - I_d)} \tag{11.44}$$

The left-hand side expresses the area contributing flow per unit contour length, and has been called the "geomorphic index of topography." It reflects the degree of convergence of flow that will be exerted by the geometry of the landscape. The right-hand side, on the other hand, collects variables that control the

ability of the landscape to pass water in the subsurface, *T* and *S*, and the climatic forcing of the system, which together with the infiltration capacity determines the water loading of the surface.

Importantly, several of these characteristics of the landscape, namely *A/c* and *S*, can be assessed at each and every point in the landscape if a good map is available. It is becoming more common now to use digital representations of the topography, assessed on a rectangular grid, called DEMs (for digital elevation models). These are usually generated from maps, although laser altimetry flown on a grid is becoming more common. Given the fields of *A/c* and *S*, and some first-hand knowledge of the hydrological character of the near-surface materials, *T* and *I*, one can assess the sites on the landscape where saturation overland flow ought to occur.

While we will not go into the details of the derivations, both the entrainment of surface materials by overland flow, and the generation of landslides by saturation of the subsurface materials, can be cast as added threshold terms in Equation 11.44. The sites of potential landslides, and of potential erosion by saturation overland flow can then be assessed within a given landscape. In a very well studied small basin in northern California known to produce many shallow landslides, it is not surprising that the locations of channel heads are best predicted using the landsliding criterion.

But the very fact that the channel head is localized by landsliding means that the channel head will not always be found at the same location. Landslides are rare events. Between events, the landscape is healing the landslide scar, and is doing so by hillslope processes that are not catastrophic – by creep, by bioturbation, and so on. A model of this interplay between hillslope and channel-forming landslides was proposed by Tom Dunne (1991), as illustrated in Figure 11.21. He prescribed a mean slope angle and a convergence angle for his hillslope element (30°). As you can see from the simulated history of channel head location reproduced in Figure 11.21, the channel head migrated back and forth over a 30 m distance as the channel first filled with colluvium and then flushed in a landslide. The time between such flushing events, which have the potential to become debris flows sluiced down the channel, is set by the recovery time of the scar, which is itself set by the rate of hillslope discharge into the hollow, and by the

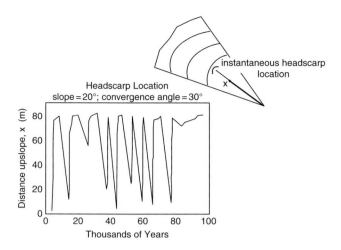

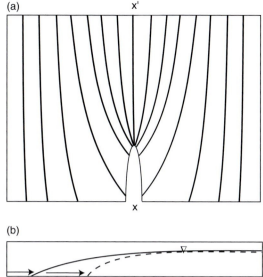

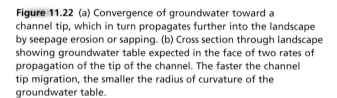

Figure 11.21 Evolution of a topographic hollow and the headscarp location that defines the tip of the channel system through time. Top: map of topographic hollow (zero order basin). Bottom: history of failure by landslide, followed by refilling by diffusive hillslope processes, in this case governed by a model regolith production rate of 0.39 mm/yr (redrawn from Dunne, 1991, Figure 10, with permission from the Japanese Geomorphological Union).

Figure 11.22 (a) Convergence of groundwater toward a channel tip, which in turn propagates further into the landscape by seepage erosion or sapping. (b) Cross section through landscape showing groundwater table expected in the face of two rates of propagation of the tip of the channel. The faster the channel tip migration, the smaller the radius of curvature of the groundwater table.

arrival of a storm with sufficient intensity and duration to trigger the slide.

Sapping and amphitheater-headed canyons

Another process proposed to form channel heads is seepage erosion. This is the entrainment of sediment from a seepage face localized by the exit of groundwater on the landscape. It is thought to result in amphitheater-headed channels, reflecting flow that converges toward the channel head. Such convergence is inevitable, as can be seen in Figure 11.22. The insertion of a channel into an otherwise uniform medium will steepen the gradients in the groundwater table that will in turn focus discharge (Dunne, 1990). The system can be seen in operation in a beach face during low tide: the groundwater table in the beach is inflated during high tide, and drains during low tide. As long as the sand is fine enough to be entrained, the face will rill into amphitheater-headed drainages (e.g., Higgins, 1982; Lobkovsky et al., 2007). At a larger scale, Abrams et al. (2009) show that the remarkable set of channels draining the old Apalachicola delta sands of the Florida panhandle, shown in Figure 11.23, can be explained by appeal to channel-head propagation by seepage erosion. Following

earlier suggestions that the channel head should propagate at a rate determined by the groundwater discharge to the channel head (e.g., Howard, 1988; Howard and McLane, 1988), they propose that the celerity of the channel head ought to go as $v = aA$, where A is the groundwater drainage area controlled by a particular channel tip. This of course will evolve (decline) as the channel tip propagates. But we also need another rule for how the channel buds off new tips. Abrams et al. (2009) propose a simple channel birth rule in which the nucleation of new tips goes as the drainage area as well: $dN/dt = bA$. These simple rules for channel growth and channel ramification, and in particular the constants a and b that control the rates, combine to govern the drainage density in such landscapes, $\Sigma L/A$, or the inverse of this, the "dissection scale" $A/\Sigma L$ (Montgomery and Dietrich, 1988).

But we should not be too cavalier in assigning process based simply on form. While it is true that we can find examples on Earth at both decimeter scales (beaches) and 1000 m scales (Apalachicola) in which it is clear that groundwater sapping rules, the export

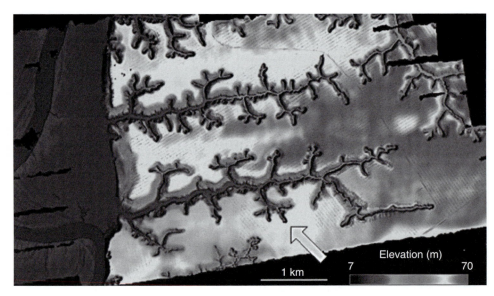

Figure 11.23 Detailed topographic DEM of the study site along the Apalachicola River, Florida panhandle. Data were collected by the National Center for Airborne Laser Mapping (NCALM) at roughly 1 m spacing. UTM is centered on 694500 easting and 337400 northing. The river runs immediately to the west of the site. Amphitheater-headed channel tips are evident throughout the channel network, which is etched into flat-lying deltaic sands (Abrams *et al.*, 2009, Figure 1, with permission from Nature Publishing Group).

of this notion to other terrestrial landscapes, and worse yet to Mars and other surfaces in the solar system, is troublesome. This is a strong theme in the work of Lamb *et al.* (2006) who discuss amphitheater-headed drainages in the layered rocks of the Colorado Plateau and the basalts of northwest United States before turning to Mars. They argue that open channel flow of water ultimately controls the rate of recession, and that the amphitheater head results from the hard-over-soft stratigraphy of the setting. The blunt-tipped drainage heads on Mars therefore should not be assigned blindly to groundwater processes.

Summary of channel head issues

We have elaborated greatly on Horton's pioneering studies of overland flow. But because overland flow is not often observed, fails to last long enough to be steady in many landscapes, and does not dominate as the runoff mechanism in vegetated landscapes, we must address alternative means by which channel tips emerge from the landscape. We see that convergence of subsurface flow is essential and can be quantified in a landscape using the metric of drainage area per contour length, A/c. In cases in which the channel tips are localized by landslides, their location at any time depends upon the time since the last landslide event. In some ideal situations on Earth, channel network extension occurs by groundwater sapping processes, which generate blunt-tipped

amphitheater heads. Similar morphology can be produced by channelized flows, however, suggesting that the geometry of a feature is insufficient to prove its origin.

Hydrology of a headwater catchment: the Coos Bay experiment

The interactions of hydrology, weathering and hillslope processes were the subject of a series of studies and experiments conducted in a small headwater catchment in the Oregon Coast Range (Anderson *et al.*, 1997a, 1997b; Montgomery *et al.*, 1997; Torres *et al.*, 1998; Anderson and Dietrich, 2001). The 860 m^2 unchanneled valley, known as CB-1 owing to its proximity to the town of Coos Bay, Oregon, is a steep, soil-mantled hollow that forms the contributing area for an ephemeral first-order channel. As we have seen, the channel head location is controlled by a combination of slope and drainage area; the small contributing area at the CB-1 channel head reflects the very steep (43–45°) slopes. The Oregon Coast Range is the exposed accretionary prism associated with subduction of the Juan de Fuca plate under North America. Moderate rock uplift rates (ca. 1 mm/yr) have brought greywacke sandstones originally deposited below sea level into subaerial exposure. The region has mild, wet winters, and relatively dry summers, with a mean annual rainfall of ~2 m. The combination of rock uplift, rain, and relatively weak rock combine to produce narrow ridges, ruffled with steep unchanneled valleys, and scored with ephemeral channels whose heads are close to the divide. The hillslopes are soil-mantled and forested.

How does water move through this landscape? The ~0.7 m-deep soils are derived from sandstone and forest duff; hence they are porous and highly permeable. Bulk densities less than $1000 \, kg/m^3$ and porosities of 60% go a long way toward explaining the high saturated hydraulic conductivities of these near-surface materials of $\sim 10^{-3} \, m/s$. The soil is well mixed by roots, tree-throw, burrowing animals, and rapid creep on the steep slopes. The high infiltration capacity of the soil ensures that rain enters the soil at the rate the storm delivers it, and then travels downward through the unsaturated zone until it encounters a zone of reduced permeability or the water table. Two key observations came out of a series of sprinkling experiments and intensive observation at the CB-1 site. First, flowpaths through the bedrock dominate runoff generation, rather than a simple perched water table at the soil/bedrock interface. Second, the dynamics of flow in the unsaturated zone is critical in determining the hydrologic response of the site to rainstorms. These hydrologic characteristics influence the chemistry of runoff from the site.

The occurrence and importance of flow through bedrock was a surprise finding in the first sprinkling experiment at CB-1. As shown in Figure 11.24, although a saturated layer did develop at the soil–bedrock interface at the site, it was patchy, and could not support a continuous flow route to the channel head. Injection of tracers directly into the saturated zone confirmed that a fast flow path linked disconnected saturated areas at the base of the soil. Nests of piezometers showed head gradients that in some areas drove water into bedrock, while in other areas head gradients pushed water out of the bedrock during long storms. The flow through bedrock appeared to be strongly associated with and controlled by fractures, which would localize points of inflow and outflow from the rock.

The dynamics of the unsaturated zone reflect the soil-water retention behavior of the sandy, high conductivity soils at CB-1. As summarized in Figure 11.25, soil-water retention (characteristic) curves

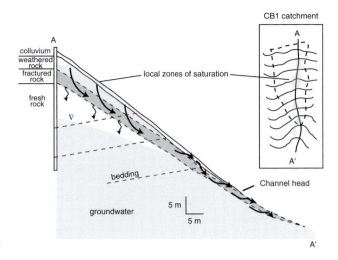

Figure 11.24 Longitudinal profile of the CB-1 catchment, an unchanneled valley in the Oregon Coast Range. Inset shows topographic map (5 m contours) of the catchment, outlined with dashed line from the channel head to the ridge crest. Layering of colluvial soil, weathered rock, fractured rock, and unweathered bedrock indicated at the ridge crest. Light shading shows the vertical extent of a perched water table during a long, steady sprinkling experiment, as well as the regional ground water table. Arrows indicate paths followed by water within the perched water table. Note that saturation in soil did not rise to the ground surface, and occurs in patches shown on inset. Flow paths cross into and out of bedrock at several locations within the catchment (after Anderson *et al.*, 1997b, Figure 14, with permission of the American Geophysical Union).

display two modes of behavior at different pressure heads. Over a large range of unsaturated pressure head, there is very little change in either water content or hydraulic conductivity (the slopes of the characteristic curves $dK/d\psi$ or $d\theta/d\psi$ are approximately zero). As the pressure head approaches zero, i.e., as the tension declines, which in the CB-1 soils (and elsewhere) occurs far from saturation, very small changes in pressure head result in very large changes in both soil moisture and hydraulic conductivity. Perhaps it is most useful to inspect the characteristic curves sideways, with pressure and conductivity being functions of water content, as in Figure 11.6. When viewed this way, we see that small additions of moisture above a threshold value have little impact on pressure head, but result in great increases in hydraulic conductivity. This nonlinear soil moisture retention behavior allows the soils when wetted to near-zero pressure heads to become highly sensitive to changes in water inputs. In this state, an increase in rainfall rate will

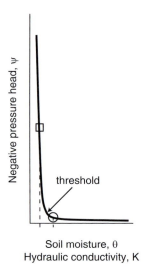

Figure 11.25 Schematic diagram of the bimodal behavior of soil water retention observed in the CB-1 soils. A threshold separates a region of low water content in which negative pressure (tension) depends strongly on water content, from a region of significant water content in which the water tension varies only weakly. A slight change in water content (e.g., from the box to the circle) greatly reduces the water tension holding water in the pores, allowing it to be released. Note that the water content at this threshold is quite low, far below saturation (after Torres, 2002, Figure 1, with permission from John Wiley & Sons).

trigger a steep rise in hydraulic conductivity, and release stored water. A pressure wave can move through the soil column much faster than water, so this hydraulic conductivity increase and water release response can rapidly mobilize water and result in discharge responses. At the CB-1 site, water table and saturated head dynamics were controlled by conditions within the unsaturated zone. Rapid transmittal of pressure head changes triggered by increases in rainfall rate may contribute to landslide initiation in sites like the CB-1 hollow.

The hydrologic characteristics of CB-1 play a role in the chemistry of runoff from the catchment. The runoff chemistry at CB-1 was remarkably steady over large variations in discharge: solute concentrations declined by 10–20% in response to increases in discharge over an order of magnitude. Solute concentrations in rivers typically vary less than discharge, but seeing this behavior in a small headwater catchment suggests either rapid reaction of water or slow transit of water through the catchment. It appears that the latter dominates, even in a setting with only ~40 m from divide to channel. A typical rainy day can displace no more than a few percent of the estimated water storage even in this tiny catchment. However, the impact of pressure wave release of stored water was noted in the first hours of sprinkling experiments. Some solutes increased in concentrations within a few hours of sprinklers being turned on, although the change in discharge was barely perceptible. The higher solute concentrations come from water being mobilized from being tightly held in pores by the first rain. The soil–bedrock flow paths through the catchment also influence water chemistry. Vadose zone soil water samples showed lower weathering-derived solute concentrations than the runoff at all times during sprinkling experiments. About half of the solute load in the channel head runoff derived from the bedrock, a consequence of most runoff flowing through bedrock along its transit through the small catchment. Weathering intensity is greater in the soil – equal solute flux is derived from a 0.7 m-thick layer of soil as from the estimated 3 m-thick hydrologically active rock layers.

Water runs through landscapes, and both are affected by the interaction. There is a chicken-and-egg aspect to water and the materials it moves through. In the short term, the water flow paths and rates are dictated by the characteristics of the soil, rocks, and topography of the landscape. Yet over long timescales, the landscape is shaped by erosion, landsliding, weathering, and other processes facilitated by water.

Summary

As in many geomorphic systems, the response of the landscape depends upon the meteorological events to which it is subjected, and upon the state of the landscape when those events occur. In hillslope hydrology, the state of the landscape can be characterized as its infiltration capacity, which depends both upon the materials comprising the surface, and upon its moisture content. Because the moisture content so strongly controls the hydraulic conductivity of the subsurface, the response will depend greatly upon how wet the land is when it rains.

Water can take several paths over and within the landscape. The fastest is overland flow, which requires that the rainfall rate exceed the rate of infiltration. Runoff dominated by this mechanism will produce flashy hydrographs. That water that does infiltrate can travel in the near subsurface in subsurface storm flow, or infiltrate toward the groundwater table. These are much slower paths, and produce runoff into the stream at times that can be long after the peak in the rainfall. Inflation of the groundwater table either due to rainfall or snowmelt will lead to its outcropping at the base of the hillslope, widening the saturated portion of the landscape. This can in turn generate rapid runoff in subsequent rainfall events, and serves as one of many examples of the feedbacks between these various hillslope hydrologic processes.

The landscape therefore acts as a filter, lagging and attenuating the water inputs to produce a hydrograph in the river. The peak in the stream hydrograph will in turn govern the ability of the river to transport sediment (Chapter 14), and if bedrock is exposed in places will dictate the rate at which the channel incises into the bedrock (Chapter 13). The paths the water takes, and the rate of water movement in the subsurface, impact the transport of material in solution – the rate of chemical denudation – from a basin. The residence time of water in the hillslope system depends greatly upon the runoff mechanism, being shortest for overland flow and greatest for groundwater. This impacts the ability of chemical reactions to alter the solute content of the water ultimately delivered to the channel. The chemistry of runoff will therefore vary with the nature of the materials, which in turn governs their hydraulic properties, but also with season and time within a storm event. The chemistry of the water reaching the channels can for the same reasons be employed to deduce the flow paths of the water in the subsurface.

While the drainage network in a landscape has been described as a fractal, this is violated at the tips of the drainage networks. The network ends at channel heads. There exists a threshold slope-area product along a flowpath in the landscape, above which the landscape channelizes, and below which it is stable and remains unchanneled. The processes that trip the system toward a channel include thickening of overland flow so that it exceeds the shear stress needed to entrain the bed material, and focused groundwater seepage that promotes saturation and hence failure as landslides. Both of these occur at sites in the landscape where flow of either surface or subsurface water converges. It is the hollows or crenulations in the landscape that contain the channel tips.

As most landscapes possess a Critical Zone characterized by a complex structure of soil overlying weathered rock overlying fresh rock (most generally with fractures), the movement of water within this subsurface architecture is complex. Importantly, it is these hydrologic paths that govern the rates and patterns of the chemical attack of the subsurface, which in turn dictate the evolution of the permeability structure of the subsurface.

Problems

1. Consider a thin flow of water on a smooth slope. The slope is $10°$, and the thickness of the flow is measured to be $4\,\text{mm}$. Calculate the mean speed of the flow and assess whether the flow will be turbulent or laminar.

2. Consider a soil dominated by fine sand. Find its porosity, field capacity, and permanent wilting point. Given these values, how much water will be available for plant growth per square meter of land surface, assuming that plants can utilize water as long as the moisture content is above the wilting point, and that the soil begins at saturation?

3. What water tension (negative pressure) must a plant be able to produce in order to utilize water in a sandy loam soil with a moisture content of 15%?

4. How fast did the front of the flash flood depicted in Figure 11.3 travel, if the distance between gages *a* and *d* is 580 m?

5. Plot the Dupuit solution for the groundwater field between two channels 200 m apart. Assume that the rainfall rate is 5 mm/hr, and that the hydraulic conductivity of the uniform substrate is 20 mm/hr. What is the maximum height of the groundwater mound between the channels?

6. Calculate the steady-state water discharge at the base of a hillslope. The hillslope is 150 m long, the rainfall rate is 7 mm/hr and the rain has been falling for long enough that the hydrology of the slope may be taken as steady, with a uniform steady infiltration rate of 1.5 mm/hr. Provide the answer both in m^3/s per m length of the bounding stream, and in cubic ft per second (cfs) per linear foot of channel.

7. In the above question, calculate the thickness of the steady-state overland flow at the base of the slope if the slope is uniform at 8° and the friction coefficient $f = 0.003$.

8. If the critical shear stress for entrainment of grains from the slope is 0.7 Pa, how far from the divide must one go to find entrainment in the case shown in Figure 11.17? This is Horton's X_C. Verify that the calculation of basal shear stress is done correctly by using the flow thickness at that location, the slope, and the acceleration due to gravity to calculate the basal shear stress.

9. Create your own fractal drainage network, using a generator function of your own design. Can you generate something that looks like the Apalachicola network shown in Figure 11.23? Why/why not?

10. *Thought question.* Describe qualitatively how the runoff from a landscape will differ under these circumstances: steady rainfall, increasing rainfall through time during the storm, and different starting moisture content.

11. *Thought question.* Discuss how paving a hillslope would alter the hydrology of the bounding stream.

12. *Thought question.* What runoff mechanisms might have dominated the Earth's surface before grasses evolved? How might their evolution have altered the runoff and sediment yields from landscapes?

13. *Thought question.* What happens in the aftermath of a fire on steep hillsides? Describe how this might affect the runoff generation mechanism, and the types of sediment yields from the source? See John McPhee's treatment of the San Gabriel Mountains debris flow sources in his essay in *Control of Nature* (McPhee, 1989).

Further reading

Dingman, S. L., 2008, *Physical Hydrology*, 2nd edition, Upper Saddle River, NJ: Prentice Hall/Waveland Press Inc., 656 pp.
A comprehensive text on physical hydrology, this is very readable and serves as an excellent textbook for an undergraduate class. That the author is well versed in cold regions hydrology is evident in the careful treatment of snow melt. The appendices are excellent.

Dunne, T. and L. B. Leopold, 1978, *Water in Environmental Planning*, San Francisco: W.H. Freeman Co., 818 pp.

This text set the bar for an environmental planning book, bringing the science of water in the landscape to the planner.

Freeze, R. A. and J. A. Cherry, 1979, *Groundwater*, Englewood Cliffs, NJ: Prentice-Hall, 604 pp.
Long considered the bible on groundwater, this remains a reference book that all those serious about groundwater should have on their shelves.

Hornberger, G. M., J. P. Raffensperger, P. L. Wiberg, and K. N. Eshleman, 1998, *Elements of Physical Hydrology*, Baltimore, MD: The Johns Hopkins University Press, 312 pp.
This modest text is a modern, very readable and well-illustrated introduction to hydrology. It comes as well with a CD of codes to work various hydrological problems.

CHAPTER 12
Rivers

What makes a river so restful to people is that it doesn't have any doubt – it is sure to get where it is going, and it doesn't want to go anywhere else.

Hal Boyle

In this chapter

In Chapter 11 we addressed how water flows across and within hillslopes. Hillslopes serve to convey both water and sediment into the river channels, whose task it is to make that ultimate delivery from the terrestrial environment to the ocean. The rivers therefore serve as the bottom boundaries to the hillslopes and, as vividly described by Leopold *et al.* (1964), as the "gutters" of continents. While the hillslopes cover the majority of the landscape, they are in important ways slaved to these thin threads of the rivers that bound them.

Here we first treat the physics of water transport in channels, and develop the fundamental equations for velocity of these turbulent flows. In an appendix to this chapter we provide for the interested reader a formal development of the Navier–Stokes equation that serves as a backdrop for these velocity equations. Several equations for the mean velocity in a river are in use, and we demonstrate how they are related. We then describe several means by which measurement of river flow speeds or the related water discharge, are made, ranging from current measurements to remote sensing methods.

Wabash River in flood, New Harmony, southern Illinois. Trees define the levee top, separating flow in the channel beyond from overbank flooding in the foreground (photograph by R. S. Anderson).

Rivers occupy self-formed channels. A channel is a topographic feature with identifiable banks created by running water. Our discussions in this chapter focus on channels formed by rivers in sediments deposited by the river. These are called alluvial channels. (Alluvium is sediment deposited by flowing streams, a word derived from the Latin – of course – *alluvius*, which means "washed against." Hence, alluvial is an adjective meaning of, related to, or derived from alluvium.) There are also bedrock channels, in which the channel is cut into solid rock, rather than loose sediment (see Chapter 13). River channels are self-formed in that they construct their channel form, their longitudinal profile, the height of the banks and the width of the floodplain. We have already described how the river network displays geometrical self-similarity, and how this feature is in common among systems designed to convey some spatially distributed property – here water. River discharge commonly increases monotonically downstream. But we must ask just how the channel adjusts in order to carry this increasing discharge – does it adjust its width, its depth, its speed? This falls under the realm of the hydraulic geometry of a channel. We also acknowledge that the river's discharge will vary in time, as it is made to carry water from precipitation and snowmelt that vary on seasonal and synoptic weather timescales. What hydraulic properties does the river change in order to accommodate the variations in discharge at a point? These timescales are too short for the entire channel geometry to change. The flow increases in stage (water depth) as discharge increases, until some point at which the water can no longer be conveyed within the channel. It then spills out of the banks, it floods, and spreads across the floodplain adjacent to the channel. The flood plain is therefore a part of the river.

We also describe the dominant plan view shapes of river channels – meandering and braided forms. We must rely upon knowledge both of the three-dimensional structure of the water flowfield, and of the mechanics of sediment transport that we discuss in detail in Chapter 14.

Another important element of the river is its longitudinal profile. In contrast to hillslopes, which generally increase in slope with distance from the divide, rivers decline in slope with distance down-valley, giving rise to the concave up profile. We discuss why rivers behave so differently, and what governs their profiles. This includes a discussion of baselevel, the elevation to which the river grades in the long term.

We close with a description of recent work on the largest river system in the world, the Amazon, using it as an example of how one can employ both first-order treatments that again go back to the principles of conservation, and marry these with modern methods of field investigation.

Theory and measurement of turbulent flows in open channels

The flow of water when it is confined within a channel belongs to a class of fluid mechanics problem termed open channel hydraulics. We first discuss the theory of open channel flow, from which we derive equations for surface and mean velocity profiles, and for discharge. This sets us up to discuss the means by which discharge is measured.

In this section we will develop one of the more famous equations in fluid mechanics, the law of the wall, which describes the theory of flow near boundaries. In our case, the most important "wall" or boundary of the flow is in fact its base, the channel bed. The law of the wall will lead the way toward understanding the "six-tenths" rule, a rule of thumb for the depth at which the flow velocity equals the mean velocity. We can then compare the formal approach with more empirically based approaches that have long been used in assessing mean flow velocities and discharge within open channels, including the Manning's, Chezy, and Darcy–Weisbach equations. Finally, we summarize methods used to measure discharge in a channel, including those based upon space-borne instruments.

In almost all geomorphically relevant situations, the flow of water over the land surface is turbulent. Experimentally, when the channel Reynolds number exceeds a critical value of about 2000, the flow in an open channel is turbulent. Recall that the Reynolds number is the ratio of inertial to viscous forces, defined for a channel as $Re = H\langle u \rangle / v$, where H is the channel depth, $\langle u \rangle$ is the mean velocity, and v is the kinematic viscosity. At high Re, inertia reigns over viscosity; momentum and all other scalar quantities are moved about within the flow by exchange of macroscopic packets of fluid, rather than by molecular collisions. We must therefore develop an equation that captures the physics of turbulent flows.

As always, our strategy for developing the velocity structure within a flow will be to start with the profile of shear stress, relate the stress to the strain rate through appeal to some rheological relationship, and then integrate the strain rate profile to yield the velocity profile, $U(z)$, where U is the velocity in the downstream (x) direction, and z is taken to be vertically upward from the bed.

Box 12.1 Steady, uniform, open channel flow assumptions

We will use the x-component of the Navier–Stokes equation (see Equation 12.69, developed in the appendix to this chapter) and solve it for the particular flow situation. We often begin by assuming that open channel flows are steady and uniform. More formally, here are the assumptions and the resulting simplifications to the Navier–Stokes equation:

- *The flow is steady.* This allows us to set all derivatives with respect to (w.r.t.) time to zero. The first term on the left-hand side of Equation 12.69 vanishes.
- *The flow is horizontally uniform.* This allows us to ignore terms with gradients in x and y. The second and third terms on the left-hand side of Equation 12.69 vanish, and the second derivatives w.r.t. x and y vanish on the right-hand side.
- *The flow is entirely driven by the downslope component of gravity*, $g_x = g \sin(\theta)$. The pressure gradient is assumed to be negligible, as the flow thickness is uniform in the channel. The second term on the right-hand side of Equation 12.69 vanishes.
- *There is no flow through the bottom of the channel, in either direction.* This requires that the vertical velocity of the flow, $w = 0$ at $z = 0$. From continuity in an incompressible medium, we know that $\partial u/\partial x + \partial v/\partial y + \partial w/\partial z = 0$. Horizontally uniform conditions require that the first two terms vanish, leaving $\partial w/\partial z = 0$. Integrating this yields $w = $ constant. The boundary condition of no flow through the boundary allows us to assess this constant to be zero. Since $w = 0$ everywhere on the bed ($z = 0$), the last term on the right-hand side of Equation 12.69 $= 0$, meaning that the entire left-hand side of the equation $= 0$. (Note, having dispensed with v and w, we will use U, rather than u, for velocity in the x direction for the rest of the chapter, as U is easier to distinguish from the u_* and v parameters we will encounter shortly.)

Given these assumptions, the x-component of the Navier–Stokes equation then becomes

$$0 = g_x + v \frac{\partial^2 u}{\partial z^2} \qquad (12.1)$$

For laminar flow in which $Re \ll 1$, the viscosity can be taken to be uniform, and we may simply integrate the equation to obtain the gradient of the velocity:

$$\frac{\partial u}{\partial z} = -\frac{\rho_f g \sin(\theta)}{\mu} z + c_1 \qquad (12.2)$$

where c_1 is a constant of integration, and the kinematic viscosity, v, is shown as the ratio of the dynamic viscosity and fluid density, μ/ρ_f. In open channel flow, we may safely assume that the velocity gradient vanishes at the top of the flow, as there is no overlying fluid exerting a shear stress on the fluid in the channel. Therefore, at $z = H$, $\partial u/\partial z = 0$ and

$$c_1 = \frac{\rho_f g \sin(\theta)}{\mu} H$$

so that Equation 12.2 becomes:

$$\frac{\partial U}{\partial z} = \frac{\rho_f g \sin(\theta)}{\mu} (H - z) \qquad (12.3)$$

Box 12.1 *(cont.)*

Integrating once more yields

$$u = \frac{\rho_f g \sin(\theta)}{\mu} \left(Hz - \frac{z^2}{2} \right) + c_2 \tag{12.4}$$

where c_2 is our second constant of integration. This we evaluate knowing that the flow speed goes to zero at the boundary (the no-slip boundary condition), in which case c_2 vanishes and we have for our final flow velocity profile:

$$u = \frac{\rho_f g \sin(\theta)}{\mu} \left(Hz - \frac{z^2}{2} \right) \tag{12.5}$$

This is the solution for a low Re, laminar flow velocity profile.

In simple flows, in particular those that are steady and uniform, the shear stress increases linearly with depth into the fluid, at a rate dictated by the density of the fluid:

$$\tau_{zx} = \rho_f g \sin(\theta)(H - z) \tag{12.6}$$

The next few steps show how one can derive this equation directly from application of Cauchy's law (developed in the appendix to this chapter). If the flow is steady, incompressible and uniform, we may neglect the left-hand side of Equation 12.63, and the second derivatives with respect to x and y on the right-hand side. The pressure gradient may be neglected relative to the gravitational force. This leaves simply

$$0 = \rho_f g_x + \frac{\partial \tau_{zx}}{\partial z} \tag{12.7}$$

which we can integrate once to obtain the stress profile:

$$\tau_{zx} = -\rho_f g_x z + c \tag{12.8}$$

The constant of integration, c, can be evaluated by appeal to the fact that at the top of the flow, at $z = H$, the shear stress vanishes. This means that

$$c = \rho_f g_x H \tag{12.9}$$

and

$$\tau_{zx} = \rho_f g_x (H - z) = \rho_f g \sin(\theta)(H - z) \tag{12.10}$$

as in Equation 12.6. Equations 12.6 or 12.10 show that the shear stress increases linearly from the surface (at $z = H$) to the bed (at $z = 0$) of the flow.

The stress reaches a maximum at the bed of

$$\tau_b = \rho_f g H \sin(\theta) \tag{12.11}$$

The subscript "b" in τ_b can be thought of as meaning bed, basal, or boundary.

It is the rheology, or, in the case of a fluid, the relationship between the strain rate and the stress, that makes this problem difficult. In the case of laminar flow (low Reynolds number, the strain is proportional to the shear stress, as first outlined in Chapter 8 (see Figure 8.10)). The proportionality constant is the inverse of the fluid viscosity, a temperature-dependent parameter you can look up in reference books. The strain rate of the fluid (equivalent to the vertical velocity profile) is then simply the product of the inverse of viscosity with the stress:

$$\dot{\varepsilon} = \frac{dU}{dz} = \frac{1}{\mu} \tau_{zx}$$

$$\frac{dU}{dz} = \frac{1}{\mu} \rho_f g (H - z) \sin(\theta) \tag{12.12}$$

In the case of turbulent flows, however, the rheology is intimately controlled by the flow itself, rather than by the material properties of the substance that is flowing. That is, the efficiency with which momentum is transported from place to place in the flow depends upon the flow structure, the vigor of the flow, and the size of the eddies doing the mixing. In a turbulent flow, the momentum is being moved about by macroscopic whorls within the flow, blobs of fluid trading places with each other, their momenta intact. The relative efficiency of momentum exchange under this mechanism vs. that associated with molecular collisions (the root of viscosity), is expressed by the Reynolds number, Re:

$$Re = \frac{\rho_f U H}{\mu} = \frac{U H}{\nu} \tag{12.13}$$

The second expression takes advantage of the definition of the kinematic viscosity of a fluid, $v = \mu/\rho$. The Reynolds number is another of those famous non-dimensional numbers we run across in fluid mechanics and geomorphology (others being the Mach number, Froude number, and Rouse number). They are named for famous mechanicians, in this case, Sir Osborne Reynolds (1843–1912), who experimented with turbulent flows in the mid-nineteenth century. In this instance the relevant length scale is the thickness of the flow, H, and the velocity is the mean vertically averaged velocity $\langle U \rangle$; this version of the Reynolds number is called the channel Reynolds number. (See our discussion of settling velocity in Chapter 10 for another application, in which we had to define a grain or particle Reynolds number.) Take for example a relatively benign river of depth 1 m, mean velocity 1 m/s. As the kinematic viscosity of water is 0.01 cm^2/s, or 1×10^{-6} m^2/s, the Reynolds number of this typical river flow is then 10^6. This is roughly three orders of magnitude above the empirically determined threshold for turbulence in an open channel (of 500–2000), meaning that we are quite safe in assuming that the flow will be turbulent.

It is clear from inspection of the measured instantaneous profiles in a flow that we haven't a prayer of writing an equation for the instantaneous flow field. Each profile varies significantly from the others, reflecting the passage of eddies past the instrument site. Calculation of two- and three-dimensional turbulent structures is a challenge even for the modern fluid mechanician, requires heavy computation, and even so depends entirely upon a choice of initial velocity structure. Instead, we will seek a theory for the *temporally* averaged flow velocity at a point in the flow. (Note the emphasis on temporal; we seek a spatial average of this temporally averaged flow later.) Symbolically, we may decompose the flow into a mean quantity and a time-varying quality, $U = \bar{U} + u'$, where the bar denotes the *time averaging* (note the distinction from $\langle U \rangle$, which is the depth-averaged velocity). We seek an expression for the profile of mean velocity with height above the bed: $\bar{U}(z)$. This decomposition of the velocity is what Reynolds accomplished in his 1895 paper (Reynolds, 1895).

We wish to know how efficient the flow is at transporting momentum, and what the spatial structure of this efficiency is. Said another way, we wish to know

the "effective viscosity" of the flow. By analogy with the viscous (low Re) case, this efficiency has been named the eddy viscosity, and the rheological equation (relating strain rate, dU/dz, to shear stress, τ) is cast as

$$\tau = \rho_f K \frac{d\bar{U}}{dz} \tag{12.14}$$

where K is the eddy viscosity, and has dimensions of L^2/T. It has been shown through experimental measurement of the strain rate as a function of the distance from a boundary, that relatively close to a boundary the eddy viscosity increases linearly with the distance from the boundary, or, in our case, with the height above the bed of the channel:

$$K = ku_*z \tag{12.15}$$

where z is the height above the bed, u_* is a characteristic velocity called the shear velocity $= \sqrt{\tau_b/\rho_f}$, and k is a non-dimensional scalar, named von Karman's constant after a famous German aeronautical engineer. This is plotted in Figure 12.1. Experiments show that $k = 0.40$. Note that this eddy viscosity is nowhere near uniform within the flow, close to the boundary. The most efficient part of the flow will be far from the boundary, reflecting the fact that here larger eddies can very efficiently exchange momentum.

We can now combine the expression for shear stress structure with the rheological rule in Equation 12.14 to derive the expected strain rate profile. There is one trick. We make use of the fact that the portion of the flow we are most interested in is very close to the bed. We approximate the shear stresses (τ) as being uniform within this region, equal to the basal shear stress, τ_b. This is shown graphically with a vertical line segment in Figure 12.1(a). The characteristic velocity we defined above, $u_* = \sqrt{\tau_b/\rho_f}$, can be rearranged to find $\tau_b = \rho_f u_*^2$. Given these assumptions, the shear strain rate becomes

$$\frac{d\bar{U}}{dz} = \frac{u_*}{kz} \tag{12.16}$$

The strain rate should diminish with height above the bed, not linearly as in the viscous case shown in Equation 12.12, but more rapidly – hyperbolically in the turbulent case. Not only is the stress diminishing with height into the flow, but the efficiency of the flow in exchanging momentum is increasing.

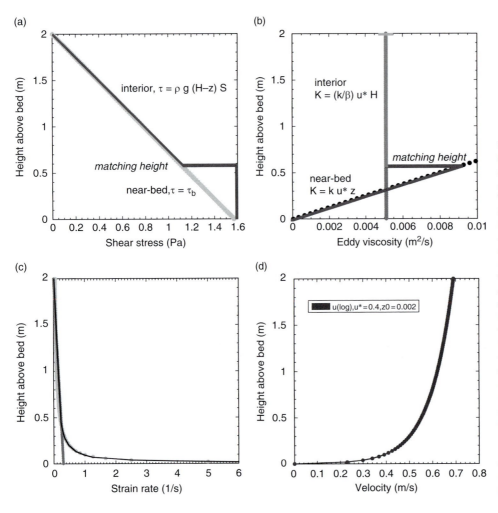

Figure 12.1 Distribution of (a) shear stress, (b) eddy viscosity, (c) strain rate, and (d) velocity in a turbulent flow of 2 m depth. The flow is broken into two regions: a near-bed boundary layer in which the shear stress is assumed to be approximately constant, and an interior region in which the shear stress declines to zero at the top. The corresponding eddy viscosity structure shown in (b) shows linearly increasing eddy viscosity with height above the bed in the boundary layer, and uniform eddy viscosity in the interior. The strain rate profile must be continuous, which determines a matching height where the strain rate from the boundary layer (black line in (c)) equals the strain rate from the interior flow (grey line in (c)). This strain rate profile is integrated to yield the expected velocity profile shown in (d).

In order to calculate the velocity profile, we must integrate this equation, yielding

$$\bar{U} = \frac{u_*}{k}\ln(z) + c \qquad (12.17)$$

where c is a constant of integration. We assess the constant of integration by appeal to a known or measurable value of the variable we are integrating, here the velocity, at some particular position within the flow. In turbulent flows, this is traditionally done by assigning the flow velocity to go to zero at a specific (but fictional) height above the bed, known as z_0. This is known as "z-nought," or "Nikuradse's roughness" (about which more later). The constant then becomes

$$c = -\frac{u_*}{k}\ln(z_0) \qquad (12.18)$$

and the final equation for the time-averaged flow velocity profile becomes

$$\bar{U} = \frac{u_*}{k}\ln\left(\frac{z}{z_0}\right) \qquad (12.19)$$

This is the famous "law of the wall," which holds true in the region of a turbulent flow that is quite close to a boundary, or wall. It is shown in Figure 12.2. The flow increases logarithmically with distance into the fluid, the greatest strain rate (or, equivalently, gradient in the mean velocity) being closest to the wall. We show an example of flow data from the Mississippi River in Figure 12.3, in which we have plotted the flow velocity as a function of the log of height above the bed. Graphically, the roughness height, z_0, is determined by the $U = 0$ intercept, while the other unknown, u_*, is reflected in the slope of the plot. Note the dependence of the profile shape on both the shear velocity, u_*, and on roughness height, z_0. The higher the shear velocity (the thicker the flow, the greater the channel slope), the higher the flow speeds at all heights; the higher the roughness, the lower the speeds at all heights.

(a)

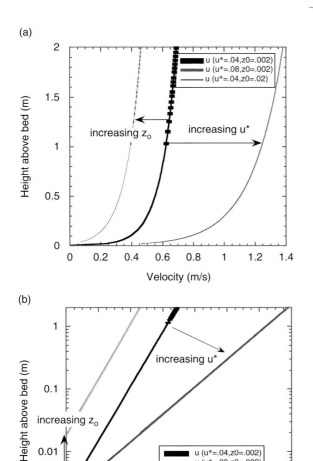

(b)

Figure 12.2 Hypothetical mean velocity profiles showing the effects of changes in shear velocity, u_*, and roughness length, z_o. (a) Linear–linear plot, (b) log–linear plot, on which straight lines represent logarithmic velocity profiles.

(a)

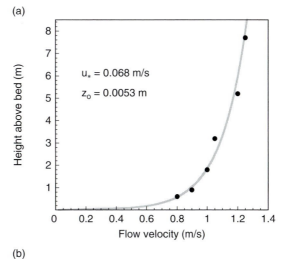

(b)

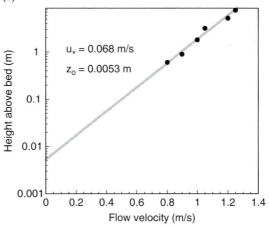

Figure 12.3 Measured flow velocity profiles plotted in both (a) linear–linear and (b) log–linear space. Data from the Mississippi River during the 8 m stage. Flow is well represented by a logarithmic profile with the shear velocity and roughness length shown.

A cautionary note: When plotted in this manner, the slope is k/u_*, so that high u_* flows give lower-looking slopes. Remember that these plots are oriented so that one can visualize the flow going left to right, and up on the plot is up in the real world. But mathematically, while the independent variable, z, is traditionally shown on the x-axis, it is shown instead on the vertical.

Why not set $U = 0$ at $z = 0$? Note that we have not used the common assumption that the flow speed goes to zero at the boundary, or the "no-slip" boundary condition that we have used in the case of viscous flow developed in Box 12.1. Mathematically, the natural log of zero is undefined, preventing us from setting $U = 0$ at $z = 0$. Physically, the turbulent fluid mechanics that we have assumed breaks down in close proximity to the boundary. In reality the flow becomes viscous as the boundary is approached, and the rheological rule we have used does not apply. But in most channels the region in which this breakdown occurs is very small, of the order of a fraction of a grain diameter.

What does the roughness height, z_o, mean? We would like to know what about the boundary sets the roughness height. We already know what sets u_*, but, unless we have an expression for z_o, we cannot predict the velocity structure. Insight into this came in the form of experiments done in the 1930s by Nikuradse, who glued sandpaper to the insides of pipes, and noted the dependence of the flow velocity (actually its integral, the pipe discharge) on the diameter of the sand grains, D. He found simply that $z_o = D/30$. Experiments in naturally occurring sedimentary beds showed a similarly simple relationship:

$$z_o = D_{84}/10 \qquad (12.20)$$

where D_{84} is the diameter of the grain in the 84th percentile of the grain size distribution (in other words, 84% of the grains are smaller) (see Whiting and Dietrich, 1990). The bigger the grains in the bed, the higher the roughness, and hence the slower the flow speeds at all heights above the bed.

The vertically averaged mean velocity

The law of the wall provides us an equation for the *temporally* averaged velocity of the fluid as a function of height above the bed. In many applications we would like an expression for the spatially (vertically) averaged velocity, $<U>$. We resort to the mean value theorem for this, evaluating the mean over the interval z_o to H:

$$\langle U \rangle = \frac{1}{H - z_o} \int_{z_o}^{H} \overline{U}(z) \, dz \qquad (12.21)$$

Using the law of the wall for $U(z)$, noting that the integral of $\ln(z) = z \ln(z) - z$, and making the simplifying assumption that $H \gg z_o$, yields

$$\langle U \rangle = \frac{u_*}{k} \left(\ln\left[\frac{H}{z_o}\right] - 1 \right) \qquad (12.22)$$

This average is shown in Figure 12.4. It occurs well below the mid-point in the flow. We now ask at what level within the flow one should measure this velocity. This would therefore be the appropriate level within the flow at which to measure the velocity, as it would immediately yield the mean velocity, simplifying the calculation of discharge (see below). We simply set the expression for the spatial mean velocity equal to the law of the wall at a specific height, call it z_m, and solve for z_m. The u_*/k cancels out, and we are left with

$$\ln(H/z_o) - \ln(z_m/z_o) = 1$$
or $\qquad (12.23)$
$$\ln(H/z_m) = 1$$

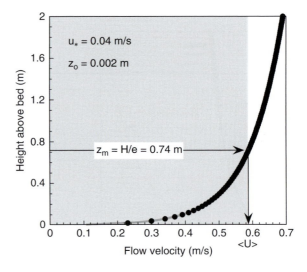

Figure 12.4 The six-tenths rule. The mean speed in a flow described by a logarithmic profile obeying the law of the wall may be found at a height of H/e, or roughly 37% of the flow height. Measured down from the top, this corresponds to roughly 63%, or, rounding, six-tenths of the distance to the bed.

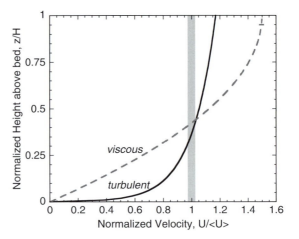

Figure 12.5 Comparison of laminar flow profile with turbulent flow profile, normalized such that they have equivalent mean velocities. The laminar profile is parabolic, while the turbulent profile is logarithmic, and shows much greater curvature near the bed. In both flows the mean speed is found approximately four-tenths of the way from the bed to the top of the flow, or six-tenths of the way from the flow top to the bed.

One more step yields

$$z_m = H/e = 0.37H \qquad (12.24)$$

This is the origin of the "six-tenths rule": the mean velocity will be found at roughly four-tenths (actually 37/100) of the way up from the bed to the surface, or six-tenths of the way down from the surface toward the bed. In Figure 12.5 we show the flow velocity profiles

Box 12.2 A caveat on the mean flow calculation

We have made several assumptions in deriving the flow velocity profile described above. One of these is that the law of the wall holds from the bed to the surface of the flow. While this was convenient for our integral embedded in the derivation of the six-tenths rule, and is not far off, the alert reader will be wondering how we can apply a rheological rule for turbulence that is designed to work "very near a boundary or wall" all the way through the flow, well into the "interior" of the flow. In fact, two assumptions made in the course of deriving the law of the wall break down. First, the approximation that the shear stress is equal to that at the boundary clearly breaks down in the interior. Second, a more appropriate rule for the momentum exchange in the interior of the flow is one in which the eddy viscosity becomes a constant, scaled not by the distance from the bed, z, but by the full flow depth, H.

Empirically, in the interior of the flow, the eddy viscosity is well characterized by $K = (k/\beta)u_*H$, where k is still von Karman's constant ($= 0.4$) and β is an empirically determined constant of about 6.27. The effective viscosity (eddy viscosity) structure is therefore more like that shown in Figure 12.1. The near-bed portion displays the linearly increasing eddy viscosity, while that in the interior becomes constant. The height at which one switches from one to the other description of the viscosity structure is set by requiring that the strain rate remain "continuous" through this region. The resulting full flow profile, plotted in Figure 12.1, becomes a patched one with the log profile of the law of the wall holding near-bed, and the interior flow being characterized by a parabolic profile (in exact analogy to the viscous flow profile derived earlier, in which viscosity is uniform through the flow). But note that the extension of the law of the wall to the surface, while inappropriate on physical grounds, still yields a fairly good approximation of the flow profile, and hence both the mean velocity and discharge based upon it are not greatly altered.

for both laminar and turbulent cases, normalized by the mean velocity. The mean velocity would be found somewhat above $0.4H$ for laminar flow, while it is somewhat below it in turbulent flows.

Other equations for the mean velocity

When attempting to estimate the discharge of a river, it is useful to have a simpler expression for the average flow velocity. These expressions approximate the mean velocity within a channel cross section, which differs from the mean velocity within a vertical profile or the time-averaged velocity at a point.

Manning's equation

One of the more commonly used equations for average velocity in a channel is Manning's equation:

$$U_a = a\frac{R_h^{2/3}S^{1/2}}{n} \tag{12.25}$$

where R_h is the hydraulic radius, S is the slope of the channel, and n is a constant called Manning's roughness coefficient, or Manning's n. The factor a is a

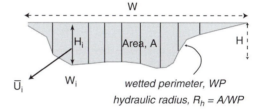

Figure 12.6 Sketch of measurement method endorsed by USGS. The cross section of the channel is broken into many segments, designed such that no more than 10% of the river discharge occurs in any segment. The discharge, Q, is the sum of the discharge in each segment, determined by the product of its depth, H_i, its width, W_i, and its mean speed, $<U_i>$. Mean flow equations described in the text utilize the wetted perimeter, and the hydraulic radius of the cross section.

dimensionless constant and takes the value of 1.0 when in SI units, and 1.48 in the English (foot–pound–second) unit system. The hydraulic radius is the cross-sectional area of the flow, A, divided by the wetted perimeter, P, as shown in Figure 12.6. For a roughly rectangular cross section the hydraulic radius is:

$$R_h = \frac{A}{P} = \frac{WH}{W + 2H} \tag{12.26}$$

where W is the channel width and H is the channel depth. For $W \gg H$ (a common geometry) the hydraulic radius is approximately H, the flow depth. The empirically determined Manning's n has been tabulated for a wide range of channel bed and bank characteristics (Barnes, 1967), with values ranging from 0.01 in smooth-walled laboratory flumes to above 0.1 in steep mountain channels. Typical values for natural channels are 0.02–0.08.

The Chezy formula

By analogy to work on flow through pipes, a formula for the mean flow velocity was first proposed by Antoine Chezy in the late 1700s, as he worked on canals in Paris:

$$U_a = C(R_h S)^{1/2} \qquad (12.27)$$

where C has become known as the Chezy coefficient, and has units of $m^{1/2}/s$.

Darcy–Weisbach equation

Yet another formula for the cross sectionally averaged flow velocity is the Darcy–Weisbach equation:

$$U_a = \frac{1}{f}(gHS)^{1/2} \qquad (12.28)$$

where f is called the Darcy–Weisbach friction factor. Note that at least this formula is dimensionally simple. As the combination $(gHS)^{1/2}$ has units of velocity, the friction factor is dimensionless. As the friction factor increases, the flow speeds at all heights should decline.

The advantage of all three channel-averaged velocity formulas is that all one needs in order to estimate the flow discharge is the cross-sectional area of the flow and this mean velocity. In other words,

$$Q = AU_a \qquad (12.29)$$

If one has the relationship between the stage of the flow, and hence its depth H, and the cross-sectional area of the flow, derived from survey information, and calibration to constrain the friction factor or Manning's roughness coefficient, then one can generate an expected rating curve between the easily measured and monitored stage and the more relevant but difficult to measure water discharge.

It is useful to know the relationships between these equations, so that one can move freely between them. Dimensional analysis suggests that we should expect that the mean velocity ought to go as the combination $(gHS)^{1/2}$. One can see that the Darcy–Weisbach formulation starts from this, and that the Chezy formulation comes close to it but neglects to include gravitational acceleration. While Chezy is fine on Earth (and most Parisian canals), this formula could not be exported to conditions on Mars, or on Titan, for example. Manning's formula comes close as well, but is both missing explicit attention to gravitational acceleration, and suggests an extra dependence on H. For very wide flows ($R_h \sim H$), we can rewrite Manning's formula to yield

$$U_a = a\frac{(HS)^{1/2}}{n}H^{1/6} \qquad (12.30)$$

to see that the added dependence on flow depth goes as $H^{1/6}$. What about the law of the wall? Recalling that the definition of $u_* = (gHS)^{1/2}$, this formula as well starts from the correct collection of variables. Like Manning's formula, however, the remaining portion of the formula implies additional dependence on H. We can reconcile the law of the wall and Manning's equation by noting that the logarithmic dependence on H/z_o only slowly increases with H. In fact, the dependence looks very much like $H^{1/6}$.

We can now compare the treatments of the roughness or friction coefficients in these formulations, meant to capture the resistance to flow imparted by the bed: z_o, n, and f. For the simple case of a very wide uniform flow, the vertically averaged flow derived from averaging the law of the wall ought to equal a cross-sectionally averaged formula, say the Darcy–Weisbach equation. Setting these expressions equal, and noting again that $(gHS)^{1/2} = u_*$, yields

$$\frac{1}{f} = \frac{1}{k}\left[\ln\left(\frac{H}{z_o}\right) - 1\right] \qquad (12.31)$$

This result suggests that we should not expect the friction factor f to be a constant, but instead that it ought to vary in a specific manner as the flow becomes deeper. In particular, f should decline as the flow thickens. What is important is the ratio of the flow thickness to the roughness height characterized by the size of the grains composing the bed.

Measurement of channel velocity and discharge

Discharge is difficult to measure. It is typically a painstaking and sometimes a dangerous (or thrilling) task. For this reason, we commonly use as a proxy for discharge the stage, or level of the river surface. This can be monitored in a variety of ways, including using a float, or an acoustic sensor. But the easily monitored time series of stage must be transformed into one of discharge through a rating curve. It is the construction of this rating curve that requires we get our feet wet and measure the discharges at a series of known stages.

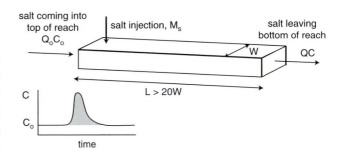

Figure 12.7 Essence of the salt discharge method. A slug of salt of known mass is added to the top of the reach, while the time series of salt concentration, $C(t)$ is measured at the base of the reach chosen such that no discharge is added over the reach, and the distance from the salt injection site is far enough downstream (roughly greater than 20 times the stream width) to ensure that the added salt solution is well mixed.

USGS stream gaging protocol

The formal USGS method of determining stream discharge (Buchanan and Somers, 1969; Wahl *et al.*, 1995), illustrated in Figure 12.6, involves measurement of flow depth and flow speed at each of many "stations" across the flow. More than 20 stations are recommended, such that in no one section is there more than 5% of the water discharge. Measurements are made from a bridge across the flow, or from a cable-car drooped across the river, or on foot by wading into the flow. A station is occupied, and the flow depth is measured (by dropping a large weight to the bed, if done from a bridge). The proper height above the bed ($0.4H$ if a single measurement is being collected) is calculated, and the current meter is placed at this height. The current is usually measured for at least 40 seconds, in order to average out the turbulent eddies that waft by. The product of this mean velocity with the spacing between stations, W_i, and the flow depth there, H_i, yields the discharge of water associated with that portion of the river. The sum of these discharges is the total discharge of the river:

$$Q = \sum_{i=1,n} W_i H_i \overline{U}_i \qquad (12.32)$$

Measurement of velocity

We measure velocity using a variety of instruments. Beyond the crudest methods, those practiced by Pooh and Piglet, tossing sticks from the bridge over the river and counting how much time it takes to reach the other side of the bridge, we use current meters that measure local flow using various technologies. The choice will depend upon the detail with which the investigator needs to know the flow field (for example, is it necessary to know the instantaneous speeds, or is the mean speed at a location sufficient?), and the monetary and logistical resources available. These methods include propellor current metering, laser velocimetry, and acoustic Doppler velocimetry. The propellor turns at a rate proportional to the flow speed, and one either counts clicks while listening on a headphone, or records the counts electronically. In the latter two methods, the speed is measured by motion of small particles or bubbles embedded in the flow.

Salt dilution method

If the flow is too gnarly, too complex geometrically, or too dangerous to measure using current meters, one can resort to a chemical tracer method. While in principal this method will work on any flow, its use is limited to relatively small discharges. And while we will describe in detail one such method, this is only one of several methods that have been employed in such situations.

Consider a reach of river chosen so that there are no tributary inputs along the reach. The discharge of water out the base of the reach, Q_1, therefore ought to equal that into the top of the reach, Q_o. As sketched in Figure 12.7, the technique entails injecting into the river at the top of the reach a known mass of salt (or any other detectable chemical tracer that does not react with the water or sediment), dissolved in water, and documenting the time series of salt concentration at the bottom of the reach. In order to understand how this could constrain the discharge of water, we will perform a balance of salt (usually NaCl) in this reach. The inputs of salt must equal the outputs of salt, provided that we measure the output for long enough to ensure that all the salt we injected has had time to make its way to the exit. The balance may be written

$$Q_o C_o T + M_s = \int_0^T Q_1 C_1 \, dt \tag{12.33}$$

where T is the time over which we have made measurements, M_s is the mass of added salt, and C is the concentration of NaCl in the water (for example, in kg/m^3). Given that we have properly chosen the reach such that $Q_o = Q_1$, we can simplify this equation to yield

$$Q_o = \frac{M_s}{\int_0^T (C_1 - C_o) \, dt} \tag{12.34}$$

where C_1 is the measured concentration of NaCl in the water at the measurement site, and C_o is its concentration in the incoming water at the top of the reach. We know how much salt we threw into the flow. Therefore all we must measure through time is the time series of salt concentration, from which we subtract a background. Commonly one measures not the salt concentration but the conductivity of the water, which is related to NaCl concentration using a calibration factor. An example of such a measurement is shown in Figure 12.8. The period of integration must extend until the conductivity has returned to the background. The background conductivity in this well-behaved case is steady, allowing simple subtraction, and integration. The final calculation of the water discharge incorporating the calibration of the conductivity probe is

$$Q_o = \frac{M_s}{\alpha \int_0^T (c_1 - c_o) \, dt} \tag{12.35}$$

where $\alpha = 2140$ (kg/m^3)/(μS/cm), and the conductivity, c, is measured in micro-Siemens per cm, μS/cm.

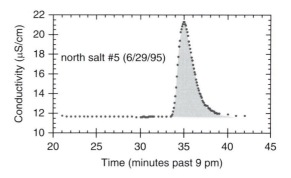

Figure 12.8 Time series of conductivity measured in a small Alaskan glacial stream, from which the water discharge can be calculated. A steady background conductivity was achieved, and measurements were taken for long enough after salt injection to assure that the entire pulse of salt had passed the measurement site. Discharge is calculated from the integral of the conductivity record that rises above the background (shaded).

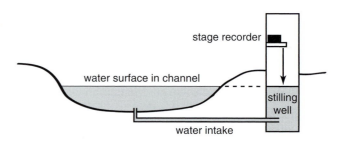

Figure 12.9 USGS stage gaging station. Housing of stage recorder serves as a stable elevation against which water surface level (stage) is measured. Distance from recorder to water surface in the stilling well is measured by a float, a pressure sensor or a sonic sensor.

Measurement of stage

While discharge itself is difficult to measure by one or another of the above methods, stage can easily be monitored continuously. Again a variety of methods have been employed. A staff gage on the bank, or painted on a bridge pier allows easy spot visual inspection. A continuous record of stage is obtained using either a float or a pressure gage in a stilling well, or a look-down sensor such as the acoustic (also called sonic) device. The design of a typical USGS gaging station is shown in Figure 12.9, in which stage is measured continuously with either a float, a pressure gage, or a sonic transducer. While the data streams from these devices were in the past collected on mechanically driven drum recorders, they can now

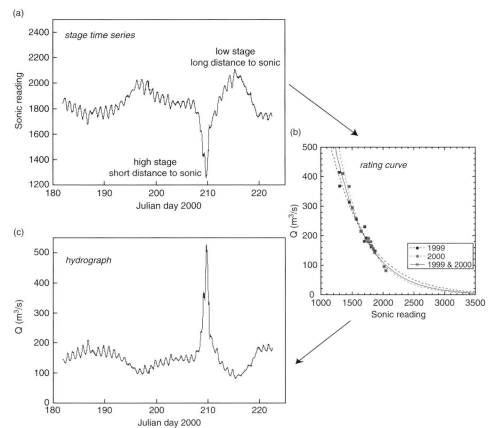

Figure 12.10 Combination of stage time series (a) and rating curve from many discharge measurements (b) to produce hydrograph (c) for Kennicott River, Alaska. Raw readings on sonic instrument are proportional to distance between sonic sensor and river surface. The high stage therefore corresponds to small sonic readings. The rating curve, here shown from measurements made over several summers, shows the expected increase in water discharge with increasing stage (smaller sonic readings). The hydrograph shows the strong pulse of discharge associated with the annual outburst flood from a side-glacier lake.

be logged with a micro-computer. The resulting time series can then be transformed into a time series of discharge using the rating curve, examples of which are shown in Figure 12.10 for the Kennicott River in Alaska.

Space-based measurement of discharge

Finally, we briefly summarize a recently developed method to measure discharge from space (Brackenridge *et al.*, 2005; Smith *et al.*, 1995; 1996; Smith, 1997; see review by Alsdorf *et al.*, 2007). This method, which relies upon orbital remote sensing, is particularly well suited to measurement of rivers in remote areas, and as we will see, is best used in braided river systems. It is straightforward to measure the width of a river from space, as water is easy to distinguish from land surface in a range of wavelengths. If the same reach of river can be measured repeatedly, one may develop a time series of river width over the runoff season. But we have just seen that a river responds to increases in discharge by widening, deepening, and increasing in

mean speed. How will measurement of width alone suffice? In this simplest application, we rely upon the fact that a braided river maintains a depth such that the shear stress exerted on the channel is roughly uniform, at nearly that necessary to transport the bedload of which the channel is made. In other words, one can safely take $\tau \approx \tau_{\text{crit}}$. In order to accommodate more discharge, another braid is accessed, meaning that the effective width of the river increases. As the depth does not increase significantly, the mean speed of the flow also does not change, and the discharge can be written: $Q(t) = H{<}U{>}W(t)$, where $W(t)$ is the time series of width. In their first application, Smith *et al.* (1995, 1996) turned to braided Alaskan glacially fed rivers to test the method. By comparing space-based width measurements and measurements of the river discharge at constrictions, they showed (Figure 12.11) that the method produced a reasonable proxy for the discharge. Ideally, one could also measure the stage of the river remotely using altimetry data from the satellite. Since then, more sophisticated measurements using other space-borne instruments have been made, allowing documentation of both river width and river stage. These can be coupled with advanced models of flow in channels and across floodplains to produce estimates of river discharge that can be validated against discharge in specific reaches (e.g., Brackenridge *et al.*, 2005; Smith, 1997). This exercise can now be performed in many areas of the world, as we can utilize the 30–90 m Shuttle Radar Topography Mission (SRTM) topographic data for all sites between 60°N and 60°S. In Figure 12.12 we show 8 years of water-level history of a large river in India. In effect, one can construct for such rivers an empirical rating curve akin to that we use to develop time series of discharge in gaged rivers. As new space-borne instruments are developed and deployed, this suite of methods will increase greatly our ability to monitor Earth's rivers (Alsdorf *et al.*, 2007).

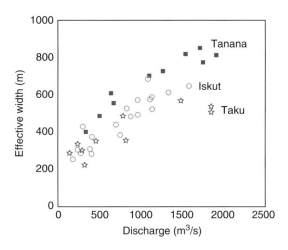

Figure 12.11 Satellite-derived effective channel width and measured river discharge for three braided Alaskan rivers. Each point is determined from a single ERS 1 SAR scene, and the daily mean discharge measured on the date of image acquisition (after Smith *et al.*, 1996, Figure 3, with permission from the American Geophysical Union).

Summary of theory and measurement of channel flow

We have developed the theory behind the law of the wall, the time-averaged turbulent velocity structure within an open channel flow. It is logarithmic with height above the bed. The detailed shape of the profile is set by two variables: u_*, the shear velocity, and z_o, the roughness height. The former is set by the channel slope and flow depth, the latter by the roughness of the bed, both the granular roughness and that associated with bedforms on the bed. This logarithmic velocity profile is then averaged in the vertical to yield the expected mean velocity, and related to other well-known equations for the mean velocity of a channel. The vertically averaged velocity is found at roughly six-tenths of the way down from the top of the flow, or at roughly $0.4H$. We discussed the means of measuring channel discharge both directly and

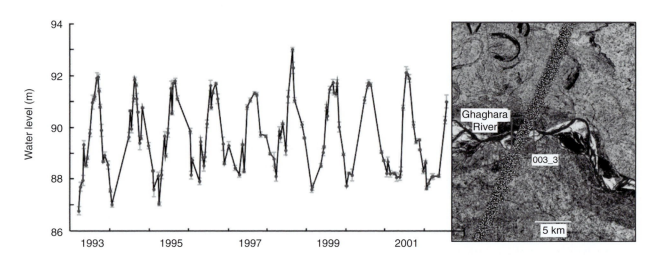

Figure 12.12 Stage history of the Ghahara River derived from satellite-based measurements (from Alsdorf *et al.*, 2007, Figure 6, with permission from the Nature Publishing Group).

indirectly. Indirect methods require a proxy for the flow, usually the flow stage, which can be measured continuously. A rating curve is required, however, which in turn requires direct measurements of discharge at a range of stages. Direct measurements are accomplished by current measurements using a variety of instruments, or by a conservative tracer (we used salt) method. In rivers that respond to increases in discharge largely through changes in width, space-borne remote sensing of river width can be used as a proxy for river discharge, enabling the monitoring of remote, large braided river systems.

Hydraulic geometry

Water discharge must increase downstream as the drainage area and the runoff from it increases. Similarly, as the hydrograph rises and falls at a given site. We have seen that water discharge in a channel cross section is the product of the mean velocity of the water with the cross-sectional area of the flow, or the width–depth product. Which of these features of the channel changes most in order to accommodate the need to increase discharge, either downstream at a given moment, or at a site through time? These relationships that reveal how the width, depth and velocity of the flow increase with discharge are called the hydraulic geometry of the river (Leopold and Maddock, 1953). The plots of width, depth and velocity as functions of water discharge, reproduced in Figure 12.13, all show roughly linear trends on log–log plots, implying power-law relationships of the form:

$$W = aQ^b$$
$$H = cQ^f \qquad (12.36)$$
$$U = kQ^m$$

Note that, since $Q = WHU$, the exponents in the relationships must sum to 1.0: $Q = WHU = ackQ^{b+f+m}$. From the graphs taken from many rivers, Leopold and Maddock deduced that b, f, and m were 0.5, 0.4, and 0.1, respectively. It appears that rivers accommodate additional discharge by equal parts of widening and deepening, with only a little increase in speed. This is perhaps why one can guess with some confidence that a river's speed is about 1 m/s.

In the course of these studies, assembled from the rich USGS gaging data collected over decades of river

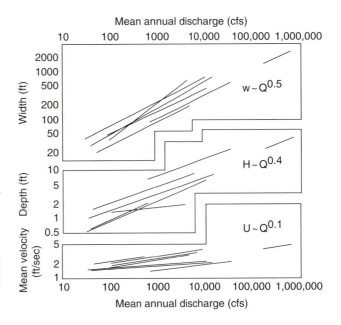

Figure 12.13 Downstream hydraulic geometry of several North American rivers, showing increase of width (top), depth (middle), and velocity (bottom), with increasing discharge downstream (redrawn from Leopold and Maddock, 1953, Figure 9).

monitoring, it was found that these quantities change significantly at a given site. This gave rise to the notion of "at a station" hydraulic geometry. The same relationships yield a different set of powers, implying that the river at a site accommodates changing discharge by changing its shape and speed differently. The values of b, f, and m are 0.26, 0.4, and 0.34. Crudely speaking, the river adjusts all of the available parameters equally. It widens, deepens and speeds up, with the deepening dominating. We have developed the theory for why the river should speed up if it deepens. Given that at any particular site the slope of a river is relatively fixed on the timescales of a single flood event, or over the monitoring period, the speed should increase as the two-third power of depth (using Manning's equation), or as $H^{1/2}\ln H$ using the law of the wall. The power on the speed exponent should therefore be less than that on the depth, as it is.

Figure 12.14 is a famous plot of channel change during the passage of a flood wave on the San Juan River in Utah. The width changes very little, while the depth of the river changes dramatically. It scours the bed deeply, here by more than 3 m, exposing and entraining sediment stored on the bed of the river. In fact in some of these rivers, including the Colorado

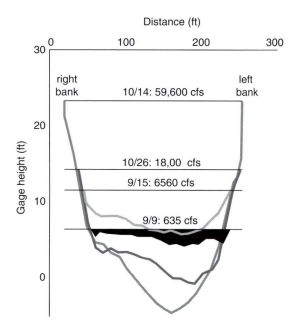

Figure 12.14 Channel scour and re-deposition through the September–December 1941 flood on the San Juan River, near Bluff Utah. Note initial slight aggradation as stage increases, followed by significant scour to depths 10 ft below low-flow bed (redrawn from Leopold and Maddock, 1953, Figure 22).

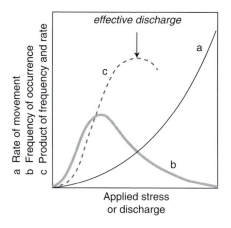

Figure 12.15 The heightened importance of rare events owing to the nonlinearity of the relationship between the applied stress and the sediment movement. The physics of the system controls the relationship between the stress and the resulting geomorphic process rate, (here the curve a), while the climate controls the frequency of occurrence, or the probability distribution of events, curve b. The resulting long-term rate of movement, and the events most important in dictating this long-term rate, are offset from the most common events toward the lower probability, larger events due to the nonlinearity of the physics (redrafted from Wolman and Miller, 1960, Figure 1, with permission from the University of Chicago Press).

through the Grand Canyon, the increase in depth due to scour of the bed can equal or exceed the rise in stage during some floods.

These at-a-station hydraulic relationships break down for a braided river, as we have already seen in our discussion of methods for measuring discharge in a braided river. These rivers accommodate additional discharge largely by widening, meaning that the exponent *m* should be much larger than the others.

Floods and floodplain sedimentation

Our formal treatment of suspended sediment transport provides a quantitative means of assessing the concentrations of various grain sizes at any level in the flow. This is highly relevant to the question of what grains succeed in escaping the channel onto the floodplain, and into the floodplain depositional system in general. When a channel is asked to convey more water than it can convey, the flow extends onto some portion of the floodplain. For planning and zoning purposes, we should therefore consider the floodplain in some sense a part of the channel. The floodplain is simply used less frequently than the rest

of the channel. Overbank flows typically occur every 2–3 years, reflecting the stochastic nature of the meteorological forcing of the fluvial system. Not all years are the same.

Flood magnitudes occur with different frequency, as is captured in Figure 12.15 by the probability distribution of flows. This is one of many examples of the tradeoff between magnitude and frequency in geomorphic systems (see Wolman and Miller, 1960; Costa and O'Connor, 1995). The probability distributions of event sizes tend to be skewed toward the small events; large events tend to be rare (thankfully, and see Chapter 17). On the other hand, the transport of sediment is nonlinearly related to properties of the flow (its depth, its speed, its shear stress). One must sum over all events and the corresponding geomorphic response to the event in order to assess what the total geomorphic response is in the long term, or to assess what might be called the geomorphically significant flow. Graphically, as shown in Figure 12.15, this means that the peak in the geomorphic response curve is shifted toward the larger events than is the peak in the frequency of the events. In the case of the fluvial system, it appears that the channel is naturally designed (meaning adjusted) to accommodate relatively frequent floods, those that occur on average every 2–3 years (Wolman and Miller, 1960).

The approach advocated by Wolman and Miller in their 1960 work has inspired many to acknowledge the non-steadiness of geomorphic processes. In systems in which flow of some fluid (water, air) incites the geomorphic activity of concern, for example bedload transport in either eolian or fluvial situations, one must marry a knowledge of the frequency with which flows of differing magnitude occur with knowledge of the relationship between the fluid motion and the sediment transport rate.

Box 12.3 Climate dependence of flood frequency

Using flows documented using the above methods over long periods of time, hydrologists can define the probability of occurrence of flows greater than or equal to a given magnitude. The details of such flood frequency analysis can be found in most hydrology texts or in USGS handbooks. Not surprisingly, the small flows are most frequent, and large flows become increasingly rare. The resulting histograms of flows have been fit with a variety of probability density functions (see Appendix B). The most commonly employed pdfs among hydrologists are the Weibull and Pearson type III distributions, which are asymmetrical distributions with one tail more elongated than the other. The USGS has prepared special probability paper on which a plot of discharge as a function of recurrence interval with a probability distribution of this sort will appear as a straight line. The average annual flood, taken as the arithmetic mean of the maximum instantaneous discharge for all years of record, will plot at a recurrence interval of 2.3 years on such a plot. For many rivers, this corresponds to bankfull flow. In other words, the channel has somehow designed itself to convey the mean annual flood within its banks.

We note that for geomorphic purposes it is the moderate to high end of the flow distribution that counts most, as depicted in Figure 12.15, as these are most capable of transporting sediment, and hence modifying the shape of the channel. The high ends of the histograms of daily flows can in fact be fit quite well with inverse power laws (power laws with negative exponents):

$$p(Q) = \beta Q^{-\alpha} \tag{12.37}$$

This can be seen by plotting not the histograms themselves but the number of days in which the flow of a given size is exceeded, so

$$N(Q) = \sum_{i}^{n} (Q_i > Q) \tag{12.38}$$

where N is the number of days the flow on that day, Q_i, has exceeded the flow of concern, Q. $N(Q)$ is the "exceedance probability" of the discharge Q. If the flows are distributed such that the pdf is a negative power law, then the integral of this distribution will also be a power law with a power differing by 1:

$$N(Q) = \int_{Q_{min}}^{Q_{max}} \beta Q^{\alpha} dQ = \left(\frac{\beta}{-\alpha + 1}\right) Q^{-\alpha + 1} \tag{12.39}$$

Visualized on a log–log plot, $N(Q)$ should be represented as a negatively sloped line (Figure 12.16) whose slope is $-\alpha + 1$. As the number of days the maximum flow is equaled or exceeded is 1, its exceedance probability is 1 ($= 10^0$). In Figure 12.16, we see that power-law fits capture the essence of the high discharge tails of the distributions, and that the power α differs significantly from catchment to catchment. In general, Molnar *et al.* (2006) showed that the power decreases with aridity, as can be seen by comparing the slopes of the exceedance probabilities from the Arizona (AZ) and Massachusetts (MA) catchments.

In rivers, progress has been made as measurement methods have evolved to document sediment transport. The Helley–Smith bedload sampler (Helley and Smith, 1971) has allowed us to document bedload sediment

Box 12.3 (*cont.*)

transport rates in a wide range of flows. This kind of data led Andrews (1980) to define the effective discharge, also called the channel-forming discharge, of a river. This can be defined as that water discharge that transports more bedload sediment than any other, averaged over a long period of time. It is formally the product of the bedload discharge rating curve, $Q_{bl}(Q)$, with the probability density function of the discharge, $p(Q)$. Working on the Yampa River in northwest Colorado, he showed (Figure 12.17), as has now been confirmed on many rivers (e.g., Nash, 1994; Emmett and Wolman, 2001), that the geomorphically effective discharge coincides remarkably well with the bankfull stage of the river. In other words, the flows that count most in the formation of a channel are those that fill the channel to its brim. In typical rivers, these occur once every 2–3 years. Those flows that exceed this size flood the floodplain, and occur too rarely to participate significantly in forming the shape of the channel. Those that are smaller, and only partially fill the channel, are too weak to cause significant bedload transport. While others have advocated the use of stream power as opposed to water discharge (Costa and O'Connor, 1995), or have worried about the incorporation of thresholds of motion for various grain sizes in the calculation of effective discharge (e.g., Lenzi *et al.*, 2006), the concept remains the same: one must acknowledge the tradeoffs between the efficiency of a flow and the frequency with which it occurs.

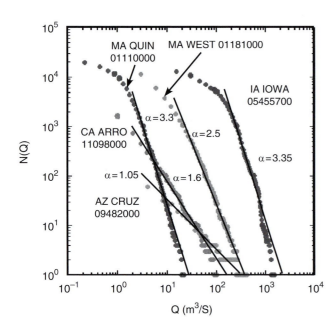

Figure 12.16 Cumulative flow exceedance plots for daily discharge from several US rivers with several decade-long records. The power-law behavior of the high discharge tails to the distributions is indicated by straight line fits to the data, with best-fitting slopes indicated by a. More arid regions (AZ = Arizona; CA = California; IA = Iowa; MA = Massachusetts) display lower slopes (after Molnar *et al.*, 2006, Figure 2, with permission from the American Geophysical Union).

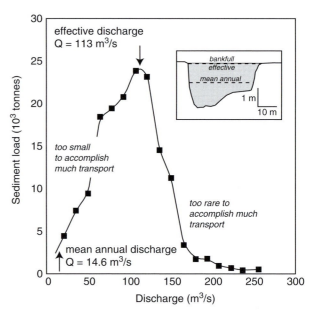

Figure 12.17 Mean annual sediment load transported by increments of water discharge in the Snake River at Dixon, Wyoming. Inset shows channel cross section, revealing how closely the effective discharge matches the bankfull discharge in this river (after Andrews, 1980, Figures 5 and 10 (inset), with permission from Elsevier).

We have seen in our discussion of bedload sediment transport the thresholded nature of the transport, and the nonlinearity of the transport rate above this threshold. This is also seen in data on suspended

sediment discharge, reported in Leopold and Maddock's USGS Professional Paper 252 on hydraulic geometry (Leopold and Maddock, 1953). They argue from assembled sediment discharge data (for example, that plotted in Figure 12.18) that the

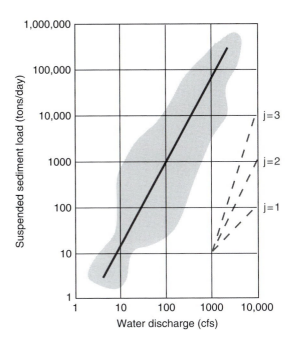

Figure 12.18 Sediment rating curve, Powder River near Arvada, Wyoming. Relationship between suspended sediment load and water discharge; gray cloud encompasses most measurement pairs. This allows calculation of sediment discharge from the more easily known water discharge. A straight line on this plot reflects a power-law relationship, or $Q_s \sim Q^j$. Here j is slightly less than 2 (after Leopold and Maddock, 1953, Figure 13).

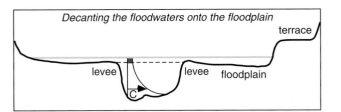

Figure 12.19 Decanting the floodwaters onto the floodplain. Concentration of suspended sediment declines toward the top of the flow, and becomes dominated by fine grains. It is these sediments that are wafted onto the floodplain during overbank conditions.

sediment load (mass of sediment transported per unit time; they report tons/day) may be related to the water discharge through another power law:

$$Q_s = pQ^j \qquad (12.40)$$

The values of the power j are typically between 2 and 3. Again, the system is nonlinear. A flood of double the discharge will transport 2^j, or 4–8 times the sediment. Such assessments for bedload tend to be even more nonlinear, i.e., have yet higher exponents.

The floodplain

Flows that exceed the ability of the channel to convey the water overtop their banks, and flood the floodplain. Overbank flow occurs, also called a flood. The question we may now ask is what sorts of grain sizes are contained in the water that makes its way over the bank? In effect, this water is decanted off the top of the river in the channel, as sketched in Figure 12.19. Consider how the escape of water onto the floodplain will affect its speed. The water slows down because the

flow is much less deep, and it experiences drag associated with roughness on the floodplain – grasses, fences, milk trucks, etc. This results in lowering of the ability of the flow to retain particles in suspension. (As discussed in Chapter 14, the relevant non-dimensional number is the Rouse number, which has the shear velocity in the denominator. As the basal shear stress drops because the flow thins, u_* drops, and the Rouse number increases.)

The first grains to drop out of suspension will be the coarser grains decanted off the channel over the banks. These will be deposited locally near the channel banks and lead to the aggradation of the levee. A natural levee is therefore built of the coarsest of the grains that escape the main channel in floods. Those grains that are yet finer will go further on into the floodplain, and will be deposited as the silts and clays that much of the world's fluvial agricultural lands depend upon.

While this is admittedly a simplified view of the floodplain depositional system, the description offered above contains the essence of the floodplain–channel connection. We note that on large rivers water escapes through a set of channels that reverse flow over the course of a flood event, rather than simply in this diffuse overbank mechanism (see the discussion of the Amazon). In addition, as we will see below, it is incorrect to assume that the floodplain is stationary, and that the style of deposition on it will be reflected deeply in the stratigraphy of the floodplain.

Channel plan views

When seen from the air, channels can have many looks, or take on many patterns. The classic subdivisions include the meandering and the braided cases. We wish to understand what it is about a river or about a tectonic setting that dictates the plan view pattern. And we wish to explore what sets the stratigraphic architecture one would see in the rock record if a deposit from one of these settings were to be preserved.

The braided case

First, consider the braided systems shown in Figure 12.20. Typically, the patterns of braided channels are explained as resulting from situations in which (1) the banks lack cohesion; (2) transport is by bedload only; (3) the flow fluctuates greatly on a frequent basis – perhaps even daily. Braided rivers are commonly found in association with glaciers, as the characteristic glacial outwash channel pattern. The flow occupies two or more channel segments simultaneously, the flow threads separated by mid-channel bars. At times, the flow system can occupy the entire valley cross section, taking up many tens of threads. The mid-channel bars themselves typically lack vegetation, and are composed of usually rather coarse bed material. Flow depths are usually not very great, being at most a few meters.

The lack of fine material and of vegetation leads to the lack of cohesion of channel banks. As the flow depth increases, the shear stress exerted on the bank becomes sufficient to entrain bank material, and the channel widens locally. Note that this is not the case if the channel banks remain intact; such a flow would instead increase in depth. The result is that as soon as the flow becomes deep enough to transport sediment, it begins to widen itself. This limits the flow depths to just greater than threshold depths, i.e., where $\tau_b = \rho g H \sin \alpha \approx \tau_c$. Given that, for a specific reach of stream, we can know the slope, this means the flow depth itself is the only variable.

Interestingly, this means that if we know the slope of a braided reach of channel, and we know the grain size of the bed material, from which we can calculate the necessary boundary shear stress to entrain it, then we can simply use the flow width as a surrogate for the flow cross-sectional area. In addition, we can use theoretical calculations to produce the mean flow velocity to be expected in the flow of critical depth. This means that we can judge the flow discharge in a braided river by

Figure 12.20 Two views of the Chitina River, Alaska, showing extensive braidplain downstream from major glaciers of the Wrangell Mountains. (a) High flow with much of the braidplain occupied by channels. (b) Medium flow, looking up-river, with abandoned train bridge segments for scale (photographs by R. S. Anderson).

summing up the width of the flow across the entire braidplain. This is a fantastic boon, as gaging such a system is essentially impossible, owing to the ephemeral nature of the flow, the mobility of the channel banks (which would have to support the instruments), etc. This method has recently been tested using satellite imagery (Smith *et al.*, 1996), and has been shown to work surprisingly well: width is indeed a very good proxy for water discharge in a braided river as seen in Figure 12.11.

The origin of the braiding behavior has been explored in numerical models (Murray and Paola, 1992), reproduced in Figure 12.21. They find that the dynamism of the system is largely due to flow expansions and contractions, or divergences and

flow

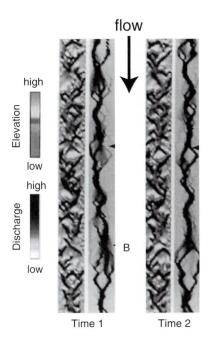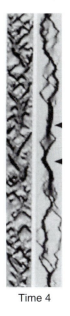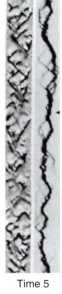

Elevation: high / low

Discharge: high / low

B

Time 1 Time 2 Time 3 Time 4 Time 5

Figure 12.21 Numerical simulation of channel braiding shown at five times, illustrating spatial patterns of both topography (left) and water discharge (right). Mid-channel bar formation at point B is especially instructive to follow through time (after Murray and Paola, 1994, Figure 2, with permission of Nature Publishing Group).

convergences. This in turn sets the divergence of sediment transport, forcing a reduction in sediment transport rate in sites of flow divergence, and an increase in sediment transport over sites of flow convergence. Recalling the erosion equation relating the rate of change of bed elevation to the spatial gradient in the sediment flux, we should expect sites of flow convergence to erode, and sites of flow divergence to deposit. This bed adjustment then changes the local flow, etc. The effect results in a very dynamic set of mid-channel bars, even in the face of essentially steady flow.

You might then ask why braided systems are so commonly associated with highly fluctuating flows. The role of flow fluctuation may relate to the efficient removal of vegetation from the system, which in turn maintains the banks as non-cohesive channel boundaries.

The meandering case

The meandering stream channel pattern is more common. An example is depicted in Figure 12.22. This sinuous snaking of a single-threaded channel is characteristic of all channels that have cohesive alluvial banks (meaning they are composed of river sediment). The sinuosity of a river is expressed as the ratio of the river channel length over the valley length, and can be many times 1. As shown in Figure 12.23, typical meander wavelengths are very tightly clustered around a dozen (actually 11) times the river width. As you can see in the three maps of meanders at the top of Figure 12.23, this means that in the absence of a scale on the map it is difficult to tell the scale of a river just by looking at a map. River meanders are self-similar.

> *meander*, n. To proceed sinuously and aimlessly. The word is the ancient name of a river about 150 miles south of Troy, which turned and twisted in the effort to get out of hearing when the Greeks and Trojans boasted of their prowess. (Ambrose Bierce (1842–1914), *The Devil's Dictionary*, 1911)

Now that we have paid particular attention to the floodplain and the development of the levee, think for a moment about the channel cross section. Here we ask what determines the stratigraphy in a cross section through the floodplain. Is it composed entirely of floodplain silts away from the channel? At depth, what would one find beneath the silts, and why?

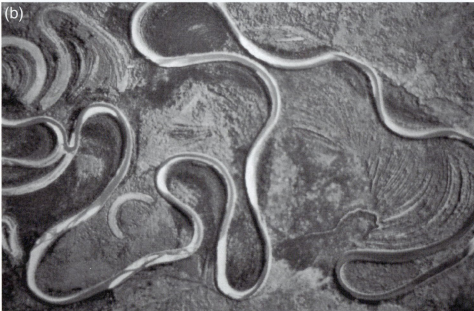

Figure 12.22 (a) Small meandering river on Yukon Flats in interior Alaska. Migration of the meanders is clear from the age (height) of the trees occupying the point bars. Abandoned meanders appear as oxbow lakes to the left of the image, and are clear water because they have no access to the fine sediment being carried by the river. (b) Scroll-bar dominated floodplain in Mississippi Valley. Point bars show as bright white deposits in the interiors of bends. Ghosts of old meander locations are shown in the pattern of the vegetation. Several cut-off meanders appear as oxbow lakes (photograph by R. S. Anderson).

In order to address this question, we have to know something about how meanders work – first how the flow works, and then how the sediment moves through it. Consider a single bend in the river shown in the sketch of Figure 12.24. The flow this time is confined within cohesive banks that are essentially non-eroding on the short timescale of a single flow event. The inside of the bend has a sandy deposit called a point bar that is an active part of the system. It is unvegetated, and dips gently into the flow. The thalweg, or the deepest part of the flow, is against the far bank on the outside of the bend. In plan view, the thalweg crosses from one side of the river to the next as it passes from one bend to a bend of opposite sign. We might expect the highest basal shear stresses to be associated with this band of highest flow depths. Importantly, the flow field is a complex one and very three-dimensional as it passes through a bend. The

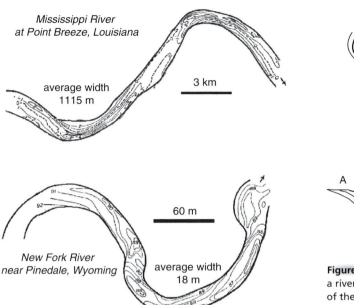

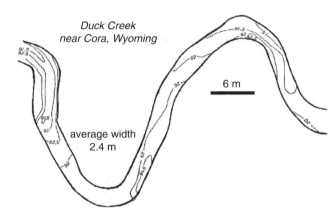

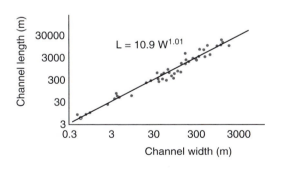

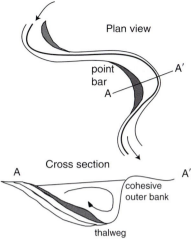

Figure 12.24 Geometry, terminology, and flowfield within a river meander. Super-elevation of the flow on the outside of the bend drives helical flow, which in turn orients the local shear stress inward toward the point bar.

Figure 12.23 Self-similarity of river meanders. Maps: representative meanders from three rivers at scales differing by three orders of magnitude look similar because the radius of curvature and the channel length scale with the channel width. Contours (in feet) show deepenings associated with outer bends in the meanders. Bottom: meander length scales linearly with channel width such that length is roughly 11 times channel width. Open circle is an ice channel (after Leopold and Wolman, 1960, Figure 3 (maps) and Figure 2 (plot)).

water experiences a centrifugal acceleration as it is made to turn a corner; recall that this goes as the square of the local velocity of the flow divided by the radius of curvature of the bend. As the water high up in the water column has the highest velocity, it experiences the highest outward acceleration. A component of its flow is therefore outward toward the outer bank. This bunches up flow there, and generates a slope on the flow back toward the inside of the bend. This slope produces a pressure gradient that drives flow inward at the bed. All told, this sets up a helical motion of the water as it passes a bend. Now consider the sediment. The shear stress at the bed mirrors the flow immediately above it. On the downstream side of the point bar, sediment is delivered to the point bar through inward motion of grains along the bed, responding to the inward basal velocity of the flow. Those grains small enough to remain in transport out of the main thread will be transported onto the point bar toe, and finer ones will be taken up higher onto the point bar slope. The net effect is a downstream and outward migration of the point bar. Over long timescales, the outer bank is also eroded, as it is there that the highest shear stresses are exerted, as illustrated in Figure 12.25 from the Muddy Creek experiment (Dietrich *et al.*, 1979, 1984; Dietrich and Smith, 1983; Dietrich, 1987). These stresses and the sediment entrainment they allow ultimately result in the erosion of the base of

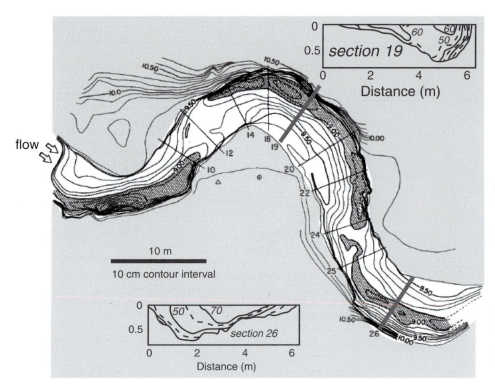

Figure 12.25 Study reach of Muddy Creek, Wyoming, showing topography of the meander bends, and stations (numbered lines) at which flow speed and sediment transport rates were documented. Insets: velocity fields at sections 19 and 26, contoured at 10 cm/s intervals. Note strong velocity gradient (strain rate) near outer channel bank (after Dietrich *et al.*, 1979, Figures 1 and 2, with permission from the University of Chicago Press).

the bank, and cause its failure (see also Hooke, 1975; Smith and McLean, 1984). Over long timescales, then, the meander migrates sideways and down-valley. The back edge of the point bar slowly becomes so inactive that it can be colonized by vegetation, and is left in the wake of the motion of the meander across and down-valley. High floods will ultimately mantle the point bar with overbank deposits. The resulting stratigraphy is a point bar migration deposit, thinly mantled by overbank deposits. Note that the deposit is normally graded – fine grains over coarse grains – as the coarser grains always travel nearest the thalweg of the channel; only the finer grains are carried high onto the point bar before the basal shear stress of the flow drops below their threshold for transport.

The geomorphic expression of a meandering system is one of a plain covered by scroll bars and oxbow lakes (Figure 12.22(b)). The scroll bars mark the migration of the meander through time, and provide strong constraint for the kinematics of pattern evolution (meaning they record the path taken, not the physics of how it happened). One can quantify the rate of meander migration in the short term by comparing aerial photographs of a site. On the ground, we can deduce migration rates by dating the vegetation that colonizes the abandoned point bar; it

should increase in age with distance from the channel. This can be seen in the photograph in Figure 12.22(a).

Models of meanders all predict the lengthening of the river through time. If this were the only process occurring, rivers would progressively increase in sinuosity through time. In fact they do not; they tend to have a well-defined sinuosity. In meandering rivers, shortening only occurs through meander cutoffs, events in which the river pinches off a bit of itself and leaves it behind as an oxbow lake. If the river is a snake, these are its snakeskins shed off at intervals, littered about in the floodplain (Figure 12.22(b)). In fact these lakes are tremendously important elements in the ecology of river systems (e.g., Ward, 1998). Only slowly do they fill in with sediment delivered to them through overbank deposition in large floods. River cutoff occurs by two mechanisms: chute cutoff in which the river forms a new channel across a meander during a large flood, and neck cutoff, in which the slow lateral migration of the planform impinges on itself. Models of meandering rivers (e.g., Howard, 1992, 1996; see also Sun *et al.*, 1996, 2001a–c) demonstrated the importance of the chute cutoff mechanism in limiting the sinuosity of a river. In Figure 12.26 we show the results of a river meander model that illustrates the extremely dynamic nature of meandering

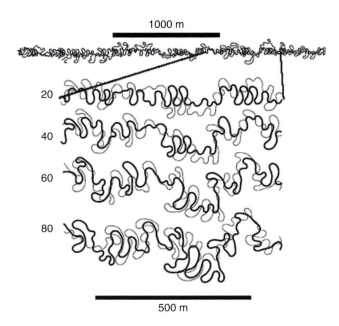

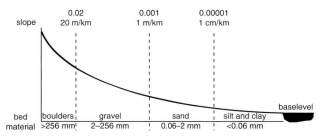

Figure 12.27 Sketch of the longitudinal profile of a river, showing concave up shape. The river here is shown graded to a baselevel that could be the level of a lake or the ocean. The caliber of the sediment comprising the bed declines downstream as the slope declines. The slopes at which the transitions between major grain size classes occur are shown.

Figure 12.26 Model of river meandering. Top: superimposed channels at 20 (solid) and 100 (dotted) time units into the simulation. Below: enlargement of an 800 m reach at several intermediate times, with the next time superimposed as dotted (after Lancaster and Bras, 2002, Figure 10, with permission from the American Geophysical Union).

systems. Here Lancaster and Bras (2002) have developed a model that is constrained to match both the small-scale geometry of a meandering system documented by Dietrich in Muddy Creek, and the large-scale geometry from a variety of Alaskan rivers.

Following work by Stolum on the Amazon (Stolum, 1996, 1998), Constantine and Dunne (2008) have mined the GoogleEarth archive to document the oxbow lake populations in many floodplains of large meandering rivers of the world. They demonstrate that lengths of oxbow lakes are log-normally distributed, with a mean that is very well predicted by

$$L = 3.0e^{0.82S} \qquad (12.41)$$

where S is the channel sinuosity. The higher the sinuosity of the river, the longer are the lakes that are pinched off by the meander cutoff process. Given this empirical rule, they go on to predict the rate of production of oxbow lakes as a function of both the river sinuosity and the rate of lengthening of the river through the meander process. For a steady sinuosity, the rate of river lengthening must balance the rate of loss of river length through meander cutoffs. Highly sinuous rivers apparently lack the

efficient chute cutoff mechanism, the physics of which is still poorly understood.

Channel profiles

Alluvial rivers display concave up longitudinal profiles, as shown in Figure 12.27. Channel slopes are steep in the headwaters, and gentle at the mouth. The elevation of the mouth of the river is the baselevel for the river, which is therefore determined by the level of the still body of water into which the river debouches. As this is either a lake or the ocean, either of which can change in level with time, river profiles are significantly affected by temporal variations in baselevel. Changes in baselevel can be either climatically driven, or in the case of the ocean, tidal, or tectonically driven. It is after all the relative sea level that counts to the river. The baselevel can therefore fall due to rock uplift of the land, or rise due to subsidence of the land. Note as well that this change in the vertical would be accompanied by lateral migration of the point at which this baselevel condition is exerted on the river profile. So the bottom boundary condition for a river is dynamic, and can move in both horizontal and vertical position.

The slope of the river must smoothly decline to zero within some small distance of encountering the water body. In turn, this lower boundary condition for the river is felt upstream, generating a "backwater effect" whose length is dependent upon the thickness of the flow and the slope of the flow.

Mackin (1948) defined a graded river as one in which the slope of the river was everywhere adjusted to allow it to transport the necessary load of sediment. John Hack (1973) defined what he called the stream gradient index to describe how the local slope of a stream depended upon distance from the divide:

$$SL = \frac{\Delta H}{\Delta L} L = \frac{\partial z}{\partial x} x \qquad (12.42)$$

where the first form is as Hack defined it, and the second is consistent with our terminology. The stream gradient index is in effect a proxy for what is now called stream power (see Chapter 13), or the rate at which energy is made available to do work on the system. If this quantity is roughly uniform throughout the fluvial profile, then we can solve the equation for slope, and integrate to arrive at the expected longitudinal profile of the river:

$$\frac{\partial z}{\partial x} = -\frac{SL}{x} \qquad (12.43)$$

$$z = -SL \ln x + C \qquad (12.44)$$

where C is a constant of integration. Asserting that the river elevation matches that of baselevel at a distance L from the divide suggests $C = SL \ln L$, which completes the full profile description:

$$z = -SL \ln \left[\frac{x}{L}\right] \qquad (12.45)$$

The longitudinal profile is therefore expected to be logarithmic in distance from the divide, as illustrated in Figure 12.28 using a range of slope gradient indices.

The alluvial river profile is concave up, declining in slope with distance downstream. This contrasts with the convexity of hillslopes. We ask here why this is so? In both cases the system must carry more sediment with distance. Recall our arguments for the convexity of hilltops: because in steady state the hillslope discharge must increase with distance, and as most hillslope transport processes increase with slope, the slope must increase. Note also that the hillslope is linked to the stream, which carries away all the sediment. If the stream system is steady as well, the stream too must carry a load of sediment that increases with distance, in fact a load that increases roughly with the drainage area of the basin. As we discuss at length in Chapter 14, the transport capacity of a stream is dependent not just upon the

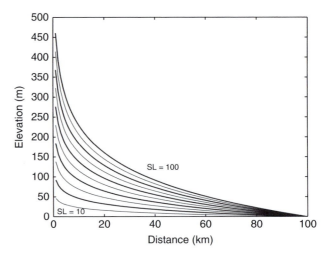

Figure 12.28 Theoretical longitudinal profiles based upon Hack's stream gradient index (*SL*). Profiles are logarithmic in distance from the divide, the concavity governed by SL, here ranging from 10 to 1000 in equal increments.

slope of the river, but upon the discharge of water. Slope is not all that matters. Another constraint is that the slope of a river surface must vanish as it enters the sea. There is no such boundary condition on the slope of a hillslope.

Let us say for simplicity that the sediment transport goes as the product of the water discharge and the slope of the river, or $Q_s \sim QS$. Let us also assume that the required sediment transport goes as the drainage area of the stream, A^m. If the water discharge also goes as A^m, then $eA^n \sim bA^mS$. Solving for the required S: $S = (e/b)A^{(n-m)}$. As long as $n < m$, the slope will decline with drainage area and hence with distance downstream.

This argument hinges on the capacity of the river to carry the sediment load delivered to it from adjacent hillslopes. Another argument hinges on the competence of a river, or the caliber of sediment it carries. Acknowledging that the mean sediment size on the bed of the river declines with distance downstream (see below), and that the competence of a fluid depends upon the shear stress it exerts on the bed (see Chapter 14), the required shear stress declines with distance. Recalling that the shear stress, τ_b, is $\rho g H S$, the required product of HS declines with distance. We know from downstream hydraulic geometry that the flow thickens with discharge. The required slope must therefore decline with distance; the profile will therefore be concave up.

Character of the bed

River beds indeed change in character from their head to their mouth. Some generalizations about changes in sediment size, bed slope, and sediment movement characteristics are outlined in Figure 12.27. The grain size transitions outlined in the figure are commonly seen in rivers, and occur at the channel slopes noted. The two main processes that have been invoked to explain this decline in sediment size are comminution or breakdown of the sediment, and selective deposition of the coarse sediment. Both clearly occur. Sternberg (1875) was the first to propose that the fining could be expressed as a weight-loss function in which the rate of loss of weight of a clast was proportional to the weight. Hence $dW/dx = -aW$, which when integrated becomes an exponential function of distance. The grain weight becomes

$$W = W_{\mathrm{o}} e^{-x/x_*} \qquad (12.46)$$

where W_{o} is the original clast weight, and x_* $(=1/a)$ is more easily interpreted as the distance downstream the clast must travel in order to decline in weight by a factor of e. This notion became so well ingrained in the engineering literature that it became known as Sternberg's law. Sediment grain size reduction during transport has long been explored experimentally. This began with the extensive work of Kuenen (1956) using drums in which the sediment was tumbled, a method modified by Kodama (1994a,b), and altered altogether by Attal *et al.* (2006) who employ instead vortex flumes. In all such experiments the rate of

decline in grain size is characterized as an exponential with a characteristic travel distance required to reduce grain size by a factor of e. (In the case of drums this was interpreted to be the product of the circumference of the drum and the total number of revolutions.) But Jones and Humphrey (1997) raise another issue. They point out that pebbles and cobbles can rest in an alluvial deposit, say a point bar, for long periods of time between discrete transport events. If during the resting period the clast weathers significantly, the weathered carapace of the clast will be quickly removed once it is again being jostled about during transport. Both physical comminution during transport and weathering during periods of rest ought to produce different rates of size reduction for grains of different rock types or minerals. This lithology-dependent effect has been explored deeply in the work of Attal and Lave (2006) on the Marsyandi River in Nepal, and is discussed further in Chapter 13.

While the grain size of the bed of rivers tends to decline downstream, the decline in grain size can be quite sharp in the case of gravel to sand. The composition of the bed is important for fish habitat, as for example salmon seek gravel beds in which to lay their eggs. This gravel–sand transition can occur over a length of channel that is far too small to be explained by comminution of the sediment. It can occur over a reach of 100 m in small rivers, and over 30 km in large rivers – in both cases a small fraction of the total length of the river. This observation has focused attention on modeling of selective transport

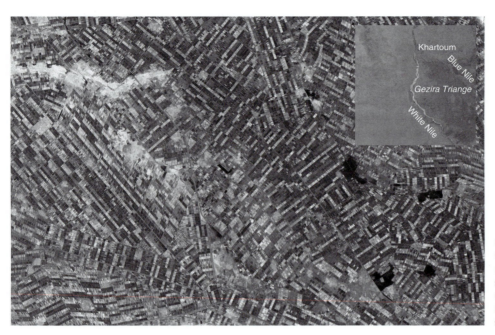

Figure 12.29 A portion of the Gezira scheme irrigation project. (NASA image PIA11079) Inset: satellite image of location of Gezira triangle between the Blue Nile and more gently sloping White Nile.

in a mixed grain size bed. Ferguson (2003) summarizes prior work in his modeling study. He simplifies the problem to two grain sizes, sand and gravel, and is able to reproduce the rapid transition by appeal to the strong dependence of the entrainment threshold on the fraction of the bed that is sand. See Chapter 14 for a detailed discussion of entrainment in sediment transport problems. For a pure gravel bed, the threshold for transport of both sand and gravel is taken to be equivalent. Any sand in the bed is hidden among the gravel particles, and cannot be transported unless the gravel moves. On a pure sand bed, any gravel that exists is highly exposed to the flow, as it sticks well above the sand bed, and therefore rolls freely. Its threshold for entrainment is reduced considerably relative to that on a pure gravel bed. The postulated dependence of entrainment thresholds for gravel and for sand on the sand fraction in the bed results in highly preferential transport of sand at the low excess shear stresses that are characteristic of gravel-bedded rivers. This means that the sand is being transported in greater concentration than it occurs in the local bed, and is therefore winnowed from the bed. This promotes the emergence of a sharp gravel–sand transition, and serves as another example of self-organization in geomorphology.

It should be clear by now that the longitudinal profile of even an alluvial river (not touching down on bedrock, and therefore immune to the direct affect of the lithology over which it is passing), is quite complex. Because the river has so many features it can adjust to accommodate the need to pass both water and sediment (recall the hydraulic geometry), and because the baselevel conditions for a river are rarely steady for long, there is no simple universal rule for what governs the profiles of rivers. We must acknowledge the details of any particular river, which will include the distribution of area with elevation which in turn governs the distribution of precipitation; the spatial distribution of the sediment supplied to the river, both in quantity and in caliber; the tectonic effects that might include altering the slopes of the river, and the baselevel of the river; and any human influences that might include ponding of water and sediment behind reservoirs, alteration of the hydrographs by land use changes, and so on.

River slopes

The slopes of rivers are quite small. The bigger the river, the smaller is its slope near the mouth. The slope of the Amazon River is of the order of 10^{-5}, or 1 cm/km. This makes large rivers more susceptible to any process that can alter the slope of the landscape. We have touched upon a couple of these in discussions of geophysical processes in Chapters 3 and 4, including isostatic rebound due to rearrangement of loads on the surface, flow of mantle across the edge of the continent due to large-scale variations in ocean volume, and other tectonic processes.

Take for example the lower Mississippi River depicted in Figure 12.30. The river elevation drops 95 m over the roughly 1000 km distance between Cairo and the Gulf of Mexico, giving it a mean slope of about 9.5×10^{-5}; the slope appears to decline to about half of this in the following 500 km to Head of Passes as it enters the Gulf of Mexico.

These low slopes also promote switching of the river from one to another course. At times of high sea level as at the present, these large rivers are aggrading near their mouths. If for any reason the river breaches its levee system in these lower reaches, a completely new course down a higher slope to the ocean may be accessed, leading to abandonment of the old channel. The Mississippi has done so several times over the course of the Holocene. In fact, the US Army Corp of Engineers spends a great deal of effort to prevent the Mississippi from occupying the Atchafalaya channel (Figure 12.30), as it is now the steeper route to the ocean. The Yellow River in China (Huang He in Chinese) has jumped from one to another channel naturally as well, entering the China Sea at locations as far apart as several hundred km. The Huang He levees now stand something like 7 m above the adjacent floodplain. Chinese generals have in the past employed the river as a weapon, making use of the low slopes and the extensive floodplains beyond the natural levees. In the latest such event, in 1938 nationalist troops led by general Chiang Kai-Shek dynamited the levees of the Huang He, flooding more than $50\,000\,\mathrm{km}^2$ of the land, and miring the invading Japanese army while killing nearly a million of his own people in the flood.

The influence of baselevel

The baselevel of a river changes with time due to a variety of processes that can change the relative elevation of the land and the sea. Alluvial rivers respond to baselevel changes by sediment infill when baselevel rises, and diffusion of the corner resulting from a baselevel drop. We can understand this by appeal to the equations of continuity of sediment

and of sediment transport. We wish to develop the equation representing the evolution of channel elevation, z. If the substrate is composed entirely of sediment (as opposed to rock), then this equation becomes simply

$$\frac{\partial z}{\partial t} = -\frac{1}{\rho_b} \frac{\partial Q_s}{\partial x} \tag{12.47}$$

where Q_s is the specific discharge of sediment, and ρ_b is the bulk density of the alluvial bed. We now seek an expression for the sediment transport rate. One approach is to use a stream power based argument for sediment transport (see Chapters 13 and 14), in which the sediment transport rate is proportional to the product of the water discharge and the local channel slope. For example, let

$$Q_s = -\alpha Q \frac{\partial z}{\partial x} \tag{12.48}$$

where Q is water discharge, and α is an efficiency coefficient. Combining these results in a diffusion equation

$$\frac{\partial z}{\partial t} = \frac{\alpha}{\rho_b} \frac{\partial (Q \partial z / \partial x)}{\partial x} = \frac{\alpha}{\rho_b} \left[Q \frac{\partial^2 z}{\partial x^2} + \frac{\partial Q}{\partial x} \frac{\partial z}{\partial x} \right] \tag{12.49}$$

In a reach of channel in which water discharge is nearly uniform, the second term in the right-hand side resulting from the chain rule of the derivative drops out, and the equation is more recognizable as a diffusion equation for the elevation of the channel:

$$\frac{\partial z}{\partial t} = \kappa \frac{\partial^2 z}{\partial x^2} \tag{12.50}$$

where the effective diffusivity is $\kappa = \alpha Q / \rho_b$. The sharp corner in the profile generated by the baselevel drop should round more quickly the greater is the sediment discharge, which is scaled by αQ. In Figure 12.31 we show an example of such a case in which the baselevel is dropped at $t = 0$ by 20 m. The corner rounds quickly at first, and more slowly thereafter, and in the long term returns to a stable concave up profile. This style of continuity-based approach is used heavily in models of the filling of basins adjacent to growing mountains, although the details of the transport rule used and the attention to grain size fractions varies between models (e.g., Paola *et al.*, 1992; Paola, 2000).

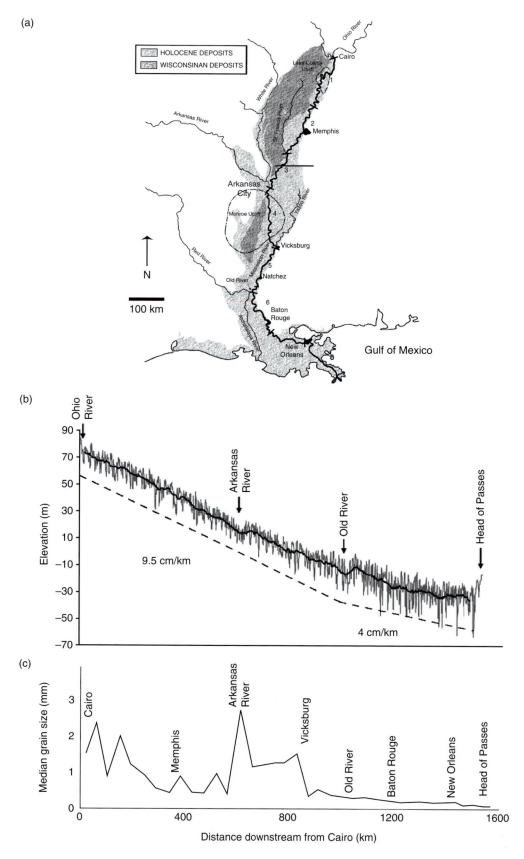

Figure 12.30 Characteristics of lower Mississippi River. (a) Map of lower 1600 km below confluence with Ohio River at Cairo, Illinois. (b) Longitudinal profile, with 40 km sliding mean (bold). (c) Grain size of the bed, showing general decline in median diameter, and influence of major inputs of the Arkansas River (after Harmar and Clifford, 2007, Figures 1, 5, and 9, with permission from Elsevier).

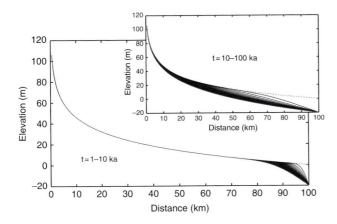

Figure 12.31 Simulations of the effect of a baselevel drop. Diffusive behavior of the system is reflected in the diffusive entrenchment of the mouth of the system. Inset shows long timescale response (1 Ma), demonstrating that the system returns to being concave-up throughout at long times.

The Amazon

We conclude this section with a discussion of the Amazon River, the largest river on Earth. The work we summarize is the product of a decade of research by a team of researchers led by Tom Dunne and Jeff Richey, and aided by many graduate students, among them Leal Mertes (e.g., Mertes, 1994, 1997, 2002; Mertes *et al.*, 1996), Liz Safran (Safran *et al.*, 2005), and Rolf Aalto (e.g., Alto *et al.*, 2003). The Amazon River is huge by any standard; to take on its study is both important and daunting. As shown in Figure 12.32, the river drains 6.15 million km^2 of the continent, delivering 6300 km^3 of water annually to the ocean (17% of the world's fresh water input) (Table 6.1; after Milliman and Syvitski, 1992). It drains a large portion of the east side of the Andes, and crosses more than 3000 km of the cratonal

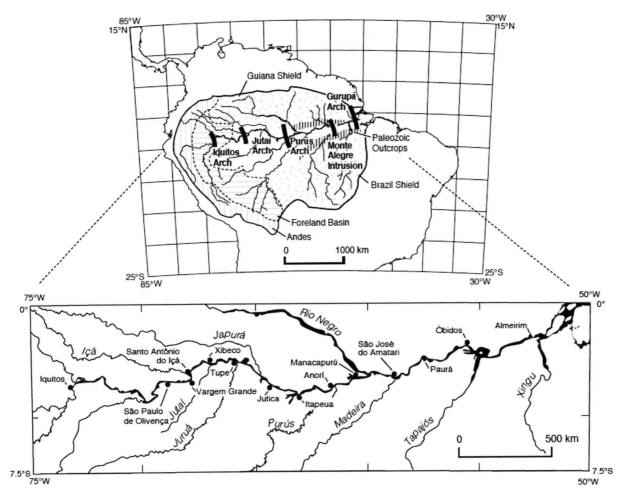

Figure 12.32 Top: map of Amazon River basin, with structural arches, and foreland basin adjacent to the Andean headwaters noted. Bottom: major towns and tributaries of the Amazon (after Dunne *et al.*, 1998, Figure 2).

shield of South America to deliver the water and sediment to the ocean. Its annual flood is so large that it depresses the crust of the Amazon basin by about 10 cm (Alsdorf *et al.*, 2007). The channel slope is so low (2 cm/km) that it is very difficult to measure accurately. Water depths are measured in tens of meters, and the channel widths in the mainstem are 2–4 km. Roughly 1200 Mt (megatonnes, or 10^9 kg) of sediment reaches the Amazon delta each year, delivering roughly 20% of the world's sediment supply to the ocean. Assuming a bulk density of sediment of roughly 1700 kg/m^3, this corresponds to 0.7 km^3 of sediment volume delivered to the ocean each year.

The questions posed in this research include: How does this sediment reach the coast? Where does the sediment come from, and how long does it linger in floodplains as it passes through the system? Is there net storage in the floodplain? How much sediment exchanges between the floodplain and the channel? The answers were obtained through construction of a sediment budget for the river, which not only required application of many of the principles we have described in this and preceding chapters, but revealed some important aspects of the river that would not be otherwise obvious. The tasks included numerous expeditions down the river, and remote sensing of the basin.

The sediment budget is illustrated in Figure 12.33. A mathematical statement that captures this budget is

$$\frac{\partial(\rho_b A z)}{\partial t} = Q_u + \sum_i Q_{\text{trib}_i} + E_{\text{bk}} - Q_d - D_{\text{bar}} - D_{\text{fpc}} \quad (12.51)$$

(Dunne *et al.*, 1998). Here the left-hand term represents the rate of change of sediment mass on the bed of the channel reach whose area is A (width times length of the reach). The first three terms on the right-hand side represent additions of mass to the reach, while the last three represent losses. The Q terms are fluvial sediment discharges as bedload and suspended load. The second and third terms capture additions of mass from tributaries, and from bank erosion. The last two loss terms represent deposition on channel bars and on the floodplain, respectively.

The documentation of down-channel transport is accomplished using measurements of water velocities, depth, width, and sediment concentration at many stations in the river at a variety of stages. From these one can construct sediment rating curves that allow

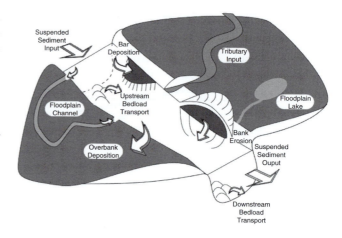

Figure 12.33 Sediment budget for a reach of channel and adjacent floodplain. The budget must account for sediment delivered to the channel from upstream, from tributaries within the reach, and from bank erosion. It may leave the reach by transport in the channel, deposition on bars, through diffuse overbank flow, and through floodplain channels (Dunne *et al.*, 1998, Figure 1).

calculation of sediment discharge from water stage, which is measured continuously. Mass delivery to the channel by bank erosion was calculated from knowledge of bank erosion rates (L/T) derived from satellite imagery, height of the bank and the bed of the river (hand surveys and Brazilian Navy bathymetric surveys), and from the sediment bulk density and grain size. Overbank delivery of sediment to the floodplain was calculated using a numerical model of flood routing through the reach, and the concentration of sediment in the water escaping the channel. The latter was estimated from detailed sediment concentration profiles in the mainstem flow, in which it was found that the average sediment concentration in the overbank flow was 0.33 of the mean concentration of sediment in the channelized water column. This ranged from 140 to 80 mg/L over the 2000 km study reach.

The plot in Figure 12.34 reveals that the dominant inputs are from the Amazon proper at Sao Paolo de Olivenca (616 MT), the Madiera (715 MT), which drains the Bolivian Andes, and the smaller tributaries (117 MT). Note that the sediment delivery from the Rio Negro is trivial, as it drains the low-relief shield north of the Amazon; it is called the Rio Negro because the water is black, largely due to its high organic content. The output at Obidos (1239 MT) is less than the sum of these inputs, reflecting a net storage of 209 MT of sediment in the system per year, most of which is on the floodplain. Roughly 14% of the sediment input to the system from the tributaries is stored in the floodplain for an unknown period of time. Another important story told by this budget exercise is that the exchange between river and bank is significant. Gain of mass by bank erosion and loss by floodplain deposition (up

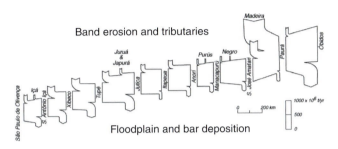

Figure 12.34 Measured sediment budget along a 2010 km reach of the Amazon mainstem. Upper left: inputs from bank erosion. Lower right: deposition on bars and floodplains. Tributary inputs are noted on top. Note huge difference between inputs from Rio Negro, draining the Guiana crystalline shield, and Rio Madeira (after Dunne *et al.*, 1998, Figure 10).

and down arrows in the diagram) is of the same order of magnitude as the channel transport. This aspect of the budget also implies that the sediment that makes its way out of the system toward the ocean at Obidos will likely have spent time in the floodplain. The pipe of the Amazon is very leaky. During storage on the floodplain, chemical weathering can act to alter the size and mineralogy of the sediment.

In fact, in a companion study, Aalto *et al.* (2003) have discovered that the delivery of sediment to the floodplain of the Amazon is quite episodic. Their data, reproduced in Figure 12.35, suggest that only about one in eight years leaves any record at all, and this record is represented by many centimeters of sediment. The authors attribute this pulsed nature of deposition to la Niña portion of El Niño/Southern Oscillation (ENSO), which brings tremendous rainfall to the headwaters of the Amazon. Once again, we see the operation of a nonlinearity of the geomorphic system, and the resulting importance of rare events.

The time series of sediment accumulation rate can be integrated to yield an accumulation rate per unit down-valley length of floodplain. The integral of an

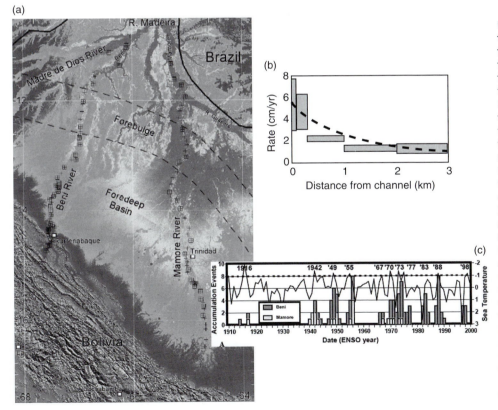

Figure 12.35 Sedimentation on the floodplains of the Amazon tributaries as they cross the foreland basin downstream of the Andes. (a) Map of the foreland basin in northern Bolivia showing major tributaries Beni and Mamore, the foredeep basin loaded by the encroaching thrust sheets, and the forebulge yet further from the load. Transects of floodplain sediment accumulation are shown in boxes. (b) Average rate of sediment accumulation determined from ^{210}Pb profiles. Falloff of sedimentation rate has a characteristic length scale of roughly 1.5 km. (c) Dates of major sedimentation events in cores, and sea surface temperatures in the eastern Pacific (la Niña threshold shown as dashed line). Events correspond to la Niña events (after Aalto *et al.*, 2003, Figures 1–3, with permission from Nature Publishing Group).

exponential is simply the value of the y-axis (here 5 cm/yr) with the characteristic length scale (here 1500 km). The authors calculate that the rate of storage on the floodplain of the Beni tributary is 100 Mt per year.

The measurements and synthesis of them that we have presented here represent roughly 20 years of effort on the part of a large group of researchers. The result is that the Amazon River now serves as an example of how one should conceptualize a geomorphic problem as a budget, here of sediment, which in turn dictates what must be measured.

Summary

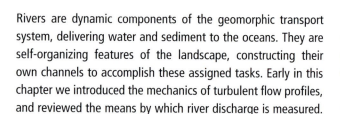

Rivers are dynamic components of the geomorphic transport system, delivering water and sediment to the oceans. They are self-organizing features of the landscape, constructing their own channels to accomplish these assigned tasks. Early in this chapter we introduced the mechanics of turbulent flow profiles, and reviewed the means by which river discharge is measured.

Rivers respond to water inputs from the landscape (hillslopes) and from glaciers in the headwaters. Given the stochastic nature of the water delivery system, which is ultimately dictated by the many scales of variability in the atmosphere discussed in Chapter 5, the peak flows transported down the river differ from one year to the next. A typical meandering river self-designs a channel capable of carrying most of these flows. Every 2–3 years, however, flows exceed the channel's capacity, causing flooding of the floodplain, and transport of the finest sediment near the top of the flow outward across the floodplain. This is one of geomorphology's greatest examples of the role of extreme events, and illustrates the need to acknowledge the full spectrum of atmospheric inputs to which a river is subjected.

The bestiary of rivers ranging from meandering to braided plan views reflects the variability of flows, the nature of the sediments involved, and the vegetation in the river corridor. Braided forms reflect high variability of flow, and little vegetation. They respond to increases in flow by simply adding braids. Meandering rivers have cohesive banks, with the cohesion commonly arising from roots of vegetation. They migrate toward their outer banks, leaving point bar deposits on the inner banks, which in turn leads to a characteristic stratigraphic architecture. Floodplains are commonly capped by the finest components of the suspended sediment as the flow goes overbank in a flood. We develop the quantitative tools to evaluate suspended sediment profiles in Chapter 14. This decanted sediment falls first on the levee, helping to construct this important natural separator between channel and floodplain.

The river profile is concave up, which contrasts with common convex hillslope profiles discussed in Chapter 9. Rivers are much more efficient at conveying water than are hillslopes. In general their slopes are very small, order tens of m/km in the headwaters and as low as cm/km near their mouths. The river comes to zero slope at its baselevel – at its junction with the bounding body of water, the ocean or a lake. Alluvial rivers are continually adjusting to changes in the baselevel. Large rivers have slopes small enough that they are vulnerable to even the broad scale geophysically driven tiltings of the land we have discussed in Chapter 3.

We have seen in the work on the Amazon that even large alluvial rivers can be treated as having sediment budgets, each component of which must be measured. In the case of the Amazon, researchers have combined classical water and sediment gaging, ^{210}Pb dating of sediments, remote sensing and flood routing models to constrain these components.

Once again, we have employed the organizing principle of continuity in several of the analyses: in the development of the Navier–Stokes equation (see the appendix to the chapter), the basis of the salt-discharge measurement of flow measurement, the erosion or deposition of an alluvial bed, and in the sediment budget of the river as a whole.

Appendix: The Navier–Stokes equation and the origin of the Reynolds and Froude numbers

The most famous and most important equation in fluid mechanics is the Navier–Stokes equation. It is application of this equation that leads to solutions for the expected velocity profiles in lava flows, for the discharge of fluid in a pipe, for the flow-field around a settling grain in a fluid. It is also the jumping off point for derivation of velocities in a turbulent fluid.

The Navier–Stokes equation is a highly embellished version of Newton's second law, which states that the rate of change of momentum of an object is the sum of the forces acting on that object. You are no doubt familiar with the form $F = ma$, where F stands for the sum of the forces, m the mass of the object and a its acceleration. Our first task is to rewrite this in a form that better reflects the words:

rate of change of momentum = sum of forces

As usual in these derivations, we expect to arrive at a differential equation reflecting these words.

The left-hand side

Recall that the momentum of an object is its mass times its velocity. The velocity is of course a vector, with x, y, and z components. The Navier–Stokes equation is therefore a vector equation. We will develop here one of the three components of this equation, the x-component. Once this is developed, the other two components can easily be written. Here as usual we will take the object to be our control volume, or box of fluid, with sides dx, dy, and dz. If ρ is the density of the fluid, and u is the velocity in the x-direction, the x-component of the momentum is then $\rho u \, dx \, dy \, dz$. We must make a strategic decision about how to proceed. The frame of reference in Newton's case is the object itself: when following along with the object, its acceleration is the sum of the forces acting upon it. This is the Lagrangian frame of reference. If we allow our box of fluid to move about, we are in this frame. If it is held still, tacked down in space, we are in the Eulerian frame of reference. One can get between the two, but one must decide which frame to operate in as we proceed with the derivation. We will choose to work in the Lagrangian frame for now. This means that the rate of change of momentum must be written as if we are following the fluid. The derivative used in this case is the total derivative, or the material derivative, or the substantial derivative, or most informatively, the derivative following the parcel. So the left-hand side becomes

$$\frac{\mathrm{D}(\rho u \, dx \, dy \, dz)}{\mathrm{D}t} \tag{12.52}$$

where $\mathrm{D}/\mathrm{D}t$ is the total derivative operator, and is a shorthand for four terms:

$$\frac{\mathrm{D}}{\mathrm{D}t} \equiv \frac{\partial}{\partial t} + u\frac{\partial}{\partial x} + v\frac{\partial}{\partial y} + w\frac{\partial}{\partial z} \tag{12.53}$$

The first of these reflects the acceleration at a point, holding x, y, and z constant. The other three terms are called advective terms, and represent the advection of momentum, here the x-component of momentum. The mathematical signature of an advective term is that it has the form of $u(dA/dx)$, where u is the component of velocity in the x-direction. The quantity A could be anything: concentration of heat (temperature), concentration of momentum, concentration of fish ...

While we can easily factor out the volume of the parcel, $dx \, dy \, dz$, from Equation 12.52, we cannot easily separate the density from the velocity. We must chain rule each of the four terms that the total derivative represents. After some rearrangement, this yields

$$dx \, dy \, dz \left[\rho\left(\frac{\partial u}{\partial t} + u\frac{\partial u}{\partial x} + v\frac{\partial u}{\partial y} + w\frac{\partial u}{\partial z} \right) \right.$$
$$\left. + u\left(\frac{\partial \rho}{\partial t} + u\frac{\partial \rho}{\partial x} + v\frac{\partial \rho}{\partial y} + w\frac{\partial \rho}{\partial z} \right) \right] \tag{12.54}$$

or

$$dx \, dy \, dz \left[\rho\frac{\mathrm{D}(u)}{\mathrm{D}t} + u\frac{\mathrm{D}(\rho)}{\mathrm{D}t} \right] \tag{12.55}$$

Note that, in the incompressible case, the total derivative of density is zero, and the second term vanishes.

The right-hand side

The forces acting on this parcel of fluid come in two flavors: the body force, F_b, and the surface tractions, F_t:

$$\sum F = F_\mathrm{b} + F_\mathrm{t} \tag{12.56}$$

The body force acts on the center of mass of the fluid, while surface tractions act on the sides of the body. The body force with which we are concerned in geomorphology is that due to gravity: the weight of the parcel. (The other body force is magnetic force.) The component of the fluid weight acting in the x-direction is simply the mass of the fluid times the component of the acceleration due to gravity in the x-direction, g_x, or

$$F_{\mathrm{b}x} = \rho \, dx \, dy \, dz \, g_x \tag{12.57}$$

It will be this component of gravity in the downslope direction that drives flow down an inclined plane.

The surface tractions are harder to deal with. These come about from the action of one element of the fluid on another,

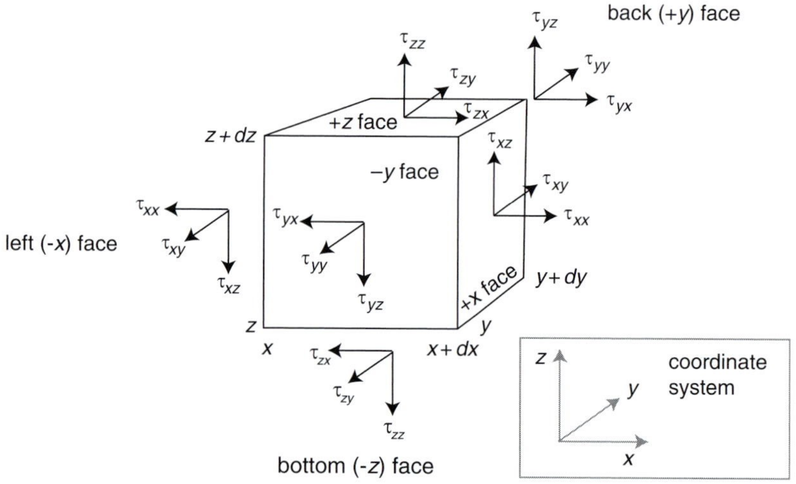

Figure 12.36 Definition sketch for system of stresses acting on a central volume. All stresses are positive as shown.

pushing or pulling it, or shearing it. The problem is that we have to recognize that these fluid forces act in directions that are both normal to a surface and tangential to it, and they act on all faces of the box. In other words, we have two directions to keep track of: the face on which the force is acting, and the direction in which it is acting. First, recall that a force is a stress times an area, in other words, a stress is a force per unit area. We are eventually going to sum up all the forces acting in the x-direction. Each of these forces will consist of the product of a stress times the surface area of the face on which the stress is acting. To be organized about this, we will obey the sign convention depicted in Figure 12.36, in which all of the stresses are depicted as positive. Normal stresses are taken to be positive in tension, in other words when they are directed away from the body. Shear stresses are taken to be positive when they act on the "downstream" face of the box, and are acting in the positive x-, y-, or z-direction. For example, the normal force acting on the left face of the box, in the x-direction, is the normal stress, $-\tau_{xx}$, where the minus sign reflects the fact that the stress as shown is acting in the negative x-direction, times the area of the side of the box, dydz.

The sum of the surface forces acting in the x-direction is then

$$F_{tx} = -\tau_{xx}(x)\mathrm{d}y\mathrm{d}z + \tau_{xx}(x+\mathrm{d}x)\mathrm{d}y\mathrm{d}z$$
$$-\tau_{yx}(y)\mathrm{d}x\mathrm{d}z + \tau_{yx}(y+\mathrm{d}y)\mathrm{d}x\mathrm{d}z \quad (12.58)$$
$$-\tau_{zx}(z)\mathrm{d}x\mathrm{d}y + \tau_{zx}(z+\mathrm{d}z)\mathrm{d}x\mathrm{d}y$$

There is one term for each of the six faces of the fluid element. Dividing this expression by the fluid volume dxdydz, and recognizing that there are three pairs of terms that reduce to derivatives in the limit as we shrink dx, dy, and dz to zero, allows us to collapse the full force balance equation to

$$\mathrm{d}x\mathrm{d}y\mathrm{d}z\left[\rho\frac{\mathrm{D}u}{\mathrm{D}t}+u\frac{\mathrm{D}\rho}{\mathrm{D}t}\right]=\mathrm{d}x\mathrm{d}y\mathrm{d}z\left[\frac{\partial\tau_{xx}}{\partial x}+\frac{\partial\tau_{yx}}{\partial y}+\frac{\partial\tau_{zx}}{\partial z}+g_x\right] \quad (12.59)$$

Dividing through by the mass of the fluid element, $\rho\mathrm{d}x\mathrm{d}y\mathrm{d}z$, and assuming that the fluid is incompressible ($\mathrm{D}\rho/\mathrm{D}t = 0$), yields a general equation for a force balance in the x-direction of an incompressible fluid:

$$\frac{\partial u}{\partial t}+u\frac{\partial u}{\partial x}+v\frac{\partial u}{\partial y}+w\frac{\partial u}{\partial z}=\frac{1}{\rho}\left[\frac{\partial\tau_{xx}}{\partial x}+\frac{\partial\tau_{yx}}{\partial y}+\frac{\partial\tau_{zx}}{\partial z}\right]+g_x \quad (12.60)$$

We must modify this slightly by recognizing that some stresses result in the distortion of the fluid, a change in its shape, while some result in changes in the size of the fluid element. The mean of the normal stresses acts from all sides and can accomplish a change in volume. We define the *pressure* as

$$P \equiv -\frac{1}{3}\left(\tau_{xx} + \tau_{yy} + \tau_{zz}\right) \quad (12.61)$$

where the minus sign represents the tradition that we think about a positive pressure as one that reduces the size of an element, while our convention for normal stresses is that they were defined to be positive in tension. Given this definition of pressure, we can then decompose the stresses into those parts that result in deformation of the fluid element, called the *deviatoric stresses*, from those that can change its volume, the *pressures*:

$$\tau_{xx} = \tau'_{xx} - P$$
$$\tau_{yy} = \tau'_{yy} - P \quad (12.62)$$
$$\tau_{zz} = \tau'_{zz} - P$$

where the primes represent the deviatoric stresses. Rewriting each of the stress terms in the force balance equation then results in

$$\frac{\partial u}{\partial t}+u\frac{\partial u}{\partial x}+v\frac{\partial u}{\partial y}+w\frac{\partial u}{\partial z}=\frac{1}{\rho}\left[\frac{\partial \tau'_{xx}}{\partial x}+\frac{\partial \tau'_{yx}}{\partial y}+\frac{\partial \tau'_{zx}}{\partial z}\right]+g_x-\frac{1}{\rho}\frac{\partial P}{\partial x} \qquad (12.63)$$

This is known as Cauchy's first law, first derived by A. L. Cauchy (1789–1857), a French mathematician (see Middleton and Wilcock, 1994, p. 307). Note that we have made no assumptions whatsoever about the nature of the materials involved. This is a very general equation reflecting conservation of momentum in an incompressible fluid. The fluid can be accelerated by either gravity or a pressure gradient.

But we are not done yet. This equation is said not to be "closed" in the sense that it is not an equation that can easily be solved for the velocity in the x-direction. The surface tractions, which come in as gradients of stresses, are not represented in terms of the flow speed, or gradients in flow speed. What is needed is a relationship between the stresses and the flow. We have some gut feeling that the flow outside of the box should push or pull or influence in some way the volume of concern. Put another way, the fluid reacts to stresses acting upon it. It does so by deforming at a given rate. This is the essence of a fluid: it responds to stresses by straining at a given rate. There is a unique relationship between stress and strain

rate in a given fluid. This relationship is called the constitutive equation, and it describes a rheology (from *rhein*, or flow). It was Newton who first described the most common of these rheologies, which he did by showing experimentally that there exists a linear relation between strain rate and stress. This may be written

$$\frac{\partial u}{\partial z}=\frac{1}{\mu}\tau_{zx} \qquad (12.64)$$

This can be solved for the stress, and re-written for each component. Most generally

$$\tau_{ij}=\mu\frac{\partial u_i}{\partial x_j} \qquad (12.65)$$

where *i* and *j* are indices representing x, y, or z. These expressions for stress can be plugged back into the force balance equation, yielding

$$\frac{\partial u}{\partial t}+u\frac{\partial u}{\partial x}+v\frac{\partial u}{\partial y}+w\frac{\partial u}{\partial z}=\frac{\partial(\mu[\partial u/\partial x])}{\partial x}+\frac{\partial(\mu[\partial u/\partial y])}{\partial y}$$
$$+\frac{\partial(\mu[\partial u/\partial z])}{\partial z}+g_x-\frac{1}{\rho}\frac{\partial P}{\partial x} \qquad (12.66)$$

If we treat first a simple system in which the viscosity is both isotropic (all 81 components representing relations between the nine components of the stress and the nine components of strain rate are equal) and does not vary in space, then we are

---

Box 12.5 The thermal analogy

We note that this equation is very similar to that describing the conservation of heat in a conducting medium:

$$\frac{\partial T}{\partial t}+u\frac{\partial T}{\partial x}+v\frac{\partial T}{\partial y}+w\frac{\partial T}{\partial z}=\kappa\frac{\partial^2 T}{\partial x^2}+\kappa\frac{\partial^2 T}{\partial y^2}+\kappa\frac{\partial^2 T}{\partial z^2}+S \qquad (12.70)$$

The gravity and pressure gradient terms serve as sources of momentum, just as radiogenic elements serve as the sources of heat, S, in the general heat equation. The gradients in stresses serve the same purpose as the gradients in heat flow, and are reflected in curvature of velocity on the one hand, and of temperature on the other hand. The analogy is made even tighter by recognizing that the position of the kinematic viscosity in the equation is analogous to the position of the thermal diffusivity. In fact, they have the same units.

Let us classify this Navier–Stokes equation. It is a partial differential equation for *u*-velocity (or speed), in that it contains derivatives of both time and all three spatial variables. It is second order, as the highest order of derivative is the second-order terms on the right-hand side. And it is nonlinear, in that the advective terms on the left-hand side contain products of the variable u and its derivatives. It is this latter quality that makes this equation particularly difficult to solve for the most general cases.

justified in pulling viscosity out of the derivatives, yielding three terms dependent upon the curvature of the velocity profile:

$$\frac{\mu}{\rho}\frac{\partial^2 u}{\partial x^2} + \frac{\mu}{\rho}\frac{\partial^2 u}{\partial y^2} + \frac{\mu}{\rho}\frac{\partial^2 u}{\partial z^2} \qquad (12.67)$$

The collection of two constants in front of each term arises so commonly in fluid mechanics problems that they have been given their own name, the kinematic viscosity:

$$\nu = \mu/\rho \qquad (12.68)$$

which has dimensions of L^2/T. Our final equation for the conservation of momentum in the x-direction is then

$$\frac{\partial u}{\partial t} + u\frac{\partial u}{\partial x} + v\frac{\partial u}{\partial y} + w\frac{\partial u}{\partial z} = \nu\frac{\partial^2 u}{\partial x^2} + \nu\frac{\partial^2 u}{\partial y^2} + \nu\frac{\partial^2 u}{\partial z^2} + g_x - \frac{1}{\rho}\frac{\partial P}{\partial x} \qquad (12.69)$$

This is the x-component of the *Navier–Stokes equation*. The French engineer Claude Navier (1785–1836) and the English mathematician George Stokes (1819–1903) derived this equation in the early nineteenth century, a step beyond Cauchy's first law taken by representing the deviatoric fluid stresses with a viscous rheological law. There are two other equations for the other two components of momentum, y and z, which we will leave to the reader to write.

The assumptions that we have used to arrive at this form of the equation are:

- the material is incompressible;
- the material behaves as a linear viscous fluid;
- the viscosity does not vary spatially.

Non-dimensionalization of the Navier–Stokes equation

It would be immensely helpful to the solution of this equation if we could simplify it by being able to ignore one or another term. One approach is to determine formally how large each of the terms is in a particular problem. Those that are large relative to others are retained, while those that are small are ignored. This is accomplished by scaling the equation, and converting it to a dimensionless form. Each of the variables can be scaled by a characteristic scale in the particular problem; this is an exercise in normalization. For example, the length scales in a grain settling problem will be the diameter (or the radius) of the particle (which we choose is in part dictated by tradition!). The velocity scale in open channel flow might be taken to be either the surface velocity or the mean velocity. This can be achieved by defining

$$T' = t/T, x' = x/L, y' = y/L, z' = z/L, \qquad (12.71)$$

$$u' = u/U, v' = v/U, w' = w/U$$

where T, L, and U are the characteristic scales chosen for time, length, and velocity, respectively, and the primes indicate new, non-dimensionalized variables. These definitions can be used in rewriting the original equation to become

$$\left[\frac{U}{T}\right]\frac{\partial u'}{\partial t} + \left[\frac{U^2}{L}\right]u'\frac{\partial u'}{\partial x} + \left[\frac{U^2}{L}\right]v'\frac{\partial u'}{\partial y'} + \left[\frac{U^2}{L}\right]w'\frac{\partial u'}{\partial z'}$$

$$= \left[\nu\frac{U}{L^2}\right]\frac{\partial^2 u'}{\partial x'^2} + \left[\nu\frac{U}{L^2}\right]\frac{\partial^2 u'}{\partial y'^2} + \left[\nu\frac{U}{L^2}\right]\frac{\partial^2 u'}{\partial z'^2} + g_x - \left[\frac{1}{\rho}\frac{P_o}{L}\right]\frac{\partial P'}{\partial x'} \qquad (12.72)$$

Here we have collected all the scales in the problem into the square brackets. The remaining variables and gradients are not only dimensionless, but should have values that are of the order of 1 to a few, if we have done our scaling correctly (and this is the art of the exercise). The magnitudes are then all carried by the scales in the brackets. It is by comparing the magnitudes of these collections of constants in brackets that one can assess the relative importance of one or another term in the equation. There is one more step, which makes this comparison even easier. If we divide all of the terms by the scales in front of, say, the inertial or advective acceleration terms, the resulting equation will be rendered dimensionless. Even more importantly, the dimensionless numbers in front of each term will reflect the importance of that term relative to the inertial terms. Let us do this:

$$\left[\frac{L}{UT}\right]\frac{\partial u'}{\partial t} + [1]u'\frac{\partial u'}{\partial x} + [1]v'\frac{\partial u'}{\partial y'} + [1]w'\frac{\partial u'}{\partial z'}$$

$$= \left[\frac{\nu}{UL}\right]\frac{\partial^2 u'}{\partial x'^2} + \left[\frac{\nu}{UL}\right]\frac{\partial^2 u'}{\partial y'^2} + \left[\frac{\nu}{UL}\right]\frac{\partial^2 u'}{\partial z'^2} + \frac{g_x L}{U} - \left[\frac{P_o}{\rho U^2}\right]\frac{\partial P'}{\partial x'} \qquad (12.73)$$

Assure yourself that each of the terms in the brackets is dimensionless. These dimensionless numbers, or slight rearrangements of them (their inverse, or one half of them) are the important dimensionless numbers of fluid mechanics. They are as follows:

Re, Reynolds number $= \frac{UL}{\nu}$

Fr, Froude number $= \frac{U}{\sqrt{g_x L}}$ (pronounced Frood)

Eu, Euler number $= \frac{P_o}{\frac{1}{2}\rho U^2}$

St, Strouhal number $= \frac{UT}{L}$

Given these dimensionless numbers, all named for famous mathematicians and fluid mechanicians, we can now re-write the last equation to yield

$$\left[\frac{1}{St}\right]\frac{\partial u'}{\partial t}+[1]u'\frac{\partial u'}{\partial x}+[1]v'\frac{\partial u'}{\partial y'}+[1]w'\frac{\partial u'}{\partial z'}$$

$$=\left[\frac{1}{Re}\right]\frac{\partial^2 u'}{\partial x'^2}+\left[\frac{1}{Re}\right]\frac{\partial^2 u'}{\partial y'^2}+\left[\frac{1}{Re}\right]\frac{\partial^2 u'}{\partial z'^2}+\frac{1}{Fr^2}-[2Eu]\frac{\partial P'}{\partial x'} \tag{12.74}$$

It should now be clear that the utility of these numbers is that they allow us to determine rapidly the importance of each term relative to the inertial term. Most important for our purposes now is the Reynolds number. Note that if Re is small, then the magnitude of the viscous terms becomes large compared to the inertial terms. The latter can then safely be ignored. As these inertial terms are the nonlinear ones, this is tremendously useful. The Reynolds number is small if the product of the length and velocity scales is small compared to the kinematic viscosity. This number was named for Osborne Reynolds (1842–1912), a civil and mechanical engineering professor at Owens College, Manchester, working in the late nineteenth century on the onset of turbulence in flow through tubes, although George Stokes (1819–1903) had pointed out the importance of this collection of constants in controlling the flow field for particular geometries 40 years earlier (e.g., what we now call Stokes law for the drag force on a sphere; see Batchelor, 1970, p. 214). Reynolds published a pair of papers that have become pillars in turbulence research (Jackson and Launder, 2007). In the first he reported his experiments on the "tendency of water to eddy" and the control of this tendency by a ratio of scales that now carries his name (Reynolds, 1883), and a theoretical approach to averaging of the Navier–Stokes equations (to yield what are now known as the Reynolds equations) as a means of exploring the switching of flows into and out of turbulence (Reynolds, 1895).

Problems

1. Estimate the time it takes for a water molecule to travel from the source of the Amazon to the ocean.

2. *Flow competence*. Assess the maximum diameter grain capable of being entrained by a flow 2 m deep, travelling down a channel with a slope of 0.002. The particles are granitic and have a density of 2700 kg/m^3. Use the simple expression presented in the text, employing the Shield's parameter of $\theta = 0.06$. Give the diameter in mm.

3. A commonly used bedload transport formula in fluvial settings is the Meyer–Peter–Muller formula. A simple version of this is $Q = A(\tau_b - \tau_c)^{3/2}$, where $A = 0.04$ (the units of A work out such that Q is in kg/(m s), and shear stresses are in N/m^2). Using this formula, and a coarse sand of diameter $D = 5$ mm in the channel, and flow conditions described in question 2, what do you expect for the discharge of sand in this channel? Give your answer in kg/s.

4. *Mississippi River velocity structure and sediment transport*. You are given a set of measurements of the flow velocity within a profile of the Mississippi River, at this place and time flowing with a flow depth of 8.5 m. The measurements are as follows:

z (m above bed)	U (m/s)
7.7	1.25
5.2	1.20
3.2	1.05
1.8	1.00
0.9	0.90
0.6	0.80

(a) Recalling the *law of the wall*, what is your best estimate of the shear velocity, u_*? What is the roughness length of the flow, z_0? (*Hint*: recall von Karman's constant, $k = 0.4$.) Plot your results in both linear–linear and log–linear fashion (linear in U as the x-axis and log of z on the y-axis), showing both the data and your predicted velocity structure, using your best estimates for u_* and z_o.

(b) Given this, what is the slope of the river in this reach? (Note that this is actually a very difficult measurement to make in these cases.) You will have to use the definition of the shear velocity for this calculation: $u_* = \sqrt{gHS}$.

(c) Would coarse quartz silt of diameter 0.125 mm be in suspension here? (*Hint*: the criterion for suspension is that the Rouse number $p < 2.5$, see Chapter 14.)

5. *Fluid discharge.* Consider a steep-walled river channel with a width of 50 m, a bankfull depth of 3 m (to the levee tops), and a slope of 0.5 m/km (i.e., note that this is the tangent of the slope, which for small angles is equivalent to the slope, in radians: 0.0005). The bed of the river is made of coarse sand, organized into ripples that result in an effective roughness length (z_o) of 3 mm. The river at the moment is flooding at a stage of 0.5 m above bankfull. Recall that the shear velocity u_* is defined as $u_* = \sqrt{\tau_b/\rho}$, or $u_* = \sqrt{gHS}$. Calculate the water discharge in the channel, in m³/s:

(a) at bankfull;

(b) at 0.5 m above bankfull (i.e., at present).

(Note that in both cases you may ignore the effects of the banks, which will be small relative to the bottom in imposing friction on the flow.) In the latter calculation, ignore the contribution from that flow that escapes over the floodplain.

6. You now obtain a water sample of the same river from a specific depth, 10 cm above the bed, and have its grain-size distribution analyzed. There are three primary grain diameters involved (all quartz; assume that they are all essentially spherical), and their sizes and concentrations at this level in the flow are as follows:

D1	0.25 mm	45 mg/L
D2	0.17 mm	30 mg/L
D3	0.01 mm	20 mg/L

(a) What is the Rouse number, p, of each of the grain sizes when the flow is 0.5 m above bankfull? (See Chapter 14.)

(b) Plot the expected concentration profiles of these grains as a function of height above the bed, using height (z) as the vertical axis on your plot. First plot the profiles on a linear–linear plot, then on a log–log plot.

(c) What is the *mean* concentration of sediment (of each grain size) in that part of the flow that goes overbank?

7. For a given discharge of a river, calculate and plot the dependence of the river stage on the roughness of the bed of the river. (This is relevant to management plans for how to modify a river to maximize the water it can pass below a given stage.)

8. Estimate the mean residence time of a sand grain on a floodplain. Assume that the sand grain was deposited in a scrollbar deposit as the river meandered within the floodplain. The floodplain is neither aggrading nor eroding. The floodplain width is 5 km, the channel width is 80 m, and aerial photos indicate that it is migrating laterally at a mean rate of 10 cm/year.

9. *Thought question.* Given your answer to this last question, contemplate how this would be reflected in the concentration of ^{10}Be in the sand contributed to the river from bank erosion.

10. *Thought question.* Discuss how rivers might differ if the fluid being transported were of a different density or viscosity, or both. (And lest you think this silly, we are learning about a landscape on Titan that appears to have been in part sculpted by liquid methane.)

Further reading

Dunne, T. and L. B. Leopold, 1978, *Water in Environmental Planning*, San Francisco, CA: W.H. Freeman Co., 818 pp.
This text sets the bar for an environmental planning book, bringing the science of water in the landscape to the planner.

Henderson, F. M., 1966, *Open Channel Flow*, New York: Macmillan, 522 pp.
This is a well-organized textbook that many turn to for clear discussion of river mechanics problems.

Knighton, D., 1998, *Fluvial Forms and Processes: A New Perspective*, 2nd edition, Oxford: Arnold, 400 pp.
This text provides a more up-to-date analysis of river mechanics and its landforms than that given in Leopold,

Wolman and Miller. This updated version of the text leans heavily on recently collected data, and draws on river studies from around the world.

Leopold, L. B., 1994, *A View of the River*, Cambridge, MA: Harvard University Press, 290 pp.
This very readable book provides a window into what Luna Leopold thought about rivers, over the many decades in which he was involved in their study and in their protection.

Leopold, L. B., M. G. Wolman, and J. P. Miller, 1964, *Fluvial Processes in Geomorphology*, San Francisco, CA: W. H. Freeman, 522 pp.
This is a classic, written by those who brought the mechanics of rivers into the geomorphic realm.

CHAPTER 13

Bedrock channels

Every river appears to consist of a main trunk, fed from a variety of branches, each running in a valley proportioned to its size, and all of them together forming a system of valleys, communicating with one another, and having such a nice adjustment of their declivities that none of them join the principal valley, either on too high or too low a level; a circumstance which would be infinitely improbable, if each of these valleys were not the work of the stream that flows in it.

Playfair, 1802, *Illustrations of the Huttonian Theory of the Earth*, p. 102

In this chapter

In some of the earliest writing on geology, Hutton and Playfair noted the tendency of rivers and their tributaries to join smoothly together. Playfair, by far the better writer of the two, stated it most elegantly in the quote with which we begin the chapter. Rivers appear to adjust themselves such that the tributaries meet their trunk streams at grade, as sketched in Figure 13.1. This simple observation has become known as Playfair's law. Note that this contrasts with glacial valleys, in which tributaries often reach the trunk stream well above grade, and hence are said to "hang." Indeed, in the glacial hanging valley cases, rivers did not do the work of forming the valleys; glaciers did. In this chapter we address how rivers erode their channels when they encounter bedrock.

View looking up-river in the Indus River middle gorge. Scalloped bedrock 15 m above mid-April low flow demonstrates the large annual swing of river stage. Note human figure crouched beside flow in middle distance (photograph by R. S. Anderson).

Uplift of rock by tectonic processes sets in motion a series of geomorphic events that are led by the action of rivers. First, the slopes of rivers draining the uplifted region are steepened, promoting river incision into bedrock. River incision subsequently promotes hillslope erosion. The pace of the response of a landscape to uplift or to tectonics in general is therefore set by the response of its rivers. Because bedrock channel incision controls landscape response, we consider this aspect of the geomorphology of rivers separately. In this chapter we will survey modern understanding of river incision into bedrock. We describe how we measure bedrock incision rates at several timescales, followed by a summary of the specific processes responsible for bedrock erosion. In this survey we introduce the concept of stream power, which is heavily utilized in the literature on river incision into bedrock. We then address the nature of the long valley profiles of bedrock rivers, meaning how they should look in simple (steady uplift, uniform rock) conditions, and in more complex (transient uplift, variable rock) conditions. In the end we will summarize the outstanding issues, as this topic remains a very active line of inquiry within the geomorphic community.

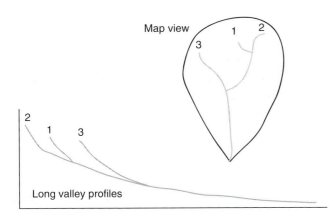

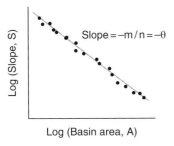

Figure 13.1 Top: map view of channel pattern in a drainage basin, and profiles of main stem and tributaries. Bottom: profile shown in log–log space. Straight line implies a power-law relationship.

Measurement techniques

River incision into bedrock is measured over several timescales using a wide range of techniques. In general, one studies this process best by choosing either very weak bedrock (clay badlands, for example), or a large river with a lot of power. Over long times, rivers leave behind markers of their former positions in the landscape. These include abandoned channel beds cut into bedrock, called strath terraces; lava flows that filled former channels, essentially fossilizing them; and certain kinds of cave passages, shown in Figure 13.2. The age of these features, T, and their height above the present channel, dz, yields the long-term incision rate, $e = dz/T$.

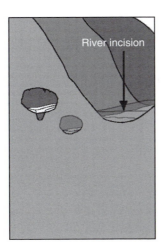
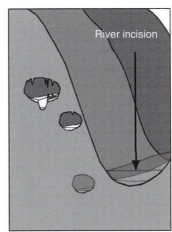

Figure 13.2 Use of caves abandoned in the walls of river canyons to deduce river incision histories. Abandoned passages can contain quartz-rich sediment that can be dated using concentrations of cosmogenic radionuclides (figure courtesy of Greg Stock).

Straths

Strath terraces are formed when a river cuts laterally into bedrock. They are abandoned when the river incises downward, leaving behind a bedrock bench that is often mantled with a thin cover of sediment (Hancock and Anderson, 2002). In some instances, especially in hard rock, the bedrock of these benches can be beautifully ornamented with potholes and flutes that clearly reflect that the bench was once the channel bed. In weaker bedrock, the bedrock morphology is less spectacular, and can be nearly planar. The mantling sediment can be coarse to fine alluvium, and up to many meters thick. They should not be confused with alluvial or fill terraces, in which the entire feature is composed of sediment.

Lava flows

Because it is a fluid, lava flows down channels. When this happens, the lava instantly buries any sediments on the channel bed. If the cooled lava flow is less erodible than the local bedrock, subsequent fluvial erosion will favor the local rock, causing the channel to migrate to the side of the lava top. The abandoned lava flow is a cast, fossilizing the river channel. One example of this is the tablelands of the western Sierra Nevada foothills, in which sinuous basalt-capped ridges stand high in the present landscape. In this case, the landscape is said to have been "inverted": what was once the channel floor is now under a volcanic veneer at the ridge top. If the lava flows can be dated, for example by Ar/Ar methods, a minimum rate of incision can be estimated by dividing the height of the base of the lava flow above the modern river channel by its age.

Caves

Some types of roughly horizontal cave passages can be identified as having been formed at or very near the baselevel imposed by the nearby surface stream (Palmer, 1991). These "phreatic" (meaning relating to underground water below the water table) passages are then abandoned as the stream incises further, sometimes developing a characteristic keyhole cross section illustrated in Figure 13.2. When sediment mantling the floors of these passages, or tucked in alcoves in them, can be identified as bedload from the adjacent river, the possibility of dating this

material exists. For example, if the sediment is granitic, and the cave is formed in marble, the sediment is clearly allochthonous (meaning it came from elsewhere). The time since the sediment was deposited in the cave can be used to calculate an incision rate. In essence, such caves serve the same role as strath terraces. However, because they are internal to the rock, they are relatively immune to removal by landsliding, and can therefore record even longer-term incision histories. Of course, carbonate bedrock is required, meaning this method is restricted to specific geologic settings.

Methods for constraining the ages of caves include paleomagnetism, dating of basal ages of speleothems, and calculation of burial ages using the concentrations of cosmogenic radionuclides (see Stock et al., 2005). All but this last method pose significant problems. Paleomagnetism is a coarse tool; the last full reversal of the field, the Bruhnes-Matayama, was roughly 700 ka. It has also been shown that the basal ages of speleothems (calcite cave deposits – usually stalagmites) are not trustworthy, and always post-date the true age of the cave derived by other more reliable methods. This reflects the fact that speleothems form very slowly from seeping water, sometimes individual drips from the ceiling, the beginning of which can significantly post-date the opening of the passage. As discussed elsewhere (Chapter 6) any quartz-bearing sediment that is buried deeply enough to be out of range of new cosmogenic production can be dated using the ratio of $^{26}Al/^{10}Be$. The ratio of these cosmogenically produced nuclides decays at a rate dictated by the decay rates of the two nuclides:

$$R = \frac{[^{26}Al]}{[^{10}Be]} = \frac{[^{26}Al]_o e^{-t/\tau_{Al}}}{[^{10}Be]_o e^{-t/\tau_{Be}}} = R_o e^{-t(1/\tau_{Al} - 1/\tau_{Be})} \quad (13.1)$$

The timescale over which this method works is several times the characteristic time, $\tau = 1/(1/\tau_{Al} - 1/\tau_{Be}) = \tau_{Al}/[1 - (\tau_{Al}/\tau_{Be})]$, which is about 2.07 Ma for the ^{10}Be–^{26}Al pair. Caves may therefore be dated back to about 5 Ma, which is a very useful time span for river evolution. In the Kings Canyon example illustrated in Figure 13.3, you can see that this long time span and the presence of numerous caves in the walls of the canyon provides unique control on the history of incision, revealing in this case a tenfold decline in incision rate.

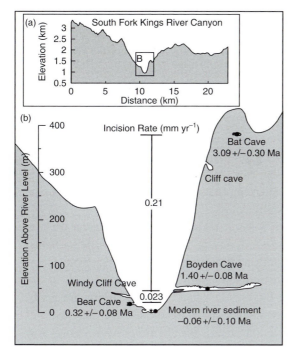

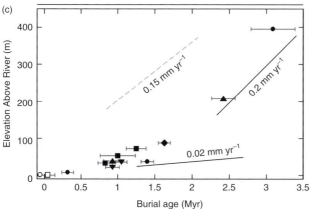

Figure 13.3 Erosion history of Kings River Canyon based upon multiple cave ages derived from ^{26}Al/^{10}Be in cave sediments. Rapid incision from 3 Ma to ~1 Ma appears to have been followed by much lower rates since then (after Stock *et al.*, 2004, Figures 2 and 3).

Cosmogenic radionuclides on the channel floor

In yet another application of cosmogenic radionuclides, erosion rates of the bedrock floor of a channel can be measured over millenial timescales. Consider sampling a piece of bedrock from the channel floor at low flow conditions. Just as for measurement of bedrock lowering rates at points on hillslopes or summits, the concentration of radionuclides will reflect the time a rock has spent within the nuclide

production zone. Ignoring decay (appropriate for rapid erosion rates), the concentration is

$$C = P_{\text{eff}}T = P_{\text{eff}}z_*/E \qquad (13.2)$$

where z_* is the production length scale (roughly 60 cm in rock), T the time the rock took to be eroded the last z_*, and E the rate of erosion. The only trick lies in calculating the effective production rate, P_{eff}. This must be adjusted not only for latitude and altitude, as we have discussed in Chapter 6, but for the shielding by water at stages above the level of the sampling site (Hancock *et al.*, 1998). This can be done with the aid of a probability distribution of flow depths, and the density of water. The instantaneous shielding due to the water column will be

$$F = e^{-((H-z)/z_{*\text{w}})} \qquad (13.3)$$

where $(H - z)$ is the thickness of the water above the sampling elevation, z, and $z_{*\text{w}}$ is the length scale for attenuation of cosmogenic radionuclide production in water (about 1.6 m). The results are average rates over a specific timescale, set by the time it took for the rock to come the last z_* to the surface. This time is $T = z_*/E$. For example, the Indus River erodes at rates of at least 1 mm/yr, making the averaging timescale roughly 600 years. This nicely fills a gap between rates deduced from real time monitoring (from one to a few years) and those deduced from dating terraces or caves.

Short-term monitoring

Short-term incision rates can be measured using erosion pins or holes sketched in Figure 13.4(a). Holes can be either hand or machine-drilled into bedrock surfaces of the bed of the channel, measured carefully with a micrometer, and then re-measured a season or a year later. A variant of this method involves drilling holes that are used to set up a monument bar, from which erosion can be measured at many places along a profile. These methods avoid the dependence on the accidents of nature needed to form a strath, or a cave, but are limited to channels with relatively high erosion rates. Repeated measurements of this sort can resolve small fractions of 1 mm of erosion. They can also be used to address more fine-scale issues relevant to the specific processes accomplishing the erosion.

(a) *Erosion hole method*

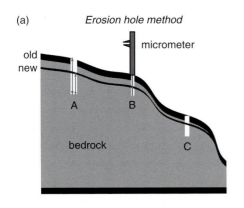

(b) *Erosion monument method*

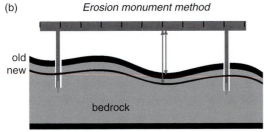

Figure 13.4 Methods for measuring short-term erosion rates in bedrock channels. (a) Erosion holes whose depths are measured with a micrometer, which is repeated to determine the change in depth of the hole. (b) Monument method in which monument mounts are drilled into bedrock of channel floor, to which a rigid bar is mounted. The latter method allows measurement at many locations (ticks on bar) between the holes.

Figure 13.5 Fluted bedrock in channel of middle gorge, Indus River, Pakistan, as it passes through the Himalayas. Flow is from left to right, pencil for scale. Lowering of the gneissic bedrock occurs by upstream propagation of small-scale flutes, which ornament larger scale bed undulations. At this and similar sites, bedrock lowering rates of several mm/yr have been documented by repeated measurement of erosion holes (photograph by R. S. Anderson).

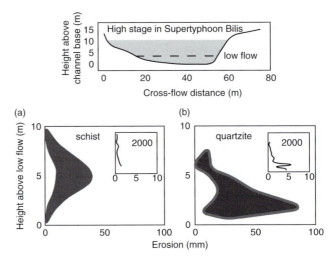

Figure 13.6 Documentation of channel erosion in a bedrock gorge in Taiwan during Typhoon Bilis. Erosion patterns in both schist (left) and quartzite (right) show maxima well above the low-flow stage, and decline smoothly to zero at the maximum stage recorded (redrawn from Hartshorn *et al.*, 2002, Figure 3, reprinted with permission from the American Association for the Advancement of Science).

We illustrate these methods with two examples. Hancock *et al.* (1998) report results from the middle gorge of the Indus River in which the river crosses a band of various rock types in an area known to be exhuming at rapid rates from fission track analysis. The depth of the Indus River in this region swings by 10–15 m annually, reflecting snowmelt in the 100,000 km^2 drainage area above it. This allowed access to portions of the bed in April that would surely be covered by >10 m of sediment-laden water for a couple of summer months, and would be exposed again when the stage falls in the winter months. Hancock used erosion holes of the sort sketched in Figure 13.4(a) to document up to 12 mm of erosion in a single year, with strong dependence on rock type and on small-scale flow dynamics. As seen in the photograph of Figure 13.5, some of the pencil-thin holes had nucleated small erosion scour flutes in the downstream direction.

Employing a variant of the erosion hole monument method, Hartshorn *et al.* (2002) documented erosion caused by the huge super-typhoon Bilis as it hammered the highlands of Taiwan in 2000. Their results reproduced in Figure 13.6 reveal that the schist bedrock channel floor was lowered by up to 100 mm in this single event, more than an order of magnitude more than the erosion in an entire year of more normal hydrologic conditions. Interestingly, they report a pattern of

erosion that displays a maximum at about 5 m above the low flow stage, falling to zero at the high-water mark (10 m). This is yet another example of the importance of large events in geomorphology, and must be captured in any theory of fluvial bedrock erosion.

Erosion processes

We now review the processes that are important in driving the erosion of bedrock in channels. We begin with a short introduction to the approach taken most commonly in modeling fluvial profile evolution, the stream power approach, as it sets the context for more specific physics.

The stream power approach

A river is water falling down a sloping channel. The water uses gravitational potential energy as it moves down. Conservation of energy demands that the total energy of the system is conserved. The energy loss in falling must therefore be transformed to some other form. Some of it may be used to perform work of erosion on the channel bed. What is the rate at which energy is made available for work? The rate of potential energy loss has been called the stream power (recall that power is defined as energy per unit time). It is expressed per unit length of channel, or as the *unit stream power*, per unit area of the bed.

Consider a channel shown in Figure 13.7 with a width W, a flow depth H, and slope S. The potential energy of a parcel of fluid of volume V is ρVzg, or $\rho Hdx Wzg$. The rate of loss of potential energy by this parcel as it moves through the channel is dictated by the rate of loss of elevation, set by the flow speed and the channel slope, US. We may now formulate the rate of potential energy loss per unit length of channel, dx, as

$$\Omega = \rho g HWUS \tag{13.4}$$

where the dx has cancelled out, and we have defined $S = \tan \theta$. Noting that the quantity HWU is equal to the water discharge, Q, this reduces to

$$\Omega = \rho g QS \tag{13.5}$$

Finally, if we formulate this as a rate of power loss per unit area of bed by dividing the stream power per unit length of channel by the width, W, the result is the unit stream power,

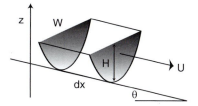

Figure 13.7 A mass of water $\rho Hdxdy$ has potential energy mgz. Since neither its mass nor gravity change as it moves, the water parcel loses potential energy at a rate dictated by the slope of the channel and the rate at which it loses elevation. The rate of loss of elevation, dz/dt, is in turn governed by the mean down-channel speed of the water parcel, U, and the slope of the bed. In other words, dz/dt = U tan(θ) = U sin(θ) for small angles.

$$\omega = \rho g QS/W \tag{13.6}$$

Noting that the shear stress at the bed of the channel is $\tau_b = \rho g HS$, an equivalent expression for unit stream power is

$$\omega = \tau_b U \tag{13.7}$$

and channel width is only slowly changing, then one can see the appeal of the stream power formulation and write

$$dz/dt = k\omega = k\rho g QS/W \tag{13.8}$$

where the constant k reflects the erodibility of the channel, among other things. This formulation suggests that the first-order control on river incision should be the local discharge of water, the local slope of the channel, and the channel width. One chief attraction of this approach is that two of these variables can be derived from maps. The channel slope may be derived from crossings of contours (or more recently from analysis of DEMs, for example using Rivertools (Peckham, 1998)). Recall the results of Seidl and Dietrich (1992), in which they deduced that erosion rates were proportional to the area–slope product, AS. If the water discharge is proportional to drainage area, then $AS \sim QS$, the proportionality constant reflecting the effective precipitation. The river discharge can at least crudely be associated with the local drainage area, which again can be assessed from maps using either a planimeter or analysis of DEMs if simplifying assumptions about the effective runoff can be justified. We note that this latter assumption breaks down if precipitation or effective runoff is highly non-uniform (see below about orographic effects). In addition, channel width cannot be derived from either

maps or commonly available DEMs with a resolution of 10–30 m, although recent LiDAR and stereo-photo methods can generate DEMs with the necessary 1 m resolution. In addition, specific processes may result in formulations that differ significantly from the stream power approach we have outlined.

The stream power formulation, or its variant, the shear stress formulation, does not identify the specific process responsible for incision. It relies upon an energy argument. A certain amount of energy is available from the water, and it may be used in a variety of ways. It could be used to heat up the water, to pick up sediment from the bed, or to generate cracks in rock that ultimately lead to erosion of the rock. The value of k may be conceived as including an efficiency factor, reflecting what fraction of this available energy goes into eroding the rock. We seek here to be more specific about the processes accomplishing the erosion. There is much debate about the efficacy of each of these processes. Their relative importance depends strongly upon the lithology (hardness, fissibility, pre-existing planes of weakness, and so on), the sediment load, and the fine-scale flow mechanics.

We can approach the problem as follows. Say lowering of the bedrock occurs in layers of thickness δ, which are connected to one another by bonds of strength E (energy required to break a bond), and of density B bonds/unit area. The energy needed to detach the slab is therefore BE. This combination is akin to what is called the work of fracture, and has units of energy per area. The mean lowering rate, averaged over many detachments, is

$$\frac{dz}{dt} = -\frac{\delta}{T} \qquad (13.9)$$

where z is the bed elevation, t is time, and T is the specific time needed to accumulate enough energy to break all these bonds. The time T may be simply calculated from

$$T = \frac{BE}{\alpha\omega} \qquad (13.10)$$

where ω is the rate of delivery of energy per unit area of bed [$= E/L^2T$], or the unit stream power, and α is the fraction of this power converted into breaking of bonds. Recall that unit stream power is the energy made available by dropping a parcel of water down the potential gradient per unit time and unit area of bed. It has units of M/T^3, or kg/s^3 in SI units.

Substituting the expression for the timescale T into the general erosion equation (Equation 13.9), yields

$$\frac{dz}{dt} = -\frac{\alpha\delta}{BE}\omega \qquad (13.11)$$

From this we can see the ingredients of k, the constant typically seen in the stream power formulation (Equation 13.8). Note that k has units of LT^2/M, or $m\ s^2/kg$ in SI units. But the way to think of it is as the product of an efficiency factor (and a pretty small one) with a slab thickness divided by the energy needed to break a unit area of such a slab free from the slab beneath it. The bigger the slabs, and the more efficiently the energy is delivered to the bed, the faster the erosion. The weaker the bonds (lower E), and the fewer bonds there are per unit area (lower B), the more rapid the erosion.

One can also accommodate a threshold in such a formulation if we acknowledge that the delivery of energy to the bed comes as packets and that the packets come in a variety of sizes (impacts of turbulent eddies, impacts of particles embedded in them). One can imagine that some of these are insufficient to break any bonds, meaning that this portion of the distribution of power does not participate in lowering the bed.

We now address those specific erosional processes that have been identified as being responsible for erosion in bedrock channels.

Abrasion

Rock can be worn away by particles directly impacting the surface, causing abrasion of the surface. The impacting particles can be of all sizes, and can approach at all angles and speeds. Each impact either promotes the growth of a micro-crack, or extends existing cracks sufficiently to separate a small piece of the impacted rock. The erosion rate depends upon the properties of the flow, the amount and caliber of the sediment, and the susceptibility of the rock to abrasion. While there is no full theory of the physics of abrasional wear, we can appeal to what has been learned in the study of ventifaction, or abrasion by particles entrained in air (see Chapter 15 on sediment transport by wind). The basic physics are parallel.

We can make much progress with two facts: (1) The particles must hit the surface, and (2) when a particle does impact, the mass of rock ejected from the impacted surface is proportional to the kinetic energy of the impacting particle. These statements can be combined by noting that the erosion or lowering rate of a rock surface is proportional to the flux of kinetic energy to that surface, q_{ke}. This may be written as

$$q_{ke} = C m_p U^3 \qquad (13.12)$$

where C is the near-bed volumetric concentration of particles in the flow, m_p is the mass of an individual particle, and U the mean near-bed flow speed (which sets the speeds of the particles). The third power of the speed comes from the product of the kinetic energy of a particle, which goes as U^2, and the flux of particles to the surface, which goes as U. If in addition the particle concentration, C, depends upon the flow speed, then the dependence on flow speed will be even stronger. For example, in eolian systems, $C \sim U^2$, implying a fifth-power dependence on wind speed. The point is that the abrasion rate ought to depend very strongly on flow speeds. The nonlinearity in q_{ke} amplifies greatly the importance of high flow events, and indeed the peak flows in a given year.

But where and when do particles impact a surface? This depends strongly upon the details of the flow field at the bed. Clearly, if the bedrock is mantled by sediment, particles entrained in the flow cannot impact the bedrock. But if the bedrock is exposed, the impacts of particles can be highly non-uniform. The trajectories of water parcels become tangent to the non-porous bedrock surface as they approach it. Hence, in order to impact the surface, the suspended sediment particles must decouple from the flow. This decoupling is more likely over highly curved surfaces: the stronger the curvature, the more likely a particle will get flung against the surface, decoupling from the fluid due to its own momentum. The more massive the particle, the greater its momentum, and the lower the curvature necessary to cause decoupling. Consider a bump in the bedrock bed, or an immobile boulder on the bed shown in Figure 13.8. The flow responds to this obstacle to the flow well ahead of the bump, and the flow lines begin to deflect in response to the pressure field set up by it. On the rear of the bump, the flow lines reconverge. Under many natural conditions of flow speed and bump size, the flow "separates" in the lee of the obstacle, creating a zone occupied by tightly wound eddies. The flow curvature on the front of the obstacle is small, while the separation eddies display strong curvature. The result is the stronger decoupling of particles from the flow in the rear of obstacles, generating what has been called the "attack from the back."

Which grain sizes do the work? Is it particles participating in bedload or suspended load that accomplish the abrasion of the bed? While suspended particles are traveling at nearly the speed of the water, and therefore possess significant kinetic energy, bedload particles re-encounter the bed periodically and travel at only small fractions of the near-bed flow speed. In addition, bedload particles will impact neither the rear of an obstacle, nor the base of overhangs. Yet ornamentation of the river channel is commonly observed at these sites. Very fine sediment, on the other hand, what we call wash load, is so small that it is very tightly coupled to the flow, and will not diverge from it even when the flow displays strong curvature. The above observations suggest that it should be the medium-size fraction, the coarse fraction of the suspended load, that accomplishes the majority of the abrasion.

This is not to say that we may ignore bedload. Indeed, laboratory experiments performed by Sklar and Dietrich (2001) in rotating abrasion mills allow us to explore the role of bedload in abrasion. These experiments, the results of which are illustrated in Figure 13.9, have suggested that the abrasion rate is strongly dependent upon the strength of the bedrock being eroded; the erosion rate goes as the inverse square of the tensile strength, and is higher when the abrading sediment is stronger. More intriguing is the relationship between sediment load and erosion rate, shown in Figure 13.10. The dependence of erosion on the tensile strength suggests a means of collapsing many experiments onto one plot by normalizing the erosion rate with the square of the tensile strength of the bed. The results imply that erosion should be most efficient under some optimum sediment load. Below this load, the number of impacting particles limits bedrock erosion, while above it, the great number of particles near the bed prevents particle impact with the bed, and effectively shields it from erosion. These experiments have greatly influenced subsequent research on bedrock river incision. Some maintain that one can craft an incision rule in which the erosion rate depends upon the local transport rate through appeal to these experiments. Others maintain that access to the bed requires that the thickness of the alluvial fill on the bed must be explicitly accounted for, and that whenever that thickness is greater than the ability of the river to scour through it in a meaningful flood, bedrock incision by whatever process is prevented. In either case, attention is focused on

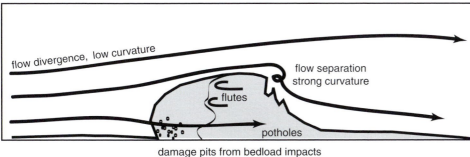

damage pits from bedload impacts

Figure 13.8 The attack from the back. Top: boulder in the channel of the middle gorge of the Indus River, Pakistan. Flow is from left to right. Person kneeling beside river in distance (photograph by R. S. Anderson). Bottom: sketch of flow separation in the lee of obstacles in the flow generates very strong curvature of the flow, efficiently decoupling suspended particles from the flow (after Hancock *et al.*, 1998, Figure 6, with permission from the American Geophysical Union).

the complexity of bedrock river incision when we acknowledge that at most times and most places bedrock beds are observed to be mantled with sediment.

Quarrying

Just as in subglacial erosion, quarrying of rock in chunks larger than single grains can be an efficient means of lowering the channel bed. There are two steps to the process. First, a crack or fracture must be generated to isolate a block of channel bedrock, and then the block must be removed. The fracturing can be accomplished by impacts of large blocks with the bed, by hydraulic wedging (see below), by frost-cracking subaerial exposure during low stages, by cavitation, and so on (e.g., Whipple *et al.*, 2000). It may also be the case that the rock arrives at the river bed already highly fractured and jointed, simply due

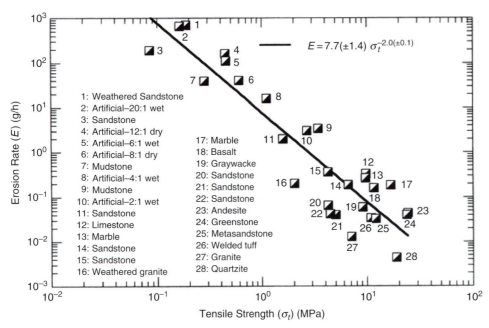

Figure 13.9 Dependence of erosion rate (in grams/hr) on the tensile strength of the rock being eroded, as determined by experiments in an abrasion mill. The power-law fit on this log–log graph suggests an inverse square relationship (after Sklar and Dietrich, 2001, Figure 2).

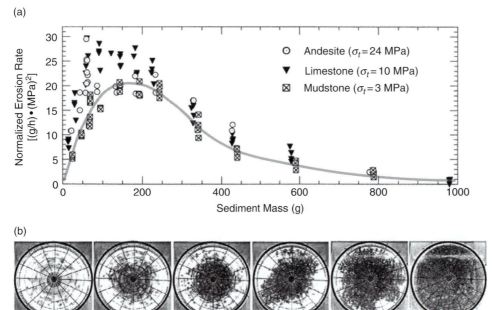

Figure 13.10 Experimentally determined dependence of erosion rate on mass of sediment in transport. Results of lab experiments in an abrasion mill. Erosion rate is normalized against square of the tensile strength of the eroding substrate. The maximum suggests a sediment loading that is most efficient at eroding the bed, above which sediment load damps erosion (after Sklar and Dietrich, 2001, Figure 4).

to the stress field associated with its position in a valley bottom (e.g., Miller and Dunne, 1996; Molnar, 2004), or the strain history to which the rock mass was subjected during faulting in the orogeny (Molnar *et al.*, 2007).

Taken most simply, the removal of bedrock chunks takes place by shoving a block across a planar surface after it has been detached from its adjoining rock mass. We sketch the situation in Figure 13.11. The force provided by the shear of the flow over the top of the block must exceed the frictional resistance against the underlying rock. The forces include the mean pressure on the top of the block, the mean stress on the base of the block, and the friction force on the

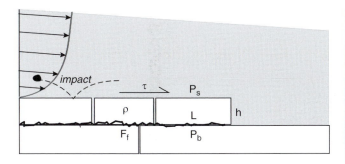

Figure 13.11 Quarrying of bedrock blocks from a river bed. Stresses imposed by the flow, and by impacting particles, must overcome the resistance to entrainment arising from frictional contact with adjacent blocks, and with the weight of the block itself. Geometry matters: a block enclosed by others on all sides is far less likely to be entrained than one at the edge of the step in the bed.

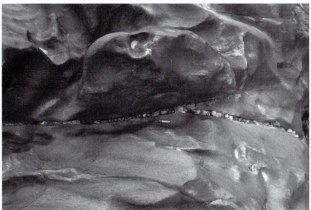

Figure 13.12 Pebbles perfectly wedged into a crack of Indus river bedrock. Lip balm for scale. Smooth surface of the outcrop reflects abrasion during immersion in moderate to high flows. Pebbles cannot be removed without breaking the pebble. Clasts are likely wedged into the crack when it is temporarily widened during pressure fluctuations associated with high flows. The phenomenon is discussed briefly by Hancock et al., 1998 (photograph by R. S. Anderson).

base and sides of the block. In general, the likelihood of extraction of a block can be cast as a shear stress that must exceed a threshold or critical stress; the erosion rate should therefore scale as the excess shear stress: $E \sim (\tau_b - \tau_c)$. But it should also be clear that edges matter. While a block in the interior of a planar bed will experience the same fluid stresses on its top, it is geometrically constrained by adjacent downstream blocks. It is therefore to be expected that bedrock beds unravel headward, essentially as a set of micro-knickpoints.

Hydraulic wedging

It is not uncommon to find cobbles, pebbles, or sand wedged into cracks in bedrock channel beds. An example of this is reproduced in Figure 13.12. These grains can be so tightly wedged that you cannot dislodge them without a crowbar. How did these rocks get into the cracks, and what role might they play in erosion of the bedrock? The obvious answers to the first question are (1) the rocks were forcefully injected into the crack, or (2) they were passively accepted into a crack that was momentarily widened while sediment was nearby. Given the specific sites, it is most likely the latter. But this merely raises the question of how cracks in the bed widen. While there exist no data to prove this, nor experiments targeting this phenomenon, the bed likely flexes due to the rapid and large pressure variations it experiences in the complex turbulence at high flows (akin to the shuddering of coastal cliffs associated with wave attack). This flexing needn't be large, as one can easily

imagine a small crack accepting a small grain, then not being able to close; some later flex allows a yet larger grain to enter, propping it open yet further, and so on. In this fashion an existing crack in the bedrock may grow. Beds with wider or more numerous initial cracks would be most susceptible. Note also that bed-load sediment is required for this process to operate. This is akin to some engineering applications for breaking down solids.

Dissolution

While not a dominant player in most settings, dissolution of a channel bed can be important in highly soluble rock types and in cave settings. The greatest testimony to this is the formation of caves in carbonate rocks. Because dissolution is generally not considered in analyses of surface channel incision, we will turn to work on cave formation to characterize the controls on channel growth by dissolution (Palmer, 1991).

The rate of wall retreat, S, in a channel due to dissolution is given by

$$S = \frac{Q}{p\rho_r}\frac{dC}{dL} \tag{13.13}$$

where Q is the water discharge, p is the wetted perimeter, ρ_r is the rock density, and dC/dL gives the change in relevant solute concentration with distance downstream. The downstream change in solute concentration is controlled by the dissolution rate of the rock, but also depends on the ratio of mineral surface area to water volume, the degree of chemical under-saturation, and the time it takes a water parcel to move through a reach. These dependences are expressed in a rate equation, which yields the rate of concentration change. In a later step, we will relate dC/dL to this rate equation:

$$\frac{dC}{dt} = \frac{A'k}{V}\left(1 - \frac{C}{C_s}\right)^n \qquad (13.14)$$

where A' is the surface area of rock exposed to water, V is the volume of water, C/C_s is the saturation ratio, k is the reaction coefficient, and n is the reaction order. We rewrite the left-hand side as being given by the change in solute concentration divided by the water residence time, or

$$\frac{dC}{dt} = \frac{\Delta C}{V/Q} \qquad (13.15)$$

where V/Q is the residence time for water within a stream reach. Using the approximation $\Delta C/L = dC/dt$ in Equation 13.15, and combining Equations 13.13–13.15, we see that the rate of wall retreat is given by

$$S = \frac{k}{\rho_r}\left(1 - \frac{C}{C_s}\right)^n \qquad (13.16)$$

Interestingly, the mineral surface area, A', in the rate equation is equivalent to the wetted perimeter and reach length, pL, in Equation 13.13 and divides out of the expression. Experimental data show that the reaction order increases as the saturation ratio approaches 1.0, which has the effect of significantly slowing dissolution rates.

Knickpoint migration

In some instances, lowering of the channel bedrock is accomplished by upriver migration of discrete steps, or knickpoints, in the profile. Here a knickpoint is defined as a steep reach sandwiched between two low-gradient reaches, as illustrated in Figure 13.13. The result is a lowering history at a point that is held at a very low rate for long periods, punctuated by rapid incision as the knickpoint migrates past. These steps occur at many scales. As a small-scale example, channels in the Blue Hills badlands of Utah, cut in the soft Cretaceous seaway shales, have profiles punctuated

(a)

(b)

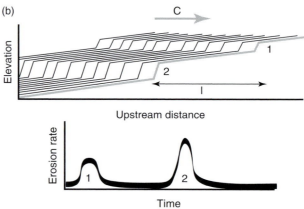

Figure 13.13 Knickpoint migration. (a) Bluehills badlands near Fremont River, Utah, with Henry Mountains in distance. The major channels in these badlands incise largely by passage of discrete knickpoints 10–50 cm tall (overview photograph by R. S. Anderson). (b) The speed, or celerity, C, of the knickpoint up the system, depends upon rock type and water discharge. Lowering of the channel bed is accomplished largely by the migration of knickpoints past a site, which occurs over a small fraction of the total time.

by knickpoints 10–30 cm tall (Figure 13.13(a)). Over the course of a single season, in which the channels were dry except during flash floods that lasted 15–30 minutes, the knickpoints migrated up-channel by about a meter. The channel floor did not experience net erosion anywhere else. The mean lowering rate of the channel may be quantified as

$$e = CH/\lambda \qquad (13.17)$$

where C is the celerity or speed of migration of the knickpoint, H its height, and λ the spacing between knickpoints. In order to estimate the long-term mean erosion rate, we must therefore know not only the characteristics of the steps, their height and spacing, but what governs their rates of migration.

At the other extreme lie some of the world's largest waterfalls, such as Victoria Falls on the Zambezi (described first by the great British explorer Livingstone), or Niagara Falls, where the river plummets many tens of meters before reaching the plunge pool at the base of the falls. The specific processes driving the lowering at the knickpoint, or headward erosion of the waterfall, are numerous. While Gilbert (1907) argued for the undermining of the head of the waterfall by backwearing induced by plunge-pool erosion, others have appealed to the weathering of the rock, groundwater sapping focused at the base of the step, stress relief of the headwall, and focusing of stress by the step in the profile itself. It is in the plunge pools at the base of waterfalls and major rapids in rivers that rivers are capable of overdeepening their bedrock floors. The topography, for example, beneath the Maid of the Mist plunge pool at the base of the present Niagara Falls is overdeepened by more than 30 m. Other similar steps in the bedrock occur further downstream, and have been interpreted as recording past locations where the rate of retreat of the falls was slow (see Figure 13.14; Philbrick, 1970).

Summary of processes

It is some combination of the processes we have described above that do the work of lowering bedrock channels. As rivers cross from one rock type to another, the dominant process may change. Clear examples of this include the tributaries to the Grand Canyon in Arizona, in which the profiles cross massive sandstone reaches bounded by shales. Erosion by quarrying of the highly fissile shale is efficient, while erosion in the massive sandstones is restricted to abrasion, which is much less efficient. The profiles reflect this: they are steep in the sandstones, and lay back in the shales. The local channel features change in accord with the dominant process. Where the rock is sufficiently fissile to allow quarrying, the bed tends to be ragged. In contrast, where the rock is massive, disallowing quarrying, abrasion dominates, and the resulting morphology is fluted, polished and/or potholed. The latter phenomenon is illustrated in Figure 13.15, in which massive Sierran granite is being eroded by potholing along the Stanislaus River.

More commonly, rivers cross complex structures that are not well bedded, and encounter three-dimensional bodies of rock with differing characteristics. These hard places may stall or steepen the profile for some time, and then let it flatten as the erosion proceeds below the hard rock. We now turn to the long-term evolution of river profiles in bedrock channels.

Stream profiles in bedrock channels

We may now ask what the longitudinal profile of a bedrock river should look like. How does it differ from that of an alluvial river? How should it reflect changes in the lithology across which the river flows? How should it reflect the uplift rate of the rock, or the pattern of uplift? How long should a bedrock river take to respond to changes in uplift, or in climate? As usual, we start with the simplest cases. Here the simplest is the steady-state case in which constant climate drives fluvial incision into a uniformly rising rock mass of uniform erodibility.

Steady uniform case

For the sake of conformity with the existing literature, we will generalize the stream power rule to acknowledge the possibility that specific processes can yield different dependences on slope and on water discharge than the pure stream power case. Starting with the unit stream power equation, we generalize to

$$E = k_1 A^m S^n \qquad (13.18)$$

where k_1 now absorbs the density of water, the acceleration due to gravity, the susceptibility of rock to erosion, the effective precipitation relating area to discharge, and the scale for the width of the channel.

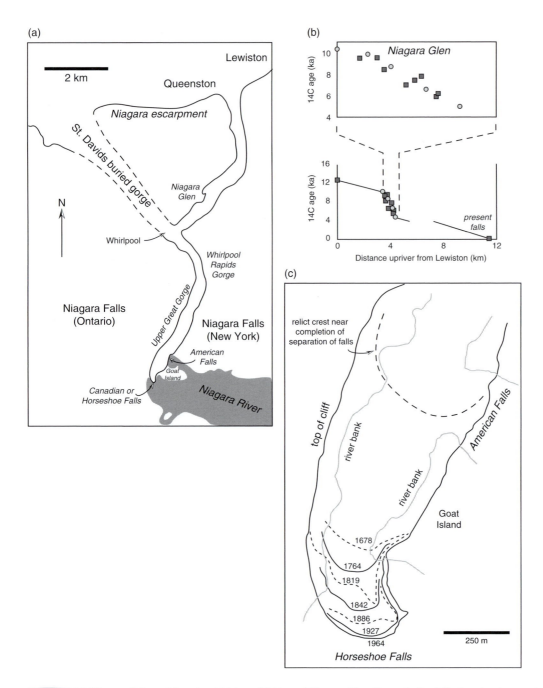

Figure 13.14 Niagara Falls and its retreat history. (a) Map of Niagara River through the falls showing two major waterfalls and several prominent narrows en route to Lewiston (after Pengelly *et al.*, 1997, Figure 2). (b) Retreat history documented using ¹⁴C dating of molluscs in riverbank (squares) and Wilson terrace deposits (circles). The retreat rate was much higher through the Niagara Glen area, leaving a narrow gorge segment. Average retreat rate has been 1 m/yr over the Holocene (after Tinkler *et al.*, 1994, Figure 5, with permission from Elsevier). (c) Historical retreat history documented by repeated surveys of the cliff top reveals continued retreat at roughly 1 m/yr (after Philbrick, 1970, Figure 1).

We return to this width problem later in the chapter. In steady state, the erosion rate, E, must balance exactly the uplift rate of the rock, U, which for simplicity we assume to be spatially uniform. This leads to

$$S = (U/k)^{1/n} A^{(-m/n)} \qquad (13.19)$$

In other words, we expect (predict) an inverse relationship between slope and drainage area that is very

Figure 13.15 Potholes beside the Indus River in its middle gorge below the Skardu basin. Here massive migmatitic granitoid rock disallows plucking of blocks, forcing the river to abrade its channel floor, resulting in ornamentation of the bed by potholes. Photo taken at April low flow conditions; high flows are roughly 14 m above this (photograph by R. S. Anderson, 1996).

specific, as is shown in Figure 13.1. The ratio m/n has been called the concavity index, $\theta = m/n$, as it dictates the rapidity with which the slope changes as area increases (and hence with distance downstream), while the collection of constants $(U/k)^{1/n}$ is called the channel steepness. Tucker and Whipple (2002) have reviewed the predictions of various stream erosion models. Below we summarize the expected profiles in several specific cases.

Steady case, but non-uniform bedrock

We can easily see from Equation 13.19 that we should expect an inverse relationship between the channel slope, S, and the erodibility of the rock, k. In fact,

over a small reach in which neither the rock uplift nor the discharge changes, the slope is a measure of the erodibility. The ratio of erodibility of adjacent reaches can be measured by the ratio of the channel slopes: $k_1/k_2 = S_2/S_1$. Note that in such cases the steep reaches of the profiles should not be interpreted to reflect a transient response to baselevel fall, as we discuss below. These sites should remain steep, and the steep section will not migrate up or downstream. This is exemplified in Figure 13.16, in which we show the profile of the Wind River as it crosses the Owl Creek Mountains. This is a granite-cored Laramide range. The basins on either side are composed of easily eroded Tertiary sedimentary rock; the river is significantly less steep as it incises through these basins.

Steady uplift, non-uniform precipitation: the orographic effect

This treatment presented so far has utilized the simplifying assumption that precipitation is uniform. This assumption allows the straightforward transformation of basin area into discharge. In general, this is not the case. In fact, a better assumption is that precipitation increases with elevation due to the orographic effect. This problem of orographic precipitation was reviewed comprehensively by Smith (1979); we have introduced this phenomenon in the Atmosphere Chapter 5. Recently, Roe *et al.* (2002) have addressed the expected changes in the steady-state stream profiles when the interaction of precipitation field with the topography (the orographic effect) is incorporated formally. In general, this requires incorporation of a function that describes the pattern of precipitation with distance from the divide, $P(x)$, so that we can assess the expected river discharge:

$$Q(x) = \int_0^x P(x') \frac{dA(x')}{dx'} dx' \tag{13.20}$$

where the prime indicates a dummy variable of integration, $A(x)$ is the distribution of drainage area as a function of distance from the drainage divide. Note that for the uniform rainfall distribution case, $P(x) = P_o$, Equation 13.20 reduces to $Q(x) = P_o A(x)$, and the drainage area may be taken as a perfect proxy for river discharge.

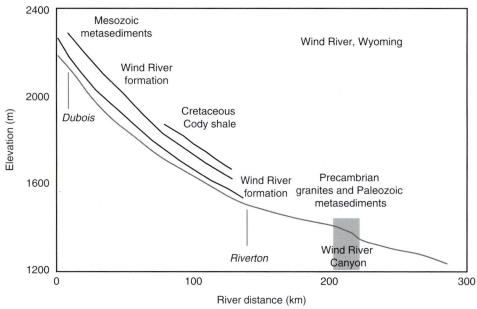

Figure 13.16 Top: DEM of the Wind River region, northern Wyoming. Bottom: profiles of Wind River and associated terraces through Wind River valley and Owl Creek Mountains (Wind River Canyon). Note the strong steepening of the river through the gorge, in which the river encounters Precambrian crystalline rocks, in contrast to the Mesozoic and Cenozoic sediments through which the river is incising up- and downstream of the gorge (after Hancock *et al.*, 1999, Figure 2, with permission from Elsevier).

But is this simple form of $P(x)$ appropriate for realistic landscapes? We have introduced the roles of topography in altering the pattern of precipitation in Chapter 5. Here assume for simplicity that the wind is steady in both speed and direction, approaching a mountain front at speed V (Figure 13.17). There are three effects (at least) that must be captured. First, droplets of water must grow from the airmass through condensation to sizes large enough to fall at rates that exceed the rate of rise of the airmass. The formation of droplets depends upon the degree of saturation of the airmass. This in turn depends upon the temperature of the air, T, and upon the rate of lifting of the air, w ($=VS$). Second, once droplets begin to fall (at which time they are dubbed by the atmospheric science community to be hydrometeors), they will drift further downwind before they reach the ground. The effects of temperature and rate of lift have opposing influences (Figure 13.17). As discussed in Chapter 5, the result is a pattern in which the maximum precipitation is shifted downwind from the maximum upwind-facing slope of the topography. For short hills, this can yield a maximum in precipitation downwind of the crest, but more commonly, at the mountain range scale, this results in a precipitation maximum upwind of the crest of the range. Roe *et al.* (2002, 2003) have explored the role of orographically induced non-uniform precipitation on the

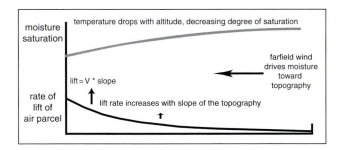

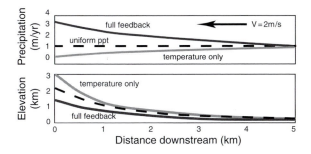

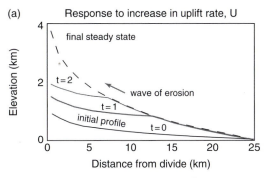

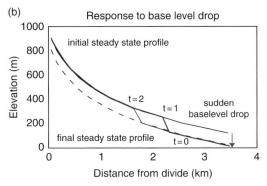

Figure 13.17 Changes in bedrock steady stream profiles caused by orographic feedbacks. Top: competing effects of slope and elevation on precipitation. The slope of the topography forces the airmass to rise, promoting precipitation. The temperature drop with elevation reduces the degree of moisture saturation. Middle: resulting change in precipitation pattern from the far-field rate for temperature effect only, and for both effects. The slope effect dominates. Uniform precipitation case shown for comparison (dashed). Bottom: resulting steady-state profiles for temperature-only and full models. The local channel slope needed to accomplish incision of rising rock mass is everywhere lower in the full model, resulting in a shallower profile and lower total relief in the channel network (redrafted from Roe *et al.*, 2002, Figure 1).

Figure 13.18 Response of bedrock river profiles to (a) an increase in rock uplift rate, and (b) a baselevel drop (from Whipple and Tucker, 1999, Figures 4 and 6, with permission from the American Geophysical Union).

profile of a bedrock river. As sketched in Figure 13.17, holding the tectonic forcing of the landscape constant (and uniform), the expected steady river profile developed in the face of more realistic orographic precipitation has a lower overall slope, resulting in a lower maximum topography (range crest) than the counterpart expected from uniform precipitation. This is a direct consequence of the enhanced precipitation in the headwaters, which results in more discharge everywhere downstream. Given that the stream power goes as the product of the slope and the discharge, less channel slope is required to drive the incision necessary to counter rock uplift.

The transient case

In reality, tectonic and climatic forces vary through time on many timescales. An important question is how rapidly a profile responds to a step change in either forcing. What is the characteristic timescale over which the change from one to the other steady-state profile occurs? And what form does the response take? Does the response occur first in the headwaters and propagate down-channel, or vice versa?

Although not definitive, one of the hallmarks of a river system in a transient state is a convexity or knickpoint in the longitudinal profile. (This is not diagnostic of transience, as a steep section of river may instead be localized by a reach of resistant bedrock, as discussed above.) As can be seen from the schematic profiles in Figure 13.18, the response to both baselevel drop and to a change in uplift rate occurs first at the mouth of the system. The information about the baselevel drop is transmitted up the river profile at a rate that is dictated by the stream discharge and the erodibility of the channel, as we saw in the discussion of knickpoints. The same occurs for a change in uplift rate. The rate of propagation of the knick, or replacement of one profile with the other, is set by the "celerity," or wave speed, *C*, of the response. This is a natural consequence of the form of the stream power rule for stream incision. One may write the erosion equation as

$$\frac{\partial z}{\partial t} = \frac{k \rho g Q}{W} \frac{\partial z}{\partial x} = C \frac{\partial z}{\partial x} \qquad (13.21)$$

which is classified as a kinematic wave equation. (As a historical aside, this equation was explored in detail in describing the flow of traffic on highways by Lighthill, 1978).

The timescale for the response to a particular site, at a distance L up the profile from the site at which the baselevel is imposed, is determined by the total distance upstream divided by the mean celerity, or

$$T = \int_0^L \frac{dx}{C(x)} \qquad (13.22)$$

(see Whipple and Tucker, 1999, equation 20 and following). Interestingly, and somewhat surprisingly at first glance, the response time is not very sensitive to drainage basin size. This is because large drainages often have high discharges, which means the response rapidly propagates up the network (because $C \sim Q$). As the discharge declines, the rate of progress of the response, be it a knickpoint or simply a break in slope, declines. This is particularly the case at tributary junctions, where the drainage area declines by a step dictated by the size of the tributary. The expected location of knickpoints in a real basin with several tributaries at some significant time after a baselevel drop is illustrated in Figure 13.19. In this instance the Colorado River serves as the local baselevel for all of the tributaries shown. The knickpoints, which here are all major dry falls on small streams exiting the thinly laminated oilshale of the Eocene Piceance Creek formation, are found at locations that differ in terms of their distance from the river. Undoubtedly these all were cast off the Colorado when this major river began incising in the late Cenozoic. That they all have experienced the same baselevel history, and are all located in uniform bedrock, makes this a fruitful natural experiment that allows one to explore the degree to which the rules we have been discussing work in a natural setting. This has indeed been done, and the predictions from the simple celerity model discussed here fit the observed locations of the knickpoints remarkably well (Berlin and Anderson, 2007; Figure 13.19).

Waterfalls

According to the above discussions, the rate of migration of a discrete knick should be governed by the discharge of the river, Q. Crudely, $C = kQ$, where k is the susceptibility of bedrock to erosion. In uniform rock, then (i.e., uniform k), the rate of propagation of the knickpoint will slow as it moves upstream into regions of lower water discharge. There are good reasons to suspect that a shear stress-based rule for channel erosion might not operate in the case of waterfalls. After all, the shear stress goes to zero for water falling vertically. Nonetheless, some fraction of the energy released by the fall of water through that height should still be made available for performing work on the underlying and adjacent bedrock. Conversion of potential to kinetic energy occurs, most of which is locally dissipated (the water exiting the plunge pool reach is not going tens of meters per second!). That the discharge of water still exerts control on the rate of recession is suggested by work on Niagara Falls.

Along the Niagara River (Figures 13.14 and 13.20), which connects Lake Erie to Lake Ontario through an 11 km channel, the rate of retreat of Niagara Falls appears to have experienced a period from roughly 10 500 to 5500 yr BP during which the recession was greatly diminished below that either prior to or after this period (Tinkler *et al.*, 1994). The story of Niagara Falls is therefore tangled intimately with the history of the Great Lakes, whose drainage has been greatly affected by the post-glacial rebound of the region as the Laurentide ice sheet vanished (e.g., Pengelly *et al.*, 1997). Tinkler *et al.* (1994) hypothesize that this 5 ka period was one in which a major fraction of the flow from the upper Great Lakes was diverted to the Ottawa River, bypassing both Lake Erie and Lake Ontario (Figure 13.14). The diminished flow resulted in a slowing of the recession of the Falls by a factor of five, accomplishing 1 km of retreat in 5 ka, compared to the roughly 1 km retreat in 1 ka, or 1 m/yr rate over the last 5–6 ka. Interestingly, the fivefold reduction of the regression rate corresponds to an estimated sevenfold diminution of the discharge. More detailed study of the recession rate in historical time shows that the retreat process is complex. The historical retreat rate of Horseshoe Falls of roughly 0.3 m/yr is accomplished by non-uniform retreat of the falls (Philbrick, 1970). It appears that generation of a deep notch in

(a)

(b)

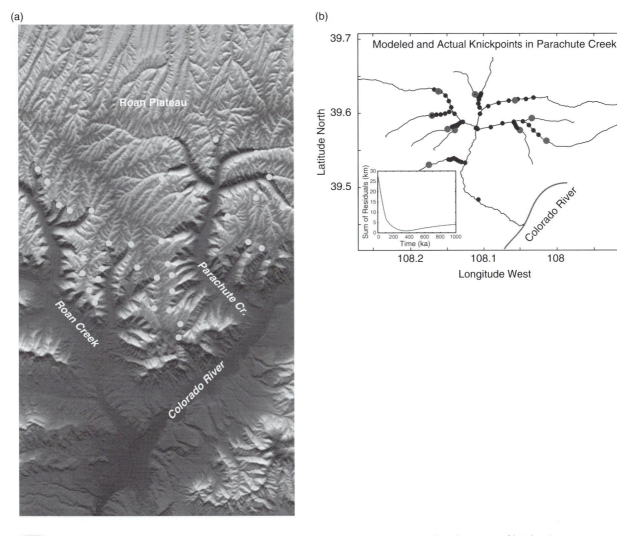

Figure 13.19 (a) Map of Parachute and Roan Creeks, Roan Plateau, Colorado, showing modern locations of knickpoints. (b) Results of 350 ka celerity model of knickpoint migration in Parachute Creek. Open circles: present positions of knickpoints. Dots: locations of model knickpoints at 20 time increments. Most knickpoints fit well with this simple model. Inset: goodness of fit measured by sum of misfits of knickpoint locations, as a function of simulation time. Best fit 350 ka for the given choice of $k = 5 \times 10^{-10}/(\text{m yr})$, and a critical area of 0.5 km^2 (after Berlin and Anderson, 2007, Figure 8, with permission from the American Geophysical Union).

plan view occurs first, followed by broadening into a new horseshoe shape.

While less information is available on the yet taller Victoria Falls on the Zambezi River, the late Pleistocene retreat rate is of the same magnitude, a fraction of a meter per year (Derricourt, 1974). The details of the processes occurring are still not clear; this issue of the mechanics of waterfall retreat remains a target for future research.

Another example of a transient profile is that of the river draining the Namche Barwa syntaxis on the eastern end of the Himalayan orogen. Finnegan

et al. (2008) have shown that the present profile of the river displays a prominent knickzone as the river punches across a zone of presumably high rock uplift rate at the western syntaxis of the Himalayas (Figure 13.21). Both the stream power and the long-term erosion rate, as revealed by cooling ages of at least two thermochronometers, peak in the knickzone.

Response to baselevel lowering

One must be careful in thinking about the role of baselevel as a boundary condition for bedrock

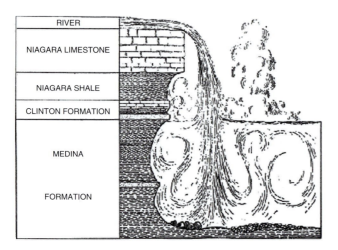

Figure 13.20 Cross section through Niagara Falls (plate VIII from G.K. Gilbert, 1907). Diagram illustrates the roles of the strong caprock of the Niagara limestone, and preferential attack of the underlying more erodible shales.

rivers. Formally, it is the level to which a stream tends, be it an ocean or a lake surface. The wrinkle is that as the stage of a lake or ocean changes, the position of the intersection between the land and the water changes, meaning that the place where the baselevel boundary condition is applied moves in the horizontal as well as in the vertical. This case was acknowledged in Rosenbloom and Anderson (1994), working on the Santa Cruz coastline, but was formally explored by Snyder *et al.* (2002, 2003a), working near the Mendocino triple junction along the northern California coast. As illustrated in Figure 13.18, rapid lowering of the baselevel (sea level fall) initiates a wave of incision that travels up the stream. Rapid sea level rise transforms the mouth to an aggrading system, disallowing further erosion of bedrock until the next sea level fall. This problem set-up has been nicely employed in the treatment of the problem of how the large circum-Mediterranean rivers responded to the dramatic drying out of the Mediterranean that has been called the Messinian salinity crisis (Hsu, 1972, 1983; Loget and Van Den Driessche, 2009).

The response to tilting of a basin due to some large-scale tectonic process can be assessed by imposing a change in slope to the entire stream profile. Once again, the response begins at the mouth of the stream, for the reason that the proportional change in slope there is much larger than in the headwaters.

Role of climatic variability: the origin of strath terraces

Finally, one must acknowledge the role of climatic variability in carving the alpine channels we see. If the stream power rule discussed above pertains, then any climate swing that yields more effective precipitation (for effective precipitation read effective discharge of the river) should result in greater rates of bedrock erosion. In fact, if we were just to consider abrasion as the dominant process, one might argue that greater water discharge should result in higher flow speeds, and hence in greater erosion rates. Couple this to the expected increase in sediment discharge that this flow increase may incite, and the effects become highly nonlinear. This argument breaks down, however, if the climate swing produces sediment in such great quantities that it clogs the channel with sediment. In other words, if the channel aggrades significantly, then the flow is unable to gain access to the bedrock bed to do any work on it. This will act as a negative feedback or a brake on the system. In rivers with glacial headwaters, this was likely the case during the glacials. Recent work on a river draining the Annapurna Himalayas has shown that the aggradational pulse associated with the last glacial maximum filled the valley to a height of many tens of meters (Pratt *et al.*, 2002).

An important consequence of this sediment loading history is the formation of strath terraces in rivers with glacial headwaters (Hancock and Anderson, 2002). These are especially well developed (broad) surfaces if the river crosses easily eroded sedimentary rock, as is the case for the Wind River as it crosses the Wind River Basin of Wyoming (Figure 13.22; see also Figure 13.16). The numerical model of strath formation is based upon the following simple reasoning, shown schematically in Figure 13.23. If the alluvial thickness is greater than a river typically scours over the course of the dominant floods, the bedrock channel floor beneath the alluvium is no longer eroded. However, the bedrock channel banks are susceptible to erosion if the river channel swings against the bedrock wall. During times of aggradation, then, the river bevels laterally, widening a braidplain or channel meander belt. Upon decline of the sediment input from the glaciers, the balance of sediment input to sediment output from the downstream reaches changes sign, and the alluvium is thinned, eventually

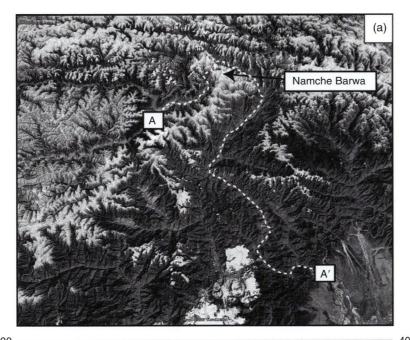

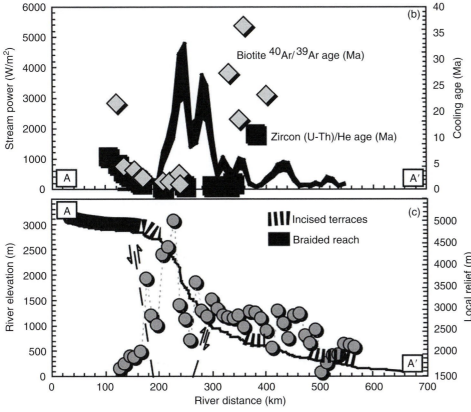

Figure 13.21 (a) Map of river draining across the eastern syntaxis of the Himalayas, anchored by Namche Barwa. (b) Stream power profile (bold line) from A–A' down the 700 km reach of the river depicted in the map. (c) Channel profile and adjacent local relief (circles) of the landscape. The steep knickzone in the channel is associated with high local relief of almost 5 km, and a peak in stream power. Cooling ages based upon biotite $^{40}Ar/^{39}Ar$ and zircon (U-Th)/He reach their minima within the knickzone (after Finnegan *et al.*, 2008, Figure 11).

allowing the river access to the bedrock floor. The river then resumes erosion into bedrock where it happens to be lowered onto the bedrock floor, and the remainder of the widened bedrock platform is abandoned as a terrace once flood flows can no longer access the surface. As shown in Figure 13.24, climate swings can therefore generate stacks of strath terraces, gravel-mantled bedrock surfaces whose

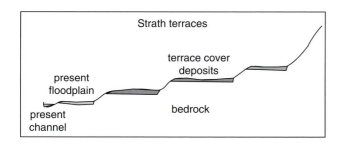

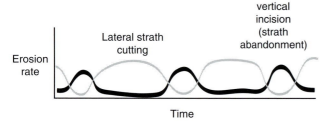

Figure 13.22 Geometry (top) and proposed origin (bottom) of strath terraces. Lateral strath cutting (grey line) occurs when the channel is blocked from vertical erosion by a thick alluvial cover. A strath is then abandoned by vertical incision (black line) when the alluvial cover thins.

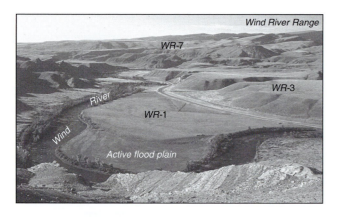

Figure 13.23 Wind River terraces as seen looking downstream near Dubois, Wyoming. Above the active flood plain, three prominent strath terraces mapped by Chadwick *et al.* (1997) are cut into the valley wall above the river: WR-1 (~20 ka); WR-3 (~120 ka); WR-7 (~600 ka) (after Hancock *et al.*, 1999, Figure 1, with permission from Elsevier).

abandonment times should correspond to the terminations of glacials.

The roles of landslide dams

Many bedrock rivers punch through steep mountainous terrain. It is in fact the incision of the rivers into

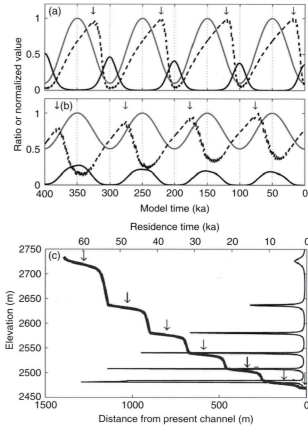

Figure 13.24 Top: history of normalized sediment inputs (gray line), normalized valley-floor width (i.e., flood-plain width) at river level (dashed line), and the ratio of vertical to lateral erosion rates (solid black line) during two simulations: (a) tenfold sediment-supply variation and (b) twofold dominant-discharge variation. Inputs and valley-floor widths are normalized by dividing the value at each time by the maximum value during each simulation. The valley floor (i.e., the "flood plain") widens most significantly when vertical erosion rates are low, producing a ratio of vertical to lateral erosion that is small or zero. Flood plains are abandoned to form terraces when vertical erosion rates and the ratio of vertical to lateral erosion rates increase, leading to renewed downcutting and narrowing of the active valley. Terraces are formed at the transition from valley-floor widening to narrowing (arrows). Terrace formation does not occur at the times of either maximum inputs or minimum inputs, but instead significantly lags the timing of the maximum (e.g., sediment-supply maxima) or the minimum (e.g., water-discharge minimum) inputs. In the simulations, all terraces generated are strath terraces. (c) The topographic profile (thick line) and channel-residence time as a function of elevation during downcutting in this simulation (thin line). Terraces are related to periods of lateral planation during long channel residence within a narrow elevation range. Residence time at individual terrace levels reaches up to many tens of thousands of years, indicating that the river spent most of the simulation forming terraces (after Hancock and Anderson, 2002, Figures 5 and 7).

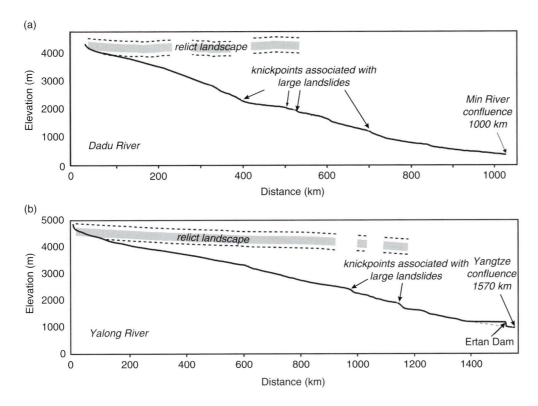

Figure 13.25 Long profiles of (a) Dadu and (b) Yalong Rivers draining the eastern edge of the Tibetan Plateau. "Relict landscape" marks the edge of the plateau into which the rivers are incising. Knickpoints or convex channel segments associated with large landslide dams are noted (after Ouimet *et al.*, 2007, Figure 4).

bedrock that lowers the baselevel for the adjacent hillslopes, which in turn steepens them. In many instances this can result in slopes that are near the angle of repose, and therefore fail occasionally as landslides. This process is inevitable, and therefore raises the issue of how such landslides create dams that should affect the long-term evolution of the river profile. There are two likely options. The landslides could serve to stall bedrock erosion by promoting sedimentation behind the dam, which prevents erosion of the bedrock by mantling it. On the other hand, if the dams fail catastrophically, the erosion downstream might be enhanced above that one would expect from non-flood discharges.

Working on the rivers draining the eastern margin of the Tibetan Plateau, Ouimet *et al.* (2007) have documented many landslides and their legacy in the channel profile. We reproduce their chief result in Figure 13.25. Clearly, the greater the number of large landslide dams that deliver large caliber sediment

to the river, the more they will slow the long-term incision of the river into rock. They argue that the longer it takes a river channel to remove a landslide dam by incision through the landslide debris, and to remove all landslide-related sediment, the stronger the influence these events have on the evolution of the river profile.

The channel width problem

In all of the analyses we have described there is an implicit assumption about channel width. Either it is held constant, or it is assumed to have some simple distribution down-valley. Investigation of this problem is one of the present frontiers in fluvial geomorphology. In reality the channel of a bedrock river can be eroded at any and all points on the channel perimeter. The rate of erosion is dictated by the erosion process, the protection of the bed by any

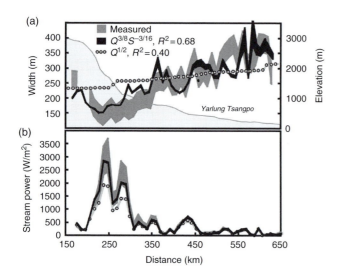

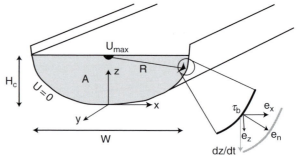

Figure 13.27 Schematic of bedrock channel cross section with flow of maximum thickness *H* and full width *W*. Local shear stress on the wall dictates local erosion rate, normal to the wall, e_n, with *x* and *z* components shown. Shear stress at the wall is taken to be proportional to the square of the shear strain rate, approximated by U_{max}/R. Erosion normal to the wall results in effective channel lowering at a rate d*z*/d*t* (after Wobus *et al.*, 2006, Figure 1, with permission from the American Geophysical Union).

Figure 13.26 (a) Channel width and elevation profile of the Yarlung Tsangpo River, and (b) associated profile of stream power per unit area. Spatial distributions of width and stream power are far better approximated assuming that width scales as $Q^{3/8}S^{-3/16}$ rather than simply as $Q^{1/2}$ (after Finnegan *et al.*, 2005, Figure 3).

sediment mantle, the availability of tools to perform the erosion, and so on – really the same issues faced in our treatment of channel lowering. But it should be clear from the first formula for stream power that a feedback exists between width evolution and channel lowering rate: the wider the channel, the lower the thickness of the flow, the lower the shear stress, and the lower the vertical erosion rate.

Empirical constraints

Data on bedrock channel width is rarely collected. Where it has been documented (e.g., Finnegan *et al.*, 2005; see also Finnegan *et al.*, 2007; see Figure 13.26), bedrock channel width appears to increase with flow discharge as $W \sim Q^{0.4}$, and to decrease with channel slope as $W \sim S^{-0.2}$. Such data also hint that a bedrock channel responds to rapid incision by narrowing (e.g., Duval *et al.*, 2004; Whittaker *et al.*, 2007a,b). Channels have been shown to narrow, for example, as they cross active folds in the Himalayan foreland (Lave and Avouac, 2000, 2001).

Theory

We summarize here recent theoretical work aimed at explaining these observations (Wobus *et al.*, 2006,

2008b). Consider a channel cross section depicted in Figure 13.27. Let us assume that the erosion rate anywhere along the periphery of the channel is set by the local shear stress (this assumption can be relaxed or replaced as we learn more about the dependence of erosion rate on flow properties). This shear stress is not set solely by the local flow depth, as there is shear within the flow in the other dimension; the flow speed goes to zero along the entire perimeter, and there must be both $\partial u/\partial z$ and $\partial u/\partial x$ shear. As a first approximation, consider the shear stress to be proportional to the square of the shear strain rate ($\partial u/\partial r$) evaluated at the wall. The erosion rate is then cast as

$$e_n = -k\left(\frac{U_{max}}{R}\right)^2 \tag{13.23}$$

where U_{max} is the flow speed in the middle of the channel, and *R* is the local distance from this core to the position on the wall. Again, this is only an approximation, as the flow field is no doubt more complex than this would imply. Indeed, slight alteration of this simplest model of the flow field can yield a very realistic map of the basal shear stress (Wobus *et al.*, 2006). The erosion rate may be decomposed into e_x and e_z components, and the location of the wall can be updated. One must also demand that a force balance on the fluid cross section exist, meaning

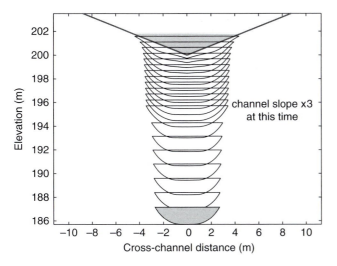

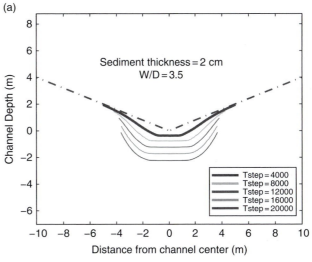

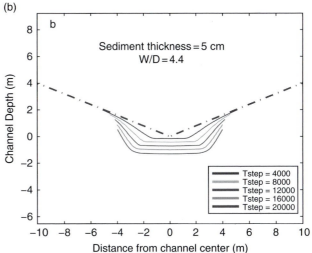

Figure 13.28 Evolution of a bedrock river cross section using model of Wobus *et al.* (2006). Local wall erosion rate is proportional to local shear stress, calculated as the core flow speed divided by the distance of the nearest core to the wall squared. Here the core of the flow is assumed to reach within 1.5 *H* of the wall. No sediment mantles the bed. The cross section evolves from the initial triangular channel to a smooth concave channel with the ratio *W/H* = 3.68 in the time it takes for channel lowering by roughly one channel depth, i.e., by *T* = *H*/(d*z*/d*t*). Tripling of the channel slope results in much faster erosion, and a narrowing of the channel.

Figure 13.29 Effect of sediment cover on channel width. The greater the sediment cover (a: 2 cm; b: 5 cm), the more the erosion is focused on the channel walls rather than the channel floor.

that the shear stress integrated along the wall must balance the down-valley weight of the fluid.

Starting with an arbitrary cross section, Figure 13.28 shows how such a channel cross section should evolve in the face of these physics. Independent of the initial shape of the channel, the channel evolves toward a smooth concave shape that eventually achieves a geometry that lowers uniformly. In other words, the channel form self-organizes. Interestingly, under these assumptions, the steady-state geometry achieves a definite width to depth ratio of *W/H* = 2.9. The detailed shape may be predicted from the demand that the channel lowers uniformly at steady state. The dependence of width on discharge and on channel slope matches well that observed in the field. Experiments may then be performed in which we link many channel cross sections to explore the interchange between channel cross-sectional shape and channel long profile.

While this is certainly not the final word on the topic, we are encouraged by the recent progress. With this theoretical edifice in place, in which channel wall elements are explicitly treated and allowed to evolve,

we can now explore the roles of specific erosional processes and how they will depend upon location in the channel cross section, the role of sediment mantle on the bed, the role of discharge variations, and so on. As a hint of things to come, consider the numerical experiment shown in Figure 13.29. We have dictated the presence of a finite alluvial mantle, and dictate that the rate of erosion depends upon the thickness of this mantle in an exponential manner. In other words, the erosion rate falls off exponentially with sediment thickness, with a length scale set by the expected distribution of scour depths. Sediment many times this thickness will prevent entirely any

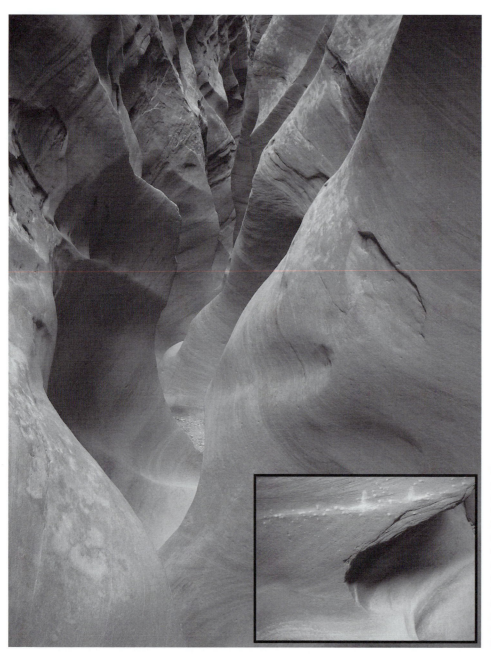

Figure 13.30 Little Wildhorse slot canyon, Utah. Smooth channel walls attest to the role of suspended sand in abrading the walls. Flow is away from observer. Inset: the bases of delicate flanges separating the smooth wall undulations are occasionally cracked, strongly suggesting a role for the larger clasts embedded in the flash floods that carve these channels. Flow is from left to right (photographs by R. S. Anderson).

erosion of the bed. The effect is to widen the channel significantly, as channel walls can be eroded while the channel floor is prevented from eroding. This effect will help to explain the significantly larger W/H ratios seen in real bedrock rivers. Indeed, as we have discussed above, this effect is the root of the generation of strath terraces, which are therefore essentially a corollary to the effect of sediment on the river bed.

Slot canyons

One cannot close a discussion of bedrock incision by rivers without at least advertising some of the most spectacular forms attributed to these processes. Slot canyons are extremely narrow canyons carved by rivers that leave as their legacy intricately carved walls reaching well above any reasonable flow depth (see Wohl and Merritts, 2001). These are especially well developed in streams tributary to the Colorado River draining terrain characterized by the massive sandstones of the American southwest. These produce the classic slot canyons that characterize this landscape, as shown in Figure 13.30. The flows themselves are rare, and are often flashfloods that have a habit of sweeping out anything in the canyon at the time, including occasionally humans. The undulations of the walls of these canyons display evidence of discrete impacts of both small and large clasts embedded in the flows. Abrasion of all sides of the channel walls smacks of the role of suspended sands. Yet, cracks in delicate flanges and flutes between wall forms near the channel floor suggest a role for the larger clasts involved as bedload (see the inset in Figure 13.30).

Summary

The dominant process of erosion by streams into bedrock varies with lithology. Highly jointed rock can be efficiently eroded by quarrying, while massive bedrock requires abrasion. As abrasion is generally less efficient than quarrying, massive bedrock reaches are typically steeper. Incision of bedrock requires that bedrock be exposed in the channel. For this reason, spatial and temporal variation of the alluvial cover of the bed, driven by sediment supply conditions on hillslopes and in the headwaters, can greatly influence the long-term rate of incision. In fact, we cannot explain the presence of strath terraces in bedrock river systems without appeal to the climate-driven variations in sediment supply.

Steady-state longitudinal bedrock river channel profiles should display concave up profiles. This reflects the inevitable increase in discharge with distance downstream; the greater the discharge, the lower the required slope to accomplish the work of erosion. For the same reason, Playfair's law holds; if small tributaries indeed join the trunk stream at grade, they must be steeper because they discharge less water.

The pattern and rate of response of a fluvial system to changes in its boundary conditions, or to changes in the climate, depends upon the size of the drainage, the lithology (loosely captured in k), and the discharge. Most changes in boundary conditions drive an upstream-propagating response, the rate of which declines with upstream distance, reflecting the decline in both drainage area and water discharge with distance upstream.

The channel cross section self-organizes into a smooth concave up form under steady conditions. It translates downward uniformly, and has a W/H ratio of roughly 3 in pure bedrock channels. The channel narrows as valley slope increases, and widens as water discharge increases. The time for accommodation of changes in these factors is scaled by the channel depth divided by the incision rate, and can be quite small for some channels. It is therefore significantly easier for a channel to adjust its width than to adjust its slope.

Appendix: Future work and research needs

While much has been accomplished in the last 15 years of research on this important geomorphic problem, there remains much to do in the field of bedrock river incision. While the community has begun to employ methods capable of documenting annual erosion rates, and has begun to exploit the available geomorphic markers to deduce erosion rates over thousand- to million-year timescales, no method of real-time (hourly to daily or weekly) measurement of bedrock erosion exists. Such a method would allow us to determine the detailed conditions of river discharge and sediment load under which erosion is most efficient.

Further exploration of the channel width issue is warranted. While a theoretical edifice is in place, we need more detailed field observations of channel width and its dependence on water discharge, rock properties, and sediment supply. Just as in the long valley profile, we also need more explicit treatment of specific erosional processes, moving beyond the simple stream power approaches. As an extreme example, the stream power rule used so heavily in this field breaks entirely on a vertical waterfall, at which the stream power should be infinite.

Models of river evolution have tended to be simplified, and lack acknowledgment of the full complexity of real systems. These include incorporation of realistic orographic effects, variation in drainage area and channel width, variation in lithology (and hence in process type), and non-uniform rock uplift patterns in both time and space. Full linkage of bedrock channel response to the history of sediment and water delivery from the headwaters of the channel system has not been accomplished.

Problems

1. Graph the expected history of retreat of a river knickpoint in which the distribution of basin area is $A = 5(L - x)^{1/2}$, where L is the length of the full channel and x is the distance upstream from the mouth of the basin. The speed of the knickpoint as it is cast off from the basin exit is $U = 10$ mm/yr. Plot the dimensionless knickpoint speed u/U as a function of the non-dimensional distance upstream x/L. (*Note*: you will find it useful to stop your plot at say $x/L = 0.9$.)

2. At the historical rate of retreat of Niagara Falls, how long will it take for the falls to reach Lake Erie?

3. Consider a strath terrace along a river etching into bedrock. We have determined its age to be 130 ka using ^{10}Be concentrations in the sediment capping the terrace. It is presently 55 m above the river. Calculate the mean rate of river incision in the time since the surface was abandoned.

4. In using the stream power formulation in bedrock incision problems, we are assuming that some fraction of the potential energy released by dropping the parcel of water in elevation is used to work on the bedrock. Here instead assume that all of that potential energy is converted to heat. Estimate how much a parcel of water will warm up as it drops from its source area in the Himalayas to the ocean. Assume that the elevation of the parcel is 6500 m when it is released by melt from a glacier. The specific heat capacity of water, c, is 4.19×10^3 J/kg-K. (*Hint*: the problem is to convert the water's initial potential energy into heat. Ignore the contribution from passage into a warmer climate; we are interested in isolating the effect of energy conversion.)

5. Calculate the mean rate of lowering of the channel floor in shale badlands in which the lowering is accomplished entirely through passage of small knickpoints. You measure the step heights to be 20 cm tall, the spacing between them to be 75 m, and repeated visits to the field reveal that the rate of propagation of the knickpoints is 40 cm/year.

6. Figure 13.3 illustrates the incision history of Kings River as deduced by dating of caves embedded in the valley walls. For the sample from Bats Cave, which is shown as having been abandoned at 3.09 Ma, calculate the ratio of ^{26}Al to ^{10}Be that must have been documented in the cave sediment.

7. Figure 13.16 shows how the Wind River in Wyoming steepens as it passes through Wind River Canyon. If we assume that the gradient of the stream changes fivefold and that the river has achieved a steady profile, meaning that it is incising at the same rates above, within, and below the canyon, how much less susceptible is the crystalline rock in the canyon to erosion than the sedimentary rocks outside the canyon? Discuss your assumptions.

8. *Thought question.* If the stream power formulation captures the essence of the bedrock river incision problem, will the effect of ponding of water behind a landslide dam, and then suddenly releasing it in a catastrophic flood, alter the expected long-term rate of incision of the river downstream of the dam?

9. *Thought question.* Again referring to Figure 13.3, discuss how different the interpretation of Kings Canyon evolution can be given that we have dated several caves rather than simply Bat Cave, or Boyden Cave.

10. *Thought question.* Can we use the stream power formulation for bedrock erosion in a vertical waterfall? Why/why not? (*Hint*: think about what the slope of the waterfall is, and calculate the expected lowering rate of the bed upstream, at, and downstream of the waterfall.)

11. *Thought question*. The river erosion rate estimates obtained from strath terraces are averages over the time since the terrace was abandoned by the river. If the erosion of the river downward into rock occurs only when the sediment supply from the headwaters is low, and this occurs during interglacials, explain why there is an apparent acceleration rate in river incision reported from dating several strath terraces on a single river. (*Hint*: you will need to acknowledge the climate history through the last several glacial–interglacial oscillations (recall the $\delta^{18}O$ record from deep sea cores).)

12. *Thought question*. Consider a simple semicircular channel in bedrock. If we assume that the erosion rate of the bedrock floor is governed by the local shear stress exerted by the flow on the bed, and we further assume that the local shear stress is simply $\tau_b = \rho g H S$, where S is the down-valley slope of the channel and H is the local depth of the water, graphically depict how the channel cross section ought to evolve through time. (It is this problem that inspired the work of Wobus *et al.* (2006, 2008b) in which they are forced to abandon this simplistic formulation for the boundary shear stress.)

13. *Thought question*. Consider a bedrock river with a longitudinal profile that displays a strong convexity well upstream of its mouth. How would you discriminate between the possibilities that this convexity is due to (i) a change in rock type, and (ii) a transient response to a baselevel drop at some time in the past?

14. *Thought question*. Discuss how a bedrock river should respond to glacial–interglacial fluctuations in baselevel of the order of 120 m. What legacy of the glacial sea level low stand might be seen in the present profile or subsurface stratigraphy, and what might govern the magnitude of these effects?

Further reading

Tinkler, K. and E. Wohl (eds.), 1998, *Rivers Over Rock*, Geophysical Monograph 107, American Geophysical Union.
This edited volume represents the state of our knowledge of bedrock rivers as of the late 1990s. It grew out of a Chapman Conference on the topic, and contains contributions from sites around the world.

Whipple, K. X., 2004, Bedrock rivers and the geomorphology of active orogens, *Annual Reviews of Earth and Planetary Science* **32**: 151–185.
An excellent review from a scientist immersed in this topic, this paper provides a convenient entrance into the theory of bedrock channel evolution as applied to rivers draining growing mountains.

Sediment transport mechanics

Even castles made of sand, fall into the sea, eventually.

Jimi Hendrix (1942–1970)

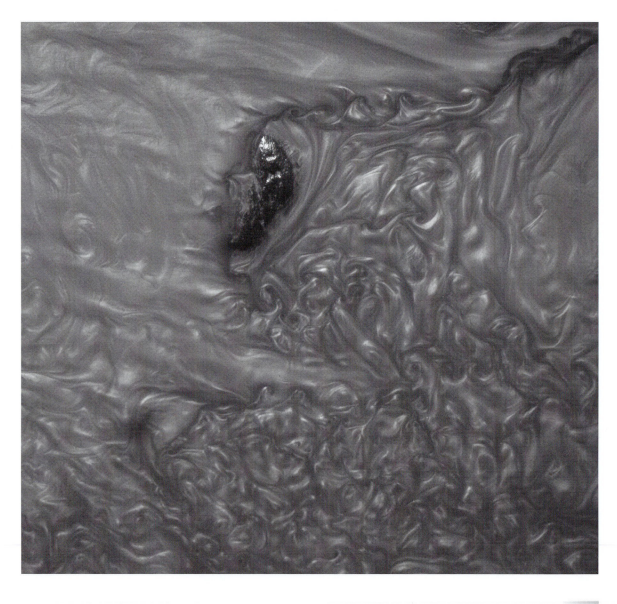

In this chapter

We focus on the physics of sediment transport. As there are many parallels between transport of sediment by water and by air, we will first focus on the physics that is in common between the systems, leaving for other chapters a fuller discussion of the application to fluvial and eolian cases. We will talk about each component of the transport system, then tie them all together to understand the behavior of the system as a whole. We hope that this will generate an understanding of why sediment transport is thresholded, why it is nonlinear above this threshold, and why different grain sizes behave differently enough during transport to force us to define different modes of sediment transport. We lay the context within which to address more specific geomorphic problems in chapters on rivers (Chapter 12) and on eolian forms (Chapter 15): how bedforms arise spontaneously and unavoidably in both fluvial and eolian settings, how the stratigraphic record of sediment transport is produced, how rocks are abraded by sediment embedded in the fluid, and so on. The physics involved includes yet more examples of force balances, this time on the grains and the fluids involved, and of conservation of mass, this time of sediment.

Turbulence in shallow stream visualized by finely ground mica glinting in the sunlight (photograph by R. S. Anderson).

The pieces of the problem

Motion of both air and water is capable of moving loose grains across beds over which the fluid is moving. These sedimentary systems display some of the more interesting examples of patterns in nature. We might ask what sets the spacing of ripples and dunes that look so intriguingly like thumbprints when seen from the air, or why sand grains have organized into ridges and bumps at all. Do these waves of sand act like waves in fluids, or do they have their own mechanics? What sets the wavelengths? Why do sand dunes come in so many shapes? Why are eolian ripples inversely graded, allowing us to distinguish them in the rock record from their fluvial counterparts?

Or consider the following very real problem. In the 1970s, and again in the late 1990s and 2000s, we managed to land packages of scientific instruments on the surface of Mars. We have known for centuries that the Martian surface is occasionally obscured from our telescopic views by what appear to be tremendous dust storms. In our design of the mission instrumentation package, how much should we worry about the abrasion of our instruments that rely on optics – like cameras! – by sand grains embedded in Martian winds?

We will approach the problem by tearing it apart into compartments whose physics we will try to understand. I (RSA) freely admit that this discussion focuses on the eolian case, as it is closest to my own experience. The physics is largely transportable between the two fluids. As in our discussion of hillslopes in Chapter 10, each of these problems combines elements that are stochastic and deterministic. For example, in our discussion of eolian saltation, we will find that trajectories of individual grains are deterministic (the simplest case), while the granular splash that occurs when a grain impacts the bed is stochastic. Put another way, the full flight of a saltating grain can be straightforwardly calculated once we know the conditions of its launch from the bed – we can know its maximum height, the distance it goes before it lands, the time it takes to travel its hop, and so on. This is not the case with the physics of the granular splash. When a grain lands, the details of the arrangement of the grains at the impact point dictate the resulting splash of grains from the surface, and since we cannot know these details at every potential point of impact, we must rely instead upon a statistical description of the process.

Finally, in order to understand the origin and migration of bedforms, ripples and dunes, we must again turn to the principle of conservation of mass. The rate of change of elevation of the bed at a point, positive if it is a depositional site, negative if it is erosional, will depend upon the spatial pattern of sediment flux – in this case of bedload material. Yet when we turn to thinking about the sediment flux over any significant period of time, we must again face the problem of the stochastic nature of the forcing of the system. While we will work to derive a simple formula to calculate the sand flux for a given flow speed, the flow speed varies in some complicated fashion over both the course of a single storm or flood, and between storms. Calculation of long-term sediment transport requires summing over the full distribution of wind (or water) speeds.

First, let us review some key observables in need of explanation. Below some threshold fluid speed, a sediment surface shows no evidence of motion. Large sand is harder to get going than fine sand. This is the "grain entrainment" problem, also at the root of what we call the "competence" of a flow. Once in motion, sand is transported very near the surface, while dust (silt) is transported at heights that can be kilometers in air, and many meters in water. The concentration of sediment above the surface appears to decline with height in any particular flow: one cannot even see the boots of the person in Figure 14.1, the cuffs of his

Figure 14.1 Tall Norwegian glaciologist on crest of a Salton Sea dune during a wind storm. His feet are invisible due to the high concentration of saltating sand near the bed (photograph by R. S. Anderson).

pants will fill with sand, his pockets will have a few grains in them, and his teeth will feel a little bit gritty with sand that is finer than that in his cuffs. The flux of sand increases tremendously with increasing speed above the threshold – nonlinearly so. We call the maximum transport rate associated with a particular flow speed its "transport capacity."

The transport mode is dependent upon the mass of the grains involved, and hence on their density and diameter. Large grains travel short smooth trajectories, while small ones travel wiggly paths. The former we call saltation, for *saltare*, the Latin for to jump or to leap, while the latter we call suspension. These are end-member behaviors; we will see that there is as usual a gray area between, which has been called "modified saltation."

Grain entrainment

While we fully recognize that the geometry of the bed is very complicated, let's start by considering the simple diagram of the bed shown in Figure 14.2. While this picture is a simplification, it ought still to capture the essence of the physics. We wish to know the dependence of grain entrainment upon characteristics of the bed and of the flow. We seek, in particular, an expression for the maximum size of grain that a particular flow can entrain, or start to move. This is called the *competence* of the flow.

In order for the grain to be entrained, it must somehow be ripped out of its geometric pocket amongst the other grains. One way to think of this is as a balance of forces in the vertical. In this case, escape from the pocket would require that the lift force imposed by the fluid be greater than the weight of the grain. While this is possible in water, it is not likely in air. Another means of treating the entrainment problem, and this more generally reflects the means of escape from a pocket in the bed, is instead as a torque problem. Although lift forces do help out, grains for the most part get torqued out of their pockets, rotating about a point of contact with the grains that comprise the pocket.

We will therefore set up a torque balance on a grain in the pocket. When the torque tending to rip the grain out of the pocket slightly exceeds the torque tending to keep the grain seated in the pocket, entrainment will occur. As illustrated in Figure 14.2,

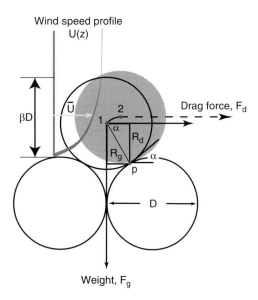

Figure 14.2 Sketch of the physics of grain entrainment from a pocket within a bed composed of similar grains. Grain depicted at two times, with centers at points 1 and 2 as it pivots out of the pocket between underlying grains.

the torque, T_g, holding the grain in place is the buoyant weight of the particle times the lever arm, R_g, about the point of contact, P. This is expressed as

$$T_g = R_g F_g = R_g(m_p - m_f)g = R_g g(\rho_p - \rho_f)\frac{\pi D^3}{6} \quad (14.1)$$

where m_p is the mass of the particle, m_f the mass of fluid it displaces, g the acceleration due to gravity, ρ_p and ρ_f the densities of the grain and the fluid, respectively, and D the particle diameter. The torque tending to move the grain out of the pocket is imposed by the drag force exerted by the fluid, F_d, about the lever arm R_d. This yields

$$T_d = R_d F_d = R_d \frac{1}{2}\rho_f \frac{\pi D^2}{4} C_d \bar{U}^2 \quad (14.2)$$

where C_d is a dimensionless drag coefficient, and the mean velocity, $\bar{U}$, represents the mean over the exposed portion of the grain, that poking up out of the pocket. Any slight mismatch of torques that results in the grain actually beginning to accelerate around its pivot point, will lead to escape. As the grain rotates up higher into the flow, the drag increases, and its associated lever arm, R_d, lengthens, while the torque arm associated with the weight of the grain, R_g, gets smaller. The drag increases because a larger portion of the particle is exposed to the flow. These feedbacks imply that once a grain has started to move, it will likely complete its pivot out of the pocket. The feedbacks are positive; the process of entrainment is irreversible.

We have ignored in the above treatment the role of lift forces on a particle. This is less well known quantitatively, but lift arises from the fact that the flow velocity over the top of the particle is greater than that experienced at the base of the particle. It was shown long ago by Bernoulli that the pressure on a surface over which flow was passing experiences a pressure that is inversely proportional to the flow velocity. This is the principle of both the flute and the airplane wing. Blow over the mouth hole of a flute (or a milk jug for that matter) and the air gets sucked out of the tube because a low pressure has been imposed over the opening, and the air responds to the pressure gradient. Flow over an airplane wing is greater over the top of the wing than over the bottom, creating a low pressure on the top and a high pressure on the bottom of the wing. The net result is a pressure gradient we call lift. The lift force exerted on a spherical particle may be represented as

$$F_\mathrm{L} = \frac{1}{2}\rho_\mathrm{f}\frac{\pi D^2}{4}C_\mathrm{L}\left[\bar{U}_\mathrm{top}^2 - \bar{U}_\mathrm{bottom}^2\right] \tag{14.3}$$

where C_L is the lift coefficient, akin to the drag coefficient, although less well defined empirically. Since the velocity at the base of the particle sitting in the pocket is essentially zero, this formula becomes very similar to that for fluid drag. That the lift force is so much smaller than the drag is reflected in the small values given to the lift coefficient, which are roughly an order of magnitude smaller than the drag coefficient.

Let us return to the torques. We can evaluate the dependence of the threshold wind velocity required for particle entrainment on the grain diameter and density. When the grain first moves, the torque inducing motion just equals the torque resisting it. Setting $T_\mathrm{g} = T_\mathrm{d}$ and rearranging results in

$$\rho_\mathrm{f}\bar{U}^2 = \frac{4}{3C_\mathrm{d}}\frac{R_\mathrm{g}}{R_\mathrm{d}}g\left(\rho_\mathrm{p} - \rho_\mathrm{f}\right)D \tag{14.4}$$

Before evaluating further the remaining variables, we can already pull some insight out of the equation. The left-hand side should look familiar from our derivation of the velocity profile in Chapters 11 and 12.

It looks something like the expression for shear stress at the base of the flow, τ_b, except that the velocity here is the mean velocity over the exposed portion of the particle, rather than the shear velocity, u_*. We expect these to be intimately related, as we will see below. Even so, inspection of the equation suggests that the threshold shear stress ought to increase linearly with grain diameter, D. In order to proceed further, we need expressions for the drag coefficient, the pivot arms, R_g and R_d, and the mean velocity $\bar{U}$.

For the simple pocket geometry shown in Figure 14.2, the torque arms may be written as $R_\mathrm{g} = \frac{D}{2}\cos(\alpha)$ and $R_\mathrm{d} = \frac{D}{2}\cos(90 - \alpha)$. As we have seen in our analysis of settling speeds of spheres in a fluid (water drops in air, Chapter 10), the drag coefficient is related to the grain Reynolds number. For the high Reynolds number case, relevant to large grains and high flow speeds, the drag coefficient is a constant: $C_\mathrm{d} = 0.4$. We will start by assuming that this applies, and re-evaluate whether this is the proper case when we turn to experimental results. Now we must assess the mean flow speed, $\bar{U}$, experienced by the exposed portion of the grain. In order to make progress, we turn to the law of the wall from Chapter 11, in which we found that the temporally averaged flow speed increases with distance from the boundary in a logarithmic fashion. As always, we use the mean value theorem in assessing averages formally:

$$\bar{U} = \frac{1}{\beta D - z_\mathrm{o}}\int_{z_\mathrm{o}}^{\beta D}\frac{u_*}{k}\ln\frac{z}{z_\mathrm{o}}\,\mathrm{d}z$$

$$\tag{14.5}$$

$$= \frac{u_*}{k}\left[\ln\left(\frac{\beta D}{z_\mathrm{o}}\right) - 1\right]$$

As expected, the mean velocity scales with the shear velocity, u_*. Here β represents the portion of the particle exposed to the flow. For the simple geometry shown in Figure 14.2, the value of β is roughly sin $(\alpha - (1/2))$, or for $\alpha = 60°$, about 0.87; about 87% of the particle is exposed to the flow. The remaining parameter, the roughness height z_o, may be approximated as $0.1D$, as has been determined by experiments of flow over natural sediments. We can now plug all these expressions back into the torque balance, Equation 14.4, to arrive at

$$\rho_\mathrm{f}u_*^2 = \tau_\mathrm{b} = \frac{4k^2}{1.2\left[\ln(\frac{0.87}{0.1}) - 1\right]^2}\frac{\cos(\alpha)}{\cos(90 - \alpha)}g\left(\rho_\mathrm{p} - \rho_\mathrm{f}\right)D \tag{14.6}$$

The messy-looking ratios on the right-hand side simply contain constants and geometric factors, and can be condensed into a single parameter, θ, that depends largely on pocket geometry (specifically α). The equation for the critical shear stress, τ_c, the basal shear stress at which entrainment occurs, can then be written more simply as

$$\tau_c = \tau_b = \theta g \left(\rho_p - \rho_f \right) D \tag{14.7}$$

Let us take stock. This expression reveals that, given the assumptions we have made, the shear stress at the bed necessary to entrain sediment must increase linearly with the diameter of that sediment. For von Karman's constant, $k = 0.4$, and $\alpha = 60°$, from our simplified geometry in Figure 14.2, we find that $\theta = 0.08$. How does this value compare with empirical, experimental information?

Relevant data are plotted in several ways. First, we can plot critical shear stress against grain size, for which Equation 14.7 predicts a linear relation. This is the case in Figure 14.3(a), at least for the larger grain sizes. Second, we can plot the critical shear velocity, u_{*c} (recall that $u_* = \sqrt{\tau_b/\rho_f}$) against grain size, for which we expect a square root dependence on grain diameter (Figure 14.3(b)). Finally, one can also plot this data in a dimensionless manner. Dividing Equation 14.7 by $g(\rho_p - \rho_f)D$, which is the normal stress imparted to the bed by the grain, yields the dimensionless equation

$$\theta = \frac{\tau_c}{g \left(\rho_p - \rho_f \right) D} \tag{14.8}$$

Because the threshold stress, τ_c, goes as D, the theory outlined above would predict that when plotted against D, this would be simply a constant, whose value is close to 0.08. In Figure 14.3(c) we see that this is the case, again for large grain sizes.

In all empirical data condensed in Figure 14.3 we see all these expected relationships for large grain sizes. In fact, the value of θ, seen best in the non-dimensionalized version of the data (Figure 14.3(c)), a modified version of a Shields diagram, is quite close but slightly lower than our predicted value of 0.08. Values of 0.03–0.06 are more common. Presumably the discrepancy reflects oversimplification of the problem in our theoretical analysis. In particular, the pocket geometry we employed was very simple; it was two-dimensional, and it assumed that all grains were the same diameter.

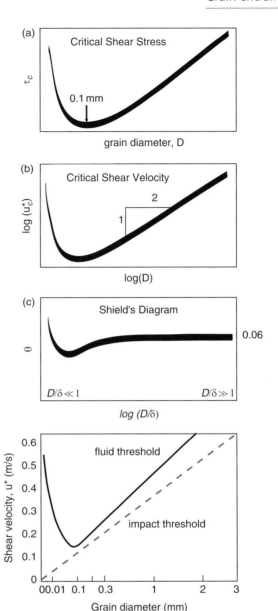

Figure 14.3 Entrainment threshold as a function of grain size. (a) Critical shear stress, showing that the most easily entrained grain is 0.1 mm. (b) Critical shear velocity, showing the power-law relationship; entrainment speed increases as square root of grain diameter. (c) Non-dimensional version of an entrainment plot (Shield's diagram). (d) Thresholds for entrainment of sediment by wind (after Bagnold, 1941, p. 88).

In addition, the instantaneous shear stress at the bed can be much larger than the mean shear stress we used to scale the drag force (through appeal to the law of the wall, which, recall, is a profile of the time-averaged velocity). Both of these effects would allow entrainment to occur at time-averaged shear stresses (hence shear velocities) that are below those we predict. That the theory works as well as it does indicates that the pocket geometry problem can be collapsed to a rescaling of the exposed diameter, and that the highest instantaneous stresses likely scale with the mean stress (see Schmeeckle *et al.*, 2007, for the discussion of the latter).

Lest we get smug about our success, we note a serious departure from the expected grain size dependence for small grain sizes. Some piece of physics must still be missing from our theory. In fact, we have ignored a couple issues altogether in our derivation, both of which become important for small particles. The first is particle–particle forces. The smaller grains become, the more important electrostatic bonds between grains become on a purely surface area to volume basis. In addition, small grains tend to be platey micas and clays, and have charged surfaces. Finally, in air, the wetness of a sandy surface can play an important role in curtailing entrainment. The water, held in thin films in pores, can exert a surface tension that serves to pull grains together. In all of these cases, the bonds between these surfaces serve as an additional force that must be exceeded before entrainment can occur. Our second oversight has to do with the flow itself. Geomorphically relevant flow of air or water at the Earth's surface transitions from fully turbulent flow high above the bed into laminar flow near the bed. Within the laminar layer the fluid is so close to grain surfaces that the transfer of momentum becomes dominated by molecular collisions – i.e., viscosity reigns. As illustrated in Figure 14.4, grains that are small relative to the thickness of this "viscous layer" can actually hide from the turbulent sweeps and bursts that characterize fully turbulent flow, diminishing the likelihood of entrainment. For all of these reasons, the entrainment diagrams, no matter how they are plotted, all display a prominent rise in the critical shear stress for small grain sizes. The result is always a trough in the diagram, meaning that there is some most easily entrained grain diameter. Surprisingly, in both air and water, that size is roughly the same, 0.1 mm (or 100 microns), assuming the mineral is common quartz. Above this size, the entrainment shear stress increases linearly with grain diameter; equivalently, the flow speed required increases as the square root of the grain diameter.

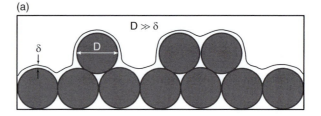

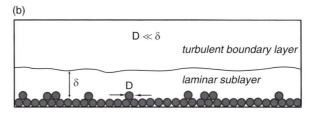

Figure 14.4 Grains on the bed beneath a turbulent flow. Grains whose diameter, *D*, is smaller than the laminar sub-layer are effectively shielded from turbulence, and will be more difficult to entrain.

You might assume that the measurement of these entrainment thresholds is easy. It is not. The problem is fraught with several complexities, which are worth pointing out as they illustrate elements of the real problem that we often have to face. The entrainment, even of well-sorted sands, occurs over a range of flow speeds, presumably owing to the range of pocket geometries in a real mixture of sand. The grains in the most susceptible pockets are ripped out first. We therefore expect experimental slop in these values. Most early measurements were made by eye – when a "significant number" of grains was seen to be in motion, the threshold had been exceeded. This meant that the measurement was subjective at some level, and would vary from observer to observer. This problem has been tackled technologically by workers in the eolian field who employ an objective measure of the number of grains up in the flow. W. Nickling, working in wind tunnels at the University of Guelph, bounced a laser beam across the flow many times between a set of mirrors on either side of the tunnel, and counted the number of interruptions of this beam per unit of time (Nickling, 1988). The results, depicted in Figure 14.5, clearly revealed an onset of saltation as the wind speed was slowly ramped up. He could then be quantitative in defining the threshold, setting the threshold for example at a fixed rate of increase in the number of particles being entrained from the bed (the strongest knee in the curves shown).

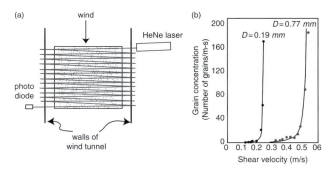

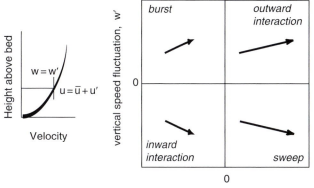

Figure 14.5 Initiation of grain motion as detected by HeNe laser system in a wind tunnel. (a) Wind tunnel set-up showing HeNe laser shone through plexiglas walls of wind tunnel in a plane just above the sediment sample tray. Laser beam bounces numerous times off parabolic mirrors with 1 m focal length before terminating at a photo diode. Interruptions of the beam corresponding to a grain leaping off the bed are recorded in a microprocessor. (b) Numbers of interruptions (per meter of beam per second) increase as the shear velocity of the flow increases, quantifying the details of the threshold of motion. As expected, small grains are easier to entrain than large (after Nickling, 1988, Figures 4 and 5, with permission from Wiley/Blackwell Publishers).

Figure 14.6 Bursts and sweeps. The classification of near-bed turbulence events. Left: mean velocity profile near a surface. At any height there is a definable mean horizontal speed, and fluctuations around it, denoted u'. While the mean vertical speed is 0, the flow still varies in its vertical speed about this mean. Right: quadrants of potential flow correlations, with established nomenclature for each quadrant. Sweeps and bursts play the greatest role in transporting momentum of the fluid, and in entrainment of particles from the bed.

But the problems don't stop there. It was recognized long ago, by the father of windblown sediment transport mechanics, R. A. Bagnold, that there is hysteresis, at least in the eolian sediment transport system. The threshold one must cross in order to initiate full blown saltation is greater than that one must cross in order to halt the process, by several tens of percent. This prompted Bagnold to define a second, lower threshold he termed the impact threshold (see Figure 14.6). This remained an unresolved problem in the physics of saltation in air for quite some time, but represented a strong constraint on any theory of the eolian saltation process.

Recent progress in the fluvial realm

Although the above treatment allows us to understand the essence of the entrainment process, we hasten to add that in the last few years there has been substantial progress made on entrainment in the fluvial world that has challenged some of the assumptions in the classical approach we have outlined. We summarize this progress here. As we will see below, the eolian case becomes so dominated by grain interactions with the bed that in full saltation the fluid entrainment problem plays a minimal role in dictating the sediment transport. This is not the case in the fluvial (or subaqueous) world. Fluid interactions with grains on the bed tell the entire story. In addition, it has been found that sediment transport can occur at fluid stresses (τ_b) far below those that would be predicted by the classical theory. Much of the progress reflects the employment of highly technical instruments, including very high-speed photography of the bed, high-definition video, laser Doppler velocimetry, and miniaturized force-sensors that allow documentation of the high-frequency history of the forces on an individual grain on the bed.

First, let's summarize some observations. High-speed filming of the granular bed in a flume showed that entrainment occurred very sporadically. Grains sit happily on the bed for long periods of time, interspersed with episodes of violent activity in which many nearby grains would become mobile. This mobilization can take place even when the bed shear stress is far below that based upon the Shields criterion, which, recall, is based upon the mean bed shear stress, $\tau_b = \rho g H \sin(\theta)$, exceeding some threshold value. This so-called "marginal transport" pointed to two potential issues. The first is the distribution of pocket geometries on the bed. Those grains in particularly favorable sites for entrainment, shallow pockets that would expose a large fraction of the grain to the flow, would tend to move before others. The second issue is the detailed turbulence of the flow. If the turbulence structure produces a history of local fluid velocities that can impart occasional high drag and/or high lift forces to a grain at the bed, this could account for entrainment events. While detailed documentation of pocket geometries in real beds has indeed shown that there exists a range of pocket geometries in which a few grains are very susceptible to entrainment, the focus has been sharpened on the role of the turbulent structure.

It has been known since the 1960s that turbulence in a boundary layer, even over a flat bed, is complex in both time and space. The relationships between horizontal and vertical velocity fluctuations (Figure 14.6) reveal a rather lopsided picture. The most likely and the most energetic flow events are those in which high-speed fluid is brought downward toward the bed (a sweep), and low horizontal speed fluid is brought vertically away from the bed (a burst or ejection). While the structures responsible for this cant to the turbulence statistics are difficult to capture, it is thought that these reflect complex horseshoe-shaped eddies. Working on the hypothesis that this structure is important in entrainment, researchers in the 1990s used high-speed film and high-frequency near-bed velocity measurements to explore whether entrainment was associated with the turbulent structures. The film speed had to be sufficient to capture individual grain movements. This required 250 frames per second, or roughly 10 times faster than typical film speeds! The velocity was documented using three-dimensional lasers in which all three components of near-bed velocity were captured. The results are summarized graphically in Figure 14.7. Entrainment was closely associated with particular types of turbulent events. In particular, entrainment and transport rates are greatly enhanced in the quadrant called the bursts, in which high horizontal speed fluid is brought downward toward the bed. This exerts the greatest shear stress on the bed particles. The data reveal that such events are indeed rare, and the greatest sediment transport rates are associated with the rarest of events; for example, transport rates are threefold enhanced over mean flow conditions during events that occur roughly 10% of the time. The implications of this observation are numerous. It implies that the sediment transport at a site can be controlled not by the *mean* shear stress at the bed, which is relatively

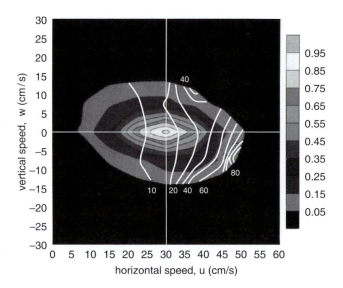

Figure 14.7 Sediment transport rates (contoured in grains/cm/s) and joint probabilities (scale on right) of streamwise (*u*) and vertical (*w*) velocities averaged over several experimental runs. Mean horizontal flow speed is 30 cm/s, while mean vertical speed is zero. Note that the greatest transport rates correspond to the lower right quadrant, in which turbulence brings high horizontal speed fluid downward toward the bed (after Nelson *et al.*, 1995, Plate 4, with permission from the American Geophysical Union).

easy to document and to handle theoretically, but upon the frequency of specific types of turbulence events at the bed. The frequency of these events is controlled by the details of the boundary layer. Their distributions and magnitudes are different in wakes behind dunes than they are on the stoss sides of dunes, and so on. In other words, outside of steady uniform flow conditions, which rarely occur in real environments, the structure of the turbulence within the boundary layer can vary strongly, and can play an important role in setting transport rates. Characterizing the spatial distribution of these turbulence events then becomes a primary target for modeling, and necessarily requires a higher order model than those used in the past – one that captures not just the mean flow, but the fluctuating structures that are the essence of turbulence in the boundary layer. We will return to this topic in the river chapter when discussing the evolution of fluvial bedforms.

Yet more recent work has documented the detailed forces on a single grain on the bed. These experiments were again carried out in a flume, but this time focused on a test particle mounted on an instrumented "stalk"

(a)

(b)

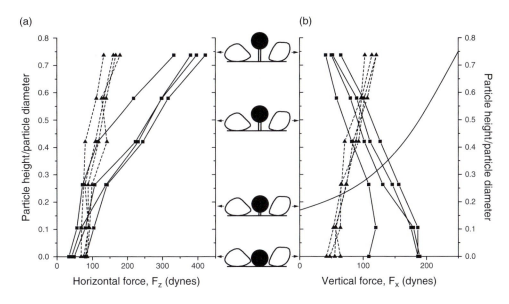

Figure 14.8 Mean horizontal and vertical force (squares) and standard deviations (triangles) for four gravel bed arrangements as a function of the height of the base of the test sphere (black) above the acrylic bed relative to the diameter of the sphere. Bed arrangements are shown in the insets. (a) Nominal drag F_x and (b) nominal lift F_z. The predicted profile of nominal lift (solid line with no symbols) in (b) was calculated using a lift coefficient of 0.2 and the measured mean horizontal velocity profile (after Schmeeckle et al., 2007, Figure 7, with permission from the American Geophysical Union).

connected to a computer through a hole in the bed. Again, the results are a little unsettling. First, by mounting a spherical particle on the stalk, researchers found that the drag coefficient used in essentially all entrainment calculations (and used above in our review of the classical approach) is off by a factor of two or so. As we have discussed, the drag coefficient has been derived from experiments in which spheres are allowed to settle in a tank of water. This new work shows that drag in the vicinity of a boundary, like the bed of a flume or a river, is different. Second, they show that the lift force can be quite high at times. The lift force is entirely ignored in the force balance we have summarized in the derivation of the Shields criterion. We focused only on torques associated with the drag force. The recent experiments reveal that there are times when the pressure difference between the base of the particle and the top can be quite high (Figure 14.8). These rare events are associated with conditions under which the flow is accelerating over the top of the particle (producing low Bernoulli pressures) while it is decelerating under the particle.

These observations of both flow structure and forces on grains on the bed point to a need to revise the classical theory based upon the mean flow. Just how much revision is needed depends upon the application. For example, if one is interested in the long-term sediment transport through a particular reach, we must know both the dependence of the sediment transport rate on the flow, and the distribution of flow speeds over the time interval of concern. If the target is instead the spatial distribution of sediment transport, which as we will see is all-important in the evolution of bedforms, then we must know how the details of the turbulence structure vary in space. It is clear from the recent work that the spatial structure is complex, and likely varies in a manner that differs from how the mean flow varies in space, making the more detailed grain-scale work highly relevant to the bedform problem.

Modes of transport

The mode of sediment transport depends largely upon the trajectories of grains once they have been entrained in the fluid. Very large grains will skitter along the bed. Grains large enough not to respond efficiently to the turbulence of the flow will travel smooth trajectories that re-impact the bed within a short distance of their launch position; these grains comprise the saltation load. These two types of motion we collectively call bedload, as the grains hug the bed. If the grains are small enough to respond to the turbulence by the time they re-impact the bed, they will carry out wiggly trajectories that can go quite far. These comprise the suspended load. The finest grains involved in the suspended load are what we call the wash load. These grains are so fine that they basically go wherever the water goes; they do not settle readily. We now describe the mechanics of the trajectory process that discriminates among these modes of transport.

The saltation trajectory

Now that we have managed to launch a grain off the bed, what path does it take before it re-encounters the bed? This trajectory problem depends sensitively upon the mass of the grain. For massive grains – and we will define this in a moment – the trajectory tends to be a smooth one, while small or low density grains tend to perform wiggly trajectories. The former we call saltation trajectories, the latter suspension trajectories. The saltation trajectory can be thought of as a deterministic process, while suspension is stochastic. Why the difference? We have to recall that the flow into which these grains are launched is turbulent. The velocities to which a grain is subjected vary widely not only because the mean flow changes through the trajectory, but because the velocity at any height above the bed actually varies widely around a mean. What becomes important is the response time of the particle to changes in the flow velocity it feels. Particles that take a long time to respond (what we called massive grains above) will effectively average or smooth out these fluctuations, while those that respond rapidly will mimic more faithfully the fluctuations – their trajectories will wiggle.

Let us break down the problem into its pieces again. The forces acting on the particle are weight, drag, and lift. In order to have something to aim at, consider first a particle with no forces acting on it except its weight. This is the pure ballistic case – the one to which gunners turn when determining the proper launch angle for cannon balls and similar ordinance. Once we specify the initial speed and angle of launch, the entire trajectory is known. As a refresher, consider

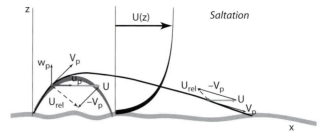

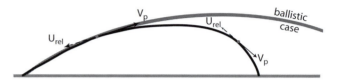

Figure 14.9 Trajectories of a saltating grain embedded in wind (top) and of a ball driven into still air (bottom). Ballistic case (grey) is shown for comparison in both cases.

Figure 14.9 in which we sketch out the ballistic trajectory problem. We wish expressions for the height, duration and length of the flight path of the particle. First: the height of the hop. We can appeal to conservation of energy: at the top of the trajectory, all of the initial kinetic energy has been converted to potential energy. Equating $PE = KE$ yields

$$mg\delta = \frac{1}{2}mw_{po}^2 \tag{14.9}$$

where w_{po} is the initial vertical component of the grain velocity. This collapses to

$$\delta = w_{po}^2/2g \tag{14.10}$$

The hop height, δ, depends sensitively upon the vertical component of the launch speed, and upon the planet we are on, which dictates g. The hop duration is simply two times the time to reach the top of the hop. Because no other forces are acting, the only acceleration acting on the particle is that due to gravity. The vertical velocity may then be determined by integrating the vertical acceleration, a, which is simply g:

$$w_p = \int a dt = -g \int dt + c = w_{po} - gt \tag{14.11}$$

where we have used the fact that at $t = 0$, $w_p = w_{po}$, the initial vertical speed of the particle, to evaluate the constant of integration, c, in the solution. Recognizing that the vertical speed must go to zero at the top of the hop means that the time to the top of the hop is simply w_{po}/g; the total hop time is therefore

$$T = 2w_{po}/g \qquad (14.12)$$

Finally, consider the hop length. Formally, the history of the horizontal position, x_p, of the particle is obtained by integrating its velocity:

$$x_p = \int u_p \mathrm{d}t = u_{po} \int \mathrm{d}t + c \qquad (14.13)$$

If we used $x_p = x_{po} = 0$ at $t = 0$ to evaluate the constant of integration, c, and integrate until the particle lands, we obtain its landing site: $x_p = u_{po}T$. Note that we have assumed the horizontal velocity does not decline, because we have assumed no drag forces are operating. The distance traveled is the time of flight times the original horizontal velocity. Employing Equation 14.12, we find that the trajectory length, L, may be written as

$$L = \frac{2}{g} u_{po} w_{po} = \frac{2}{g} v_{po}^2 \sin(\alpha)\cos(\alpha) \qquad (14.14)$$

Where α is the launch angle. Note a couple of features of these ballistic trajectories. The longest trajectory length occurs if the particle is launched at 45°. Its trajectory length and height are scaled by the square of the launch speed, while the time of flight is scaled linearly by the launch speed. Finally, ballistic trajectories would be longer on Mars than on Earth, given that Martian gravitational acceleration is roughly 0.38 times that on Earth.

Of course, this simple description of the trajectory falls short (so to speak) of reality in several important ways. Saltation lengths are longer than this ballistic case suggests, because the drag force acts to accelerate grains downwind. Typical saltation paths are very asymmetrical: a typical grain is launched at 45°, and lands at 10–15°. (Note that this is the opposite asymmetry from a baseball or golf ball trajectory, because the drag acts in the opposite direction.) Importantly, acceleration by the wind allows grains to land with significantly more speed and hence kinetic energy than they had at launch. The drag force is a reflection of the extraction of momentum from the wind (I call them ventivores). This momentum is then available for damage at the impact site. Note also the low angle of impact. This will emerge as an important factor in the generation of ripples on a sandy surface.

To be more concrete about it, we show in Figure 14.9 the differences between purely ballistic

(without drag), saltation, and baseball trajectories. The drag force acting on the grain, which we neglected in the ballistic case, is dependent upon the relative velocity between the grain and the fluid: $\overrightarrow{U_{rel}} = \overrightarrow{U_{fluid}} - \overrightarrow{V_p}$.

In the case of sediment transport, the fluid is in motion, while on a still day in the ballpark, the air into which a baseball is launched is not in motion. As we have shown in the calculation of settling speeds (Chapter 9), the drag force, which is a vector, may be written

$$\overrightarrow{F_d} = \frac{1}{2}\rho_f \frac{\pi D^2}{4} C_d |U_{rel}| \overrightarrow{U_{rel}} \qquad (14.15)$$

Here the absolute value symbols denote the magnitude of the quantity, while the arrow symbol represents a vector quantity. Note that the drag goes as the square of the magnitude of the relative velocity, and acts in the direction of the relative velocity. In Figure 14.9 we see that the relative velocity of the grain and the fluid in saltation are such that the grain is accelerated in the direction of the flow for the majority of its path. Only as it nears its re-encounter with the bed does its speed exceed that of the fluid, resulting in a slight deceleration. On the other hand, at least on a windless day, the baseball (or golf ball) experiences a drag that acts to slow it in the horizontal at all points in its path. The symmetry of its path will therefore be opposite that of the sand grain. Sadly, at least for the batter or the golfer, the path traveled will be far shorter than that one would estimate with a ballistic calculation.

The granular splash

In eolian sediment transport, the interaction of a saltating grain with the bed is a forceful one. This is not the case in sediment transport under water, and represents perhaps the largest difference between the two cases. When a saltating grain impacts the bed, it imparts its kinetic energy, including that extracted from the wind, to the bed. The resulting mini-explosion is called the granular splash. That this does not occur in the subaqueous case reflects the density of the fluid. A grain approaching the bed under water must displace the near-bed water before it can impact the bed. Because water is roughly a third as dense as the typical grain, this uses up a substantial fraction of its momentum, and the grain tends to land rather softly, at least in comparison with the air case in which the density contrast is a thousand-fold.

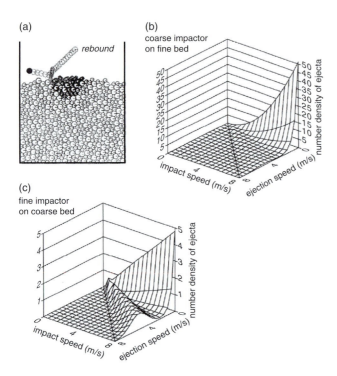

(a)

rebound

(b)

coarse impactor
on fine bed

impact speed (m/s)   ejection speed (m/s)   number density of ejecta

(c)

fine impactor
on coarse bed

impact speed (m/s)   ejection speed (m/s)   number density of ejecta

Figure 14.10 (a) Example of grain splash calculation.
Large grain impacts from left at 1 m/s and is shown in 20 frames.
The impact disturbs a large volume of the mixed grain-size bed
in front of the impact point, and rebounds efficiently from
the surface. Disturbance of the bed is indicated by initially
horizontal tick marks on the grains. (b) and (c): the splash function,
a statistical description of the granular splash process. Results from
(b) a coarse impactor on a fine bed, and (c) a fine impactor on
a coarse bed. Total number of ejecta increases roughly linearly
with impact speed. Low-speed ejecta are splashed grains, while
the high-speed bump represents the rebound of the impacting
grain at roughly half of its impact speed. Rebounds are more
likely and more energetic when fine grains impact a coarse bed,
while splashed grains are more numerous and energetic when
a coarse grain impacts a fine bed (after Haff and Anderson,
1993, Figures 10 and 13, with permission from John
Wiley & Sons).

The granular splash resulting from the impact
of a massive sand grain is quite complicated, and
certainly stochastic, owing to the variability in the
micro-packing of the local grains, and the local geometry of the surface. The key elements of the splash,
however, are that the impacting grain tends to
rebound off the surface with about half its original
impact speed, and it ejects up to several grains, all at
velocities that are quite low – perhaps a tenth of its
impact speed. These are summarized in Figure 14.10.
Most of the energy, however, is dissipated in the
frictional rearrangement of grains within the bed.
Impacts of fine grains into coarse beds (or portions

of beds) result in very different statistics of the splash
(what we call the splash function) than those of coarse
grains impacting a fine bed. We will see that this
becomes relevant to the generation of sorting patterns
in eolian ripples. This saltation impact phenomenon
can be thought of as a micro-tamping of the bed. It
compacts the local surface. On the stoss (upwind) side
of the dune, where saltation occurs, the surface is
tamped and jostled many times per second. This firms
up the surface. On the lee side, the grains fall out of the
air column, accumulate in a grainfall deposit until it
steepens to failure, and then move downslope in
granular flows that require dilatation of the pores.
Bagnold knew this (learned it fast, that is) from his
travels around the Egyptian and Libyan deserts early
in the twentieth century in modified Model Ts:
you can drive on the stoss side of dunes, but avoid
the lee at all costs (see his descriptions in *Libyan Sands*
(Bagnold, 1935)).

We know these things about the saltation process
in the eolian world because they have been filmed at
very high speeds. Since a typical saltation hop time is
of the order of 0.1 seconds, and we need many points
along the trajectory to be visualized (say 100), the film
must be exposing frames at a rate of about 1000
frames per second (fps). This is about two orders of
magnitude faster than most movie film (16 fps), even
filmed in slow motion (32 fps). The impact event itself
can also be explored using another trick from the
physics class – the strobe. In experiments run at
Caltech by B. T. Werner, individual sand grains were
shot at a sand surface, and a single frame of film
was exposed under stroboscopic lighting. While these
sorts of techniques work well in air, they can also
be employed – although with much more technical
complexity – in water. It is just this sort of experimentation that is revealing a much more complex picture
of the bedload transport process under water – but
more on that later.

Mass flux: transport "laws"

Now let us turn back to the major problem of what
sets the mass flux of sand in bedload. We have
explored the entrainment problem, and found that
there is a distinct threshold that is dependent upon
grain size, below which there is no transport. Above
this, grains are entrained, are accelerated downwind
once they are up into the airstream, and finally impact

the bed with more momentum than they started with. At the impact site, energetic impacting grains rebound from the surface at a higher angle and a lower speed than that at which they landed, and they eject up to several other grains, at much lower speeds. These grains then are accelerated downwind, impact the bed, and so on. A chain reaction has been set up, fed by the energy from the wind. More and more grains are splashed off the bed into the wind. One might rightfully ask why the entire world doesn't start saltating! The answer lies in the conservation of momentum equation,

$$\rho_a \frac{\partial U}{\partial t} + \rho_a U \frac{\partial U}{\partial x} = -\frac{\partial p}{\partial x} + \frac{\partial \tau_{zx}}{\partial z} - F_x \qquad (14.16)$$

where ρ_a is the density of air, U the mean horizontal speed, g the acceleration due to gravity, τ_{zx} the turbulent Reynolds stress, and F_x the horizontal body force exerted by the grains on the air. In the turbulent boundary layer under conditions of a steady horizontally uniform flow of air, this reduces to

$$\frac{\partial \tau_{zx}}{\partial z} = F_x(z) \qquad (14.17)$$

The momentum extracted from the wind by the grains as they are accelerated forward by the wind does not come for free. The wind loses momentum to the grains – and as it does so the wind speed near the bed, within which the sand grains are flying about, declines. This is documented in wind tunnel experiments shown in Figure 14.11. The horizontal force acting on the fluid, F_x, is the sum of all the drag forces acting on all the grains at any level z in the fluid (Anderson and Haff, 1991). The reduced wind speed results in less acceleration of grains, lower impact speeds, fewer splashed grains, and fewer grains being entrained directly by fluid stresses. This is an example of self-regulation. The integrated set of feedbacks is captured in the saltation system diagram of Figure 14.12, in which all of the components are illustrated. The sediment flux at any wind speed will represent a balance between the positive feedback of each grain producing a cascade of splashed grains, and the negative feedback of momentum extraction from the wind by the grains.

Proper linkage of all of these components, entrainment, trajectory, granular splash, and momentum extraction from the wind, into a simulation of the

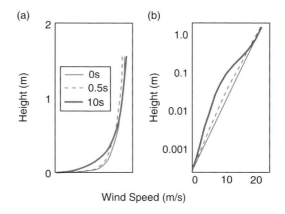

Figure 14.11 Wind profiles showing departure from log profiles during sediment transport at a position 10 m downwind from the edge of the erodible surface in the experiment. Plots are for three times in the experiment: 0 s (no sediment transport), 0.5 s, and 10 s (full transport). (a) Linear–linear, (b) log–linear plot (after Shao and Li, 1999, Figure 9, with permission from Springer).

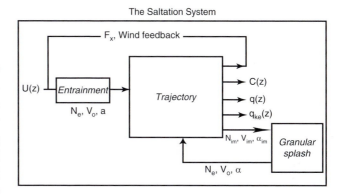

Figure 14.12 The saltation system, depicting processes and feedbacks (after Anderson and Haff, 1988, with permission from the American Association for the Advancement of Science).

entire eolian saltation process should be capable of reproducing the many properties of the system we have discussed. These include the thresholded nature of the system, the hysteresis involved in a second lower threshold, a nonlinear relationship between the wind speed and mass flux above this threshold, concentration and mass flux profiles that decline with height above the bed, and an altered wind speed profile. Recently such models have been constructed, and pass this test with flying colors, as seen in Figure 14.13. Interestingly, the models predicted in addition certain

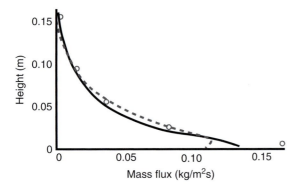

Figure 14.13 Profile of mass flux during eolian saltation. Circles: measurements. Lines: calculated using individual trajectories with full range of initial conditions (from Anderson and Hallet, 1986, Figure 11).

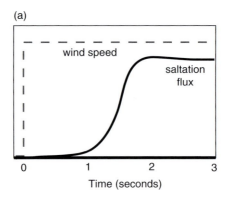

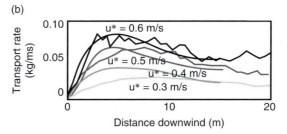

Figure 14.14 Expected history of saltation flux at inception of transport by wind. The response time of the saltation system is a few seconds. (a) Numerical simulation (after Anderson and Haff, 1988, with permission from the American Association for the Advancement of Science). (b) Downwind pattern of sediment transport by wind in three different winds, from wind tunnel measurements (after Shao and Raupach, 1992, Figure 4) and model (smooth lines, from Shao and Li, 1999, Figure 10, with permission from Springer).

behaviors that had not been documented before. In the simulations, the time to reach a steady number of grains in transport, or mass flux, is of the order of 1–2 seconds (see Figure 14.14(a)). This timescale has important implications for how one goes about estimating long-term sediment transport. In experiments in a long wind tunnel, it has subsequently been shown that this temporal evolution of the saltation system translates into a downwind evolution. It takes several meters for the saltation cloud to evolve to its full carrying capacity; similarly, it takes several seconds for the wind speed profile at any location to settle down to one in equilibrium with the sand that it is carrying. It also implies that the variations in wind speed over short periods of time (seconds) are important in setting the total transport rates.

For any given wind speed, there is a given maximum mass flux of grains that it can support in transport. The relation is illustrated in Figure 14.15, and is called the *transport capacity* of the wind. You can easily see both the thresholded nature of the problem, and importantly, the very strongly nonlinear increase in transport rate once the threshold is exceeded. While there are about as many transport laws as there are data sets to document such a relationship (see a table of such relationships in Greeley and Iverson, 1985), the relation is essentially a cubic one between eolian sediment transport and shear velocity:

$$Q = C\frac{\rho_a}{g}\sqrt{\frac{D}{D_o}}(u_* - u_{*c})^3 \qquad (14.18)$$

where ρ_a is the density of air, g the acceleration due to gravity, D the grain diameter, D_o a reference grain diameter (0.25 mm here) and C is a dimensional constant equal to 0.18 if all dimensions are in SI units. Bagnold documented this kind of relationship from his wind tunnel work as early as the 1930s. It is noteworthy that both effects, the threshold and the cubic increase above it, enhance the importance of the rare high fluid speeds. This is the case in both eolian and fluvial systems.

One might ask why the system is so nonlinear once transport has begun. Why is there this cubic dependence on the shear velocity? The basic argument is as follows. The flux of mass goes as the product of grain mass with the number in transport per unit area of bed, N, and the transport or hop length of the grains, L:

$$Q = m_p NL \qquad (14.19)$$

which has the proper units of a specific discharge [M/LT]. The hop length goes as the product of

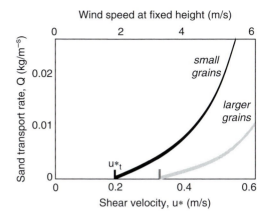

Figure 14.15 Sediment transport rate as a function of flow speed, for two different grain sizes with different entrainment thresholds. Transport is thresholded, above which the transport rate increases nonlinearly with wind speed.

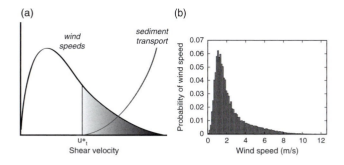

Figure 14.16 (a) Hypothetical probability distribution of shear velocities. The transport rate of sediment is shown as well with its threshold and a nonlinear increase in transport rate above the threshold. The shaded portion of the wind distribution will contribute to transport, despite the fact that the mean wind speed is below the threshold. This emphasizes the importance of wind statistics, and the problems associated with the use of only long-term mean winds in calculations of sand transport. (b) Probability distribution of wind speeds. Measurements made over a seven-year interval at Crucero site, USGS field site SW07. Winds at 3 m above the ground are measured every 4 seconds, with the average recorded every hour. Sediment transport is measured to occur above wind speeds of roughly 2.5 m/s.

the horizontal speed of the hop and its duration. If the launch speeds of the grains are scaled by u_* in both the horizontal and vertical, then the duration goes as $2u_*/g$ (from Equation 14.12) and the hop lengths should therefore go as u_*^2. If the number of grains launched per unit area of bed goes roughly as u_*, then the product in Equation 14.19 must go as u_*^3.

Now let us consider the problem of how to estimate the amount of sand that might be transported over the course of a year. You might want to know this for calculating how much a particular sand dune might migrate in a year – a very real problem in the Pampa la Jolla dunefield in Peru (e.g., Hastenrath, 1987), and in the extensive arid regions of China. Recall that the response time for the sand saltation system to changes in wind speed is of the order of 2 seconds. Ideally, you would therefore want to have or to estimate wind data with 2-second resolution for the year. The calculation would entail about 30 million time steps (roughly the number of seconds in a year). On the other hand, one could approach it from a statistical angle. From some nearby meteorological station, a wind record can be obtained, from which a probability distribution of wind speeds can be extracted. This is nothing but a histogram of wind speeds: the wind speeds are discretized into a set of bins, and the time spent in each bin is tabulated. Consider the distribution of wind speeds shown in Figure 14.16. From the transport law, we can calculate the total sand transport in each wind speed bin, and then sum to arrive at a discharge of sand for the year.

We emphasize that because the transport rate is nonlinearly related to the wind speeds, one cannot simply apply the sand transport equation using the mean wind speed. The essence of the problem is the following inequality:

$$\overline{[u_*^3]} \neq (\overline{u_*})^3 \tag{14.20}$$

The problem is made even worse if the function is thresholded. If we take the end-member case that the mean wind speed averaged over a year is below the threshold, we would calculate that there ought not to be any sand transport at all! It is very unusual that wind speeds averaged over even a day exceed the threshold of entrainment, much less so averaged over a year. It should be clear, then, that the proper data set with which to perform this calculation is one with temporal resolution of the same order as the response time of the system – a couple of seconds in this case. It is surprisingly rare that such wind data are available, and even less common that they are properly used.

We apply our knowledge of sediment transport mechanics to bedforms that arise when a surface is subjected to wind, and erosion caused by wind in Chapter 15.

Suspended sediment transport

Grains so small that they are affected by the turbulent velocity fluctuations of a flow are said to be in suspension. These grains cause the muddy look of a river. They are the important contributors to overbank deposits. They represent anywhere from 50 to 90% of all sediment in transport in a river. Within the eolian system, these silt and clay grains are the dust of dust storms. They are fine enough to be carried in the atmosphere from the Sahara to Florida, and from the Gobi desert to Hawaii.

Here we develop a theoretical basis for this important component in these transport systems. While the treatment will not be exhaustive, we will provide enough detail that the appropriate measurements and analysis can be accomplished. We take as our first goal the documentation of the mass discharge of suspended sediment, taking here the fluvial system as our example. We will see that some aspects of suspension complicate the problem, while others allow us to make assumptions that greatly simplify the problem.

There are two ways to view the problem, one at the grain level, the other at the continuum level. At the grain level, each grain in the flow is simply experiencing a set of forces that are always of two flavors: weight and fluid drag. We could calculate the trajectories of a large number of grains, each one of which experiences a history of drag forces that depend upon the turbulent eddies it encounters, and then calculate the concentration of these grains as a function of the height above the bed. This laborious calculation is time consuming, although it is faithful to the physics at the grain level. The alternative approach is to view the problem as a continuum, a mixture of grains and fluid. We review both approaches.

The suspension trajectory

First consider the trajectory of a small grain. Less massive grains respond much more rapidly to turbulent eddies and fluctuations in flow speed through the course of their trajectory. Of course, the smallest grains, motes of dust, will travel at essentially the speed of the air. They will follow faithfully a parcel of air as it wafts down to the surface, trades some of its high momentum with another parcel, is entrained in an upward eddy, and moves onward. The suspension

trajectory is therefore a stochastic one. That is to say that we can no longer determine with certainty the hop height, the hop length, or the travel time of the grain in the fluid, even if we know perfectly the speed and angle of ejection from the bed; we are as far from the ballistic case as one can get. Rather, we can describe the results *statistically*. While we will turn to the problem of the resulting concentration profile of suspended grains in a later section, know that the physics is still the same – the grains respond to imbalances of forces on them, their weight and fluid drag, by accelerating. The big ones, those in pure saltation, are simply big enough to filter out all the small fluctuations that force the small ones to travel wiggly paths. To illustrate this, in Figure 14.17 we reproduce simulations of trajectories of grains small

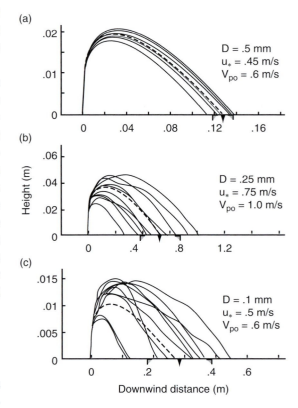

Figure 14.17 Saltation trajectories with turbulent wind component included. All trajectories are given identical initial conditions of liftoff angle (90°), and spin (0). Initial vertical speed is noted. Mean hop length of 20 calculated trajectories is marked with an arrow; brackets indicate one standard deviation. Note exaggerated vertical scales. (a) $D = 0.5$ mm, $u_* = 0.45$ m/s, $V_{po} = 0.6$ m/s: $P_s = 16.4$. (b) $D = 0.25$ mm, $u_* = 0.75$ m/s, $V_{po} = 1.0$ m/s: $P_s = 2.46$. (c) $D = 0.1$ mm, $u_* = 0.5$ m/s, $V_{po} = 0.6$ m/s: $P_s = 0.66$ (after Anderson, 1987a, Figure 6, with permission from the University of Chicago Press).

enough to react rapidly to wind speed fluctuations. All grains in the calculations were launched at the same speed and angle. Each grain then encounters a different set of wind speed fluctuations, resulting in trajectories that are greatly different from one another. As the grains get more massive, the trajectories become more alike, and approach the saltation end-member in which the trajectories are essentially identical. In the saltation case, the trajectories are fully characterized by the mean flow speed, and can be considered deterministic, while the suspended trajectories are clearly stochastic. One could simply tally such quantities as concentration, mass flux as functions of height above the bed, as was done in the saltation simulations, to generate the expected concentration and mass flux profiles. On the other hand, it is useful to employ a continuum approach, treating the grains as an ensemble, and track bulk properties of the mixture of grains and fluid. This approach we show below. Happily, the two approaches yield the same profiles of concentration and mass flux.

The continuum approach

The continuum approach presumes that the mixture of grains and fluid varies in its properties in a smooth and continuous way. In general, as in the saltation case treated earlier, we must know the velocity of the grains and their concentration in order to calculate the specific sediment mass discharge:

$$Q_{\mathrm{sus}} = \int_0^H CU_{\mathrm{sed}}\mathrm{d}z \qquad (14.21)$$

where U_{sed} is the horizontal velocity of the sediment, and C the concentration of sediment, here taken to be the mass concentration [$=M/L^3$]. Note that the units on specific discharge are M/LT, or the mass discharged per unit width of the flow per unit time. In the fluvial case, in order to calculate the full sediment discharge of a river, one must then also multiply by the channel width. Because suspended grains are traveling at roughly the horizontal velocity of the fluid, the sediment velocity may be replaced with the horizontal velocity of the fluid, and we may easily calculate the total discharge of sediment once we know the concentration profile, simply by integrating the product of the concentration with the mean flow velocity, U:

$$Q_{\mathrm{sus}} = \int_0^H CU\mathrm{d}z \qquad (14.22)$$

In order to perform this integral, we need expressions for the concentration profile of suspended sediment concentration, $C(z)$, and of the flow velocity profile, $U(z)$. We have already seen (Equation 12.19) that the flow velocity profile is well captured by the classic "law of the wall," or logarithmic velocity profile:

$$U = \frac{u_*}{k}\ln\left(\frac{z}{z_o}\right) \qquad (14.23)$$

What do we expect the sediment concentration profile to look like? The problem is complicated. The flow is highly turbulent (high Reynolds number), and the grains are embedded in turbulent eddies within the flow. We will make a set of simplifying assumptions that lead to a description of the mean concentration profile. Our first simplification is to restrict our attention to situations in which the flow is steady over significant timescales (i.e., long times compared to the average eddy time).

Consider a horizontal plane across which grains can move, as sketched in Figure 14.18. Grains will always want to slip downward in the fluid due to their negative buoyancy. This imposes a negative (downward) flux of grains across this plane, set by the

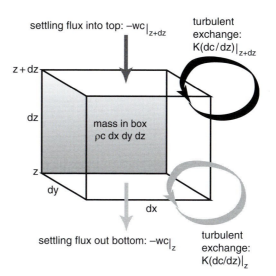

Figure 14.18 Sketch for conservation of mass of sediment in suspension, in which the upward diffusion of sediment is balanced by settling.

product of the local concentration C and their settling velocity, w_{sett}:

$$Q_{sett} = w_{sett}C \tag{14.24}$$

In steady state, this must be balanced by an upward flux of the same magnitude, but how can grains pass upward through the plane? This is the role of turbulence. In the same manner that turbulence passes momentum from one place to another in a flow, it passes very small particles and any other passive marker in the flow – temperature, for instance. One may think of this as an eddy forcing a packet of fluid from one place in the flow to trade places with another packet of fluid elsewhere in the flow. If the concentration of particles is different between the two packets, the turbulence will transport sediment from places of high concentration to places of low concentration.

The transport of sediment by turbulent eddies is represented by a turbulent diffusive flux,

$$Q_{diff} = -K\frac{\partial C}{\partial z} = -ku_*z\frac{\partial C}{\partial z} \tag{14.25}$$

where here the big K, called the eddy diffusivity, represents the efficiency of turbulence in passing fluid from one place to another, and $\partial C/\partial z$ is the local concentration gradient. As for momentum, the mass diffusivity can be represented by the eddy diffusivity approximation relevant to near-surface conditions: it linearly increases with distance from the bed, z, scaled by von Karman's constant, k ($= 0.4$) and the shear velocity u_*.

If the sediment concentration profile is steady in time, then the downward flux due to settling (Equation 14.24) must be balanced by the upward flux allowed by turbulent mixing (Equation 14.25). This results in

$$w_{sett}C = -ku_*z\frac{\partial C}{\partial z} \tag{14.26}$$

which can be rearranged to obtain

$$\frac{dC}{C} = -\frac{w_{sett}}{ku_*}\frac{dz}{z} \tag{14.27}$$

This can now be integrated to obtain the concentration profile. We dodge a bullet by imposing a boundary condition near the bed that the concentration at a reference level z_a be C_a. Hence,

$$\int_{C_a}^{C} \frac{dC}{C} = -\frac{w_{sett}}{ku_*} \int_{z_a}^{z} \frac{dz}{z} \tag{14.28}$$

(a)

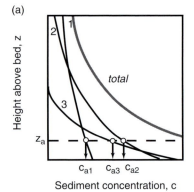

(b)

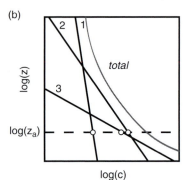

Figure 14.19 Concentration profiles expected during suspended sediment transport. (a) Linear–linear plot, (b) log–log plot. Profiles for three grain sizes are shown, increasing in grain diameter from 1–3. Reference level concentrations for each grain size are depicted. Total profile is the sum of the three individual profiles.

which results in the concentration profile for any given grain size with settling velocity w_{sett}:

$$C = C_a\left(\frac{z}{z_a}\right)^{-\frac{w_{sett}}{ku_*}}$$
$$\frac{C}{C_a} = \left(\frac{z}{z_a}\right)^{-p} \tag{14.29}$$

It is important to understand this equation. Note that because the equation includes a particular settling velocity, it is written for a single grain size. Figure 14.19 shows profiles for three different grain sizes. The reference level concentration, C_a, defines where the profile crosses the reference level height, z_a, and is determined by the local bed conditions: the availability of a particular grain size in the bed, and the degree to which the flow is in excess of the entrainment threshold for this grain size. The combination of constants that goes into the exponent of the height above the bed sets the profile shape. This dimensionless power, p, is called the Rouse

number, in honor of Hunter Rouse who performed detailed sediment transport experiments (e.g., Rouse, 1939; see also the work of Vanoni, 1941, 1946; Vanoni *et al.*, 1966). Note how p varies with grain size, for a fixed flow velocity (hence fixed u_*). For larger grains, such as grain 3 in Figure 14.19, the profile is much steeper, the gradient much higher. Why? Larger grains settle more rapidly, producing a larger settling flux for a given concentration. In order to counter the settling flux, which is a prerequisite for a steady concentration profile, the concentrations of the sediment in fluid parcels being traded by turbulent eddies must be very different. Translated, this means the concentration gradient must be high. The opposite is the case for small grains. In fact, in the end-member case of vanishingly small (or neutrally buoyant) particles, $p \sim 0$, and a uniform concentration results. The very fine sediment that approaches this behavior constitutes the "wash load" (grain 1 in Figure 14.19). The other limit, in which grains are expected to be in saltation rather than suspension, corresponds to $p > 2.5$, or $w_{sett} > u_*$.

The suspension trajectory and continuum approaches result in the same predicted power-law profiles (Figure 14.20). Summing over all grains in the simulation their probability of being found at each level in the fluid, one can arrive at a concentration profile. When plotted in log–log space, the resulting concentration profile has a characteristic slope, which corresponds to the expected Rouse number for the settling velocity and the shear velocity of the simulation. As long as the turbulence structure of the near-bed wind is well enough characterized in the simulation, these approaches are equivalent.

Let us illustrate with a real-world example. While most of our attention is on sedimentary grains, the same physics pertains to the transport of snow. In very cold climates the blowing of snow will quickly round the edges of original snowflakes and the grains become well rounded. They also get very hard. At temperatures well below freezing, snow grains are very elastic. As anyone who has tried to climb mountains into the wind will know, the snow grains are no longer the friendly floating flakes, but hard bullets. The howling, persistent winds of Antarctica about which Mawson wrote so eloquently in his tome *Home of the Blizzard* (Mawson, 1930), lead to low-visibility ground blizzards that make travel yet less palatable. In Figure 14.21 we show the measured profiles of snow concentration in a wind storm in

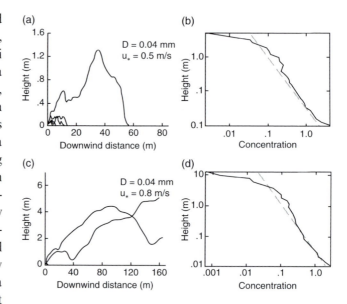

Figure 14.20 (a and c) Suspension trajectories for (a) $D = 0.04$ mm, $u_* = 0.5$ m/s, and (c) $D = 0.04$ mm, $u_* = 0.8$ m/s. (b and d) Concentration profiles corresponding to suspension trajectories in (a) and (c), respectively, with corresponding power-law fits using the Rouse numbers, (b) $p = 0.71$, and (d) $p = 0.44$, shown as dashed straight lines. Concentrations are normalized with respect to a reference level concentration at $z = 0.04$ m (see Anderson and Hallet 1986) (after Anderson, 1987a, Figure 5, with permission from the University of Chicago Press).

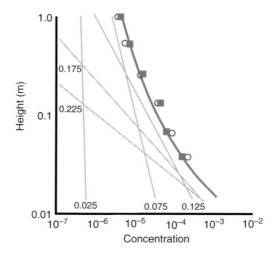

Figure 14.21 Concentration profile in blowing snow, showing application of suspension theory for a range of grain sizes. The total concentration profile is the sum of profiles from each grain diameter (labeled in mm). The snow concentration data shown as squares and circles are from two profiles with $u_* = 0.67$ m/s (from Anderson and Hallet, 1986, Figure 16a).

Antarctica. While the profile of any particular grain size is a power law, hence a straight line on a log–log plot, the total concentration profile is concave upward due to the addition of these power-law profiles.

Finally, let us explore the character of this new dimensionless number, p. The Rouse number can be interpreted in terms of a ratio of timescales. It is reasonable that if the grain is up in the fluid for long enough to respond "significantly" to the fluid velocities to which it is subjected, then its trajectory will be modified considerably from the ballistic one it would otherwise travel. These we would consider suspended, while those trajectories that are not greatly affected by the fluid will be saltation trajectories. Consider the eolian case. The time a grain launched from the bed with a vertical velocity w_o is in the air should be $T = 2w_o/g$. It takes w_o/g on the way up, and w_o/g on the way back down. But what is the timescale for a grain to respond to a fluid moving at a different speed?

We can assess this from the following calculation. In the grain (or raindrop) settling problem, we have already developed the equation governing a grain's acceleration when simply dropped in a fluid of a different velocity (zero) (see Chapter 10). The acceleration equals the sum of the forces divided by the mass of the particle. For small particles, with $\mathrm{Re} < 1$,

$$\frac{\mathrm{d}w_\mathrm{p}}{\mathrm{d}t} = -g + AU_\mathrm{rel} = -g + A(w_\mathrm{p} - w) \tag{14.30}$$

where A is a collection of constants with units of inverse time. For a still fluid, $w = 0$, this results in a solution for the asymptotic approach to the settling velocity of the particle:

$$w_\mathrm{p} = w_\mathrm{sett}\left(1 - \mathrm{e}^{-At}\right) \tag{14.31}$$

The particle achieves $(1 - 1/e)$, or roughly two-thirds of its settling velocity, in a timescale of $t_\mathrm{r} = 1/A = D^2\rho_\mathrm{p}/(18\rho_\mathrm{f}v)$. For quartz in air, this becomes $8 \times 10^6 D^2$,

---

Box 14.1 Suspended sediment discharge in a river

Given Equation 14.29 for the concentration profile of one grain size, we sum the profiles for all the grain sizes present to obtain a total concentration profile for suspended particles. This sum can be inserted in Equation 14.22 to obtain the total suspended sediment discharge. Performing the integral in Equation 14.22 is not a simple task, in general. Both the concentration of sediment, C, and the flow velocity, U, vary with height above the bed, and with opposite sign – concentration decreasing with distance from the bed, and velocity increasing. This is shown in Figure 14.22. Neither C nor U is uniform enough with height above the bed to be taken as a constant and is removed from the integral. Either we must actually integrate the equation, or we must perform the integral numerically on the computer. The approximation is as follows: $Q_\mathrm{sus} = \sum_i (C_i U_i \mathrm{d}z_i)$, where the index i corresponds to an interval in the vertical. If one uses a consistent vertical spacing, $\mathrm{d}z$, this becomes more simply: $Q_\mathrm{sus} = \mathrm{d}z \sum_i (C_i U_i)$. Summing over all vertical slices results in the final estimate of suspended sediment discharge per unit width of flow. If we multiply this "lane discharge" by the width of the flow, W, we obtain the total suspended discharge of the river, in units of M/T.

To review, one may document the sediment discharge in a channel by taking, at a minimum, the following measurements:

• width of the channel;
• sediment concentration at a measured height above the bed, z_a;
• a velocity profile that will allow assessment of u_* and z_o;
• back in the lab, size analysis of this sample to break it into several size classes with differing settling velocities.

In Chapter 12, we apply this theory to the evolution of floodplains, and in Chapter 15 we address the eolian equivalent, the widely dispersed eolian silts and clays that accumulate in important loess deposits around the world.

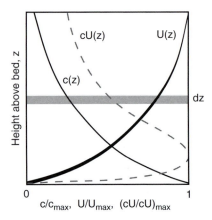

Figure 14.22 Sediment concentration, flow velocity, and mass flux (dashed) profiles during transport of suspended sediment. All profiles are normalized by their maximum value. Total suspended sediment discharge requires summing the sediment mass flux, cU, over all elements, dz.

where D is in meters. For example, if $D = 0.1\,\text{mm}$, or very fine sand, $t_r = 0.08\,\text{s}$. Recalling the equation for settling velocity, note that $w_\text{sett} = t_r g$, or, turning this around, $t_r = w_\text{sett}/g$.

Suspended grains can be defined as those for which the time of flight in the fluid, T, exceeds this response time, t_r: equivalently, suspension occurs when $T > t_r$, or $t_r/T < 1$. Replacing the response time with w_sett/g, this ratio becomes $(w_\text{sett}/g)/(2w_o/g) = w_\text{sett}/2w_o$. So the ratio of timescales is approximated by a ratio of velocities. At least in the eolian case, the vertical velocity component at liftoff can be expected to scale with the intensity of the turbulence, and hence $w_o \sim u_*$, in which case we have come full circle to the Rouse number. Given that $p = w_\text{sett}/(ku_*)$, that $p < 2.5$ is required for suspension, and that $k = 0.4$, this condition reduces to $w_\text{sett} < u_*$. Similarly, in the subaqueous case, where fluid stresses reign even under conditions of full saltation, the relevant velocity scale for establishing the particle trajectory is certainly u_*, and the same argument holds. You may think of the Rouse number, then, as either a ratio of velocity scales, or a ratio of timescales.

Summary

Sediment transport is thresholded, the *competence* of a flow being strongly dependent upon flow speed. For grains larger than about 0.1 mm, theory and empirical data alike suggest that the critical shear stress increases linearly with grain size. Departures from this result from electrostatic forces between small platey grains, and the ability of small grains to hide within the thin laminar layer near the bed.

Sediment transport occurs in two dominant modes. Large grains saltate; they travel smooth, deterministic paths that reflect response to the mean wind profile. Small grains travel in suspension; they are small enough to respond to turbulent velocity fluctuations of the flow, and take wiggly paths that can only be characterized statistically. The difference in modes can be characterized by the dimensionless Rouse number, p, which can be thought of as a ratio of timescales or of velocity scales.

The transport *capacity* of a flow is nonlinearly related to flow speed because both the number of grains in transport and their trajectory lengths increase with flow speed. Strong feedbacks in the saltation system allow a steady transport to occur, as the acceleration of grains acts as a brake on the flow. The timescale for achieving a steady transport is a few seconds, at least in the eolian case. Many 0.1 s trajectories and re-impacts with the bed must occur before the chain reaction allows a steady population of saltators. This timescale has strong implications for how one should calculate long-term transport rates.

Suspended sediment is distributed as a power law with height above the bed. The power is determined by the Rouse number, a ratio of the settling speed of particles to the vertical speed fluctuations of the turbulent flow. The Rouse number controls the rate at which concentration declines with height above the bed. Very low Rouse numbers, corresponding to small grains and strong flows, result in nearly uniform concentrations we call wash load. Knowledge of this vertical distribution of sediment (1) explains why finer grains will dominate the upper levels of a flow, and (2) provides a quantitative theory with which to calculate the total discharge of sediment from only a few measurements.

Problems

1. Calculate the depth of water flow in a wide channel (the stage) that would be necessary to entrain 2 mm quartz grains from the bed, given that the slope of the river is 2 m/km. Take the Shield's parameter, $\theta = 0.06$.

2. How much time is a grain in the air if it is launched at a speed of 0.5 m/s at an angle of 45° with respect to the horizontal? Repeat the calculation for a grain launched at a speed of 1 m/s at 60°.

3. Ignoring air drag, calculate the speeds (in m/s) and angles at which a baseball must leave the bat in order to clear an outfield wall 15 ft high and 400 feet away. The ball is launched from 2 feet above the home plate. (Pardon the English units – they are still used in this sport.) There are many possible solutions. Now assess the minimum speed (hence launched at the optimal angle) to achieve this homerun. Discuss how this will differ when one incorporates drag into the calculation.

4. *Long-term transport.* We wish to know the total transport of sand in an eolian system over the course of a year. You know that the threshold shear velocity for entrainment is $u_* = 0.3$ m/s for the sand involved. For the simplest case assume that the wind speed distribution is uniform, so that the probability of a given shear velocity is the same as that of any other over an interval [0, 1 m/s]. Calculate the expected total transport of sand, given that the instantaneous transport rate is given by $Q = B(u_* - u_{*c})^3$. Contrast this against the figure for the total transport obtained using instead the mean shear velocity over the entire year. (If you calculate the ratio of the two answers, the B cancels out.) Now perform the same comparison using a more realistic

distribution in which the probabilities fall off exponentially with higher shear velocities: $p(u_*) = (1/\bar{u}_*)e^{-u_*/\bar{u}_*}$, where $\bar{u}_*$ is the mean shear velocity.

5. Calculate the response time, t_r, for quartz in air for the following grain sizes: 10, 100, 300 microns. What would the response time be for the same grains in water?

6. How much higher would the saltation curtain be on Mars than on Earth? Assume that the distribution of particle launch speeds and angles would be the same as on Earth; for simplicity, take the range of speeds from 0.1 m/s to 2 m/s, and assume that all grains are launched at 45°. The acceleration of gravity on Mars is 3.75 m/s². Assume that the trajectories are ballistic – i.e., ignore drag in the vertical.

7. What flow depth is required in a turbulent river to support quartz grains of diameter 0.25 mm in suspension if the slope of the river is 1 m/km? Assume that the critical Rouse number for suspension is 2.5. Perform the same calculation for a 0.25 mm grain of gold. (*Hint:* you should find that gold is much less likely to be in suspension. This is the essence of the gold panning method.)

8. *Thought question.* Given the theory of suspended sediment transport we have discussed, develop a means by which one could measure the total suspended sediment discharge at a given position on a river given only one measurement of sediment concentration (say from a sediment sample collected at a given height in the water column). Outline all of the information one would need to gather from the sample and from the river to accomplish this task.

Further reading

Bagnold, R. A., 1941, *Physics of Blown Sand and Desert Dunes*, Methuen: London, 265 pp.
This remains the classic book to which we turn when first learning about wind transport of sediment. In this book Bagnold describes both his field observations and careful experiments.

Middleton, G. V. and J. B. Southard, 1984, *Mechanics of Sediment Movement*, Society of Economic Paleontologists and Mineralogists.
In a delightfully informal presentation, these authors summarize the field at the time of writing, nicely marrying experimental data sets and theoretical treatment of sediment transport problems.

Yalin, M. S., 1972, *Mechanics of Sediment Transport*, New York: Pergamon Press, 360 pp.
This is another classic text. Yalin ends each of his chapters with suggested future research themes. Some of these were well ahead of their time.

CHAPTER 15

Eolian forms and deposits

Sea waves are green and wet,
But up from where they die
Rise others vaster yet,
And those are brown and dry.

Robert Frost (1874–1963), "Sand Dunes"

In this chapter

We employ the knowledge of sediment transport mechanics to develop an understanding of a variety of eolian features. In treating the origins and styles of eolian bedforms, we will rely upon the discussion of the saltation process. In describing the processes involved in sand dune migration, and in the origin of the unique stratigraphic signature of dunes, we add to the list of processes to include grainflows that dominate on the lee side of dunes. We describe models of how both dunes and ripples evolve, and pay particular attention to why dunes differ so markedly in shape to generate everything from barchan crescents to star dunes. Some of the largest accumulations of eolian sediments in the world, called loess, are fine-grained silts that travel not in saltation but in suspension. Here we describe their origin and distribution, and lay the foundation for interpretation of the climate history that is embedded in the classic loess sequences around the world. Finally, we treat some unique features associated with wind transport of sediment, those associated not with deposition of sediment, but with the erosion of obstacles subjected to sand-blasting: ventifacts. While at some level these features are curiosities, they still raise important questions that we must consistently face as geomorphologists. For example, what is the role of rare wind events in driving dune motion, or in carving facets in ventifacts? Is it the saltating (large) grains that are largely responsible, or the smaller (but faster) suspended grains?

Detail of eolian stratigraphy exposed on a wall of Buckskin Gulch, Utah. Cross-sets reveal individual grainflows bounded by thin grainfall deposits. Vertical lines are staining from modern surface water draining down the vertical face of the exposure (photograph by R. S. Anderson).

Bedforms

That sand grains organize themselves into ridges and bumps has long intrigued both artists and scientists. Despite the similarity in shape pointed out by Robert Frost in his poem "Sand Dunes," water waves and sand dunes or sand ripples have little in common except that they naturally arise from the motion of a fluid over some substrate. We discuss wind-driven waves in water in Chapter 16. These "bedforms" in sediment come in a hierarchy of sizes, from the smallest ripples, to dunes, and sometimes to larger yet mega-dunes or complex dunes – dunes on the backs of larger dunes. While ripples and dunes display a similar asymmetrical shape, with shallow stoss (upwind) and steep lee (downwind) faces, the distinction between these two classes of bedforms is best reflected in the fact that the ripples do not have a slip face. While individual grains may roll down the lee of a ripple, the means by which sand moves down the lee face of a dune is by grainflows that transport billions of grains at a time.

A bedform may be characterized by a height, H, a length, L, and a velocity, c (also called the celerity, or wavespeed) as depicted in Figure 15.1. In all but antidunes (about which more later), the motion of the bedform is down-wind. Think for a moment about the required pattern of sand transport that will allow downflow motion, and retain the shape of the bedform. This "steady form" translation is in fact common. A sand dune this year can look much the same as it did last year, but tens of meters downwind. As can be seen in Figure 15.1, the stoss side of the form must be experiencing erosion, while the lee must be experiencing deposition. We can understand the required pattern of sand transport by appealing to the "erosion equation" we developed in thinking about hillslopes (Chapter 10). To refresh,

$$\frac{\partial z}{\partial t} = -\frac{1}{\rho_b} \frac{\partial Q}{\partial x}$$

(15.1)

where z is the elevation of the bed, Q is the down-flow transport rate [$= M/LT$], ρ_b is the bulk density of the sand, and x is the down-flow distance. (For the record, this is often called the Exner equation in sediment transport circles.) The left-hand side is negative in the case of erosion, which requires that the gradient in sand transport be positive; on the other hand, deposition requires a negative gradient in sand transport.

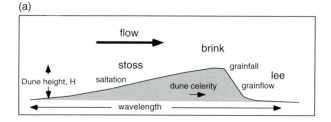

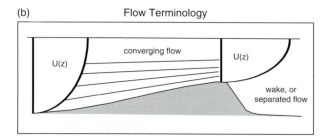

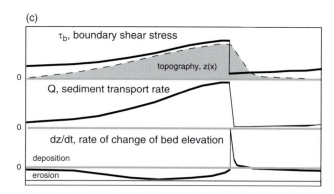

Figure 15.1 (a) Dune terminology. (b) Flow field over a dune. (c) Patterns of shear stress, sediment transport rate, and erosion/deposition rate over a dune.

This is simply a mathematical statement that the local sand account, reflected in the elevation of the local column of sand, is dictated by the difference between what is passed forward and what arrives from behind. We can therefore at least qualitatively graph the pattern of mass flux (sand transport) required for a steady bedform (Figure 15.1). The increasing rate of sand transport up the stoss face is allowed by the speed-up of the fluid flow over the top of the form. Why does this happen? Consider the profiles illustrated in Figure 15.1. In order to maintain the same discharge of water over the form, the mean flow speed over the crest must be greater than that over the trough. Note that the flow decelerates after it goes over the top. A wake is formed in which the flow speeds are very low, and even reverse in direction. The flow takes a long while to recover from its encounter with the bedform, typically a distance

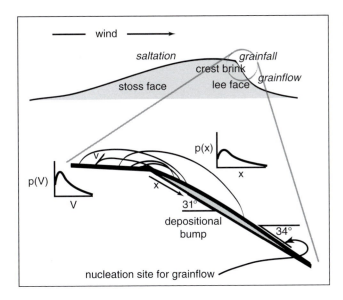

Figure 15.2 Dune anatomy and detail of saltation trajectories near the brink of the dune.

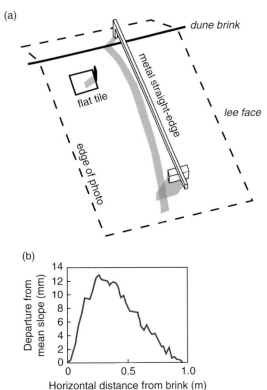

Figure 15.3 (a) Shadow of a straight-edge on the top of a dune lee face, showing bulge topography. (b) Difference in elevation between the measured profile and a straight line, showing a 12 mm bulge centered approximately 0.3 m downslope from the brink (from McDonald and Anderson, 1995, Figure 2, with permission from John Wiley & Sons).

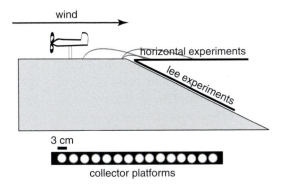

Figure 15.4 Experiments to test distribution of deposition on the lee face of a dune (after McDonald and Anderson, 1995, Figure 1, with permission from John Wiley & Sons).

equal to several dune heights. As the transport rate is dependent upon the flow speed, this requires that the sand transport rate decline as well. This is what allows deposition on the lee side of the bedform.

In the case of eolian dunes, the deposition takes place quite near the top of the lee face. Most saltation lengths are quite small, millimeters to centimeters in length, the rare ones being perhaps a meter in length. For all but the smallest of grains, the top of the lee face acts as a perfect trap for the sand in saltation. As shown in Figure 15.2, the result is a "depositional bump" at the top of the lee face, a bump that continues to grow in amplitude until some portion of it fails as a small landslide. You can easily see this at the top of a lee face by looking parallel to the face. The bulge can be easily quantified by photographing the shadow cast by a straight-edge (Figure 15.3). The detailed shape of the bulge has been used to verify a model of saltation that predicts the number of grains in motion in a particular wind speed, with a distribution of launch speeds that map into a distribution of deposition sites on the lee face. As sketched in Figure 15.4, grains can reach a particular point on the lee face by launching from far upwind with a high velocity, or from nearby with a low velocity. Formally, we would predict the number of grains of sand caught in a particular cup, or landing at a particular place on the lee face, by integrating, from the brink upwind, the number of grains launched with the proper velocity.

The ensuing failure of the bump then transports the grains comprising the depositional bump downslope as a grainflow, spreading the deposition out over the lee face. The dune therefore ratchets itself forward. Even in the face of steady winds, the movement

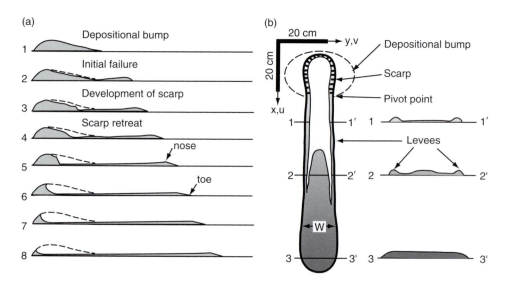

Figure 15.5 Grainflow terminology and evolution. (a) Sketch of the failure of the depositional bump (view rotated into parallel with the mean surface). Initially the bump fails as a roughly equi-dimensional slab or thin dribble on the steepest part of the depositional bump (2). As the slab spills away onto the platform, a steep scarp, extending laterally across the bump, develops (3). This migrates headward toward the top of the depositional bump (4–8). After this scarp retreat occurs, and the grain flow moves as a solitary wave down the platform. (b) General geometrical features of a typical grain flow and definition of the various features of grain flows Cross sections 1–3 show typical cross-flow profiles, with distinct levees found after passage of the bulk of the flow (after McDonald and Anderson, 1996, Figures 2 and 3, with permission from SEPM Society of Sedimentary Geology).

of the lee face will be a lurching one, reflecting steady accumulation and periodic tapping of this reservoir of sand at the top of the lee face. The dune acts as a little catastrophe machine.

The grain avalanches are interesting in their own right, as they display almost fluid-like behavior. The failure typically occurs at the steepest portion of the bump, which is downslope from the peak deposition rate. As this material moves away, a small headscarp propagates upslope to tap the main mass of sand, adding this sand to the flow. This is illustrated in Figure 15.5. The flow behavior is allowed by a slight dilation of the mass. The toe of the flow moves downslope at speeds of perhaps 0.2 m/s, and typically stops when it encounters a decrease in the slope near the base of the dune (except on truly huge dunes). Interestingly, this stoppage initiates at the toe of the flow and then zips rapidly upslope, as sand in motion feels that the sand immediately downslope of it has stopped. The effect has been likened to the stopping of a train. Couplings between cars are not perfect, and are a little bit extended while the train is in motion. When the engine stops, you can hear the couplings between all the cars close up in a wave that rapidly progresses toward the caboose. By analogy,

all the little gaps between grains, which had allowed the mass to behave as a granular fluid, close from the toe upslope. We have used the speed of this wave, and the diameter of the grains involved, to deduce the mean spacing of grains during flow – it is of the order of a few percent of a grain diameter.

The resulting stratigraphic package in a simple downwind translating duneform is dominated near the dune crest by a set of grainflows dipping at about 30°, which interfinger with grainfall deposits that become more prevalent lower in the dune. The portion of the bedform that is most likely to be preserved is the base of the form, as shown in a photograph in Figure 15.6, from the Navajo sandstone of the western United States. Just how much is preserved depends upon the net deposition rate of the entire field of forms, which is dictated by local subsidence rates, and by the eustatic sea level history. In the Navajo formation, these fractions of dunes, sometimes still tens of meters thick, add up to a kilometer thick package of dune sand. The base of a dune is dominated by the toes of the grainflow avalanches. These flows interfinger with finer-grained deposits that either have arrived at the base of the dune by wafting over the top in suspension in high winds (grainfall), or have

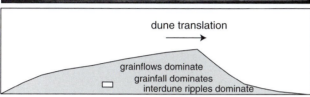

Figure 15.6 Top: Navajo sandstone stratigraphy in western Utah, showing intercalation of grainflows and grainfall at the base of a Jurassic dune. The toes of grainflows are darker because they are more permeable and have been cemented differently than the lighter grainfall deposits with which they interfinger. Bottom: this stratigraphy likely represents a record of the base of a much larger dune (white box). The remainder of the dune is not recorded in the rock record, as it is truncated by passage of the next dune. That any record is left at all requires that there be net deposition in the system. The package of sand at the top of the photo represents an interdune flat, dominated by inversely graded ripple laminae.

snuck around the dune in winds from other directions. These portions of the deposit are typically finer grained and are more tightly laminated than the grainflows, which results in slight changes in color, cementation, and groundwater flow behavior. Rubin has addressed the stratigraphic architecture of complex dunes in a very accessible study (Rubin and Carter, 2006).

The details and the periodicity of the stratigraphy of the eolian deposits of the Navajo sandstone have been studied extensively. For example, close inspection of beautifully regular cross-sets in northern Arizona have been used to suggest the presence of monsoonal rains. The interfingered deposits show roughly a 1 m advance of the dune between interruptions by rippled surfaces that reflect wind reversals. From the size of the cross sets, it is suggested that the dunes were of the order of at least 30 m in height, and that this 1 m advance therefore corresponds to an annual cycle. Common at the top of the grain avalanche sets are slumps with attendant thrust faults that have been interpreted to be failures of the slip face under wet conditions.

Classification of dune types

Sand dunes can be classified most straightforwardly by plan view pattern as sketched in Figure 15.7. Isolated forms include the barchan, the parabolic and the

---

Box 15.1 Self-organized criticality

The stable slope of a pile of dry granular materials is remarkably constant in a variety of materials. This slope, called the angle of repose, is considered a threshold because steeper slopes will fail while shallower ones will not. For all this, the angle of repose is actually ill-defined. There are in fact two angles, one static, and the other dynamic. One can pile grains one by one up to an angle of about 34° before they fail. The ensuing flow comes to a stop on a slope that is about 31°. The physics community is greatly interested in this phenomenon, using these piles of sand grains as an example of "self-organized criticality." Bak, Tang, and Wiesenfeld's 1988 paper showed that a simple cellular automaton of a sand pile could produce several characteristic features observed in nature, such as the power-law distribution of the sizes of avalanches, and that the emergent complexity was robust in that it did not depend on the details of the system. Such systems are said to display self-organized criticality in that a wide range of possible initial conditions and variable settings will result in emergence of this critical behavior. The emergence of complexity from simple local interactions (in the models, and in the sand piles) suggested that this could indeed occur in nature as well. Perhaps because Bak *et al.* (the team is known as BTW) used as their simple illustration a sand pile on which slow addition of new sand grains caused avalanches, many experiments were subsequently done using real avalanches in granular matter, including rice.

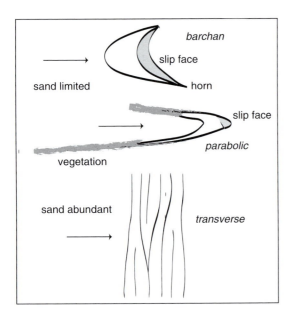

Figure 15.7 Plan view shapes of dunes in unidirectional winds. Pattern depends upon the supply of sand, and the abundance of vegetation.

star dune, while continuous fields of dunes include transverse and linear dunes. There are of course transitional forms, for example the seif dune, and there are compound forms, with little dunes atop larger forms. It has long been a challenge to the geomorphologist to determine the controls on dune shape. To interpret the eolian rock record, it is important that we be able to distinguish dune forms and their stratigraphic signature (see Rubin and Carter, 2006). The pattern is dictated by a combination of sand supply and by the degree to which the sand-transporting winds vary in direction.

Barchan dunes

Perhaps the most distinctive is the barchan dune, an isolated crescentic form with arms that stretch downwind. Barchans are not huge, often with heights of only a few meters. They are often found in well-organized fields, and are commonly separated by hard rock-strewn surfaces on which desert pavement can form. Classic examples include the barchan dunefields of the Pampa la Jolla desert near Arequipa, Peru (Lettau and Lettau, 1969), and the small field of dunes on the west edge of the Salton Sea, California (Haff and Presti, 1995), although they have now been discovered on Mars, as seen in the NASA image in

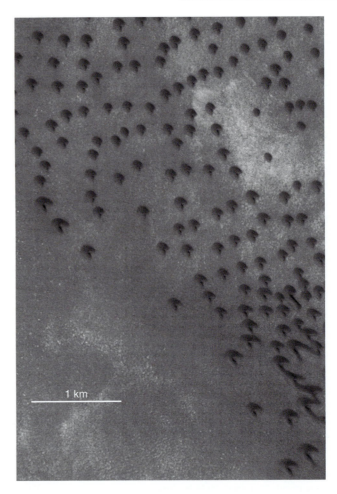

Figure 15.8 Barchan dunes on Mars at 71°N. Dunes are dark because they are likely composed of darker minerals. Slip faces at lower right of each dune reveal that the dominant wind direction is from the upper left. Scale bar is only approximate (NASA/JPL/Malin Space Science Systems, 2004, Mars Global Surveyor Mission, Image PIA05920).

Figure 15.8. In a field of barchan dunes, sand leaving one or another "horn" of one dune typically saltates rapidly across a desert pavement, to be trapped by the next dune downwind. This may be several kilometers downwind. Sand is then steered by the local pattern of wind across the stoss surface of the dune, some of it being trapped by the lee avalanching face, other grains saltating out of the horns, to be lost to yet another transit across the desert. Barchan dunes can move many tens of meters per year. It has long been recognized that the crescentic shape of the barchan is attributable to the uni-directionality of the winds, or more precisely to a probability distribution of wind directions that is tightly clustered about a mean.

Box 15.2 How fast can a barchan dune move?

One may estimate the eolian sand discharge (volume per unit length along the crest per time, m^3/(m yr), or m^2/yr) by the product of the dune height with the dune speed (also called its celerity): $Q = Hc$. While measurement of dune speed has traditionally been accomplished by surveying methods, or by repeat aerial photography, a more recent method allows simultaneous measurement of an entire dune field. Using a correlation method to locate features that can be found on each of two or more digital images, one can assess the movement of objects between images. Because barchan dunes translate without much change in form, such an automatic correlation method is well suited to the task. Vermeesch and Drake (2008) use this method to determine the speeds of dozens of barchan dunes in a desert depression in Chad over intervals of up to six years. As you can see in Figure 15.9, they then plot their data as the inverse of dune speed vs. dune height. This has the feature that the slope of the line connecting dunes in a particular region can be used to assess the sand discharge. The barchans in this region of the Chad desert are not only some of the tallest barchans, reaching 50 m, but are perhaps the fastest in the world.

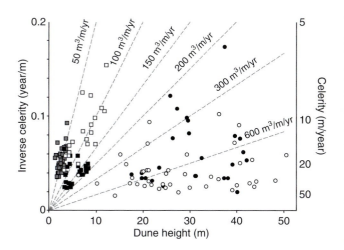

Figure 15.9 Inverse of dune celerity (speed) vs. dune height, showing results reported for many dune fields on Earth. We expect a linear relationship in a field in which the sand flux, Q, may be assumed to be relatively uniform. Just as the sand flux may be calculated from the speed and height of a dune ($Q = Hc$), we may rearrange this to yield $(1/c) = (1/Q) H$. The slope on the plot is therefore the inverse of the sand flux. Several sand fluxes are labeled, increasing as the slope on the plot declines (after Vermeesch and Drake, 2008, Figure 4, with permission from the American Geophysical Union).

Parabolic dunes

Dunefields can be significantly affected by minor amounts of vegetation. Parabolic dunes result from vegetation of the low-lying flanks of dunes. The arms of parabolic dunes point upwind, are vegetated, and can become very long (of the order of kilometers).

The end-member case is a set of vegetated ridges left in the wake of passing dunes as the sand in the ridges becomes trapped by vegetation. Some of these forms have active tips to the parabolas, while others are completely dormant. Vegetation will preferentially colonize the thin sand at the dune edges, where it has most access to water, and where the impacts of saltating sand are minimal. Once a site is colonized, the wind speeds drop significantly, and entrainment of sand is no longer possible (see Nield and Baas, 2008, for an example of modeling of such forms).

Transverse dunes

Dunes with their crest lines oriented normal to the dominant wind that occur in continuous fields with little non-sandy surface between them are called transverse dunes. Such dunefields and the dunes within them can vary widely in scale. Those in Death Valley, California, are a few meters in height, while those in Namibia are more than 100 meters tall. Again, the empirical evidence has supported an interpretation of these dunes as resulting from a tightly clustered wind direction, and a significant sand supply.

Linear dunes

A very common duneform in the world's largest deserts is the linear dune. Like transverse dunes, these occur in fields with sandy surfaces between the dunes, but their crest lines appear to be aligned *with* the mean wind direction rather than normal to it. These dunes can become very large, with crests more than 100 meters above the desert floor, with spacing between the dunes of the order of several kilometers. Of all the dune types, the linear dune has engendered the most debate. Why are there two types of dunes that are orthogonal to one another? What determines which type emerges? Several hypotheses have been advanced. Some have taken the 1–2 km wavelength between dunes as the key feature in need of explanation. That this corresponds roughly to the height of the atmospheric boundary layer – the layer of the atmosphere within which wind velocity changes significantly – and that other wind-parallel patterns emerge in clouds and in "windrows" of seaweed (for example), has led some to believe that they reflect the long-term operation of pairs of roller vortices in the lower atmosphere. Others have called upon significant variation in the wind direction to cause these dunes first to elongate in the direction of the mean wind, and second to maintain them in that orientation. Which is it? The answers lie in the collection of detailed field evidence, both stratigraphic and meteorological, and in the generation of models.

Star dunes

The star dune has several arms pointing in all directions. Star dunes can be very large, reaching many tens of meters in height. While sometimes alone, they often occur as fields of dunes. Excellent examples of such fields are found in the Gran Desierto of northern Mexico, and many places in the Sahara. The common assumption has been that these reflect highly variable winds. In the case of very large star dunes, the center of the dune may not change much from year to year, while the position of the arms changes more rapidly, much like a stationary starfish. This is not surprising, as the rate of movement of any bedform is scaled by the sand flux divided by the cross-sectional area of the dune normal to the flow: the larger the cross section, the slower the movement. The center, with its larger cross section, therefore moves slowly while the arms with their tapering cross sections can move rapidly.

Models of dunes and their stratigraphy

Only recently has the issue of what controls a dune pattern been addressed with a numerical model. Any model worth its saltation must be able to produce the entire zoo of dune types. How does one approach the problem? Dunes, after all, contain many millions of grains, preventing modeling using a grain-by-grain strategy.

The modeling strategy, then, must abstract the saltation and grain flow processes sufficiently to be able to explore the evolution of large-scale features. Any rule set used in such a model must hold firmly to

---

Box 15.3 How many grains does a sand dune contain?

Think for a moment about how many grains comprise a typical barchan dune – even roughly. Let's say it is 50 m wide, 10 m tall, and 100 m long. The sand involved is typical fine-medium sand, of 0.25 mm diameter. In a cubic meter of sand there are therefore $4000 \times 4000 \times 4000$ grains (actually about 65% of this, accounting for pore space), or about 4×10^{10}. That's already a pretty big number. You can hold in your hand several million grains! Approximating the dune as a triangle in cross section, whose effective width is about half the dune width, suggests a dune volume of about 10^4 m^3. Even this small barchan dune will have of the order of a few times 10^{14} grains. Standing on a large star dune and performing the same calculation can result in numbers as high as a few times 10^{18}. That's a lot. Clearly, it would be fruitless to design a computer code to track the movements of all of these grains in the wind. Flipping this calculation on its head, it is astonishing to stand on a large dune and realize that the number of grains is still a few orders of magnitude short of a mole, or 6.023×10^{22}; in fact, it is not far off to say that the number of grains in such a dune is approximately the number of atoms in a single grain of sand!

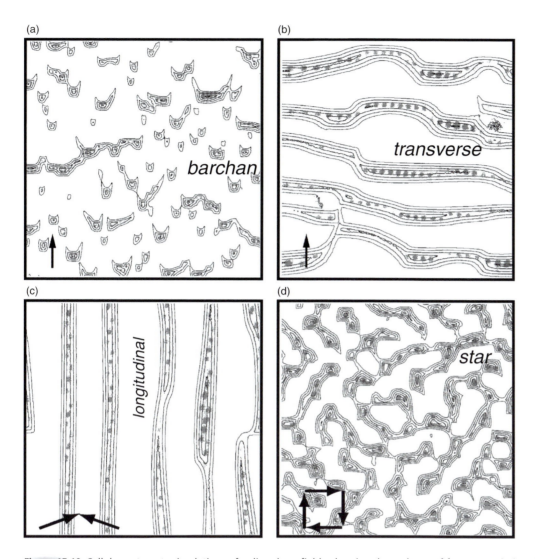

Figure 15.10 Cellular automata simulations of eolian dune fields, showing dependence of form on variations in wind speed (shown with arrows) and abundance of sediment supply. (a) Barchan dunes, unidirectional wind with little sand. (b) Transverse dunes, unidirectional winds, and abundant sand. (c) Longitudinal dunes, bidirectional wind, and abundant sand. (d) Star dunes, multidirectional wind, and abundant sand (after Werner, 1995, Figures 2 and 3).

the principle of conservation of mass. The model must capture the essence of the problem, and not violate any physics we know. As we have seen, the model must be able to address the roles of sand supply and of variation in wind direction. The timescale over which the model operates is dictated by the natural timescale for significant movement of dunes. This timescale can be estimated from knowledge of the cross-sectional area of a dune divided by the sand discharge. For example, if the dune has a cross-sectional area of $500\,m^2$, and the annual sand discharge per unit width of dune crest is $5\,m^3/(m\ yr)$, the timescale for moving the dune forward by one wavelength is $500/5 = 100$ years. Given that there are

roughly $\pi \times 10^7$ seconds in a year, it is neither feasible, nor is it necessary, to calculate the sand flux at the timescale over which the saltation clouds respond to the wind (a few seconds – recall our discussion of this in Chapter 14). Werner (1995) employed a cellular automaton approach to the problem, in which the amount of sand in each cell of the model was tracked. The pixels, representing slabs of sand in the model, were $1 \times 1\,m$. In Figure 15.10 we reproduce several of his simulations revealing that with a simple rule set he could capture the essence of the four main dune forms: barchan, transverse, linear and star. The chief determinants of the forms were two variables: the sand supply and variability of wind direction. Narteau *et al.*

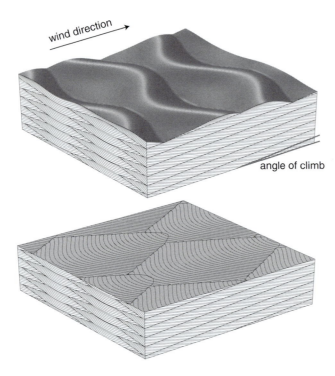

wind direction →

angle of climb

Figure 15.11 Top: transverse bedforms with curved, out-of-phase crestlines, and sine-shaped plan forms. The angle of climb is dictated by the ratio of the rate of net deposition and the rate of migration of the duneform. Bottom: the animation block is then cut to reveal the internal structure (from Rubin and Carter, 2006, Figure 34a, with permission from SEPM Society of Sedimentary Geology).

Figure 15.12 Eolian ripple field. Ripple wavelength roughly 10 cm. Note the long crests of the ripples, and the fingerprint-like pattern in plan view. This pattern requires that ripples both terminate and bifurcate. Wind toward lower left (photograph by R. S. Anderson).

whether it is a large dune moving with small dunes on its back, etc. The form is then cut open to reveal the internal architecture, as reproduced for example in Figure 15.11. This is instructive for anyone wishing to interpret the rock record for the types of dunes involved, the deposition rate and the complexity of the wind field.

Eolian ripples

The crests of ripples, the smallest bedforms (see photo in Figure 15.12) are aligned perpendicular to the wind direction, and translate in the direction of the wind. They are asymmetrical in cross section, with gentle stoss slopes reaching perhaps 10°, and steeper lee slopes about twice this. Aspect ratios between eolian ripple wavelength and height are of the order of 10–20, significantly higher than subaqueous ripples. In plan view, ripple fields are not perfect. Crests both terminate and bifurcate, like the ridges on your fingerprint. These imperfections have been called dislocations, in analogy with the imperfections in crystal lattices (see box on ripple dislocations). Interestingly, eolian ripple crests tend to be coarser than the troughs, again in contrast with the subaqueous case. Ripples preserved in the rock record appear as thin, inversely graded pinstripes that reflect the passage of one ripple after another over a site. These traits raise the following questions: Why do ripples form at all? What sets their shape: their wavelength, cross section, and plan view? How does the stratigraphic signature of eolian ripples, inverse grading, evolve?

(2009) have employed a cellular automaton approach as well, although their wind field is more realistic, allowing for example recirculating eddies in the lees of evolving forms. They show not only the nucleation and amalgamation of dunes, but the emergence of superimposed dunes, and use these forms to set the time and length scales of the model that are otherwise difficult to determine.

If the question being asked of the system involves an even longer timescale, associated with the development of the stratigraphic architecture of a dunefield one might expect to see in the rock record, an even simpler geometrical approach can be taken. In David Rubin's work, most recently made available as movies in Rubin and Carter (2006), dunes of increasing complexity are allowed to move through the calculation space simply as geometric entities with associated internal stratigraphy reflecting slip faces. They explore the dependence of the resulting stratigraphy on the rate of net deposition (climb rate), the one- or two-dimensional topography of the form being moved,

Box 15.4 Make your own ripples

Spread a patch of sand on any flat surface, preferably outside. Use your hand or a straight-edge to smooth the surface. Arm yourself with a pile of loose sand next to the smooth surface. Now kneel down and repeatedly throw handfuls of this sand at this patch such that the sand impacts the surface at a very low angle (at say 10°). Here you are mimicking the wind, which accelerates the sand grains that then impact the sand surface at small angles. Ripples will emerge in the otherwise smooth patch. They will grow in size, and they will migrate in the direction in which you are throwing the sand. Wind is not required. What is required is the low impact angle, which results in the strong variation in impact intensity between the windward and lee sides of the emerging form.

We can recast the first question more formally: Is a flat sandy bed unstable with respect to small perturbations in its topography when subjected to saltation? Do these small perturbations grow into finite size waves that translate downwind? Or are there conditions under which a wrinkle in the bed can heal itself? Let's take a counter example first. We ask the same question but subject the surface not to saltation but to rainsplash. This process acts to remove sand from high places, and moves it into adjacent lows. Any original roughness of the surface is damped; the process is diffusive. On the other hand, turning to a little experiment anyone can do among the dunes, if we scrape a sandy surface flat with a straight-edge, and then allow the wind to blow across it, the original flat surface magically organizes itself into ripples. It appears that a sand bed is indeed unstable to any little perturbation. But why?

Bagnold (1941) was the first to propose a solution. As soon as there is some topography on the bed, the very low incidence angle (order 10°) of the saltation trajectories results in more intense bombardment of the stoss slopes than of the lee slopes. In essence, the lee is shadowed from the beam of incoming grains as sketched in Figure 15.13. Given that most grains are entrained from the bed by impacts once eolian saltation begins, this means that the ejection rate from the stoss sides will exceed that from the lee. The majority of the ejected grains hop only a small distance (a few millimeters to a few centimeters) downwind before landing. The deposition pattern is then a downwind-translated version of the ejection pattern. One can easily imagine that if the translation distance is about a quarter of the wavelength of the original bump, the grains blasted from the middle of the stoss face will land on the crest. Similarly, the low ejection rate in

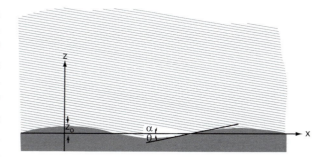

Figure 15.13 Schematic bed topography with initial amplitude, z_o, and descending portions of the impacting successive saltation trajectories. The angle α between the descending saltation paths and the horizontal is taken as positive. The angle θ between the local bed slope and the horizontal is positive on the stoss sides of bed perturbations, and negative in the lee (after Anderson, 1987b, Figure 4, with permission from John Wiley & Sons).

the lee shadow translates into a low arrival rate of hopping grains in the trough. That the arrival rate of new grains on the crest is greater than the ejection rate from the crest allows it to grow in height. Likewise, the arrival rate of new grains in the trough is smaller than the rate of entrainment from the trough, allowing it to erode. This enhances the amplitude of the ripple. When cast more formally, we show in Figure 15.14 that all perturbations grow in amplitude, and that the fastest growing perturbations will be those forms with wavelengths that are about four times the hop length of the ejected grains. This theory therefore succeeds in explaining that flat beds are unstable to perturbations, and predicts that the fastest growing ripples ought to be a few times the dominant hop length of grains in saltation. But the theory breaks down for large amplitude forms, when the shape of the ripples can no longer be viewed as simple sinusoids.

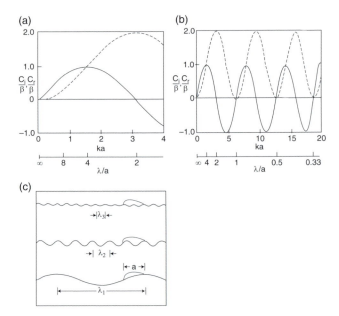

Figure 15.14 (a) Dimensionless translation speed (dashed line), and growth rate (solid line) for the uniform reptation length case, plotted against the product of the wave number $k = 2\pi/\lambda$ with the reptation length, a. The peak in growth rate – the fastest growing wave – at $ka = \pi/2$ corresponds to wavelengths four times greater than the reptation length. (b) Same as (a) except for the expanded ka scale, showing many maxima in the growth rate, corresponding to shorter and shorter wavelengths. (c) Schematic diagram illustrating the origin of the higher frequency peaks in the growth rate arising from the uniform reptation length analysis (after Anderson, 1987b, Figures 7 and 8, with permission from John Wiley & Sons).

The stability analysis therefore does not answer the question of what governs the ripple wavelength. From the work of R. A. Bagnold (1941) and R. P. Sharp (1963), we know that ripples tend to be more widely spaced when the wind speed is higher. Simulations of ripple evolution have shown that ripples subjected to a steady wind grow rapidly at first from small mottles in the bed, and continue to grow in amplitude and wavelength for as long as the simulation is run (e.g., R. S. Anderson, 1991; see also Werner and Gillespie, 1993). Their rate of growth slows down, but they continue to grow. This has raised a debate about what is the ultimate control on ripple wavelength. The question is whether ripple size is controlled by wind speed alone, or by some combination of wind speed and duration of the wind. After all, winds do not blow forever; they switch direction, and they stop.

Ripple stratigraphy

To approach the problem of grain size sorting in ripples, which results in both the coarse crests of ripples in the field, and the inverse grading of ripple laminations in the rock record, we must understand the behavior of different grain sizes during saltation. Coarse and fine grains behave differently in both the granular splash, and their trajectories. Simulations of the grain impact process reveal that a grain impact into a bed composed of a mixture of coarse and fine grains will result in the preferential ejection of fine

---

Box 15.5 Ripple dislocations

Eolian ripples always contain imperfections. While mature ripples are long-crested normal to the prevailing wind, they do not wrap around the world. They terminate. As you can see from the photograph in Figure 15.12, they also bifurcate. These features have also been called ripple dislocations, in honor of their similarity with lattice dislocations in a crystal. Dislocations are an inevitable consequence of the means by which ripples evolve. Ripples begin as a set of small short-crested mottles that then must link. The ripple lattice slowly rids itself of dislocations that are trapped in the ripple field by this early linkage. Dislocations move more rapidly than the ripples themselves. This too we can understand from our knowledge of the dependence of bedform speed on cross-sectional area. At a ripple termination, the ripple cross section declines, allowing it to move more rapidly than the remainder of the ripple. It therefore quickly catches up with and attaches to the next downwind ripple. The portion of this new ripple to which the termination is now attached (making a new bifurcation) is now shadowed from sand impacts, affecting its rate of migration. The result is that dislocations (terminations and associated bifurcations) rapidly sweep downwind through the ripple lattice.

grains, and at higher launch speeds for the fine grains than the coarse. In the absence of any other effect, this will leave the highly impacted stoss face of the ripple coarser, as the fines get removed. Because they have less mass, fine grains will be accelerated more by the wind than coarse grains, and hence travel longer paths before reencountering the bed. This and the low ejection speeds of the coarse grains means that the coarse grains creep up the stoss face in tiny hops, and barely make it over the crest, into the impact-shadowed lee. As the fine grains are ejected at higher speeds, they travel longer paths before landing. They can therefore waft over the crest of the ripple and make it well down the lee face, even into the next trough. Statistically speaking, then, the lee face of the ripple is populated by coarse grains near the crest and finer grains near the trough. As the ripple migrates, this pattern is sequestered in the subsurface, only to be re-exhumed on the stoss face. Then the process of impact sorting runs again. The net result is that coarse grains tend to get trapped near the crest of the ripple. In a slightly aggrading system, just as in the case of dunes, it is the bottom fraction of each bedform that makes it into the rock record. As shown in the simulation in Figure 15.15, the result is a set of thin pinstripes, each of which is inversely graded (coarser over finer). This micro-sorting results as well in the slightly different cementation, which is revealed in the sharpened color contrast in the resulting sandstone.

Summary of bedforms

We have discussed the origin and evolution of eolian bedforms and their stratigraphic signatures. In both dunes and ripples, saltation dominates. On dunes the very different grainflow process, driven not by wind but by gravity alone, takes over on the lee side, completing the transport down the lee face. The thresholded nature of the failure process yields discrete failures that are recorded both because the grainflows sort themselves into inversely sorted packets and because they are interfingered with grainfall deposits at the toe of the dune. In contrast, it is the details of the splash process within the saltation system that lead to the inverse grading of ripple deposits. We now turn to deposits of fine-grained eolian sediments whose emplacement in the landscape is through suspension rather than saltation. These far-traveled sediments have become

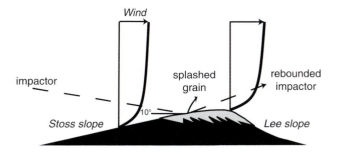

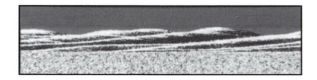

Figure 15.15 Modeling of eolian ripple stratigraphy. Top: model set-up. Energetic grains impact the granular surface at low angles, resulting in rebound of the impactor and splash of other grains from near the impact site dictated by the splash function for mixed grainsize material. The grayed crest of the ripple represents dominance of coarse particles there. The subsequent paths of the splashed grains are calculated explicitly using a wind field that honors the ripple shape. Bottom: resulting ripple stratigraphy when net deposition is allowed to take place. Reverse grading occurs, with coarse grains over fine grains within the body of an individual lamina, with abrupt bounding surfaces between lamina. Base of the simulation shows unsorted mixture of fine and coarse grains that formed the initial bed (after Anderson and Bunas, 1993, Figures 1 and 4, with permission from Nature Publishing Group).

well removed from their sources in deserts and river corridors, and now mantle large parts of the terrestrial and oceanic landscapes.

Loess

One of the best terrestrial records of climate history lies in huge deposits of dust called loess. The record of paleoclimate in the Chinese loess extends back 8 Ma. These deposits accumulate downwind of major dust sources at long-term average rates that can be significant fractions of a millimeter per year. Grain sizes involved are typically of the order of 20–40 microns and smaller, and can therefore be wafted great distances in the turbulent atmosphere. While the thickest loess deposits are found in China's loess plateau, other significant deposits occur in eastern Europe, Siberia, the Pampas of Argentina, and in several sites

Box 15.6 Corrugated or washboard roads

Anyone who has driven significantly on dirt roads has likely experienced the shuddering of the vehicle as its tires bounce over ripples in the road. Even a short time after the road has been smoothed by re-grading, this washboard re-emerges. Just as in the formation of bedforms beneath flows of fluid, these arise naturally. In the case of the dirt road, the ripples form by the passage of tires bouncing over the road. In many ways the problem is analogous to eolian ripples. The car tires provide impacts on the stoss side of the ripples just as wind-accelerated grains provide the impulses on the stoss sides of eolian ripples. The problem has been studied by road engineers for decades. As a child, I (RSA) recall a *Scientific American* article on the topic. Recently, physicists have become intrigued by the problem (Taberlet *et al.*, 2007), and have performed experiments designed to elucidate the origin and evolution of the ripples. The experiment consists of a 1 m-diameter circular bed of sand that is rotated beneath a rubber wheel attached to an arm. The wheel bounces and rolls across the surface of the sand. In Figure 15.16 we reproduce the chief result showing the vertical position of the tire through time. The figure shows how the initially flat bed develops bumps that force the wheel to bounce. The first bumps to form are small and force only small vertical bounce. These grow in amplitude and in wavelength, and they migrate in the direction of the motion of the wheel. They go on to simulate the process in numerical models, but the point is really a simple one: the pattern emerges from the asymmetrical distribution of the impulse provided by the tire (on the stoss side of the form) and the net transport of the disturbed sediment in the direction of motion.

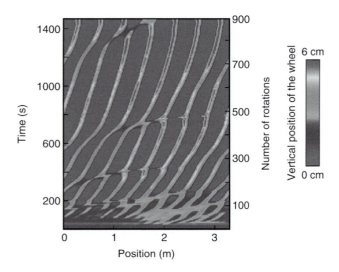

Figure 15.16 Space–time diagram of the vertical position of a tire as it passes over the surface of sand in an experiment designed to explore the evolution of washboard roads. As the diameter of the rotating circular sand box is 1 m, each horizontal line in the diagram (πD in length) represents one full revolution of the box. The pattern migrates in the direction of the relative motion of the tire. The amplitude and wavelength of the bounce, and hence that of the sand on which it is bouncing, increase through time (after Taberlet *et al.*, 2007, Figure 3, reprinted with permission. Copyright 2007 by the American Physical Society).

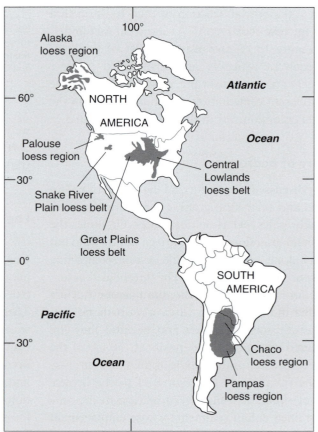

Figure 15.17 Loess regions of the western hemisphere (after Muhs and Bettis, 2003, Figure 1).

in North American. In the map of Figure 15.17, we can see that these include Alaska, eastern Washington (the Palouse Prairie), and a band along the Mississippi that stretches hundreds of kilometers in either direction.

Very long-traveled dust contributes strongly to sedimentation in the deep sea. Quartz dust has been fingered as a likely contributor of quartz to Hawaiian soils. Being built of basalt, with no free quartz content, any quartz in the soils must be allochthonous. And the dust content in ice cores in the middle of ice sheets (Antarctica, Greenland) is used to assess the windiness of the climate through the glacial ages.

But all this is in the ancient record. Modern dust storms provide the analog for this far-traveled sediment. Their impact on civilization provides the motivation for our study of the processes involved. Concentrations of dust in the atmosphere can be large enough to impair visibility, and to cause respiratory illnesses. In some instances, the chemistry of the dust exacerbates the problem. For example, in the Owens Valley of eastern California, a major playa was dried up by diversion of Owens River water to Los Angeles in the early 1900s. Wind storms now periodically engulf local towns in dust that contains a significant component of playa salts. A major effort has been launched to find a solution for the problem.

Recent analysis of sediment in small lakes in the San Juan Mountains of southwest Colorado (Neff et al., 2008) has shown that the deposition rate of fine dust greatly increased in the 1800s. The chemical and isotopic signature of the dust reveals that its source is not local, but comes from upwind semi-arid landscapes that were heavily grazed in the 1800s and early 1900s. Not only does grazing alter the vegetative cover of the landscape, but the trampling of delicate soil crusts (called cryptogam) breaks up the soil between clumps of vegetation, making it more susceptible to entrainment in the wind.

Major questions raised by both modern dust storms and the ancient record that we have not yet treated in this chapter include the details of the entrainment mechanism for dust, what sets how far the dust will travel in the atmosphere, and what sets the spatial pattern of deposition? While we have treated entrainment of large particles with a theory that appears to be able to explain well the threshold of entrainment, we found departures at small grain sizes that would all make it more difficult to entrain smaller grains. All of the theory presented, and most of the empirical data used to check the theory, comes from single grain size beds: all large grains … all small grains. It turns out that mixed grain size beds are important in the entrainment of dust. While it is difficult to entrain dust from a bed composed entirely of dust (for reasons already discussed: it can hide in the laminar layer; electrostatic forces reign), large saltating grains can effectively disrupt the laminar sub-layer, and they can forcefully eject small grains from the bed. This implies that once saltation is operating, dust can be mobilized. This fits better with our observations in the real world. It has also been found that while some surfaces may be entirely composed of dust, this dust often forms larger particles as either pellets, or pieces of crust, that are entrained as large grains. Upon impact with the bed at the end of their trajectories, these particles then disintegrate into their original finer grains.

The area affected by the dust bowl is shown on the map in Figure 15.18. The period of the dust bowl of the 1930s in the USA was punctuated with occasional dust storms in which the visibility went to nearly zero. These events go by many names, one of the more dramatic being Black Rollers. You could see the front of the cloud approaching from the west, an opaque wall over a kilometer high. In some years, there were as many as 7–10 dust storms per month, dominantly in the early months of the year. Some of these events resulted in reduction of visibility and deposition of fine dust as far downwind from the source as the east coast (see extensive first-hand accounts of the dust bowl in Egan, 2006). During the Dust Bowl, the dust source was fields from which the original bunch grass cover had been removed, but which remained unplanted because of severe drought.

While there are many natural sources of dust for eolian entrainment, the most common are alluvial rivers that experience significant swings in stage, playas, and alluvial fans at the edges of the deserts (see Muhs, 2007). Glacial outwash plains are a perfect example. Fine grains produced subglacially are carried in suspension in the river, which on both daily and seasonal timescales rises and falls in stage. The remainder of the sediment includes everything from sand to boulders. The suspended sediments are left high and dry on poorly vegetated mid-channel bars and on floodplains. Entrainment occurs when the common down-valley winds entrain the mud-cracked fines directly, or cause saltating sand to impact the surfaces of the fine-grained deposits. Dust plumes are common in

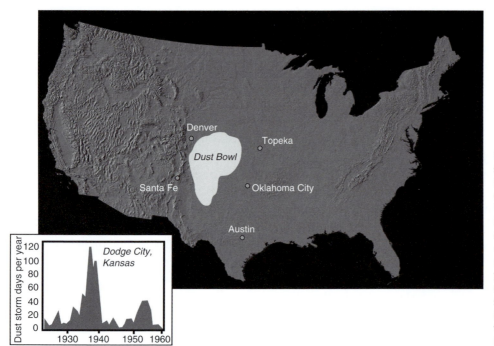

Figure 15.18 Distribution of severely impacted sites on the High Plains of North America during the Dust Bowl of the 1930s. While this depicts the lands most greatly degraded, the dust from the Dust Bowl wafted over the entire eastern US. Inset: number of dust storm days per year through the dust bowl decades of 1930s through 1950s, shown for Dodge City, Kansas (after Pye, 1987, Figure 4.21, with permission of Academic Press).

such settings; dust is steered down the river corridors, and is wafted into the nearby landscape, coating the grass and the trees on the surrounding hills.

The thickness of a loess deposit commonly falls off with the distance from its source. Examples abound, but we illustrate it here with the pattern around the Mississippi (Figure 15.19). The pattern can be fit well with an exponential, the length scale for the decline in thickness being of the order of ten kilometers (see Bettis *et al.*, 2003). The cliffs of Hannibal, Missouri, in which Tom Sawyer and Huck Finn played, were cliffs cut into the massive loess deposits that border the river.

It is the loess deposit in the $640\,000\,km^2$ loess plateau of China, however, that has seen the greatest investigation. Here the loess is up to 300 m thick and declines to the east with distance from the primary source. The loess plateau is closely associated with the great bend in the Huang He (Yellow River). The primary sources lie to the northwest in the several deserts of interior China and Tibet. Thickness and grain size of the loess systematically decline to the southwest. The climate becomes wetter in the same direction. It is the loess of the Chinese loess plateau that has been used most successfully to correlate with the deep sea record of climate variability in the Pleistocene. As shown in Figure 15.20, the loess section is characterized by a series of soils separated by monotonous unweathered silt. Dating the profile, at first based

simply on the location of the Bruhness-Matuyama magnetic reversal boundary, and later enhanced by profiles of ^{10}Be concentrations, has demonstrated that the soils can be correlated to interglacial periods (odd marine isotope stages). This suggests the model of loess accumulation in which deposition rates in the glacial periods outpaced the rate at which pedogenic processes could weather the incoming silt, while in interglacial periods lower rates of dust accumulation allowed soils to form. These extensive piles of dust remain some of the most promising terrestrial records of climate change whose quantitative interpretation requires simultaneous treatment of soil-forming processes (e.g., Anderson and Hallet, 1996).

Cave dwellings are numerous in the easily carved loess. The greatest death toll in any earthquake in history came from the great Senshi earthquake of 1556, in China, in which untold thousands (probably more than $800\,000$) of residents in loess caves died when their caves collapsed. The loess is easily terraced, leading to great agricultural production in the region. But for the same reasons the loess is also easily eroded. The silt is transported down tributaries to the Yellow River (Huang He), the extremely high suspended sediment load giving the river its name. But it is the record of climate contained in the loess that has most excited the scientific community. The loess contains tens of buried soil profiles that reflect variation in climate

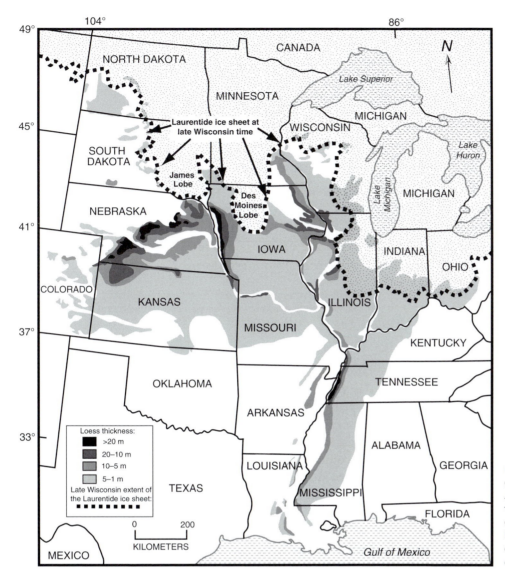

Figure 15.19 Distribution of last glacial age loess in the North American mid-continent. Loess thickness declines regularly with distance from the fluvial sources. Maximum Laurentide ice sheet extent is also shown (after Muhs *et al.*, 2008, Figure 1).

through time. The degree of development of these soils varies with position on the plateau, being redder and more highly developed in the wetter southwest, and more subtle in the arid northwest. Several variables are involved in the generation of this classic alternating sequence of loess–soil–loess. The picture is not one of high rates of dust deposition during glacial times, followed by generation of a soil profile into the stationary top of the deposit in the interglacial. Instead, it is now thought that accumulation rate varies but does not go to zero, while precipitation varies out of phase with accumulation.

Erosion by windblown particles

One of the more striking phenomena associated with sediment transport is the erosion of obstacles that happen to be in the way of a sediment-laden flow. In the case of air, the forms produced are called *ventifacts*, literally, wind-faceted rocks (see Knight, 2008). These are common in some desert settings, and in places that have either now or in the past experienced occasional very high winds. These stones vary in size, and the signature of windblown abrasion

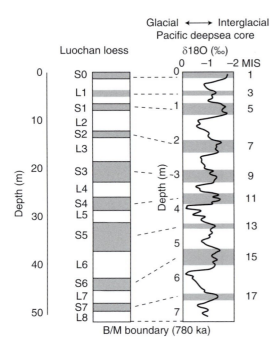

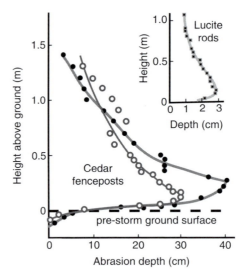

Figure 15.21 Pattern of erosion of fence posts and lucite rods (inset). Note maximum erosion well above the ground surface (after Anderson, 1986, Figure 2).

Figure 15.20 Correlation of loess stratigraphy in Luochan, China (e.g., Kukla and An, 1989), with Pacific deepsea core V-28–239 (Shackleton and Opdyke, 1976). Periods of loess deposition (L-units) correspond to glacial periods, while periods of soil formation (S-units) correspond to interglacials (odd marine isotope stages (MIS)). The Bruhnes-Matuyama (B/M) magnetic reversal at 780 ka is the most robust tie-line between the two records (after Muhs, 2007, Figure 6, with permission from Elsevier).

varies with location on the rock, and with the lithology. The most spectacular of these rocks display flutes and grooves imprinted by the local wind field. Some of these grooves can be many tens of centimeters long, and have meandering paths. The explanation of these features took the geomorphic community through some circuitous paths toward the answers, a path that included field experiments, lab experiments, and theory.

In an attempt to explain ventifacts, Bob Sharp (1964) placed a set of Lucite (plastic) rods in a region of the desert of southern California that is famous for high winds and ventifacts – San Gorgonio Pass. This is now the site of a major wind farm with hundreds of wind turbines – see Box 15.7. After many years of exposure to the occasional windstorms in this valley, the rods all showed profiles of erosion that were similar. In particular, as shown in the inset to Figure 15.21, they all showed a maximum of erosion a few centimeters above the ground surface, above which the erosion fell off with increasing height. This

raises the question of what grains are responsible for this pattern – is it those in saltation, or those in suspension? And which winds are most effective in causing erosion? Put more formally, what is the functional relation between wind speed and erosion rate?

Around this time, in fact on December 21, 1977, a huge dust storm enveloped the southern end of California's San Joaquin Valley for a significant portion of a day (Wilshire *et al.*, 1981). Wind speeds exceeded 300 km/hr, and broke every anemometer in the path of the storm. Fields that had been overgrazed by cattle were scoured by 20 cm. The sediments entrained in the wind filled the California aqueduct, the finer grains reaching the Santa Barbara channel. The dust plume reached many kilometers in height. Human-made objects were highly abraded: paint was stripped off cars and houses, 10 × 10 cm cedar fence posts were abraded entirely away, and centimeter-size grains were embedded in cracks in telephone poles many meters above the ground. The southern San Joaquin valley was not a place to be that day. Upon closer examination of the pattern of abrasion (Figure 15.21), it was found again that abrasion fell off with height above a maximum – but this time the maximum erosion occurred at several tens of centimeters above the ground.

This issue of wind abrasion was important to NASA as well, as NASA at that time (in the late 1970s) was designing the first spacecraft to land on the surface of

Box 15.7 Wind energy as a power source

Wind is one of several natural sources of energy that are being harnessed increasingly in the world. Wind has for centuries been employed to pump water, and to grind grain, resulting in the icons of the American West and of Holland, respectively. Until the 1930s, wind power was used to generate electricity on individual rural farms, until the rural electrification effort replaced these with the power grid that now supplies electricity to the vast majority of the population. Recently, however, large wind farms have been established at particularly windy sites in the American West, some of them containing thousands of turbines. Wind energy varies as the cube of the wind speed. Of course, one of the characteristics of wind is its erratic nature. While one cannot count on the wind as a source of peak power, the power produced can be put directly onto the grid to reduce the need for power generation from fossil fuels. Two principal designs for wind turbines are vertical axis and horizontal axis machines, distinguished obviously by the orientation of the axis of rotation. The Darius rotor, which looks like a giant egg-beater, is the dominant vertical axis machine, while the horizontal axis machines differ markedly from one another, some oriented upwind, some downwind. Most wind turbines produce power in the 100 kW range. Major sites in the USA include several passes in California, and near Cheyenne in Wyoming. The total wind energy produced in the USA is 11.6 GW, which accounts for more than 1% of US energy consumed, with turbine farms spanning 34 states.

Mars – a place that was known for some time to produce planet-engulfing dust storms. What damage might sand or dust grains embedded in these winds do to such landers, and especially how likely is it that sand would frost the lenses of cameras placed on or near the surface? NASA-funded researchers attacked this problem empirically, that is, with experiments. It was determined that particles flung rapidly against plates removed material from the plates at a rate that was proportional to the particle diameter cubed, and to the square of the velocity – that is, to the kinetic energy of the particle.

This gave researchers a place to start. They had two targets: explain the vertical pattern of erosion, and deduce which wind events are most important. This latter question could be cast as how strongly does erosion scale with wind speed. Erosion is proportional to the kinetic energy of a particle. Therefore, the erosion rate must be proportional to the flux of kinetic energy in the particles embedded in the wind. How might such a flux vary with height above the ground, and how does it depend upon wind speed? Several factors compete for control over the problem. First, it is the speed (and kinetic energy) of the particles that counts – not of the wind. In any given wind, particles that are suspended, for which it is a safe approximation that they are traveling at the speed of the wind, will be traveling faster. The smaller

the particle, the more faithfully it follows the wind in both speed and direction. On the other hand, as the wind encounters an obstacle, it flows around it. In fact at the surface of the obstacle the speed of the wind is zero – recall the no-slip boundary condition on the velocity profile. This means that in order to impact a surface, the particles must decouple from the wind. Here the smaller the particle the less likely it will decouple from the wind (for the same reason that it is traveling at the wind speed). The more massive the particle, the more likely it will be flung from the flow and impact the obstacle. Interestingly, this portion of the problem was tackled by scientists interested in the ice riming of ship masts and rigging during ice storms. The buildup of ice above deck can so greatly change the center of gravity of a ship that it can capsize. The fraction of those particles (or ice droplets) in the wind that actually collect on the mast is called the collection efficiency. As illustrated in Figure 15.22, the larger the droplet, the more likely it is to impact the mast.

Putting this collection efficiency issue aside for the moment, it is actually rather simple to envision how the kinetic energy flux might vary with height in the case of suspension. The kinetic energy flux is the product of the particle concentration (mass per unit volume) and the wind speed cubed (two powers to get kinetic energy, and another to get flux). Ignoring

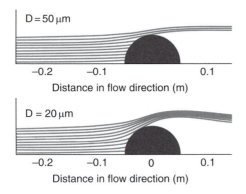

Figure 15.22 Collection efficiency determined by deflection of particle trajectories in flowfield around a cylinder. The smaller the particle, the more tightly coupled the particle and flowlines, and therefore the less likely the particle will impact the cylinder.

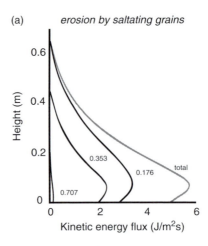

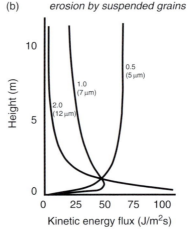

Figure 15.23 Wind erosion caused by (a) saltating and (b) suspended grains (after Anderson, 1986, Figures 5 and 6).

effects of density gradients on the wind speed profile, we have viable theories for both the concentration profile and the wind speed profile. They go in opposite directions, concentration declining monotonically with height, wind speed (and hence its cube) increasing monotonically with height. The product of the two must therefore cross at some height above the ground. While this is easy to envision, a complete theory of abrasion requires knowledge of the contribution from saltating grains as well. Here too we have seen that the concentration profile declines with height above the bed. The complications arise from the fact that the particles are not traveling at the speed of the fluid. Their response times are so long that they are still being accelerated for the majority of their trajectories. One can still proceed with numerical calculations. As expected, the calculations all reveal that the erosion associated with saltation ought to generate a maximum at some height above the bed. The height of the maximum increases with the mean of the launch velocity of the saltating grains, as this dictates the heights of the saltation paths. The sum of the contributions from saltation and suspension is shown in Figure 15.23, and can explain the general pattern of erosion preserved in the fence posts of the southern San Joaquin Valley.

How strongly does erosion depend on the wind speed? In both the saltation and suspension cases, it has been shown that the flux of kinetic energy will scale as the fifth power of the wind speed, and hence as u_*^5. We get two powers of velocity from the kinetic energy of a particle, another power from the flux of these particles. The other two powers come from the fact that the concentration of grains in the airstream tends to be controlled by the shear stress (actually excess shear stress, but at the high shear stresses relevant to wind erosion, subtracting off the threshold does not reduce the value much). Recalling the definition of the shear velocity as $u_* = \sqrt{\tau_b/\rho_a}$, this dependence introduces yet another two powers of the wind speed. The long-term erosion at a site is therefore controlled by the very few highest wind speed events – even more sensitively than the mass flux that controls for example the rate of sand dune migration, which varies as the third power of u_*. This means that places with well-developed ventifacts experience *extremely* strong winds. While the erosion may only have taken a few days of very high winds, it may take centuries to millennia to accumulate that many days worth of the extreme winds necessary.

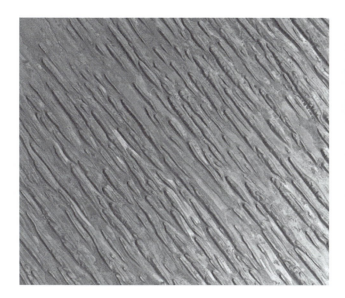

Figure 15.24 Satellite image of yardang field in Lut Desert, Iran. Dominant wind is from upper left. Forms are elongated and streamlined downwind; lanes between high ridges are sand-filled with sand en route to a sand sea to the south. Suggestion for this location courtesy of John Whitney, USGS (Digital Globe image from Google Earth: 30° 35′ 38.78″ N, 58° 16′ 0.25″ E.)

See Sharp (1949) for a discussion of ventifacts likely to have been wind-blasted in the Pleistocene.

While this discussion is relevant to the generation of some small-scale enigmatic features to be found by the sharp-eyed tourist in the desert on our planet or a neighboring one (for example, see Laity and Bridges, 2009), there are indeed larger scale forms that are attributable to wind erosion. These include yardangs, deflation basins, and in the case of snow, sastrugi. Yardangs are streamlined ridges carved from any substrate by particles entrained in the wind (Blackwelder, 1934). They can reach up to kilometers in length, and tens of meters tall. Their plan view shapes have been likened to inverted boat hulls with the highest and widest end facing upwind. While abrasion dominates the windward edges of yardangs, deflation dominates the leeward side. The best known of these in North America are carved in soft lake deposits bounding playa basins (e.g., Blackwelder, 1934; Ward and Greeley, 1984). Globally, it is the deep deserts of Africa and Asia, and in particular the great deflation basins, that display the most prominent of these relatively rare forms. A prominent example from the Lut desert in Iran is shown in the satellite image reproduced in Figure 15.24.

Figure 15.25 Small-scale sastrugi in snow on windblown Niwot Ridge, Colorado. View looking east, downwind, with lee clouds in the distance. Amplitude of the roughness in this image is roughly 20 cm. Truncation of the snow layering is visible in the foreground, demonstrating that the forms are indeed erosional, cut into the stratigraphy (photograph by R. S. Anderson).

Windblown snow

Just as mineral grains are transported by wind, so too is snow. The loss of snow from the Antarctic continent by entrainment in wind is second only to calving of icebergs. Snow too travels in both saltation and suspension, depending upon its grain size and the wind speed involved. At some level one needs only to alter the density of the grains from that of silicate minerals to that of ice in the calculations we have already illustrated in order to make progress on a theory of snow transport by wind. The high concentrations of snow in the lower atmosphere can reduce

visibility to near zero. Yet in these same ground blizzards that make travel dangerous to impossible, one
can often look straight up to see blue sky. This reflects
the strong concentration gradients we have already
discussed: concentrations fall off as power laws with
height above the bed. While we often think of snow as
consisting of delicate flakes, both the metamorphism
that takes place during residence in a snow pack, and
the jolting transport as grains re-encounter the bed
while in saltation, can rapidly round off those delicate
edges. At temperatures well below 0 °C snow is in fact
very hard, approaching the hardness of quartz at
−40 °C. This hardness allows snow to ornament the
surfaces of nunataks that poke above glacier surfaces.
Every climber knows how stinging the impacts of
snow grains on the face can be when forced to turn
into the wind.

At temperatures closer to the melting point, snow
is more likely to sinter with other grains upon impact
with the surface. The sintering produces a cohesion
among the grains of the snow surface that is lacking
in all but damp sediment surfaces. This can transform a snow surface into one in which erosional
sculpting of the form dominates, rather than deposition. The result is *sastrugi*, a Russian word for a set
of erosional forms in snow; the singular is *sastruga*.
A small-scale example is shown in Figure 15.25.
Just as yardangs can come in clusters and fields, so
too can sastrugi. Like yardangs, they are aligned
parallel to the direction of the prevailing wind.
Indeed, their orientation is used to map the wind
field from aerial photos and space imagery. Because
the snow is softer than rock, however, these sastrugi
fields can evolve more rapidly than yardangs. They
are indeed the bane of the polar traveler, who must
traverse these features with sled or ski, as they represent a surface roughness that can reach several
meters in amplitude.

Eolian evidence on Mars

Given that free water on the Martian surface was
likely lost at least two billion years ago, it is not
surprising that the surface of Mars is dotted with
eolian forms. We have known for decades that the
Martian surface is occasionally shrouded by global
dust storms (e.g., Sagan and Pollack, 1969),

Figure 15.26 Sliding stone on Racetrack Playa, Death Valley
National Park, California. This roughly 10 cm diameter rock is one
of many tens of stones that have found their way from an alluvial
fan source onto the playa surface. They have subsequently taken
mysterious paths out toward the center of the playa. The tracks
they leave record the most recent increment in their journey. Some
are straight, most are meandering, and many appear to be quite
parallel to one another (photograph by R. S. Anderson).

presumably reflecting both high winds and the availability of fine-grained material on the surface. In a
prescient paper, Carl Sagan and Ralph Bagnold
(1975) collaborated to discuss the relevance of sediment transport physics to other planetary surfaces.
Early landers such as in the Viking mission photographed the surface littered with rocks that appear
to have been ventifacted. Since the arrival of high-
resolution cameras in Martian orbit in the late 1990s,
we have been treated with image after image of a
wide variety of eolian features, from barchan dunes
to yardangs, and even dust devils and their wiggly
vacuumed tracks.

Box 15.8 The sliding rocks of Racetrack Playa

Among the great mysteries in geomorphology, the sliding stones of Racetrack Playa rank among the top. An example of a track left by a small stone is shown in the photograph in Figure 15.26. No one has witnessed the movement of these rocks, which range up to several tens of centimeters in diameter. Yet the tracks differ year to year, implying that some event or combination of events allows the rocks to skitter about on this flat playa surface. Some of the displacements from year to year are tens of meters long. Some of the tracks are long and relatively straight, as in the photo. Others are zig-zagged. The sliding rocks of Racetrack Playa have repeatedly been studied (e.g., Shelton, 1953; Sharp and Carey, 1976; Sharp *et al.*, 1996). (They do occur elsewhere, but not as spectacularly and not as consistently.) It is agreed that high wind speeds are required. The rocks present a sail to the wind, allowing a high drag force that exceeds the resisting friction force with the underlying substrate. It is also agreed that the substrate must be wetted; attempts to move the rocks when the surface is dry are fruitless. So, in this scenario, the sliding events apparently require a conspiracy of rare events: wind must occur while the surface is wet. How high the wind must be depends upon how low the friction coefficient with the playa can go. In any case, both the high winds and the wetted surface are rare events. But researchers differ as to whether the motion requires that the rocks be embedded in a sheet of ice. The issue here is the parallelism of many of the tracks. Some of the tracks are indeed remarkably parallel when closely analyzed, and one way to accomplish this is to embed the rocks in a rigid substance. It is difficult however to account for the following observation (Sharp and Carey, 1976). Knowing that the ice sheet hypothesis was being entertained by others, Sharp and Carey drove stakes deeply into the playa surface on four sides of one of the large rocks. When they returned, the rock had escaped from the stake perimeter, leaving a track recording its motion. Any sheet of ice in which the rock had been embedded would have been pinned by the stakes. Still, the parallelism remains to be explained. If ice is indeed required, the meteorological conspiracy required is threefold: rain, freezing conditions, and high winds, in that order. The tracks are defined by a smoothed center as wide as the rock, and by narrow levees on either side. The cross sections of the pair of levees constrain how much material was shoved aside to allow the rock to move. The thickness of material scraped aside is always quite small, implying that the rock is gliding on a very thin layer of very slippery stuff. Indeed, when one wets the surface, it is very slippery! Some of this slipperiness is clearly due to the fact that the surface is composed of clay. Breaking open an individual mudcrack always reveals fibers that are apparently algae. Perhaps some of the slippery nature of the wetted playa surface results from these algae. It is our sense that the mysteries will remain until we actually witness the motion of these rocks.

Summary

Bedforms arise inevitably and spontaneously from a flat sandy bed. Dominant forms are ripples and dunes. Eolian ripples grow from small three-dimensional mottles whose initial wavelength is set by the hop lengths of grains ballistically ejected from the bed. Ripple fields self-organize into long-crested asymmetric ridges, a pattern interrupted by ripple terminations and bifurcations that migrate more rapidly through the ripple "lattice" than the ripples themselves move. Eolian ripple crests are coarser than the troughs. This results from divergence of coarse and fine grain behavior in both the granular splash and trajectory processes.

Dunes have slip faces down which transport occurs by grain flows. Saltating grains bounce up the stoss slope, and accumulate as a depositional bump at the top of the lee face. When some portion of this bump achieves a failure angle of about $34°$, the sand slope fails locally. The ensuing grainflow taps the depositional bump and accomplishes the transport of sand down the lee face. Interfingering of these grainflows with grainfall deposits of the finer grains, and with ripple laminations on the lee slope dominate the rock record of eolian dunes.

Dune types vary strongly with both the variability of wind direction and the availability of sand. Unidirectional winds generate barchan dunes, seif dunes, and then transverse dunes as sand supply increases. Linear dunes require two dominant wind directions with less than 90° variability. Star dunes require significant wind direction variability. Parabolic dunes require that vegetation stabilize the thinnest portions of the dune, effectively turning barchan dunes inside out. Simple modeling strategies that include saltation and that allow for a slip face can reproduce this full spectrum of dune forms.

Transport by suspension in wind can re-distribute dust on continental scales. The primary entrainment mechanism is through bombardment of mixed grain size surfaces with saltating grains. Major dust storms, which are given different names in different societies, can transport dust hundreds to thousands of kilometers. The resulting deposits, called loess, typically show a pattern of thickness that falls off with distance from the source with a characteristic length scale of tens of kilometers. In the dust bowl of the Midwest USA in the 1930s, the late winter months were punctuated with many days in which visibility declined to near zero. Similar problems exist in China, Korea, and the Sahel of western Africa at present, related not only to climate but to land use histories. In the terrestrial realm, the continuity of deposition through time in the loess sequences of the world records major swings in both deposition rate and pedogenic digestion of the loess associated with glacial–interglacial climate swings. In addition, the dust concentration profile in ice cores from Greenland and Antarctica speaks to great variation in the wind strength and the availability of dust from unvegetated sources through time.

Particles embedded in the wind can also abrade obstacles in their paths, creating such desert curiosities as ventifacts and yardangs (or, in the case of snow, sastrugi). Abrasion rates peak at some height, typically a few centimeters, above the ground, where the flux of kinetic energy is greatest. That the abrasion rate varies as roughly the fifth power of the wind speed implies that these forms record the effects of the very highest and rarest of winds.

Problems

1. Calculate the expected celerity of an eolian sand dune, in m/yr, if the dune is 5 m tall (a typical barchan) and the annual average sand discharge is 2.5×10^{-6} m^3/(m s).

2. Calculate the expected celerity of an eolian ripple, in cm/hr, if the ripple is 1 cm tall (a typical ripple) and the average sand discharge over the hour of measurement is 5×10^{-6} m^3/(m s).

3. *Wind erosion*. Calculate the ratio of the wind erosion that might be expected in the same site in winds that differ by a factor of two. Assume that the lower wind condition is twice that needed to entrain grains, meaning that the higher wind is four times the threshold.

4. Analyze the washboard road figure (Figure 15.16) for the speed of translation of the ripples.

5. Calculate the wind speed required to move one of the Racetrack Playa rocks. Assume that the rock is rectangular, with sides of 30 cm and a height of 20 cm, and that it is a limestone. Assume that the friction coefficient when the playa is wet is as low as 0.15.

6. Calculate the residence time of a sand grain in a barchan dune. The footprint of the dune is 75 m long in the wind direction, and repeat photos have shown that the dune is moving at 15 m/year. For the same dunefield, calculate the mean speed of a sand grain over the desert floor if the spacing between barchan dunes in the wind direction, as deduced from air photo patterns, is 500 m. (*Hint*: it is safe to say that there is no storage of sand between the dunes on these rocky desert floors.)

7. How many sand grains are there in a one-grain diameter thick cross section of a simple 10 cm wavelength ripple? Assume that the ripple is 5 mm high, 10 cm long, and can be approximated as a triangle. The sand is 0.2 mm in diameter, and its porosity is 35%.

8. *Thought question.* Develop a strategy for documenting the conditions under which the rocks of Racetrack Playa move. As we need to capture them in motion, concentrate on proposals for instrumentation or remote sensing that would accomplish this.

Further reading

Bagnold, R. A., 1935, *Libyan Sands: Travel in a Dead World*, London: Travel Book Club, 351 pp.
For one wishing to know how Bagnold got involved in his work on the transport and deposition of sand, this provides access. It reveals how he and companions toured into the western Egyptian deserts on military leave, modifying their Model Ts and learning the hard way how to traverse dunes.

Bagnold, R. A., 1941, *The Physics of Blown Sand and Desert Dunes*, London: Methuen, 265 pp.
This is the classic text on the topic, providing both insight into the physics of the motion of sand and the origin of sand bedforms, and insight into how science works. These were the musings of an independently wealthy man whose passion for science led him to construct wind tunnel facilities on his own estate.

Egan, T., 2006, *The Worst Hard Time: The Untold Story of Those Who Survived the Great American Dust Bowl*, New York: Houghton Mifflin, 340 pp.
Rather than focus on the people who left the Dust Bowl for California, as Steinbeck did in Grapes of Wrath, Egan sticks with interviews of those who stayed behind. He builds a tale of a land transformed from one used by buffalo to one used by cattle to one used by farmers, becoming increasingly susceptible to drought.

Lancaster, N., 1995, *Geomorphology of Desert Dunes*, London: Routledge, 312 pp.

Banking on his deep knowledge of the deserts of both North America and South Africa, Lancaster provides a very useful entrance point to the study of sand dunes in the field.

Pye, K., 1987, *Aeolian Dust and Dust Deposits*, London: Academic Press, 334 pp.
This text is a compilation of information about dust in all its aspects, from the characterization of the grain size and mineralogy, to the accumulation in the great loess deposits of the world.

Pye, K. and H. Tsoar, 1990, *Aeolian Sand and Sand Dunes*, London: Unwin Hyman, 458 pp.
This book nicely complements Pye's tome on dust from a few years earlier, providing an encyclopedic view of the topic with parallel attention to the grains involved and to their accumulation.

Rubin, D. M. and C. L. Carter, 2006, *Bedforms and Cross-Bedding in Animation: SEPM Atlas Series* No. 2, http://www.sepm.org (see also http://walrus.wr.usgs.gov/seds/bedforms/animation.html).
This is a wonderful guide to the interpretation of eolian sand dune deposits. They employ simple geometric models of dune translation, but acknowledge the variability in both the wind direction and the complexity of the dunes. The animations available show the power of the moving image to allow insight into complex features.

CHAPTER 16

Coastal geomorphology

In every outthrust headland, in every curving beach, in every grain of sand there is the story of the earth.

Rachel Carson

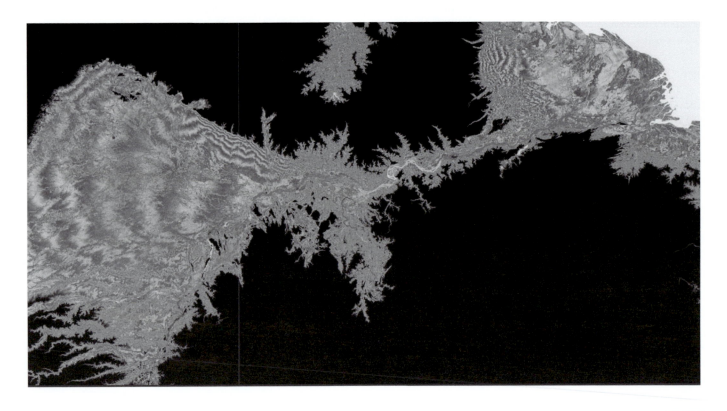

In this chapter

Coasts are the most dynamic elements of the world's landscapes. This reflects the arrival of energy in the form of waves that are generated by winds beneath storms at sea. Millions of waves arrive each year to every site along the ocean perimeter. Most of the energy from these waves is dissipated in the coastal zone, some of it converted to work of both sediment transport and erosion of rock. The coasts serve as the bottom or exterior boundary condition for the terrestrial landscape. The evolving coastline influences both the gradients of the rivers and the heights of the cliffs in rocky coastlines. The oceans are the final sinks for sediment and water from the terrestrial landscape.

Delta of the Orinoco River, Venezuela. Contours colored using cyclic bands at 1 m contour intervals, from SRTM 90 m DEM. Landscape above 100 m is black (image courtesy of James Syvitski, INSTAAR, University of Colorado).

As waves are all-important in the coastal zone, we will address both their origin in the deeper ocean, and their transformation as they approach shore. Winds at sea generate a broad spectrum of waves. At any particular coastal site, we can characterize the arriving spectrum of waves, called the wave climate. We will characterize the concentration of energy in a wave, and the rate at which that energy is delivered to the coast as it moves. It is the large waves that accomplish much of the work.

Sediment delivered to the coast by rivers generates deltas that are some of the most productive agricultural sites in the world and house tens of millions of people, but are also, for the same reasons, vulnerable to large-scale disasters both from floods from upstream and from the ocean. In order to understand the scope of the resources and the risks inherent in deltas, we must understand their dynamics. We will see the strength of the influence on deltas from human activities well upstream of the deltas themselves.

The coarse-grained sediment in the coastal zone is smeared along the coast in littoral drift, generating map-view patterns from capes to spiral bays to beach cusps that reflect the ability of the system to self-organize. We wish to be able to read such landscapes for the dominant processes, and to understand the feedbacks that allow this self-organization.

As in other chapters, we will utilize the concepts of continuity and of transport to explore several of the elements of coastal geomorphology. We will see that the sediment transport in the littoral zone is non-linearly related to both wave height and wave angle. It is in part this nonlinearity that promotes the feedbacks involved in the self-organization of coastal landforms.

The relative movement of land and sea

We start by discriminating between rocky and sandy coastlines. This is essentially equivalent to organizing coasts into tectonically active and passive coasts. Where rock is moving upward faster than sea level, coasts are typically rocky, and are ornamented by highly crenulated lines of cliffs, interspersed with pockets of sand. In contrast, where sea level is rising relative to the land, the coastlines are sandy, the sand is smeared along the coast by littoral (beach-driven) processes, and organized into a completely different set of geomorphic forms including barrier islands, capes, cusps and so on.

We must therefore acknowledge the importance of relative sea level change. As in many examples in geomorphology, more than one interface is moving. Here sea level, the interface between the sea and the atmosphere, can and is changing, while the landmass at the coastline, the interface between land and air, may also be moving. Both interfaces can go in both directions. We define the rate of relative sea level change to be $RSL = SL - R$, where SL is the rate of sea level change, and R is the rate of change of elevation of the landmass ($R > 0$ implies rock uplift, $R < 0$ implies subsidence). Importantly, the reference frame for both interfaces is the center of the Earth, or more measurably, the geoid. The ocean is made to attack rock only when $RSL < 0$, or when the rate of rise of sea level is insufficient to outpace rock mass uplift.

These statements force us to come to grips with both sea level history and tectonic history on our coasts. Changes occur on several timescales, reflecting myriad processes involving climate, tectonics, and even human-caused land subsidence. Below we treat these one by one.

The Pleistocene record

The primary medium-term (many ka) driver of sea level change is the growth and decay of the world's ice masses. These have constituted huge changes in global ocean volume within the last few million years. The record is best documented using the isotopic concentration of benthic forams found in the stratigraphy of deep-sea cores depicted in Figures 1.3 and 1.4. Although the history derived from these tests is contaminated to some degree by the changing temperature of the deep ocean waters, it remains our best continuous record. But these are changes in isotope concentration. How do we relate these to changes in sea level? The amplitude of sea level change is documented by the depth of the Last Glacial Maximum (LGM) shorelines. On tectonically quiet coastlines, these shorelines are found at depths of $-120\,m$ (relative to modern sea level). The crudest approximation for sea level history is simply obtained by rescaling the $\delta^{18}O$ time series into one of sea level as shown in Figure 16.1. Unfortunately the transformation is not linear. The world's ocean is not a perfect bathtub with vertical walls. This is the only case in which the level of water in the basin is a

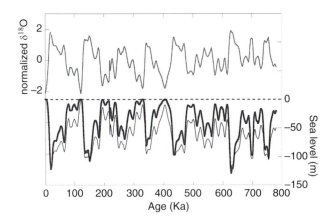

Figure 16.1 Sea level histories predicted from marine isotope history. Top: $\delta^{18}O$ curve from Imbrie *et al.* (1984), normalized by subtracting mean and dividing by standard deviation over this 800 ka interval. Bottom: sea level histories calculated using linear (light line) and nonlinear (bold line) algorithms. Constraints include 0 level at present, + 6 m at 120 ka maximum (MIS 5e), and −120 m at LGM. While such estimates are straightforward and may be our only recourse for long-term sea level histories covering the full Quaternary, the true sea level history as constrained by corals in the last two glacial cycles is more complicated.

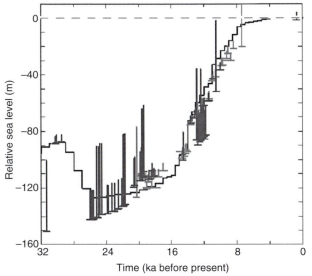

Figure 16.2 Sea level curve derived from corals in Barbados, showing sea level history that includes the LGM at roughly 26 ka. Each symbol is a dated coral sample. Horizontal error bars in age reflect age resolution. Depth of sample is well known, but the upward pointed bar reflects the depth range for the particular species of coral sampled. Smallest depth range of 5 m corresponds to *Acropora palmata*. Continuous black line is modeled sea level history derived from ICE 5G eustatic sea level results (from Peltier and Fairbanks, 2006, Figure 2, with permission from Elsevier).

perfect linear proxy for the volume of water in the basin. The oceans have a definable hypsometry just as a terrestrial drainage basin:

$$\frac{dV}{dt} = \frac{d(Az)}{dt} \tag{16.1}$$

Only if the area of the basin, A, is not a function of bathymetric level, z, can one remove A from the derivative to result in the simple scaling

$$\frac{dz}{dt} = \frac{1}{A}\frac{dV}{dt} \tag{16.2}$$

allowing the sea level change to be directly scaled by the isotopically derived rate of change of global ice volume. This effect, however, is not large. While the continental shelves, whose outer edges correspond remarkably well with the 120 m bathymetric contour, are perhaps on average about 100 km wide, this is still quite small relative to the radius of the major ocean basins (which is of the order of 3000 km). The ocean area, in other words, does not change that much during sea level swings of 120 m, and we can indeed use Equation 16.2. However, the problem is complicated by the fact that the water delivered to the oceans from melting ice is not all the same isotopic composition. Indeed, it is the isotopic stratigraphy of ice that we use to extract a climate signal from ice cores. Our best opportunity to

derive a sea level history comes within the last glacial cycle during which we have independent means of documenting sea level. These probes at least allow us to assess the dangers in attempting to scale sea level from the marine isotopic record.

Sea level change in the Holocene

As the last great ice sheets melted and calved icebergs into the sea, sea level rose. The history of the demise of these ice sheets is therefore recorded not only in the moraine record (see Dyke, 2004), but also in sea level records. The best of these are obtained from coral records on stable shelves. In the Caribbean, a particular species of coral, *Acropora palmata,* dominates the architecture of the reefs, lives within 5 m of sea level, and is capable of growing fast enough to keep pace with sea level rise. Dating of corals taken from cores extracted by drilling through these coral reefs can be used to document the deglacial record of sea level rise. The key here is that the corals can be dated with high precision, using both U/Th and ^{14}C methods (see Chapter 6). The record reproduced in Figure 16.2

reveals that sea level began at 26 ka (about the LGM) at −120 m. Sea level rose at an increasing rate from 18 ka to 10 ka, and then flattened out at about −6 m at 10 ka. Thereafter it has risen at very slow rates. This history is punctuated with one to several abrupt rises. The largest and most agreed upon of these is melt-water pulse (MWP) 1A, during which a rise of roughly 25 m occurred in 500 years between 14.7–14.2 ka (Weaver *et al.*, 2003). The rate of rise, which reached 5 cm/yr, can only be attributed to the very large rate of demise of one or more ice sheets.

Let us calculate what rate of addition of water volume this 5 cm/yr requires. The rate of volume change of ocean water is given by Equation 16.1. Recalling that the oceans cover about three-quarters of the globe, $A = 0.75 \times (4\pi R^2)$, where R is the radius of the Earth (about 6370 km). Then the rate of addition of water must be $dV/dt = 3.8 \times 10^8 \, (0.05 \times 10^{-3}) = 18\,000 \, \mathrm{km}^3/\mathrm{yr}$. Dividing by the number of seconds in a year, we obtain $6 \times 10^{-4} \, \mathrm{km}^3/\mathrm{s}$, or $6 \times 10^5 \, \mathrm{m}^3/\mathrm{s}$. This discharge is oceanic in scale, more than half a Sverdrup ($= 10^6 \, \mathrm{m}^3/\mathrm{s}$), and this discharge must have been maintained for the entire 500 years to achieve the total rise of 25 m. This is not a typical river discharge. Even the Amazon fails to reach this scale of discharge, although it comes close with its discharge of about $10^5 \, \mathrm{m}^3/\mathrm{s}$. So the rate of sea level rise recorded in the corals in these pulses requires the equivalent of several additional Amazon Rivers turning on for several centuries, and then shutting off.

The bottom line, therefore, is that sea level was lowered by ∼120 m during the LGM, and has risen since then, rapidly at first, and more slowly through time, with a couple of periods of very rapid rise that likely reflect extraordinary discharges of fresh water to the sea. Since 10 ka, and certainly since 6 ka, the contributions from ice sheets have been relatively minor. This sets the scene for the record within the last few hundred years, within which time the sea level history is recorded instrumentally.

The last century of sea level change and its causes

We live at a time during which sea level is rising again. We measure sea level using tide gages; a typical setup is sketched in Figure 16.3. These are ideally attached to rock, and record the rate of rise or drop of the sea relative to this substrate. Of course, since they are

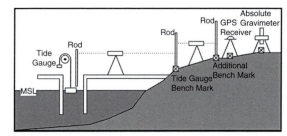

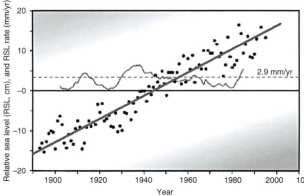

Figure 16.3 Tide gages record sea level rise. (a) Typical modern tide gage set-up utilizing a float in a stilling well to isolate the sea surface from the wavefield, tied back to benchmarks on land. GPS receiver measures rate of change of land surface relative to the satellite constellation (courtesy of S. Nerem, CU Boulder). (b) Example tide gage record of relative sea level (RSL) from New York City over more than a century. Dots show annual mean sea level, and continuous line shows running average rate of sea level rise. Mean rate over the century is 2.9 mm/yr. Approximately 1 mm/yr, or about 30% of the signal, is attributable to glacial isostatic adjustment (GIA) (after Douglas and Peltier, 2002, Figure 1, with permission from the American Geophysical Union).

attached to the rock, what they really measure is relative sea level rise, as the rock could be moving relative to the center of the Earth as well. If we go to "stable" margins, however, we presumably take this tectonic signal out of the picture, although it is not really that simple (for example, see Tushingham and Peltier, 1991). There has been much attention within the last few decades to the rates and causes of sea level rise, and modern instrumentation has allowed the various contributions to be quantified (see review in Cazenave and Nerem, 2004). The present rate of sea level rise is about 2.9 mm/yr. While this does not sound very significant, we note that the potential impact on low-lying landscapes could be large. Within the last century this amounted to almost 30 cm, and the next century could see more. Atolls and deltaic landscapes are most at risk, where the elevations are very low to begin with, and subsidence of the rock and sediment that comprise the landscape exacerbate any true (global) sea level rise.

A list of possible contributions to sea level rise must include the following:

- melting of glaciers and ice sheets;
- thermal expansion of the oceans;
- change in storage of fresh water on land and in groundwater.

Thermal expansion, also called the steric effect (*steric* = pertaining to the arrangement of atoms), reflects the expansion of water when warmed. Let us calculate how much the column of seawater would expand if it were to warm by 1 °C. Just as in our calculation of the density of the lithosphere, the one-dimensional strain of the water column associated with a change in temperature may be calculated from

$$\Delta z = \alpha H \Delta T \qquad (16.3)$$

where α is the coefficient of thermal expansion (strain per °C), ΔT is the change in temperature (°C), and H is the height of the oceanic column that is warmed. Let $H = 3000$ m, $\Delta T = 1$ °C, and $\alpha = 1$–2×10^{-4}/°C. The resulting change in the height of the water column (the sea level rise) is 0.3–0.6 m. This is large because the ocean is so deep. In reality, of course, the ocean does not uniformly warm over its entire depth, but warms first in the upper well-mixed surface layer (see problem 1 in this chapter). Importantly, it may take thousands of years for the full column to warm, or to come into equilibrium with the surface temperature of the ocean, as the mixing time of the ocean is very long. This means that even if warming were to cease now, it may take many lifetimes for the full effect of the warming to be felt within the ocean; this steric effect will therefore continue to raise global sea level for centuries to come. Present estimates of the steric contribution to sea level rise are about 1.5 mm/yr (Cazenave and Nerem, 2004).

Of the roughly 3 mm/yr of present day sea level rise, about half of it comes from new water being added to the oceans, most of it from the decline in ice masses on land. The contribution of glaciers to global sea level rise is surprisingly hard to constrain. As there are tens of thousands of glaciers in the world, one cannot measure them all. Even the contribution from a single glacier is difficult to measure. This can be obtained by documentation of changes

in the glacier volume over time, but even this is difficult. Glacier area, on the other hand, can be obtained readily from space imagery. Conversion of area to volume, however, requires a relationship between glacier area and mean glacier thickness. Meier *et al.* (2007) have assembled mass balance information for many of the world's glaciers, and claim that their shrinkage represents 60% of the global sea level rise rate within the last century that is attributable to new water (non-steric effects). The rate of contribution is also continuing to increase. One can also document changes in ice thickness on individual glaciers using laser altimetry. By flying down the centerlines of many glaciers in Alaska, and comparing the resulting profiles with those derived from 1950s maps of the same glaciers, Arendt *et al.* (2002, 2006) have documented the thinning of Alaskan glaciers. Multiplying the change in cross-sectional area of the glacier by the width of the glacier results in a change in volume. They have demonstrated that the major glaciers in Alaska represent a significant fraction of the sea level rise rate attributable to small glaciers. These particular glaciers are apparently losing volume much more rapidly than their counterparts in other parts of the world.

Rock uplift

We have seen that relative sea level rise requires knowledge of both changes in sea level and changes in the level of the rock mass with respect to the geoid with time. While we have discussed tectonically driven rock uplift in Chapter 3, we emphasize here the utility of coastal geomorphic markers as some of the best available. Marine terraces and beaches are particularly useful because they are scribed as horizontal lines whose initial altitude we know from sea level history. They are better than their fluvial counterparts because river-formed features are not initially horizontal and have initial slopes we do not necessarily know. We know that marine terraces are originally horizontal.

We briefly summarize three cases: a rocky coastline, a coral coastline, and a sandy coastline. In rocky coasts, a line of cliffs bounds the sea. As the ocean erodes the base of the cliff, the cliff line recedes into the landmass. The base of the cliff marks the mean level of the sea; this is called the wave-cut angle, as seen in Figure 16.4. Documentation of the present

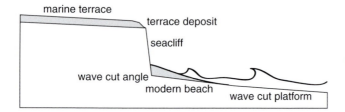

Figure 16.4 Anatomy of a modern rocky coastline. The intersection of the seacliff with the wave-cut platform, often mantled with modern beach sands, defines the wave-cut angle. This is a proxy for modern mean sea level, as wave-cut angles on paleo-seacliffs are proxies for past sea levels.

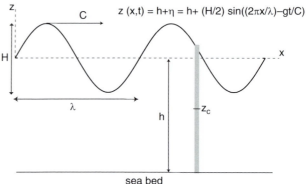

Figure 16.5 Definition sketch of moving sinusoidal wave with wavelength, λ, full amplitude (wave height), H, and celerity, C. The waves modulate the mean water depth, h. The elevation of the center of mass of a local water column, z_c, is shown, which enters the calculation of the potential energy of the wave.

elevations of wave-cut angles from old abandoned cliff lines reveals the pattern of relative sea level change since the time of formation and abandonment of the seacliffs. To derive a long-term mean uplift rate, one must also date the form, which is accomplished by dating sediment atop the adjacent wave-cut platform. We can either date fossils in the deposits, or use the cosmogenic radionuclide concentration in the sediments themselves (see Chapter 6, also Chapter 18 on Santa Cruz).

In low latitudes, the geomorphologist is provided with another tool. Corals grow in a band of roughly 30° north and south from the equator; the major reef-forming coral *Acropora*, for example, is presently restricted to 31 °S to 31 °N (Wallace and Rosen, 2006). Corals grow by photosynthesis of the polyps that construct their $CaCO_3$ homes, and therefore grow within the photic zone. Those that are restricted to very shallow depths serve as excellent markers of sea level. In addition, they are dateable using the U/Th clock. This has made tectonically rising coastlines within this latitudinal belt, like Papua New Guinea's Huon Peninsula, into world-class laboratories in which to document both past sea levels and the pattern of rock uplift (for example, Ota and Chappell, 1996).

Sandy coastlines are typically less active tectonically. They can however be either rising or falling with respect to sea level due to larger scale geophysical phenomena. For example, the east coast of the United States is still responding to the removal of the load of the LGM ice sheet. Some of the coast is therefore subsiding due to collapse of the forebulge, while more northern portions are rising due to rebound from direct loading by ice. Deposits in these post-glacial settings can be dated using ^{14}C on organic matter. In rising coastlines, old strandlines (sandy deposits reflecting old beaches) can record past sea levels. In subsiding coastlines, the information about past sea level is extracted from depth in cores. In both settings it is often peats that are dated. We will see below their use in documenting the subsidence history of the Mississippi delta.

Waves

Most of the work done by the geomorphic system at the intersection of the land, sea and air is accomplished by waves. We first address their origin, then their transformation as they move from deep water into the shallow water adjacent to the coast.

Origin of waves

In all but the exceptional case of tsunamis, which are generated by earthquakes and landslides, waves are generated by wind. To first order, the original waves may be considered sinusoidal, as illustrated in Figure 16.5. The winds of a storm at sea transform a still water surface into one with a broad spectrum of waves. As these waves are broadcast from beneath the storm, the spectrum evolves due to dispersion. Upon encountering the edge of a basin, they begin to be influenced by the bottom topography. The subsequent evolution involves steering of the wave fronts or refraction, and changes in the height and speed of

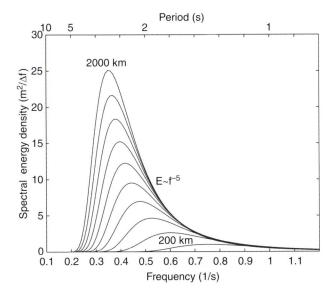

Figure 16.6 Spectrum of wave energy for varying fetch lengths (in 200 km intervals), calculated from Komar (1998, Equation 5-5) after Liu (1971) using $u_* = 0.7$ m/s. Note that the high frequency (short period) tail behaves as a $1/f^5$ distribution, and that the dominant wave grows in period with increase in fetch.

the wave. This eventually causes steeping of the wave to the point of breaking. It is the resulting complex turbulent surf that accomplishes the sediment transport and erosion of rock. We will briefly treat each of these steps in the generation, translation, and transformation of the waves.

A flat sea surface is unstable to perturbations when wind blows across it. Stated less mathematically, waves are inevitable when water is subjected to wind. The streamlines of the air are compressed over the top of an incipient wave, and diverge over the trough. The Bernoulli principle suggests that this generates low pressure over the crest and high over the trough, pulling up the crest and pushing down the trough. This feedback promotes the growth of the waveform. The form translates downwind.

But what dictates the wavelength and the height of the waves? Wave height grows with the speed and duration of the wind, and the distance over which the wind blows unobstructed, or the fetch (Figure 16.6). While the predictive empirical relationships are complicated (see for example Komar, 1998), they suggest that the wave height grows as the wind speed squared and as the square root of the fetch.

It is also important that wave speeds are dependent upon the wavelength or period of the wave, and upon the depth of the water. If the water is very deep relative to the wavelength ($D \gg \lambda$), the wave is said to be a deep-water wave and translates at a speed that is dependent upon wave period:

$$C = \frac{gT}{2\pi} \ (\text{for } D > 0.5\lambda) \tag{16.4}$$

where T is the wave period. The longer period waves travel more rapidly. Taking $g = 10 \text{ m/s}^2$, the celerity of a 20 s wave should be 30 m/s. In contrast, if the wave is long relative to the water depth the wave is said to be a shallow water wave and travels at a speed that is simply

$$c = \sqrt{gD} \ (\text{for } D < 0.05\lambda) \tag{16.5}$$

where D is the water depth. Recall that we encountered this formula in calculating ocean depths from the translation speed of tsunamis in Chapter 3.

Returning to the deep-water case, the longer period waves that are generated beneath a storm will travel most rapidly away from the storm. Given that wave speed and wavelength are related through $C = \lambda/T$, we can rearrange Equation 16.4 to arrive at

$$C = \left[\frac{g}{2\pi}\lambda\right]^{1/2} \tag{16.6}$$

The longer wavelength waves travel more rapidly. This suggests that from a storm well out in the Pacific that generates a wide spectrum of waves, the waves that will arrive first at a distant shore will be those that traveled fastest, and hence will be the longest period, longest wavelength waves. This is the ocean swell, and it can have a very clean, narrow spectrum and a clear dominant wavelength. This can be seen in wave buoys off the coast of California as a strong swell arrives. At 30 m/s, these waves can travel about 2400 km in a day. The surfers love these waves. As time since the storm generated the waves goes on, the waves arriving at the shoreline should decline in period and wavelength.

The energy per unit length of wave, or the energy density of the wave, may be written

$$E = \frac{\rho g H^2}{8} \tag{16.7}$$

This is the sum of the kinetic and potential energy associated with the wave. While we do not derive this here (a full derivation is provided for example in Dean and Dalrymple, 1984), the principle is straightforward. We will evaluate the mean potential energy of the wave; one integrates over the wavelength the product of the mass of the column of water with the height of its center of mass (Figure 16.5). Once again we employ the mean value theorem to arrive at a mean potential energy:

$$\overline{PE} = \frac{1}{\lambda} \int_0^\lambda [\rho g z] \left[\frac{z}{2}\right] dz \qquad (16.8)$$

where

$$z = h + \eta = h + \frac{H}{2}\sin(\frac{2\pi x}{\lambda}) \qquad (16.9)$$

This reduces to

$$\overline{PE} = \frac{\rho g}{2\lambda} \int_0^\lambda (h + \eta)^2 dx = \frac{\rho g}{2\lambda} \int_0^\lambda (h^2 + 2h\eta + \eta^2) dx \quad (16.10)$$

and finally

$$\overline{PE} = \frac{\rho g}{2\lambda} \left[h^2\lambda - \frac{hH\lambda}{2\pi} \left[\cos\frac{2\pi x}{\lambda}\right]_0^\lambda \right. \\ \left. + \frac{H^2}{4} \int_0^\lambda \sin^2\frac{2\pi x}{\lambda} dx \right] \qquad (16.11)$$

When evaluated in the limits, the cos term vanishes. The $\sin^2$ integral leaves $\lambda/2$. The resulting final form for the mean potential energy:

$$\overline{PE} = \frac{\rho g h^2}{2} + \frac{\rho g H^2}{16} \qquad (16.12)$$

The first term represents the mean potential energy associated with the mean water column height above the ocean floor. The second term represents that associated with waves. Combining this latter term with an equal contribution from kinetic energy of the wave results in Equation 16.7 for the total energy due to waves.

The flux of energy, or the wave power, is the product of the energy density with the group wave speed:

$$\omega = EC = \frac{\rho g^{3/2}}{8} H^2 h^{1/2} \qquad (16.13)$$

in the shallow water case. It is this latter quantity that is conserved upon refraction during shoaling. This

quantity is analogous with stream power, and will be important in quantifying both transport of sediment and erosion of rock. The strong dependence of the wave power on wave height, H, suggests that yet again the geomorphic transport process is non-linearly related to the size of the event – here a wave. The big waves will do a lot of the work.

Transformation of waves

Now let us explore how a deep-water wave is transformed as it approaches a coastline. Both the speed and the angle of the wave evolve. As a wave approaches the shore, it will slow as it shoals, as a direct manifestation of the dependence of shallow water wave speed on water depth. Consider the most common case of wave approach from an angle other than orthogonal to the coast. The coastward edge of the wave will experience shallower water before its seaward counterpart. The coastal edge of the wave will slow, forcing the wave front to bend, and the wave rays likewise to bend. The wave fronts tend to become more parallel to the shore as they approach, and the wave rays indicating the direction of water motion become more normal to the shore. The wave refracts, in a process that is perfectly analogous to light refraction of light, as shown in Figure 16.7. Snell's law holds in a water wave field as well.

In general, the wave rays will therefore impact the shore at some obtuse angle. We will see that this drives motion of sediment along the coastline, as longshore drift. In situations other than planar shorelines, the bathymetry over which the wave shoals will steer the wave. Wave rays will focus on headlands, and defocus away from the heads of submarine canyons. This is sketched in Figure 16.7.

As a deep-water wave shoals, its velocity and wavelength decline, its height increases, and its period is retained. The orbital trajectories of the water parcels transform from circular to elliptical. As the wave evolves, the growth in height and decline in wavelength steepens the wave as it approaches the shore. At some threshold steepness, the wave breaks. The plunging wave then thrusts water up onto the shoreface, which generates a swash upward and a backwash back down the shoreface. If the angle of wave approach upon impacting the shoreline is not perpendicular, this will drive sediment transport down the coast.

The delivery of water to the shoreface causes a setup of the water surface above the mean water level.

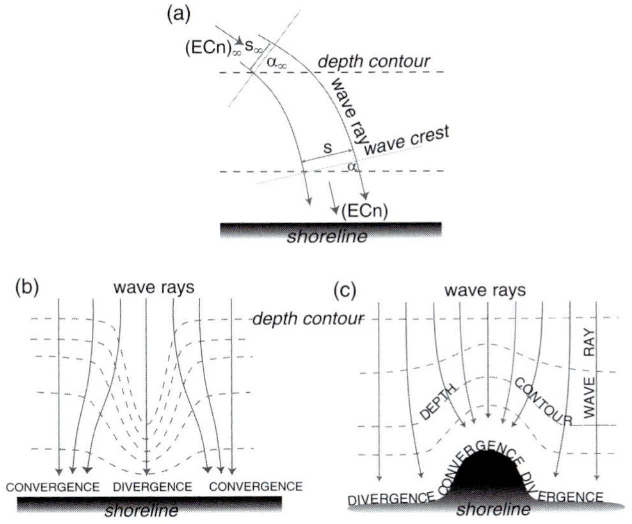

decays smoothly to zero over several such length scales. Typical reported peak current speeds are large fractions of 1 m/s. This current is responsible for transportation of sediment that has been thrown up into suspension by the intense turbulence of the waves.

Because mass must be conserved, what runs up onto the shoreface must run off. This is accomplished by both undertow and rip currents. In the vertical plane, slow seaward velocities at depth (undertow) balance the high shoreward speeds at the water surface. In the shore-parallel dimension, rip currents accomplish the return flow in discrete high-speed jets that interrupts or deflects the breaker line. Its jet-like nature is the key to how swimmers must escape – swim along shore, rather than fight back toward shore into the core of the jet.

The mean position of the breaker line can be documented by stacking video images. Because the breaker generates white foam, which contrasts strongly with the dark water, such stacking results in a gray-scale image that illuminates well the mean position of the breakers (for example, Holland and Holman, 1993).

Waves deposit energy along a line where the sea meets the land. We have talked about how rock uplift affects the location of this relative sea level. On a shorter timescale, however, tides and storm surges also influence the location of this important interface. As we show below and in Chapter 18, much more wave energy can be delivered to seacliffs when the tide is high. Less energy is dissipated across the shelf as the wave shoals. The same is true with storm surge, which can cause a significant rise in local sea level at times when the waves are also high due to the local winds. This double whammy is yet another example of the extreme roles of large events in geomorphology.

Hurricane storm surge

The ocean surface is mounded up in a dome centered on the eye of the storm for two reasons (see Kerry Emanuel's (2005) wonderful book on hurricanes). First, the low pressure of the atmosphere in the storm center, which itself generates the winds of the storm, creates a pressure gradient in the ocean that drives water toward the storm center. Second, the winds generate a current that piles water up as the storm encounters shallow water. The latter effect generates the majority of the storm surge and produces the

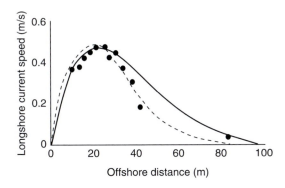

Figure 16.7 Refraction of waves as they approach a shoreline. Refraction of waves obeys the same Snell's law that controls refraction of light. (a) Definition sketch of refraction showing conservation of wave energy flux as the wave approaches a shoreline. (b) Refraction results in defocusing of waves as they traverse a submarine canyon, and (c) focusing as they approach a headland (after Komar, 1998, Figures 5-42 and 5-43, with permission from Prentice-Hall).

Figure 16.8 Cross-shore profile of longshore current. Curves from profiles using (dashed) linear (Airy) theory and nonlinear theory (solid line) (after Wu et al., 1985, Figure 9, with permission of the American Geophysical Union).

Wave arrival at an angle will also result in a current of water parallel to the shoreline called the longshore current. The confinement of this current to the near-shore zone demonstrates its causal link to the breaking of waves on a sloping beach. The cross-shore pattern of this velocity revealed in Figure 16.8 shows a distinct peak at roughly halfway between the breaker line and the top of the swash excursion, and

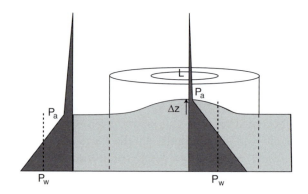

Figure 16.9 Mounding of the ocean surface beneath the eye of a hurricane. Low atmospheric pressure centered on the eye of the storm produces a pressure gradient in the ocean that pushes water from outside the storm toward the eye. Conservation of water requires that the amount of water increase within the cylinder (dashed margins) beneath the storm, resulting in rise of the water surface. The magnitude of the rise Δz is set by the requirement that the mean pressure in the water columns (vertical thin-dashed lines) inside and outside the storm be equal. Under these conditions the pressure gradient driving flow across the cylinder vanishes. Here the pressure profiles in the atmosphere and the water are sketched illustrating this condition of equilibrium. Note that the gradient in pressure in the air column differs, due to the different density of the hot rising air in the storm center vs. outside the storm, while that within the water column is the same.

asymmetry of the surge (to the right of where the eye hits the land in the northern hemisphere). These surges can be up to 20 meters!

As demonstrated in Figure 16.9, we can estimate the magnitude of the pure pressure effect. The mean atmospheric pressure is 1013 millibars, or 1.013×10^5 Pa. Given that the formula for pressure beneath a fluid of a given uniform density, ρ, is $P = \rho g H$, we can estimate the thickness of this layer, H. For atmosphere at surface density of $\rho = 1.22$ kg/m^3, this corresponds to a layer 8464 m thick. To generate the same pressure, water at $\rho = 1000$ kg/m^3 would have to be $H = 10.3$ m thick. It is pure coincidence that the pressure exerted by the atmosphere is worth about a 10 m column of water. But it sure is convenient. Given that the lowest pressures associated with category 5 hurricanes are about 900 millibars, or about 9/10th of an atmosphere, the pressure drop corresponds to 1/10th of the column. This translates into 1/10th of 10 m, or about 1 m of water. So, to balance the pressures in the column of water beneath the eye vs. those in a comparable column exterior to a large storm, the water must dome up by about 1 m. That this is only a small fraction of the

10 m typical of a storm surge associated with such monster storms implies that the remainder and indeed the majority of the storm surge must come from the wind-driven currents.

Physics of sand movement in the littoral system

Longshore drift, or the mass discharge of sand along the coast, has been closely studied theoretically, in the field and in the lab. The most often used equation for littoral or longshore drift is (Komar, 1998, eqn. 10.8)

$$Q_l = 1.1 \rho g^{3/2} H_{br}{}^{5/2} \sin \alpha_{br} \cos \alpha_{br} \qquad (16.14)$$

where the breaker height H_{br} is in meters, water density and the acceleration due to gravity are in SI units, and the sediment transport rate is in m^3/day. The angle α_{br} is the angle between the breaker line and the local shoreline. Computation of the net littoral drift over a year or several years, which is relevant to the assessment of filling rates behind human-made structures on the coast (for example jetties and groins), then requires information about the distribution of wave heights and wave approach angles. This is reported in deepwater conditions off the coastline. Transformation of the deepwater wavefield must then be done to transform both wave height from H to H_{br}, and the wave angle from α to α_{br}.

Measurement of this sediment transport in real time is extremely difficult. Over short periods of time, tracer methods have been employed (e.g., Komar and Inman, 1970). Here colored or otherwise distinct sand is injected as a line across the surf zone. One then measures the downdrift field of tracer concentrations in the sand to establish the distance of transport over a fixed period of time, and hence its mean speed. The product of the mean speed with the width of the breaker zone, L, and the thickness of the moving sand, δ, then yields the volumetric sediment flux (see Figure 16.10):

$$Q_{litt} = u \delta L \qquad (16.15)$$

Measurement of long-term rates can be accomplished by either measuring the accumulation of sediment behind natural or human-made structures such as seawalls, jetties, groynes, and breakwaters (an example of which is shown in Figure 16.11) or by

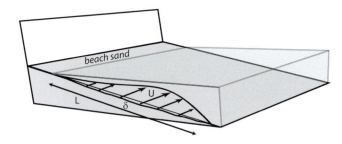

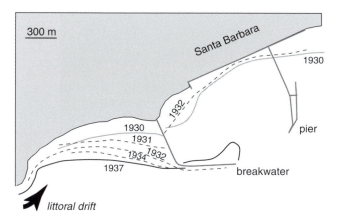

Figure 16.10 Littoral drift is down-beach transport of sand by waves impacting the beach at an angle. The total discharge of sand depends upon the thickness of the transport layer, δ, the speed of the sand averaged over many waves, u, and the cross-beach distance over which waves influence the bed, L.

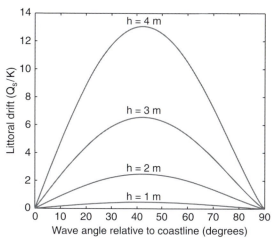

Figure 16.12 Dependence of littoral drift rate on angle of far-field wave attack. Sediment flux vanishes at both very shallow and normal incident waves, and is maximum at roughly 45°. Drift rate increases with wave height, h, to the power 12/5.

Figure 16.11 Accumulation of sand from littoral drift after construction of the Santa Barbara, California, breakwater (after Komar, 1998, Figure 9-4, with permission from Prentice-Hall).

compiling dredging records at harbor mouths (for example, Griggs *et al.*, 1994).

Sandy coasts

We now ask how the littoral sand is arranged on sandy coastlines. Beach morphology includes organization at two very different scales, including beach cusps at the scale of several tens of meters, to capes that can be many tens of kilometers in spacing.

Capes and spits

On a large scale, spits and capes punctuate sandy coastlines. The east coast of North America provides an excellent example, with Cape Hatteras one of

several Carolina capes that are spaced at roughly 150 km intervals. These represent one of the largest scale examples of self-organization in geomorphology. The physical basis of their origin and evolution has been illuminated by numerical simulations reported by Ashton *et al.* (2001). The essence of the problem lies in the dependence of longshore sediment transport rate on the angle of wave approach. This is sketched in Figure 16.12. At 0° incidence, the transport rate vanishes because the wave has no onshore momentum. At 90° there is no component of momentum in the alongshore direction. Transport is maximized at roughly 45°. For wave approach angles less than this maximum, slight deviations in the shoreline will enhance sediment transport, the pattern of transport rate that will cause erosion; the perturbation will be damped, meaning that it will not grow. But for approach angles that exceed this maximum, deviations or perturbation in the shoreline will cause a decline in transport rate, which in turn causes downdrift decline in transport rate, which in turn results in deposition. The form will grow. Under these conditions the shore is said to be unstable to perturbations. Ashton *et al.* (2001) explored this instability, and assessed the roles of dominant wave angle, and of a long-term distribution of wave angles in the evolution of sandy shorelines. In Figure 16.13 we reproduce some of their results in which they showed that over long timescales the evolving coastal landforms

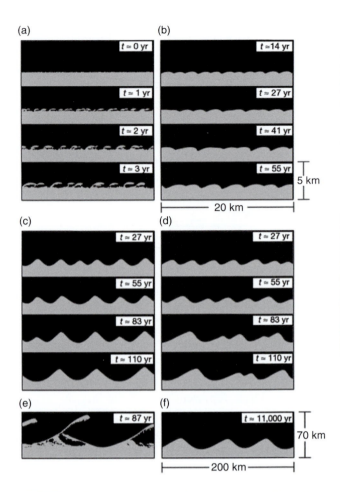

Figure 16.13 Simulated evolution of sandy coastline subjected to wave attack from a variety of angles with dominant angle from upper left (after Ashton *et al.*, 2001, Figure 2, with permission from the Nature Publishing Group).

Figure 16.14 Well-developed beach cusps on a beach in Santa Cruz, California. Approximate spacing of cusps is 25 m. Notice the backwash focusing in the troughs of the cusps (photograph by Suzanne Anderson).

interact with one another. This occurs largely by shadowing of one form by another, preventing the oncoming waves from causing sediment transport on the downdrift form. This is most astonishingly seen in the growth of cuspate spits shown in Figure 16.13(e). The only waves that can enter the bay protected by the spit are the rare waves coming from the upper right. These drive sediment transport in the opposite direction, into the bay. It is remarkable that such a broad array of forms can result from the simple feedbacks in the system. One need not appeal to separate physical mechanisms for capes and cuspate spits. They result from the same physics.

Beach cusps

At a much finer scale, one you are more likely to observe from the ground rather than from the air,

beaches are often graced with a set of beach cusps that are tens of meters in wavelength. These are characterized by offshore pointing ridges or horns that bound small bays (Figure 16.14). They form within a few days, and can be quite persistent. They are another example of nature's ability to self-organize. In other words, like ripples and other bedforms, rather than reflecting some pattern in the fluid itself, they result from a set of feedbacks involving the topography. The topography steers the waves as they both rush upward in the swash, and as they recede in the backwash off the beach face. This in turn dictates the pattern of sediment transport, the gradients in which control the sites of erosion and deposition (Figure 16.15). Several researchers discussed the basic elements of the phenomenon, including Bagnold (1940). The self-organizational nature of the forms have been simulated by Werner and Fink (1993), and embellished by Coco *et al.* (2000). The wavelength

(a)

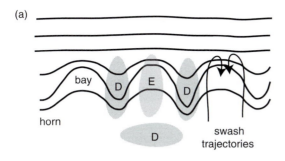

(b)

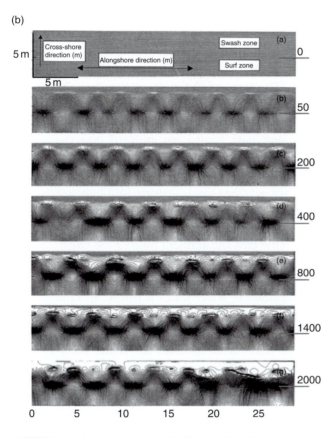

Figure 16.15 (a) Schematic map of self-organizing beach cusps. Evolving topography steers water parcels as they rush upslope as swash, and return down the beach face. The return backwash runs down the fall line. Convergence of water in the bays leads to erosion of the bay (site labeled E), and deposition of eroded sand at the mouth of the bay (site labeled D). Deceleration of the onrushing swash as it encounters the horn leads to deposition on the horns. Simulated cusp spacing is controlled by the swash excursion (after Werner and Fink, 1993, Figure 1, with permission from the American Association for the Advancement of Science). (b) Modeled evolution of swash zone topography showing emergence of beach cusps. Number of cycles is labeled; initial beach slope = 10°, mean swash excursion 1.8 m (after Coco et al., 2000, Figure 9, with permission of the American Geophysical Union).

of the features appears to be linearly proportional to the swash excursion, or the height to which the swash is pushed onto the beach face. The evolution of above-water topography is mirrored by bathymetric evolution below water. Sediment from the bays appears to generate a small bar, while offshore of the horns the topography is low. This in turn promotes "tripping" of the wave as it approaches the bay, and acceleration of the wave as it approaches the horn. As the upslope speed of the water must decline to zero before the backwash, the backwash is entirely controlled by the local topography – water simply flows down the fall line. This results in convergence of flow of water in the embayment, which in turn promotes erosion there. While these topography–wave–steering–sediment transport feedback loops might seem complicated, they appear to be robust: the results are not sensitive to the details of the rules one uses.

Deltas

Where a river encounters a still body of water, it drops its sediment. The resulting landform is a delta, a term that was introduced by Herodotus to describe the delta of the Nile River, on which five million people now live. In map view, the shape is indeed a capital Greek delta, Δ, as long as we take up to be south (see the Nile satellite image in Figure 16.16). We are interested in the pattern of deposition of the sediment, and its fate within the coastal system. The broader geological interest is stimulated by the petroleum potential of these systems, especially on the Mississippi and the Niger deltas. In addition, we emphasize here that (1) the slopes of these landforms are very low, (2) they are by definition at sea level, (3) they are heavily occupied by civilization, and (4) they are therefore very susceptible to flooding of devastating magnitudes, both by floods from upstream, and from storm surges and intense rainfall associated with tropical storms.

Deltas are complex, highly variable features that defy simple classification. The basic architecture, however, was described first by G. K. Gilbert (1890) while he mapped and described shorelines in his remarkable study of the Bonneville lake basin. That the origin of coastal geomorphology was through the study of a basin in the arid west reflects not only the intellectual breadth of Gilbert, but the complete exposure of the

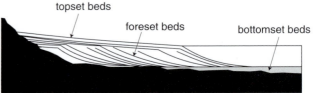

Figure 16.17 Cross section of a Gilbert-type delta, as first described by Gilbert in the Bonneville basin, Utah. The basic architecture of deltas is revealed (after Gilbert, 1890, Figure 14).

Figure 16.16 MODIS image of Nile delta, Egypt, acquired June 3, 2002. Red Sea in right margin of image, Mediterranean Sea at top. Cairo is at the base of the delta. The aridity of the region through which the Nile flows demonstrates the allochthonous nature of the Nile: all of its water is derived in headwaters of the Blue and White Nile (image generated by Jacques Descloitres, MODIS Land Rapid Response Team, NASA/GSFC).

landforms there, as they have been left high and dry by the Bonneville flood (see Chapter 17), and have subsequently been little modified due to the region's aridity. The Gilbert delta, as we now know it, has a flat top, and a steeply dipping front that then gradually merges with the bottom topography of the basin (Figure 16.17). Gilbert called the beds on the flat top of the delta the topset beds, those on the leading edge the foreset beds, and those at the base of the delta bottomset. The deltas of the Bonneville basin are well preserved because the arid climate has prevented significant modification of the forms since their formation only a few thousand years ago, and their internal structure is well exposed because drawdown of the lake caused incision through them.

We will first describe the flowfield of water from a river interacting with a basin of still water. Then we will address how the sediment in the river water decouples from it to produce the sedimentary feature. Excellent treatment of this problem is available in Allen (1997). The flow of the river as it enters the sea or lake is best described as a free turbulent jet. You are familiar with this general class of flows in smokestacks that pass heated gas upward into the atmosphere. Much progress has been made in volcanology by treatment of the violent volcanic eruptions as axi-symmetric jets that can transform into buoyant plumes. In the river case, the flow is not axi-symmetric, but is planar. Turbulence is generated by shearing of the river plume past the ambient fluid of the basin, and against the bed of the basin. This shearing entrains some of the basin water, which both expands the plume and slows it. River water is forced into the still basin by either the momentum of the river, or by the buoyancy of the river water. If the momentum dominates, we call it a momentum jet. If the buoyancy dominates we call it a plume (see for example Slingerland *et al.*, 1994). The architecture of a plane turbulent jet, illustrated in Figure 16.18, consists of a near-shore region in which the core of the incoming river momentum is eroded by entrainment of ambient fluid, and a far-field region in which there is no remaining signature of the river core. The flow expands linearly in both regions. In the fully established flow, the basin-ward velocities decline as the inverse of the square root of distance from the river mouth, and are normally distributed around the axis of the flow.

It is the deceleration of the jet that promotes sedimentation. The competence (the maximum grain size the flow is capable of entraining) and the capacity of the flow (the total sediment discharge per unit width of flow) decline with distance from the river mouth. Bedload is no longer transported as the flow declines. Finer particles delivered to the basin in the river are no longer supported by turbulence. This fine

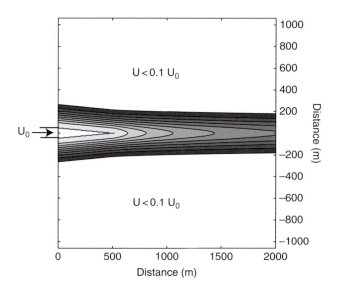

Figure 16.18 Contour plot of seaward velocity in the plume from a river entering a still basin from the left. Contours are 0.1 fractions of the incoming speed. Full incoming horizontal speed is retained in the triangular plume core.

suspended particulate matter, which includes fine sands, silts, clays, and the remains of organisms, is collectively called hemipelagic sediment (*hemi* = Greek for half, *pelagos* = Greek for sea). The resulting pattern of deposition can result in oversteepening of the slopes that promote redistribution of sediment by slumping and gravity flows.

We may appeal to the continuity equation for sediment, which must include terms for both the rain of suspended material and the gradient of discharge of bedload:

$$\frac{\partial z}{\partial t} = -\frac{1}{\rho_b}\frac{\partial Q}{\partial x} + \frac{\partial S}{\partial t} \qquad (16.16)$$

where Q is the rate of sediment transport by bedload, and S is the total suspended sediment in the column. Its rate of decline will depend upon the settling speed of the grains in suspension. In general, the rain of suspended material cannot easily be calculated by appeal to the settling speeds of the fine particles. This is because the interaction of fine grains with seawater results in flocculation of sediment to generate clusters that become effectively larger grains with higher settling speeds than one would assume from the constituent grains (see for example Syvitski, 1991).

In general, the shape of the resulting deposit should reflect the bedload deposition at the flow mouth, the

trajectories of parcels of fluid within the jet, the declining rate of deposition of hemipelagic sediment with time a parcel spends in this flowfield, and the efficiencies of redistribution by slope-related processes on the prograding feature.

Once deposited on the subaqueous front of a delta, slope processes involving bulk sediment motion, including slumping, and granular creep under the influence of waves, can redistribute sediment. To acknowledge the effect of such processes, Kenyon and Turcotte (1985) appeal to an analog with diffusive, slope-driven hillslope processes to derive an expression for the evolution of the delta front, taking the transport rate to be proportional to local slope:

$$Q = -K\frac{\partial z}{\partial x} \qquad (16.17)$$

The resulting equation for the delta form is a translating exponential:

$$z = z_0 e^{-\frac{u_0}{K}(x - u_0 t)} \qquad (16.18)$$

where x is seaward distance, z elevation above the farfield sea floor, z_0 is the depth of the farfield seafloor below sea level, u_0 is the rate of progradation of the delta front, and K is the efficiency of slope-dependent transport. Inspection of this equation reveals that the combination K/u_0 is a characteristic length scale in the problem; it determines the distance over which the topography decays to $1/e$ of the total thickness of the delta. The faster the progradation speed, and the lower the efficiency of subaqueous transport, the shorter the length scale over which the topography decays. They use the rate of advance of delta fronts and bathymetric profiles of several deltas to estimate the efficiency of this subaqueous slope process; the estimates range from $2.4 \times 10^4 \, \mathrm{m^2/yr}$ (Fraser) to $5.6 \times 10^5 \, \mathrm{m^2/yr}$ (Mississippi). They also show that the height of the delta above the basal plain may indeed be expressed as an exponential, as seen in the fit of a straight line on a semilog plot of elevation shown in Figure 16.19.

The stratigraphic arrangement of the sediment is of great interest as well, and has been the target of research in both academic and industry settings. In general, coarser grains are deposited more proximal to the delta top than finer. As the settling rates of the finer grains are smaller, they remain in the broadening plume for longer and are taken further from the delta top before they rain to the seafloor. The resulting

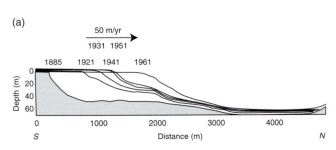

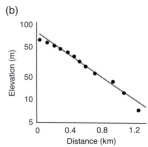

Figure 16.19 (a) SN cross section of the delta of the Rhine River over 80 years of progradation (after Muller, 1966) showing steady translation of the form through time at a rate of about 50 m/yr. Dates mark the slope-break at the delta front. (b) Semi-logarithmic profile of the delta face (dots) and the shape of the delta predicted by Kenyon and Turcotte (1985, Figure 13) assuming that slope-processes dominate the form.

deposit is therefore reverse-graded, with finer grains at its base, coarser near its top. This is visible in the simulations of Fagherazzi and Overeem (2007) reproduced in Figure 16.20.

We have so far assumed that the basin water is still. Its motion by both waves and currents serve to redistribute the sediment both down-drift and down-current. This has led to the classification of deltas as being wave-dominated, river-dominated, or tidally dominated.

The behavior of the plume is strongly controlled by any contrast in density between the river and the water into which it is debouching (seawater or ocean water). When the bulk density of the fluid from the river is less than that of the ocean, the plume spreads over a saltwater wedge and is called hypopycnal (*hypo* = under or below; *pycnal* = density). In contrast, when the concentration of sediment is so large that the bulk density of the river water exceeds that of the basin water, the flow dives to the bed, and is rapidly transported down the face of the delta. Such flows are termed hyperpycnal (*hyper* = above or higher). This is most common in lakes, where the density of seawater need not be overcome by sediment concentration, and the incoming water is denser than the ambient fluid of the reservoir. This situation results in underflows or turbidity currents that efficiently transport sediment down the face of the delta. In such cases the foreset slopes can be far less than the angle of repose seen in Gilbert-type deltas. In Lake Mead, for example, the delta from the entrance of the Colorado River has a foreslope that is of the order of 1° (Kostic *et al.*, 2002).

Deltas around the world are home to roughly one in ten humans, having roughly an order of magnitude higher population density than the global average.

The footprint of a delta scales with the size of the basin, comprising one-hundredth to one-thousandth of the drainage basin (Ericson *et al.*, 2006). The slopes of the subaerial portions of deltas can be as low as $1\,\text{m}/10\,\text{km}$ (10^{-4}), meaning that large tracts of land are within one to a few m of sea level. (For example, a 5 m storm surge can inundate $5/10^{-4} = 50\,000\,\text{m}$ or 50 km of land.) These areas and the population on them are therefore vulnerable from both sides: to storm surges associated with tropical cyclones, and to tsunamis from the oceanic side, and to floods from the terrestrial side. More than most other landforms, the land itself is also subsiding, in places at rates of several centimeters/year. This combined with eustatic sea level rise conspires to cause rapid relative sea level rise, both causing reduction of wetland area, and accentuating the risk of flood inundation through time.

Briefly, let us summarize the causes of land subsidence. First, on a regional scale the viscous mantle of the Earth is responding to the unloading of the northern continents from their LGM ice load, and the reloading of the global oceans with the 120 m of water (e.g., Peltier, 2004). Second, the huge load of sediment delivered to the margin by the river causes subsidence. Third, the compaction of the sediment column as pore water is squeezed out of the sediments causes subsidence. Compaction rates are especially high when the sediments are organic-rich, which is characteristic of wetlands. Under natural (non-anthropogenically enhanced) conditions, the subsidence due to basin flexing and compaction serves to make room for the new sediment. In other words, room is made to accommodate the new sediment – so-called accommodation space. Humans can perturb this system in several ways: (1) the extraction of hydrocarbons (e.g., Niger, Mississippi, Po) and/or water from the subsurface

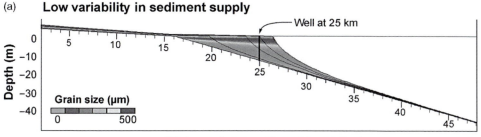

(a) **Low variability in sediment supply**

Distance from sediment input (km)

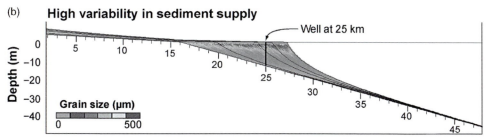

(b) **High variability in sediment supply**

Distance from sediment input (km)

(c)

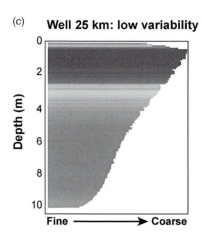

Well 25 km: low variability

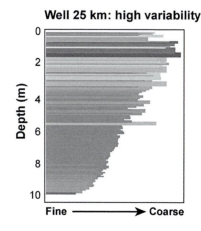

Well 25 km: high variability

Figure 16.20 Model of prograding delta including grain size distribution. This allows prediction of the resulting stratigraphy. Two cases are shown in which the sediment supply is taken to be slowly varying (a), with resulting stratigraphy (c), and highly varying (b and d) (Fagherazzi and Overeem, 2007, Figure 3, reprinted, with permission, from the *Annual Review of Earth and Planetary Sciences*, vol. 35 © 2007 by Annual Reviews www.annualreviews.org).

causes very local subsidence; and (2) a reduction in sediment delivery to the river mouth reduces the sediment available to fill the growing accommodation space. Recent compilations of the reduction of sediment supply by the construction of dams within the river system have allowed an assessment of this effect, and it is shown to dominate on many deltas (Ericson *et al.*, 2006), often following a period of enhanced sediment supply associated with deforestation and farmland expansion (Syvitski, 2008).

More than ten million people are flooded from the sea each year on the world's deltas. The lessons from Hurricane Katrina, and from Cyclone Nagris should be strong ones. Large regions of the Mississippi delta, including a fraction of New Orleans, are below sea level, prevented from inundation only by artificial human-built structures such as levees. This is not uncommon on modern deltas where relatively well-to-do countries have the wherewithal to manipulate the system (for example the Po; see Syvitski, 2008). It is well documented using GPS and inSAR on permanent scatterers (Dixon *et al.*, 2006) that the levees along the Mississippi are subsiding along with the rest of the landscape of the delta, making them less capable of restraining floodwaters through time. This subsidence inevitable and ongoing, as recorded in the salt marsh peat deposits of the coastal wetlands, where Törnqvist *et al.* (2008) have shown that the compaction of these Holocene organic-rich sediments dominates the subsidence signal, reaching 5–10 mm/yr.

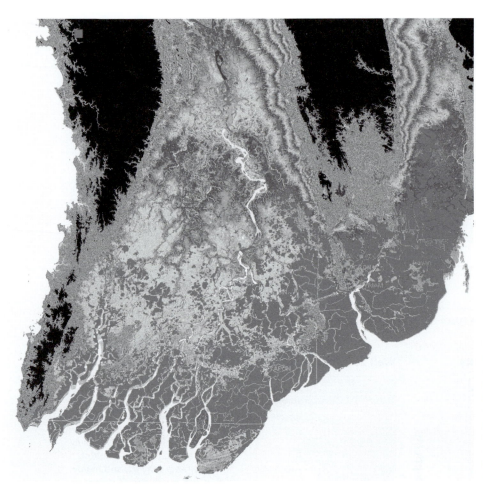

Figure 16.21 DEM of the Irrawaddy River delta (cycles of color repeat at 10 m intervals; black above 100 m elevation). Near-coastal areas shaded dark gray were flooded during the passage of Tropical Cyclone Nargis in May 2008 (image courtesy of James Syvitski, INSTAAR, University of Colorado).

Cyclone Nagris delivered high winds, high rainfall and a storm surge to the Irrawaddy delta of Myanmar (Burma) over a two-day period in early May 2008. There, no such artificial structures held back the waters, and thousands of square kilometers of this low-lying landscape were flooded. This is dramatically evident in the shaded DEM of Figure 16.21. The storm surge was meters in amplitude, and wind-driven waves riding atop the surge will have efficiently churned the water. Once on the landscape, the water takes a long time to drain to the river or the ocean on this gently sloping landscape.

Rocky coasts

Rocky coastlines are highly crenulated, with many abrupt headlands separated by coves that are home to pocket beaches. The sand of the littoral zone is not continuous and should not be considered the littoral super-highway it is on sandy coastlines. Understanding the evolution of these complicated landscapes requires understanding erosion of rock by waves. In general, rocky coastlines bound landmasses that are rising relative to the sea. In other words, these landscapes are experiencing rock uplift, due either to local tectonics, as in California and Japan, or to glacio-isostatic rebound, as in Maine, for example. The combination of rock uplift and glacial–interglacial oscillation of sea level generates the marine terraces that are so useful in documentation of uplift rates and patterns (see Chapter 3).

Just as in the fluvial and glacial environments, erosion of rock by waves occurs by three processes: abrasion, quarrying and dissolution. The last of these is much less viable in the case of coastal erosion, as seawater already has a high ionic concentration, slowing chemical reactions. (An exception however is salt weathering in the sub-aerial environment above the waves, promoted by the growth of salt crystals in the wetted and dried surface of the rocks (see Chapter 7)). Again, quarrying or plucking of rock is most efficient. The process of hydraulic wedging can enhance this as described in the fluvial environment. The source of the vibration of the rock mass is even easier to imagine in the coastal setting, as the impact of individual waves can be quite high.

While we are far from a detailed understanding of the processes of rock erosion, we can cast the problem as one in which the erosion is associated with the delivery of wave power to the rock. Here we bank on the analogy with stream power, which is similarly distant from specific processes. This allows decomposition of the problem into the following pieces: (1) the power in the wave per unit length of wave is determined by winds in the source area of the waves, which sets the wave heights emerging from the storm. This wave energy is retained as the wave traverses the ocean toward a coastline. (2) The distribution of power evolves as the wave shoals, which dissipates energy, and the wave front refracts as it experiences variable bottom topography. The bottom topography will be shifted according to the level of the sea, which in turn is controlled by both the local tide and any storm surge. (3) Shoaling eventually topples the wave into a breaker, which very efficiently dissipates energy, depositing some fraction of it as work on the underlying substrate. If sandy, this energy is available to transport the sediment. If rock, it can abrade or pluck the rock.

A typical ocean swell has a period of 10–20 seconds. This means that in a day the coast will see about 6000 waves, or in a year 1–3 million waves. Anyone who has been to the beach knows that waves differ from day to day and season to season. Just as in the fluvial case, then, we must acknowledge the stochastic nature of the system, and describe it statistically. The coastal community relies heavily on statistics reported from wave buoys. These are anchored in deep water well away from shoaling effects of the nearshore. A typical report from a buoy at Monterey California (Figure 18.20) shows the distribution of wave heights from each direction to characterize the incoming wavefield.

The question is which waves accomplish the majority of the geomorphic work in the coastal setting. In the discussion of sandy coastlines we found a strong dependence on the angle of wave approach. In the case of rocky coastlines, one might surmise that the waves must actually have access to the rock in order to damage it. We will see that this is both right and wrong. In order to assess the power delivered to the coastline, Adams et al. (2002, 2005) installed seismometers on the coastal cliffs near Santa Cruz (see also Chapter 18). They found that (1) the cliff was jolted at high frequency by each impacting wave, showing a response amplitude that depended upon the far-field wave height and the tide (Adams et al., 2003), and (2) the cliff swayed with an amplitude of about 10 microns with each passing wave (Adams et al., 2004). The former jolting would likely allow blocks to be loosened. If the wave doing the impact is charged with sandy sediment, the sand can cause abrasion. And any cracks in the bedrock can become filled with sand that will act as a hydraulic wedge to help pry new blocks from the wave-cut platform. If beneath too much sand to allow the wave direct access to the bedrock, only the last of these three processes will be active. The problem is closely akin to that in the fluvial realm where bedrock erosion can be accomplished when the bedrock is exposed to the erosional processes of the flow itself. In contrast the rocking and swaying of the cliff occurs whether there is a sandy beach or not. The length scale over which the swaying decays inland appears to be several cliff-heights, and is set by the elastic properties of the substrate. The edge of the terrestrial landscape is being elastically loaded and unloaded at 10–20 second intervals. This can act as a fatigue mechanism. If for example the cliff erosion rate is 10 cm/yr, and the elastic effects occur as far inland as 30 m, the rock mass arriving at the coast will have been processed through 3 million cycles per year × 300 years, or roughly 1 billion cycles. Adams et al. hypothesize that this cycling will fatigue the rock mass, effectively preparing it for removal upon exposure to direct wave attack.

Coastal littoral sand budget

A littoral cell is defined as being a self-contained unit of coastline within which the sand sources and sinks are contained. It is sketched in Figure 16.22. The top of a cell is defined as a line across the coast across which sand is not transported. The bottom of a cell is commonly a site where sand is lost from the littoral system down a submarine canyon. In other words,

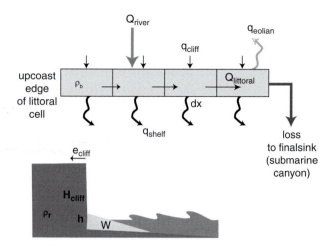

Figure 16.22 Schematic diagram of sediment budget within a littoral cell, seen in plan view (top) and in cross section showing beach wedge (bottom). Sources of sediment include contributions from cliff retreat and rivers, while sinks include losses to the shelf in suspension, and loss to submarine canyons. In the case illustrated, the submarine canyon dictates the downcoast edge of the littoral cell. At steady state the volume of sand in the beach in each portion of the cell will remain the same, meaning that inputs and outputs of sand must balance. While not shown here, one could also cast this problem as a balance of each of many grain sizes, or of particular minerals.

again sand does not cross the line normal to the coastline, but its instead lost from the system. Sand moves along the beach or littoral system by longshore drift, being pushed by waves one little hop at a time. We may formalize the sediment balance in a portion of the cell just as on a segment of hillslope. We must consider both the sources and sinks of sediment, and the rate at which it is passed from one point to another within the system (the fluxes).

As usual, the art lies in choosing what to balance or conserve: we will balance the volume of sand-sized sediment, that large enough to be retained within the beach or littoral system. Grains finer than a particular "cutoff diameter," which may depend upon the wave energy available to cause significant suspension, will be lost to the shelf and will no longer count in the littoral sediment balance. For simplicity of discussion, let us call the grains large enough to be retained "sand," although in for example the Santa Cruz coastline of California, the cutoff diameter is about 0.2 mm. The sources of sediment include both rivers and sand from cliff retreat. Sinks include any loss to the shelf (we could call this leakage), or to eolian

dunes onshore, and the ultimate sinks are submarine canyons that intercept the littoral drift.

In words, we may write the balance as

rate of change of mass of sand in the box = rate of inputs of sand to the box – rate of loss of sand from the box

Translation into mathematics yields

$$\frac{\partial M}{\partial t} = Q_x - Q_{x+dx} + q_{cliff} dx - q_{shelf} \, dx$$
$$- q_{eolian} dx + Q_{river} \tag{16.19}$$

where the terms in Q are mass discharges [= M/T], and those in q are specific discharges [= M/LT]. Reference to the definition in Figure 16.22 shows that the mass of sand in the cell of length dx is $M = \rho_{litt} A dx = \rho_{litt} W \bar{h}_{litt} dx$. This serves to organize our thoughts about what we must know about a specific field example. A river contributes sand to the littoral cell at a point, at a rate that reflects the basin area, A_b, the mean erosion rate in the basin, $\dot{e}_b$, and the fraction of the products that are sand, f_r:

$$Q_r = \rho_r f_r \dot{e}_b A_b \tag{16.20}$$

Cliffs contribute sand at a rate dictated by the cliff recession rate, the height of the cliff, and the proportion of that cliff material that is above the grain size threshold, also shown in Figure 16.22:

$$q_{cliff} = \dot{e}_{cliff} \, \rho_{cliff} \, f_{cliff} \, H_{cliff} \tag{16.21}$$

All of these terms can now be inserted into the final equation for the sediment budget in the littoral cell. Doing so, and dividing by both the distance dx and the bulk density of littoral sand, ρ_{litt}, yields

$$\frac{\partial A}{\partial t} = -\frac{1}{\rho_{litt}} \frac{\partial Q}{\partial x} + \varepsilon_{cliff} \left(f_{bed} H_{cliff} \frac{\rho_{cliff}}{\rho_{litt}} \right)$$
$$+ \frac{\rho_{river}}{\rho_{litt}} \left[\sum_i f_{riv,i} \dot{e}_{b,i} A_{b,i} \delta(x - x_i) \right] \tag{16.22}$$
$$- \frac{\rho_{com}}{\rho_{shelf}} q_{shelf}$$

Note that this is an evolution equation for the cross-sectional area of sand in a segment of beach, A, and that we have ignored for the moment the potential loss to eolian dunes. While the equation looks complicated, it explicitly accounts for all of the processes involved, and the relevant erosion rates of both cliffs and river basins. As in our treatments of hillslopes and glaciers, for example, if we assume that the

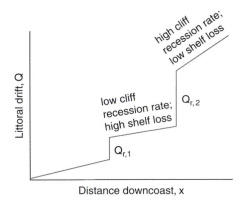

Figure 16.23 Expected down-cell pattern of the steady-state sand discharge. In this hypothetical case, two rivers contribute point-sources of sediment, the second stronger than the first. The sand contributed from cliff recession is likewise stronger in the second segment than in the first.

system is steady, we may ignore the time-varying term on the left-hand side, and integrate the remaining expression to obtain the expected spatial pattern of beach sand discharge. Between point inputs of sand from the rivers, the discharge of sand down-cell must simply be the integral of the contributions for the cliff back-wearing and losses to the shelf. If the back-wearing and shelf loss are uniform over this interval, this leads to a linear increase in discharge: $Q = (\dot{e}_{cliff} f_{cliff} H_{cliff} - q_{shelf})x$. At the rivers, the discharge must experience a jump or step comparable to the fluvial input, Q_r. The resulting pattern of down-coast discharge, displayed in Figure 16.23, is a linear increase punctuated by steps at the river mouths. This kind of pattern has been observed in the Santa Cruz coastline, as documented by Perg *et al.* (2003) who combined a mixing model with concentrations of cosmogenic radionuclides in the littoral sand and its sources to document the rates involved (see Chapter 18).

While to our knowledge this has never been done, one *could* write a similar balance for each of several grain sizes, and for each of several minerals. This would allow a formal framework in which to address the issues of grain size and mineralogical evolution in a littoral system. As these characteristics of the beach are commonly and easily documented, they would provide strong constraints on models of provenance, and strong tests of our understanding of grain size evolution in beaches.

Pocket beaches and headlands

Rocky coastlines are quite complicated, and display highly crenulated map forms. Rocky headlands punctuate the coastline, isolating stretches of coastline that are mantled instead by sand. The coastline therefore alternates between smooth and rough.

One of the more spectacular features of such coastlines is spiral or hooked bays. An example from Half Moon Bay, California, is shown in Figure 16.24. These have been a curiosity for decades, and have been shown to obey a geometric form called a logarithmic spiral. This is defined by the simple equation

$$r = r_o e^{\theta \cot \alpha} \tag{16.23}$$

where r_o is the length of a ray from the center of the spiral near the head of the bay to the shoreline, θ is the angle from the initial ray, and α is the angle between the shoreline and the ray from the bay center and sets the tightness of the spiral (Figure 16.24). The upcoast, y, and inland, x, distances of points along the spiral from the bay center are

$$y = r \cos \theta$$
$$x = r \sin \theta \tag{16.24}$$

Fitting for the best parameters r_o and α, and allowing the entire pattern to rotate yields remarkably good fits for these features, as can be seen for the Half Moon Bay case in Figure 16.24. Such hooked bays have subsequently been modeled (an early example is Rea and Komar, 1975) using wave refraction and sediment transport algorithms.

While these spiral bays may be one of the most elegant features on rocky coasts, there is significant order to the other patterns produced by rocky headlands. The beaches (small isolated ones are called pocket beaches) are commonly asymmetric. The headlands serve to refract waves in their lee that then produce predictable patterns of sand transport. The asymmetry of sediment transport produces a grain to the coastline that can easily be read for the dominant wave approach angle.

We can understand these features as equilibrium forms, just as we can understand parabolic hillslopes as equilibrium or steady-state forms. In steady state there can be no net erosion or deposition at any site along the coastline. In other words, $\partial A/\partial t = 0$, where A is the cross-sectional area of the littoral sand at a point along the coast. This in turn requires that the

Figure 16.24 Half Moon Bay, California, localized by the headland at Seal Point, and a set of log spirals with varying characteristic angles, α (see inset for definitions of parameters that define the spiral). The curve with $\alpha = 41°$ appears to be the best fit to the coastline within the bay, in accord with the results of Yasso, 1965 (image from Google Earth, with credit to NASA).

longshore discharge of sand must everywhere be the same; if there are no strong sources of sand in the reach (no rivers, no cliff-generated littoral sediment), then $dQ_{litt}/dx = 0$ and $Q_{litt} = $ constant. We have seen that the littoral drift rate is controlled by the breaker height and the angle of the breaker with respect to the coastline. This implies either that the equilibrium coastline is adjusted so that the angle of the breakers with respect to the local shoreline is uniform along the feature, or that any variation in breaker height is perfectly compensated by variation in the incidence angle.

Icy coasts

Icy coasts such as those that ring the Arctic Ocean represent a special case of rocky coastlines. In many instances the coastal bluffs consist of ice-cemented sands and silts. These deposits represent old coastal plains that prograded onto the shelf during glacial times. They are underlain by permafrost that is hundreds to more than 1 km thick. Their bluffs point north, so do not experience direct solar radiation. Yet coastal bluff retreat as revealed by repeat air and satellite photography (e.g., Figure 16.25) has now reached astonishing rates of tens of meters per year in places.

While considerable ongoing research is focused on assessing the rates of retreat as these coasts unravel, there are two potential causes of the acceleration, both associated with the recent demise of the sea ice cover of the Arctic basin illustrated in Figure 16.26. Loss of sea ice allows waves to grow over a longer fetch, promoting stronger wave attack during storms over the Arctic Ocean. It also removes the reflective lid from the ocean, effectively replacing white with black. This change in albedo influences the radiation heat budget in the lower atmosphere. It also allows more solar heat to be deposited in the ocean itself. While one might expect the Arctic Ocean to be a perfect ice bath, at roughly $0\,°C$, summer

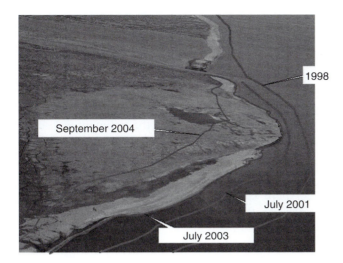

Figure 16.25 Coastal retreat of the order of many tens of meters per year along the north coast of Alaska, documented from repeat photography (courtesy Gary Clow, USGS, via Susan Flores, BLM).

temperatures near coastal Alaska can reach above 10 °C. The greatest recent temperature change in the Arctic climate is in the Fall, when the summer-warmed ocean water releases its heat to the atmosphere; in the past it did not warm much in summer and was fully lidded by sea ice in the Fall.

While we have discussed wave attack of rock earlier in the chapter, we have not discussed melting of permafrost by waves. If the problem is chiefly thermal, the essence of the problem is that the top or surface boundary condition has been changed by raising the temperature of the surface. Efficient mixing of the water column by the turbulence associated with waves assures that the cool water released by melt of the ice cement is removed rapidly and replaced by warm sea-water. The rate of melting also depends upon the temperature of the ice, as the ice must first be warmed to the melting point. The problem is analogous to the melting of icebergs once they are calved into the ocean from tidewater glaciers (Andrews, 2000). As ocean currents waft them into warmer water, their melt rate increases.

The geometry of the cliff recession is in places strongly controlled by the periglacial setting. The cliff retreat is quantized by the presence of ice-wedge polygons. As sketched in Figure 16.27, growth of a notch first undermines them as the ice cement is melted. As this occurs a lengthening mat of tundra vegetation hangs over the edge, demonstrating the strength of

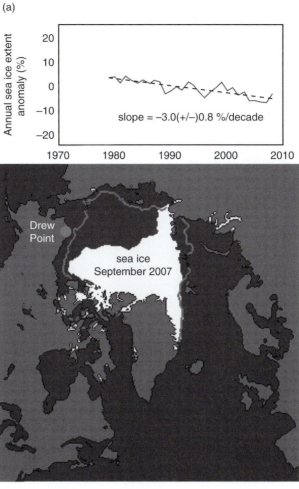

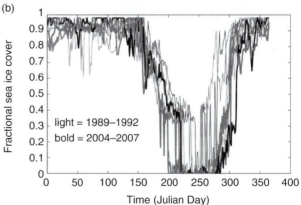

Figure 16.26 Sea ice context for acceleration of coastal erosion around the Arctic. Sea ice shown in map (middle) at minimum extent September 2007. Bold line depicts median sea ice minimum. Dramatic loss of ice in western Artic is evident. (a) Decline in sea ice extent averages 3% loss/decade (data and image from National Snow and Ice Data Center, NSIDC), (b) fractional sea ice cover along Drew Point, Alaska, coastline (dot in map). Early 1990s time series suggests 20–30 ice-free days, in contrast with the 50–70 ice-free days in the mid-2000s, during which coastal erosion can proceed.

the tangle of roots in this vegetation. Finally, the entire polygon tilts in a block failure, and subsequently melts rapidly as it is bathed in warm seawater. One such topple along the Beaufort Sea coast is shown in Figure 16.28.

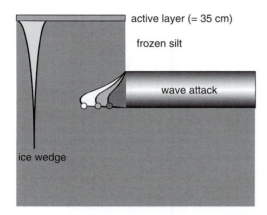

Figure 16.27 Undercutting of coastal bluff by melt of permafrost notch. Subaerial melt undermines the turf carpet in the active layer, lengthening the tundra carpet overhang. Much more efficient erosion occurs when sea ice leaves in mid-summer, allowing warm ocean water to melt ice in the permafrost. Inland movement of the pivot point (dots) as the notch grows eventually allows topple of a block of permafrost. These usually fail on ice wedges, suggesting that the bond between ice wedge and frozen silt is weak.

Several potential feedbacks exist that could damp the erosion rate. First among these is the deposition of sediment freed from the cliffs as the ice cement melts. If this sediment is coarse enough to remain in the littoral system as a beach, its growth through time could protect the cliff from attack by warm waves. In many places, however, it appears that the sediment is instead quite fine – silt likely deposited as loess. Upon release, this fine sediment is simply wafted in suspension well offshore to be deposited on the widening shelf. Hence there is no negative feedback to stall erosion of such coastlines.

The continental shelf

While today's coastlines are highly dynamic, we must also acknowledge that over the Quaternary all of the world's continental shelves have been repeatedly subjected to coastal processes as the coastline has oscillated between interglacial highstands and glacial lowstands. The shelf edge is easily seen in the global bathymetric map of Figure 16.29. It is not a coincidence that the edge of the continental shelves follows closely the 120–140 m contour, averaging

Figure 16.28 Coastline of northern Alaska near Drew Point, Smith Bay. Coastal bluffs topple when undercut by melting of a notch into the ice-rich permafrost. Sea ice is present at the time this photograph was taken, preventing further erosion until mid-summer. Tundra turf forms a strong mat that lengthens and droops over the block edge as the ice is removed by subaerial melting (photograph by R. S. Anderson).

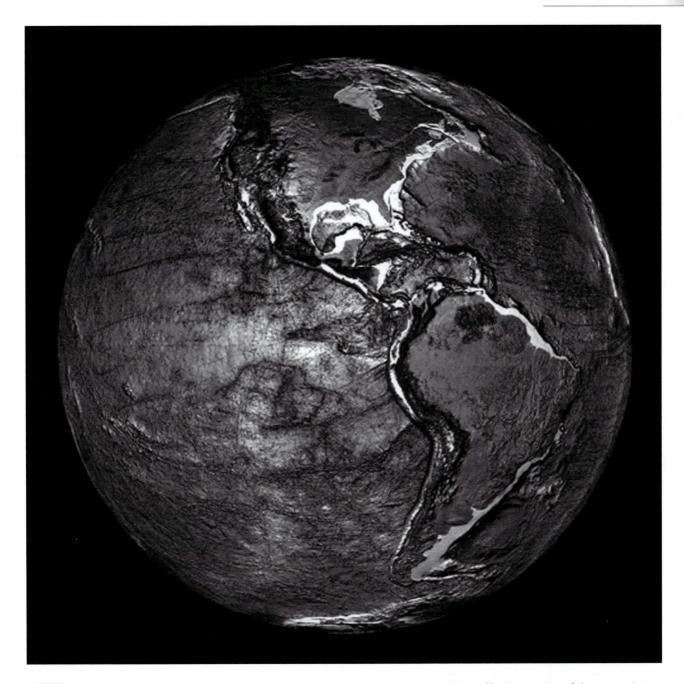

Figure 16.29 NOAA DEM of the globe highlighting the continental shelves. Note narrow shelves off active margins of the west coasts of the Americas, and wider shelves on the passive margins (from NOAA website, http://www.ngdc.noaa.gov/mgg/image/images/ etopo2v2-modis-globes/JPEGfull/).

135 m. The continental shelves average roughly 100 km wide, and comprise 7% of the present ocean area. But these are very productive seas, with high nutrient loads from the rivers, and the shallow waters assuring that sediment quickly settles to the bottom.

In the case of continental margins at which the rock is rising due to tectonic activity, the Quaternary oscillations of sea level have served to grate off a slice of rock with each cycle. At the inner margin of such shelves the record is left in a set of marine terraces, whereas on the shelf itself the record is lost. The opposite is the case on passive margins. There the long-term subsidence of the shelf is recorded in shelf deposits in which the littoral sediments walk back and forth across the shelf, burying sub-aerially deposited fluvial sediments that were laid on during sea level lowstands. The shelves and their inner and outer margins therefore represent the response of the geomorphic system to the oscillation of sea level in the last glacial cycles. The long-term evolution of the shelf has recently emerged as a research topic in geomorphology (e.g., see Fagherazzi *et al.*, 2003).

While our attention in this book must stop somewhere, and we have chosen to focus on terrestrial landscapes, it is nonetheless interesting to ask what the continental margins must have looked like prior to the great sea level oscillations of the Plio-Pleistocene. Just as the deep fjords outside of Antarctica did not exist prior to say 3 Ma, so too the continental shelves of the world must have been very different, in all likelihood much narrower.

Summary

Sea level varies at several timescales, from tidal to glacial–interglacial. We focused on the longer timescale variations in sea level that reflect variations in the volume of terrestrial ice. Sea level has bounced by 120 m over the last couple of million years, and was last at its 120 m lowstand at roughly 25–20 ka. All of the coastal processes we have discussed have therefore exerted their influence over the landscape of the continental shelves. They have repeatedly beveled and filled this important boundary to our continents. The recent rise in sea level, now at a little more than 3 mm/yr, is attributable to equal portions of warming of the oceans' surface layer, and further loss of terrestrial ice. The new water from ice comes from decay of both the great ice sheets, but importantly from the smaller glaciers, which now contribute 60% to that sea level rise attributable to new water.

We have described coastal forms on both sandy and rocky shorelines. We have shown examples in which one can make considerable progress by combining equations of conservation of mass with those for transport, in this case of sand. The transport of sand and the erosion of rock are driven by the delivery of power to the coast by waves. As in many systems we have addressed in this book, the geomorphic process is highly nonlinear: large waves carry much more of the power

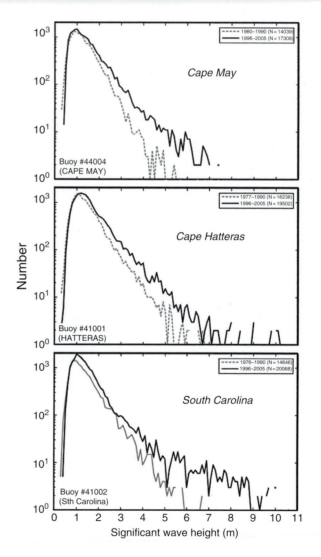

Figure 16.30 Distributions of significant wave height at three wave buoys on the east coast of the United States. Comparison of two decades-long measurement periods reveals a dramatic increase in large waves in the most recent decade (after Allan and Komar, 2006, Figure 6, with permission from the Coastal Education and Research Foundation).

than do small waves. This requires that we acknowledge the probability distribution of waves. Once again, as in our discussion of floods and rivers, and as in the discussion of rainsplash transport, a fair portion of the work is likely accomplished by the rare large events, here waves produced by cyclones either locally or elsewhere in the ocean.

The coastal system is packed with examples of self-organization, including beach cusps, flying spits, and capes in which the evolving shape of the coast feeds back on the pattern of wave energy delivered to it, which in turn fixes the pattern of deposition and erosion.

A large fraction of the world population of humans lives within a few meters of sea level. We therefore interact intimately with the ocean's edge. Many of these people live on the deltas of the great rivers of the world, whose low slopes and subsidence rates translate into high risk. Ten million people per year are inundated from the sea on these landscapes. Activities on land, including trapping of sediments in reservoirs, levee construction, and extraction of both water and hydrocarbons from the subsurface make these landscapes even more vulnerable to both floods from upstream and the arrival of water from the ocean. The human cost of these events is huge. In addition, the expense of maintaining infrastructure on coastal landscapes, both through sand nourishment and riprap on rocky coastlines, is great. These expenses directly reflect the dynamism of the coastal landscape, and its contrast with the permanence and inflexibility of human structures. Coasts change rapidly, at rates of meters/year that are not seen elsewhere in the terrestrial landscape. This is not surprising, given that (1) the energy available to accomplish geomorphic work on the coast arrives from all portions of the world's oceans in the form of waves that have not lost much energy since they were formed beneath the oceanic storms, and (2), as anyone who has been thrown about by a wave knows, water is heavy.

Finally, we have touched upon several aspects of the coastal zone that are sensitive to climate change. Not only is sea level rising, which affects the locus of wave energy deposition on the landscape, but the waves appear to be changing as well. As you can see in Figure 16.30, the spectra of waves impacting the coastlines of North America are changing. Large waves are becoming more common, presumably reflecting the increased likelihood of major storms.

Problems

1. *The steric effect.* Calculate the expected rise in sea level if the global ocean were to warm by $1\,^{\circ}$C. Take the mean depth of the ocean to be 3800 m, and the coefficient of thermal expansion to be 1–$2 \times 10^{-4}/^{\circ}$C (the coefficient increases with water temperature). Repeat the calculation assuming that only the well-mixed layer of 100 m thickness has warmed by $3\,^{\circ}$C.

2. Calculate the total volume of water (in km^3) that must have been added to the global oceans to cause the sea level rise associated with meltwater pulse I.

3. *Hyperpycnal flows.* Calculate the concentration of quartz sediment necessary to make the bulk density of river water equal to that of seawater ($1003\,kg/m^3$). (Concentrations exceeding this will generate hyperpycnal flows.)

4. Predict the spatial down-coast pattern of littoral drift on a 50 km reach of coast, assuming that the littoral cell is in steady state. Assume that the backwearing rate of the bounding 10-m tall seacliffs is 20 cm/year, and that they are composed of materials that are 20% sand, 80% silt and clay. The coastline is punctuated by two rivers, at 10 and 35 km down coast from the top of the cell, which deliver sediment that is 80% sand, at rates of $1.2 \times 10^5\,m^3/yr$, and $1.8 \times 10^5\,m^3/yr$ respectively. Assume that no sand is lost to the shelf.

5. Contrast the power arriving in waves generated from two storms at sea, one in which the significant wave height is $H = 3$ m, traveling at a group speed of 20 m/s, the other with wave height of $H = 4$ m, traveling at 30 m/s.

6. Predict the storm surge associated with a Category 4 hurricane, due to the following:
   (a) the low pressure of the eye of the storm;
   (b) sustained winds of 100 mph ($= 160$ km/hr). How far up the Mississippi river would this surge be felt, assuming that the slope of the river at its mouth is about 1 cm/km?

   Given the 20 km/hr speed of the storm normal to the coastline, sketch the expected pattern of storm surge along the coastline.

7. How would the heights of tides differ if the Moon were closer to the Earth? At the current recession rate of 3.8 cm/yr, how might the tides differ between now and 2 Ga? (The assumption of a steady rate of recession of the Moon is of course only a simplification.)

8. On a shelf sloping at 10 m/km, calculate the greatest rate of shoreline transgression associated with post-LGM sea level rise.

9. *Thought question.* Discuss how the changes in the distribution of waves revealed in Figure 16.30 might alter the coastlines they impact.

10. *Thought question.* Consider a changing world in which the coverage of the Arctic Ocean by sea ice is shrinking, as revealed in Figure 16.26. Discuss several feedbacks by which this should accelerate coastal erosion, given that the bluffs along the coast are simply frozen silt.

Further reading

Bascom, W., 1980, *Waves and Beaches*, Garden City, NY: Anchor Press/Doubleday, 366 pp.
This is a classic largely descriptive treatment of waves and how they form beaches. It is very readable and accessible to introductory students.

Komar, P. D., 1998, *Beach Processes and Sedimentation*, 2nd edition, Upper Saddle River, NJ: Prentice-Hall, Inc., 544 pp.
Written by one of the luminaries in the field, this textbook provides a modern highly quantitative synthesis of coastal geomorphology.

Sunamura, T., 1992, *Geomorphology of Rocky Coasts*, Chichester: Wiley and Sons, 302 pp.

Given how the Japanese coastline is dominated by bedrock forms, it is not surprising that one of the first syntheses of their origin and evolution comes from a Japanese scientist. This is well illustrated, quantitative, and provides both theory and examples of both scientific and engineering interest.

Trenhaile, A. S., 1987, *The Geomorphology of Rock Coasts*, Oxford: Clarendon Press, 384 pp.
This synthesis of rocky coastline processes and forms draws on more Canadian examples, and is perhaps more accessible to introductory students than is Sunamura.

CHAPTER 17
The geomorphology of big floods

What I have always loved best about the history of the world is that it is true. That all the extraordinary things we read were no less real than you and I are today. What is more, what did happen is often far more exciting and amazing than anything we could invent.

E. H. Gombrich, *A Little History of the World*, 2005

In this chapter

It is a favorite occupation of geomorphologists to sit around the campfire on a starry night in the field and contemplate geomorphic events that would have been fun to see – in other words, to argue about when and where it would have been interesting to be alive. One common answer to this question is during large floods. We focus in this chapter on a few case examples of landscape-altering floods. We first review why these events are worthy of study, besides their intrinsic excitement, and lay the historical framework within which such study has occurred. The largest of floods occur when dams fail, releasing ponded water in torrents. We discuss scaling arguments for the prediction of flood magnitude from the geometric properties of the lake and of the floodway. Within the case studies, we outline the techniques used when trying to recreate the floods, a subdiscipline known as paleoflood hydrology. The keys of the chapter will remain the stories of the largest floods and how we have come to know them. It is only when armed with a strong array of evidence that the scientific community has accepted most of these floods as having happened; they do indeed screech loudly against the hum of everyday experience to which our imagination is tuned.

Hidden Creek Lake immediately before (top) and after (bottom) draining through a tunnel beneath the Kennicott Glacier, Alaska. Ice of Kennicott Glacier is in foreground of top photo. Icebergs once floating in the lake are now grounded, providing evidence of the shoreline height. Maximum lake level drop is approximately 100 m (photographs by R. S. Anderson).

Why should we study large floods?

We have stressed in many places throughout this text that geomorphic processes are stochastic in one sense or another. Floods are a classic example. The study of flood frequency is an ancient one, and the terminology of the 100-year flood, the 50-year flood, and so on, is well embedded even in the public's lexicon. Yet most of this field of study focuses on meteorologically driven hydrographs, associated with spring snowmelt or with the sequence, duration, and intensity of rainfall events. These flood magnitudes generally define a well-behaved distribution, smoothly decreasing in frequency (and hence likelihood of occurrence) with flood size. In contrast, floods caused by failure of dams usually fall outside of this distribution, or at least fall in the rarefied tails of the distribution. So, happily, these are rare events. The reason they incite and demand interest on our part is that these very rare events can induce tremendous change in the channel system, and in some instances in the entire landscape downstream. Their effects can be so large that they dominate the look of the landscape for centuries and even tens of thousands of years. The lasting impact of large floods reflects the thresholded, nonlinear relationships between sediment transport and channel incision processes to the shear stresses at the boundary. Material transported during truly large floods may be immune to all but the largest floods (and weathering, of course). As a corollary to this, as we have noted elsewhere in this book, there are many landscapes in which even continuous monitoring of sediment discharge out of a basin for many decades fails to explain the long-term erosion rates of the basin, which we can now document using cosmogenic radionuclides within stream sediments (e.g., Kirchner *et al.*, 2001).

Finally, it will be seen that many of these events occur in the context of glaciation. This is not an accident, and should indeed be expected. Ice serves as an excellent dam (at least up to the point where the weight of the dam is buoyantly compensated by the impounded water). Glaciers and ice sheets disrupt drainages and pond water on the landscape. This occurs on scales that range from small ice-dammed lakes (Alaska has hundreds of these lakes; Post and Mayo, 1971), to continental scale paleolakes like Lakes Agassiz and Ojibway that were ponded by the retreating Laurentide Ice sheet. Moraine-dammed glacial lakes pose a serious hazard in deglaciating mountainous terrain (for example, see Clague and Evans, 2000). In addition, landslide dams serve to pond large lakes along some of the high mountains of the world (see Hewitt, 1998). The failure of these earthen dams can lead to major floods downstream. Furthermore, the transfer of water from the ocean into ice sheets on land lowers sea level dramatically. We will see that the 120 m sea level lowering during the Pleistocene led to the disconnection of the Black Sea from the Mediterranean Sea. The Black Sea refilled upon subsequent sea level rise during deglaciation, an event recently postulated to have taken place in a catastrophic flood.

While we focus on floods associated with glaciers and glacial times, we will also address cases in which the dams are landslides. In these cases one calamity can lead to another, a landslide acting to dam water that is then released rapidly downstream. The May 2008 earthquake in Sechuan province of China generated huge numbers of landslides, illustrating a yet longer chain of events: earthquake, landslide, flood.

A historical backdrop

From the first days of geology as a formal scientific line of inquiry, geologists have been on guard against any tendency to appeal to the rare event in explaining a geological or geomorphological feature. This was in large part a response to the prevailing thought in the seventeeth and eighteenth centuries in which Noah's flood, The Great Deluge, was commonly trotted out as an explanation for features such as large isolated rocks in alpine meadows. Louis Agassiz (1840) replaced this explanation with his glacial hypothesis, that these are glacial erratics that signify occupation of the landscape not by floodwater but by ice. The root of Lyell's approach (1833) was that observable forces and processes can explain geologic history, that the present is the key to the past. This uniformitarian view has a couple of different manifestations: uniformity of process, and uniformity of process rate. The former is essentially a restatement that physics is the same today as it was yesterday, and that physics, chemistry, and biological principles have not changed through time. The uniformitarian view of process rates is less easy to define, and is less defensible. Over what time period must one average a process rate in order to characterize it well? We have seen many examples and can imagine many more in which the averaging time must be very long in order to characterize well a stochastic process, even one that is purely meteorologically driven. This year's flood is certainly not the same as last year's, and hence the sedimentation rates in deltas, for example, are going to vary accordingly. On top of this, acknowledgement that the climate of the Earth has changed radically over timescales of tens to hundreds of thousands of years in the late Pleistocene ice ages requires that we ponder how such climatic shifts have changed the probability distribution of weather events that force the geomorphic system.

But the story we tell in this chapter goes a step further, and forces us to think about the role of events that we will likely not have a chance to witness in our lifetime. This had been the intellectual ground of the catastrophists in the past, those who would call upon Noah's flood or some other calamity that may be described in mythology but not documented in written human history. It is only within the last few tens of years that the Earth sciences community has been forced to think hard about the profound roles of truly rare and truly large events. The largest wake-up call was of course the Alvarez hypothesis, put forth in 1980, that a 10 km diameter bolide impacted the Earth to wreak environmental havoc sufficient to cause the demise not only of the dinosaurs but also of 75% of the species on Earth (Alvarez *et al.*, 1980). This event falls outside of human experience altogether, unless you wish to count the remotely observed machine-gun like pummeling of Jupiter by the strung out pieces of comet Shoemaker–Levy 9 in July 1994. This served to confirm that large impacts still occur, and that they have the capability of leaving their marks on large planets. So also do the large floods we will introduce here. The difference lies in the fact that smaller scale dam-burst floods do occur here on Earth. Their study has allowed us to explore the physical processes involved, so that we can scale up to the expected effects from the Bonneville flood, the Missoula floods, and now the proposed flooding of the Black Sea by the Mediterranean – a flood that may in fact be the origin of the Noachian flood myth. The acknowledgement of the roles of these events has forced the geological community to shift toward a mindset in which both the everyday events and the truly rare events that we never actually witness must

be mixed to yield the landscapes we observe both on Earth and on other planetary objects.

A recipe for truly big floods: a bunch of water, a breach of the dam

Extraordinarily big floods occur when water impounded by some sort of dam is released by a failure or breach of the dam. Rain-driven flood discharges are limited by the rainfall rate integrated across the watershed, and in most settings are damped by slow transmission through subsurface flow paths. As we have seen, storm discharge in channels usually is dominated by old water impelled out of storage during rainstorms, a process that releases water at a rate no greater than the input (rainfall) rate. Snowmelt and rain-on-snow floods can be larger than rain-driven floods, but are limited by the rate of snowmelt. That this rate is restricted to a few tens of centimeters per day, that the snow is distributed widely in the basin, and that the melt water so produced must travel at least some of its path down snow-laden hillslopes, means that the peak in the hydrograph from snowmelt cannot be as high as that created by the failure of a dam. In the case of dam failure, discharge is limited by characteristics of the lake and dam, rather than by meteorological forcing.

Dams come in several forms: glacier ice, landslides and moraines (some ice-cored), and log jams. Humans have engineered earthen and concrete dams, and human activities can influence the formation of log jams. Summaries of the largest floods in the Quaternary reveal that their discharges have reached tens of millions of cubic meters per second (O'Connor *et al.*, 2002). We will see that it is no coincidence that these have occurred during times of continental glaciation, as either the drawdown of sea level by 120 m, and its subsequent rise, or the blockage of river drainages by the ice itself have generated the largest of the floods.

Given the failure of a dam, what sets the magnitude of the flood? What sets its peak discharge? Or, more generally, what sets the shape and magnitude of the flood hydrograph? Is it symmetric? How long after the beginning of the flood is the peak? How drawn out is the receding limb? Imagine being an official in a town downstream from a dam. It is most useful to know the maximum flood discharge in the event of a dam failure, and how much warning one might have of such a catastrophe. Joe Walder and colleagues, of the US Geological Survey Cascade Volcano Observatory, tackled this problem recently (Walder and Costa, 1996; Walder and O'Connor, 1997). Using data sets assembled for both earthen and ice dam failures, they found a simple power-law relation between flood peak discharge, Q_{peak}, and the volume of the lake at the time of failure, V_{lake}:

$$Q_{peak} = aV_{lake}^{p} \qquad (17.1)$$

The power, p, in the relationship can be determined simply from the slope of the line fit through the data on a log–log plot shown in Figure 17.1. This empirical relationship is valuable, but still leaves us lacking a physical understanding of the phenomenon. What sets the power in this relationship? Can it be predicted from the physics of the dam failure process?

Let us first think about the earthen dam case. Lakes impounded by earthen dams drain by either the overtopping of the dam, and the scouring of a channel down into the dam surface, or by catastrophic failure of the dam by landsliding, perhaps induced by the seepage of water through the dam. In either case, the worst-case scenario, which is certainly the one of interest to the official in the town downstream, is one in which the dam is very rapidly removed. In this scenario, the physics of the enlargement of the breach is sidestepped, and the problem is simply one of calculating how fast water can get out of the opening.

First, as always, let us set up the problem as one of conservation of the volume of water in the lake. The word picture is as follows:

rate of change of water in storage = rate of water inputs – rate of water outputs

Or, put mathematically,

$$\frac{dV}{dt} = Q_{in} - Q_{out} \qquad (17.2)$$

where V is the volume of the lake, Q_{in} is the volumetric rate of inputs from streams upstream from the dam, and Q_{out} is the output through the breach. In the case of a catastrophic flood, the outputs are much greater than the inputs and we can simplify the equation accordingly by ignoring the first term on the right-hand side. The quantity of greater interest than the lake volume, and the one more easily measured, is the height of the lake. We can convert between lake

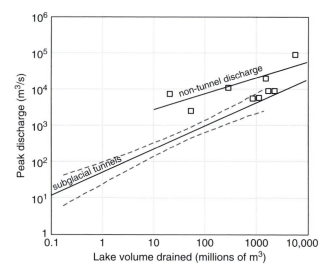

Figure 17.1 Power-law relationship of peak flood discharge to lake volume for glacially dammed lakes. Those that drain through subglacial tunnels appear to produce slightly lower peak discharge than those that flow in open channels (after Walder and Costa, 1996, Figure 2, with permission from John Wiley & Sons).

volume and lake height knowing the hypsometry of the lake, or the lake area as a function of altitude. For example, if the lake basin were vertically walled, the volume is simply $V = A \times H$, and the simplified equation becomes

$$\frac{dH}{dt} = -\left[\frac{1}{A}\right]Q_{out} \qquad (17.3)$$

If on the other hand the lake basin were a cone, with a mean slope of the lake floor of α, then the lake area and lake volume become

$$A = \frac{\pi H^2}{\tan^2 \alpha} \qquad (17.4)$$

$$V = \frac{\pi H^3}{3 \tan^2 \alpha} \qquad (17.5)$$

and the differential equation becomes

$$\frac{dH}{dt} = -\frac{\tan^2 \alpha}{H^2}Q_{out} \qquad (17.6)$$

As long as the hypsometry is known, an expression for conservation of water can be crafted along the lines of the one shown above for the cone-shaped lake. What we need to know now is what controls the discharge of water through the breach in the dam.

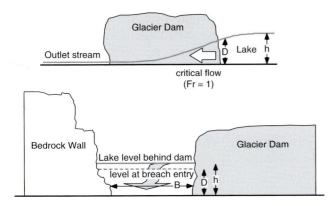

Figure 17.2 Geometry of glacial outburst seen in cross-valley view (top) and up-valley view (bottom). Flow through the breach in the ice dam is presumed to go critical (Fr = 1), allowing simplification of the calculation of peak discharge (after Walder and Costa, 1996, Figure 5, with permission from John Wiley & Sons).

The discharge is the product of the cross-sectional area of the breach and the mean velocity through the breach. If the breach cross section is fixed at the outset and does not change, then the problem becomes simply one of calculating the expected mean velocity. One may turn to the physics of flow through a weir, as the breach serves effectively as a weir to the flow out of the lake. Within a weir, the flow must go through the condition we call "critical." Equivalently, somewhere within the weir the Froude number, Fr = 1, where $Fr = \bar{U}/\sqrt{gh}$, or the ratio of the mean velocity of the flow, with the velocity of a shallow water wave in the flow. Here g is the acceleration due to gravity, and h is the flow depth, which in general need not be the full lake depth (see Figure 17.2). At critical flow, Fr = 1, and the mean velocity $\bar{U} = \sqrt{gh}$, with only minor adjustments associated with the details of the geometry of the weir. If the mean width of the breach is W, and the water depth in the lake above the base of the breach is h ($= H - z$, z being the altitude of the base of the breach), the cross-sectional area of the breach is Wh, and the discharge of water through the breach becomes

$$Q = Wg^{1/2}(H - z)^{3/2} \tag{17.7}$$

Using H_{max} as the maximum lake elevation, one can calculate the worst-case scenario discharge to which the river downstream would be subjected. As the lake level drops, the discharge will also decline in a predictable way. We can now fill in our equation for lake volume, and arrive at an equation for the evolution

of the lake surface elevation. For the simplest case of a vertical walled lake basin, this becomes

$$\frac{dH}{dt} = -\frac{W}{A}g^{1/2}(H - z)^{3/2} \tag{17.8}$$

We can learn a fair bit from simple inspection of these two equations. First, the greatest rate of loss of water from the lake will occur just after the failure of the dam, when the value of $H - z$, the head driving the flow, is greatest. In this simple basin, the drawdown rate of the lake will be greatest at the beginning of the flood, and will taper off with time. If we insert a slightly more realistic lake basin geometry, the equation for the drawdown rate becomes

$$\frac{dH}{dt} = -\left[\frac{\tan^2 \alpha}{\pi H^2}\right] Wg^{1/2}(H - z)^{3/2} \tag{17.9}$$

Note the tradeoff between the rapid drawdown driven by the great head difference in the early going, which varies as $H^{3/2}$, and the H^{-2} dependence that comes from the lake geometry. As the lake declines in elevation, the same volume loss leads to greater elevation drop. In real basins, one would document the relationship between lake surface area and water elevation in the basin (the hypsometry of the basin) and insert this $A(z)$ dependence into Equation 17.8.

Paleoflood analysis

Since megafloods are extremely rare, much of our understanding comes from interpreting the mark of these huge events on the landscape. In this context, it is appropriate to ask what happens to the water down the flood path. Paleoflood analysis (see Kochel and Baker, 1992) uses erosional and depositional features to deduce the magnitude, peak discharge, and duration of past floods. The first task is to assemble the clues about the width, depth and slope of the flow. This type of analysis has been applied to many events from the past. For instance, working along much of the flood path along the Snake River, O'Connor (1993) mapped sites of significant scour, sites of deposition, and quantified the grain size of the debris in the deposits. Using these data, it is straightforward to calculate the peak discharge of the river at points along its path. With these constraints in hand, one can use flood routing methods to calculate the expected shape of the hydrograph as it travels down the Snake River.

Slackwater and separation eddy deposits

The flood height can be reconstructed from a variety of evidence. Scour lines, where regolith has been stripped from the landscape, can be very persistent and are easily observed. Although usually found deeper in the flow, anomalous boulders deposited by floodwaters can be identified either from anomalous grain size or lithology. But in the ancient record, one of the most useful markers of paleoflood altitude is slackwater deposits. These form where the flood has either backed up water in a tributary, or the flood itself has traveled up a tributary. In either case, the slowing of the flow results in deposition of the suspended sediment. These distinctive deposits often fine upward, reflecting the waning of the flood and the raining out of finer and finer sediment from the water column. In addition, and very tellingly, these deposits often show up-tributary migration of bedforms such as ripples. Other useful indicators of flow height are separation eddy deposits found behind obstacles at the edges of the flow. A bedrock protrusion, coarse debris flow fan, or similar object jutting out into the current cause separation of the flow, and a downstream eddy forms. The slower velocities in the eddy promote deposition of the suspended material near the top of the flow, which act as a record of the flow height.

Estimates of flow competence

By estimating the flow depth at many cross sections, the slope of the floodwater surface can be estimated. With these in hand, it is possible to estimate the shear stress imposed by the flow on the bed of the channel: $\tau_b = \rho g H \sin(\theta)$. Using the entrainment condition discussed elsewhere, we can then estimate the largest clasts that the flood should have been able to transport. Of course, one can also turn this on its head, and invert the observed sizes of clasts that were clearly in transport to calculate the available shear stress. Since the slope of the channel can be surveyed, this flow competence can then be used to estimate the flow thickness, H, even in the absence of flow indicators such as scour lines or slackwater deposits.

Paleodischarge estimates

Using this and the cross-sectional area of the channel at a station, A, one can estimate the flood discharge using any of several flow equations. The discharge

will be $Q = A \times U$, where U is the mean velocity. This latter can be obtained using the Manning's, Darcy–Weisbach, or Chezy equations we discussed in Chapter 10. A slightly more sophisticated approach entails use of a numerical model to capture the evolving flow. For example, using the flood depth deduced from high water marks, O'Connor and colleagues have employed the step-backwater model to calculate peak flood discharges in both the Bonneville and Missoula floods.

The Bonneville flood

Let us now consider an example in which a lake flooded over an earthen spillway. We have already encountered Lake Bonneville several times in this text. It has been a focus of much geomorphic and paleoclimatic work since G. K. Gilbert wrote about the system in the first monograph of the United States Geological Survey in 1890. Gilbert identified two major shorelines from a Pleistocene lake. The shorelines are plastered onto the Wasatch Front, and ring many of the ranges in the eastern Basin and Range province, scribing the margins of a 47 000 km² lake. He named the higher of these the Bonneville and the lower the Provo shoreline. The difference in altitude between the two is about 100 m. Gilbert hypothesized that the lake catastrophically dropped from the Bonneville to the Provo shoreline, the water escaping the basin through Red Rocks Pass northward toward the present city of Pocatello, Idaho, there feeding into the Snake River. Over the last century, geomorphic and paleoclimatic investigations have illuminated the character and timing of the flood. Within the basin, dating of the shorelines and of the lake depositional sequences has constrained the lake level history in the Bonneville basin, shown in Figure 17.3 extraordinarily well, using ^{14}C methods. We know that the lake slowly rose through roughly 15 ka to reach the Bonneville level about 15 ka. Around 14.5 ka the Bonneville shoreline was abandoned and the Provo shoreline was formed; this was the time of the flood. (As an aside, because the date of this flood is so well known, and because the Bonneville deposits are so tightly constrained in time, it has led to the use of the Bonneville shoreline deposits and shoreline notches as calibration tools for a variety of dating techniques; among these are cosmogenic nuclide

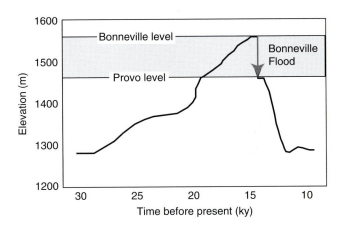

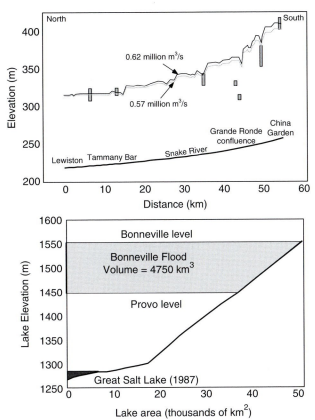

Figure 17.3 History of Bonneville lake level as constrained by ^{14}C dating of shorelines and lake deposits. Abrupt drop in lake level from Bonneville to Provo shorelines corresponds to the Bonneville Flood. Subsequent lake level drop is caused by changes in the water balance within the basin (redrawn from Currey, 1990, Figure 11, with permission from Elsevier).

Figure 17.4 (a) Paleoflood levels of the Bonneville flood down a 60 km reach of the Snake River, along with calculations of discharge. Note roughly 100 m flood depth (after O'Connor, 1993, Figure 49). (b) Elevation of Bonneville and Provo shorelines and associated lake areas. The 100 m drop from Bonneville to Provo shorelines corresponds to a lake volume loss of 4750 km^3 (O'Connor, 1993, Figure 50, following Currey, 1990).

production rates.) Detailed mapping of the shorelines and knowledge of the topography of the basin reveals that roughly 4750 km^3 of water was drained from Lake Bonneville in the flood.

Harold Malde (1968) was the first to study the flood path from the Red Rocks Spillway in any detail. His report opened the eyes of the geological community to the magnitude of the floods that can be generated by such events. O'Connor (1993) is the latest to have assembled the paleoflood information, providing the most recent picture of the Bonneville flood. As seen in Figure 17.4, the flood depth was of the order of 100 m as far downstream as the present Oregon – Idaho border, more than 100 km downstream from the Red Rocks Pass. The various flow depth data can be accommodated by a model in which the peak discharge is roughly 0.6 million m^3/s, or 0.6 Sv (a Sverdrup, Sv, is an oceanographic flow unit, equal to 1 million m^3/s, a scale appropriate to ocean currents such as the Gulf Stream). This is a boggling discharge for a river flood. A minimum duration of the flood can be determined by dividing the total flood volume (4750 km^3) by this peak discharge. Recalling that 1 yr $\approx \pi \times 10^7$ s, we see that the estimated duration of 8×10^6 s is almost a quarter of a year. Flood routing calculations provide a more complete picture of the flood wave as it progressed down the Snake River. According to these calculations, the flood crest arrived at Farewell Bend, about

800 km from Red Rock Pass, about 12 days after the initiation of the flood.

The stream power (Figure 17.5) and the shear stresses exerted on the bed were tremendous, reaching up to 3000 N/m^2. Given this, what is the likely competence of the flow? What grains might be suspended, and what should be traveling as wash load? As the shear velocity can be calculated from the shear stress through $u_* = \sqrt{\tau_b/\rho}$, and the condition for suspension is that the shear velocity exceed the settling velocity, one can solve for the grain whose settling velocity is just small enough to be suspended. For the maximum Bonneville flood conditions, O'Connor estimates that grains smaller than about 100 to 200 mm ought to have been suspended (!) and that 1 mm grains ought to constitute the wash load. These inferences are corroborated by inspection of the slackwater and eddy deposits.

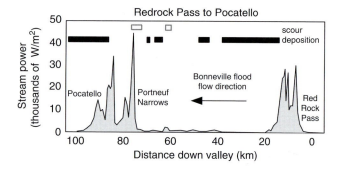

Figure 17.5 Estimated stream power along 100 km reach of Bonneville flood path down the Snake River. Sites of scour correspond well with reaches in which stream power is increasing, while sites of deposition correspond to reaches of decreasing stream power (after O'Connor 1993, Figure 55a).

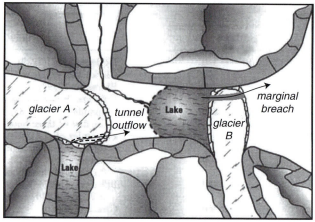

Figure 17.6 Geometries of glacially ponded lakes and the flood paths from them. Trunk valley glacier, A, blocks tributary valley, forcing breach through a subglacial tunnel. Glacier in tributary valley, B, blocks trunk valley stream, allowing breach of dam against the far wall in a marginal breach (after Walder and Costa, 1996, Figure 3, with permission from John Wiley & Sons).

It is no wonder, then, that this sediment-laden flow highly scoured the flood path. It has left a record that is easily read today in the landscape. The aridity of the local climate so greatly limits the weathering rates that scoured surfaces are still evident, and regolith stripped in the flow has not been replaced in 14.5 ka. This and the fact that prominent many-meter diameter boulders, known as the Melon gravels, sit atop these scoured surfaces in some places, and are amalgamated into coarse boulder bars in others, are testimony to a flood that must have been a sight to behold.

Glacial floods: Jökulhlaups

In alpine settings, there are several geometrical arrangements of glaciers in valley systems that can lead to the damming of either trunk or tributary streams, and hence to the generation of lakes capable of producing glacial outburst floods. These are sketched in Figure 17.6. These floods are also called jökulhlaups, an Icelandic word for glacier (*jökul*) flood (*hlaup*). The geometry of any particular lake depends upon the timing of the occupation of tributary vs. trunk valley with glaciers. Walder and Costa (1996) have shown that both styles of blockage are common. The mode of failure of the ice dam varies as well, having two modes. One is analogous to the earth dam failure mechanism, in which overtopping of the dam leads to a feedback process that causes increasing size of the breach, the flood flowing in a subaerial channel. This is most common when the trunk valley is dammed by a tributary, as the low spot on the ice

dam is against the far wall. While the mechanics of the erosion of the breach are different, the ice melting in this case while in the earthen dam case the sediments comprising the dam are entrained, the flow mechanics are similar. One can appeal to the weir calculation, and again the worst-case scenario is one in which the breach very rapidly attains a maximum depth and width.

The second failure mechanism is more particular to glaciers. It involves the breakage of a subglacial seal, and the ensuing growth of a subglacial conduit, or tunnel, through which the flood water is delivered directly to the terminus of the glacier (see Nye, 1976; Clarke and Matthews, 1991). This is a more common mechanism for the case in which the trunk valley glacier dams up a tributary, out of which the tributary glacier has already receded. Here again a strong feedback mechanism operates to allow the rapid dumping of the lake water. The key to the erosion of ice channels is the fact that the water can melt the channel walls, leading to enlargement of the channel or tunnel. The larger the channel is, the greater the discharge of water it can accomplish. Melt is promoted by two factors: (1) water coming from the lake may be above the melting point, and (2) potential energy in the water is convertible to thermal energy, which is delivered to the walls by turbulence of the flow. The flow in the pipe is driven by a pressure gradient, with atmospheric pressure at the exit to the tunnel, and pressure associated with the depth of water in the lake at the entrance to the tunnel. One other phenomenon must be accounted for in this mechanism. The glacier ice creeps toward the tunnel at a rate dictated by the difference between the tunnel pressure and the ice overburden. When the pressure gradient driving the flow through the tunnel backs off as the lake begins to draw down, the tunnel walls creep into the conduit, countering some of the melt feedback, and ultimately shutting off the flood. It is this creep closure of the tunnel walls that allows a glacier to generate flood after flood. It can re-plug the leak and do it all over again, in contrast to the one-time nature of the failure of an earthen dam.

The Lake Missoula floods and the channeled scablands

Now we are in a position to explore the famous Missoula floods responsible for carving the "channeled scablands" of eastern Washington (Figure 17.7).

The problem of interpreting the scablands was tackled as a life-long quest by J. Harlen Bretz, who published a series of papers in the 1920s through 1960s on the topic (Bretz, 1923, 1925, 1928, 1969; Bretz *et al.*, 1956). Any geomorphologist who travels through this country of dry canyons cut into the Columbia River basalts and of smooth wheat-farmed hills cannot fail to ask the question of their origin. Bretz brought a vast array of field observations to the table as evidence that there had to have been a very large flood across this landscape. He mapped the anastamosing set of scoured channels that dissects this wheat country. He described megaripples, standing 30 m high with wavelengths of the order of 100 m, composed of basalt cobbles. He mapped the shapes of hills that look to be streamlined by a flood from northeast to southwest. He argued that the dry box canyons known locally as coulees – of which the Grand Coulee, dammed in the 1930s WPA project, is one – were fluvially carved, and that even if the Columbia River had been diverted temporarily it would not have cut these canyons. He traced the evidence of the flood through Wallula Gap near the town of Walla Walla, through the Dalles of the Columbia River, to

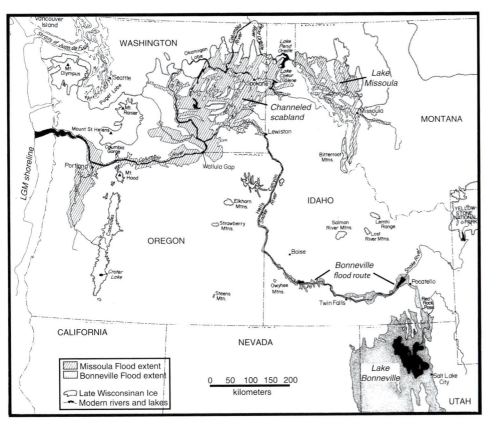

Figure 17.7 Map of major Late Pleistocene Lakes and floodpaths to the Pacific. Lake Bonneville shown at its highstand lake area dominates northwest Utah, while Lake Missoula occupies valleys in western Montana. The Bonneville flood was a single event, and traveled down the Snake River. The Missoula floods (also called the Spokane Floods) spread out over eastern Washington, through the Channeled Scabland, before recombining to flow down the Columbia to the Pacific. These were numerous floods, separated by decades of lake-filling behind an ice dam formed by a lobe of the Cordilleran ice sheet (after O'Connor, 1993, Figure 1).

Figure 17.8 Multiple Lake Missoula shorelines above the present city of Missoula, Montana, nicely revealed by incomplete snow cover. Shorelines on Mt. Jumbo. Highest shorelines are 290 m above the city (photograph by Don Hyndman, courtesy of the University of Montana, Missoula, Montana).

a fluvial delta at the mouth of Oregon's Willamette Valley. Indeed, these patterns are remarkably evident from space, a vantage point that Bretz did not have.

Yet in a classic example of enforced uniformitarianism, he was loudly and summarily dismissed by some of the most prominent geologists of the day. The kinds of flows Bretz called for were huge. They smacked of the kinds of catastrophes the geological community had held at arm's length since Lyell. And he had not identified the source of the water. The debates raged at professional meetings and in the scientific literature, arguments against the Bretz megaflood hypothesis often reaching the absurd, and seldom being based upon field investigation. Bretz only asked for the ideas to be tested against the field evidence. Although he too was befuddled by the lack of an obvious origin for sufficient water, Bretz nonetheless argued that an "enormous volume, existing for a very short time, alone will account for their existence" (Bretz, 1925).

It took work in the Missoula area, performed largely by Pardee in the 1920s through 1940s (Pardee, 1942), to identify a possible source of water. There he found obvious signs of a large lake in shorelines on the valley walls above the town of Missoula, nicely depicted in low-snow conditions in Figure 17.8. When

mapped, these demarcated a very large lake indeed, one that could hold more than 2000 cubic km of water. Glacial Lake Missoula appeared to have resulted from the damming of the Clarke's Fork River by the Purcell Lobe of the Cordilleran ice sheet. Bedforms within the lake basin, for example those in the Camas Prairie in westernmost Montana and shown in Figure 17.9, spoke of rapid, deep flow of the water toward the exit. That the lake outlet path led to the channeled scablands was supported by the discovery of a further set of megaripples.

Only when a field conference took geologists to the field in August 1965, a conference that Bretz was unable to attend due to poor health, was he vindicated. The years since have seen the publication of significant observational and theoretical work on the more frequent modern glacial outburst floods from Iceland's Vatnajokul, bringing this process into the mainstream. Within a few years, too, Malde (1968) was to publish about the dramatic effects of the Bonneville flood down the Snake River.

The scale of the proposed Missoula Lake flood is even more staggering than the Bonneville flood. High-water evidence indicates that the water was over 150 m deep at the dam site, and more than 200 m thick as it passed through the Wallula Gap (O'Connor and

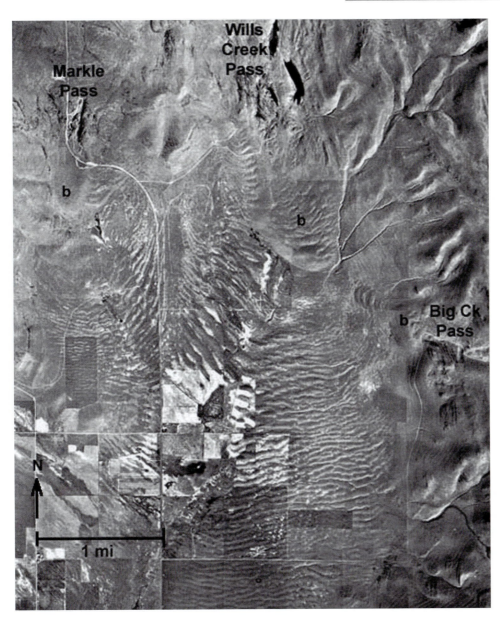

Figure 17.9 Camas Prairie and its giant ripples, used as evidence of the Spokane megaflood.

Baker, 1992). Peak discharges of the order of 15–17 Sv just downstream of the ice dam site, and about 10 Sv at the Wallula Gap constriction, have been proposed (see Figure 17.10). This is roughly an order of magnitude greater than the peak discharge proposed for the Bonneville flood. One can also estimate the duration of the flood by dividing the lake volume by this peak discharge:

$$T_{\min} = \frac{V}{Q_{\text{peak}}} \tag{17.10}$$

This yields a minimum duration, as the rising and falling limbs of the hydrograph would have lower

discharge, but it nonetheless establishes the timescale. The lake volume is roughly 2700 km^3. If we take an average of the various estimates of peak discharge to be 13×10^6 m^3/s, this yields an estimate for the flood of about 2×10^5 seconds, or about 3 days. Call it a week for the whole event to take place. That would have been one spectacular week to be alive in what is now eastern Washington.

But the story does not end there. A curiosity remained. Bretz called upon one or perhaps two floods to carve the scablands. Others working in the Wallula Gap area were left pondering the meaning of the odd stratified silts and sands called the Touchee beds.

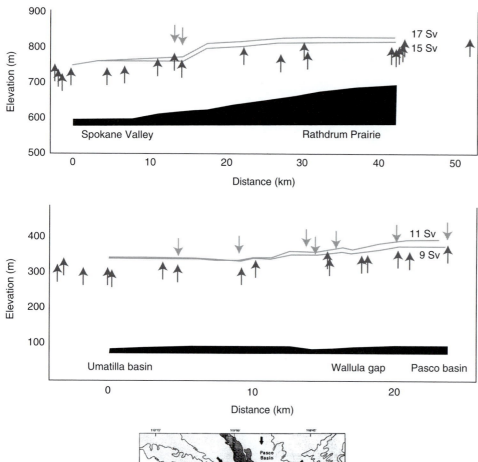

Figure 17.10 Mapped cross sections across the Missoula flood channels (bottom, through Wallula Gap – through cross sections noted on map), and peak discharges (in Sverdrups, Sv) during a Missoula Flood, calculated using step-backwater methods. Note the flood depth of ∼150 m in Spokane Valley (top) and 200 m depth through Wallula Gap (bottom). Arrows represent upper and lower bounds on peak stage (after O'Connor and Baker, 1992, Figures 2, 3 and 7).

Something like 50 beds, with consistent enough bedding features to be called rhythmites, had been identified; all are about the same thickness of a few tens of centimeters, and all fine upwards. Early workers called upon oscillations or sloshing of a single Lake Missoula flood to cause the rhythmicity. The reinterpretation of these beds, accomplished by Waitt (1985), relies upon not only these rhythmite slackwater deposits in the scablands region, but upon a fuller record from within the Missoula basin. The essence of

this record is reproduced in Figure 17.11. Waitt interprets each of these upward fining beds to represent a single outburst flood, the fining occurring as the flow wanes, the coarser particles settling out first. The Lake Missoula record comes not from shorelines on the valley walls but more recently discovered varved lake deposits in the depo-center of the basin. Here too there are tens of packets, each consisting of 40–60 varves. The packets fine upward as well, and then abruptly coarsen as a new packet begins. Waitt interprets each

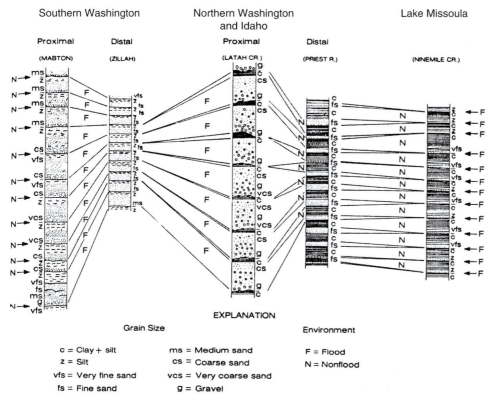

Southern Washington | Northern Washington and Idaho | Lake Missoula

Proximal (MABTON) | Distal (ZILLAH) | Proximal (LATAH CR.) | Distal (PRIEST R.) | (NINEMILE CR.)

EXPLANATION

Grain Size

c = Clay + silt
z = Silt
vfs = Very fine sand
fs = Fine sand

ms = Medium sand
cs = Coarse sand
vcs = Very coarse sand
g = Gravel

Environment

F = Flood
N = Nonflood

Figure 17.11 Correlation of rhythmites of Lake Missoula lakebeds (right) with slackwater deposits of northern Washington (middle) and southern Washington (left). Each set grades from coarse to fine, and is separated from the next by an abrupt unconformity back to coarse. This is primary evidence for there having been many such floods (after Waitt, 1985, Figure 17).

of these packets to represent a lake-filling episode. As the lake shoreline becomes more distant from the sample site on the lake floor, the sediments capable of reaching the site decline in size. The abrupt coarsening as the next set begins reflects the fact that the flood has occurred and the shoreline is again much closer to the site. The combination of these two data sets, and their seeming corroboration of many tens of floods, stand as strong evidence for repeated outburst flooding from a major glacial lake whose dam is capable of both catastrophically releasing water, and of repairing itself. The number of varves in the lake-filling sequences can be used to assess the interval between floods: 40–60 years. This story of multiple Missoula floods is further supported by the discovery of a well-preserved Mt. St. Helens ash bed (dating to 13 ka) in the midst of the Touchee beds; it lies below 11 and above 28 of the rhythmic beds. Waitt argues that the likelihood of preservation of an intact ashbed in the midst of a single flood is negligible. The ash must have fallen between flood events, when the landscape was subaerial. These events therefore dominated the last glacial maximum in this region, between roughly 15.3 ka and 12.7 ka, with a

pacing that would have allowed most generations of any humans living in this region to witness at least one flood.

More recent research on deposits found in the subsurface beneath Wallula Gap demonstrates that this entire ice dam configuration may have existed in earlier glaciations as well, generating an earlier set of floods (see Pluhar *et al.*, 2006). So the story continues to be extracted from this interesting landscape.

The further record of these events, both Bonneville and Missoula floods, is found offshore in the Astoria submarine fan as gravel packages. While these floods were huge, and are probably the largest floods that western North America has witnessed in the last few hundreds of thousands of years, the processes by which they operate are reproduced in many other smaller sites. For example, the Alsek River of westernmost Canada has left a legacy of flood scars along its path, and a record in the Gulf of Alaska (Clarke and Matthews, 1981). Peak discharge could potentially have been somewhere between Bonneville and Missoula in size. Megafloods have also been hypothesized to have occurred on the Tsangpo River, Asia

(Montgomery *et al.*, 2004), where again glaciers are invoked to have stored the waters.

Lakes Agassiz and Ojibway

One other glacial outburst megaflood must be addressed, this time with implications for the climate of at least the landmasses adjoining the northern Atlantic. This flood is associated with the very extensive proglacial lakes Agassiz and Ojibway (called "superlakes," see Clarke *et al.*, 2003) that fronted the Laurentide ice sheet as it retreated from its Last Glacial Maximum extent (Figure 17.12). The story begins however in Greenland. In ice cores extracted from Greenland in the late 1990s, the detailed record of climate on the summit of Greenland was documented in detail. While the Holocene record was remarkably stable (i.e., boring), it was interrupted by a single large cold spike that is clearly evident in Figure 17.13. Alley *et al.* (1997) interpreted this ice core record as a 4–8 °C drop in temperature in Greenland that lasted roughly 200 years, and corresponded to dry, dusty conditions, and an increase in forest fire frequency that has been documented in many sites around the northern Atlantic (e.g., see review in Alley and Agustdottir, 2005). This "8.2 ka climate event" is now thought to have resulted from the catastrophic draining of Glacial Lakes Agassiz and Ojibway through the Hudson Strait into the Labrador Sea. Until this time, drainage from these proglacial lakes had been routed through the Great Lakes region to the east, and then out to the North Atlantic through the St. Lawrence River, and/or through the Mississippi River to the Gulf of Mexico. Barber *et al.* (1999) documented a marine record of this outburst flood in deep sea cores from the Hudson and Labrador Straits as a marker bed consisting of distinctive hematite-rich sediment (hence red) observed in all cores over 700 km away, whose source is presumably a hematite-rich till in the Hudson Bay region. This sediment layer, up to 80 cm thick in the western Hudson strait, and thinning to 5 cm well into the Labrador Sea, is thought to represent the sediment-laden plume associated with the flood. Using maps of the lake shorelines that date to that time, and corresponding maps of the LIS margin, they also crudely estimated the volume of water involved in the flood to be 1–$2 \times 10^{14}\, m^3$, or 1–$2 \times 10^5\, km^3$. At the time of the proposed

(a)

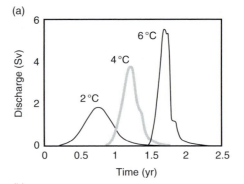

(b)

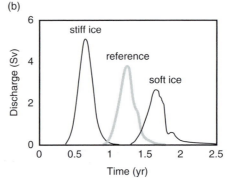

Figure 17.12 Sensitivity of simulated hydrographs from Lake Agassiz flood to (a) lake temperature and (b) the flow-law parameter for the ice (after Clarke *et al.*, 2004, Figure 2 and Figure 9a,c, with permission from Elsevier). See also Figure 17.14.

catastrophic drainage, the shoreline was more than 175 m above the contemporaneous sea level. Modeling of the dispersion of the sediment plume suggested that the outburst lasted less than 1 year, corresponding to $6 \times 10^6\, m^3/s$ mean discharge, or 6 Sv. This is of the same magnitude as the estimated peak discharges of the Missoula floods. They also tightly constrained the timing of the event by carefully dating sediment above and below the marker horizon, properly accounting for local variations in the reservoir effect needed to correct ^{14}C ages (see Chapter 6). They arrived at an age of 8470 years BP, corresponding nicely to the beginning of the 200-year long 8.2 ka climate event in the Greenland ice cores.

The importance of this release of water is that it was fresh water, and would have formed a significant fresh water cap over the portion of the Labrador Sea that is responsible for forming North Atlantic intermediate deep water. This formation of deep water is one of the cylinders of the big engine that drives the thermohaline conveyor belt. The weakening of the driving of the conveyor belt would have reduced the northward delivery of warm surface water to the North Atlantic, which in turn would have allowed

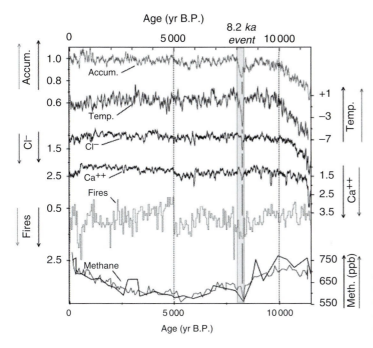

Figure 17.13 Summary plot of Holocene record in the Greenland ice core from GRIP. The 8.2 ka event is clearly revealed in the accumulation rate, temperature, and Cl and Ca time series (after Alley and Agustdottir, 2005, Figure 2, with permission from Elsevier).

the cooling of the adjacent landmasses. The resulting climate cold spike is by far the largest cold event in the Holocene record.

But the details of how such a flood could occur had not been addressed. Was a flood of such magnitude plausible? Clarke *et al.* (2004) have modeled the outburst flood from these proglacial lakes. In fact, they argue that such a flood is not only plausible, but is inevitable, given the geometry of the decaying ice sheet over the landscape of northern North America. Using a physics-based model of evolving flow through a subglacial conduit that pulls from our best analog for these events in the Icelandic jökulhlaups, they assessed the potential flood hydrographs from six different potential floodways beneath the decaying ice sheet. As seen in Figure 17.14, most of these hydrographs peak at 5–7 Sv. Not surprisingly, the peak discharge is sensitive to the assumed temperature of the lake water (higher temperature water allows the conduit to grow most rapidly) and to the assumed stiffness of the ice – stiffer ice (the higher flow-law parameter in Glen's flow law) reduces the rate of ice creep toward the tunnel allowing larger conduits. As we have seen in other estimates in this chapter, the duration of the flood can be scaled by $T = V/Q$. Given their calculated lake volumes of $100\,000\,\text{km}^3$, 5 Sv results in an estimated flood duration of about two-thirds of a year. Interestingly,

most of the simulated hydrographs displayed complex multi-peak behavior corresponding to growth and decay of constrictions along the flood path. In addition, most of the calculated floods ended before the lake had fully drained, meaning that estimates of sea level rise from the lake volume alone are overestimates.

Such estimates of abrupt sea level rise have sent the geological community on a search for independent confirmation of sea level rise associated with this flood. The resulting estimates vary, which has led to estimates of the lake volume that range from $1\text{–}5 \times 10^{14}\,\text{m}^3$. Kendall *et al.* (2008) have shown why these estimates might vary so greatly. Using a detailed model of viscous response of the Earth to spatially and temporally varying glacial unloading (e.g., Peltier, 2004), corresponding water reloading of the oceans, and this addition of new water from lakes Agassiz and Ojibway, they predict how spatially variable the sea level rise might be in this interval. Because the northern hemispheric landmasses are rebounding so steeply due to deglaciation at that time, the sea level "fingerprint" of the 8.2 ka climate event is highly contaminated by the glacial isostatic rebound signal in most northern continental edges. This is evident in the global map of Figure 17.15. Instead, they advocate hunting for a less contaminated record on the edges of the southern landmasses.

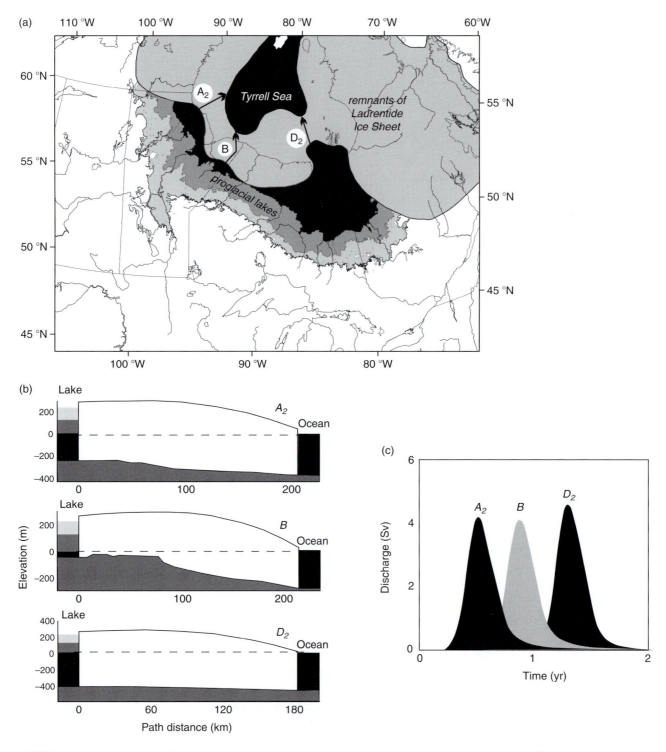

Figure 17.14 Lakes Agassiz and Ojibway amalgamated, shown at time of proposed megaflood at roughly 7700 ^{14}C year BP. (a) Map shows configuration of remains of the Laurentide ice sheet at that time, with three possible flood routes draining the lake to the south of the ice to the Tyrrell Sea to the north. Lake filled to Fidler and Kinojevis levels is shown in dark and medium gray; that part of the lake below sea level at that time is painted black. (b) Cross sections through the ice along these flood paths, showing lake levels, up to 220 m above the sea level at the time, the ice surface and basal topography. (c) Calculated hydrographs of the floods along these paths. Each hydrograph is shifted by 0.5 years to allow distinction between them. Floods along these paths would have lasted about 1 year with a peak discharge of between 4–5 Sv (after Clarke *et al.*, 2004, Figures 3, 6, and 7, with permission from Elsevier).

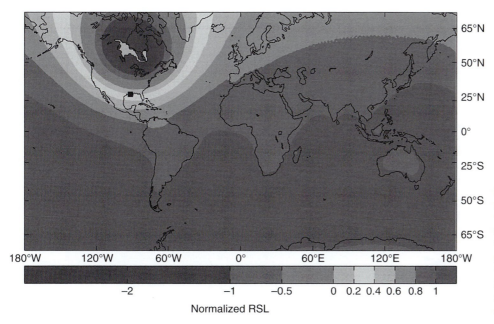

Figure 17.15 Modeled relative sea level rise due to flooding from Lake Ojibway-Agassiz at 8.2 ka. Values normalized by the expected RSL due to ongoing glacial isostatic rebound at that time. RSL signal from the flood is masked by rebound signal in northern continents (Kendall *et al.*, 2008, Figure 2).

The English Channel reinterpreted

Collection of detailed bathymetry in the English Channel separating England from France has revealed what appear to be fluvial channels carved by large floods (Gupta *et al.*, 2007; Figure 17.16). The scour features include elongated mid-channel hills akin to those Bretz used to argue for the Missoula Floods. The authors argue that the source of water was a lake formed by portions of the Fenno-scandian ice sheet to the north and east, and an arch of land that stands high because it reflects the structure of an anticline noted in Figure 17.17. It is this Weald-Artois arch that was overtopped by the lake waters, resulting in the flood through what we now call the Straits of Dover. While the timing for the carving of the canyon is less well constrained, it is believed that the first flood to scour the channel was likely during a major glaciation between 450–400 ka, when the Fennoscandian ice sheet extended well past its LGM moraines and into the north-central European plain. An inset channel cut into an abandoned bench, clearly visible in Figure 17.16, is potential evidence for a second major flood.

As in the case of the Lake Agassiz-Ojibway mega-flood, this event is triggered directly by an ice sheet, and its ability to block drainage of water from the land in its path toward the ultimate baselevel of the ocean. The hydraulic head that drives the flow is established by the difference in elevation of the lake and the ocean. When blocked, the lake can lose water only by evaporation, while it continues to receive runoff from both northern central Europe and the ice sheet itself. The ocean, on the other hand, will be drawn down well below modern sea level because so much water is temporarily stored as ice on the continents.

Noah's flood

The next example is another corollary to the situation of a lake and an ocean during a glacial cycle. While it is of only minor local human interest that the floods in the sparsely and newly populated American West occurred, it is of great interest that a huge flood could have occurred in the eastern Mediterranean region. Not only does the Bible have a flood "myth" in Noah's flood, but so also do many of the civilizations that cropped up around Europe and the Middle East at the beginning of the agricultural revolution, some 7000 years ago. Through a combination of coincidences and both hard and creative scientific work, William Ryan and Walter Pitman, of Lamont Doherty, have pieced together knowledge of an event about that time that may have given rise to all of these

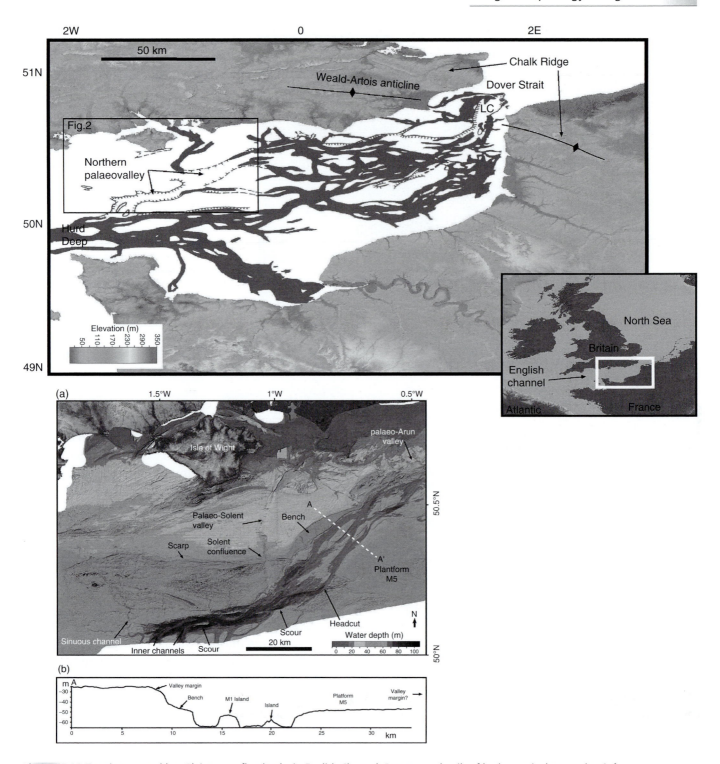

Figure 17.16 Top: interpreted late Pleistocene floodpaths in English Channel. Bottom, a: details of bathymetric data used to infer floodpaths. Bottom, b: cross section across floodpath showing inset bench and deeper channel cut into it (after Gupta *et al.*, 2007, Figures 1 and 2, with permission from Nature Publishing Group).

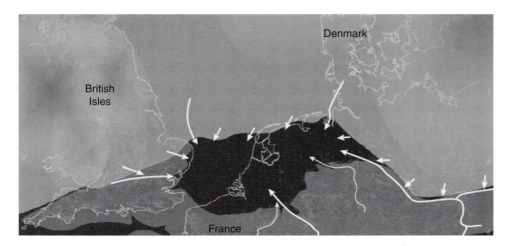

Figure 17.17 Proposed lake whose drainage through the Dover Straits initiated the English Channel. Ice from the Fennoscandian ice sheet extensions into eastern England and into central western Europe diverted rivers into the lake. The arch of land that acted as the southwestern dam for this lake is supported by the Weald-Artois anticlinal ridge. This is proposed to have been breached first during a glaciation between 450–400 ka (after Gibbard, 2007, Figure 2, modified from Cohen *et al.*, 2005, with permission from Nature Publishing Group).

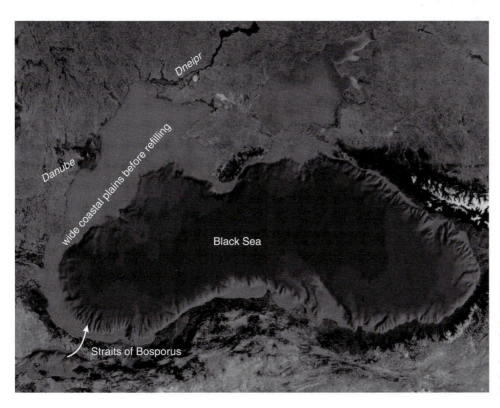

Figure 17.18 Image of the Black Sea and surrounding landmasses. It is connected to the Mediterranean Sea through the Bosporus Strait. Note the broad shallow shelves on the west and north that would be exposed at lowstands of the sea (NASA image).

stories. It involves the Mediterranean Sea and the Black Sea, which are presently separated by first the Dardanelles Strait, and then the Bosporus Strait, separating Greece from the Turkish peninsula. This is shown in the map in Figure 17.18. If we turn the clock back 7000 years, one of the chief differences around the world's oceanic edge is the location of the shoreline. At sea level lowstands associated with glacial maxima, the eustatic sea level stood 120–150 m below present. As the base of the strait between the Mediterranean Sea and the Black Sea is above this, the two would have been separated during major glacial epochs. The Black Sea would have become a land-locked freshwater lake fed only by the local rivers such as the Danube, the Dneiper, and the Don. Its level would have become dynamically controlled not by the world's oceans and hence global ice volume, but by the local water balance. Like any other lake, its level would have risen during times of more runoff input than evaporation, and, conversely, would have fallen when the evaporation exceeded runoff from the contributing rivers. What happens when sea level rises upon the ebbing of the continental scale ice sheets depends upon the integrity of the pathway that the Mediterranean water would take through the Bosporus gateway.

Ryan and Pitman claim that the sea level rise resulted not in a gentle refilling of the Black Sea, but a catastrophic flooding of the Black Sea, at about 7150 BP (Ryan *et al.*, 1997, 2003). The story they flesh out in their book *Noah's Flood* (1999) is not only geologic, but ends in educated speculation about the effects this flood might have had on the region's civilization. Using geophysical techniques, they document a major erosional unconformity through the strait. They also show through detailed bathymetric mapping and dredging of deposits that the filling of the lake from about 150 m below present to within 15 m of the present sea level had to be very rapid. Nowhere is there a transgressive "systems tract" indicative of slow onlap, or slow upward and outward motion of the coastal zone. Everywhere, both in the deep basin and on the shelf (above −150 m), there is an abrupt transition of sediment type, fauna, and isotopic signature. The sediment abruptly becomes a sapropel, jelly-like organic rich ooze. On the shelf this overlies sediments that indicate an arid landscape, with loess plains crossed by fluvial systems, and evidence of desiccation. The fauna change abruptly from freshwater dwellers to those tolerant of Mediterranean salinity. And the timing is everywhere the same; when datable shells are found, they do not range widely from 7150 ^{14}C years. At this time, sea level would have been about 15 m below present, as shown by the work on Barbados corals by Fairbanks (1989), as discussed in Chapter 16. The ancient shoreline of the Black Sea is found at 150 m below present.

Let us consider the problem quantitatively. We would like to know what the peak discharges might have been through the Bosporus Strait, how fast the level of the Caspian Sea must have been rising at its greatest rate, and how long did this flood last? The problem is only slightly different from those we have talked about earlier in this chapter. We are considering a flood from one basin to another, through a

breach whose geometry we have to specify. We can use the same hydraulic physics. What differs is that the water supply is so large that the filling of the Black Sea will not significantly draw down the Mediterranean Sea, which is linked to the global oceans. The flood will rage at first as the water level difference in the two water bodies is then at it greatest, and will tail off as the water level difference, the slope of which is driving the flow, declines toward zero. We again appeal to the water balance Equation 17.1, the volume V now being that of the Black Sea. The outputs from the Black Sea will be zero, while the inputs will include both the rivers and the flood through the Bosporus Strait. For now we will ignore the river discharge during the flood, an assumption we can revisit. In order to proceed, we need a means of converting the water volume to water elevation, and we need an expression for the input by flood discharge. As the lake area at pre-flood conditions is not that greatly different from the present lake area, we can consider the volume to be simply the pre-flood area, A, times the water thickness above this pre-flood altitude, h_0. Then the differential dV/dt becomes $A\, dh/dt$, and both sides can be divided by A to leave dh/dt, the rate of Black Sea level rise. We again appeal to the physics of water passing through a weir of width W (Equation 17.7) to produce a final expression for the expected rate of gain of water level during the flood:

$$\frac{dh}{dt} = \frac{1}{A} W (H - h)^{3/2} g^{1/2} \qquad (17.11)$$

where $(H-h)$ is the maximum difference between the levels of the Mediterranean Sea and the Black Sea. Remember that this is a worst-case scenario: the breach is assumed to open instantaneously to its maximum depth. By inspection, the flood magnitude, and hence the rate of rise of the Black Sea, should be greatest at the outset, when $(H-h)$ is greatest. This rate should decline as the water level difference $(H - h)$ declines. We can scale the peak discharge by estimating the width of the Bosporus Strait (700 m, the narrowest reported reach), and the height difference driving the flow (about 120 m). The peak discharge becomes about 3×10^6 m^3/s, a discharge that lies between that of the Missoula and Bonneville floods.

The duration is a little harder to estimate without a full calculation, but as we did for the Missoula floods, we can crudely estimate the timescale by dividing the total volume necessary to fill up the Black Sea at this

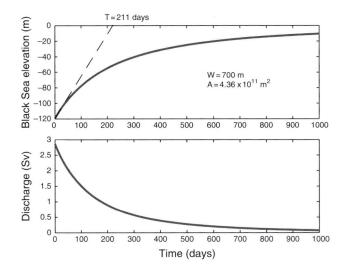

Figure 17.19 Calculated discharge through the Bosporus Strait (bottom) and the elevation of the Black Sea as it is refilled by the Mediterranean in the proposed megaflood. Assumed dimensions of the breach and the area of the Black Sea are listed. The initial rate of rise of the shoreline would have been 120/211 m/day, or a little more than 0.5 m per day.

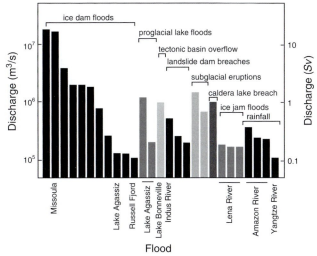

Figure 17.20 Floods within the Quaternary with peak discharges exceeding 100 000 m³/s. Floods are arranged by mechanism, with major floods discussed in the text labeled (after O'Connor et al., 2002, Figure 3, with permission from the American Geophysical Union).

peak discharge. We know that this will be a lower bound on the true timescale, as the discharge will decline through time. If the area, A, is about 4.36×10^{11} m² and the height difference is 120 m, then the volume change is about 5.2×10^{13} m³, and the timescale for the flood (Equation 17.10) becomes $T = 5.2 \times 10^{13}/2.8 \times 10^6$ m³/s $= 1.8 \times 10^7$ s. Given that a day is about 0.8×10^5 s, this timescale is equivalent to about 220 days. The full calculation of Black Sea level rise history, based upon Equation 17.8, is shown in Figure 17.19. One can see that this scaling has done well to establish the basic scale of the graph. The timescale T is shown on the plot as well; the actual time to achieve almost full parity of sea levels will be several times this timescale.

We emphasize that this sort of calculation is an end-member case. In other words, the assumption that the Bosporus Strait instantaneously breached to 120 m water depth is undoubtedly not true. The importance of the calculation is that it bounds the answer, in this case on the high discharge side, and the short duration side. In any case, it looks like the Black Sea could be filled in a matter of several months, or perhaps a very few years, depending upon the rapidity of the erosion of the sill.

Note that the initial rate of rise, at least in this end-member calculation, is a little more than 0.5 m/day.

While not in itself horrifying (even a sloth can climb a tree at that rate), it is easy to calculate that, on a very low-sloping coastal plain adjacent to the ancient pre-flood shoreline, this 0.5 m/day rise rate translates into about a kilometer of horizontal travel per day that would have been necessary to keep one's self and one's flocks from drowning. We will leave it up to the curious reader to follow up on the consequences of this flood, as sketched by Ryan and Pitman in their very readable and intriguing book. Let it simply be said that the rise of early Neolithic farming culture within Europe appears to have occurred at about this time, and that a story bearing striking similarities with the Noachian flood myth is to be found in many cultures in Europe and the Middle East.

Floods from the failure of landslide dams

While many natural dams are made of ice, most are constructed of landslide debris. Landslides fall from the adjacent valley sides to form a temporary dam. They are common, and they range widely in scale (see for example Cenderelli, 2000; Clague and Evans, 2000; and Hewlitt, 1998). Ouimet and Whipple (2008) report that the rivers draining the eastern edge of the Tibetan Plateau are pummeled by landslides, the long profiles of the rivers being punctuated by the boulder-strewn remnants of the landslides. The longevity of such a dam depends upon its size and upon the nature of the debris. Similarly, the ultimate release of the natural reservoir from behind the dam can be either slow or rapid. The hazard associated with the landslide itself is far smaller than the hazard posed by the potential flood, if for no other reason than that the footprint of the flood greatly exceeds the footprint of the landslide. As one example, a great flood on the Indus River, caused by the catastrophic breaching of a landslide dam in the middle Indus gorge, wiped out the entire Raj army encamped on the open floodplain at Attock, several hundred kilometers downstream.

Summary

We recount these tales of huge floods (many of which are listed in Figure 17.20) for several reasons. In some instances these floods have etched indelibly into the landscape patterns that have not been erased in several thousand years. Nothing in our lifetime will rival them in scale; no meteorological setting can produce such floods. Instead it takes the ponding of water behind some breakable dam. Within the past couple of tens of thousands of years, there were several settings in which this phenomenon occurred: high lake levels associated with a different water balance in glacial times overtopped their alluvial earthen enclosures; lobes of continental scale ice sheets dammed off glacial drainage; and sea level drop isolated basins that used to be connected, which were reconnected upon sea level rise. These events not only have left their marks on the landscapes of the world, but also in some cases were potentially of sufficient magnitude to alter global climate or to cause significant displacement of human cultures.

In modern times the only rivals to the magnitudes of these floods will be failures of human-made dams, of which there are presently tens of thousands in the world. The means of calculating the maximum possible discharges from hypothetical dam failures parallel those we have employed in this chapter. Note that once again we have employed a conservation equation in setting up the flood problem.

These lake outburst floods force the modern geomorphologist, and for that matter the modern Earth scientist, to acknowledge the role played by very rare events. A lesson from Bretz's experience is that one must listen hard to the clues whispered by the landscape. As we approach new worlds in our travels through the solar system, this open mindset will be even more crucial to maintain. Odd and spectacular things do happen. It is indeed the case, as Gombrich pointed out in his *Little History of the World*, that "what did happen is often far more exciting and amazing than anything we could invent." And what fun it would have been to see one of these floods!

Problems

1. You are the mayor of a town immediately downstream from a glacier with a side-glacier lake that could catastrophically drain through a subglacial tunnel. The geometry of the lake basin is such that failure will most likely occur when the lake has achieved a level corresponding to a lake volume of 100 million m^3.

   (a) What is your estimate of the peak discharge of the flood?

   (b) What is your estimate of the duration of the flood, given the lake volume and peak discharge? Report your answer in days.

2. Estimate the duration of the Lake Bonneville flood, given the volume available between the Bonneville and Provo shorelines, and the estimated peak discharge. Report your answer in years.

(a) How much would sea level rise due to the Bonneville flood?

(b) How much would sea level rise if a flood of 120 000 km³ occurred (one estimate for the size of the Lake Agassiz flood)?

3. Given the map of the expected response to the Lake Ojibway flood reproduced in Figure 17.15, where would be a good site to measure the true sea level rise associated with the flood?

4. Calculate the peak discharge to be expected from the proposed catastrophic refilling of the Black Sea through the Bosporus Strait. Assume that the minimum width of the strait is 700 m, and the difference in level between the Mediterranean and Black Sea was 120 m (see Figure 17.18).

(a) At this discharge, what is the expected rate of rise of the Black Sea?

(b) Given a shelf slope of 1m/km, how rapidly would the shoreline advance across the shelf?

(c) How much would global sea level drop as this water drains into the Black Sea (in other words, is the assumption that sea level is not affected a valid one?)

5. *Thought question.* Imagine that you are a manager in a town downstream from a moraine-ponded lake. Discuss the strategy you would use to assess the height above the river that you would deem "safe" in the advent of an outburst flood from the dam.

6. *Thought question.* What is your favorite flood tale? Summarize its main features, how we know about it, and review the evidence we use to deduce its peak discharge.

Further reading

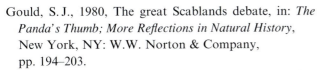

Gould, S. J., 1980, The great Scablands debate, in: *The Panda's Thumb; More Reflections in Natural History,* New York, NY: W.W. Norton & Company, pp. 194–203.
Gould tells the story of J. Harlen Bretz's attempts to explain the evidence for megafloods in the eastern Washington scablands, and how it was met with resistance from the broader geological community.

Hsu, K. J., 1983, *The Mediterranean Was a Desert: A Voyage of the Glomar Challenger,* Princeton, NJ: Princeton University Press, 197 pp.
This is a very accessible account of the discovery of the Messinian salinity crisis, when the Mediterranean Sea became disconnected from the Atlantic and dried out.

Ryan, W. and W. Pitman, 1999, *Noah's Flood: The New Scientific Discoveries about the Event that Changed History,* New York: Simon and Schuster, 319 pp.
Starting with a retelling of the discovery of the Messinian salinity crisis of the latest Miocene, this very accessible narrative focuses on the evidence for a rapid re-filling of the Black Sea, the politics behind these discoveries in the midst of the Cold War, and speculation about the role this major flood may have had on the birth of modern civilizations.

CHAPTER 18

Whole landscapes

Today's scientists have substituted mathematics for experiments, and they wander off through equation after equation, and eventually build a structure which has no relation to reality.

Nikola Tesla (1934)

In this chapter

Landscapes ground geomorphologists in reality. While theories usually address one process at a time, whole landscapes represent a large-scale natural experiment that demands that we acknowledge the linkages among the components. Here we illustrate how the processes and landform elements we have been discussing throughout this book are linked in a landscape. We choose a coastal landscape on an active margin, as it is one in which deep Earth and atmospheric and oceanic processes conspire to shape the landscape. We touch upon several themes introduced in the first chapter of the book, including the importance of climate history (here manifested most explicitly in the huge swings of sea level), acknowledgement of the roles of individual events (here earthquakes, storms, waves), and the utility of new dating tools to unlock the timing in the landscape.

We choose to illustrate using the Santa Cruz landscape that was our back yard for over 15 years. It is a landscape in which the marine terraces of the coast document the pattern of rock uplift associated with the tectonic setting. These same terraces can be used to assess the rates of geomorphic processes acting to modify them. Here we summarize what we have learned so far from the Santa Cruz experiment, and demonstrate how the use of modern geomorphic methods can be used to tell the story of a whole landscape.

Along-coast view of the decayed seacliff of one of the Santa Cruz marine terraces. Note the sigmoidal but asymmetric slope profile; sharp curvature at the top of the slope gives way to broad curvature at the base. Bedrock occasionally outcrops on upper slopes (photograph by R. S. Anderson).

The Santa Cruz landscape: introduction

The rocky coastline north of Santa Cruz is famous for its well-developed marine terraces, its surf, and its numerous pocket beaches isolated by bedrock cliffs tens of meters high. Within the deeply indented Monterey Bay revealed in the topography in Figure 18.1, the beach becomes more continuous, and more hospitable for volleyball. The marine terraces are for the most part flat, and have sandy soils. They are used for growing crops dominated by Brussel sprouts and artichokes, and for houses.

As seen in Figure 18.2, the flat terraces slowly disappear from the landscape as one ascends the hills toward the crest of the Santa Cruz Mountains. The vegetation changes from the grasses that dominate the terraces into redwoods that first line the river valleys and then cloak the entire landscape; all but small plots are second growth forest, as the landscape was extensively logged in the late nineteenth century. The hillslopes are very steep, many near 30°, and the river channels are occasionally choked with coarse debris, suggesting a large role for debris flows in the upper catchments.

Tides are 1–2 m. Kelp forests grow in a zone just offshore, their fronds damping the higher frequency waves. That kelp requires a rocky substrate for its holdfasts demonstrates that the shelf platform just offshore is already rocky. The shelf here is several kilometers wide, and is defined by an abrupt break in slope at 120 m depth. Modern deposits covering the mid-Tertiary sedimentary rock are thin where they exist at all.

The present climate is Mediterranean, in which all precipitation arrives between October and May in discrete storms lasting several days. More than nine storms in a year leads to floods; fewer than five storms

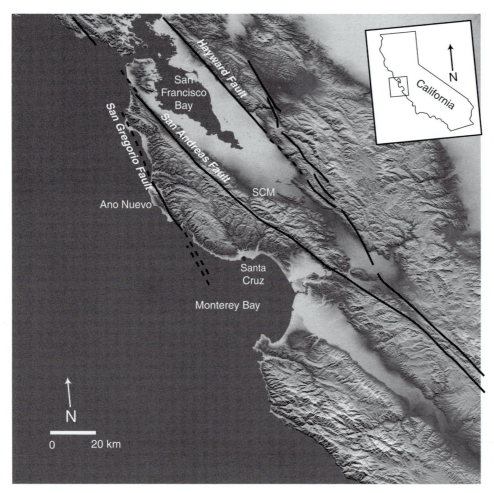

Figure 18.1 Santa Cruz, California and surrounding landscape. The San Francisco Bay area is cut by various strike-slip faults of the Pacific–North American plate boundary. The Santa Cruz Mountains (SCM) occur along a bend on the San Andreas Fault, the rock uplift occurring at least in part on subsidiary faults such as the Loma Prieta structure that ruptured in the 1989 earthquake. Rock uplift of the coast is marked by the famous marine terraces along the Santa Cruz coastline. Seacliffs along the coast north of Santa Cruz, and rivers draining the Santa Cruz Mountains contribute sediments that comprise the littoral sands of the coast. The Santa Cruz littoral cell moves sand southward along the coast to its southern margin at the Monterey submarine canyon, where Monterey Bay most deeply indents the coastline.

Figure 18.2 Air photo of coastline just northwest of Santa Cruz. Three main marine terraces and their backing seacliffs are depicted, dissected by a major stream system. Highway 1 runs along the first terrace platform.

a year leads to a drought. Many streams all but dry up in the summer season, in which fog dominates the mornings. River discharge picks up only after significant rains have altered the infiltration capacity of the soils; the first rains are not very productive of runoff. It rarely freezes, snow at the crest of the Santa Cruz Mountains being a front-page news event. The precipitation gradient is extreme, the strong orographic effect driving a threefold increase within only 10 km of the coast. The El Niño cycle plays a significant role in both precipitation and wave impacts on the landscape. The feedbacks between the terrestrial and marine processes are interesting. Logs are flushed from river channels during El Niño-driven floods, and are then bashed against the shoreline by associated high waves.

How has this landscape evolved? What has driven the uplift of the rock – at what rate and in what pattern? Which processes remove the marine terraces from the landscape? What allows the deep indentation into the coastline that forms Monterey Bay? And what is the signature of climate change in this landscape?

Happily, we can appeal to the very features that best represent this landscape – the marine terraces themselves, shown in Figure 18.2 – to answer some of these questions. We will pivot our discussion of the landscape about these forms that fringe the land. Marine terraces are wonderful geomorphic markers. They can be used to infer rock uplift rates if we know their ages and the sea level at which they were formed. In addition, that they form with a characteristic shape, chiefly defined by a uniform gentle offshore slope of the bedrock platform, a thin mantle of terrace sands, and an abrupt edge at the 80–90° seacliff, allows us to compare one terrace with another older terrace to deduce patterns and rates of evolution of this landscape. It is rare that we know how a landscape started, or, put mathematically, that we know its initial condition.

Box 18.1 Historical aside

It is on this reach of coast that the modern hypothesis for how marine terraces form on an active coastline was first proposed. C. S. Alexander (1953) proposed that the terraces require that the landmass be uplifted while sea level fluctuates over a wide range. Each terrace flat (wave-cut platform) is formed while sea level is high, beveled into the landscape by wave action, then abandoned during sea level fall and never again reoccupied, because by the time sea level is again high the terrace has been uplifted. Alexander proposed neither a mechanism for uplift, nor a mechanism for sea level fluctuation.

Rock uplift: advection around a fault bend

Let us first explore the tectonic story that can be extracted. At the largest scale, a glance at the topographic map in Figure 18.1 reveals the dominance of the landscape by the Santa Cruz Mountain range that forms the spine of the SF Peninsula, and extends south to Watsonville. The San Andreas Fault (SAF) can be easily traced along the northwest side of the mountain range until just north of Santa Cruz, where it steps over to the southwest side of the range. Note that the fault performs a left step as it crosses the range, which on a right lateral fault is a restraining bend. It should restrict motion on the fault, and generate compression within the crust. The highest mountains within the range lie near this fault bend: Mt Umunhum, Loma Prieta; the height of the range crest dies off in either direction, toward San Francisco and toward Watsonville. While others have tended to explain this and all other coast ranges within coastal California as resulting from compressional features associated with a slight component of convergence between Pacific and North American plates, the specific geometry of the Santa Cruz mountains has inspired a different view (Anderson, 1990, 1994). If the bend in the San Andreas Fault serves as a knot around which crust must move, and if the crust is largely incompressible, then the arrival of crust in the bend must result in thickening of the crust. This will take place along one or several structures, and should give rise to uplift surrounding the bend, the specific pattern of which will be determined by the geometries of the ancillary faults. The thickened crust and the topography associated with it on either side will then make its way down-fault by dextral slip on the San Andreas Fault. This serves to advect topography generated within the fault bend toward regions in which there is no new source of rock uplift (Figure 18.1). The topography then simply decays through geomorphic processes that include river incision into the uplifted rock mass, and erosion of hillslopes that bound the streams.

The details of such an uplift pattern are difficult to discern in all but very special circumstances. The Santa Cruz side of the fault is one such example, the marine terraces being the tool of the trade. There are five terraces where best developed, and more commonly three. They vary in width from about a

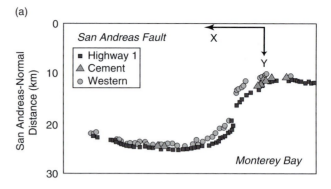

(a)

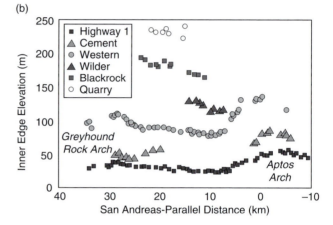

(b)

Figure 18.3 Pattern of marine terraces in vicinity of Santa Cruz, CA. (a) Map view, rotated into coordinate system of the San Andreas Fault. (b) Elevation of terrace inner edge locations. Note two maxima, at Aptos Arch and Greyhound Rock arch (redrawn from Anderson and Menking, 1994, Figure 2).

kilometer to only a few tens of meters, the widest being the youngest, the Highway 1 platform (Bradley and Griggs, 1976). Most importantly, the elevation of the inner edge of any particular wave-cut platform varies from place to place. In Figure 18.3 we show both a map of the inner edge locations, and their elevations as a function of distance along the coast. Note that a distinctive elevation pattern recurs in each platform, and that the older the platform the greater the amplitude of the warps in the surface. The pattern has two high points, one corresponding to the deepest indentation of Monterey Bay into the coastline, the other to a shallower high that occurs up the coast toward Año Nuevo. Given that these geomorphic features must have formed at sea level, which is an equipotential that on this length scale should be uniform, the warping of the platforms must have occurred subsequent to their formation.

These are relatively short wavelength features, of the order of 10–20 km.

Immediately following the M7.1 Loma Prieta earthquake of October 17, 1989, several investigators began to explore what pattern of coseismic rock uplift had been generated, and whether repeating such a pattern could explain the pattern of terrace elevations. One of the big surprises of the Loma Prieta earthquake was the large component of dip-slip on the fault, or for that matter that this fault existed at all. We now know from aftershock patterns that the Loma Prieta structure dips offshore at 60–70°. It slipped over a region extending from 5–18 km in depth, and 37 km in length. The coseismic slip of 2.3 m was broken into roughly equal amounts of dextral strike-slip and thrust dip-slip; the coastal side came up relative to the San Francisco Bay side. The expected pattern of uplift from such an event is dominated by the dip-slip component (Figure 18.4). This general pattern was subsequently measured using GPS, and re-leveling surveys along roads and railroad tracks in the area. It has since been shown that one can explain very well the pattern of the terrace elevations in a 10–12 km reach around and downcoast of Santa Cruz by simply repeating this coseismic pattern (Figure 18.4). Given the coseismic uplift at a point on the terrace, and the difference between its present elevation and the sea level at which it had formed (its total uplift since formation), one can then calculate the number of earthquakes, n, necessary to generate this uplift. This number increases with the height and hence the age of the terrace. Of course, if we are in the business of estimating the hazard posed by the Loma Prieta structure, we require a repeat time for such events, τ. This requires knowledge of the age of the terrace, T, as $\tau = T/n$.

While the terrace elevation pattern is well explained in the reach of coast immediately surrounding Santa Cruz, it departs dramatically from the expected pattern toward Año Nuevo. The second bump in the terrace elevations requires another source of rock uplift. The most likely culprit is a slight component of vertical motion on the San Gregorio fault, which intersects the shoreline just south of Año Nuevo. If this is the case, then the Santa Cruz reach of coastline can be viewed as being bounded by faults that dip beneath it, on each of which thrust motion occurs to drive it upward.

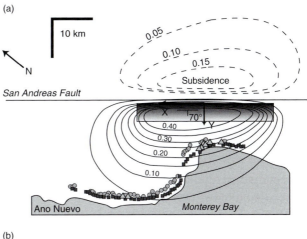

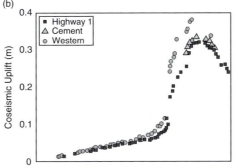

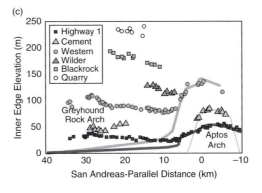

Figure 18.4 Deformation pattern associated with 1989 Loma Prieta earthquake. (a) Map rotated into the coordinate system parallel to the San Andreas Fault, showing uplift on Monterey Bay side, subsidence on San Jose side. Marine terrace inner edge locations are shown along the coastline. Note the coastline most deeply indents the coseismic uplift pattern at the corner of Monterey Bay. (b) Pattern of expected uplift of terrace inner edge locations. (c) Pattern of present inner edge elevations (symbols), with predicted patterns of inner edge elevations (gray lines) after an integer number of Loma Prieta coseismic uplift events. Pattern is well matched in Monterey Bay, and falls well short along the coast further north (redrawn from Anderson and Menking, 1994, Figures 3, 4 and 7).

Evolution of the terraces

We have identified the uplift pattern and its cause. Now let us turn to the degradation of the landscape. While the decay of the entire Santa Cruz mountain range might be one target, we will appeal again to the marine terraced portion of the landscape for insight and for constraint on process rates. Backwearing of seacliffs produced the wave-cut platforms that are ultimately recorded in the landscape as the marine terraces. The seacliffs that back each terrace appear to decay with age (Figure 18.5). The seacliffs appear to round off; the curvature, and the maximum slope on the paleocliff decay monotonically with age. This observation was used in the early 1980s as an example of landscape diffusion (Hanks *et al.*, 1984). Assuming a set of ages for the terraces, these authors determined the best fitting landscape diffusivity, $\kappa = 10 \ m^2/ka$, or $0.010 \ m^2/yr$. Implicit in their analysis is that the rate-limiting process is the transport of debris.

Two observations led to further work on the problem. First, the profiles are not symmetrical. The curvature of the cliff-tops is sharper than that at the cliff-base. Pure linear diffusion should produce symmetrical profiles. Second, even on cliffs that are highly decayed, occasional outcrops of the Santa Cruz mudstone bedrock are scattered over the top half of the paleocliff profile (see Figure 18.0). The weathering of bedrock, or the production of transportable debris from immobile bedrock, may therefore play an important role in limiting the rate of this landscape's evolution. Numerical models that account for both regolith production and its transport can better explain the evolution of the seacliff profiles (Rosenbloom and Anderson, 1994). These models required explicit rules for regolith production, and for regolith transport. The regolith production rule was constrained by the fact that under the several meters thick terrace sands that mantle the marine terrace flats there are intact *pholad* borings, holes several centimeters deep drilled by bivalves into the soft bedrock while it was in the littoral zone. The weathering rate must therefore decline to near-zero under such thick cover in order to allow preservation of these features. The modeling results suggest that weathering rates on bare bedrock approach 0.1 mm/ year. But what hillslope processes are active in this landscape, and how do we acknowledge these

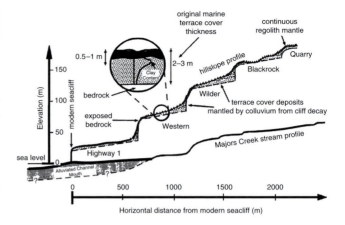

Figure 18.5 Profiles of hillslopes and of adjacent channel normal to the Santa Cruz coastline at Majors Creek. Seacliffs decay with elevation and hence with age. The stream profile displays several knickpoints, and is alluviated at its mouth, reflecting significant erosion during the last sea level drop associated with the Last Glacial Maximum (after Rosenbloom and Anderson, 1994, Figure 4, with permission from the American Geophysical Union).

processes in the models? Until recently, it has simply been assumed that the transport process is a diffusive one, in other words that the regolith transport rate is proportional to the local slope. As we have seen, this indeed captures the essence of rainsplash, frost heave, and regolith creep processes. But these are not the dominant hillslope processes acting to modify the Santa Cruz landscape. The landscape is almost entirely clothed in vegetation, with the only exceptions being the minor bedrock exposures on the crests of the paleocliffs, and the numerous piles of bare dirt produced by the burrowing of rodents. At least at present, it is the activity of these animals that dominate in transporting material in the landscape. That these animals tend to place the material downslope of their burrows leads to a diffusive character to their activities. Indeed, the landscape diffusivity of roughly $10 \ m^2/ka$ we employed in our models of the seacliff degradation (Rosenbloom and Anderson, 1994) likely captures broadly the diffusivity of the landscape (Figure 18.6). But the animals responsible for the transport do not live uniformly on the landscape. They do not dig in bedrock, and they tend not to live on perfectly flat terrace tops, as these become saturated in the winter rains (we will see why below: it is a consequence of the weathering of the terrace deposits). Only when we have a proper, biologically

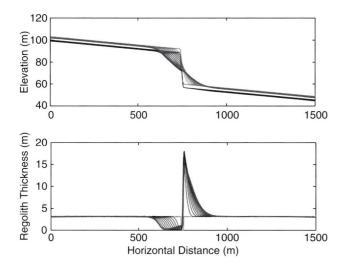

Figure 18.6 Numerical simulation of a degrading seacliff (top) and of the thickness of regolith on the landscape. Initial condition is a vertical cliff (riser) separating two gently sloping marine terrace surfaces, each mantled with 3 m of terrace sands. Final profile shown in bold. Landscape is shown in 10 ka snapshots over the 200 ka simulation. Note very slight erosion of the marine terrace bedrock tread. Top of cliff retains bare bedrock character. High curvature at the crest of the form contrasts with the gentle curvature on the depositional base.

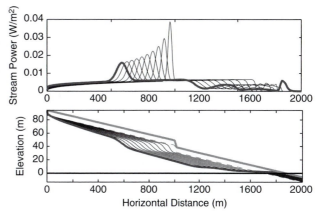

Figure 18.7 Numerical simulation of the evolution of a coastal channel. Initial condition meant to mimic a riser separating two marine terraces (dashed) is slowly uplifted at 0.5 mm/yr, reaching the final position (bold) after 15 ka. The channel profile evolves by eroding headward, at a rate dictated by the local pattern of unit stream power (top), producing a final profile (bold) with a knickpoint considerably upstream of its origin. Stream power is highest where the channel slope is greatest, but the peak decays as the knickpoint propagates toward the headwaters where the water discharge is small.

defensible rule for the geomorphic role of these animals will we have a full description of the processes involved in decay of the seacliffs. Note again, however, that the calibration of any such model of landscape evolution will require assignment of ages for each of the cliffs. This requires dating of the terraces.

Stream channels

The Santa Cruz landscape is drained by a number of streams that head along the crest of the mountains, and cross the emerging terraces. The baselevel experienced by any stream is slaved to the ~ 100 m oscillations of sea level that reflect variation in global ice volume. Just as a seacliff is abandoned by sea level drop, so the mouth of any stream channel chases the sea across the shelf as it drops to its minimum during the next glacial maximum. This drop in baselevel incites stream incision into rock as we have discussed in Chapter 13. This drives bedrock incision across an old terrace and its bounding seacliff. It also serves to incise a new bedrock river channel across the now

subaerial wave-cut platform. A simple illustration of the expected upstream migration of a knickzone associated with an initial step created as a terrace is abandoned is shown in Figure 18.7. The longitudinal channel profile resulting from repeated drops in baselevel can show vestiges of this complex history as a set of knickzones found well above present sea level (Rosenbloom and Anderson, 1994).

Interestingly, if the notch cut into the old wave-cut platform by a stream during the sea level lowstand is not entirely erased by wave action as sea level rises to its new interglacial level, a linear bedrock trough should remain in the bathymetry. Indeed, these troughs exist in the present Santa Cruz platform (Anima *et al.*, 1997; see also Storlazzi and Field, 2000). They serve to steer sand in the offshore environment. They are made visible as well in aerial photos of the shoreline because kelp requires a bedrock substrate for its anchors (its "holdfasts"). No such anchors exist in the sand-filled troughs, which therefore generate gaps in the kelp forest.

Streams also contribute to the ability of a landscape to "forget" that it was ever terraced. Consider the following problem. Rock uplift has presumably been going on for several million years in the

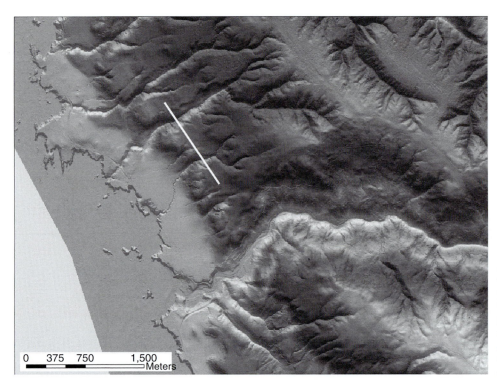

(a)

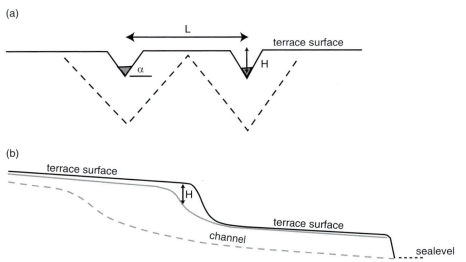

(b)

Figure 18.8 How marine terraces are removed or forgotten from the landscape. Top: DEM of a portion of the terraced landscape, showing progressive loss of terraces with elevation. (a) Cross section across two channels and intervening terrace tread (line in DEM), at early (solid) and late (dashed) times. (b) Long-profile of terrace and incising channel, the latter shown at early (solid) and later (dashed) times. Deepening of the stream channel below the terrace surfaces incites broadening of the adjacent valley walls, which in turn bite into the remnants of the flat marine terrace treads. The resulting timescale for removal of the terrace flat is dependent on the distance between adjacent channels, L, and is inversely dependent upon the rock uplift rate, which scales the stream incision rate (diagrams after Anderson *et al.*, 1999, Figure 10, with permission from Wiley-Blackwell Publishing).

Santa Cruz landscape, yet remnants of only five marine terraces exist in the landscape, and none above a couple of hundred meters above sea level. Why is this? While the terrace edges representing old seacliffs round off through time by a set of processes we have discussed, this decay of seacliffs is inefficient as a means of erasing a terrace. Given the diffusivity of the landscape, it would take a very long time for a wide flat to decay enough to be unrecognizable as an old terrace. However, stream channels serve to chop up these flats into little pieces. We have seen that streams seek a baselevel that is here the ocean. As the land rises, that baselevel effectively falls, which in turn inspires the stream to erode through rock. The stream can chop through a terrace surface, and the next and the next as it seeks an ever-changing sea level. In contrast, once abandoned as an isolated line of cliffs during sea level drop, a seacliff is ignorant of sea level. A seacliff decays in the same fashion and at the same rate whether it is 10 m, 100 m, or 400 m above sea level. Each old terrace is therefore incised by one to a few master streams (visible in Figure 18.2).

But that is not all. As we discussed in Chapter 11, streams bifurcate to form dendritic networks. They branch. And each of the branches serves to chop up the old terrace flats into yet smaller pieces. In the end, as depicted in Figure 18.8, the pieces are small enough that the steep slopes that bound the incising stream channels intersect to become ridges, and the flat is removed entirely from the memory in the landscape. Terrace surfaces are therefore lost from the record above some elevation. In the Santa Cruz case, this leaves us with only small isolated remnants of the 5th terrace, and none that have been identified above this.

Terrace ages

It has proven difficult to date these terraces absolutely. Until recently, the common practice, as shown in Figure 18.9, has been to assign an age to the lowest terrace, and then use knowledge of sea level history and the heights of terrace inner edges to arrive at the best correlation. Each inner edge should correspond to a sea level highstand. The alternative to this method is to obtain absolute ages for the terraces. Unfortunately, although they are plentiful in southern California, the fossil solitary coral (*Balanophilia elegans*) on which U/Th dating could be performed has been difficult to find in the terrace deposits along this reach of coast, as it is just at the northern latitudinal edge of their range. On the other hand, the setting is appropriate for application of cosmogenic radionuclide (CRN) surface exposure dating. Recall that this method relies upon the accumulation of CRNs in surface materials, in which the production rate profile falls off exponentially with depth. These surfaces pose two problems for the CRN method. First, the sands in which the CRNs accumulate spent some time within the near-surface production zone prior to arrival on the beach that ultimately is abandoned to become the terrace cover deposit. In this system, this inheritance includes exhumation on some hillslope upcoast of the sampling site, travel down the hillslope and through the fluvial system to the coast, and finally transport in the littoral system. As we have seen in applications of CRNs to date fluvial terraces, this results in the shift in the CRN concentration profile shown in Figure 18.10. The second complication arises from soil processes subsequent to formation of the terrace. These include pedogenic processes, but most likely in this setting are

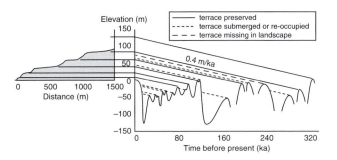

Figure **18.9** Terrace prediction diagram. Left: surveyed elevation profile of marine terraces near Mendocino triple junction, Bruhel point, California, with inner edge elevations noted as horizontal lines. Right: sea level history. If rock uplift rate has been steady, each sea level highstand, corresponding to an interglacial period, will result in generation of a marine terrace. Parallel solid lines represent predicted steady uplift histories of inner edges. Dashed lines represent highstands that generated terraces that are either submerged, or have been re-occupied by later terraces. Implied steady uplift rate at this location on the coast is 0.4 mm/year (redrawn from Merritts and Bull, 1989, Figure 2).

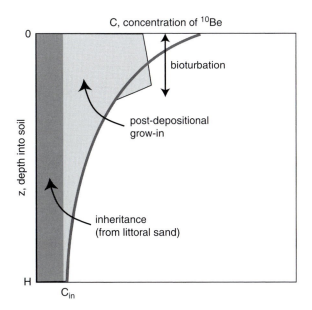

Figure **18.10** Expected profile of CRNs, for example ^{10}Be, in bioturbated terrace sands. Inherited nuclides serve to shift the profile with a uniform concentration. Subsequent grow-in of nuclides produces the exponential, the top of which is blunted by bioturbation. The integral of the profile remains a clock that will increase linearly with time.

dominated by rodent burrowing activity – bioturbation. This should effectively mix the top portion of the profile, homogenizing and blunting the exponential shape (Figure 18.10). The method that can correct for this, called the integral method, relies upon the

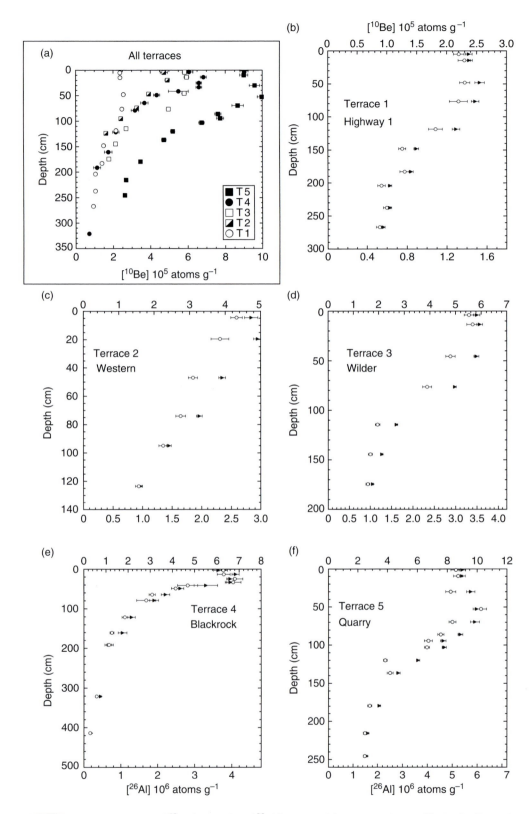

Figure 18.11 Measurements of ^{10}Be (top axis) and ^{26}Al (bottom axis) concentration profiles in the five main marine terraces of Santa Cruz, California. (a) All profiles. (b–f) Individual terrace profiles. Note in all cases that the top 50–150 cm shows relatively uniform concentration, reflecting turbation of the soils. Below this the profile is roughly exponential, and asymptotes to a finite value reflecting the inherited radionuclide concentration (after Perg *et al.*, 2001, Figures 2 and 4; see also Perg *et al.*, 2002).

fact that the integral of the CRN production profile acts as a clock just as well as the concentration at any depth in an undisturbed profile would. As long as the inherited component can be identified and subtracted off, the remaining depth-integrated concentration can be inverted for a deposition age. Constraining the inherited component requires samples from at least a couple of meter depths. Constraining the integral then requires many samples to define the full shape of the profile. This technique was first applied to the five terraces in the swath of coastline between 10–20 km on Figure 18.2. As shown in Figure 18.11, the concentrations increase monotonically with the height of the terrace, as expected (hoped). Modeling of the profile evolution and direct integration of the profiles yield the ages of all five terraces. In general, they correspond well with sea level highstands. It was, however, a surprise that the youngest of the terraces appears to correspond to the MIS 3, at roughly 60 ka, rather than one or another of the MIS 5 highstands, as previously thought. The general age pattern also translates into a nearly steady rock uplift rate along that reach of coast, of slightly higher than 1 mm/yr.

One could also ask of these CRN profiles what sets the CRN inheritance. As revealed in Figure 18.11, the inheritance appears to vary from one terrace to another, and the inherited ^{10}Be concentration is typically of the order of 10^5 atoms/g. Can the inheritance be used to assign rates to geomorphic processes involved in the exhumation and transport of the sand? This pointed toward the modern littoral system as an analog, the thinking being that an understanding of the concentrations in the modern system would

allow better interpretation of the inheritance in the older terrace cover deposits.

Evolution of soils on the terraces

The prominent terraces of the Santa Cruz coastline also represent a natural experiment with which to explore the evolution of soils. If the terrace sands that cap each terrace are compositionally uniform from one to another terrace, then the terraces of different ages allow researchers to tease out the temporal evolution of soils developed on them. Soils that form in similar parent materials, and have seen similar climates and biological activity, are called a chronosequence. Such sites are rare and hence precious to soil scientists. It was in fact the father of modern soil science who first identified the importance of sets of marine terraces as chronosequences (Jenny, 1941).

White *et al.* (2008) have made use of the absolute dates on the Santa Cruz terraces to explore the roles of water flux, mineral reaction rates, and degree of saturation in dictating the weathering of the terrace sands. They first documented both the elemental and mineralogical profiles in the five terraces. These in turn constrained the profiles of strain, and of the mass transfer coefficients, or the missing mass of each element in the five soils.

The elemental profiles reveal a deepening weathering profile through time that is well represented by the evolution of the Na and Ca profiles plotted in Figure 18.12. These reflect the weathering of plagioclase ($CaAl_2Si_2O_8$ to $NaAlSi_3O_8$), as is nicely

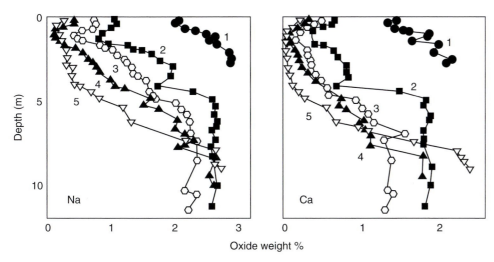

Figure 18.12 Profiles of Na and Ca in the Santa Cruz terraces (labeled by number). Progressive depletion of both Na and Ca occurs; Ca is depleted most rapidly (after White *et al.*, 2008, Figure 2, with permission from Elsevier).

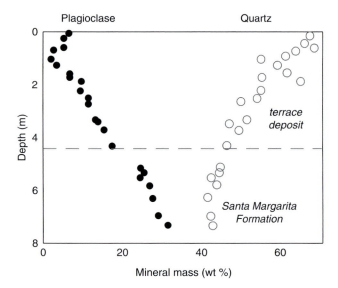

Figure 18.13 Profile of quartz and plagioclase weight percent on the third Santa Cruz terrace. On this terrace the weathering profile appears to extend smoothly into the underlying sands of the Santa Margarita Formation (after White *et al.*, 2008, Figure 3, with permission from Elsevier).

corroborated by the profiles of plagioclase and quartz in the third terrace reproduced in Figure 18.13. SEM images of the soils reveal that the quartz remains intact while the plagioclase grains become highly pitted. Clay contents in the profiles increase with time, as do both MgO and FeO components. This reflects the production of smectite, and the conversion of original and pedogenic smectite to kaolinite + iron oxides.

In these settings the parent material is terrace sand, which is derived ultimately from weathering elsewhere in drainages up-coast of the sampled terrace site. These sands are already enriched in quartz relative to the parent rocks, and are reduced in concentrations of heavy minerals and plagioclase. Given the variability of the trace mineralogy of the terrace sands from one site to the next, the authors decided to use Si as their inert component in the strain calculations, rather than the more commonly used zirconium or titanium. These latter elements are concentrated in trace minerals that may have been winnowed, or concentrated in thin layers by the action of waves. Profiles of mass transfer of Na and Ca deepen with time, in places extending below the base of the terrace sands into sandy bedrock. Al and Fe profiles show accumulation in shallow argillic (clay-rich)

horizons that broaden and intensify through time. These observations motivate a model, sketched in Figure 18.14, of the progressive weathering of the soils, in which the dominant variables are the water flux through the soils, the kinetic reaction rates, and the dependence of reaction rate on the degree of saturation of the soil solutions. The authors argue that the most important of these variables is the water flux through the soil, which is within a factor of two of the modern fluid flux as documented using a Cl balance.

Implications of the weathering of soils for the hydrology

Note that the clay content is significant even on the second and third marine terraces. These clays greatly alter the hydrology of the terraces, which in turn dominate the hydrology of the bottom portion of the landscape. Once hydraulically very conductive beach sands, these deposits have evolved to become poor conductors of water. Recall that water is delivered to the coast by a very few rainstorms in the winter months. The first of these rains fall upon a landscape that has baked in the sun for 6–8 months. The clays have cracked, and happily accept the first rains of the fall. In the early season, the runoff from the landscape is therefore extremely small, yielding of the order of 7% of the runoff that we can deduce from Figure 18.15. But as further rains swell the clays, the cracks disappear and the flatter portions of the terraces can even become waterlogged. The same intensity and duration of rainstorm occurring in January or February will generate up to 70% runoff, a tenfold greater yield than its early season counterpart. It is therefore in part the weathered nature of these terrace deposits that produces this great importance of antecedent moisture in the hydrology of the Santa Cruz landscape.

Littoral system

The terraced nature of the coastline provides yet another opportunity to tease out the geomorphic process rates in this setting. We will make use of the fact that there are two sources of sand for the beach (littoral) system: rivers, and back-wearing of cliffs. The terrace cover sands atop the cliffs are simply

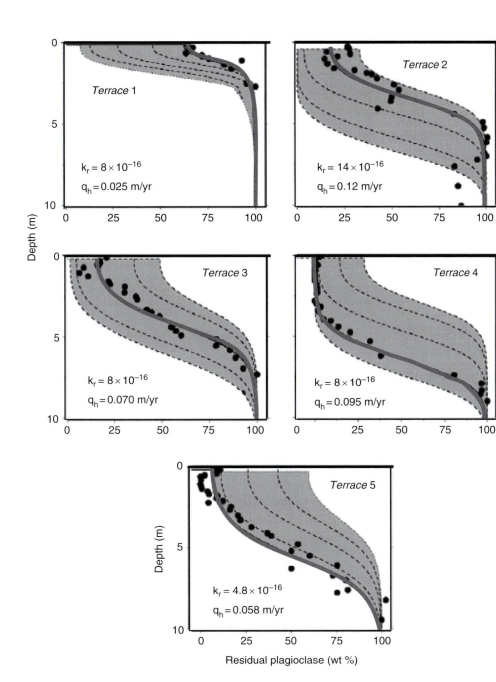

Figure 18.14 Modeled profiles of residual plagioclase, shown against mineral data from the five Santa Cruz terraces (after White *et al.*, 2008, Figure 16, with permission from Elsevier).

recycled into the beach system. In order to be quantitative about the problem, we must write two balances, the first a sediment balance in the littoral system, the second a CRN balance. These can then be linked to obtain a mixing model in which both components to the system can be constrained.

Consider the littoral sediment balance shown in Figure 18.16. Just as on a hillslope, we must consider both the sources and sinks of sediment, and the rate at which it is passed from one point to another within

the system (the fluxes). As the littoral system rapidly leaks grains to the shelf that are finer than about 0.2 mm, we will concern ourselves only with larger grains, which we will call sand. We can capture the mass balance of sand in the littoral system:

$$\frac{\partial A}{\partial t} = -\frac{1}{\rho_{\text{litt}}}\frac{\partial Q}{\partial x} + \varepsilon_{\text{cliff}}\left(f_{\text{bed}}H_{\text{bed}}\frac{\rho_{\text{bed}}}{\rho_{\text{litt}}} + f_{\text{terr}}H_{\text{terr}}\frac{\rho_{\text{terr}}}{\rho_{\text{litt}}}\right)$$

$$+ \frac{1}{\rho_{\text{river}}}\left[\sum_i f_{\text{river},i}Q_{\text{river},i}\delta(x-x_i)\right] - \frac{\rho_{\text{com}}}{\rho_{\text{litt}}}q_{\text{com}} \quad (18.1)$$

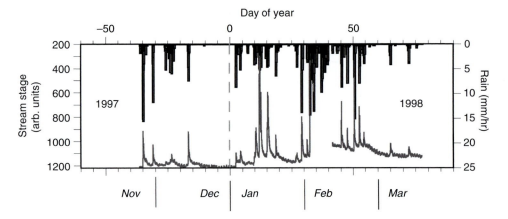

Figure 18.15 Hydrograph (bottom, left axis), and hyetograph (top, right axis) of Wilder Creek catchment during the 1997–98 winter. This shows many typical features of the rainfall and hydrologic response in such systems. The rainfall is delivered in a few storms, all in the winter months. Early storms result in only small stream discharge, here depicted by the stage of this small creek (in units measured downward from a sonic look-down sensor). Later storms of the same intensity and duration, say in mid-January, result in much greater stage and hence discharge, demonstrating the role of antecedent moisture. The intense and enduring storms at the end of January 1998 knocked out the stream gage until it was repaired 2 weeks later.

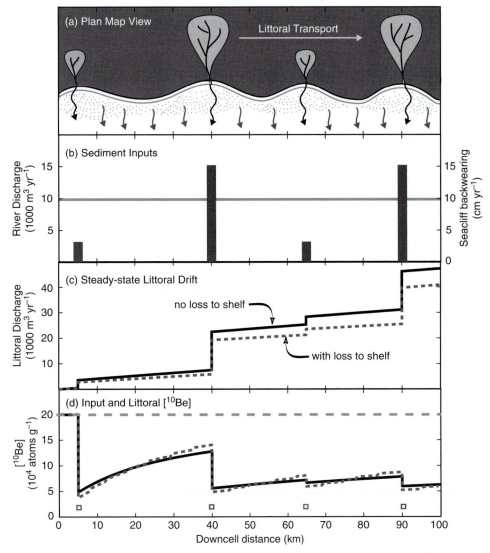

Figure 18.16 Conceptual model for sediment budget in a littoral system in which both rivers and cliff back-wearing contribute sand (after Perg *et al.*, 2003, Figure 1).

where the terms on the right-hand side of the equation represent the gradient in the discharge of sand by littoral drift, contributions from cliff back-wearing and from rivers, and losses to the shelf, respectively. Here A [m^2] is the cross-sectional area of the beach, ρ_{litt} [kg/m^3] is the littoral bulk sediment density, Q [kg/yr] is the littoral mass discharge, ϵ_{cliff} [m/yr] is the rate of cliff back-wearing, f_{bed} and f_{terr} are, respectively, the fraction of bedrock and capping regressive marine terrace sediment above the littoral grain size cut-off diameter, H_{bed} and H_{terr} [m] are, respectively, the heights of the bedrock cliff and capping terrace sediment, ρ_{bed} and ρ_{terr} [kg/m^3] are, respectively, the density of bedrock and terrace sediment, ρ_{river} [kg/m^3] is the fluvial sediment density, f_{river} is the fraction of fluvial sediment above the cut-off diameter, Q_{river} [kg/yr] is the fluvial mass discharge inserted at $x = x_i$, q_{com} [m^3/(m yr)] is the sediment flux of those grain sizes that have been comminuted to below D_{litt}, and ρ_{com} [kg/m^3] is the bulk density of this comminuted sediment.

A river contributes sand to the littoral cell at a point, with a rate that reflects the basin area, A_b, the mean erosion rate in the basin, $\dot{e}$, and the fraction of the products that are sand, f_r:

$$Q_r = f_r \dot{e} A_b \tag{18.2}$$

Cliffs, a continuous source, contribute sand at a rate dictated by the cliff recession rate, the height of the cliff, and the proportion of that cliff material that is above the grain size threshold. On the Santa Cruz coastline this can be broken into contributions from the bedrock of the cliff, and the terrace cover deposits:

$$Q_{cliff} = \dot{e}_{cliff}[f_{terr}H_{terr}\rho_{terr} + f_{br}H_{br}\rho_{br}] \tag{18.3}$$

In a steady system, we may ignore the time-varying term, and integrate the remaining expression to obtain the expected pattern of sand discharges, as seen in Figure 18.16. If the leakage of sand to the shelf is negligible, then between point inputs of sand from the rivers, the discharge of sand down-cell must simply be the integral of the contributions for the cliff back-wearing. If the back-wearing is uniform, this leads to a linear increase in discharge with distance: $Q = f_{terr}H_{terr}\dot{e}_{cliff}X$. At the rivers, the discharge must experience a jump or step comparable to the fluvial input, Q_r. The resulting pattern of down-cell discharge is a linear increase punctuated by steps at the rivers. Leakage of sand to long-term storage on the shelf simply reduces the necessary discharge down-cell.

While it is simple to construct such a theory, it is far more difficult to quantify the components of the littoral system in the field, including cliff back-wearing, littoral drift rates, and river inputs. In the Santa Cruz area, repeat photographs of the cliff-line have been used to quantify the rate of cliff back-wearing, which is relevant to the prediction of cliff retreat in the future and hence to the maintenance of existing infrastructure and planning of new construction within the coastal zone. As cliff failure is a very spotty process, with short periods of activity (sometimes seismically triggered, sometimes storm-associated) and long periods of stasis, it is difficult to establish the long-term rates relevant to the littoral cell. Nonetheless, historical data show rates of the order of tens of centimeters per year for much of the Santa Cruz coastline (Best and Griggs, 1991; Moore and Griggs, 1998). Rates of littoral drift are perhaps even more difficult to measure. The most robust measurements come from construction and maintenance of the Santa Cruz boat harbor. The jetty was built to protect the harbor entrance. For many years the littoral drift simply accumulated upcoast of the jetty, generating a broadening beach whose volume history could be used to constrain the sand discharge. Subsequent to the infill of the space up-coast of the jetty, the harbor entrance has been maintained through active dredging. The dredge records can also be used to constrain sand discharge. Both methods yield sand discharge at this point along the cell of roughly 1.5×10^5 m^3 sand per year. The rivers contributing to the Santa Cruz cell are for the most part ungaged, especially for sediment. The patchy USGS gaging records have been augmented by modeling of the expected sand discharges using a rating curve relating sediment discharge to water discharge (Best and Griggs, 1991).

Happily, we can use cosmogenic radionuclides to estimate the long-term rates of sand discharge from these small basins. Such estimates can be tested against existing data where available, and can be used in the absence of any such data on ungaged streams. The basin-averaged erosion rate can be constrained using a single sample of sand from a point bar near the base of the fluvial system. Sampling even a handful of sand yields roughly a million sand grains that have come from a wide variety of sites within the basin. The CRN concentration of this sand can therefore be used to constrain an average of the exhumation rate over the basin. Note that we need this concentration in the mixing model for assessment of

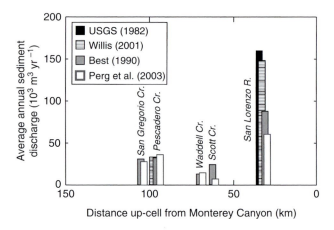

Figure 18.17 Comparison of USGS-style stream-gaged sediment discharge with CRN-based measurements (from Perg *et al.*, 2003). Comparison is excellent except in the San Lorenzo watershed, which shows major urbanization (after Perg *et al.*, 2003, Figure 4, and references therein).

the littoral budget. Ignoring decay, the equation for the concentration in the point bar sample is

$$\bar{C} = \frac{\bar{P}_{o} z_*}{\dot{\bar{e}}} \tag{18.4}$$

which can easily be solved for the mean basin erosion rate. In order to justify ignoring decay, the erosion rate must be high enough that the time spent in passing through the CRN production zone is short enough that not much decay will have occurred. This requires formally that $z_*/\dot{e} < \tau$, where τ is the decay constant for the radionuclide. For erosion rate fractions of 1 mm/year, this is certainly the case. In Figure 18.17 we show that the CRN-based rates are comparable to those obtained from the existing gaged records. All of the fluvial contributions from major streams along the Santa Cruz cell have

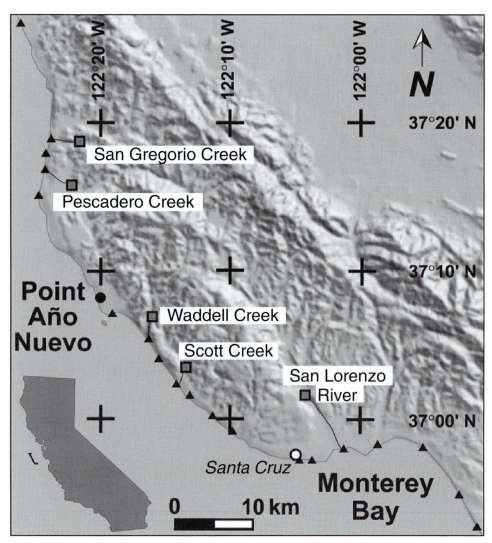

Figure 18.18 Map of Santa Cruz coastline showing sampling sites for ^{10}Be in beach sediments (triangles), and the main drainages contributing sand to the Santa Cruz littoral cell (boxes) (after Perg *et al.*, 2003, Figure 2).

been assessed in this manner, and fall within the 0.1–0.3 mm/yr range.

Using Equation 18.4 we can turn these mean erosion rates into long-term river sediment discharges to the littoral system. All that remains in the littoral budget are the contributions from the back-wearing of cliffs. For these we rely upon the CRN concentrations of the littoral sands themselves, and the mixing model outlined above that allows us to constrain relative contributions from cliffs and rivers. Here we can take advantage of the fact that the CRN concentrations from the cliff sediments (the terrace sands) are very different from those derived from the river basins. Sands on the cliffs have been irradiated for tens of thousands of years, resulting in high mean CRN concentrations. In contrast, at erosion rates of fractions of a mm/year, sands from the rivers spend only a couple of thousand years being exhumed through the roughly 0.7 m-thick production zone, resulting in low CRN concentrations. (The timescale is simply $T = z_*/\dot{e}$; if $z_* = 0.7$ m, and $\dot{e} = 0.1$ mm/yr, $T = 7000$ years.) The littoral CRN data collected from sites shown in Figure 18.18 are depicted in Figure 18.19. The CRN concentrations are everywhere intermediate between those of the terrace sands, and those from the river basins. Using the mixing model, we can quantitatively estimate the cliff erosion rates necessary to produce these concentrations. The best-fitting results of the model shown as the solid lines in Figure 18.19 suggest that cliff erosion rates of about 10 cm/year are sufficient to produce the CRN concentrations.

To summarize, the sand in transport in the littoral system is a mixture of sand from terrace cover deposits atop seacliffs eroding at about 10 cm/year, and hillslopes delivering sand to the rivers eroding at about 0.3 mm/year. If the system is indeed at roughly steady state, this sand must all be transported down-cell; the discharge must monotonically increase from the top of the cell to its end, increasing linearly between rivers, and undergoing step increases at each of these point sources. The CRN concentrations of the littoral sand support this story. The proportions of seacliff- to river-delivered sand vary, but in this coastline, they are roughly equal partners. All this sand is delivered to the tip of the Monterey Canyon, which is found within spitting distance from shore in the center of Monterey Bay.

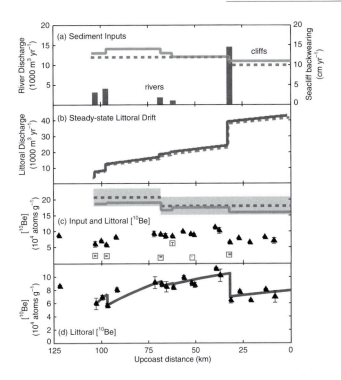

Figure 18.19 Use of cosmogenic radionuclides to document the relative contributions of rivers and seacliffs in the littoral sediment budget. (a) Best-fitting sediment input pattern to the littoral cell. (b) Calculated steady-state longshore drift required to accommodate the pattern of sediment sources. (c) [10]Be concentrations of river sediment (boxes), littoral sand (triangles) and seacliff-capping terrace sands (gray bars). (d) Spatial pattern of calculated [10]Be in littoral sands based upon the source pattern in (a) and the concentrations in (c) (after Perg et al., 2003, Figure 3).

Seacliff evolution

Note the discrepancy in the erosion rates discussed above. The fastest rate at which bedrock is attacked in this environment is that at which the seacliff is eroding; the difference between seacliff and hillslope erosion rates is several orders of magnitude (0.1 m/yr vs. 0.3 mm/yr). The energy delivered to the coastline by waves is enormous. It not only drives the littoral sediment downcoast, toward Monterey canyon, but drives bedrock erosion at the base of the cliff that ultimately results in cliff failure and removal of a chunk of terrace. Every 6–15 seconds, a wave breaks along the entire coastline, ten thousand per day, two to three million per year. Here we summarize recent work on the problem of coastline evolution in the face of such energetic attack.

(a)

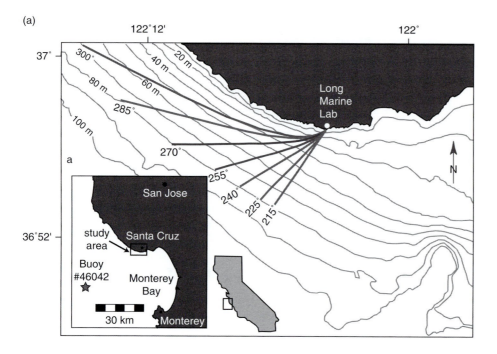

(b)

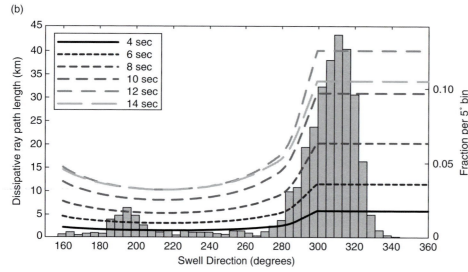

Figure 18.20 (a) Map of study area with bathymetry at 10 m contour intervals. Wave rays are shown refracting across the shelf to arrive at the instrument site at the Long Marine Lab (LML). Inset shows buoy from which wave data are collected. (b) Probability distribution of swell direction, and the ray path length for all possible swell directions arriving at the instrument site (after Adams *et al.*, 2002, Figure 2).

The problems one might address with such an investigation include the formation of the stair-stepping terraces we have been discussing, or the embayment of the coastline to generate Monterey Bay and similar indentations on actively uplifting coastlines. The first involves the cross-shore pattern of the coastline, while the second involves the along-shore pattern. The first problem requires understanding the long-term variation in sea level, which modulates the location on the rock mass that the waves will

deliver their energy. The terrace ages require that such a model extend through several glacial–interglacial swings in sea level, covering several hundred thousand years. We might ask under what conditions a young terrace might erase an older terrace from the coastline altogether, leaving no memory of its prior existence. Are there feedbacks that allow the seacliff to erode more quickly early in a sea level highstand than late? On the coast-parallel problem, we might ask what it is that allows a coastline to become embayed. Is there

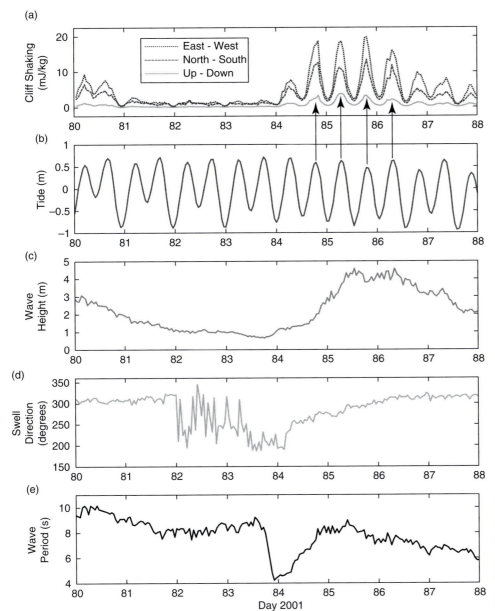

Figure 18.21 (a) Cliff shaking averaged over 1-hour periods in an 8-day experiment in 2001, as measured by three components of the seismometer record. (b–e) Wave and tide climate records for the corresponding period (after Adams *et al.*, 2002, Figure 3).

any limit to the depth of embayment if sea level were to remain high forever; in other words, is there a steady geometry toward which a coastline tends?

It is clear in any of these problems that we must understand the power delivered by waves to the coastline. Anyone who has stood on the cliff-edge during a big storm has felt the vibration of the cliffs as each wave arrives. This inspired a measurement campaign in which we monitored the vibration by installing a seismometer at the cliff top over parts of two recent winters. The wave setting is shown in

Figure 18.20. The seismometer records ground velocities in north–south, east–west, and up–down directions at 50 Hz (50 measurements per second). Not surprisingly, each wave could be detected. In order to convert the time series of ground motion to something more compatible with the oceanic variables of waves and tides, the trace was first squared and then summed over 15-minute intervals. The resulting time series of energy (velocity squared is kinetic energy per unit mass, the mass here being the mass of the cliff) is reproduced in Figure 18.21 along with the

(a)

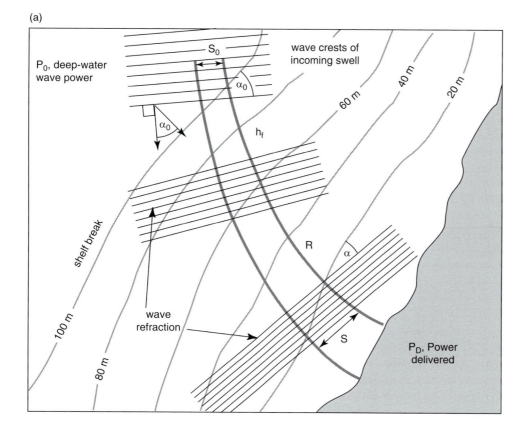

(b)

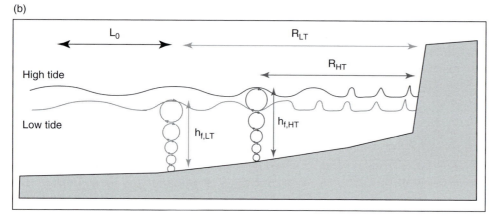

Figure 18.22 Transformation of waves as they approach the coast over a shoaling shelf. (a) Map view, showing stretching of wave crests as they refract. (b) Change in wave shape as they interact with the shelf topography. Waves shown at two tidal levels. Waves at high tide do not interact until they are considerably closer to the seacliff (after Adams *et al.*, 2002, Figure 1).

time series of wave height, swell direction, and tide for a week-long interval. As expected, the energy delivered to the coast varies strongly with wave height, but the strength of the tidal signal in the energy delivery was surprising. More energy arrives at the cliff during high tide than at low tide, despite the low (1–2 m) tidal range.

We have seen in Chapter 16 that the energy in the far-field waves before they interact with the coastline is simply a function of their height; it varies as H^2.

That the energy actually delivered to the cliffs is so strongly affected by tide can be explained by considering where waves lose, or dissipate, their energy. At low tide, the waves interact with the shelf over a longer distance (Figure 18.22), and most importantly break further offshore, dissipating most of their energy on the sandy beach. This energy is used to drive littoral transport, to comminute the sand, and is lost to radiation of sound and generation of heat in the water. At high tide, it is more

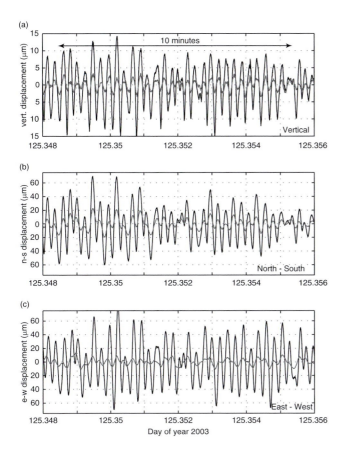

Figure 18.23 Comparison of inland (gray) and shoreward (black) time series of displacements in the (a) vertical, (b) north–south, and (c) east–west directions over a 12-minute period. The amplitude of deflection at the inland site is always a small fraction of that at the seaward site, and perfectly in phase with it (after Adams *et al.*, 2005, Figure 11, with permission from the American Geophysical Union).

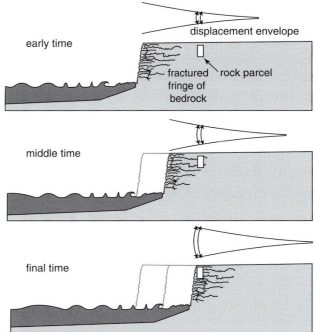

Figure 18.24 Schematic of the zone of rock fatigue caused by repeated wave-induced flexing of the coastal cliffs. A rock parcel (white box) begins at a great distance from the coast (top), becomes increasingly affected by wave loading, and emerges at the cliff edge in the final image. It will experience increasing amplitude of deflection through time over tens to hundreds of millions of cycles (after Adams *et al.*, 2004, Figure 16, with permission from the American Geophysical Union).

likely that the energy in the wave will be retained until it impacts the cliff.

The seismometer experiments also revealed a longer period signal. We did not understand this signal at first and simply filtered it out so that we could see more clearly the short-period shaking of the seacliffs. Later examination of this long period signal was allowed by removing instead the high-frequency shaking, resulting in the time series shown in Figure 18.23. This revealed a very consistent pattern that cried out for interpretation. Each wave appeared to cause a rocking of the cliff, downward and toward the wave as it arrived, followed by recovery. The amplitude of both the horizontal and vertical deflection was of the order of a few to 20 microns. While this does not seem like very much, the fact that it

happens two to three million times per year implies that the rock mass of the cliff may eventually suffer from fatigue, as sketched in Figure 18.24. As we stated in Chapter 16, the rate of back-wearing of the cliff, and the length scale to which this deflection occurs, and the number of waves per year can be used to assess the number of displacement cycles to which a volume of rock is subjected prior to the time at which it becomes the face of the cliff. Given these numbers, the rocks of the cliffs of the Santa Cruz coastline will have felt roughly 1 billion cycles over the 300 years that are within range of the cliff edge.

We should ask what sets the length scale for the decay of displacement amplitude. This is not a problem in which the loading of the weight of the wave is accommodated by flexure in the sense we have encountered in ice and lake loads. Rather, the short timescales and length scales involved require that the loading be accommodated by purely elastic displacements. The problem of deflection of the near-surface rocks by local loads is common in engineering, as buildings and other human infrastructure are supported by an elastic response of the crust. Boussinesq (1885) developed the equations of deflection beneath a point load on an otherwise stress-free half-space. These equations have been summarized nicely in Farrell (1972), who presents solutions for vertical and radial deflection beneath point loads (a force P at $[0,0]$):

$$w = -\frac{P}{4\pi\mu R}\left(\frac{\sigma}{\eta} + \frac{z^2}{R^2}\right) \tag{18.5}$$

and

$$v = \frac{P}{4\pi\mu R}\left(1 + \frac{z}{R} + \frac{\eta r^2 z}{\mu R^3}\right) \tag{18.6}$$

where $R = \sqrt{r^2 + z^2}$ and the parameters $\sigma = \lambda + 2\mu$ and $\eta = \lambda + \mu$, λ and μ being the two Lamé elastic constants. The pattern is shown in Figure 18.25. For the deflection of the surface itself ($z = 0$), and noting that in this case $R = r$, the vertical and horizontal (radial) deflections become more simply:

$$w = -\left(\frac{\sigma P}{4\pi\mu\eta}\right)\frac{1}{r} \tag{18.7}$$

and

$$v = \left(\frac{P}{4\pi\mu}\right)\frac{1}{r} \tag{18.8}$$

Given that the deflections have dimensions of lengths, and the load P has units of force, the collections of constants in parentheses must have units of length squared. Both vertical and horizontal deflections fall off as the inverse of distance from the load, as $1/r$. The collections of constants in the parentheses dictate the length scale over which the elastic response decays with distance.

One can extend this analysis to line loads by summing the deflections associated with a number of point (or disk) loads aligned in a line. In order to assess the expected temporal history of deflection associated with an arriving wave, one must also allow this deflection pattern to move with the wave. As the system is elastic, there will be no lag between the loading and the deflection. Indeed, one must acknowledge the fact that there is an array of waves canted at some approach angle to the cliff on which the seismometer sits, each of which constitutes a line load. When inserting reasonable values of the elastic

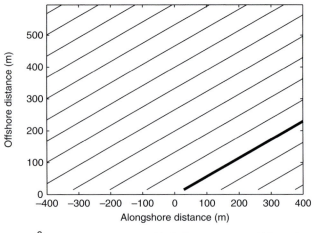

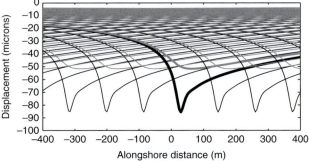

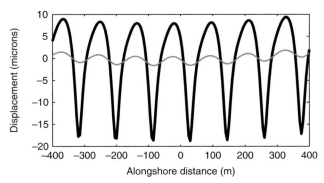

Figure 18.25 Modeled vertical displacement associated with waves approaching the coast (top) at a fixed angle (60°). Middle panel displays deflection measured at the coast (black) and inland by 31 m (gray) caused by each wave. Bold profiles correspond to the bold wave in top panel. Bottom panel displays sum of deflections from all waves, with mean displacement subtracted. Calculated inland displacements (thin line) are a small fraction of those at the coastal site (bold line).

constants, Young's modulus and Poisson's ratio, the calculated maximum deflection associated with this lattice of waves is of the order of 10–20 microns, as can be seen in Figure 18.25. In addition, the calculated deflection of the surface falls off with distance inland, such that the amplitude of the wave-induced deflection is roughly 30% of that at the coast by 30 m

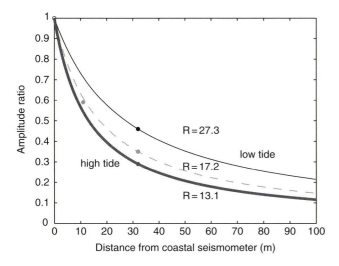

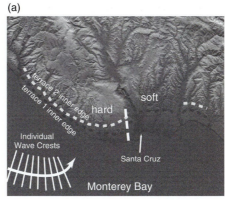

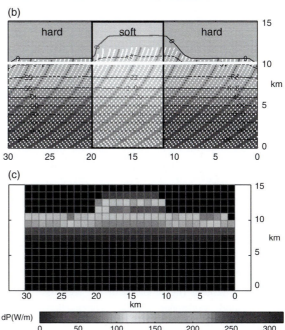

Figure 18.26 Measured and predicted patterns of decay of wave-induced deflection amplitude with distance inland from the first seismometer. Measured amplitude ratios at inland station (32 m) are used to predict the center of wave loading offshore (R). The distance increases as the tide drops. The amplitude ratio at an intermediate site (11 m), measured at medium tide (dashed curve), is well predicted (after Adams *et al.*, 2005, Figure 13, with permission from the American Geophysical Union).

Figure 18.27 Two-dimensional numerical simulation of a coastal embayment. (a) Santa Cruz coastline showing deeper indentation of marine terrace inner edges 1 and 2 where the lithology is a softer sandstone. (b) Simulated coastal erosion of a coastline in which a rock type five times more susceptible to erosion is embedded in harder rock. Waves are taken to come from 60° orientation; every fifth wave ray is black. An embayment develops in the softer rock. (c) Map of coastline and wave power dissipation as a proxy for erosion of the seafloor, reported at the final timestep of the model (after Peter N. Adams, Figures 1 and 5 from unpublished Chapter 4 of his PhD dissertation, UC Santa Cruz, with permission).

from the seacliff. That both of these calculations are in accord with the measured deflections (20 microns at the coast, 30% of this at 30 m inland) support the interpretation of the swaying of the cliff as being caused by the elastic response to passage of the moving set of line loads associated with the wave field. Indeed, we can go one step further with this data. The simple point load theory suggests that the deflection ought to fall off as $1/r$, the distance from the load. We can use the amplitudes of the deflections measured with the seismometers at the three distances from the cliff to predict where the locus of the load must have been. Graphically, this is visible in Figure 18.26. The resulting projections to the load suggest that it was 15 m from the cliff-edge seismometer at high tide, and 27 m at low tide. These are reasonable values, given the geometry of the cliff.

Long-term evolution of the coastal plan view

Now consider the longer-term problem of evolving the shape of the coastline. The northwestern edge of Monterey Bay coincides with a break in lithology, being Miocene Santa Cruz mudstone west of Santa Cruz point, and Plio-Pleistocene Purisima formation east of it, within the bay. We sketch these in Figure 18.27 as "hard" and "soft" lithologies, respectively. The eastern edge of Monterey Bay is similarly defined by a lithology contrast, the Monterey peninsula famous for its golf courses with rocky coastal vistas being composed of Mesozoic granite, while the interior is Tertiary sediments. This suggests that lithology might control the location of such embayments, and inspires the following simple numerical experiment. Consider an initially straight coastline subjected to steady sea level and steady distribution of waves. The susceptibility of the coast to erosion is allowed to vary in a stepwise manner, being hard in the middle of the domain, soft (more susceptible)

elsewhere. Assume an initially planar shelf out to an edge at say 120 m depth. In the face of this steady forcing, how does the topography evolve? We must have rules for both the back-wearing of the cliff, and for any wave-induced erosion of the shelf. Wave refraction and dissipation of wave energy must be done appropriately, as the shelf topography across which they pass evolves. Results of such an experiment are shown in Figure 18.27, revealing the development of an embayment in the coastline where softer rock appears.

Summary

The Santa Cruz landscape is remarkably suited to geomorphic study. Its marine terraces have aided in documenting the rate of rock uplift, the rate of seacliff degradation (hillslopes, weathering, and channels), and the rate of chemical weathering of the terrace cover sands. Each of these cases has been ground-breaking, from the original recognition by Alexander (1953) that marine terraces reflect rock uplift, to the early work of Hanks and Andrews (1989) on cliff degradation, to the most recent use of the terraces as a chronosequence by White *et al.* (2008). The simplicity of the seacliffs composed of fine-grained mudstones allowed construction of a littoral sand budget that was amenable to a simple two-component cosmogenic mixing model. Finally, the infrastructure of the Monterey Bay region, with its buoys and marine institutes, has promoted oceanic research into the realm of the Monterey submarine canyon, and has enabled sensing of the shaking and swaying of the coastal cliffs as waves arrive from far-flung oceanic storms. The landscape has therefore contributed to our understanding of most elements of geomorphology, from the raising of rock into harm's way by geophysical processes, to its attack by both atmospheric and oceanic processes. It is a landscape that will no doubt continue to act as a natural experiment from which we will learn.

Problems

1. *Advection and uplift.* Given the slip rate, V, of the San Andreas Fault, and a stationary rock uplift pattern tied to a bend in the fault represented by $U = U_{\max}e^{-(x/x_*)^2}$, what is the total uplift one would expect by translation of the crust past the entire uplift welt? (*Hint*: first generate a mathematical expression, and then perform a specific calculation by letting $U_{\max} = 1$ mm/yr, $x_* = 15$ km, and the SAF slip rate to be 35 mm/yr.)

2. Using the plot in Figure 18.4, estimate how many coseismic uplift events the Highway 1 terrace must have experienced in order to reach its present elevation in the middle of the Aptos Arch. (*Hint*: you will have to assert an age and a sea level at which the terrace platform formed – information provided elsewhere in the chapter.)

3. Estimate the background "inheritance" of ^{10}Be in the Highway 1 terrace profile (Figure 18.11), and

use this to estimate the mean erosion rate in the catchments that once supplied beach sand to this terrace. (*Hint*: assume that the majority of the inheritance arises from exposure during exhumation on hillslopes rather than during translation in the fluvial and littoral systems.)

4. If the waves arriving at the coast have a 10 second period, the coastal retreat rate is 20 cm/yr, and the horizontal exponential length scale for decay of flex-swaying of the cliffs by the load of the waves is 15 m, through how many wave cycles will a rock mass experience significant strain before the cliff wears back to it? (*Hint*: here we define "significant" strain associated with the swaying when the amplitude is 10% of that at the cliff.

5. If the ^{10}Be concentration in littoral sand is 1×10^5 atoms/gram quartz, that in river sand is 1×10^4 atoms/gram quartz, and the mean ^{10}Be in terrace deposits atop the back-wearing cliffs is

2×10^5 atoms/gram quartz, what proportion of the littoral sand is derived from the rivers and the cliffs, respectively? Assume that the only contribution to the littoral sand from cliff backwearing is due to the terrace deposits themselves – in other words, the bedrock of the cliffs produces no sand.

6. If the CRN-derived age of the terrace is 80 ka, and the terrace sand is 1.5 m thick, and has a bulk density of 2000 kg/m^3, what is the mean concentration of ^{10}Be in the terrace sands?

7. *Thought question.* Discuss why the Santa Cruz Mountains, which rise more than 1 km above sea level, are not terraced from bottom to top? In other words, why do the terraces end in the landscape at some elevation despite the fact that the processes causing the landmass to rise have been the same for several millions of years, much longer than the oldest terrace (∼250 ka)?

Further reading

Anderson, R. S., C. A. Riihimaki, E. B. Safran, and K. R. MacGregor, 2006b, Facing reality: late Cenozoic evolution of smooth peaks, glacially ornamented valleys and deep river gorges of Colorado's Front Range, in: S. D. Willett, N. Hovius, M. T. Brandon, and D. M. Fisher (eds.), *Tectonics, Climate and Landscape Evolution*, Geological Society of America Special Paper 398, Boulder, CO: Geological Society of America, pp. 397–418.
Although this pales in comparison with Gilbert's analysis of the Henry Mountains more than a century earlier, the authors attempt to address the origin of several key elements of the landscape of the Front Range of the Rockies, and by analogy other Laramide Ranges in the western United States.

Gilbert, G. K., 1877, *Report on the Geology of the Henry Mountains*, US Geographical and Geological Survey of the Rocky Mountain Region, Washington, DC: Government Printing Office, 160 pp.
This is Gilbert's tour de force. His chapter on landscape development in this isolated arid mountain range just north of Glen Canyon is most likely the first compact analysis of the entire geomorphic system. From the generation of regolith by weathering to its transport down hillslopes to its transport in rivers and the erosion of the riverbed by its passage, Gilbert's treatment of each component in the system anticipates our modern understanding of geomorphology. It is required reading for any geomorphologist.

APPENDIX A: PHYSICS

Primary units

length, L	meter	m
mass, M	kilogram	kg
time, T	second	s
temperature, °T	kelvin	K

Key definitions

Quantity	Equation	Units	Name, symbol, comment
Force	$F = ma$ (Newton's second law)	ML/T^2	newton, N
Momentum	$P = mU$	ML/T	
Energy (or work)	$i =$ force $\times$ distance (work)		joule, J ($=$ N-m)
Stress (normal, shear)	force per unit area	M/LT^2	pascal, Pa ($=$ N/m^2)
Pressure (same as stress)	force per unit area	M/LT^2	pascal, Pa

Power	energy per unit time	ML^2/T^3	watt, W ($=$ J/s)
Kinetic energy	$KE = \frac{1}{2}MV^2$	ML^2/T^2	joule, J
Potential energy	$PE = mgh$	ML^2/T^2	joule, J h, height above an arbitrary datum
Heat energy	$E = \rho c T_k dx dy dz$	ML^2/T^2	joule
Flux	$q = A/L^2 T$	A/L^2T	where A is any quantity
Discharge	$Q =$ flux $\times$ cross-sectional area	A/T	cumecs ($=$ m^3/s) if water discharge
Unit discharge (or specific discharge)	$Q = A$ per unit width per unit time	A/LT	
Strain	$\varepsilon = \Delta L/L_o$ (linear strain) or $\varepsilon = \Delta V/V_o$ (volumetric strain)	dimensionless	
Strain rate	change in size per unit time, normalized by original size	$1/T$	

Heat transport mechanisms

Radiation

$Q = \sigma T_s^4$, where σ is the Stefan-Boltzmann constant, and T_s is the surface temperature (K)

Wien's law: $\lambda_{peak} = a/T_s$, where λ_{peak} is the peak in the spectrum and $a = 2898$ μm-K

$Q \sim 1/R^2$, where R is the distance from the emitting object: an inverse square law

Conduction

Fourier's law: $Q = -k(dT/dz)$

Advection

Described formally by terms like UdA/dx, where U is the component of velocity in the x-direction, and A is the quantity being advected

Convection

Motion of a material (and heat and other scalar quantities with it) driven by density contrasts, themselves caused by either thermal or chemical gradients

Rheologies

Fluids: strain rate ~ stress

Viscous linear (Newtonian viscous): $\tau = \mu\,du/dz$ or $du/dz = (1/\mu)\tau$

Nonlinear viscous: $du/dz = A\tau^n$, e.g., cubic ($n = 3$, Glen's flow law)

Plastic: $du/dz = 0$, $\tau < \tau_o$, $du/dz = \infty$, $\tau > \tau_o$, where τ_o is the threshold or yield stress

Coulomb granular solids – frictional behavior in bulk: $\tau = C + \sigma \tan\phi$ everywhere

Solids: strain ~ stress

Elastic: $\varepsilon = E\sigma$

$\varepsilon_{yy} = -\nu\varepsilon_{xx}$

Important dimensionless numbers

Name	Dimensions	Ratio of what to what	End-member behaviors
Fr, Froude number	$U/\sqrt{gH}$	inertial–gravity	tranquil–supercritical
Ma, Mach number	U/U_{sound}		subsonic–supersonic
Pr, Prandtl number	ν/κ	diffusion of heat–diffusion of momentum	
Re, Reynolds number	$= LU/\nu$	inertial–viscous forces	laminar–turbulent
Ra, Rayleigh number		buoyancy–viscous	stable–convecting
Ro, Rouse number	$w_{sett}/ku*$	response time–hop time	suspended–bedload

Important natural constants

c	speed of light	2.998×10^8 m/s
σ	Stefan–Boltzmann constant	5.67×10^{-8} W/(m^2 K^4)
R	universal gas constant	8.3144 J/(mol K)
A	Avagadro's number	6.022×10^{23} mol^{-1}
G	gravitational constant	6.6732×10^{-11} N m^2/kg^2

Physical properties

Symbol	Name	Units (SI)
a	albedo	dimensionless [0,1]
c	heat capacity, or specific heat	J/(kg K)
C_d	drag coefficient	dimensionless
D	flexural rigidity	N-m
E	Young's modulus	Pa
E_a	activation energy	J/mole
f	coefficient of friction	dimensionless
k	thermal conductivity	W/(m K)
K	hydraulic conductivity	m/s
P	permeability	m^2
$t_{1/2}$	half-life	s
α	thermal expansion coefficient	K^{-1}
ϕ	porosity	dimensionless [0,1]
κ	thermal diffusivity $= k/\rho c$	m^2/s
μ	dynamic viscosity	Pa-s
ν	kinematic viscosity $= \mu/\rho$	m^2/s
ν	Poisson's ratio	dimensionless
ρ	density	kg/m^3

APPENDIX B: MATHEMATICS

In this appendix we list important numbers, formulas, rules, and functions, and show what they look like graphically.

Numbers worth memorizing

$\pi = 3.14159 \ldots$
$e = 2.7183 \ldots$

We often normalize a function or distribution in order to compare it with others. This is shown in many of the graphs below. Normalization is accomplished by dividing both the x-values and the y-values by constants, and plotting the resulting (now non-dimensional) values on the new axes. This process is something of an art, which comes in the choice of what scales to use in normalizing. An obvious choice for the constant is the maximum value. Upon dividing by the maximum, the resulting ratios must lie between 0 and 1. Often, however, there is no clear maximum value (especially on the x-axis), and we must choose something that is characteristic of the problem. Examples include the standard deviation of a distribution, or the period of the oscillation, or the length scale over which the value changes by a factor of 2 or of e. (See graphs of exponentials, Figures B.2–B.3.) The result is always a graph that has values that go from 0 to 1 or from 0 to a few. It also allows us to compare the shapes of functions

to one another, as this normalization removes the role of the scales themselves. Note for example how all of the Gaussian curves plot on top of one another when normalized.

Important functions

1. *Straight lines:* $y = mx + b$ (Figure B.1). Here the slope on the plot is m, while the y-intercept is b. Lines are everywhere in geomorphology. They define the straight slopes of landslide-prone hillslopes. They relate the flux of regolith and slope angle for rainsplash and frost-creep processes.
2. *Negative exponentials* (Figure B.2) are encountered in radioactive decay, in production profiles of cosmogenic nuclides: $y = Ae^{-x/x^*}$. These are characterized by two constants, A and x_*. The first is the maximum value, found at $x = 0$. The second is the scale over which the function falls by a factor of e. We call this scale the e-folding scale. It is found graphically by finding the place at which the value of the function is $1/e$ of A, or A/e. Recalling that e is about 3 (actually 2.7183...), this is roughly a third of the value of A, which is easy to estimate on the graph. These functions are encountered in the decay of radioactive nuclides.
3. *Positive exponentials* (Figure B.3) are found in the unchecked growth of populations: $y = Ae^{x/x^*}$. In this case, x stands for time. Exponential growth

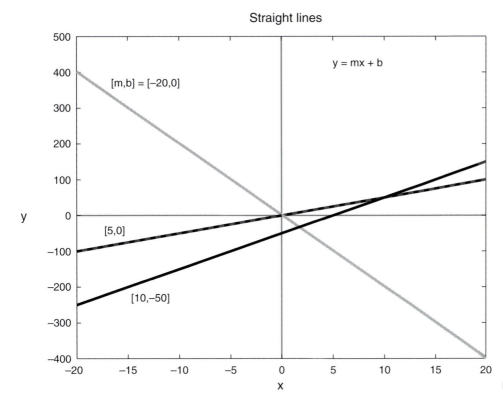

Figure B.1 The straight line.

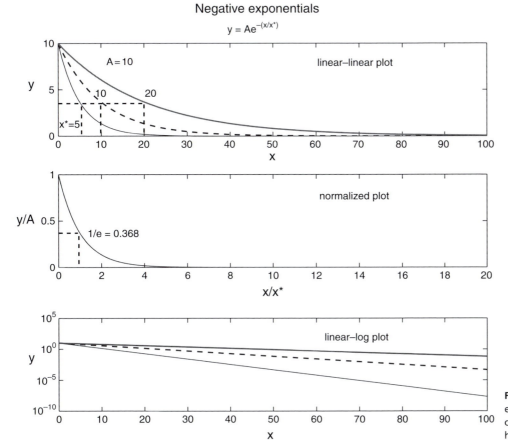

Figure B.2 The negative exponential function. This is characterized by two constants, here denoted A, and x^*.

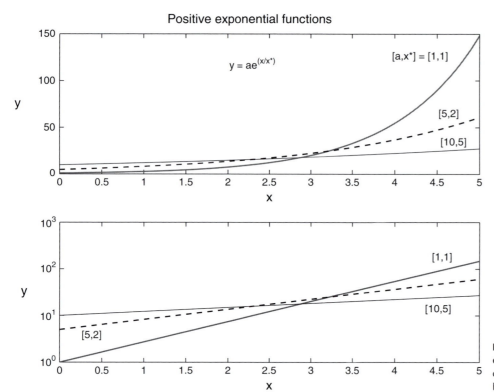

Positive exponential functions

$$y = ae^{(x/x^*)}$$

$[a,x^*] = [1,1]$

$[5,2]$

$[10,5]$

$[1,1]$

$[5,2]$

$[10,5]$

Figure B.3 The positive exponential function. This too is characterized by two constants, here denoted a and x^*.

is what we expect in a population that grows at a rate dictated by the number of individuals in the population at any time. Just as for negative exponentials, it too is characterized by two constants, A and x_*. The first is again the initial value, found at $x = 0$. The second is the scale over which the function changes (this time *increases*) by a factor of e (the e-folding scale).

4. Closely related to exponentials is a function that approaches an *asymptote* as an exponential (Figure B.4): $y = A(1 - e^{-x/x^*})$. Here the function is defined by the value of the asymptote, A, and by the rate at which the asymptote is approached. Again, this is set by a scale, we use x_*. Note the values of the function at three places: at $x = 0$, $e(0) = 1$, implying $y = 0$. At $x = \infty$, $e^{(-\infty)} = 0$, implying $y = A$. Finally, at $x = x^*$, $y = A(1 - e^{-1}) = A(1-(1/e)) = 0.63A$. This approach toward an asymptote is found in systems in which both growth and decay occur, the asymptote reflecting a balance of growth and decay (also called secular equilibrium in radioactive decay series).

5. *Power-law functions* (Figure B.5) are very common in geomorphology: $y = Ax^p$. They arise in drainage

basin characteristics. They have the important property that they become straight lines when plotted on log–log graphs. The slope of the line is the power. You can see this by logging both sides of the equation: $\log(y) = \log(A) + p\log(x)$. This has the form of $y = b + mx$, which we all recognize as a straight line with intercept b and slope m. This means that an easy way to evaluate the power in a power-law function is by plotting it in this manner. Power-law functions are everywhere in geomorphology. To name two examples, power laws describe the relationship between the number of streams of one order with respect to the number in the next order (one of Horton's laws), and the relationship between slope and drainage area in a bedrock stream profile.

6. The *parabola* (Figure B.6) is of course a special case of a power law: $y = y_0 + A(x - x_0)^2$, but as it is so commonly seen in geomorphology it is worth breaking out separately. Sand grains splashed up by raindrops carry out parabolic trajectories. Steady-state hilltops have parabolic topographic profiles. Flow of a viscous fluid between two plates (like magma in a dike) has a

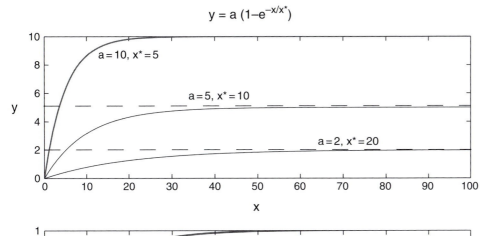

$$y = a\,(1-e^{-x/x^*})$$

a = 10, x* = 5

a = 5, x* = 10

a = 2, x* = 20

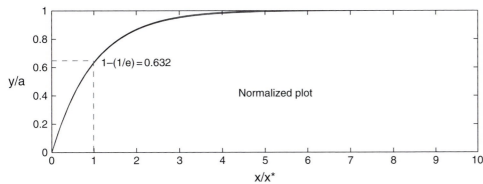

1−(1/e) = 0.632

Normalized plot

Figure B.4 An example of an asymptotic function.

Power law

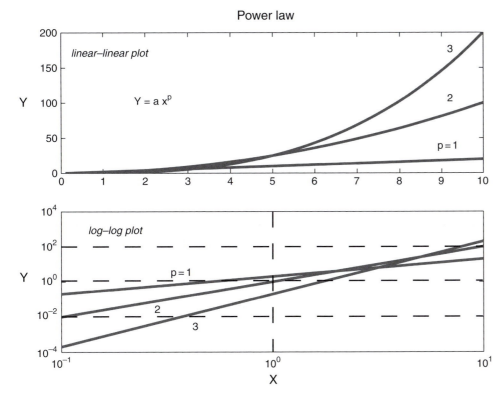

linear–linear plot

$$Y = a\,x^p$$

3

2

p = 1

log–log plot

p = 1

2

3

Figure B.5 Power-law functions, here shown with positive powers. Note that power laws on log–log plots are straight lines.

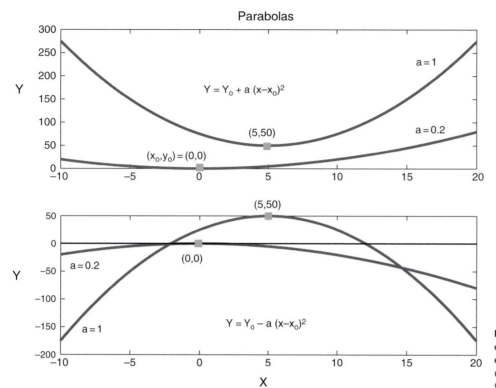

Figure B.6 Parabolas. Here examples are shown that are either downward convex or upward convex, and shifted from (0,0).

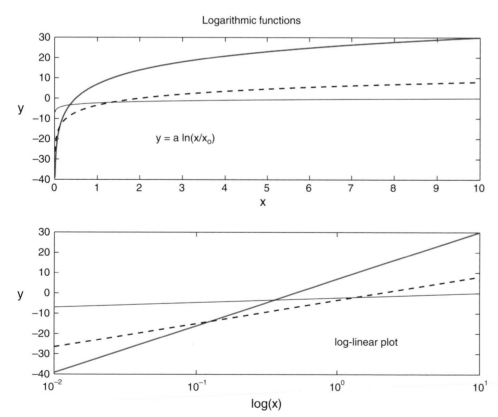

Figure B.7 Logarithmic functions. These plot as straight lines on log–linear graphs.

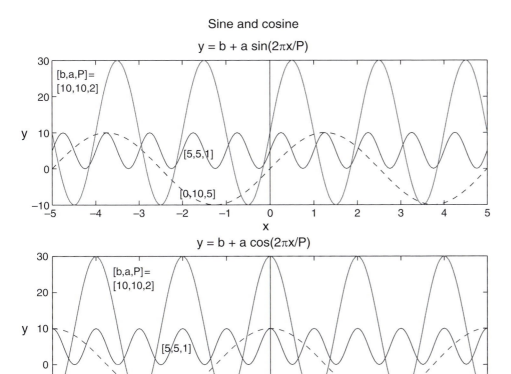

Sine and cosine

$y = b + a\,\sin(2\pi x/P)$

[b,a,P]=
[10,10,2]

[5,5,1]

[0,10,5]

$y = b + a\,\cos(2\pi x/P)$

[b,a,P]=
[10,10,2]

[5,5,1]

[0,10,5]

Figure B.8 Trigonometric functions sine and cosine. Effects of changing shift, b, amplitude, a, and period, P, are shown.

parabolic velocity profile. The general form here accommodates a parabola centered not on [0,0] but on $[x_o, y_o]$. This can be seen by setting $x = x_o$. The value of $y = y_o$ then follows.

7. *Logarithmic functions* (Figure B.7): $y = A\,\log(x/x_o)$. These are encountered in fluid mechanics. For example, the flow speed increases as a logarithm of height above the bed in an open channel flow. These functions have the property of rapidly increasing at first and increasing much more slowly thereafter.

8. Plots of two major *trigonometric functions* (Figure B.8) are shown in order to emphasize the role of scaling in both the x and y dimensions. Recall that $\sin(0) = 0$ and that $\cos(0) = 1$. Quite complicated looking graphs can be constructed by combining sin and cosine curves – this is the essence of the Fourier transform. The surface temperature of the Earth carries out sinusoidal swings on both daily and annual timescales.

9. The *hyperbolic tangent* (Figure B.9). We include this function because it serves to step smoothly from one value (here $b - a$) to another $(b + a)$ over a specified distance, scaled by x_* and centered at x_o. The formula is $y = b + a\,\tanh((x - x_o)/x_*)$.

10. *Hyperbolic sine* and *hyperbolic cosine* (Figure B.10) can also be scaled as shown on their plots. One encounters these functions in solutions for the displacement profile around a fault.

11. The *Gaussian* (Figure B.11) is often encountered in error analysis, as errors are supposed to be normally, or Gaussianly distributed: $y = A e^{-((x-x_o)/x_*)^2}$. The function is named for Karl Freidrich Gauss (1777–1855), a German mathematician and astronomer. It yields the classic bell-shaped curve. As written, it is centered on $x = x_o$. The value of x_* (also known as the standard deviation if this is a probability density function – see below), sets how sharply the curve falls off away from the peak value of A.

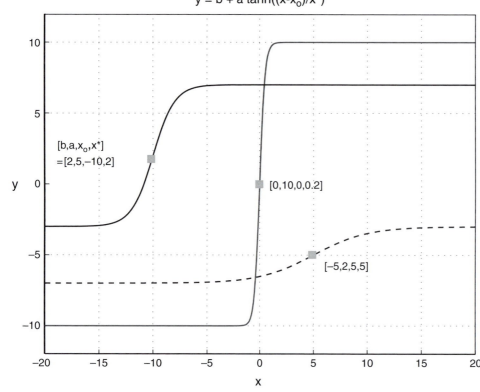

Figure B.9 Hyperbolic tangent. Effects of changing shift, *b*, amplitude, *a*, center, x_0, and horizontal scale over which the step occurs, *x**, are shown.

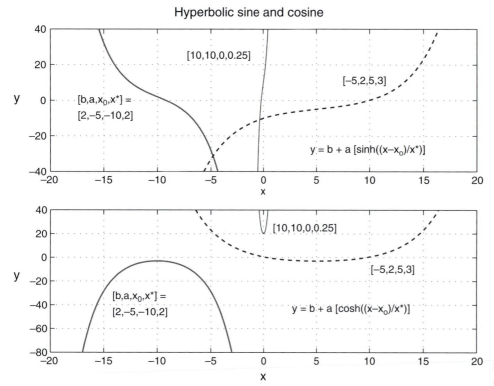

Figure B.10 Hyperbolic sine, sinh, and cosine, cosh.

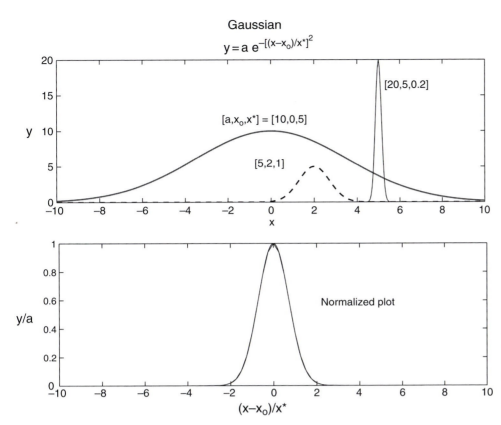

Figure B.11 Gaussian function. All plots in the top panel collapse to that on the bottom panel when y is normalized using a, and when the x-axis is shifted by x_0 and then scaled by $x*$.

Basic rules of thumb for manipulation of expressions

Logs, powers and exponentials

Laws of exponents

$$a^m a^n = a^{(m+n)}$$

$$\frac{a^m}{a^n} = a^{(m-n)}$$

$$a^0 = 1$$

$$a^{p/q} = \sqrt[q]{a^p}$$

$$a^{-t} = \frac{1}{a^t}$$

Logarithms

Definition: if $y = \log_a(x)$, then $a^y = x$, where a is called the base of the logarithm. The most important bases for us are 10, 2, and e.

Terminology: $\log_e(x) = \ln(x)$ or the log base e, is also called the *natural logarithm*.

Conversion from $\log_{10}$ to ln: $\log_{10}(a) = \ln(a)/\ln(10) = \ln(a)/2.303$

Laws of logarithms

$$\log(xy) = \log(x) + \log(y)$$

$$\log(x/y) = \log(x) - \log(y)$$

$$\log(x^b) = b \log(x)$$

Trigonometry

In Figure B.12 we show a right-angled triangle, one with a perpendicular (90° or $\pi/2$ radians) angle. The definitions of sin, cos, and tangent are as follows:

$$\sin = \text{opp}/\text{hyp}$$

$$\cos = \text{adj}/\text{hyp}$$

$$\tan = \text{opp}/\text{adj} = \sin/\cos$$

$$\sin^{-1}(x) = \text{invsin}(x) = \text{inverse}\sin(x)$$

$$\text{if}\ \sin^{-1}(x) = a,\ \text{then}\ \sin(a) = x$$

$$1\ \text{radian} = 360/2\pi = 57.3°$$

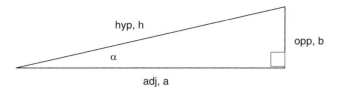

Right-angled triangle. Definition sketch for trigonometric functions
sin = opp/hyp; cos = adj/hyp; tan = opp/adj

Figure B.12 Right-angled triangle. Definitions of sine, cosine, and tangent use sides labeled opposite, adjacent, and hypotenuse.

Small-angle approximation: $\sin(\alpha) = \tan(\alpha) = \alpha$ for very small angles, α (where α is taken to be in radians!). For the same reasons, $\cos(\alpha) = 1$ for very small angles.

Pythagorean theorem: for a right-angled triangle with hypotenuse h and sides a and b, $h^2 = a^2 + b^2$. This can be manipulated to solve for the length of any unknown side given the other two. This is named for Pythagoras, the great Greek mathematician and philosopher who died around 497 B.C.

Angle formulas

$\sin^2\theta + \cos^2\theta = 1$

$\sin(\theta \pm \phi) = \sin\theta\cos\phi \pm \cos\theta\sin\phi$

$\sin 2\theta = 2\sin\theta\cos\theta$

$\cos 2\theta = \cos^2\theta - \sin^2\theta$

Law of sines : $\dfrac{a}{\sin A} = \dfrac{b}{\sin B} = \dfrac{c}{\sin C}$

Geometry

Volume, area, and circumference

Circle: $C = 2\pi R = \pi D$

$A = \pi R^2 = \frac{1}{4}\pi D^2$

Sphere: $A = 4\pi R^2 = \pi D^2$

$V = \frac{4}{3}\pi R^3 = \frac{1}{6}\pi D^3$

Algebra

The *quadratic formula*: for an equation that may be written

$ax^2 + bx + c = 0,$

the roots (the values of x where the equation is 0) may be found using the formula:

$$x = \frac{-b \pm \sqrt{b^2 - 4ac}}{2a}.$$

Calculus

Derivatives

The *first derivative* is the local slope of a function, say $y(x)$. It is defined as follows:

$$\frac{dy}{dx} \equiv \lim_{dx \to 0} \frac{y(x + dx) - y(x)}{dx}$$

so that as the interval over which the slope is being evaluated shrinks, the local slope is better and better approximated (see Figure B.13). The derivative is variously expressed as dy/dx, $y'(x)$, y', or even y_x. When you see the notation $\partial z/\partial x$, it signifies a *partial derivative*, and is read the partial derivative of z with respect to x. This occurs when there are several variables involved, for example, when z is a function of x and y, $z(x,y)$, and means the derivative of z taken with respect to the variable x, while holding all other variables constant. The dimensions of a derivative are those of y over those of x.

The *second derivative* of a function, d^2y/dx^2 is the curvature of the function, or the slope of the slope. It is positive if the slope is increasing with distance, x, and negative if the slope is declining with distance (Figure B.14). Comparable notation for the second derivative is d^2y/dx^2, $y''(x)$, y'', and y_{xx}. The dimensions of a second derivative are those of y over those of x squared.

While massive tables exist of both derivatives and integrals (e.g., in the front of any CRC Handbook of Chemistry and Physics), we list important examples here for convenience.

In the formulas to follow, a is a real constant, x is a variable, u is a function of x, $u(x)$, and all trigonometric functions are measured in radians.

1. $\dfrac{d}{dx}(a) = 0$

2. $\dfrac{d}{dx}(x) = 1$

Definition sketch for the derivative

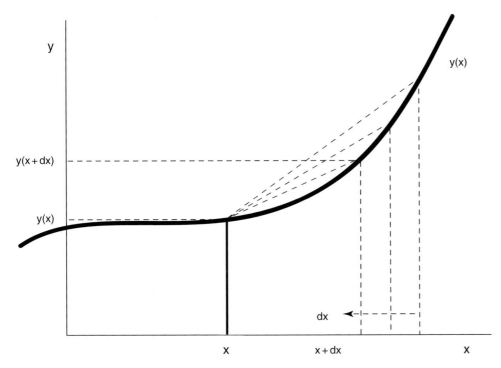

Figure B.13 Definition of the derivative, dy/dx.

Slope and curvature

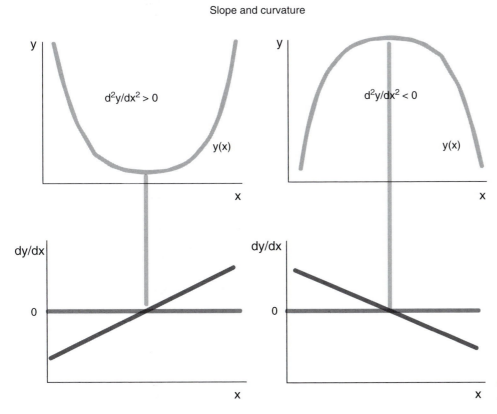

Figure B.14 Slope and curvature of the function $y(x)$.

3. $\dfrac{d}{dx}(u^n) = nu^{n-1}\dfrac{du}{dx}$

4. $\dfrac{d}{dx}(uv) = v\dfrac{du}{dx} + u\dfrac{dv}{dx}$ – the product rule

5. $\dfrac{d}{dx}\ln(u) = \dfrac{1}{u}\dfrac{du}{dx}$

6. $\dfrac{d}{dx}(e^u) = e^u\dfrac{du}{dx}$

7. $\dfrac{d}{dx}(\sin(u)) = \cos(u)\dfrac{du}{dx}$

8. $\dfrac{d}{dx}(\cos(u)) = -\sin(u)\dfrac{du}{dx}$

9. $\dfrac{d}{dx}(\tan(u)) = \sec^2(u)\dfrac{du}{dx}$

The *integral* of a function is the area under the function, lying between the function and the *x*-axis (see Figure B.15). A *definite integral* is taken over a specified interval, [*a,b*], while for an *indefinite integral* this interval is not specified.

Integrals

Indefinite integrals: to each of the solutions shown, you must add a "constant of integration":

1. $\displaystyle\int a\,dx = ax$

2. $\displaystyle\int x^n\,dx = \dfrac{x^{n+1}}{n+1}$ except when $n = -1$

Definition sketch for the integral

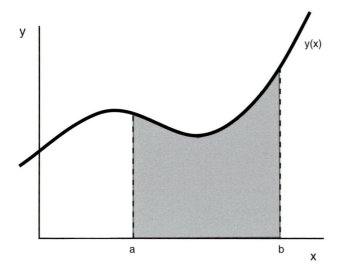

Figure B.15 Definition of the definite integral of *y(x)* as the area under the curve between $x = a$ and $x = b$.

3. $\displaystyle\int \dfrac{1}{x}\,dx = \log(x)$

4. $\displaystyle\int e^x\,dx = e^x$

5. $\displaystyle\int e^{ax}\,dx = \dfrac{e^{ax}}{a}$

6. $\displaystyle\int \log(x)\,dx = x\log(x) - x$

7. $\displaystyle\int \sin(ax)\,dx = -\dfrac{1}{a}\cos(ax)$

8. $\displaystyle\int \cos(ax)\,dx = \dfrac{1}{a}\sin(ax)$

A couple of useful definite integrals are as follows:

1. $\displaystyle\int_0^\infty e^{-ax}\,dx = \dfrac{1}{a}, \quad a > 0$

2. $\displaystyle\int_0^\infty e^{-a^2x^2}\,dx = \dfrac{1}{2a}\sqrt{\pi}, \quad a > 0$

Mean value theorem

The *mean value theorem* is used to assess formally the mean value (or average) of a function over a specified interval. This is shown graphically in Figure B.16, and mathematically as:

$$\bar{y} = \dfrac{1}{b-a}\int_a^b y(x)\,dx$$

Note that if we rearrange this by multiplying by $(b - a)$, it becomes:

$$\bar{y}(b - a) = \int_a^b y(x)\,dx$$

The left-hand side corresponds to the area of the box $\bar{y}$ tall and $(b - a)$ wide, while the right-hand side corresponds to the area under the curve, $y(x)$. The mean value of the function is therefore formally the value of y for which these areas are equal. We make use of this, for example, in evaluating the mean flow velocity, given a flow velocity profile.

Taylor series expansion

The Taylor series expansion, named for the English mathematician Brook Taylor, is used to estimate the

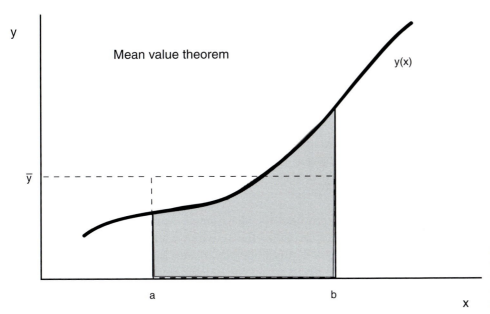

Figure B.16 Graphical representation of the mean value theorem used to obtain the mean of the function $y(x)$ over the interval $[a,b]$.

value of a function at a location nearby one at which we know the value of the function and its derivatives. Of course one could obtain a first-order estimate by assuming that the value is the same as that at the known location. But that wouldn't honor the fact that the function is changing. For example, it may have a finite slope at the known location, and a curvature, and the curvature may be changing, and so on. We obtain better and better estimates by taking into account these additional pieces of information. The Taylor series is represented as a sum of terms in an infinite series:

$$y(x + dx) = y(x) + \frac{1}{1!}\frac{dy}{dx}(x)dx + \frac{1}{2!}\frac{d^2y}{dx^2}(x)dx^2$$

$$+ \frac{1}{3!}\frac{d^3y}{dx^3}(x)dx^3 \cdots$$

Recalling that one factorial (1!) is simply 1, the first two terms in this series collapse to:

$$y(x + dx) \approx y(x) + \frac{dy}{dx}(x)dx$$

where the $\approx$ sign acknowledges that we are ignoring higher-order terms (those with higher-order derivatives) in the approximation. We employ this approximation in deriving several differential equations in this book.

Ordinary differential equations (ODEs)

These contain a variable and its derivatives; they are limited to one variable – if more, then partial differential equations (PDEs).

They are classified according to several criteria:

- the highest degree of derivative in the equation (called the order of the equation);
- whether they contain variables as multiples or powers, or products of variables with derivatives (linear if not, nonlinear if so);
- whether coefficients are constants or not.

Examples are as follows:

$$\frac{dN}{dt} + aN = 0 \text{ first-order, linear homogeneous ODE}$$
$$N = N_0 e^{-at}$$

We encounter this equation in the decay of radio-nuclides such as ^{14}C used to date geomorphic surfaces.

$$\frac{dN}{dt} + aN = P \text{ first-order, linear, nonhomogeneous ODE}$$
$$N = \frac{P}{a}(1 - e^{-at})$$

This describes a population with both new production or growth, represented by P, and decay or death, represented by aN. The solution is a function that approaches an asymptote, called secular equilibrium, in which growth and decay are perfectly balanced.

$$\frac{d^2 z}{dx^2} = A \quad \text{second-order, linear, nonhomogeneous ODE}$$

$$z = \frac{1}{2} A x^2 + c_1 x + c_2$$

where c_1 and c_2 are constants of integration, for which one must appeal to boundary conditions. In this example, if z is topographic elevation, the equation would represent a system in which topographic curvature is uniform (at A). This occurs on steady hilltops, and leads to a parabolic form if $c_1 = 0$, reflecting the condition that the slope at the hillcrest is zero.

Partial differential equations (PDEs)

$$\frac{\partial T}{\partial t} = A \frac{\partial^2 T}{\partial x^2} \quad \text{second-order, linear PDE}$$

This is the classic diffusion equation, encountered in the study of conducting systems. It has strong analogs in the flow of viscous fluids, and the evolution of topography in the face of diffusive hillslope processes.

$$\frac{\partial u}{\partial t} + u \frac{\partial u}{\partial x} = A \frac{\partial^2 u}{\partial x^2} \quad \text{second-order, nonlinear PDE}$$

These are a few terms of the Navier–Stokes equation describing the conservation of momentum in a one-dimensional flow dominated by viscous forces. The nonlinear term in this case is the second term, with a product of the variable and its derivative.

Statistics

For a discrete set of values, x_i, the *mean* of the population is:

$$\bar{x} = \frac{1}{n} \sum_{i=1}^{n} x_i$$

or for a random variable, x, with the probability of occurrence of the variable x_i taken to be $f(x_i)$,

$$\bar{x} = \sum_{i=1}^{n} x_i f(x_i)$$

Note that the probabilities $f(x)$ are constrained to have $\sum_n f(x_i) = 1$.

The *variance* is:

$$Var = \sum_{i=1}^{n} x_i^2 f(x_i) - \bar{x}^2$$

and the *standard deviation* is the square root of the variance:

$$\sigma = \sqrt{Var}$$

For continuously distributed variables, however, we may formalize the statistics using a continuously distributed probability of occurrence. A *probability density function*, or *pdf*, is a function that identifies the probability of occurrence of some event, x. The pdf, $f(x)$ is defined as:

$$P(a, b) = \int_a^b f(x) dx$$

where $P(a, b)$ signifies the probability of finding the value between a and b.

All pdfs are constrained by $\int_{-\infty}^{\infty} f(x) dx = 1$; in other words, the probability that the variable will fall between $-$ and $+$ infinity is unity. In addition, one might want to know the cumulative probability density function, $F(x)$, defined as:

$$F(a) = P(x \le a) = \int_{-\infty}^{a} f(x') dx'$$

We may now be precise in defining several statistical measures, as shown in Figure B.17.

The *mode* of a distribution is that value of x corresponding to the peak in the probability density function, i.e., the most probable value.

The *mean* of a distribution, $\bar{x}$, is defined formally to be:

$$\bar{x} = \int_{-\infty}^{\infty} x' f(x') dx'$$

The *median* is that value of x at which the cumulative probability distribution function is 0.5, i.e.:

$$F(x) = \int_{-\infty}^{x} f(x') dx' = 0.5$$

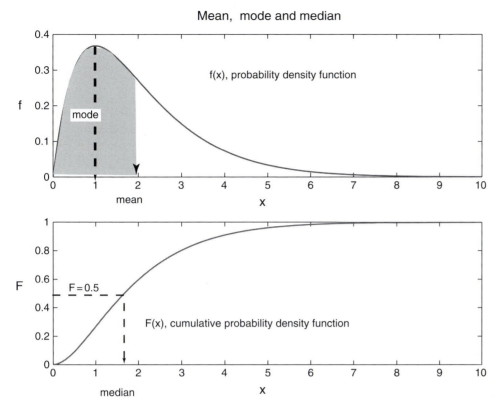

Figure B.17 Definitions of mean, mode, and median.

The spread in the distribution, or its width, is often desired as well. Formally, the *standard deviation* of a distribution is

$$\sigma = \left[\int_{-\infty}^{\infty} (x - \bar{x})^2 f(x)\mathrm{d}x \right]^{1/2}$$

and the *variance* is the square of this. Note that the standard deviation has the same units as the variable, while the variance has units of the variable squared. This is also called the second moment of the distribution, named for the second power in the formula. (In this terminology, the first moment is the mean, as x is taken to the first power.)

Probability density functions (PDFs)

We encounter a wide range of pdfs in geomorphology, most of which are captured in the following few examples (some shown in Figure B.18). Note that in each case there is an easily identified function with which you are likely already familiar, multiplied by some collection of constants out in front. The role of these constants is to assure that the integral over the full range is 1.

The *uniform distribution*. Here there is an equal, or uniform, probability of finding the function over a specified interval:

$$f(x) = \frac{1}{b - a}; \quad a \leq x \leq b$$
$$f(x) = 0; \quad x < a, x < b$$

The *exponential* distribution. The probability is maximum at $x = 0$ and falls off exponentially to zero at infinity:

$$f(x) = \frac{1}{x_*}\mathrm{e}^{-x/x_*}; \quad x \geq 0$$
$$f(x) = 0; \quad x < 0$$

Note that the distribution is defined by a single parameter, x_*, which sets how rapidly the distribution falls off.

The *Gaussian* or *normal* distribution (already illustrated in Figure B.11). This is often encountered in

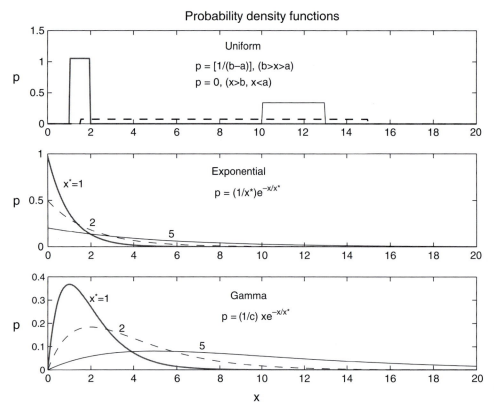

Figure B.18 Commonly used probability density functions, pdfs.

error analysis, as errors are supposed to be normally, or Gaussianly distributed:

$$f(x) = \frac{1}{\sigma\sqrt{2\pi}} e^{-\left(\frac{x - x_0}{\sigma}\right)^2}$$

The two parameters in the distribution as defined here are the mean, x_o, and the standard deviation, σ. This defines the classic bell-shaped curve.

The cumulative Gaussian pdf, centered on 0, is so commonly encountered that it has been given its own name – the *error function*:

$$\mathrm{erf}(\eta) = \frac{2}{\sqrt{\pi}} \int_0^{\eta} e^{-\eta^2} d\eta$$

Similarly, for problems of opposite symmetry, a *complementary error function* is defined as:

$$\mathrm{erfc}(\eta) = 1 - \mathrm{erf}(\eta) = 1 - \frac{2}{\sqrt{\pi}} \int_0^{\eta} e^{-\eta^2} d\eta$$

See our discussion of thermal diffusion problems.

The *gamma* distribution. Note that the probability goes to zero at both $x = 0$, and $x = $ infinity:

$$f(x) = \frac{1}{c} x e^{-x/a}$$

A more general gamma distribution allows x to be taken to some power.

The *log-normal* distribution. This is often encountered in describing grain size distributions (see Ole Barnsdorff-Neilsen's work):

$$f(x) = \frac{1}{x\sigma\sqrt{2\pi}} e^{-\frac{1}{2}\left(\frac{\log(x) - m}{\sigma}\right)^b}; \quad x \geq 0$$
$$f(x) = 0; \quad x < 0$$

where m is the mean and s is the standard deviation of the distribution. This distribution is flexible in that its shape can vary from essentially an exponential shape to a gamma-function like shape.

The *Weibull* distribution is defined by three parameters. This and variations of it are commonly used in describing meteorological data.

$$f(x) = \frac{b}{a}\left(\frac{x-c}{a}\right)^{b-1} e^{-\left(\frac{x-c}{a}\right)^b}; \quad x \geq c$$

$$f(x) = 0; \quad x < c$$

The Weibull distribution is similar to the gamma function, in that it collapses to the exponential (when $b = 1$), but is more flexible (i.e., has more parameters) to allow non-zero centered distributions (i.e., $c \neq 0$), and greater than exponential declines in probability at high x.

The *Binomial* distribution:

$$f(x) = [n!/x!(n-x)!)]p^x q^{n-x}, \quad \text{for } x = 0, 1, 2, \ldots, n$$

where p is the probability that the respective event will occur, q is equal to $1 - p$, and n is the maximum number of independent trials.

The *Poisson* distribution:

$$f(k; \lambda) = \frac{\lambda^k e^{-\lambda}}{k!}$$

Both the mean and the variance of this distribution equal λ, meaning that the standard deviation is $\sqrt{\lambda}$.

Goodness of fit

We often need to evaluate how well our model fits a particular set of data. One common means employs the chi-square statistic (see Bevington, 1969):

$$\chi^2 = \sum_{i=1}^{n} \left(\frac{x_{mi} - x_{oi}}{\sigma_i}\right)^2$$

in which x_m corresponds to a modeled value, while x_o is the corresponding data value. The standard deviation, σ, reflects the expected range or error in our knowledge of the real data, and acts to weight the data. Modeling strategies reduce to a search for the model that best minimizes this statistic.

REFERENCES

Aalto, R., L. Maurice-Bourgoin, T. Dunne, D. R. Montgomery, C. A. Nittrouer, and J. L. Guyot, 2003, Episodic sediment accumulation on Amazonian floodplains influenced by El Niño/Southern Oscillation, *Nature* **425**: 493–497.

Abad, J. D. and M. H. Garcia, 2006, RVR Meander: a toolbox for re-meandering of channelized streams, *Computers & Geosciences* **32**: 92–101.

Abbott, L. D., E. A. Silver, R. S. Anderson *et al.*, 1997, Measurement of tectonic surface uplift rate in a young collisional mountain belt, *Nature* **385**: 501–507.

Abramowitz, M. and I. A. Stegun (eds.), 1981, *Handbook of Mathematical Functions with Formulas, Graphs, and Mathematical Tables*, New York: Dover Publications.

Abrams, D. M., A. E. Lobkovsky, A. P. Petroff *et al.*, 2009, Growth laws for channel networks incised by groundwater flow, *Nature Geoscience* **2**, doi: 10.1038/NGEO432.

ACIA, 2005, *Arctic Climate Impact Assessment*, Cambridge: Cambridge University Press.

Adams, P. N., R. S. Anderson, and J. S. Revenaugh, 2002, Microseismic measurement of wave energy delivery to a rocky coast. *Geology* **30**: 895–898.

Adams, P. N., C. D. Storlazzi, and R. S. Anderson, 2005, Nearshore wave-induced cyclical flexing of seacliffs. *Journal of Geophysical Research – Earth Surface* **110**, No. F2: F02002, 10.1029/2004JF000217.

Adler, R. F., G. J. Huffman, A. Chang *et al.*, 2003, The version-2 global precipitation climatology project (GPCP) monthly precipitation analysis (1979–present), *Journal of Hydrometeorology* **4**: 1147–1167.

Agassiz, L., 1840, *Études sur les Glaciers*, Neuchâtel.

Ahnert, F., 1970, A comparison of theoretical slope models with slopes in the field, *Zeitschrift für Geomorphologie* **9**: 88–101.

Aitken, M. J., 1998, *An Introduction to Optical Dating: The Dating of Quaternary Sediments by the Use of Photon-stimulated Luminescence*, New York: Oxford University Press.

Alexander, C. S., 1953, The marine and stream terraces of the Capitola-Watsonville area, *University of California Publications in Geography* **10**: 1–44.

Allan, J. C. and P. D. Komar, 2006, Climate controls on US West Coast erosion processes, *Journal of Coastal Research* **22** (3): 511–529.

Allen, J. R. L., 1965, Late Quaternary Niger delta, and adjacent areas; sedimentary environments and lithofacies, *AAPG Bulletin* **49**: 547–600.

Allen, P. A., 1997, *Earth Surface Processes*, Oxford: Blackwell Science.

Alley, R. B. and A. M. Agustdottir, 2005, The 8 k event: cause and consequences of a major Holocene abrupt climate change, *Quaternary Science Reviews* **24**: 1123–1149, doi: 10.1016/j.quascirev.2004.12.004.

Alley, R. B., P. A. Mayewski, T. Sowers, M. Stuiver, K. C. Taylor, and P. U. Clark, 1997, Holocene climatic instability: a prominent, widespread event 8200 yr ago, *Geology* **25**: 483–486.

Alsdorf, D. E. and D. P. Lettenmaier, 2003, Tracking fresh water from space, *Science* **301**: 1485–148.

Alsdorf, D. E., E. Rodríguez, and D. P. Lettenmaier, 2007, Measuring surface water from space, *Reviews in Geophysics* **45** (2): RG2002, doi: 10.1029/2006RG000197.

Alsdorf, D. E., J. M. Melack, T. Dunne, L. A. K. Mertes, L. L. Hess, and L. C. Smith, 2000, Interferometric radar measurements of water level changes on the Amazon flood plain, *Nature* **404**: 174–177, doi: 10.1038/35004560.

Alvarez, L. W., W. Alvarez, F. Asaro, and H. V. Michel, 1980, Extraterrestrial cause for the Cretaceous-Tertiary extinction, Science, *New Series* **208**: 1095–1108.

Amorese *et al.*, 2006, Physical modeling of fault scarp degradation under freeze-thaw cycles, *Earth Surfaces Processes and Landforms* **31** (14): 1731.

Amos, C. B. and D. Burbank, 2007, Channel width response to differential uplift, *Journal of Geophysical Research* **112**: F02010, doi:02010.01029/02006JF000672.

Amundson, J. M. and Iverson, N. R., 2006, Testing a glacial erosion rule using hang heights of hanging valleys, Jasper National Park, Alberta, Canada. *Journal of Geophysical Research* **111**: F01020, doi:10.1029/2005JF000359.

Anders, A. M., G. H. Roe, D. R. Durran, and J. R. Minder, 2007, Small-scale spatial gradients in climatological precipitation on the Olympic Peninsula, *Journal of Hydrometeorology* **8**: 1068–1081.

Anders, A. M., G. H. Roe, B. Hallet, D. R. Montgomery, N. Finnegan, and J. Putkonen, 2006, Spatial patterns of precipitation and topography in the Himalaya, in: S. D. Willett, N. Hovius, M. Brandon, and D. M. Fisher, (eds.), *Tectonics, Climate, and Landscape Evolution*, Geological Society of America Special Paper 398, Boulder, CO: Geological Society of America, pp. 39–53.

Anders, A. M., G. H. Roe, D. R. Montgomery, and B. Hallet, 2008, Influence of precipitation phase on the form of mountain ranges, *Geology* **36**: 479–482.

Anderson, M., 2007, Introducing groundwater physics, *Physics Today* May.

Anderson, R. K., G. H. Miller, J. P. Briner, N. A. Lifton, and S. B. DeVogel, 2007, A millennial perspective on Arctic warming from ^{14}C in quartz and plants emerging beneath ice caps, *Geophysical Research Letters* **35**, doi:10.1029/2007GL032057.

Anderson, R. S., 1986, Erosion profiles due to particles entrained by wind: Application of an eolian sediment transport model, *Geological Society of America Bulletin* **97**: 1270–1278.

Anderson, R. S., 1987a, Eolian sediment transport as a stochastic process: the effects of a fluctuating wind on particle trajectories, *Journal of Geology* **95**: 497–512.

Anderson, R. S., 1987b, A theoretical model for aeolian impact ripples, *Sedimentology* **34**: 943–956.

Anderson, R. S., 1988, The pattern of grainfall deposition in the lee of aeolian dunes, *Sedimentology* **35**: 175–188.

Anderson, R. S., 1990, Evolution of the northern Santa Cruz Mountains by advection of crust past a San Andreas Fault bend, *Science* **249** (4967): 397–401.

Anderson, R. S., 1991, Eolian ripples as examples of self-organization in geomorphological systems, *Earth Science Reviews* **29**: 77–96.

Anderson, R. S., 1994, Evolution of the Santa Cruz Mountains, California, through tectonic growth and geomorphic decay, *Journal of Geophysical Research* **99**: 20 161–20 179.

Anderson, R. S., 1998, Near-surface thermal profiles in alpine bedrock: Implications for the frost-weathering of rock, *Arctic and Alpine Research* **30**: 362–372.

Anderson, R. S., 2000, A model of ablation-dominated medial moraines and the generation of debris-mantled glacier snouts, *Journal of Glaciology* **46** (154): 459–469.

Anderson, R. S., 2002, Modeling of tor-dotted crests, bedrock edges and parabolic profiles of the high alpine surfaces of the Wind River Range, Wyoming, *Geomorphology* **46**: 35–58.

Anderson, R. S., S. P. Anderson, K. R. MacGregor *et al.*, 2004c, Strong feedbacks between hydrology and sliding of a small alpine glacier, *Journal of Geophysical Research – Earth Surface* **109**: F1, doi:10.1029/2004JF000120.

Anderson, R. S., A. L. Densmore, and M. A. Ellis, 1999, The generation and degradation of marine terraces, *Basin Research* **11** (1): 7–19.

Anderson, R. S., S. P. Anderson, and S. Riggins, 2006c, Modeling chemical weathering rates at the bedrock-regolith interface. *AGU Spring Meeting Abstract.*

Anderson, R. S. and K. L. Bunas, 1993, The mechanics of aeolian ripple sorting and stratigraphy as visualized through a cellular automaton model, *Nature* **365**: 740–743.

Anderson, R. S. and P. K. Haff, 1988, Simulation of eolian saltation, *Science* **241**: 820–823.

Anderson, R. S. and P. K. Haff, 1991, Wind modification and bed response during saltation of sand in air, *Acta Mechanica Supplement* **1**: 21–51.

Anderson, R. S. and B. Hallet, 1986, Sediment transport by wind: Toward a general model, *Geological Society of America Bulletin* **97**: 523–535.

Anderson, R. S. and B. Hallet, 1996, Simulating magnetic susceptibility profiles in loess as an aid in quantifying rates of dust deposition and pedogenic development, *Quaternary Research* **45**: 1–16.

Anderson, R. S., B. Hallet, B. Aubry, and J. Walder, 1982, Observations in a cavity beneath the Grinnell Glacier, Montana, *Earth Surface Processes and Landforms* **7**: 63–70.

Anderson, R. S. and Humphrey, N. F., 1989, Interaction of weathering and transport processes in the evolution of arid landscapes, in: T. Cross, (ed.), *Quantitative Dynamic Stratigraphy*, Englewood Cliff, NJ: Prentice- Hall, pp. 349–361.

Anderson, R. S. and K. M. Menking, 1994, The Quaternary marine terraces of Santa Cruz, California: evidence for coseismic uplift on two faults, *Geological Society of America Bulletin* **106**, 649–664.

Anderson, R. S., P. Molnar, and M. A. Kessler, 2006a, Features of glacial valley profiles simply explained, *Journal of Geophysical Research* **111**: F01004, doi:10.1029/2005JF000344.

Anderson, R. S., D. L. Orange, and S. Y. Schwartz, 1990, Implications of the October 17th 1989 Loma Prieta earthquake for the emergence of marine terraces along the Santa Cruz coast, and for the long term evolution of the Santa Cruz Mountains, in R. E. Garrison, *et al.* (eds.), *Geology and Tectonics of Coastal California, San Francisco to Monterey*, Bakersfield, CA: Pacific Section of the AAPG, pp. 205–224.

Anderson, R. S., C. A. Riihimaki, E. B. Safran, and K. R. MacGregor, 2006b, Facing reality: late Cenozoic evolution of smooth peaks, glacially ornamented valleys and deep river gorges of Colorado's Front Range, in: S. D. Willett, N. Hovius, M. T. Brandon, and D. M. Fisher (eds.), *Tectonics, Climate and Landscape Evolution;* Geological Society of America Special Paper 398, Boulder, CO: Geological Society of America, pp. 397–418.

Anderson, R. S., J. L. Repka, and G. S. Dick, 1996, Dating depositional surfaces using in situ produced cosmogenic radionuclides, *Geology* **24**: 47–51.

Anderson, R. S., M. L. Sorenson, and B. B. Willetts, 1991, A review of recent progress in the understanding of aeolian sediment transport. *Acta Mechanica* Supplement **1**: 1–20.

Anderson, S. P., 1988a, The upfreezing process: experiments with a single clast, *Geological Society of America Bulletin* **100**: 609–621.

Anderson, S. P., 1988b, Upfreezing in sorted circles, western Spitzbergen, in: *Permafrost, Fifth International Conference, Proceedings*, vol. **1**, Trondheim, Norway: Tapir, pp. 666–671.

Anderson, S. P., 2005, Glaciers show direct linkage between erosion rates and chemical weathering fluxes, *Geomorphology* **67** (1–2): 147–157.

Anderson, S. P., 2007, Biogeochemistry of glacial landscape systems, *Annual Reviews of Earth and Planetary Sciences* **35**: 375–399.

Anderson, S. P., A. E. Blum, J. Brantley *et al.*, 2004a, Proposed initiative would study Earth's weathering engine, *Eos, Transactions, AGU* **85** (28): 265, 269.

Anderson, S. P. and W. E. Dietrich, 2001, Chemical weathering and runoff chemistry in a steep, headwater catchment, *Hydrological Processes* **15**: 1791–1815.

Anderson, S. P., W. E. Dietrich, and G. H. Brimhall, Jr., 2002, Weathering profiles, mass balance analysis, and rates of solute loss: linkages between weathering and erosion in a small, steep catchment, *Geological Society of America Bulletin* **114** (9): 1143–1158.

Anderson, S. P., W. E. Dietrich, D. R. Montgomery, R. Torres, M. E. Conrad, and K. Loague, 1997b, Subsurface flow paths in a steep, unchanneled catchment, *Water Resources Research* **33** (12): 2637–2653.

Anderson, S. P., W. E. Dietrich, R. Torres, D. R. Montgomery, and K. Loague, 1997a, Concentration-discharge relationships in a steep, unchanneled valley, *Water Resources Research* **33** (1): 211–225.

Anderson, S. P., J. I. Drever, C. D. Frost, and P. Holden, 2000, Chemical weathering in the foreland of a retreating glacier, *Geochimica et Cosmochimica Acta* **64** (7), 1173–1189.

Anderson, S. P., D. H. Mann, and A. E. Blum, 2004b, Chemical weathering along a deposequence of glacial loess-derived soils in Alaska, in: R. Wanty and R. Seal (eds.) *Eleventh International Symposium on Water–Rock Interaction*, vol. **1**, Rotterdam: Balkema, pp. 797–799.

Anderson, S. P., F. von Blanckenburg, and A. F. White, 2007, Physical and chemical controls on Critical Zone form and function, *Elements* **3**: 315–319.

Andrews, D. J. and T. C. Hanks, 1985, Scarp degraded by linear diffusion: inverse solution for age, *Journal of Geophysical Research* **90**: 10 193–10 208.

Andrews, E. D., 1980, Effective and bankfull discharges of streams in the Yampa River basin, Colorado and Wyoming, *Journal of Hydrology* **46**: 311–330.

Andrews, J. T., 1968, Postglacial rebound in Arctic Canada: similarity and prediction of uplift curves, *Canadian Journal of Earth Sciences* **5**: 39.

Andrews, J. T., 2000, Icebergs and iceberg-rafted detritus (IRD) in the North Atlantic: facts and assumptions, *Oceanography* **13** (3): 100–108.

Anikouchine, W. A. and R. W. Sternberg, 1973, *The World Ocean. An Introduction to Oceanography*, Englewood Cliffs, NJ: Prentice-Hall.

Anima, R. J., Y. W. Rodriguez, and G. B. Griggs, 1997, Seafloor morphology map between Año Nuevo and Rio del Mar reveal offshore extensions of onshore geology, *Monterey Bay National Marine Sanctuary Symposium, Sanctuary Currents '97, Facets of Biodiversity*, Santa Cruz, California.

Arendt, A., K. Echelmeyer, W. Harrison, C. Lingle, and V. Valentine, 2002, Rapid wastage of Alaska glaciers and their contribution to rising sea level, *Science* **297**: 382–386, doi:10.1126/science.1072497.

Arendt, A., K. Echelmeyer, W. Harrison *et al.*, 2006, Updated estimates of glacier volume changes in the Western Chugach Mountains, Alaska, USA and a comparison of regional extrapolation methods, *Journal of Geophysical Research* **111**, doi:10.1029/2005JF000436.

Ashton, A., A. B. Murray, and O. Arnoult, 2001, Formation of coastline features by large-scale instabilities induced by high-angle waves, *Nature* **414**: 296–300.

Attal, M. and J. Lavé, 2006, Changes of bedload characteristics along the Marsyandi River (central Nepal): implications for understanding hillslope sediment supply, sediment load evolution along fluvial networks, and denudation in active orogenic belts, in: S. D. Willett, N. Hovius, M. T. Brandon, and D. Fisher (eds.), *Tectonics, Climate, and Landscape Evolution*, Geological Society of America Special Paper 398, Boulder, CO: Geological Society of America, pp. 143–171, doi:10.1130/2006.2398(09).

Attal, M., J. Lavé, and J. P. Masson, 2006, A new experimental device to study pebble abrasion and transpose to natural systems, *Journal of Hydraulic Engineering* **132**: 624–628, doi:10.1061/(ASCE)0733-9429(2006)132:6(624).

Atwater, B. F., S. Musumi-Rokkaku, K. Satake, Y. Tsuji, K. Ueda, and D. K. Yamaguchi, 2005, *The Orphan Tsunami of 1700: Japanese Clues to a Parent Earthquake in North America*, US Geological Survey Professional Paper 1707, Reston, VA/Seattle, WA: US Geological Survey/University of Washington Press.

Avouac, J-P., 2003, Mountain building, erosion and the seismic cycle in the Nepal Himalaya, *Advances in Geophyics* **46**, doi:10.1016/S0065–2687(03)46001–9.

Avouac, J-P., P. Tapponnier, M. Bai, H. You and G. Wang, 1993, Active thrusting and folding along the northeastern Tien Shan and late Cenozoic rotation of Tarim with respect to Dzungaria and Kazakhstan, *Journal of Geophysical Research* **98**, 6755–6804.

Bagnold, R. A., 1935, *Libyan Sands: Travel in a Dead World*, London: Travel Book Club.

Bagnold, R. A., 1940, Beach formation by waves: some model experiments in a wave tank, *Journal of the Institute of Civil Engineers* **15**: 27–52.

Bagnold, R. A., 1941, *The Physics of Blown Sand and Desert Dunes*, London: Methuen.

Bagnold, R. A., 1973, The nature of saltation and of "bedload" transport in water, *Proceedings of the Royal Society of London* **332a**: 473–504.

Bak, P., C. Tang, and K. Wiesenfeld, 1988, Self-organized criticality, *Physical Review A* **38**: 364–374.

Baker, V. R., 1973, *Paleohydrology and Sedimentology of Lake Missoula Flooding in Eastern Washington*, Geological Society of America Special Paper 144, Boulder, CO: Geological Society of America.

Baker, V. R. and D. Nummedal (eds.), 1978, *The Channeled Scabland; A Guide to the Geomorphology of the Columbia Basin*, Washington: Washington, DC: NASA.

Baker, V. R., 1995, Joseph Thomas Pardee and the Spokane flood controversy, *GSA Today* **5** (9).

Baker, V. R., 1981, *Catastrophic Flooding: The Origin of the Channeled Scabland*, Stroudsburg, PA: Dowden Hutchinson and Ross Inc.

Balco, G., C. W. Rovey, II, and J. O. H. Stone, 2005a, *The First Glacial Maximum in North America, Science* **307** (5707): 222, doi:10.1126/science.1103406.

Balco, G., J. O. H. Stone, and J. Mason, 2005c, Numerical ages for Plio-Pleistocene glacial sediment sequences by ^{26}Al/^{10}Be dating of quartz in buried paleosols, *Earth and Planetary Science Letters* **232**: 179–191.

Balco, G., J. O. H. Stone, and C. Jennings, 2005b, Dating Plio-Pleistocene glacial sediments using the cosmic-ray-produced radionuclides ^{10}Be and ^{26}Al, *American Journal of Science* **305**: 1–41.

Baldwin, J. A., K. X. Whipple, and G. E. Tucker, 2003, Implications of the shear-stress river incision model for the timescale of post-orogenic decay of topography, *Journal of Geophysical Research* **108** (B3), doi:10.1029/2001JB000550.

Barber, D. C., Dyke, A., Hillaire-Marcel, C. *et al.*, 1999, Forcing of the cold event of 8,200 years ago by catastrophic drainage of Laurentide lakes, *Nature* **400**: 344–348.

Bard, E., B. Hamelin, R. G. Fairbanks, and A. Zindler, 1990, Calibration of the 14-C timescale over the past 30,000 years using mass spectrometric U-Th ages from Barbados corals, *Nature* **345**: 405–409.

Barnes, H. H., Jr., 1967, *Roughness Characteristics of Natural Channels*, US Geological Survey Water-Supply Paper 1849.

Bartholomaus, T. C., R. S. Anderson, and S. P. Anderson, 2007, Response of glacier basal motion to transient water storage, *Nature Geoscience* **1** (1): 33–37, doi:10.1038/ngeo.2007.5298.

Barton, N. R., R. Lien, and J. Lunde, 1974, Engineering classification of rock masses for the design of tunnel support, *Rock Mechanics* **6** (4): 189–239.

Bascom, W., 1980, *Waves and Beaches*, Garden City, NY: Anchor Press/Doubleday.

Batchelor, G. K., 1970, *An Introduction to Fluid Dynamics*, Cambridge: Cambridge University Press.

Bear, J., 1979, *Hydraulics of Groundwater*, New York: Dover Press.

Bear, J., 1988, *Dynamics of Fluids in Porous Media*, New York: Dover Press.

Beatty, J. K., C. C. Petersen, and A. Chaikin (eds.), 1998, *The New Solar System*, 4th edition, Cambridge: Cambridge University Press.

Begonha, A. and M. A. Sequeira Braga, 2002, Weathering of the Oporto granite: geotechnical and physical properties, *Catena* **49**: 57–76.

Bell, R. E., 2008, The role of subglacial water in ice-sheet mass balance, *Nature Geoscience* **1**: 297–304, doi:10.1038/ngeo186.

Benda, L. and T. Dunne, 1997, Stochastic forcing of sediment supply to channel networks from landsliding and debris flow, *Water Resources Research*, **33** (12): 2849–2863.

Benedict, J. B., 1970, Downslope soil movement in a Colorado Alpine region: Rates, processes and climatic significance, *Arctic and Alpine Research* **2** (3): 165–226.

Benito, G. and J. E. O'Connor, 2003, Number and size of last-glacial Missoula floods in the Columbia River valley between the Pasco Basin, Washington, and Portland, Oregon, *Geological Society of America Bulletin* **115**: 624–638.

Benn, D. I. and D. J. A. Evans, 1998, *Glaciers and Glaciation*, London: Arnold.

Bennett, M. R. and N. F. Glasser, 1996, *Glacial Geology: Ice Sheets and Landforms*, London: Wiley.

Berger, G. W., 1995, Progress in the luminescence dating methods for Quaternary sediments, in: N. W. Rutter and N. R. Catto (eds.), *Dating Methods for Quaternary Deposits*, St John's, Newfoundland: Geological Association of Canada, pp. 81–104.

Berlin, M. M. and R. S. Anderson, 2007, Modeling of knickpoint retreat on the Roan Plateau, western Colorado, *Journal of Geophysical Research* **112**: F03S06, doi:10.1029/2006JF000553.

Berner, E. K. and R. A. Berner, 1996, *The Global Environment: Water, Air and Geochemical Cycles*, Upper Saddle River, NJ: Prentice-Hall.

Berner, E. K., R. A. Berner, and K. L. Moulton, 2003, Plants and mineral weathering: present and past, in: J. I. Drever (ed.), *Surface and Ground Water, Weathering, and Soils;*

Treatise on Geochemistry, vol. **5**, Amsterdam: Elsevier, pp. 169–188.

Berner, R. A., 2004, *The Phanerozoic Carbon Cycle: CO$_2$ and O$_2$*, Oxford: Oxford University Press.

Berner, R. A. and E. K. Berner, 1997, Silicate weathering and climate, in: W. Ruddiman, (ed.) *Tectonic Uplift and Climate Change*, New York: Plenum Press, pp. 353–365.

Berner, R. A., A. C. Lasaga, and R. M. Garrels, 1983, The carbonate-silicate geochemical cycle and its effect on atmospheric carbon dioxide over the past 100 million years, *American Journal of Science* **283**: 641–683.

Beskow, G., 1930, Erdfliessen und Strukturböden der Hochgebirge im Licht der Frosthebung, *Geologiska Föreningens Stockholm Förhandlinger* **52**: 622–638.

Best, T. and G. B. Griggs, 1991, The Santa Cruz littoral cell: difficulties in quantifying a coastal sediment budget, *Proceedings of Coastal Sediments* **91**: 2262–2276.

Bettis, E. A., III, D. R. Muhs, H. M. Roberts, and A. G. Wintle, 2003, Last glacial loess in the conterminous U.S.A., *Quaternary Science Reviews* **22**: 1907–1946.

Bevington, P. R., 1969, *Data Reduction and Error Analysis for the Physical Sciences*, New York: McGraw-Hill.

Bieniawski, Z. T., 1973, Engineering classification of jointed rock masses. *Transactions of the South African Institute of Civil Engineers* **15**: 335–344.

Bierman, P. R., 1994, Using in situ produced cosmogenic isotopes to estimate rates of landscape: evolution: a review from the geomorphic perspective, *Journal of Geophysical Research* **99**: 13 885–13 896.

Bierman, P. R., 2007, Cosmogenic glacial dating, 20 years and counting, *Geology* **35** (6): 575–576.

Bierman, P. R. and M. Caffee, 2002, Cosmogenic exposure and erosion history of Australian bedrock landforms, *Geological Society of America Bulletin* **114**: 787–803, doi:10.1130/0016–7606(2002)114.

Bierman, P. R. and K. K. Nichols, 2004, Rock to sediment – slope to sea with ^{10}Be – rates of landscape change, *Annual Review of Earth and Planetary Sciences* **32**: 215–255.

Bilham, R. and N. Ambraseys, 2004, Apparent Himalayan slip deficit from the summation of seismic moments for Himalayan earthquakes, 1500–2000, *Current Science* **88**.

Bilham, R., V. K. Gaur, and P. Molnar, 2001, Himalayan seismic hazard, *Science* **293**: 1442–1444.

Bilham, R. and G. King, 1989, The morphology of strike slip faults: examples from the San Andreas fault, California, *Journal of Geophysical Research* **94**: 10 204–10 226.

Bills, B. G., D. R. Currey, and G. A. Marshall, 1994, Viscosity estimates for the crust and upper mantle from patterns of lacustrine shoreline deformation in the Eastern Great Basin, *Journal of Geophysical Research* **99** (B11): 22 059–22 086.

Birkeland, P., 1999, *Soils and Geomorphology*, 3rd edition, New York: Oxford University Press.

Biswas, S., I. Coutand, D. Grujic, C. Hager, D. Stöckli, and B. Grasemann, 2007, Exhumation and uplift of the Shillong plateau and its influence on the eastern Himalayas: New constraints from apatite and zircon (U-Th-[Sm])/He and apatite fission track analyses. *Tectonics* **26**: TC6013, doi:10.1029/2007TC002125.

Bjornstad, B. N., K. R. Fecht, and C. J. Pluhar, 2001, Long history of Pre-Wisconsin, Ice-Age cataclysmic floods: evidence from Southeastern Washington State, *Journal of Geology* **109** (6): 695–713.

Black, R. F. and W. L. Barksdale, 1949, Oriented lakes of northern Alaska, *Journal of Geology* **57**: 105–118.

Black, T. A. and D. R. Montgomery, 1991, Sediment transport by burrowing mammals, Marin County, California, *Earth Surface Processes and Landforms* **16**: 163–72.

Blackwelder, E., 1934, Yardangs, *Geological Society of America Bulletin* **45**: 149–156.

Blum, J. D. and Y. Erel, 1997, Rb-Sr isotope systematics of a granitic soil chronosequence: the importance of biotite weathering, *Geochimica et Cosmochimica Acta* **61** (15): 3193–3204.

Bookhagen, B. and D. W. Burbank, 2006, Topography, relief, and TRMM-derived rainfall variations along the Himalaya, *Geophysical Research Letters* **33** (8): L08405, doi:10.1029/2006GL026037.

Bookhagen, B., R. C. Thiede, and M. R. Strecker, 2005a, Abnormal monsoon years and their control on erosion and sediment flux in the high, arid northwest Himalaya, *Earth and Planetary Science Letters* **231**: 131–146.

Bookhagen, B., R. C. Thiede, and M. R. Strecker, 2005b, Late Quaternary intensified monsoon phases control landscape evolution in the northwest Himalaya, *Geology* **33**: 149–152.

Boorstein, D., 1983, *The Discoverers: A History of Man's Search to Know His World and Himself*, New York: Random House.

Bougamont, M., S. Tulaczyk, and I. Joughin, 2003, Response of subglacial sediments to basal freeze-on: II. Application to the stoppage of Ice Stream C, West Antarctica, *Journal of Geophysical Research* **108**, ETG **20**: 1–16.

Boussinesq, J., 1885, *Application des Potentiels a l'Etude de l'Equilibre et du Mouvement des Solides Elastiques*, Paris: Gauthier-Villars.

Brackenridge, G. R., S. V. Nghiem, E. Anderson, and S. Chien, 2005, Space-based measurement of river runoff, *Eos, Transactions, AGU* **86** (19): 185.

Bradley, W. C. and G. B. Griggs, 1976, Form, genesis, and deformation of central California wave-cut platforms, *Geological Society of America Bulletin* **87** (3): 433–449.

Brantley, S. L., 2008, Kinetics of mineral dissolution, in: S. L. Brantley, J. D. Kubicki, and A. F. White (eds.), *Kinetics of Water–Rock Interaction*, New York and Heidelberg: Springer-Verlag, pp. 151–210.

Brantley, S. L. and C. F. Conrad, 2008, Analysis of rates of geochemical reactions, in: S. L. Brantley, J. D. Kubicki, and A. F. White (eds.), *Kinetics of Water–Rock Interaction*, New York and Heidelberg: Springer-Verlag, pp. 1–37.

Brantley, S. L., M. B. Goldhaber, and K. V. Ragnarsdottir, 2007, Crossing disciplines and scales to understand the Critical Zone, *Elements* **3**: 307–314.

Brantley, S. L., T. S. White, A. F. White *et al.*, 2006. *Frontiers in Exploration of the Critical Zone: Report of a workshop sponsored by the National Science Foundation (NSF)*, October 24–25, 2005, Newark, DE.

Braun, J., 2005, Quantitative constraints on the rate of landform evolution derived from low-temperature thermochronology, *Reviews in Mineral Geochemistry* **58**: 351–374.

Bretz, J. H., 1923, The channeled scabland of the Columbia Plateau, 1, *Geology* **31**: 617–649.

Bretz, J. H., 1925, The Spokane Flood beyond the channeled scablands, *Journal of Geology* **33**: 259.

Bretz, J. H., 1928, The channeled scabland of Eastern Washington, *Geographical Review* **18** (3): 446–477.

Bretz, J. H., 1969, The Lake Missoula floods and the channeled scabland, *Journal of Geology* **77**: 505–543.

Bretz, J. H., H. T. U. Smith, and G. E. Neff, 1956, Channeled scabland of Washington; new data and interpretations, *Geological Society of America Bulletin* **67** (8): 957–1049.

Brewer, I. D., D. W. Burbank, and K. V. Hodges, 2003, Modelling detrital cooling-age populations: insights from two Himalayan catchments, *Basin Research* **15**: 305–320.

Brimhall, G. H., C. N. Alpers, and A. B. Cunningham, 1985, Analysis of supergene ore-forming processes and ground water solute transport using mass balance principles, *Economic Geology* **80**: 1227–1256.

Brimhall, G. H. and Dietrich, W. E., 1987, Constitutive mass balance relations between chemical composition, volume, density, porosity, and strain in metasomatic hydrochemical systems: results on weathering and pedogenesis, *Geochimica et Cosmochimica Acta* **51**: 567–587.

Briner, J. P. and T. W. Swanson, 1998, Using inherited cosmogenic ^{36}Cl to constrain glacial erosion rates of the Cordilleran ice sheet, *Geology* **26**: 3–6.

Briner, J. P., D. S. Kaufman, W. F. Manley, R. C. Finkel, and M. W. Caffee, 2005, Cosmogenic exposure dating of late Pleistocene moraine stabilization in Alaska, *Geological Society of America Bulletin* **117**: 1108–1120.

Briner, J. P., G. H. Miller, P. T. Davis, and R. C. Finkel, 2006, Cosmogenic radionuclides from fiord landscapes support differential erosion by overriding ice sheets, *Geological Society of America Bulletin* **118**: 406–420, doi:10.1130/B25716.1.

Brocklehurst, S. H. and K. X. Whipple, 2002, Glacial erosion and relief production in the Eastern Sierra Nevada, California, *Geomorphology* **42**: 1–24.

Brocklehurst, S. H. and K. X. Whipple, 2004, Hypsometry of glaciated landscapes, *Earth Surface Processes and Landforms* **29**: 907–926.

Brocklehurst, S. H. and K. X. Whipple, 2007, The response of glacial landscapes to spatial variations in rock uplift rate, *Journal of Geophysical Research-Earth Surface* **112**(F2).

Brown, C. S., M. F. Meier, and A. Post, 1982, *Calving Speed of Alaska Tidewater Glaciers, with Application to Columbia Glacier*, US Geological Survey Professional Paper 1044–9612, Reston, VA: US Geological Survey, pp. C1–C13.

Brown, E. T., R. F. Stallard, M. C. Larsen, G. M. Raisbeck, and F. Yiou, 1995, Denudation rates determined from the accumulation of in situ-produced ^{10}Be in the Luquillo Experimental Forest, Puerto Rico, *Earth and Planetary Science Letters* **129**: 193–202.

Brown, G. C. and A. E. Mussett, 1981, *The Inaccessible Earth*, London: Allen & Unwin.

Brown, R. J. E., 1970, *Permafrost in Canada*, Toronto, Ontario: University of Toronto Press.

Brunauer, S., P. H. Emmett, and E. Teller, 1938, Adsorption of gases in multimolecular layers, *Journal of the American Chemical Society* **60**: 309–319.

Brunner, C. A., W. R. Normark, G. G. Zuffa, and F. Serra, 1999, Deep-sea sedimentary record of the late Wisconsin cataclysmic floods from the Columbia River, *Geology* **27** (5): 463–466.

Buchanan, T. J. and W. P. Somers, 1969, *Discharge Measurements at Gaging Stations: US Geological Survey, Techniques of Water-Resources Investigations, Book 3*, Washington DC: US Government Printing Office, chapter A8.

Bull, W., 1991, *Geomorphic Response to Climatic Change*, New York: Oxford University Press.

Burbank, D. W. and R. S. Anderson, 2000, *Tectonic Geomorphology*, Oxford: Blackwell Science.

Burbank, D. W., A. E. Blythe, J. Putkonen *et al.*, 2003, Decoupling of erosion and precipitation in the Himalayas, *Nature* **426**: 652–655.

Burbank, D. W., J. Leland, E. Fielding *et al.*, 1996, Bedrock incision, uplift, and threshold hillslopes in the northwest Himalaya, *Nature* **379**: 505–510.

Burn, C., 2002, Tundra lakes and permafrost, Richards Island, western Arctic coast, Canada, *Canadian Journal of Earth Sciences* **39**: 1281–1298, doi:10.1139/E02–035.

Buss, H. L., P. B. Sak, S. M. Webb, and S. L. Brantley, 2008, Weathering of the Rio Blanco quartz diorite, Luquillo Mountains, Puerto Rico: coupling oxidation, dissolution, and fracturing, *Geochimica et Cosmochimica Acta* **72**: 4488–4507.

Butler, D., 1995, *Zoogeomorphology*, Cambridge University Press. 239 p.

Campbell, C. S., P. W. Cleary, and M. Hopkins, 1995, Large-scale landslide simulations: global deformation, velocities and basal friction, *Journal of Geophysical Research* **100** (B5): 8267–8284.

Cande, S. C. and D. V. Kent, 1995, Revised calibration of the geomagnetic polarity time scale for the Late Cretaceous, *Journal of Geophysical Research* **100**: 6093–6095.

Carroll, A. R., L. M. Chetel, and M. E. Smith, 2006, Feast to famine: Sediment supply control on Laramide basin fill, *Geology* **34**: 197–200.

Carslaw, H. S. and J. C. Jaeger, 1967, *Conduction of Heat in Solids*. New York: Oxford University Press.

Carson, M. A. and M. J. Kirkby, 1972, *Hillslope Form and Process*, New York: Cambridge University Press.

Carver, S., N. Mikkelsen, and J. Woodward, 2002, Long-term rates of mass wasting in Mesters Vig, Northeast Greenland: notes on a re-survey, *Permafrost and Periglacial Processes* **13**: 243–249.

Carver, G. and G. Plafker, 2008, Paleoseismicity and neotectonics of the Aleutian subduction zone – an overview, in: J. T. Freymueller, P. J. Haeussler, R. Wesson, and G. Ekstrom (eds.), *Active Tectonics and Seismic Potential of Alaska*, Geophysical Monograph Series, vol. **179**, pp. 43–63, 10.1029/GM1794443.

Cathles, L. M., 1975, *The Viscosity of the Earth's Mantle*, Princeton, NJ: Princeton University Press.

Cazenave, A. and R. S. Nerem, 2004, Present-day sea level change: observations and causes, *Reviews in Geophysics* **42**: RG3001, doi:10.1029/2003RG000139.

Cenderelli, D. A., 2000, Floods from natural and artificial dam failures, in: E. E. Wohl (ed.), *Inland Flood Hazards*, New York: Cambridge University Press, pp. 73–103.

Cerling, T. E. and H. Craig, 1994, Geomorphology and in-situ cosmogenic isotopes, *Annual Review of Earth and Planetary Sciences* **22**: 273–317.

Chadwick, O. A., R. D. Hall and F. M. Philips, 1997, Chronology of Pleistocene glacial advances in the central Rocky Mountains, *Geological Society of America Bulletin* **109**: 1443–1452.

Champagnac, J. D., P. Molnar, R. S. Anderson, C. Sue, and B. Delacou, 2007, Quaternary erosion-induced isostatic rebound in the Western Alps, *Geology* **35** (3): 195–198.

Chappell, J. M., 1974, Upper mantle rheology in a tectonic region: evidence from New Guinea, *Journal of Geophysical Research* **79**: 390–398.

Chepil, W. S. and N. P. Woodruff, 1963, The physics of wind erosion and its control, *Advances in Agronomics* **15**: 211–302.

Chinnery, M. A., 1961, The deformation of the ground around surface faults, *Bulletin of the Seismological Society of America* **51**: 355.

Chmeleff, J., F. von Blanckenburg, K. Kossert, and D. Jakob, 2009, Determination of the ^{10}Be half-life by multicollector ICP-MS and liquid scintillation counting, *Nuclear Instruments and Methods in Physics Research Section B: Beam Interactions with Materials and Atoms*, doi:10.1016/j.nimb.2009.09.012.

Chow, V. T., 1959, *Open Channel Hydraulics*, McGraw-Hill.

Chowdhury, A. M. R., 2004, Arsenic crises in Bangladesh, *Scientific American* **291**: 86–91.

Clague, J. J. and S. G. Evans, 2000, A review of catastrophic drainage of moraine-dammed lakes in British Columbia, *Quaternary Science Reviews* **19** (17–18): 1763–1783, doi:10.1016/S0277–3791(00)00090–1.

Clark, M. K. and L. H. Royden, 2000, Topographic ooze: building the eastern margin of Tibet by lower crustal flow, *Geology* **28**: 703–706.

Clarke, G. K. C., D. W. Leverington, J. T. Teller, and A. S. Dyke, 2003, Superlakes, megafloods and abrupt climate change, *Science* **301**: 922–923.

Clarke, G. K. C., D. W. Leverington, J. T. Teller, and A. S. Dyke, 2004, Paleohydraulics of the last outburst flood from glacial Lake Agassiz and the 8200 BP cold event, *Quaternary Science Reviews* **23**: 389–407.

Clarke, G. K. C., D. W. Leverington, J. T. Teller, A. S. Dyke, and S. J. Marshall, 2005, Fresh arguments against the Shaw megaflood hypothesis, *Quaternary Science Reviews* **24**: 1533–1541.

Clarke, G. K. C. and W. H. Mathews, 1981, Estimates of the magnitude of glacier outburst floods from Lake Donjek, Yukon Territory, Canada, *Canadian Journal of Earth Sciences* **18**, 1452–1463.

Clow, G. D. and F. E. Urban, 2002, Largest permafrost warming in northern Alaska during the 1990s determined from GTN-P borehole temperature measurements, *Eos, Transactions, AGU* **83** (47), Fall Meeting, Supplementary Abstract B11E-04, p. F258.

Coco, G., D. A. Huntley, and T. J. O'Hare, 2000, Investigation of a self-organization model for beach cusp formation and development, *Journal of Geophysical Research* **105** (C9): 21 991–22 002.

Cohen, K. M., P. L. Gibbard, and F. S. Busschers, 2005, in: A. Dehnert and F. Preusser (eds.), *INQUA-SEQS Meeting Volume of Abstracts*, vol. **4**, Bern: INQUA-SEQS.

Cohen, D., R. LeB. Hooke, N. R. Iverson, and J. Kohler, 2000, Sliding of ice past an obstacle at Engabreen, Norway, *Journal of Glaciology* **46**: 599–610.

Cohen, D., T. S. Hooyer, N. R. Iverson, J. F. Thomason, and M. Jackson, 2006, Role of transient water pressure in quarrying: A subglacial experiment using acoustic emissions, *Journal of Geophysical Research* **111**: F03006, doi:10.1029/2005JF000439.

Cohen, D., N. R. Iverson, T. S. Hooyer, U. H. Fischer, M. Jackson, and P. Moore, 2005, Debris-bed friction of hard-bedded glaciers, *Journal of Geophysical Research* **110**: F02007, doi:10.10297/2004JF000228.

Colgan, J. P., D. L. Shuster, and P. W. Reiners, 2008, Two-phase Neogene extension in the northwestern Basin and Range recorded in a single thermochronology sample, *Geology* **36**(8): 631–634.

Colman, S. M. and D. P. Dethier, eds., 1986, *Rates of Chemical Weathering of Rocks and Minerals*, New York: Academic Press.

Colman, S. M. and K. L. Pierce, 1986, The glacial sequence near McCall, Idaho – weathering rinds, soil development, morphology, and other relative-age criteria, *Quaternary Research* **25**: 25–42.

Constantine, J. A. and T. Dunne, 2008, Meander cutoff and the controls on the production of oxbow lakes, *Geology* **36** (1): 23–26, doi:10.1130/G24130A.1.

Costa, J. and R. D. Jarrett, 1981, Debris flows in small mountain stream channels of Colorado and their hydrologic implications, *Bull. Assoc. Engineering Geol. XVIII* **3**: 309–322.

Costa, J. E. and J. E. O'Connor, 1995, Geomorphically effective floods, in: J. E. Costa, A. J. Miller, K. W. Potter, and P. R. Wilcock (eds.), *Natural and Anthropogenic Influences in Fluvial Geomorphology*, American Geophysical Union Monograph **89**, pp. 45–56.

Costa, J. E. and R. L. Schuster, 1988, The formation and failure of natural dams, *Geological Society of America Bulletin* **100**: 1054–1068.

Costa, J. E. and R. L. Schuster, 1991, *Documented Historical Landslide Dams from Around the World*, US Geological Survey Open-File Report 91–239, Reston, VA: US Geological Survey.

Costa, J. E. and G. P. Williams, 1984, *Debris Flow Dynamics* (Video), USGS Open File Report 84–606.

Costard, F., L. Dupeyrat, E. Gautier, and E. Carey-Gailhardis, 2003, Fluvial thermal erosion investigations along a rapidly eroding river bank: application to the Lena river (central Yakutia), *Earth Surface Processes and Landforms* **28**: 1349–1359.

Costard, F., E. Gautier, D. Brunstein *et al.*, 2007, Impact of the global warming on the fluvial thermal erosion over the Lena River in Central Siberia, *Geophysical Research Letters* **34**: L14501, doi:10.1029/2007GL030212.

Cowgill, E., 2007, Impact of riser reconstructions on estimation of secular variation in rates of strike–slip faulting: Revisiting the Cherchen River site along the Altyn Tagh Fault, NW China, *Earth and Planetary Science Letters* **254** (3–4): 239–255.

Cowie, P. A. and Z. K. Shipton, 1998, Fault tip displacement gradients and process zone dimensions, *Journal of Structural Geology* **20**: 983–997.

CRC *Handbook of Chemistry and Physics*, D. R. Lide, editor in chief, CRC Press.

Crosby, B. T. and K. X. Whipple, 2006, Knickpoint initiation and distribution within fluvial networks: 236 waterfalls in the Waipaoa River, North Island, New Zealand, *Geomorphology* **82**: 16–38, doi:10.1016/j.geomorph.2005. 08.023.

Crowley, T., 2005, Raising the ante on the climate debate, *Eos Forum*, July **12**, p. 262.

Currey, D. R., 1990, Quaternary paleolakes in the evolution of semidesert basins, with special emphasis on Lake Bonneville and the Great Basin, U.S.A., *Palaeogeography, Palaeoclimatology, Palaeoecology*, **76**: 189–214.

Dade, B. W. and H. E. Huppert, 1998, Long-runout rockfalls, *Geology*, **26** (9): 803–806.

Danjon, F., D. H. Barker, M. Drexhage, and A. Stokes, 2008, Using three-dimensional plant root architecture in models of shallow-slope stability, *Annals of Botany* **101**: 1281–1293.

Darwin, C., 1881, *The Formation of Vegetable Mould Through the Action of Worms With Observation of Their Habits*, John Murray.

Davies, G., 1999, *Dynamic Earth: Plates, Plumes and Mantle Convection*, Cambridge: Cambridge University Press.

Dean, R. G. and R. A. Dalrymple, 1984, *Water Wave Mechanics for Engineers and Scientists*, Prentice Hall.

DeMestre, N., 1990, *The Mathematics of Projectiles in Sport*, Australian Mathematical Society Lecture Series, vol. **6**, Cambridge: Cambridge University Press.

Denmark Kommissionen, 1983, *Meddelelser Om Grønland*, Reports on Greenland.

Denny, C. S. and J. C. Goodlett, 1956, Microrelief resulting from fallen trees, in: *Surficial Geology and Geomorphology of Potter County*, Pennsylvania, USGS Professional Paper 288, pp. 59–68.

Densmore, A. L., R. S. Anderson, B. McAdoo, and M. E. Ellis, 1997, Hillslope evolution by bedrock landslides, *Science* **275**: 369–372.

Densmore, A. L., M. E. Ellis, and R. S. Anderson, 1998, A numerical model of landscape evolution by bedrock landslides, *Journal of Geophysical Research* **103**: 15 203–15 220.

Derricourt, R. M., 1974, Retrogression rate of the Victoria Falls and the Batoka Gorge, *Nature* **264**: 23–25.

Dethier, D. P. and E. D. Lazarus, 2006, Geomorphic inferences from regolith thickness, chemical denudation and CRN erosion rates near the glacial limit, Boulder Creek catchment and vicinity, Colorado, *Geomorphology* **75**: 384–399.

Dick, G. S., R. S. Anderson, and D. Sampson, 1997, Controls on flash flood magnitude and hydrograph shape, Upper Blue Hills badlands, Utah: Application of an acoustic sensor for stream gauging, *Geology* **25**: 45–48.

Dietrich, W. E., 1987, Mechanics of flow and sediment transport in river bends, in: K. Richards (ed.), *River Channels Environment and Process*, Oxford: Basil Blackwell, pp. 179–224.

Dietrich, W. E. and T. Dunne, 1993, The channel head, in: K. Beven and M. J. Kirby (eds.), *Channel Network Hydrology*, Hoboken, NJ: Wiley, pp. 175–219.

Dietrich, W. E., T. Dunne, N. F. Humphrey, and L. M. Reid, 1982, Construction of sediment budgets for drainage basins: sediment budgets and routing in forested drainage basins, *US Forest Service General Technical Report PNW-141, Pacific Northwest Forest and Range Experiment Station*, pp. 5–23.

Dietrich, W. E. and J. T. Perron, 2006, The search for a topographic signature of life, *Nature* **439**: 411–418.

Dietrich, W. E., R. Reiss, M. Hsu, and D. R. Montgomery, 1995, A process-based model for colluvial soil depth and shallow landsliding using digital elevation data, *Hydrological Processes* **9**: 383–400.

Dietrich, W. E. and J. D. Smith, 1983, Influence of the point bar on flow through curved channels, *Water Resources Research* **19**: 1173–1192.

Dietrich, W. E., J. D. Smith, and T. Dunne, 1979, Flow and sediment transport in a sand bedded meander, *Journal of Geology* **87**: 305–315.

Dietrich, W. E., J. D. Smith, and T. Dunne, 1984, Boundary shear stress and sediment transport in river meanders of sand and gravel, in: C. M. Elliot (ed.), *River Meandering: Proceeding of the Conference Rivers 1983*, New York: American Society of Civil Engineers, pp. 632–639.

Dietrich, W. E., C. J. Wilson, D. R. Montgomery, J. McKean, and R. Bauer, 1992, Erosion thresholds and land surface morphology, *Geology* **20**: 675–679.

Dietrich, W. E., C. J. Wilson, D. R. Montgomery, and J. McKean, 1993, Analysis of erosion thresholds, channel networks, and landscape morphology using a digital terrain model, *Journal of Geology* **101**: 259–278.

Dingman, S. L., 2008, *Physical Hydrology*, 2nd edition, Prentice Hall Waveland Press Inc.

Dixon, T. H., F. Amelung, A. Ferretti *et al.*, 2006, New Orleans subsidence and relation to flooding after Hurricane Katrina as measured by space geodesy, *Nature* **441**, 587–588.

Dodson, M. H., 1973, Closure temperature in cooling geochronological and petrological systems, *Contributions to Mineralogy and Petrology* **40**: 259–274.

Douglas, B. C. and W. R. Peltier, 2002, The puzzle of global sea-level rise, *Physics Today* **55**: 35–40.

Drever, J. I., 1985 (ed.), *The Chemistry of Weathering*, New York: Springer-Verlag.

Drever, J. I., 1988, *The Geochemistry of Natural Waters*, Upper Saddle River, NJ: Prentice Hall.

Drever, J. I., 1997, *The Geochemistry of Natural Waters: Surface and Groundwater Environments*, 3rd edition, Upper Saddle River, NJ: Prentice Hall.

Drever, J. I., editor, 2005, *Surface and Ground Water, Weathering, and Soils*, Treatise on Geochemistry, vol. **5**, Elsevier Science & Technology.

Drever, J. I. and D. W. Clow, 1995, Weathering rates in catchments, in: A. F. White and S. L. Brantley (eds.), *Chemical Weathering Rates of Silicate Minerals, Reviews in Mineralogy* **31**: 463–483, Washington, DC: Mineralogical Society of America.

Drewry, D., 1986, *Glacial Geologic Processes*, London: Edward Arnold Ltd.

Driscoll, N. W. and G. D. Karner, 1994, Flexural deformation due to Amazon Fan loading: a feedback mechanism affecting sediment delivery to margins, *Geology* **22** (11): 1015–1019.

Duhnforth, M., R. S. Anderson, D. J. Ward, and G. M. Stock, 2008, Importance of Joint Spacing and Rock Hardness on the Pattern and Efficiency of Glacial Erosion in Alpine Settings: An Example From Yosemite National Park, California, AGU Abstracts.

Duhnforth, M., A. L. Densmore, S. Ivy-Ochs, P. A. Allen, and P. W. Kubik, 2007, Timing and patterns of debris flow deposition on Shepherd and Symmes creek fans, Owens Valley, California, deduced from cosmogenic ^{10}Be, *Journal of Geophysical Research* **112**, F03S15, doi:10.1029/2006JF000562.

Dunne, T., 1980, Formation and controls of channel networks, *Progress in Physical Geography* **4**: 211–239.

Dunne, T., 1990, Hydrology, mechanics, and geomorphic implications of erosion by subsurface flow, in: C. G. Higgins and D. R. Coates (eds.), *Groundwater Geomorphology: The Role of Subsurface Water in Earth-Surface Processes and Landforms*, Geology Society of America Special Paper **252**, Geological Society of America, pp. 1–28.

Dunne, T., 1991, Stochastic aspects of the relations between climate, hydrology, and landform evolution, *Transactions, Japanese Geomorphological Union* **12** (1): 1–24.

Dunne, T. and B. F. Aubry, 1986, Evaluation of Horton's theory of sheetwash and rill erosion on the basis of field experiments, in: A. D. Abrahams (ed.), *Hillslope Processes*, London: Allen and Unwin, pp. 31–53.

Dunne, T. and R. D. Black, 1970a, An experimental investigation of runoff production in permeable soils, *Water Resources Research* **6**: 478–499.

Dunne, T. and R. D. Black, 1970b, Partial area contributions to storm runoff in a small New England watershed, *Water Resources Research*, **6**: 1296–1311.

Dunne, T., and W. E. Dietrich, 1980, Experimental investigation of Horton overland flow on tropical hillslopes, 2. Hydraulic characteristics and hillslope hydrographs, *Zeitschrift für Geomorphologie Supplement* **35**: 60–80.

Dunne, T. and L. B. Leopold, 1978, *Water in Environmental Planning*, San Francisco, CA: W.H. Freeman.

Dunne, T., L. A. K. Mertes, R. H. Meade, J. E. Richey, and B. R. Forsberg, 1998, Exchanges of sediment between the flood plain and channel of the Amazon River in Brazil, *Geological Society of America Bulletin* **110** (4): 450–467.

Dunne, T., K. X. Whipple, and B. F. Aubry, 1995, Microtopography and hillslopes and initiation of channels by Horton overland flow, in: *Natural and Anthropogenic Influences in Fluvial Geomorphology*, Washington, DC: American Geophysical Union, pp. 27–44.

Dunne, T., W. Zhang, and B. Aubry, 1991, Effects of rainfall, vegetation, and microtopography on infiltration and runoff, *Water Resources Research* **27** (9): 2271–2285.

Dutton, C. E., 1882, *The Tertiary History of the Grand Canyon*, USGS Monograph 2, with atlas, Washington, DC: US Government Printing Office.

Duvall, A., E. Kirby, and D. Burbank, 2004, Tectonic and lithologic controls on bedrock channel profiles and processes in coastal California, *Journal of Geophysical Research Earth Surface* **109**: F03002, doi:10.1029/2003JF000086.

Dyke, A. S., 2004, An outline of North American deglaciation with emphasis on central and northern Canada, in: J. Ehlers and P. Gibbard, *Quaternary Glaciations – Extent and Chronology. Part II: North America*, Amsterdam: Elsevier.

Edmond, J., M. R. Palmer, C. I. Measures, B. Grant, and R. F. Stallard, 1995, The fluvial geochemistry and denudation rate of the Guayana Shield in Venezuela, Colombia, and Brazil, *Geochimica et Cosmochimica Acta* **59**: 3301–3325.

Egan, T., 2006, *The Worst Hard Time: The Untold Story of Those Who Survived the Great American Dust Bowl*, New York: Houghton Mifflin.

Eggler, D. H., Larson, E. E., and Bradley, W. C., 1969, Granites, grusses, and the Sherman erosion surface, southern Laramie Range, Colorado-Wyoming, *American Journal of Science* **267**: 510–522.

Ehlers, T. A., 2005, Crustal thermal processes and the interpretation of thermochronometer data, *Reviews in Mineral Geochemistry* **58**: 315–350.

Ehlers, T. A. and K. A. Farley, 2003, Apatite (U-Th)/ He thermochronometry: methods and applications to problems in tectonic and surface processes, *Earth and Planetary Science Letters* **206**: 1–14.

Ehlers, T. A., S. D. Willett, P. A. Armstrong, and D. S. Chapman, 2003, Exhumation of the central Wasatch Mountains, Utah: 2. Thermokinematic model of exhumation, erosion, and thermochronometry interpretation, *Journal of Geophysical Research* **108**: 2173, doi:10.1029/2001JB001723.

Einstein, A., 1905, On the motion – required by the molecular kinetic theory of heat – of small particles suspended in a stationary liquid, *Annalen der Physik* **17**: 549–560.

Einstein, H. A., 1950, *The Bed-load Function for Sediment Transportation in Open Channel Flows*, Technical Bulletin no. 1026, Washington, DC: US Department of Agriculture, Soil Conservation Service.

El-Baz, F., C. S. Breed, M. J. Grolier, and J. F. McCauley, 1979, Eolian features in the western Desert of Egypt and some applications to Mars, *Journal of Geophysical Research* **84**: 8205–8221.

Elder, J., 1976, *The Bowels of the Earth*, Oxford: Oxford University Press.

Ellison, W. D., 1947, Soil erosion studies, 1, *Agricultural Engineering* **28** (4): 145–146.

Emanuel, K. A., 2005, *Divine Wind: The History and Science of Hurricanes*, New York: Oxford University Press.

Emmett, W. W. and M. G. Wolman, 2001, Effective discharge and gravel-bed rivers, *Earth Surface Processes and Landforms* **26**: 1369–1380.

Ericson J. P., C. J. Vörösmarty, S. L. Dingman, L. G. Ward, and M. Meybeck, 2006, Effective sea-level rise and deltas: causes of change and human dimension implications, *Global and Planetary Change* **50**: 63–82.

Evans, D. L, T. G. Farr, and J. J. van Zyl, 1992, Estimates of surface roughness derived from synthetic aperture radar (SAR) data, *IEEE Transactions on Geoscience and Remote Sensing* **30**: 382–389.

Fagherazzi, S., A. D. Howard, and P. L. Wiberg, 2004, Modeling fluvial erosion and deposition on continental shelves during sea level cycles, *Journal of Geophysical Research* **109**: F03010, doi:10.1029/2003JF000091.

Fagherazzi, S. and I. Overeem, 2007, Models of deltaic and inner continental shelf landform evolution, *Annual Review of Earth and Planetary Sciences* **35**: 685–715, doi:10.1146/ annurev.earth.35.031306.140128.

Fahnestock, M., S. Ekholm, K. Knowles, R. Kwok, and T. Scambos, 1997, *Digital SAR Mosaic and Elevation Map of the Greenland Ice Sheet* (CD-ROM), Boulder, CO: National Snow and Ice Data Center.

Fairbanks, R. G., 1989, A 17,000 year glacio-eustatic sea level record: influence of glacial melting rates on the Younger Dryas event and deep ocean circulation, *Nature* **342**: 637–642.

Farley, K. A., 2000, Helium diffusion from apatite: General behavior as illustrated by Durango fluorapatite, *Journal of Geophysical Research* **105**: 2903–2914.

Farley, K. A., 2002, (U-Th)/He dating: techniques, calibrations, and applications. *Reviews in Mineral Geochemistry* **47**: 819–844.

Farr, T. G., 1992, Microtopographic evolution of lava flows at Cima volcanic field, Mojave Desert, California, *Journal of Geophysical Research* **97**: 15171–15179.

Farr, T. G. and M. Kobrick, 2000, Shuttle Radar Topography Mission produces a wealth of data, *Eos, Transactions, AGU* **81**: 583–585.

Farr, T. G., *et al.*, 2007, The Shuttle Radar Topography Mission, *Reviews in Geophysics*, **45**: RG2004, doi:10.1029/ 2005RG000183.

Farrell, W. E., 1972, Deformation of the Earth by surface loads, *Reviews of Geophysics and Space Physics* **10** (3): 761–797.

Feldl, N. and R. Bilham, 2006, Great Himalayan Earthquakes and the Tibetan Plateau, *Nature* **444**: 165–170, doi:10.1038/ nature05199.

Ferguson, R. I., 2003, Emergence of abrupt gravel to sand transitions along rivers through sorting processes, *Geology* **31**: 159–164.

Fetter, C. W., 2001, *Applied Hydrogeology*, 4th edition, New Jersey: Prentice Hall.

Fielding, E. J., B. L. Isacks, M. Barazangi, and C. Duncan, 1994, How flat is Tibet?, *Geology* **22**: 163–167.

Finkel, H. J., 1959, The barchans of southern Peru, *Journal of Geology* **67**: 614–647.

Finnegan, N. J., B. Hallet, D. R. Montgomery *et al.*, 2008, Coupling of rock uplift and river incision in the Namche Barwa Gyala Peri massif, Tibet, *Geological Society of America Bulletin* **120** (1–2): 142–155.

Finnegan, N. J., G. Roe, D. R. Montgomery, and B. Hallet, 2005, Controls on the channel width of rivers: implications for modeling fluvial incision of bedrock, *Geology* **33**: 229–232.

Finnegan, N. J., L. S. Sklar, and T. K. Fuller, 2007, Interplay of sediment supply, river incision, and channel morphology revealed by the transient evolution of an experimental bedrock channel, *Journal of Geophysical Research* **112**, F03S11, doi:10.1029/2006JF000569.

Fleagle, R. G. and J. A. Bussinger, 1980, *An Introduction to Atmospheric Physics*, 2nd edition, New York: Academic Press.

Fleischer, R. L., P. B. Price, and R. M. Walker (eds.), 1975, *Nuclear Tracks in Solids: Principles and Applications*, Berkeley, CA: University of California Press.

Fleming, K., P. Johnston, D. Zwartz, Y. Yokoyama, K. Lambeck, and J. Chappell, 1998, Refining the eustatic sea-level curve since the Last Glacial Maximum using far- and intermediate-field sites, *Earth and Planetary Science Letters* **163** (1–4): 327–342, doi:10.1016/ S0012–821X(98)00198–8.

Fletcher, R. C., H. L. Buss, and S. L. Brantley, 2006, A spheroidal weathering model coupling porewater chemistry to soil thicknesses during steady state erosion, *Earth and Planetary Science Letters* **244**: 444–457.

Font, M., J.-L. Lagarde, D. Amorese, J.-P. *et al.*, 2006, Physical modelling of fault scarp degradation under freeze/thaw cycles, *Earth Surface Processes and Landforms* **31** (14): 1731–1745.

Forte, A. M., J. X. Mitrovica, R. Moucha, N. A. Simmons, and S. P. Grand, 2007, Descent of the ancient Farallon slab drives localized mantle flow below the New Madrid seismic zone, *Geophysical Research Letters* **34**: L04308, doi:10.1029/2006GL027895.

Fountain, A. G. and J. S. Walder, 1998, Water flow through temperate glaciers, *Reviews of Geophysics* **36**: 299–328.

Freeze, A. R. and J. A. Cherry, 1979, *Groundwater*, Englewood Cliffs, NJ: Prentice-Hall.

Freeze, A. R., 1974, Streamflow generation, *Reviews of Geophysics and Space Physics* **12** (4): 627–647.

French, H. M., 2007, *The Periglacial Environment*, 3rd edition, Chichester: Wiley.

Freymueller, J. T., N. E. King, and P. Segall, 1994, The co-seismic slip distribution of the Landers earthquake, *Bulletin of the Seismological Society of America* **84** (3): 646–659.

Frohn, R. C., K. M. Hinkel, and W. R. Eisner, 2005, Satellite remote sensing classification of thaw lakes and drained thaw lake basins on the North Slope of Alaska, *Remote Sensing of the Environment* **97**: 116–126.

Furbish, D. J., 1997, *Fluid Physics in Geology: An Introduction to Fluid Motions on Earth's Surface and Within its Crust*, New York: Oxford University Press.

Furbish, D. J., E. Childs, P. K. Haff, and M. W. Schmeeckle, 2009, Rainsplash of soil grains as a stochastic advection-dispersion process, with implications for desert plant–soil interactions and land-surface evolution, *Journal of Geophysical Research – Earth Surface* **114**, F00A03, doi:10.1029/2009JF001265.

Furbish, D. J., K. K. Hamner, M. W. Schmeeckle, M. N. Borosund, and S. M. Mudd, 2007, Rainsplash of dry sand revealed by high-speed imaging and sticky-paper splash targets. *Journal of Geophysical Research – Earth Surface* **112**, F01001, doi:10.1029/2006JF000498.

Gabet, E. J., 2000, Gopher bioturbation: field evidence for nonlinear hillslope diffusion, *Earth Surface Processes and Landforms* **25** (13): 1419–1428.

Gabet, E. J., 2003, Sediment transport by dry ravel, *Journal of Geophysical Research* **108** (B1): 2050, 10.1029/2001J B001686, 2003.

Gabet, E. J., 2007, A coupled chemical weathering and physical erosion model for a landslide-dominated landscape, *Earth and Planetary Science Letters* **264**: 259–265.

Gabet, E. J. and T. Dunne, 2003, Sediment detachment by rain power, *Water Resources Research* **39** (1): 1002, doi:10.1029/2001WR000656.

Gabet, E. J. and T. Dunne, 2004, Correction to "Sediment detachment by rain power," *Water Resources Research* **40**: W08901, doi:10.1029/2004WR003422.

Gabet, E. J., Edelman, R., and Langner, H., 2006, Hydrological controls on chemical weathering rates at the soil–bedrock interface, *Geology* **34**: 1065–1068.

Gabet, E. J., O. J. Reichman, and E. Seablooom, 2003, The effects of bioturbation on soil processes and hillslope evolution, *Annual Review of Earth and Planetary Sciences* **31**: 249–273.

Gaillardet, J., B. Dupré, P. Louvat, and C. J. Allègre, 1999, Global silicate weathering and CO_2 consumption rates deduced from the chemistry of large rivers, *Chemical Geology* **159**: 3–30.

Galewsky, J., 1998, The dynamics of foreland basin carbonate platforms: tectonic and eustatic controls. *Basin Research* **10** (4): 409–416.

Galewsky, J., 2008, Orographic clouds in terrain-blocked flows: an idealized modeling study, *Journal of the Atmospheric Sciences* **65**: 3460–3478.

Galewsky, J., 2009, Rain shadow development during the growth of mountain ranges: an atmospheric dynamics perspective, *Journal of Geophysical Research* **114**: F01018, doi:10.1029/2008JF001085.

Galewsky, J., E. A. Silver, C. D. Gallup, R. L. Edwards, D. C. Potts, 1996, Foredeep tectonics and carbonate platform dynamics in the Huon Gulf, Papua New Guinea, *Geology* **24**(9): 819–822.

Galewsky, J., C. P. Stark, S. J. Dadson, C.-C. Wu, A. H. Sobel, and M.-J. Horng, 2006, Tropical cyclone triggering of sediment discharge in Taiwan, *Journal of Geophysical Research* **111**: F03014, doi:10:1029/2005JF000428.

Gallagher, K., 1994, Genetic algorithms: a powerful new method for modelling fission-track data and thermal histories, in: M. A. Lanphere, G. B. Dalrymple, and B. D. Turrin (eds.), *Proceedings International Conference of Geochronology, Cosmochronology and Isotope Geology*, Reston, VA: US Geological Survey.

Gallagher, K., R. Brown, and C. Johnson, 1998, Fission track analysis and its applications to geological problems, *Annual Review of Earth and Planetary Sciences* **26**: 519–572.

Galy, V., O. Beyssac, C. France-Lanord, and T. Eglinton, 2008, Recycling of graphite during Himalayan erosion: a geological stabilisation of carbon in the crust, *Science* **322**: 943–945.

Galy, V., C. France-Lanord, O. Beyssac, P. Faure, H. Kudrass, and F. Palhol, 2007, Efficient organic carbon burial in the Bengal fan sustained by the Himalayan erosional system, *Nature* **450**: 407–410.

Galy, V., C. France-Lanord, and B. Lartiges, 2008, Loading and fate of particulate organic carbon from the Himalaya to the Ganga-Brahmaputra delta, *Geochimica et Cosmochimica Acta* **72**: 1767–1787.

Gardner, W., 1919, The movement of moisture in soil by capillarity, *Soil Science* **7**: 313–317.

Garrels, R. M. and Christ C., 1965, *Solutions, Minerals and Equilibria*, San Francisco, CA: Freeman-Cooper.

Gee, J. S., S. C. Cande, D. V. Kent, R. Partner, and K. Heckman, 2008, Mapping geomagnetic field variations with unmanned airborne vehicles, *Eos, Transactions, AGU* **89** (19): 178–179.

Gelfenbaum, G. and J. D. Smith, 1986, Experimental evaluation of a generalized suspended-sediment transport theory, in: R. J. Knight and J. R. McClean (eds.), *Shelf Sands and Sandstones*, Memoir II, Calgary: Canadian Society of Petroleum Geologists, pp. 133–144.

Gibbard, P., 2007, Paleogeography: Europe cut adrift, *Nature* **448**: 259–260.

Gilbert, G. K., 1877, *Report on the Geology of the Henry Mountains*, US Geographical and Geological Survey of the Rocky Mountain Region, Washington, DC: Government Printing Office.

Gilbert, G. K., 1890, Lake Bonneville, *US Geological Survey Monograph* **1**, Reston, VA: US Geological Survey.

Gilbert, G. K., 1907, Rate of recession of Niagara falls, *USGS Bulletin* **306**.

Gilbert, G. K., 1909, Convexity of hilltops, *Journal of Geology* **17**: 344–350.

Gilbert, G. K., 1914, *The Transportation of Debris by Running Water*, US Geological Survey Professional Paper 86, Reston, VA: US Geological Survey.

Gilpin, L. M., 1995, Holocene Paleoseismicity and Coastal Tectonics of Kodiak Islands, Alaska. PhD dissertation. UC-Santa Cruz. University Microfilms International, Inc., Ann Arbor.

Gilpin, L. M., G. A. Carver, S. Ward, and R. S. Anderson, 1994, Tidal benchmark readings and post-seismic rebound of the Kodiak Islands, SW extent of the 1964 Great Alaskan Earthquake rupture, *Seismological Research Letters* **65** (1): 68.

Glen, J. W., 1952, Experiments on the deformation of ice, *Journal of Glaciology* **2** (12): 111–114.

Gold, L. W. and A. H. Lachenbruch, 1973, Thermal conditions in permafrost – a review of North American literature, in: *Proceedings Second International Conference on Permafrost, Yakutsk, U.S.S.R., North American Contribution*, Washington, DC: National Academy of Sciences, pp. 3–25.

Goldich, S. S., 1938, A study in rock-weathering, *Journal of Geology* **46**: 17–58.

Gomberg, J. and M. Ellis, 1994, Topography and tectonics of the central New Madrid seismic zone: results of numerical experiments using a three-dimensional boundary-element program, *Journal of Geophysical Research* **99**: 20 299–20 310.

Gombrich, E. H., 2005, *A Little History of the World*, New Haven, CT: Yale University Press (English translation by Caroline Mustill; original publication in 1935 in German).

Gomez, B. and M. Church, 1989, An assessment of bed load sediment transport formulae for gravel bed rivers, *Water Resources Research* **25**: 1161–1186.

Gómez-Heras, M., B. J. Smith, and R. Fort, (2006), Surface temperature differences between minerals in crystalline rocks: implications for granular disaggregation of granites through thermal fatigue, *Geomorphology* **78**: 236–249.

Goodfriend, G. A., M. J. Collins, M. L. Fogel, S. A. Macko, and J. F. Wehmiller (eds.), 2000, *Perspectives in Amino Acid and Protein Geochemistry*, New York: Oxford University Press.

Gooseff, M. N., A. Balser, W. B. Bowden, and J. B. Jones, 2009, Effects of hillslope thermokarst in northern Alaska, *Eos, Transactions, AGU* **90**: 29–30.

Gosse, J. C. and F. M. Phillips, 2001, Terrestrial in situ cosmogenic nuclides: theory and application, *Quaternary Science Reviews* **20**: 1475–1560.

Gosse, J. C., J. Klein, E. B. Evenson, B. Lawn, and R. Middleton, 1995, BE-10 dating of the duration and retreat of the last Pinedale glacial sequence, *Science* **268**: 1329–1333.

Goudie, A. S., 2006, The Schmidt Hammer in geomorphological research, *Progress in Physical Geography* **30**: 703–718.

Gould, S. J., 1980, The great Scablands debate, in: *The Panda's Thumb; More Reflections in Natural History*, New York, NY: W. W. Norton & Company, pp. 194–203.

Govers, R., 2009, Choking the Mediterranean to dehydration: the Messinian salinity crisis, *Geology* **37**: 167–170.

Granger, D. E. and P. Muzikar, 2001, Dating sediment burial with cosmogenic nuclides: theory, techniques, and limitations, *Earth and Planetary Science Letters* **188** (1–2): 269–281.

Granger, D. E. and A. L. Smith, 2000, Dating buried sediments using radioactive decay and muogenic production of ^{26}Al and ^{10}Be, *Nuclear Instruments and Methods in Physics Research, B: Beam Interactions with Materials* **172**: 822–826.

Granger, D. E., D. Fabel, and A. N. Palmer, 2001, Pliocene-Pleistocene incision of the Green River, Kentucky, determined from radioactive decay of cosmogenic ^{26}Al and ^{10}Be in Mammoth Cave sediments, *Geological Society of America Bulletin* **113**: 825–836.

Granger, D. E., J. W. Kirchner, and R. C. Finkel, 1996, Spatially averaged long-term erosion rates measured from *in situ* cosmogenic nuclides in alluvial sediment, *Journal of Geology* **104**: 249–257.

Granger, D. E., J. W. Kirchner, and R. C. Finkel, 1997, Quaternary downcutting rate of the New River, Virginia, measured from differential decay of cosmogenic ^{26}Al and ^{10}Be in cave-deposited alluvium, *Geology* **25**: 107–110.

Granger, D. E., C. S. Riebe, J. W. Kirchner, and R. C. Finkel, 2001, Modulation of erosion on steep granitic slopes by boulder armoring, as revealed by cosmogenic ^{26}Al and ^{10}Be, *Earth and Planetary Science Letters* **186** (2): 269–281.

Greeley, R. and J. D. Iversen, 1985, *Wind as a geological process on Earth, Mars, Venus and Titan*, Cambridge: Cambridge University Press.

Green, W. H. and G. A. Ampt, 1911, Studies in soil physics. I. The flow of air and water through soils, *Journal of Agricultural Science* **4** (1): 1–24.

Greene, H. G., N. M. Maher, and C. K. Paull, 2002, Physiography of the Monterey Bay Marine National Marine Sanctuary and implications about continental margin development, *Marine Geology* **181**: 55–82.

Griggs, G. B., 1992, Flooding and slope failure during the January 1982 storm, Santa Cruz County, California, *California Geology* **35**: 158–163.

Griggs, G. B., J. F. Tait, and W. Corona, 1994, The interaction of seawalls and beaches: seven years of monitoring, Monterey Bay, California, *Shore and Beach* **62** (3): 21–28.

Griggs, G. B., K. B. Patsch, and L. E. Savoy, 2005, *Living with the Changing Coast of California*, University of California Press.

Grinnell, J., 1923, The burrowing rodents of California as agents in soil formation, *Journal of Mammalogy* **4**: 137–149.

Guido, Z. S., D. J. Ward, and R. S. Anderson, 2007, Pacing the post-LGM demise of the Animas Valley glacier and the San Juan Mountain Icecap, Colorado, *Geology* **35** (8): 739–742, doi:10.1130/G23596A.

Gupta, S., J. S. Collier, A. Palmer-Felgate, and G. Potter, 2007, Catastrophic flooding origin of shelf valley systems in the English Channel, *Nature* **448**: 342–345, doi:10.1038/nature06018.

Gurnis, M., 2001, *Sculpting Earth from Inside Out: Our Ever Changing Earth*, Special Editions, *Scientific American*.

Hachinohe, S., N. Hiraki, and T. Suzuki, 2000, Rates of weathering and temporal changes in strength of bedrock of marine terraces in Boso Peninsula, Japan, *Engineering Geology* **55** (1): 29–43(15).

Hack, J. T., 1973, Stream profile analysis and stream-gradient index, *US Geological Survey Journal of Research* **1** (4): 421–429.

Haff, P. K., 2001, Desert pavement: an environmental canary?, *Journal of Geology* **109**: 661–668.

Haff, P. K. and R. S. Anderson, 1993, Grain-scale simulations of loose sedimentary beds: the example of grain-bed impacts in aeolian saltation, *Sedimentology* **40**: 175–189.

Haff, P. K. and D. E. Presti, 1995, Seven barchan dunes of the Salton Sea region, California, in: V. P. Tchakerian (ed.), *Desert Aeolian Processes*, London: Chapman & Hall.

Haff, P. K. and Werner, B. T., 1996, Dynamical processes on desert pavements and the healing of surficial disturbances, *Quaternary Research* **45**: 38.

Hales, T. C. and J. J. Roering, 2005, Climate controlled variations in scree production, Southern Alps, New Zealand, *Geology* **33**: 701–704.

Hales, T. C. and J. J. Roering, 2007, Climate controls on frost cracking and implications for the evolution of bedrock landscapes, *Journal of Geophysical Research-Earth Surface* **112**: F02033, doi:10.1029/ 2006JF000616.

Hales, T. C. and J. J. Roering, 2009, A frost "buzzsaw" mechanism for erosion of the Eastern Southern Alps, New Zealand, *Geomorphology* **107**: 241–253.

Hall, K., M. Guglielmin, and A. Strini, 2008, Weathering of granite in Antarctica: II. Thermal stress at the grain scale, *Earth Surface Processes and Landforms* **33**: 465–493.

Hallet, B., 1976, Deposits formed by subglacial precipitation of $CaCO_3$, *Geological Society of America Bulletin* **87** (7): 1003–1015.

Hallet, B., 1979, A theoretical model of glacial abrasion, *Journal of Glaciology* **23**: 39–50.

Hallet, B., 1990, Self-organization in freezing soils: from microscopic ice lenses to patterned ground, *Canadian Journal of Physics* **68**: 842–852.

Hallet, B., 1996, Glacial quarrying: a simple theoretical model, *Annals of Glaciology* **22**: 1–8.

Hallet, B. and R. S. Anderson, 1981, Detailed glacial geomorphology of a proglacial bedrock area at Castleguard Glacier, Alberta, Canada, *Zeitschrift fur Gletscherkunde und Glazialgeologie* **16**: 171–184.

Hallet, B. and S. Prestrud, 1986, Dynamics of periglacial sorted circles in western Spitzbergen, *Quaternary Research* **26**: 81–99.

Hallet, B. and E. D. Waddington, 1992, Buoyancy forces induced by freeze-thaw in the active layer: implications for diapirism and soil circulation, in: J. C. Dixon and A. D. Abrahams (eds.), *Periglacial Geomorphology*, Chichester: Wiley, pp. 251–79.

Hallet, B., L. Hunter, and J. Bogen, 1996, Rates of erosion and sediment evacuation by glaciers: a review of field data and their implications, *Global and Planetary Change* **12**: 213–235.

Hallet, B., S. P. Anderson, C. W. Stubbs, and E. C. Gregory, 1988, Surface soil displacements in sorted circles, western Spitsbergen, in: *Permafrost, Fifth International Conference, Proceedings*, vol. **1**, Trondheim, Norway: Tapir, pp. 770–775.

Hallet, B., J. S. Walder, and C. W. Stubbs, 1991, Weathering by segregation ice growth in microcracks at sustained sub-zero temperatures: verification from an experimental study using acoustic emissions, *Permafrost and Periglacial Processes* **2**: 283–300.

Hancock, G. S. and R. S. Anderson, 2002, Numerical modeling of fluvial strath-terrace formation in response to oscillating climate, *Geological Society of America Bulletin* **114**: 1131–1142.

Hancock, G. S., R. S. Anderson, and K. X. Whipple, 1998, Bedrock erosion by streams: beyond stream power, in: K. Tinkler and E. Wohl (eds.), *Rivers Over Rock*, Geophysical Monograph **107**, American Geophysical Union, pp. 35–60.

Hancock, G. S., R. S. Anderson, O. A. Chadwick, and R. C. Finkel, 1999, Dating fluvial terraces with ¹⁰Be and ²⁶Al profiles, Wind River, Wyoming, *Geomorphology* **27**: 41–60.

Handy, R. L., 1976, Loess distribution by variable winds, *Geological Society of America Bulletin* **87**: 915–927.

Hanks, T. C. and D. J. Andrews, 1989, Effect of the far-field slope on morphologic dating of scarplike landforms. *Journal of Geophysical Research* **94** (B1): 565–573.

Hanks, T. C., R. C. Bucknam, K. R. Lajoie, and R. E. Wallace, 1984, Modification of wave-cut and faulting-controlled landforms, *Journal of Geophysical Research* **89**: 5771–5790.

Hanks, T. C. and R. E. Wallace, 1985, Morphological analysis of the lake Lahontan shoreline and beachfront fault scarps, Persching County, Nevada, *Bulletin of the Seismological Society of America* **75**: 835–846.

Harbor, J. M., 1992, Numerical modeling of the development of U-shaped valleys by glacial erosion, *Geological Society of America Bulletin* **104**: 1364–1375.

Harbor, J. M., 1995, Development of glacial-valley cross-sections under conditions of spatially variable resistance to erosion, *Geomorphology* **14** (2): 99–107.

Harmar, O. P. and N. J. Clifford, 2007, Geomorphological explanation of the long profile of the Lower Mississippi River, *Geomorphology* **84** (3–4): 222–240, doi:10.1016/j.geomorph.2006.01.045.

Harper, J. T., N. F. Humphrey, and W. T. Pfeffer, 1998, Three-dimensional deformation measured in an Alaskan Glacier, *Science* **281** (5381): 1340–1342.

Harper, J. T., N. F. Humphrey, W. T. Pfeffer, S. V. Huzurbazar, D. B. Bahr, and B. C. Welch, 2001, Spatial variability in the flow of a valley glacier: Deformation of a large array of boreholes, *Journal of Geophysical Research* **106** (B5): 8547–8562.

Harris, C. and M. C. R. Davies, 2000, Gelifluction: observations from large-scale laboratory simulations, *Arctic, Antarctic and Alpine Research* **32** (2): 202–207.

Harris, C., M. Luetschg, M. C. R. Davies, F. W. Smith, H. H. Christiansen, and K. Isaksen, 2007, Field instrumentation for real-time monitoring of periglacial solifluction, *Permafrost and Periglacial Processes* **18**: 105–114.

Harris, R. A., 1998, Introduction to special section: Stress triggers, stress shadows, and implications for seismic hazard, *Journal of Geophysical Research* **103**, 24 347–24 358.

Harris, R. A., J. F. Dolan, R. Hartleb, and S. M. Day, 2002, The 1999 Izmit, Turkey, Earthquake: a 3D dynamic stress transfer model of intraearthquake triggering, *Bulletin of the Seismological Society of America* **92**: 245–255.

Harris, R. A. and R. W. Simpson, 1998, Suppression of large earthquakes by stress shadows: A comparison of Coulomb and rate-and-state failure, *Journal of Geophysical Research* **103**, 24 439–24 451.

Hartmann, D. L., 1994, *Global Physical Climatology*, San Diago, CA: Academic Press.

Hartmann, W. K., 1999, *Moons and Planets*, 4th edition, Brooks-Cole.

Hartshorn, K., N. Hovius, W. B. Dade *et al.*, 2002, Climate-driven bedrock incision in an active mountain belt, *Science* **297**: 2036–2038.

Hastenrath, S. L., 1987, The barchan dunes of southern Peru revisited, *Zeitschrift fur Geomorphologie* **31**: 167–178.

Hay, W. W., J. L. Sloan II, and C. N. Wold, 1988, The mass/age distribution of sediments on the ocean floor and the global rate of loss of sediment, *Journal of Geophysical Research* **93**: 14 933–14 940.

Hearty, P. J., G. H. Miller, C. Stearns, and B. J. Szabo, 1986, Aminostratigraphy of Quaternary shorelines around the Mediterranean basin, *GSA Bulletin* **97**: 850–858.

Heffern, E. L., P. W. Reiners, C. W. Naeser, and D. A. Coates, 2008, Geochronology of clinker and implications for evolution of the Powder River Basin landscape, Wyoming and Montana, *Geological Society of America Reviews in Engineering Geology* **18**: 155–175.

Heimsath, A. M., Chappell, J., Dietrich, W. E., Nishiizumi, K., and Finkel, R. C., 2000, Soil production on a retreating escarpment in southeastern Australia, *Geology* **28**: 788–790.

Heimsath, A. M., J. Chappell, N. A. Spooner, and D. G. Questiaux, 2002, Creeping soil, Geology **30**: 111–114.

Heimsath, A. M., W. E. Dietrich, K. Nishiizumi, and R. C. Finkel, 1997, The soil production function and landscape equilibrium, *Nature* **388**: 358–361.

Heimsath, A. M., W. E. Dietrich, K. Nishiizumi, and R. C. Finkel, 1999, Cosmogenic nuclides, topography, and the spatial variation of soil depth, *Geomorphology* **27** (1/2): 151–172.

Heimsath, A. M., D. J. Furbish, and W. E. Dietrich, 2005, The illusion of diffusion: field evidence for depth-dependent sediment transport, *Geology* **33**, 949–952.

Heller, P. L., D. L. Anderson, and C. L. Angevine, 1996, Is the Middle Cretaceous pulse of rapid sea-floor spreading real or necessary?, *Geology* **24**: 491–494.

Helley, E. J. and W. Smith, 1971, *Development and Calibration of a Pressure-difference Bedload Sampler*, Open-File Report, Washington, DC: United States Geological Survey.

Henderson, F. M., 1966, *Open Channel Flow*, New York: Macmillan.

Hewitt, K., 1998, Catastrophic landslides and their effects on the Upper Indus streams, Karakoram Himalaya, northern Pakistan, *Geomorphology* **26**: 47–80.

Hewlett, J. D. and W. L. Nutter, 1970, The varying source area of streamflow from upland basins, in: *Proceedings of the Symposium on Interdisciplinary Aspects of Watershed Management*, New York: American Society of Civil Engineers, pp. 65–83.

Higgins, C. G., 1982, Drainage systems developed by sapping on Earth and Mars, *Geology* **10**: 147–152.

Hildes, D. H. D., G. K. C. Clarke, G. E. Flowers, and S. J. Marshall, 2004, Subglacial erosion and englacial sediment transport modelled for North American ice sheets, *Quaternary Science Reviews* **23**: 409–430.

Hinkel, K., R. Frohn, F. Nelson, W. Eisner, and R. Beck 2005, Morphometric and spatial analysis of thaw lakes and drained thaw lake basins in the Western Arctic Coastal Plain, Alaska, *Permafrost and Periglacial Processes* **16**: 327–341.

Hinzman, L. D., N. D. Bettez, W. R. Bolton *et al.*, 2005, Evidence and implications of recent climate change in northern Alaska and other Arctic regions, *Climatic Change* **72**: 251–298.

Hodges, K., C. Wobus, K. Ruhl, T. Schildgen, and K. Whipple, 2004, Quaternary deformation, river steepening, and heavy precipitation at the front of the Higher Himalayan ranges, *Earth and Planetary Science Letters* **220**: 379–389.

Hodges, K. V., K. Ruhl, C. Wobus, and M. Pringle, 2005, ^{40}Ar/^{39}Ar geochronology of detrital minerals, in: P. W. Reiners and T. A. Ehlers (eds.), *Low-temperature Thermochronology: Techniques, Interpretations and Applications*, vol. **58**, Washington, DC: Mineralogical Society of America, pp. 239–257.

Hoek, E. and E. T. Brown, 1997, Practical estimates of rock mass strength, *International Journal of Rock Mechanics and Mining Sciences* **34** (8): 1165–1186.

Hoke, G. D. and D. L. Turcotte, 2002, Weathering and damage, *Journal of Geophysical Research* **107** (B10): 2210, doi:10.1029/2001JB001573.

Hoke, G. D. and D. L. Turcotte, 2004, The weathering of stones due to dissolution, *Environmental Geology* **46**: 305–310.

Hole, F. D., 1981, Effects of animals on soils, *Geoderma* **25**: 75–212.

Holland, H. D., 1978, *The Chemistry of the Atmosphere and Oceans*, New York: Wiley.

Holland, K. T. and R. A. Holman, 1993, The statistical distribution of swash maxima on natural beaches, *Journal of Geophysical Research* **98**: 10 271–10 278.

Hooke, R. LeB., 1975, Distribution of sediment transport and shear stress in a meander bend, *Journal of Geology* **83**: 543–566.

Hooke, R. LeB., 1991, Positive feedbacks associated with erosion of glacial cirques and overdeepenings, *Geological Society of America Bulletin* **103**: 1104–1108.

Hooke, R. LeB., 2005, *Principles of Glacier Mechanics*, 2nd edition, Cambridge: Cambridge University Press.

Hooke, R. LeB., P. Calla, P. Holmlund, M. Nilsson, and A. Stroeven, 1989, A 3 year record of seasonal variations in surface velocity, Storglaciaren, Sweden, *Journal of Glaciology* **35**: 235–247.

Hopkins, D. M., 1949, Thaw lakes and thaw sinks in the Imuruk Lake area, Seward Peninsula, Alaska, *Journal of Geology* **57**: 119–131.

Hornberger, G. M., J. P. Raffensperger, P. L. Wiberg, and K. N. Eshleman, 1998, *Elements of Physical Hydrology*, Baltimore, MD: The Johns Hopkins University Press.

Horton, R. E., 1945, Erosional development of streams and their drainage basins, hydrophysical approach to quantitative morphology, *Geological Society of America Bulletin* **56** (3): 275–370.

Houghton, H. G., 1985, *Physical Meteorology*, The MIT Press.

Hovius, N., C. P. Stark, and P. A. Allen, 1997, Sediment flux from a mountain belt derived by landslide mapping, *Geology* **25** (3): 231–234.

Howard, A. D., 1988a, Groundwater sapping on Earth and Mars, in: A. D. Howard, R. C. Kochel, and H. R. Holt (eds.), *Sapping Features of the Colorado Plateau, a Comparative Planetary Geology Field Guide*, Washington, DC: NASA Scientific and Technical Information Division, pp. 1–4.

Howard, A. D., 1988b, Groundwater sapping experiments and modelling, in: A. D. Howard, R. C. Kochel, and H. E. Holt (eds.), *Sapping Features of the Colorado Plateau, a Comparative Planetary Geology Field Guide*, Washington, DC: NASA Scientific and Technical Information Division, pp. 71–83.

Howard, A. D., 1992, Modelling channel migration and floodplain development in meandering streams, in: P. A. Carling and G. E. Petts (eds.), *Lowland Floodplain Rivers*, Chichester: Wiley, pp. 1–42.

Howard, A. D., 1994, Rockslopes, in: A. D. Abrahams and A. J. Parsons (eds.), *Geomorphology of Desert Environments*, Boca Raton, FL: CRC Press, pp. 123–172.

Howard, A. D., 1996, Modelling channel evolution and floodplain morphology, in: M. G. Anderson, D. E. Walling, and P. D. Bates (eds.), *Floodplain Processes*, New York: Wiley, pp. 15–62.

Howard, A. D., W. E. Dietrich, and M. A. Seidl, 1994, Modeling fluvial erosion on regional to continental scales, *Journal of Geophysical Research* **99**: 13 791–13 986.

Howard, A. D. and G. Kerby, 1983, Channel changes in badlands, *Geological Society of America Bulletin* **94**: 739–752.

Howard, A. D. and T. R. Knutson, 1984, Sufficient conditions for river meandering: a simulation approach, *Water Resources Research* **20**: 1659–1667.

Howard, A. D. and C. F. McLane, 1988, Erosion of cohesionless sediment by groundwater seepage, *Water Resources Research* **24** (10): 1659–1674.

Howat, I. M., I. Joughin, and T. Scambos, 2007, Rapid changes in ice discharge from Greenland outlet glaciers, *Science* **315**: 1559, doi:10.1126/science.1138478, advanced publication in *Science Express* on February 8, 2007.

Howat, I. M., B. E. Smith, I. Joughin, and T. A. Scambos, 2008, Rates of southeast Greenland ice volume loss from combined ICESAT and ASTER observations, *Geophysical Research Letters* **35**: L17505, doi:10.1029/2008GL03449.

Hsu, K. J., 1972, Origin of saline giants: a critical review after the discovery of the Mediterranean evaporite, *Earth-Sciences Review* **8**: 371–396.

Hsu, K. J., 1983, *The Mediterranean was a Desert; a Voyage of the Glomar Challenger*, Princeton, NJ: Princeton University Press.

Hughen, K., J. Overpeck, S. J. Lehman, *et al.*, 1998, Deglacial changes in ocean circulation from an extended ^{14}C calibration, *Nature* **391**: 65–68.

Hughen, K., S. Lehman, J. Southon, *et al.*, 2004, ^{14}C activity and global carbon cycle changes over the past 50,000 years, *Science* **303**: 202–207, doi:10.1126/science. 1090300.

Hughen, K. A., J. R. Southon, S. J. Lehman, and J. T. Overpeck, 2000, Synchronous radiocarbon and climate shifts during deglaciation, *Science* **290**: 1951–1954.

Hulbe, C., T. A. Scambos, T. Youngberg, and A. K. Lamb, 2008, Patterns of glacier response to disintegration of the Larsen B Ice Shelf, Antarctic Peninsula, *Earth and Planetary Change* **63** (1): 1–8, doi:10.1016/j.gloplacha. 2008.04.001.

Humphrey, N. F. and C. F. Raymond, 1994, Hydrology, erosion and sediment production in a surging glacier: Variegated Glacier, Alaska, 1982–83, *Journal of Glaciology* **40** (136): 539–552.

Humphrey, N. F., C. F. Raymond, and W. D. Harrison, 1986, Discharges of turbid water during mini-surges of the Variegated Glacier, Alaska, *Journal of Glaciology* **32** (111): 195–207.

Humphreys, G. S. and M. T. Wilkinson, 2007, The soil production function: a brief history and its rediscovery, *Geoderma* **139**: 73–78.

Hyndman, R. D., 1995, Giant earthquakes of the Pacific Northwest, *Scientific American*, December.

Hyndman, R. D. and K. Wang, 1995, The rupture zone of Cascadia great earthquakes from current deformation and the thermal regime, *Journal of Geophysical Research* **100**: 22 133–22 154.

Hyndman, R. D., K. Wang, and M. Yamano, 1995, Thermal constraints on the seismogenic portion of the southwestern Japan subduction thrust, *Journal of Geophysical Research* **100**: 15 373–15 392.

Ikeda, S., G. Parker, and K. Sawai, 1981, Bend theory of river meanders, part 1: linear development, *Journal of Fluid Mechanics* **112**: 363–377.

Iken, A., 1981, The effect of subglacial water pressure on the sliding velocity of a glacier in an idealized numerical model, *Journal of Glaciology* **27**: 407–422.

Iken, A., Flotron, A., Haeberli, W., and Rothlisberger, H., 1983, The uplift of Unteraargletscher at the beginning of the melt season; a consequence of water storage at the bed?, *Journal of Glaciology* **9**: 28–47.

Iken, A. and M. Truffer, 1997, The relationship between subglacial water pressure and the velocity of Findelengletscher, Switzerland, during its advance and retreat, *Journal of Glaciology* **43** (144): 328–338.

Imbrie, J. J., J. Hays, D. Martinson *et al.*, 1984, The orbital theory of Pleistocene climate: support from a revised chronology of the marine [18]O record, in: A. L. Berger, J. Imnrie, J. Hays, G. Kukla, and B. Saltzman (eds.),

Milankovitch and Climate, Part 1, Dordrecht: D. Reidel, pp. 269–305.

Isherwood, D. and A. Street, 1976, Biotite-induced grussification of the Boulder Creek granodiorite, Boulder County, Colorado, *Geological Society of America Bulletin* **87**: 366–370.

Iverson, N. R., 1990, Laboratory simulation of glacial abrasion: Comparison with theory, *Journal of Glaciology* **36**: 304–314.

Iverson, N. R., 1991, Potential effects of subglacial water-pressure fluctuations on quarrying, *Journal of Glaciology* **37** (125): 27–36.

Iverson, N. R., 1995, Processes of erosion, in: J. Menzies (ed.), *Modern Glacial Environments, Processes, Dynamics and Sediments*, Oxford: Butterworth-Heinemann, pp. 241–260.

Iverson, R., 1997, The physics of debris flows, *Reviews of Geophysics* **35**, 3: 245–296.

Iverson, N. R., 2002, Processes of glacial erosion, in: J. Menzies (ed.), *Modern and Past Glacial Environments*, New York: Elsevier, pp. 131–145.

Iverson, N. R., D. Cohen, T. S. Hooyer *et al.*, 2003, Effects of basal debris on glacier flow, *Science* **301**: 81–83.

Iverson, N. R., T. S. Hooyer, U. H. Fischer *et al.*, 2007, Soft-bed experiments beneath Engabreen, Norway: regelation infiltration, basal slip, and bed deformation, *Journal of Glaciology* **53** (182): 323–340.

Iverson, R. M., J. Costa, and R. G. LaHusen, 1992, *Debris-flow Flume at H. J. Andrews Experimental Forest, Oregon*, USGS Open-File Report 92–483.

Iverson, R. M., M. E. Reid, N. R. Iverson *et al.*, 2000, Acute sensitivity of landslide rates to initial soil porosity, *Science* **290**: 513–516.

Iverson, R. M., M. E. Reid, and R. G. LaHusen, 1997, Debris-flow mobilization from landslides, *Annual Review of Earth and Planetary Sciences* **25**: 85–138.

Iverson, R. M. and J. W. Vallance, 2001, New views of granular mass flows, *Geology* **29** (2): 115–118.

Jackson, D. and B. Launder, 2007, Osborne Reynolds and the publication of his papers on turbulent flow, *Annual Review of Fluid Mechanics* **39**: 19–35.

Jahns, R. H., 1943, Sheet structure in granites, its origin and use as a measure of glacial erosion in New England, *Journal of Geology* **51**: 71–98.

Jenny, H., 1941, *Factors of Soil Formation*, New York: McGraw-Hill.

Johnson, A. M., 1970, *Physical Processes in Geology*, San Francisco, CA: Freeman, Cooper and Co.

Johnson, A. M., 1979, Field methods for estimating rheological properties of debris flows (m.s.). Unpublished.

Johnson, D. W., 1915, The nature and origin of fjords, *Science* **41**: 537–543.

Joly, J., 1898, An estimate of the geological age of the Earth, *Scientific Transactions of the Royal Dublin Society* **7** (Ser. 2): 22–36.

Jones, C. H., G. L. Farmer, and J. R. Unruh, 2004, Tectonics of Pliocene removal of lithosphere of the Sierra Nevada,

California, *Geological Society of America Bulletin* **116** (11/12): 1408–1422.

Jones, L. S. and N. F. Humphrey, 1997, Weathering-controlled abrasion in a coarse-grained, meandering reach of the Rio Grande: Implications for the rock record, *Geological Society of America Bulletin* **109** (9): 1080–1088.

Jones, P. D. and M. E. Mann, 2004, Climate over past millennia, *Reviews of Geophysics* **42** (2): RG2002, doi:10.1029/2003RG000143.

Jorgenson, M. T. and T. E. Osterkamp, 2005, Response of boreal ecosystems to varying modes of permafrost degradation, *Canadian Journal of Forestry Research* **35**: 2100–2111.

Jorgenson, M. T., Y. L. Shur, and E. R. Pullman, 2006, Abrupt increase in permafrost degradation in Arctic Alaska, *Geophysical Research Letters* **33**: L02503, doi:10.1029/ 2005GL024960.

Joughin, I., I. M. Howat, M. Fahnestock *et al.*, 2008, Continued evolution of Jakobshavn Isbrae following its rapid speedup, *Journal of Geophysical Research* **113**: F04006, doi:10.1029/2008JF001023.

Jyotsna, R. and P. K. Haff, 1997, Microtopography as an indicator of modern hillslope diffusivity in arid terrain, *Geology* **25**: 695–698.

Kale, V. S., 2007, Geomorphic effectiveness of extraordinary floods on three large rivers of the Indian Peninsula, *Geomorphology* **85** (3–4): 306–316.

Kamb, B., C. F. Raymond, W. D. Harrison, *et al.*, 1985, Glacier surge mechanism: 1982–83 surge of Variegated Glacier, Alaska, *Science* **227**: 469–479.

Karner, G. D., 2004, *Rheology and Deformation of the Lithosphere at Continental Margins*, Columbia Press.

Kaufman, D. S., 2003, Amino acid paleothermometry of Quaternary ostracodes from the Bonneville Basin, Utah, *Quaternary Science Reviews* **22**: 899–914.

Kaufman, D. S. and G. H. Miller, 1992, Overview of amino acid geochronology, *Comparative Biochemistry and Physiology* **102B**: 199–204.

Keeling, C. D., S. C. Piper, R. B. Bacastow *et al.*, 2001, Exchanges of atmospheric CO_2 and $^{13}CO_2$ with the terrestrial biosphere and oceans from 1978 to 2000. I. Global aspects, *SIO Reference Series*, no. 01–06, San Diego, CA: Scripps Institution of Oceanography.

Keller, E. A. and N. Pinter, 2002, *Active Tectonics (Earthquakes, Uplift and Landscape)*, New Jersey: Prentice Hall.

Kendall, R. A., J. X. Mitrovica, G. A. Milne, T. E. Tornqvist, and Y. Li, 2008, The sea-level fingerprint of the 8.2 ka climate event, *Geology* **36**: 423–442.

Kenyon, P. M. and D. L. Turcotte, 1985, Morphology of a delta prograding by bulk sediment transport, *Geological Society of America Bulletin* **96**: 1457–1465.

Kesel, R. H., E. G. Yodis, and D. J. McCraw, 1992, An approximation of the sediment budget of the lower Mississippi River prior to major human modification, *Earth Surface Processes and Landforms* **17**: 711–722.

Kessler, M. A. and R. S. Anderson, 2004, Testing a numerical glacial hydrological model using spring speed-up events and outburst floods, *Geophysical Review Letters* **31**: L18503, doi:10.1029/2004GL020622.

Kessler, M. A., A. B. Murray, B. T. Werner, and B. Hallet, 2001, A model for sorted circles as self-organized patterns, *Journal of Geophysical Research* **106** (B7): 13 287–13 306.

Kessler, M. A., R. S. Anderson, and G. S. Stock, 2006, Modeling topographic and climatic control of east–west asymmetry in Sierra Nevada Glacier length during the Last Glacial Maximum, *Journal of Geophysical Research* **111**: F2, F02002, doi:10.1029/2005JF000365.

Kessler, M. A., R. S. Anderson, and J. P. Briner, 2008, The insertion of fjords into continental margins, *Nature Geoscience* **1** (6): 365–369, doi:10.1038/ngeo201.

King, G. P., R. S. Stein, and J. B. Rundle, 1988a, The growth of geological structures by repeated earthquakes – 1. Conceptual framework, *Journal of Geophysical Research* **93** (B11): 13 307–13 318.

King, G. P., R. S. Stein, and J. B. Rundle, 1988b, The growth of geological structures by repeated earthquakes – 2. Field examples of continental dip-slip faults, *Journal of Geophysical Research* **93** (B11): 13 319–13 331.

King, G. C. P., R. S. Stein, and J. Lin, 1994, Static stress changes and the triggering of earthquakes, *Bulletin of the Seismological Society of America* **84** (3): 935–953.

Kirby, E., P. W. Reiners, M. A. Krol *et al.*, 2002, Late Cenozoic evolution of the eastern margin of the Tibetan Plateau: inferences from $^{40}Ar/^{39}Ar$ and (U-Th)/He thermochronology, *Tectonics* **21** (1): 10.1029/ 2000TC001246.

Kirchner, J. W., R. C. Finkel, C. S. Riebe, D. E. Granger, J. L. Clayton, and W. F. Megahan, 2001, Episodic mountain erosion inferred from sediment yields over 10-year and 10,000-year timescales, *Geology* **29** (7): 591–594.

Kirkby, M. J., 1967, Measurement and theory of soil creep, *Journal of Geology*, **75**: 359–378.

Kirkby, M. J., 1971, Hillslope process-response models based on the continuity equation, *Trans IBG, Special Pub* **3**: 15–30.

Kirkby, M. J., 1976, Deterministic continuous slope models, *Zeitschrift für Geomorphologie, Supplement* **25**: 1–19.

Kirkby, M. J., 1977, Soil development models as a component of slope models, *Earth Surface Processes* **2**: 203–30.

Kirkby, M. J., 1985, A model for the evolution of regolith-mantled slopes, in: M. J. Woldenburg (ed.), *Models in Geomorphology*, London: Allen & Unwin, pp. 213–237.

Kirkby, M. J., 1988, Hillslope runoff processes and models, *Journal of Hydrology* **100**: 315–339.

Kite, G., 1993, Computerized streamflow measurement using slug injection, *Hydrological Processes* **7**: 227–233.

Kjøllmoen, B. (ed.), 1999, *Glasiologiske undersøkelser i Norge 1998*, NVE rapport 5.

Kneale, W. R., 1982, Field measurements of rainfall drop-size distribution and the relationship between rainfall parameters and soil movement by rain splash, *Earth Surface Processes and Landforms* 7: 499–502.

Knight, J., 2008, The environmental significance of ventifacts: a critical review, *Earth-Science Reviews* **86** (1–4): 89–105.

Knighton, D., 1998, *Fluvial Forms and Processes*, Oxford: Edward Arnold.

Kochel, R. C. and V. R. Baker, 1982, Paleoflood hydrology, *Science* **215**: 353–361.

Kocurek, G., M. Carr, R. Ewing, K. G. Havholm, Y. C. Nagar, and A. K. Singhvi, 2006, White Sands Dune Field, New Mexico: age, dune dynamics and recent accumulations, *Sedimentary Geology*, doi:10.1016/j.sedgeo.2006.10.006.

Kodama, Y., 1994a, Downstream changes in the lithology and grain size of fluvial gravels, the Watarase River, Japan: evidence of the role of abrasion in downstream fining, *Journal of Sediment Research* **A64**: 68–75.

Kodama, Y., 1994b, Experimental study of abrasion and its role in producing downstream fining in gravel-bed rivers, *Journal of Sediment Research* **A64**: 76–85.

Komar, P. D., 1971, The mechanics of sand transport on beaches. *Journal of Geophysical Research* **76** (3): 713–721.

Komar, P. D., 1998, *Beach Processes and Sedimentation*, 2nd edition, Englewood Cliffs, NJ: Prentice-Hall, Inc.

Komar, P. D. and D. L. Inman, 1970, Longshore sand transport on beaches, *Journal of Geophysical Research* **75**: 5914–5927.

Koons, P. O., 1989, The topographic evolution of collisional mountain belts: a numerical look at the Southern Alps, N. Z., *American Journal of Science* **289**: 1041–1069.

Koons, P. O., 1990, The two sided wedge in orogeny: erosion and collision from the sand box to the Southern Alps, New Zealand, *Geology* **18**: 679–682.

Korschinek, G. *et al.*, 2009, A new value for the half-life of [10]Be by heavy-ion elastic recoil detection and liquid scintillation counting, *Nuclear Instruments and Methods in Physics Research Section B: Beam Interactions with Materials and Atoms*, doi:10.1016/j.nimb.2009.09.020.

Korup, O., A. L. Strom, and J. T. Weidinger, 2006, Fluvial response to large rock-slope failures: examples from the Himalayas, the Tien Shan, and the Southern Alps in New Zealand, *Geomorphology* **78**: 3–21.

Kostic, S., G. Parker, and J. G. Marr, 2002, Role of turbidity currents in setting the foreset slope of clinoforms prograding into standing fresh water, *Journal of Sedimentary Research* **72**: 353–362.

Kranz, R. L., 1983, Microcracks in rocks: a review, *Tectonophysics* **100**: 449–480.

Kuenen, Ph. H., 1956, Experimental abrasion of pebbles. II: rolling by current, *Journal of Geology* **64**: 336–368.

Kuhlemann, J., 2001, Post-collisional sediment budget of circum-Alpine basins (Central Europe), *Mem. Sci. Geol. Padova* **52**: 1–91.

Kuhlemann, J., W. Frisch, I. Dunkl, and B. Székely, 2001, Quantifying tectonic versus erosive denudation by the sediment budget: the Miocene core complexes of the Alps, *Tectonophysics* **330**: 1–23.

Kuhlemann, J., W. Frisch, B. Székely, I. Dunkl, and M. Kázmér, 2002, Post-collisional sediment budget history of the Alps: tectonic versus climatic control, *International Journal of Earth Science (Geolische Rundschau)* **91**: 818–837.

Kutzbach, J. E., W. L. Prell, and W. F. Ruddiman, 1993, Sensitivity of Eurasian climate to uplift of the Tibetan Plateau, *Journal of Geology* **101**: 177–190.

Lachenbruch, A. H., 1962, Mechanics of thermal contraction cracks and ice-wedge polygons in permafrost, *Geology Society of America Special Paper* **70**.

Lachenbruch, A. H., 1970, Some estimates of the thermal effects of a heated pipeline in permafrost, *US Geology Survey Circular* **632**.

Lachenbruch, A. H. and B. V. Marshall, 1986, Changing climate: geothermal evidence from permafrost in the Alaskan Arctic, *Science* **234**: 689–696.

Laity, J. E. and N. T. Bridges, 2009, Ventifacts on Earth and Mars: Analytical, field, and laboratory studies supporting sand abrasion and windward feature development, *Geomorphology* **105**: 202–217.

Laity, J. E. and M. C. Malin, 1985, Sapping processes and the development of theater-headed valley networks on the Colorado plateau, *Geological Society of America Bulletin* **96** (2): 203–217.

Lal, D., 1988, In situ produced cosmogenic isotopes in terrestrial rocks, *Annual Reviews of Earth and Planetary Science* **16**: 355–388.

Lal, D., 1991, Cosmic ray labeling of erosion surfaces: in situ nuclide production rates and erosion models, *Earth and Planetary Science Letters* **104**: 424–439.

Lal, D. and B. Peters, 1967, Cosmic ray produced radioactivity on the earth, in: K. Sitte (ed.), *Handbuch der Physik*, Berlin: Springer-Verlag, pp. 551–612.

Lamb, M. P., A. D. Howard, J. Johnson, K. X. Whipple, W. E. Dietrich, and J. T. Perron, 2006, Can springs cut canyons into rock?, *Journal of Geophysical Research* **111**: E07002, doi:10.1029/2005JE002663.

Lambeck, K. and J. Chappell, 2001, Sea level change through the last glacial cycle, *Science* **292**: 679–686.

Lancaster, N., 1995, *Geomorphology of Desert Dunes*, London: Routledge.

Lancaster, S. T. and R. L. Bras, 2002, A simple model of river meandering and its comparison to natural channels, *Hydrological Processes* **16** (1): 1–26.

Landry, W. and B. T. Werner, 1994, Computer simulations of self-organized wind ripple patterns, *Physics D* **77**: 238–260.

Lane, D. S., 1994, Thermal properties of aggregates, in: P. Klieger and J. F. Lamond (eds.), *Significance of Tests and Properties of Concrete and Concrete-making Materials*, ASTM Special Technical Publication **169-C**, pp. 438–445.

Lantz, T. C. and S. V. Kokelj, 2008, Increasing rates of retrogressive thaw slump activity in the Mackenzie Delta region, N.W.T., Canada, *Geophysical Research Letters* **35**: L06502, doi:10.1029/2007GL032433.

Lasaga, A. C., 1984, Chemical kinetics of water–rock interactions, *Journal of Geophysical Research* **89**: 4009–4025.

Lasaga, A. C., J. M. Soler, J. Ganor, T. E. Burch, and K. L. Nagy, 1994, Chemical weathering rate laws and global geochemical cycles, *Geochimica et Cosmochimica Acta* **58**: 2361–2386.

Lavé, J. and J. P. Avouac, 2000, Active folding of fluvial terraces across the Siwalik Hills Himalaya of central Nepal, *Journal of Geophysical Research* **105**: 5735–5770.

Lavé, J. and J. P. Avouac, 2001, Fluvial incision and tectonic uplift across the Himalayas of central Nepal. *Journal of Geophysical Research* **106** (B11): 26 561–26 591.

Laws, J. O., 1941, Measurements of the fall velocity of water-drops and raindrops, *Eos, Transactions, AGU* **21**: 709–721.

Laws, J. O. and D. A. Parsons, 1943, The relation of raindrop size to intensity, *Eos, Transactions, AGU* **24**: 452–459.

Leeder, M., 1999, *Sedimentology and Sedimentary Basins: From Turbulence to Tectonics*, Oxford: Blackwell Science.

Lehre, A. K., 1987, Rates of soil creep on colluvium-mantled hillslopes in north-central California, in: *Erosion and Sedimentation in the Pacific Rim (Proceedings of the Corvalis Symposium, August 1987)*, IAHS Publications No. **165**, pp. 91–100.

Lenzi, M. A., L. Mao, and F. Comiti, 2006, Effective discharge for sediment transport in a mountain river: computational approaches and geomorphic effectiveness, *Journal of Hydrology* **326**: 257–276.

Leopold, L. B., 1974, *Water – A Primer*, San Francisco, CA: W. H. Freeman.

Leopold, L. B., 1994, *A View of the River*, Cambridge, MA: Harvard University Press.

Leopold, L. B. and W. W. Emmett, 1977, 1976 bedload measurements, East Fork River, Wyoming, *National Academy of Science Proceedings* **74** (7): 2644–2648.

Leopold, L. B. and W. B. Langbein, 1966, River meanders, *Scientific American*, June, pp. 60–70.

Leopold, L. B. and T. Maddock, 1953, The hydraulic geometry of stream channels and some physiographic implications, *USGS Professional Paper* **252**, pp. 1–57.

Leopold, L. B. and M. G. Wolman, 1960, River meanders, *Geological Society of America Bulletin* **71**: 769–794.

Leopold, L. B., M. G. Wolman, and J. P. Miller, 1964, *Fluvial Processes in Geomorphology*, San Francisco, CA: Freeman.

Leprince, S., E. Berthier, F. Ayoub, C. Delacourt, and J. P. Avouac, 2008, Monitoring Earth surface dynamics with optical imagery, *Eos*, Transactions, AGU **89**, January, 1.

Lettau, K. and H. Lettau, 1969, Bulk transport of sand by the barchans of the Pampa La Joya in Southern Peru, *Zeitschrift für Geomorphologie* **13**: 182–195.

Light, M., 2002, *Full Moon*, New York: Knopf.

Lighthill, M. J., 1978, *Waves in Fluids*, New York: Cambridge University Press.

Ling, F. and T. Zhang, 2003, Numerical simulation of permafrost thermal regime and talik development under shallow thaw lakes on the Alaskan Arctic coastal plain, *Journal of Geophysical Research* **108**, doi:10.1029/2002JD003014.

Liu, F., M. W. Williams, and N. Caine, 2004, Source waters and flowpaths in a seasonally snow-covered catchment, *Water Resources Research* **40**: W09401.

Liu, J., Y. Klinger, K. Sieh, and C. Rubin, 2004, Six similar sequential ruptures of the San Andreas fault, Carrizo Plain, California, *Geology* **32**: 649–652, doi:10.1130/G20478.1.

Liu-Zeng, J., Y. Klinger, K. Sieh, C. Rubin, and G. Seitz, 2006, Serial ruptures of the San Andreas fault, Carrizo Plain, California, revealed by three-dimensional excavations, *Journal of Geophysical Research* **111**: B02306, doi:10.1029/2004JB003601.

Lobkovsky, A. E., B. Jensen, A. Kudrolli, and D. H. Rothman, 2004, Threshold phenomena in erosion driven by subsurface flow, *Journal of Geophysical Research – Earth Surface* **109**, Art. No. F04010.

Lobkovsky, A. E., B. E. Smith, A. Kudrolli, D. C. Mohrig, and D. H. Rothman, 2007, Erosive dynamics of channels incised by subsurface water flow, *Journal of Geophysical Research* **112**: F03S12, doi:10.1029/2006JF000517.

Lock, G. S. H., 1990, *The Growth and Decay of Ice*, Cambridge: Cambridge University Press.

Loewenherz-Lawrence, D. S., 1991, Stability and the initiation of channelized surface drainage: a reassessment of the short wavelength limit, *Journal of Geophysical Research* **96** (B5): 8453–8464.

Loewenherz-Lawrence, D. S., 1994, Hydrodynamic description for advective sediment transport processes and rill initiation, *Water Resources Research* **30** (11): 3203–3321.

Loget, N. and J. Van Den Driessche, 2009, Wave train model for knickpoint migration, *Geomorphology* **106**: 376–382, doi:10.1016/j.geomorph.2008.10.017.

Løken, O. H. and D. A. Hodgson, 1971, On the submarine geomorphology along the east coast of Baffin Island, *Canadian Journal of Earth Sciences* **8**: 185–195.

Long, J. T. and R. P. Sharp, 1964, Barchan-dune movement in Imperial Valley, California. *Geological Society of America Bulletin* **75**: 149–156.

Loso, M. G., Anderson, R. S., and Anderson, S. P., 2004, Post Little Ice Age record of fine and coarse clastic sedimentation in an Alaskan proglacial lake, *Geology* **32** (12): 1065–1068, doi:10.1130/G20839.1.

Loughnan, F., 1969, *Chemical Weathering of the Silicate Minerals*, New York.

Lozinski, W. von, 1909, Über die mechanische Verwitterung der Sandsteine im gemässigten klima, *Bulletin International de l'Academie des Sciences de Cracovie Class des Sciences Mathematique et Naturalles* **1**: 1–25 (English translation by

T. Mrozek (1992), On the mechanical weathering of sandstones in temperate climates. in: D. J. E. Evans (ed.), *Cold Climate Landforms*, Chichester: Wiley, pp. 119–134).

Lundstrom, S. C., 1992, The budget and effect of superglacial debris on Eliot Glacier, Mount Hood, Oregon, unpublished Doctoral thesis, University of Colorado, Boulder, CO.

Lyell, C., 1833, *The Principles of Geology*, London: Murray.

Lyle, M., J. Barron, T. J. Bralower *et al.*, 2008, Pacific Ocean and Cenozoic evolution of climate, *Reviews in Geophysics* **46**: RG2002, doi:10.1029/2005RG000190.

MacGregor, K. C., R. S. Anderson, S. P. Anderson, and E. D. Waddington, 2000, Numerical simulations of longitudinal profile evolution of glacial valleys, *Geology* **28** (11): 1031–1034.

MacGregor, K. R., C. A. Riihimaki, and R. S. Anderson, 2005, Spatial and temporal evolution of rapid basal sliding on Bench Glacier, Alaska, USA, *Journal of Glaciology* **51**: 49–63.

Machete, M. N., S. F. Personius, A. R. Nelson, D. P. Schwartz, and W. R. Lund, 1991, The Wasatch fault zone, Utah, segmentation and history of Holocene earthquakes, *Journal of Structural Geology* **13**: 137–149.

Mackay, J. R., 1987, Some mechanical aspects of pingo growth and failure, western Arctic coast, Canada, *Canadian Journal of Earth Sciences* **24**: 1108–1119.

Mackay, J. R., 1993, Air temperature, snow cover, creep of frozen ground, and the time of ice-wedge cracking, western Arctic coast, *Canadian Journal of Earth Sciences* **30**: 1720–1729.

Mackay, J. R., 1998, Pingo growth and collapse, Tuktoyaktuk Peninsula area, western Arctic coast, Canada: a long-term field study, *Géographie Physique et Quaternaire* **52**: 1–53.

Mackin, J. H., 1948, Concept of the graded river, *Bulletin of the Geological Society of America* **59**: 463–512.

Magee, J. W., G. W. Miller, N. A. Spooner, and D. Questiaux, 2004, Continuous 150 k.y. monsoon record from Lake Eyre, Australia: insolation-forcing implications and unexpected Holocene failure, *Geology* **32** (10): 885–888, doi:10.1130/G20672.1.

Malde, H. E., 1968, *The Catastrophic Late Pleistocene Bonneville Flood in the Snake River Plain, Idaho*, US Geological Survey Professional Paper 596, Reston, VA: US Geological Survey.

Mancktelow, N. S. and B. Grasemann, 1997, Time-dependent effects of heat advection and topography on cooling histories during erosion, *Tectonophysics* **270**: 167–195.

Mandelbrot, B., 1982, *The Fractal Geometry of Nature*, San Francisco, CA: W H Freeman & Co.

Mann, M. E., 2000, Climate change: lessons for a new millennium, *Science* **289** (5477): 253–254, doi:10.1126/science.289.5477.253.

Mann, M. E. and P. D. Jones, 2003, Global surface temperatures over the past two millennia, *Geophysical Research Letters* **30** (15): 1820, doi:10.1029/2003GL017814.

Mann, M. E., R. S. Bradley, and M. K. Hughes, 1998, Global-scale temperature patterns and climate forcing over the past six centuries, *Nature* **392**: 779–787.

Mann, M. E., R. S. Bradley, and M. K. Hughes, 1999, Northern hemisphere temperatures during the past millennium: inferences, uncertainties, and limitations, *Geophysical Research Letters* **26**: 759–762.

Mann, M. E., R. S. Bradley, M. K. Hughes, and P. D. Jones, 1998, Global temperature patterns, *Science* **280**: 2029–2030.

Mann, M. E., Z. Zhang, M. K. Hughes *et al.*, 2009, Proxy-based reconstructions of hemispheric and global surface temperature variations over the past two millennia, *Proceedings of the National Academy of Sciences* **2009** (105): 13252–13257, doi:10.1073/pnas.0805721105.

Mansinha, L. and D. E. Smylie, 1971, The displacement fields of inclined faults, *Bulletin of the Seismological Society of America* **61**: 1433–1440.

Marshall, G. A., R. S. Stein, and W. Thatcher, 1991, Faulting geometry and slip from co-seismic elevation changes: the 18 October 1989, Loma Prieta, California, earthquake, *Bulletin of the Seismological Society of America* **81**: 1660–1693.

Marshall, J. and Plumb, R. A., 2008, *Atmosphere, Ocean and Climate Dynamics, an Introductory Text*, Amsterdam: Elsevier Academic Press.

Martel, S. J., 2006, Effect of topographic curvature on near-surface stresses and application to sheeting joints, *Geophysical Research Letters* **33**: L01308, doi:10.1029/2005GL024710.

Matsuoka, N., 1990, The rate of bedrock weathering by frost action: field measurements and a predictive model, *Earth Surface Processes and Landforms* **15**: 73–90.

Matsuoka, N., 1994, Diurnal freeze–thaw depth in rockwalls: field measurements and theoretical considerations, *Earth Surface Processes and Landforms* **19**: 423–435.

Matsuoka, N., 1995, Rock weathering processes and landform development in the Sor Rondane Mountains, Antarctica, *Geomorphology* **12**: 323–339.

Matsuoka, N. and Moriwaki, K., 1992, Frost heave and creep in the Sor Rondane Mountains, Antarctica, *Arctic and Alpine Research* **24**: 271–280.

Matsuoka, N. and J. Murton, 2008, Frost weathering: recent advances and future directions, *Permafrost and Periglacial Processes* **19**: 195–210, doi:10.1002/ppp.620.

Matsuoka, N. and Sakai, H., 1999, Rockfall activity from an alpine cliff during thawing periods, *Geomorphology* **29**: 309–328.

Mawson, D., 1930, *The Home of the Blizzard, Being the Story of the Australasian Antarctic Expedition, 1911–1914*, 3rd edition, Birlinn Ltd.

Mayo, L. R., 1988, Advance of Hubbard Glacier and closure of Russell Fiord, Alaska – Environmental effects and hazards in the Yakutat area, in: *US Geological Survey Orc. /0/6*, Reston, VA: US Geological Survey pp. 4–16.

Mayo, L. R., 1989, Advance of Hubbard Glacier and 1986 outburst of Russell Fiord, Alaska, USA, *Annals of Glaciology* **13**: 189–194.

McCarthy, C. J., 1980, Sediment transport by rainsplash, unpublished PhD thesis, University of Washington, Seattle.

McCoy, W. D., 1987, The precision of amino acid chronology and paleothermometry, *Quaternary Science Reviews* **6** (1): 43–54.

McDonald, R. R. and R. S. Anderson, 1992, The morphology and dynamics of natural and laboratory grain flows, ASCE, *Engineering Mechanics, Proceedings of the Ninth Conference*, pp. 748–751.

McDonald, R. R. and R. S. Anderson, 1995, Experimental verification of aeolian saltation and lee side deposition models, *Sedimentology* **42**: 39–56.

McDonald, R. R. and R. S. Anderson, 1996, Constraints on eolian grain flow dynamics through laboratory experiments on sand slopes, *Journal of Sedimentary Research* **66**: 642–653.

McDougall, I. and T. M. Harrison, 1998, *Geochronology and Thermochronology by the $^{40}Ar/^{39}Ar$ Method*, New York: Oxford University Press.

McFadden, L., Eppes, M., Gillespie, A., and Hallet, B., 2005. Physical weathering in landscapes due to variation in the direction of solar heating, *Geological Society of America Bulletin* **117**: 161–173.

McKean, J., W. Dietrich, R. Finkel, J. Southon, I. Proctor, and M. Caffee, 1993, Quantification of soil production and downslope creep rates from cosmogenic ^{10}Be accumulations on a hillslope profile, *Geology* **21**: 343–346.

McKee, E., 1979, An introduction to the study of global sand seas, in: E. McKee (ed.), *A Study of Global Sand Seas*, US Geological Survey Paper 1052, Washington, DC: US Geological Survey, pp. 1–20.

McPhee, J., 1989, *The Control of Nature*, Farrar, Straus & Giroux.

Meade, R. H., 1985, *Suspended Sediment in the Amazon River and its Tributaries in Brazil during 1982–1984*, US Geological Survey Open-File Report 85–492, Reston, VA: US Geological Survey.

Meier, M. F., M. B. Durgerov, U. K. Rick *et al.*, 2007, Glaciers dominate eustatic sea level rise in the 21st century, *Science 24* **317** (5841): 1064–1067, doi:10.1126/science.1143906 (originally published in Science Express, July 19).

Menzies, J., 1996 (ed.), *Modern Glacial Environments: Processes, Dynamics and Sediments, Volume 1: Glacial Environments*, Oxford: Butterworth-Heinemann.

Merritts, D. and W. B. Bull, 1989, Interpreting Quaternary uplift rates at the Mendocino triple junction, Northern California, from uplifted marine terraces, *Geology* **17**: 1020–1024.

Mertes, L. A. K., 1994, Rates of flood-plain sedimentation on the central Amazon River, *Geology* **22** (2): 171–174.

Mertes, L. A. K., 1997, Documentation and significance of the perirheic zone on inundated floodplains, *Water Resources Research* **33**: 1749–1762.

Mertes, L. A. K., 2002, Remote sensing of riverine landscapes, *Freshwater Biology* **47**: 799–816.

Mertes, L. A. K., T. Dunne, and L. A. Martinelli, 1996, Channel-floodplain geomorphology along the Solimões-Amazon River, Brazil, *Geological Society of America Bulletin* **108**: 1089–1107.

Mesolella, K. J., R. K. Matthews, W. S. Broeker, and D. L. Thurber, 1969, The astronomical theory of climatic change: Barbados data, *Journal of Geology* **77**: 250–274.

Meybeck, M., 1976, Total mineral transport by world major rivers, *Hydrological Sciences Bulletin* **21**: 265–284.

Meyer-Peter, E. and R. Müller, 1948, Formulas for bed-load transport, *Proceedings of the International Association for Hydraulic Research*, 2nd meeting, Stockholm, Sweden.

Meysman, F. J. R., J. J. Middelburg, and C. H. Heip, 2006, Bioturbation: a fresh look at Darwin's last idea, *Trends in Ecology and Evolution* **12**: 688–695.

Meysman, F. J. R., V. S. Malyuga, B. P. Boudreau, and J. J. Middelburg, 2003, Relations between local, nonlocal, discrete and continuous models of bioturbation, *Journal of Marine Research* **61**: 391–410.

Michel, R., J. P. Avouac, and J. Taboury, 1999a, Measuring ground displacements from SAR amplitude images: application to the Landers earthquake, *Geophysical Research Letters* **26**, 875–878.

Michel, R., J. P. Avouac, and J. Taboury, 1999b, Measuring near-field coseismic displacements from SAR images: application to the Landers earthquake, *Geophysical Research Letters* **26**, 3017–3020.

Middleton, G. V. and J. B. Southard, 1984, *Mechanics of Sediment Movement*, Society of Economic Paleontologists and Mineralogists.

Middleton, G. V. and P. R. Wilcock, 1994, *Mechanics in the Earth and Environmental Sciences*, Cambridge: Cambridge University Press.

Miller, D. J. and T. Dunne, 1996, Topographic perturbations of regional stresses and consequent bedrock fracturing, *Journal of Geophysical Research* **101**: 25 523–25 536, doi:10.1029/96JB02531.

Miller, G. H., Fogel, M. L., Magee, J. W., Gagan, M. K., Clarke, S., and Johnson, B. J., 2005, Ecosystem collapse in Pleistocene Australia and a human role in megafaunal extinction, *Science* **309**: 287–290.

Miller, G. H., A. J. T. Jull, T. Linick *et al.*, 1987, Racemization-derived late Devensian temperature reduction in Scotland, *Nature* **326**: 593–595.

Miller, G. H., J. W. Magee, and A. J. T. Jull, 1997, Low-latitude glacial cooling in the Southern Hemisphere from amino acid racemization in emu eggshells, *Nature* **385**: 241–244.

Miller, G. H., J. W. Magee, B. J. Johnson *et al.*, 1999, Pleistocene extinction of *Genyornis newtoni*: human impact on Australian megafauna, *Science* **283**: 205–208.

Milliman, J. D. and R. H. Meade, 1983, World-wide delivery of river sediment to the oceans, *Journal of Geology* **91**: 1–21.

Milliman, J. D. and J. Syvitski, 1992, Geomorphic/tectonic control of sediment discharge to the ocean: the importance of small mountainous rivers, *Journal of Geology* **100**: 525–544.

Mitchell, S. G. and P. W. Reiners, 2003, Influence of wildfires on apatite and zircon (U-Th)/He ages, *Geology* **31**: 1025–1028.

Mitra, A. K., M. D. Gupta, S. V. Singh, and T. N. Krishnamurti, 2003, Daily rainfall for the Indian Monsoon region from merged satellite and rain gauge values: large-scale analysis from real-time data, *Journal of Hydrometeorology* **4**: 769–781.

Mitrovica, J. X., C. Beaumont, and G. T. Jarvis, 1989, Tilting of continental interiors by the dynamical effects of subduction, *Tectonics* **8**: 1079–1094.

Mitrovica, J. X. and A. M. Forte, 2004, A new inference of mantle viscosity based upon joint inversion of convection and glacial isostatic adjustment data, *Earth and Planetary Science Letters* **225** (1–2): 177–189.

Molnar, P., 2004a, Interactions among topographically induced elastic stress, static fatigue, and valley incision, *Journal of Geophysical Research* **109**: F02010, doi:10.1029/2003JF000097.

Molnar, P., 2004b, Late Cenozoic increase in accumulation rates of terrestrial sediment: how might climate change have affected erosion rates?, *Annual Review of Earth and Planetary Sciences* **32**: 67–89, doi:10.1146/annurev.earth.32.091003.143456.

Molnar, P. H., R. S. Anderson, and S. P. Anderson, 2007, Tectonics, fracturing of rock, and erosion, *Journal of Geophysical Research* **112**: F03014, doi:10.1029/2005JF000433.

Molnar, P., R. S. Anderson, G. Kier, and J. Rose, 2006, Relationships among probability distributions of stream discharges in floods, climate, bed load transport, and river incision, *Journal of Geophysical Research* **111**: F02001, doi:10.1029/2005JF000310.

Molnar, P. and P. England, 1990, Late Cenozoic uplift of mountain ranges and global climate change: chicken or egg?, *Nature*, **346** (6279), 29–34.

Montgomery, D. R. and W. E. Dietrich, 1988, Where do channels begin?, *Nature* **336** (6196): 232–234.

Montgomery, D. R. and W. E. Dietrich, 1992, Channel initiation and the problem of landscape scale, *Science* **255** (5046): 826, doi:10.1126/science.255.5046.826.

Montgomery, D. R. and W. E. Dietrich, 1994, A physically based model for the topographic control on shallow landsliding, *Water Resources Research* **30**: 1153–1171.

Montgomery, D. R., B. Hallet, L. Yuping *et al.*, 2004, Evidence for Holocene megafloods down the Tsangpo River gorge, southeastern Tibet, *Quaternary Research* **62**: 201–207.

Montgomery, D. R., G. Balco, and S. Willett, 2001, Climate, tectonics, and the morphology of the Andes, *Geology* **29**: 579–582.

Montgomery, D. R., W. E. Dietrich, R. Torres, S. P. Anderson, J. T. Heffner, and K. Loague, 1997, Hydrologic response of a steep unchanneled valley to natural and applied rainfall, *Water Resources Research* **33** (1): 91–109.

Moon, V. and J. Jayawardane, 2004, Geomechanical and geochemical changes during early stages of weathering of Karamu Basalt, New Zealand, *Engineering Geology* **74**: 57–72.

Moore, J. G., 1964, Giant submarine landslides on the Hawaiian Ridge, *U S Geological Survey Professional Paper* **501D**, pp. D95–D98.

Moore, J. G., D. A. Clague, R. T. Holcomb, P. W. Lipman, W. R. Normark, and M. E. Torresan, 1989, Prodigious submarine landslides on the Hawaiian Ridge, *Journal of Geophysical Research* **94** (B12): 17 465–17 484.

Moore, J. R., J. W. Sanders, W. E. Dietrich, and S. D. Glaser, 2009, Influence of rock mass strength on the erosion rate of alpine cliffs, *Earth Surface Processes and Landforms*, doi: 10.1002/esp.1821.

Moore, L. and G. B. Griggs, 1998, Measuring shoreline erosion rates: strategy, techniques and accuracy, in: *Proceedings of California and the World Ocean '97*, pp. 719–730.

Moore, L. J. and G. B. Griggs, 2002, Long term cliff retreat and erosion hot spots along the central shores of the Monterey Bay National Marine Sanctuary, *Marine Geology* **181**: 265–283.

Moore, P. L. and N. R. Iverson, 2002, Slow episodic shear of granular materials regulated by dilatant strengthening, *Geology* **30** (9): 843–846.

Morsi, S. A. and A. J. Alexander, 1972, An investigation of particle trajectories in two-phase flow systems, *Journal of Fluid Mechanics* **55**: 193–200.

Motyka, R. J. and M. Truffer, 2007, Hubbard Glacier, Alaska: 2002 closure and outburst of Russell Fjord and postflood conditions at Gilbert Point, *Journal of Geophysical Research* **112**: F02004, doi:10.1029/2006JF000475.

Mudd, S. M. and D. J. Furbish, 2004, Influence of chemical denudation on hillslope morphology, *Journal of Geophysical Research* **109**: F02001, doi:10.1029/2003JF000087.

Mudd, S. M. and D. J. Furbish, 2006, Using chemical tracers in hillslope soils to estimate the importance of chemical denudation under conditions of downslope sediment transport, *Journal of Geophysical Research* **111**: F02021, doi:10.1029/2005JF000343.

Mudd, S. M. and D. J. Furbish, 2007, Responses of soil-mantled hillslopes to transient channel incision rates, *Journal of Geophysical Research – Earth Surface* **112**: F03S18, doi:10.1029/2006JF000516.

Muhs, D. R., 2007, Loess deposits, origins and properties, in: *Encyclopedia of Quaternary Science*, Amsterdam: Elsevier, pp. 1405–1418.

Muhs, D. R. and E. A. Bettis, III, 2003, *Quaternary Loess-paleosol Sequences as Examples of Climate-driven Sedimentary Extremes*, Geological Society of America

Special Paper 370, Boulder, CO: Geological Society of America, pp. 53–74.

Muhs, D. R., E. A. Bettis, J. N. Aleinikoff *et al.*, 2008, Origin and paleoclimatic significance of late Quaternary loess in Nebraska; evidence from stratigraphy, chronology, sedimentology, and geochemistry, *Geological Society of America Bulletin* **120** (11–12): 1378–1407.

Muhs, D. R., K. R. Simmons, G. L. Kennedy, and T. K. Rockwell, 2002, The last interglacial period on the Pacific Coast of North America: Timing and paleoclimate, *Geological Society of America Bulletin* **114**: 569–592.

Müller, R. D., M. Sdrolias, C. Gaina, and W. R. Roest, 2008, Age, spreading rates, and spreading asymmetry of the world's ocean crust, *Geochemistry, Geophysics, Geosystems* **9**: Q04006, doi:10.1029/2007GC001743.

Murray, A. B. and C. Paola, 1994, A cellular model of braided rivers, *Nature* **371**: 54–57.

Murray, A. B. and C. Paola, 1996, A new quantitative test of geomorphic models, applied to a model of braided streams, *Water Resources Research* **32** (8): 2579–2587.

Murton, J. B., 1996, Near-surface brecciation of chalk, Isle of Thanet, southeast England: a comparison with ice-rich brecciated bedrocks in Canada and Spitsbergen, *Permafrost and Periglacial Processes* **7** (2): 153–164.

Murton, J. B., R. Peterson, and J.-C. Ozouf, 2006, Bedrock fracture by ice segregation in cold regions, *Science* **314** (5802): 1127, doi:10.1126/science.1132127.

Naeser, C. W., 1967, The use of apatite and sphene for fission track age determinations, *Bulletin of the Geological Society of America* **78**: 1523–1526.

Naeser, C. W. and H. Faul, 1969, Fission track annealing in apatite and sphene, *Journal of Geophysical Research* **74**: 705–710.

Nakada, M. and H. Inoue, 2005, Rates and causes of recent global sea-level rise inferred from long tide gauge data records, *Quaternary Science Reviews* **24** (10–11): 1217–1222, doi:10.1016/j.quascirev.2004.11.006.

Narteau, C., D. Zhang, O. Rozier, and P. Claudin, 2009, Setting the length and time scales of a cellular automaton model from the analysis of superimposed bedforms, *Journal of Geophysical Research* **114**: F03006, doi:10.1029/2008JF001127.

Nash, D. B., 1994, Effective sediment transporting discharge from magnitude–frequency analysis, *Journal of Geology* **102**: 79–95.

Natawidjaja, D. H., K. Sieh, S. N. Ward *et al.*, 2004, Paleogeodetic records of seismic and aseismic subduction from central Sumatran microatolls, Indonesia, *Journal of Geophysical Research* **109**: B04306, doi:10.1029/2003JB002398.

Natawidjaja, D. H., K. Sieh, M. Chlieh *et al.*, 2006, Source parameters of the great Sumatran megathrust earthquakes of 1797 and 1833 inferred from coral microatolls, *Journal of Geophysical Research* **111**.

National Research Council, 2001, *Basic Research Opportunities in Earth Science*, Washington, DC: National Academy Press.

Neff, J. C., A. P. Ballantyne, G. L. Farmer *et al.*, 2008, Increasing eolian dust deposition in the western United States linked to human activity, *Nature Geoscience* **1**: 189–195, doi:10.1038/ngeo133.

Nelson, F. E., O. E. Anisimov, and O. I. Shiklomanov, 2001, Subsidence risk from thawing permafrost, *Nature* **410**: 889–890.

Nelson, J. M. and J. D. Smith, 1989, Flow in meandering channels with natural topography, in: S. Ikeda and G. Parker (eds.), *River Meandering*, Geophysical Monographs Series, vol. **12**, Washington, DC: American Geophysical Union, pp. 69–126.

Nelson, J. M., R. L. Shreve, S. R. McLean, and T. G. Drake, 1995, Role of near-bed turbulence structure in bed load transport and bed form mechanics, *Water Resources Research* **31** (8): 2071–2086.

Nerem, R. S., E. Leuliette, and A. Cazenave, 2006, Present-day sea-level change: a review, *Comptes Rendus Geosciences* **338** (14–15): 1077–1083.

Nick, F. M., A. Vieli, I. M. Howat, and I. Joughin, 2009, Large-scale changes in Greenland outlet glacier dynamics triggered at the terminus, *Nature Geoscience* **2**: 110–114, doi:10.1038/NGEO394.

Nickling, W. G., 1988, The initiation of particle movement by wind, *Sedimentology* **35** (3): 499–511.

Nickling, W. G. and M. Ecclestone, 1980, A technique for detecting grain motion in wind tunnels and flumes, *Journal of Sedimentary Research* **50**: 652–654.

Nield, J. M. and A. C. W. Baas, 2008, Investigating parabolic and nebkha dune formation using a cellular automaton modelling approach, *Earth Surface Processes and Landforms* **33**, doi:10.1002/esp.1571.

Nienow, P., M. Sharp, and I. Willis, 1998, Seasonal changes in the morphology of the subglacial drainage system, Haut Glacier D'Arolla, Switzerland, *Earth Surface Processes and Landforms* **23**: 825–843.

Nishiizumi, K., E. L. Winterer, C. P. Kohl *et al.*, 1989, Cosmic ray production rates of ^{26}Al and ^{10}Be in quartz from glacially polished rocks, *Journal of Geophysical Research* **94**: 17 907–17 915.

Nye, J., 1965, The flow of a glacier in rectangular, elliptic and parabolic cross-section, *Journal of Glaciology* **5**: 661–690.

Nye, J. F., 1976, Water flow in glaciers: jokulhlaups, tunnels and veins, *Journal of Glaciology* **17**: 181–207.

O'Loughlin, E. M., 1986, Prediction of surface saturation zones in natural catchments by topographic analysis, *Water Resources Research* **22** (5): 794–804.

O'Connor, J. E., 1993, *Hydrology, Hydraulics, and Geomorphology of the Bonneville Flood*, Geological Society of America Special Paper 274, Boulder, CO: Geological Society of America.

O'Connor, J. E. and V. R. Baker, 1992, Magnitudes and implications of peak discharges from glacial Lake

Missoula, *Geological Society of America Bulletin* **104**: 267–279.

O'Connor, J. E., G. E. Grant, and J. E. Costa, 2002, The geology and geography of floods, in *Ancient Floods, Modern Hazards: Principles and Applications of Paleoflood Hydrology*, Water Science and Application vol. **5**, Washington, DC: American Geophysical Union, pp. 359–385.

Oerlemans, J., 1984, Numerical experiments on large-scale glacial erosion, *Zeitschrift für Gletscherkunde und Glazialgeologie* **20**: 107–126.

Oerlemans, J., 1994, Quantifying global warming from the retreat of glaciers, *Science* **264**: 243–245.

Oerlemans, J., 2001, *Glaciers and Climate Change*, Rotterdam, Netherlands: A. A. Balkema Publishers.

Oerlemans, J., 2005, Extracting a climate signal from 169 glacier records, *Science* **308**: 675–677, doi:10.1126/science.1107046.

Oerlemans, J. and J. P. F. Fortuin, 1992, Sensitivity of glaciers and small ice caps to greenhouse warming, *Science* **258**: 115–117.

Okada, Y., 1992, Internal deformation due to shear and tensile faults in a half-space, *Bulletin of the Seismological Society of America* **82** (2): 1018–1040.

Oremland, R. S. and J. F. Stolz, 2005, Arsenic, microbes and contaminated aquifers, *Trends in Microbiology* **13** (2): 45–49.

Ota, Y. and J. Chappell, 1996, Late Quaternary coseismic uplift events on the Huon Peninsula, Papua New Guinea, deduced from coral terrace data, *Journal of Geophysical Research* **101** (B3): 6071–6082.

Ouimet, W., K. Whipple, L. Royden, S. Zhiming, and Z. Chen, 2007, The influence of large landslides on river incision in a transient landscape: eastern margin of the Tibetan plateau (Sichuan, China), *Geological Society of America Bulletin* **119** (11/12): 1462–1476, doi:10.1130/B26136.1.

Owen, P. R., 1964, Saltation of uniform grains in air, *Journal of Fluid Mechanics* **20**: 225–242.

Palmer, A. N., 1991, Origin and morphology of limestone caves, *Geological Society of America Bulletin* **103**: 1–21.

Palumbo, L., L. Benedetti, D. Bourles, A. Cinque, and R. Finkel, 2004, Slip history of the Magnola fault (Apennines, Central Italy) from Cl-36 surface exposure dating: evidence for strong earthquakes over the Holocene, *Earth and Planetary Science Letters* **225**: 163–176.

Paola, C., 2000, Quantitative models of sedimentary basin filling, *Sedimentology* **47**: 121–178.

Paola, C., P. L. Heller, and C. L. Angevine, 1992, The large-scale dynamics of grain-size variation in alluvial basins, 1: theory, *Basin Research* **4** (2): 73–90.

Papoulis, A., 1990. *Probability and Statistics*, Englewood Cliffs, NJ: Prentice Hall.

Pardee, J. T., 1910, The Glacial Lake Missoula, Montana, *Journal of Geology* **18**: 376–386.

Pardee, J. T., 1942, Unusual currents in Glacial Lake Missoula, Montana, *Geological Society of America Bulletin* **53**: 1569–1600.

Parker, G., 1976, On the cause and characteristic scales of meandering and braiding in rivers, *Journal of Fluid Mechanics* **76**: 457–480.

Parker, G., 1978, Self-formed straight rivers with equilibrium banks and mobile bed, 2, The gravel river, *Journal of Fluid Mechanics* **89**: 127–146.

Parkhurst, D. L. and C. A. J. Appelo, 1999, User's guide to PHREEQC (Version 2) – A computer program for speciation, batch-reaction, one-dimensional transport, and inverse geochemical calculations, *US Geological Survey Water-Resources Investigations Report* 99–4259.

Parsons, B. and J. G. Sclater, 1977, An analysis of the variation of ocean floor bathymetry and heat flow with age, *Journal of Geophysical Research* **82**: 803–827.

Paterson, W. S. B., 1994, *The Physics of Glaciers*, 3rd edition, Oxford: Butterworth-Heinemann.

Paull, C. K., P. Mitts, W. Ussler, III, R. Keaten, and H. G. Greene, 2005, Trail of sand in upper Monterey Canyon: Offshore California, *Geological Society of America Bulletin* **117**: 1134–1145.

Paull, C. K., W. Ussler, III, H. G. Greene, P. Mitts, R. Keaten, and J. Barry, 2003, Caught in the act: 20 December 2001 gravity flow event in Monterey Canyon, *Geo-Marine Letters* **22**: 227–232.

Paulson, A., S. Zhong, and J. Wahr, 2005, Modelling post-glacial rebound with lateral viscosity variations, *Geophysical Journal International* **163** (1): 357–371.

Paulson, A., S. Zhong, and J. Wahr, 2007, Inference of mantle viscosity from GRACE and relative sea level data, *Geophysical Journal International* **168**: 1195–1209.

Pavich, M. J., 1986, Processes and rates of saprolite production and erosion on a foliated granitic rock of the Virginia Piedmont, in: S. M. Colman and D. P. Dethier (eds.), *Rates of Chemical Weathering of Rocks and Minerals*, Orlando, FL: Academic Press, pp. 551–590.

Pavich, M. J., L. Brown, J. Harden, J. Klein, and R. Middleton, 1986, ^{10}Be distribution in soils from Merced River terraces, California, *Geochimica et Cosmochimica Acta* **50**: 1727–1735.

Pavich, M. J., L. Brown, J. N. Valette-Silver, J. Klein, and R. Middleton, 1985, ^{10}Be analysis of a Quaternary weathering profile in the Virginia Piedmont, *Geology* **13**: 39–41.

Peckham, S. D., 1998, Efficient extraction of river networks and hydrologic measurements from digital elevation data, in: O. E. Barndorff-Nielsen *et al.* (eds.), *Stochastic Methods in Hydrology: Rain, Landforms and Floods*, Singapore: World Scientific Publishing, pp. 173–204.

Pelletier, J. D., 2005, Formation of oriented thaw lakes by thaw slumping, *Journal of Geophysical Research* **110**: F02018, doi:10.1029/2004JF000158.

Peltier, W. R., 1981, Ice age geodynamics, *Annual Review of Earth and Planetary Sciences* **9**: 199–225, doi:10.1146/annurev.ea.09.050181.001215.

Peltier, W. R., 2004, Global glacial isostasy and the surface of the Ice-age Earth: the ICE-5G (VM2) model and GRACE, *Annual Review of Earth and Planetary Sciences* **32**: 111–149, doi:10.1146/annurev.earth.32. 082503.144359.

Peltier, W. R. and J. T. Andrews, 1976, Glacial-isostatic adjustment I. The forward problem, *Geophysical Journal of the Royal Astronomical Society* **46**: 605–646.

Peltier, W. R. and R. G. Fairbanks, 2006, Global glacial ice volume and Last Glacial Maximum duration from an extended Barbados sea level record, *Quaternary Science Reviews* **25**: 3322–3337.

Peltier, W. R., R. A. Drummond, and A. M. Tushingham, 1986, Post-glacial rebound and transient lower mantle rheology, *Geophysical Journal International* **87** (1): 79–116.

Pengelly, J. W., K. J. Tinkler, W. G. Parkins, and F. M. G. McCarthy, 1997, 12600 years of lake level changes, changing sills, ephemeral lakes and Niagara Gorge erosion in the Niagara Peninsula and eastern Lake Erie basin, *Journal of Paleolimnology* **17** (4): 377–402.

Perg, L. A., R. S. Anderson, and R. C. Finkel, 2001, Young ages of the Santa Cruz marine terraces determined using ^{10}Be and ^{26}Al, *Geology* **29** (10): 879–882.

Perg, L. A., R. S. Anderson, and R. C. Finkel, 2002, Young ages of the Santa Cruz marine terraces determined using ^{10}Be and ^{26}Al, Reply to comments by Brown and Bourles, *Geology* **30** (12): 1148, doi:10.1130/0091–7613(2002)030 1148: 2.0.CO;267.

Perg, L. A., R. S. Anderson, and R. C. Finkel, 2003, Use of cosmogenic radionuclides as a sediment tracer in the Santa Cruz littoral cell, California, United States, *Geology* **31** (4): 299–302.

Perron, J. T., M. P. Lamb, C. D. Koven, I. Y. Fung, E. Yager, and M. Adamkovics, 2006, Valley formation and methane precipitation rates on Titan, *Journal of Geophysical Research* **111**: E11001, doi:10.1029/2005JE002602.

Péwé, T. L. (ed), *Desert Dust: Origin, Characteristics, and Effect on Man*, Geological Society of America Special Paper 186, Boulder, CO: Geological Society of America.

Pfeffer, W. T., 2007, *The Opening of a New Landscape: Columbia Glacier at Mid-Retreat*, AGU Special Publications Series, 59.

Philbrick, S. S., 1970, Horizontal configuration and the rate of erosion of Niagara Falls, *Geological Society of America Bulletin* **81**: 3723–3732.

Phillips, F. M., B. D. Leavy, N. O. Jannik, D. Elmore, and P. W. Kubik, 1986, The accumulation of cosmogenic chlorine-36 in rocks: A method for surface exposure dating, *Science* **231**: 41–43.

Phillips, F., M. Zreda, S. Smith, D. Elmore, P. Kubik, and P. Sharma, 1990, Cosmogenic chlorine-36 chronology for glacial deposits at Bloody Canyon, eastern Sierra Nevada, *Science* **248**: 1529–1532, doi:10.1126/science.248.4962.1529.

Pierce, K. L., 1986, Dating methods, in: *Active Tectonics*, National Research Council, Geophysics Study Committee, Washington, DC: National Academy Press, pp. 195–214.

Pierce, K. L., J. D. Obradovich, and I. Friedman, 1976, Obsidian hydration dating and correlation of Bull Lake and Pinedale glaciations near West Yellowstone, Montana, *Geological Society of America Bulletin* **87**: 703–710.

Piexoto, J. P. and A. H. Oort, 1992, *Physics of Climate*, College Park, MD: American Institute of Physics.

Pizzuto, J. E., 1987, Sediment diffusion during overbank flows, *Sedimentology* **34**: 301–317.

Playfair, J., 1802, *Illustrations of the Huttonian Theory of the Earth*, Edinburgh: W. Creech; London: Cadell and Davies.

Plug, L. J. and B. T. Werner, 2001, Fracture networks in frozen ground, *Journal of Geophysical Research* **106** (B5): 8599–8613.

Plug, L. J. and B. T. Werner, 2002, Nonlinear dynamics of ice-wedge networks and resulting sensitivity to severe cooling events, *Nature* **417**: 929–933.

Plug, L. J. and J. J. West, 2009, Thaw lake expansion in a two-dimensional coupled model of heat transfer, thaw subsidence, and mass movement, *Journal of Geophysical Research* **114**: F01002, doi:10.1029/2006JF000740.

Pluhar, C. J., B. C. Bjornstad, S. P. Reidel, R. S. Coe, and P. B. Nelson, 2006, Magnetostratigraphic evidence from the Cold Creek bar for onset of ice-age cataclysmic floods in eastern Washington during the early Pleistocene, *Quaternary Research* **65**: 123–135.

Plummer, M. A. and F. M. Phillips, 2003, A 2-D numerical model of snow/ice energy balance and ice flow for paleoclimatic interpretation of glacial geomorphic features, *Quaternary Science Reviews* **22**: 1389–1406.

Pollard, D. D. and A. M. Johnson, 1973, Mechanics of growth of some laccolithic intrusions in the Henry Mountains, Utah, II. Bending and failure of overburden layers and sill formation, *Tectonophysics* **18**: 311–354.

Pollard, W. H. and H. M. French, 1980, A first approximation of the volume of ground ice, Richards Island, Pleistocene Mackenzie Delta, Northwest Territories, Canada, *Canadian Geotechnical Journal* **17**: 509–516.

Porter, S. C. and G. Orombelli, 1980, Catastrophic rockfall of September 12, 1717 on the Italian flank of the Mont Blanc massif, *Zeitschrift für Geomorphologie* **24**: 200–218.

Post, A. and L. R. Mayo, 1971, *Glacier-dammed Lakes and Outburst Floods in Alaska*, US Geological Survey Atlas HA-455, Reston, VA: US Geological Survey.

Pouliquen, O., J. Delour, and S. B. Savage, 1997, Fingering in granular flows, *Nature* **386**: 816–817.

Powell, R. D. and B. F. Molnia, 1989, Glacimarine sedimentary processes, facies and morphology of the south–southeast Alaska shelf and fjords, *Marine Geology* **85**: 359–390.

Pratt, B., D. W. Burbank, A. Heimsath, and T. Ojha, 2002, Impulsive alluviation during early Holocene strengthened monsoons, central Nepal Himalaya, *Geology* **30**: 911–914.

Prentice, C., T. Niemi, and T. Hall, 1991, Quaternary tectonics of the Northern San Andreas Fault, geologic excursions in Northern California, *California Division of Mines and Geology Special Publication* **109**: 25–34.

Price, E. J. and R. Bürgmann, 2002, Interactions between the Landers and Hector Mine, California, Earthquakes from space geodesy, boundary element modeling, and time-dependent friction, *Bulletin of the Seismological Society of America* **92** (4): 1450–1469, doi:10.1785/0120000924.

Pullman, E. K., M. T. Jorgenson, and Y. Shur, 2007, Thaw settlement in soils of the Arctic coastal plain, Alaska, *Arctic, Antarctic and Alpine Research* **39**: 468–476.

Putkonen, J. and B. Hallet, 1994, Surface dating of dynamic landforms: Young boulders on aging moraines, *Science* **265**: 937–940, doi:10.1126/science.265.5174.937.

Putkonen, J. and T. Swanson, 2003, Accuracy of cosmogenic ages for moraines, *Quaternary Research* **59**: 255–261.

Pye, K., 1987, *Aeolian Dust and Dust Deposits*, London: Academic Press.

Pye, K. and H. Tsoar, 1990, *Aeolian Sand and Sand Dunes*, London: Unwin Hyman.

Pysklywec, R. and J. X. Mitrovica, 1998, Mantle flow mechanisms for the large-scale subsidence of continental interiors, *Geology* **26**: 687–690.

Raats, P. A. C. and M. T. van Genuchten, 2006, Milestones in soil physics, *Soil Science* **171**: S21–S28.

Randriamazaoro, R., L. Dupeyrat, F. Costard, and E. C. Gailhardis, 2007, Fluvial thermal erosion: heat balance integral method, *Earth Surface Processes and Landforms* **32** (12): 1828–1840, doi:10.1002/esp.1489.

Rasmussen, K. R. and H. E. Mikkelsen, 1991, Wind tunnel observations of aeolian transport rates, *Acta Mechanica Supplement* **1**: 135–144.

Ravelo, A. C., D. Andreasen, M. Lyle, A. Olivarez Lyle, and M. W. Wara, 2004, Regional climate shifts caused by gradual global cooling in the Pliocene epoch, *Nature* **429**: 263–267.

Raymo, M. E. and W. F. Ruddiman, 1992, Tectonic forcing of late Cenozoic climate, *Nature* **359**: 117–122.

Raymo, M. E., W. F. Ruddiman, and P. N. Froelich, 1988, Influence of late Cenozoic mountain building on ocean geochemical cycles, *Geology* **16**: 649–653.

Raymond, C. F., 1971, Flow in a transverse section of Athabasca Glacier, Alberta, Canada, *Journal of Glaciology* **10**: 55–84.

Raymond, C. F. and W. D. Harrison, 1988, Evolution of Variegated Glacier, Alaska, U.S.A., prior to its surge, *Journal of Glaciology* **34** (117): 154–169.

Rea, C. C. and P. D. Komar, 1975, Computer simulation models of a hooked beach shoreline configuration, *Journal of Sedimentary Research* **45**, doi: 10.1306/212F6E6A-2B24–11D7–8648000102C1865D.

Reasenberg, P. A. and Simpson, R. W., 1992, Response of regional seismicity to the static stress change produced by the Loma Prieta earthquake, *Science* **255**: 1687–1690.

Reichman, O. J. and E. W. Seabloom, 2002, The role of pocket gophers as subterranean ecosystem engineers, *Trends in Ecology and Evolution* **17**: 44–49.

Reid, J. B., E. P. Bucklin, L. Copenagle *et al.*, 1995, Sliding rocks at the Racetrack, Death Valley: what makes them move? *Geology* **23** (9): 819–822.

Reid, L. M. and T. Dunne, 1996, *Rapid Evaluation of Sediment Budgets*, Reiskirchen, Germany: Catena.

Reiners, P. W., 2002, (U-Th)/He chronometry experiences a renaissance, *Eos, Transactions, AGU* **83**: 21–27.

Reiners, P. W., 2005, Zircon (U-Th)/He Thermochronometry, in: P. W. Reiners and T. A. Ehlers (eds.), *Thermochronology, Reviews in Mineralogy and Geochemistry* **58**: 151–176.

Reiners, P. W., 2007, Thermochronologic approaches to paleotopography, in: Kohn, M. J. (ed.), *Paleoaltimetry: Geochemical and Thermodynamic Approaches, Reviews in Mineralogy and Geochemistry* **66**: 243–267.

Reiners, P. W. and M. T. Brandon, 2006, Using thermochronology to understand orogenic erosion, *Annual Reviews of Earth and Planetary Science* **34**: 419–466.

Reiners, P. W. and K. A. Farley, 1999, He diffusion and (U-Th)/He thermochronometry of titanite, *Geochimica et Cosmochimica Acta* **63**: 3845–3859.

Reiners, P. W., T. A. Ehlers, and P. K. Zeitler, 2005, Past, present, and future of thermochronology, in: P. W. Reiners and T. A. Ehlers (eds.), *Thermochronology, Reviews in Mineralogy and Geochemistry* **58**: 1–18.

Rempel, A. W., 2007, Formation of ice lenses and frost heave, *Journal of Geophysical Research* **112**: F02S21, doi:10.1029/2006JF000525.

Rempel, A. W., J. S. Wettlaufer, and M. G. Worster, 2004, Premelting dynamics in a continuum model of frost heave, *Journal of Fluid Mechanics* **498**: 227–244.

Repka, J. L., R. S. Anderson, and R. C. Finkel, 1997, Cosmogenic dating of fluvial terraces, Fremont River, Utah, *Earth and Planetary Science Letters* **152**: 59–73.

Reynolds, O., 1883, An experimental investigation of the circumstances which determine whether the motion of water in parallel channels shall be direct or sinuous and of the law of resistance in parallel channels, *Philosophical Transactions of the Royal Society* **174**: 935–982.

Reynolds, O., 1895, On the dynamical theory of incompressible viscous fluids and the determination of the criterion, *Philosophical Transactions of the Royal Society* **186**: 123–164.

Rice, M. A., B. B. Willetts, and I. K. McEwan, 1996, Observations of collisions of saltating grains with a granular bed from high-speed cine-film, *Sedimentology* **43** (1): 21–31.

Richards, L. A., 1931, Capillary conduction of liquids through porous mediums, *Physics* **1**: 318–333.

Richards, P. L. and L. R. Kump, 2003, Soil pore-water distributions and the temperature feedback of weathering in soils, *Geochimica et Cosmochimica Acta* **67**: 3803–3815.

Richey, J. E., L. A. K. Mertes, T. Dunne *et al.*, 1989, Sources and routing of the Amazon River flood wave, *Global Biogeochemical Cycles* **3**: 191–204.

Richter, D. D. and D. Markewitz, 2001, *Understanding Soil Change: Soil Sustainability Over Millennia, Centuries and Decades*, Cambridge: Cambridge University Press.

Riebe, C. S., J. W. Kirchner, D. E. Granger, and R. C. Finkel, 2000, Erosional equilibrium and disequilibrium in the Sierra Nevada mountains, inferred from cosmogenic ^{26}Al and ^{10}Be in alluvial sediment, *Geology* **28**: 803–806.

Riebe, C. S., J. W. Kirchner, D. E. Granger, and R. C. Finkel, 2001a, Strong tectonic and weak climatic control of long-term chemical weathering rates, *Geology* **29** (6): 511–514.

Riebe, C. S., J. W. Kirchner, D. E. Granger, and R. C. Finkel, 2001b, Minimal climatic control on erosion rates in the Sierra Nevada, California, *Geology* **29** (5): 447–450.

Riebe, C. S., J. W. Kirchner, and D. E. Granger, 2001c, Quantifying quartz enrichment and its consequences for cosmogenic measurements of erosion rates from alluvial sediment and regolith, *Geomorphology* **40** (1–2): 15–19.

Riebe, C. S., J. W. Kirchner, and R. C. Finkel, 2004, Erosional and climatic effects on long-term chemical weathering rates in granitic landscapes spanning diverse climate regimes, *Earth and Planetary Science Letters* **224**: 547–562.

Rignot, E. and P. Kanagaratnam, 2006, Changes in the velocity structure of the Greenland Ice Sheet, *Science* **311** (5763): 986–990, doi:10.1126/science.1121381.

Riihimaki, C. A., R. S. Anderson, E. B. Safran, D. P. Dethier, and R. Finkel, 2006, Longevity and progressive abandonment of the Rocky Flats surface, Front Range, Colorado, *Geomorphology* **78**: 265–278.

Roberts, G. P. and A. M. Michetti, 2004. Spatial and temporal variations in growth rates along active normal fault systems: an example from Lazio-Abruzzo, central Italy, *Journal of Structural Geology* **26**: 339–376.

Rodbell, D. T., 1993, The timing of the last deglaciation in Cordillera Oriental, northern Peru based on glacial geology and lake sedimentology, *Geological Society of America Bulletin* **105**: 923–934.

Roe, G. H., 2005, Orographic precipitation, *Annual Review of Earth and Planetary Science* **33**: 645–671.

Roe, G. H. and M. Baker, 2006, Microphysical and geometrical controls on the pattern of orographic precipitation, *Journal of Atmospheric Science* **63**: 861–880.

Roe, G. H., D. R. Montgomery, and B. Hallet, 2002, Effects of orographic precipitation variations on the concavity of steady-state river profiles, *Geology* **30**: 143–146.

Roe, G. H., D. R. Montgomery, and B. Hallet, 2003, Orographic climate feedbacks on the relief of mountain ranges, *Journal of Geophysical Research, Solid Earth* **108** (B6): 2315, doi:10.1029/2001JB001521

Roering, J. J., J. W. Kirchner, L. S. Sklar, and W. E. Dietrich, 2001, Experimental hillslope evolution by nonlinear creep and landsliding, *Geology* **29**: 143–146.

Rosenbloom, N. A. and R. S. Anderson, 1994, Evolution of the marine terraced landscape, Santa Cruz, California, *Journal of Geophysical Research* **99**: 14 013–14 030.

Rouse, H., 1939, *An Analysis of Sediment Transport in the Light of Turbulence*, Technical Paper SCS-TP-25, Washington, DC: US Soil Conservation Service.

Rouse, H. and S. Ince, 1957, *History of Hydraulics*, State University of Iowa.

Rubin, D. M. and C. L. Carter, 2006, *Bedforms and Cross-Bedding in Animation*, SEPM Atlas Series No. 2, http://www.sepm.org/.

Ruddiman, W. F., 1997, *Tectonic Uplift and Climate Change*, New York and Heidelberg: Springer-Verlag.

Ruddiman, W. F., 2008, *Earth's Climate: Past and Future*, 2nd edition, New York: W. H. Freeman.

Ruddiman, W. F., M. E. Raymo, W. L. Prell, and J. E. Kutzbach, 1997, The uplift-climate connection: a synthesis, in: W. F. Ruddiman (ed.), *Tectonic Uplift and Climate Change*, New York: Plenum, pp. 471–515.

Ruhl, K. and K. V. Hodges, 2005, The use of detrital mineral cooling ages to evaluate steady state assumptions in active orogens: an example from the central Nepalese Himalayas, *Tectonics* **24**: TC4015, doi:10.1029/2004TC001712.

Ryan, W. and W. Pitman, 1999, *Noah's Flood: The New Scientific Discoveries about the Event that Changed History*, New York: Simon and Schuster.

Ryan, W. B. F., C. O. Major, G. Lericolais, and S. L. Goldstein, 2003, Catastrophic flooding of the Black Sea, *Annual Review of Earth and Planetary Sciences* **31**: 525–554.

Ryan, W. B. F., W. C. Pitman, III, C. O. Major *et al.*, 1997, An abrupt drowning of the Black Sea shelf, *Marine Geology* **138**: 119–126.

Saffer, D. M. and B. A. Bekins, 1998, Episodic fluid flow in the Nankai accretionary complex: timescale, geochemistry, flow rates, and fluid budget, *Journal of Geophysical Research* **103** (B12): 30 351–30 370.

Saffer, D. M. and B. A. Bekins, 1999, Fluid budgets at convergent plate margins: implications for the extent and duration of fault zone dilation, *Geology* **27** (12): 1095–1098.

Safran, E. B., 2003, Geomorphic interpretation of low-temperature thermochronologic data: insights from two-dimensional thermal modeling, *Journal of Geophysical Research* **108** (B4): 2189, doi:10.1029/2002JB001870.

Safran, E. B., P. R. Bierman, R. Aalto, T. Dunne, K. X. Whipple, and M. Caffee, 2005, Erosion rates driven by channel network incision in the Bolivian Andes, *Earth Surface Processes and Landforms* **30** (8): 1007–1024.

Sagan, C. and R. A. Bagnold, 1975, Fluid transport on Earth and eolian transport on Mars, *Icarus* **26** (2): 209–218.

Sagan, C. and J. B. Pollack, 1969, Windblown dust on Mars, *Nature* **223** (5208): 791–794.

Satake, K. and B. K. Atwater, 2007, Long-term perspectives on giant earthquakes and tsunamis at subduction zones, *Annual Reviews of Earth and Planetary Sciences* **35**: 349–374.

Savage, J. C. and W. S. B. Paterson, 1963, Borehole measurements in the Athabasca Glacier, *Journal of Geophysical Research* **68**: 4521–4536.

Schaetzl, R. J. and L. R. Follmer, 1990, Longevity of treethrow microtopography: implications for mass wasting, *Geomorphology* **3**: 113–123.

Schaller, M., F. von Blanckenburg, N. Hovius, and P. W. Kubik, 2001, Large-scale erosion rates from in situ-produced cosmogenic nuclides in European river sediments, *Earth and Planetary Science Letters* **188**: 441–458.

Schaller, M., F. von Blanckenburg, A. Veldkamp, L. A. Tebbens, N. Hovius, and P. W. Kubik, 2002, A 30,000 year record of erosion rates from cosmogenic ^{10}Be in Middle European river terraces, *Earth and Planetary Science Letters* **204**: 307–320.

Schaller, M., F. von Blanckenburg, A. Veldkamp, M. W. van den Berg, N. Hovius, and P. W. Kubik, 2004, Paleo-erosion rates from cosmogenic ^{10}Be in a 1.3 Ma terrace sequence: River Meuse, the Netherlands, *Journal of Geology* **112**: 127–144.

Scherler, D., S. Leprince, and M. R. Strecker, 2008, Glacier-surface velocities in alpine terrain from optical satellite imagery – accuracy improvement and quality assessment, *Remote Sensing of Environment* **112**: 3806–3819.

Schlichting, H., 1979, *Boundary Layer Theory*, New York: McGraw-Hill.

Schmeeckle, M. W. and J. M. Nelson, 2003, Direct numerical simulation of bedload transport using a local, dynamic boundary condition, *Sedimentology* **50** (2): 279–301, doi:10.1046/j.1365–3091.2003.00555.x.

Schmeeckle, M. W., J. M. Nelson, J. Pitlick, and J. P. Bennett, 2001, Interparticle collision of natural sediment grains in water, *Water Resources Research* **37** (9): 2377–2392.

Schmeeckle, M. W., J. M. Nelson, and R. L. Shreve, 2007, Forces on stationary particles in near-bed turbulent flows, *Journal of Geophysical Research* **112**: F02003, doi:10.1029/2006JF000536.

Schoenbohm, L., B. C. Burchfiel, and L. Chen, 2006, Propagation of surface uplift, lower crustal flow, and Cenozoic tectonics of the southeast margin of the Tibetan Plateau, *Geology* **34**: 813–816.

Schoonmaker, D., 1998, Jokulhlaup, *American Scientist* **86**: 426–427.

Schott, J. and R. A. Berner, 1984, X-ray photoelectron studies of the mechanism of iron silicate dissolution during weathering, *Geochimica et Cosmochimica Acta* **47**: 2233–2240.

Schuster, R. L., 2000, A worldwide perspective on landslide dams, in: D. Alford and R. Schuster (eds.), *Usoi Landslide Dam and Lake Sarez: An Assessment of Hazard and Risk in the Pamir Mountains, Tajikistan*, International Strategy for Disaster Reduction Series No. I, New York: United Nations, pp. J9–22.

Schuur, E. A. G., J. Bockheim, J. G. Canadell *et al.*, 2008, Vulnerability of permafrost carbon to climate change: implications for the global carbon cycle, *BioScience* **58**: 701–714.

Schwartz, D. P. and K. J. Coppersmith, 1984, Fault behavior and characteristic earthquakes: examples from the Wasatch and San Andreas Fault Zones, *Journal of Geophysical Research* **89**: 5681.

Schwartz, S. Y., D. L. Orange, and R. S. Anderson, 1990, Complex fault interactions in a restraining bend on the San Andreas Fault, southern Santa Cruz Mountains, California, *Geophysical Research Letters* **17**: 1207–1210.

Seidl, M. A. and W. E. Dietrich, 1992, The problem of channel erosion into bedrock, *Catena Supplement* **23**: 101–124.

Selby, M. J., 1980, A rock mass strength classification for geomorphic purposes: with tests from Antarctica and New Zealand, *Zeitschrift für Geomorphologie* **24**: 31–51.

Selby, M. J., 1982, *Hillslope Materials and Processes*, Oxford: Oxford University Press.

Serreze, M. C., J. E. Walsh, F. S. Chapin III *et al.*, 2000, Observational evidence of recent change in the northern high-latitude environment, *Climate Change* **46** (1–2): 159–207.

Shackleton, N. J. and N. D. Opdyke, 1973, Oxygen isotope and palaeomagnetic stratigraphy of equatorial Pacific core V28–238: oxygen isotope temperatures and ice volumes on a 10^5 and 10^6 year scale, *Quaternary Research* **3**: 39–55.

Shackleton, N. J. and N. D. Opdyke, 1977, Oxygen isotope and paleomagnetic evidence for early Northern Hemisphere glaciation, *Nature* **270** (5634): 216–219.

Shakesby, R. A. and S. H. Doerr, 2006, Wildfire as a hydrological and geomorphological agent, *Earth-Science Reviews* **74**: 269–307.

Shao, Y. and A. Li, 1999, Numerical modeling of saltation in the atmospheric surface layer, *Boundary-Layer Meteorology* **91**: 199–225.

Shao, Y. and M. R. Raupach, 1992, The overshoot and equilibration of saltation, *Journal of Geophysical Research* **97**: 20 559–20 564.

Sharma, P. P. and S. C. Gupta, 1989, Sand detachment by single raindrops of varying kinetic energy and momentum, *Soil Science Society of America Journal* **53**: 1005–1010.

Sharma, P., M. Bourgeois, D. Elmore *et al.*, 2000, PRIME Lab performance, upgrades, and research applications, *Nuclear Instrumentation and Methods in Physics Research, B: Beam Interactions with Materials* **172**: 112–123.

Sharp, M., W. Lawson, and R. S. Anderson, 1988, Tectonic processes in a surge-type glacier – an analogue for the emplacement of thrust sheets by gravity tectonics, *Journal of Structural Geology* **10**: 499–515.

Sharp, R. P., 1949, Pleistocene ventifacts east of the Big Horn Mountains, Wyoming, *Journal of Geology* **57**: 175–195.

Sharp, R. P., 1964, Wind-driven sand in Coachella Valley, California. *Geological Society of America Bulletin* **74**: 785–804.

Sharp, R. P., 1963, Wind ripples, *Journal of Geology* **71** (5): 617–636.

Sharp, R. P., 1966, Kelso dunes, Mojave desert, California, *Geological Society of America Bulletin* **77** (10): 1045–1073.

Sharp, R. P., 1991, *Living Ice: Understanding Glaciers and Glaciation*, Cambridge: Cambridge University Press.

Sharp, R. P. and D. L. Carey, 1976, Sliding stones, Racetrack Playa, California. *Bulletin of the Geological Society of America* **87** (12): 1704–1717.

Sharp, R. P., D. L. Carey, J. B. Reid Jr., P. J. Polissar, and M. L. Williams, 1996, Sliding rocks at the Racetrack, Death Valley: what makes them move? Discussion and Reply, *Geology* **25**: 766–767.

Shelton, J. S., 1953, Can wind move rocks on Racetrack Playa, *Science* **117** (3042): 438–439, doi:10.1126/science.117.3042.438-a.

Shen, Z.-K., C. Zhao, A. Yin *et al.*, 2000, Contemporary crustal deformation in East Asia constrained by Global Positioning System measurement, *Journal of Geophysical Research* **105**: 5721–5734.

Shields, A., 1936, Anwendung der Aehnlichkeitsmechanik und der Turbulenzforschung auf die Geschiebebewegung, *Mitt. Preuss. Versuchsanst. Wasserbau Schiffbau*, **26**: 26 (English translation by W. P. Ott and J. C. van Uchelen, US Department of Agriculture, Soil Conservation Service, Coop. Laboratory, California, Institute of Technology, Pasadena, 1936).

Shreve, R. L., 1968a, Leakage and fluidization in air-layer lubricated avalanches, *Geological Society of America Bulletin* **79** (5): 653–657.

Shreve, R. L., 1968b, *The Blackhawk Landslide*, Geological Society of America Special Paper 108, Boulder, CO: Geological Society of America.

Shreve, R. L., 1972, Movement of water in glaciers, *Journal of Glaciology* **11**: 205–214.

Shreve, R. L., 1985a, Esker characteristics in terms of glacier physics, Katahdin esker system, Maine, *Bulletin of the Geological Society of America* **96**: 639–646.

Shreve, R. L., 1985b, Late-Wisconsin ice surface profile calculated from esker paths and types, Katahdin esker system, Maine, *Quaternary Research* **23**: 27–37.

Shuster, D. L. and K. A. Farley, 2003, 4He/3He thermochronometry. *Earth and Planetary Science Letter* **217**: 1–17.

Shuster, D. L. and K. A. Farley, 2005, 4He/3He thermochronometry: theory, practice, and potential complications, *Reviews in Mineralogy and Geochemistry* **58**: 181–203.

Shuster, D. L., T. A. Ehlers, M. R. Rusmore, and K. A. Farley, 2005, Rapid glacial erosion at 1.8 Ma revealed by ^{4}He/^{3}He thermochronometry, *Science* **310**: 1668–1670.

Shuster, D. L., P. M. Vasconcelos, J. A. Heim, and K. A. Farley, 2005, Weathering geochronology by (U-Th)/He dating of goethite, *Geochimica et Cosmochimica Acta* **69**: 659–673, doi:10.1016/j.gca.2004.07.028.

Sidle, R., A. D. Ziegler, J. N. Negishi, A. Rahim Nik, R. Siew, and F. Turkelboom, 2006, Erosion processes in steep terrain – truths, myths, and uncertainties related to forest management in Southeast Asia, *Forest Ecology and Management* **224**: 199–225.

Siegesmund, S., S. Mosch, Ch. Scheffzük, and D. I. Nikolayev, 2008, The bowing potential of granitic rocks: rock fabrics, thermal properties and residual strain, *Environmental Geology* **55**: 1437–1448.

Sieh, K., 1978, Pre-historic large earthquakes produced by slip on the San Andreas fault at Pallett Creek, California, *Journal of Geophysical Research* **83**: 3907–3939.

Sieh, K., 2006, Sumatran megathrust earthquakes: from science to saving lives, *Philosophical Transactions of the Royal Society A* **364**: 1946–1963.

Sieh, K. E. and R. H. Jahns, 1984, Holocene activity of the San Andreas Fault at Wallace Creek, California, *Geological Society of America Bulletin* **5**: 883–896.

Skinner, B. J., S. C. Porter, and D. B. Botkin, 1999, *The Blue Planet: An Introduction to Earth System Science*, 2nd edition, New York: Wiley.

Sklar, L. S. and W. E. Dietrich, 2001, Sediment and rock strength controls on river incision into bedrock, *Geology* **29**: 1087–1090.

Sklar, L. S. and W. E. Dietrich, 2004, A mechanistic model for river incision into bedrock by saltating bedload, *Water Resources Research* **40**, W06301, doi:10.1029/2003WR002496.

Sklar, L. S. and W. E. Dietrich, 2008, Implications of the saltation-abrasion bedrock incision model for steady state river longitudinal profile relief and concavity, *Earth Surface Processes and Landforms*, doi:10.1002/esp.1689.

Slingerland, R. L., K. Furlong, and J. Harbaugh, 1994, *Simulating Clastic Sedimentary Basins/Physical Fundamentals and Computing Procedures*, Englewood Cliffs, NJ: Prentice Hall.

Small, E. E. and R. S. Anderson, 1995, Geomorphically driven late Cenozoic rock uplift in the Sierra Nevada, California, *Science* **270**: 277–280.

Small, E. E. and R. S. Anderson, 1998, Pleistocene relief production in Laramide mountain ranges, western United States, *Geology* **26**: 123–126.

Small, E. E., R. S. Anderson, R. C. Finkel, and J. Repka, 1997, Erosion rates of summit flats using cosmogenic radionuclides, *Earth and Planetary Science Letters* **150**: 413–425.

Small, E. E., R. S. Anderson, G. S. Hancock, and R. C. Finkel, 1999, Estimates of regolith production from ^{10}Be and ^{26}Al: evidence for steady state alpine hillslopes, *Geomorphology* **27**: 131–150.

Smallwood, K. S., M. L. Morrison, and J. Beyea, 1998, Animal burrowing attributes affecting hazardous waste management, *Environmental Management* **22** (6): 831–847.

Smith, J. D. and S. R. McLean, 1984, A model for flow in meandering streams, *Water Resources Research* **20** (9): 1301–1315.

Smith, L. C., 1997, Satellite remote sensing of river inundation area, stage, and discharge: a review, *Hydrological Processes* **11**: 1427–1439.

Smith, L. C., B. L. Isacks, R. R. Forster, A. L. Bloom, and I. Preuss, 1995, Estimation of discharge from braided glacial rivers using ERS I synthetic aperture radar: first results, *Water Resources Research* **31** (5): 1325–1329.

Smith, L. C., B. L. Isacks, A. L. Bloom, and A. B. Murray, 1996, Estimation of discharge from three braided rivers using synthetic aperture radar (SAR) satellite imagery: potential application to ungaged basins, *Water Resources Research* **32** (7): 2021–2037.

Smith, L. C., Y. Sheng, G. M. MacDonald, and L. D. Hinzman, 2005, Disappearing Arctic lakes, *Science* **308**: 1429, doi:10.1126/science.1108142.

Smith, R. B., 1979, The influence of mountains on the atmosphere, *Advances in Geophysics* **21**: 187–230.

Smith, T. R. and F. P. Bretherton, 1972, Stability and the conservation of mass in drainage basin evolution, *Water Resources Research* **8** (6): 1506–1529.

Snyder, N. P., K. X. Whipple, G. T. Tucker, and D. J. Merritts, 2002, Interactions between onshore bedrock-channel incision and nearshore wave-base erosion forced by eustasy and tectonics, *Basin Research* **14**: 105–127.

Snyder, N. P., K. X. Whipple, G. T. Tucker, and D. J. Merritts, 2003a, Channel response to tectonic forcing: analysis of stream morphology and hydrology in the Mendocino triple junction region, northern California, *Geomorphology* **53**: 97–127.

Snyder, N. P., K. X. Whipple, G. E. Tucker, and D. J. Merritts, 2003b, Importance of a stochastic distribution of floods and erosion thresholds in the bedrock river incision problem, *Journal of Geophysical Research* **108**: 2117, doi:10.1029/2001JB001655.

Soliva, R., A. Benedicto, R. A. Schultz, L. Maerten, and L. Micarelli, 2008, Displacement and interaction of normal fault segments branched at depth: Implications for fault growth and potential earthquake rupture size, *Journal of Structural Geology* **30** (10): 1288–1299, doi:10.1016/j.jsg.2008.07.005.

Solomon, S., D. Qin, M. Manning *et al.* (eds), 2007, *Fourth Assessment Report of the Intergovernmental Panel on Climate Change*, Cambridge, UK and New York: Cambridge University Press.

Sørensen, M., 1985, Estimation of some aeolian saltation transport parameters from transport rate profiles, in: O. E. Barndorff-Nielsen, J. T. Møller, K. R. Rasmussen, and B. B. Willetts (eds.), *Proceedings of International Workshop on the Physics of Blown Sand*, Denmark: Aarhus University, pp. 141–190.

Stallard, R. F., 1995a, Relating chemical and physical erosion, in: A. F. White and S. L. Brantley (eds.), *Chemical Weathering Rates of Silicate Minerals, Reviews in Mineralogy* **31**: 543–564.

Stallard, R. F., 1995b, Tectonic, environmental, and human aspects of weathering and erosion: a global review using a steady-state perspective, *Annual Review of Earth and Planetary Sciences* **23**: 11–39.

Stanley, G. M., 1955, Origin of playa stone tracks, Racetrack Playa, Inyo County, California, *Bulletin of the Geological Society of America* **66** (11): 1329–1350.

Stein, R. S., 2003, Earthquake conversations, *Scientific American*, **288** (1): 72–79.

Stein, R. S., G. C. P. King, and J. Lin, 1992, Change in failure stress on the southern San Andreas fault system caused by the 1992 Magnitude = 7.4 Landers earthquake, *Science* **258**: 1328–1332.

Stern, T. A., A. K. Baxter, and P. J. Barrett, 2005, Isostatic rebound due to glacial erosion within the Transantarctic Mountains, *Geology* **33**: 221–224.

Sternberg, H., 1875, Untersuchungen über Längen- und Querprofile geschiebeführender Flüsse, *Zeitschrift für Bauwesen* **25**: 483–506.

Stock, G. M., R. S. Anderson, and R. C. Finkel, 2004, Cave sediments reveal pace of landscape evolution in the Sierra Nevada, California, *Geology* **32** (3): 193–196; doi:10.1130/G20197.1.

Stock, G. M., R. S. Anderson, and R. C. Finkel, 2005, Late Cenozoic topographic evolution of the Sierra Nevada, California, inferred from cosmogenic ^{26}Al and ^{10}Be concentrations, *Earth Surface Processes and Landforms* **30**: 985–1006, doi:10.1002/esp.1258.

Stock, G. M., T. A. Ehlers, and K. A. Farley, 2006, Where does sediment come from? Quantifying catchment erosion with detrital apatite (U-Th)/He thermochronometry, *Geology* **34** (9): 725–728.

Stock, G. M., D. E. Granger, R. S. Anderson, I. D. Sasowsky, and R. C. Finkel, 2005, Dating cave deposits for use in landscape evolution studies: insights from caves in the Sierra Nevada, California, *Earth and Planetary Science Letters* **236**: 388–403.

Stock, J. D. and D. R. Montgomery, 1996, Estimating paleorelief from detrital mineral age ranges, *Basin Research* **8**: 317–327.

Stolum, H., 1996, River meandering as a self-organization process, *Science* **271**: 1710–1713.

Stolum, H., 1998, Planform geometry and dynamics of meandering rivers, *Geological Society of America Bulletin* **110**: 1485–1498.

Storlazzi, C. D. and M. E. Field, 2000, Sediment distribution and transport along a rocky, embayed coast: Monterey Peninsula and Carmel Bay, California, *Marine Geology* **170**: 289–316.

Strahler, A. N., 1952, Hypsometric (area altitude) analysis of erosional topology, *Geological Society of America Bulletin* **63**: 1117–1142.

Strudley, M. W., A. B. Murray, and P. K. Haff, 2006, Emergence of pediments, tors, and piedmont junctions from a bedrock weathering-regolith thickness feedback, *Geology* **34** (10): 805–808.

Stumm, W. and J. J. Morgan, 1996, *Aquatic Chemistry: Chemical Equilibria and Rates in Natural Waters*, 3rd edition, New York: Wiley-Interscience.

Sturm, M., C. Racine, and K. Tape, 2001, Climate change: increasing shrub abundance in the Arctic, *Nature* **411**: 546–547.

Sugden, D. E., 1978, Glacial erosion by the Laurentide Ice Sheet, *Journal of Glaciology* **83**: 367–391.

Sugden, D. E. and B. S. John, 1976, *Glaciers and Landscape: A Geomorphological Approach*, London: Edward Arnold.

Summerfield, M., 1991, *Global Geomorphology: An Introduction to the Study of Landforms*, Longman Scientific & Technical.

Sun, T., P. Meakin, and T. Jossang, 1996, A simulation model for meandering rivers, *Water Resources Research* **32**: 2937–2954, doi:10.1029/96WR00998.

Sun, T., P. Meakin, and T. Jossang, 2001a, Meander migration and the lateral tilting of floodplains, *Water Resources Research* **37** (5), 1485–1502.

Sun, T., P. Meakin, and T. Jossang, 2001b, A computer model for meandering rivers with multiple bed load sediment sizes. 1. Theory, *Water Resources Research* **37** (8): 2227–2241.

Sun, T., P. Meakin, and T. Jossang, 2001c. A computer model for meandering rivers with multiple bed load sediment sizes. 2. Computer simulations, *Water Resources Research* **37** (8): 2243–2258.

Sunamura, T., 1992, *Geomorphology of Rocky Coasts*, Chichester: Wiley.

Sverdrup, H. U., M. W. Johnson, and R. W. Fleming, 1942, *The Oceans: Their Physics, Chemistry and General Biology*, Englewood Cliffs, NJ: Prentice-Hall.

Swoboda-Colberg, N. G. and J. I. Drever, 1993, Mineral dissolution rates in plot-scale field and laboratory experiments, *Chemical Geology* **105**: 51–69.

Syvitski, J. P. M. (ed.), 1991, *Principles, Methods and Applications of Particle Size Analysis*, New York: Cambridge University Press.

Syvitski, J. P. M., 2008, Deltas at risk, *Sustainable Science*, doi:10.1007/s11625–008–0043–3.

Syvitski, J. P. M. and Y. Saito, 2007, Morphodynamics of deltas under the influence of humans, *Global and Planetary Change* **57**: 261–282.

Syvitski, J. P. M., D. C. Burrell, and J. M. Skei, 1987, *Fjords: Processes and Products*, New York: Springer-Verlag.

Taber, S., 1930, The mechanics of frost heaving, *Journal of Geology* **38**: 303–317.

Taberlet, N., S. W. Morris, and J. N. MacElwaine, 2007, Washboard road: the dynamics of granular ripples formed by rolling wheels, *Physical Review Letters* **99**: 068003.

Tagami, T. and P. B. O'Sullivan, 2005, Fundamentals of fission-track thermochronology, in: *Low-temperature Thermochronology: Techniques, Interpretations, and Applications, Reviews in Mineralogy and Geochemistry* **58** (1): 19–47.

Tagami, T., Galbraith, R. F., Yamada, R., and Laslett, G. M., 1998, Revised annealing kinetics of fission tracks in zircon and geological implications, in: P. Van den Haute and F. de Corte, (eds.), *Advances in Fission-Track Geochronology*, Solid Earth Science Library, **10**, Norwell, MA: Kluwer, pp. 99–112.

Taylor, G. and R. A. Eggleton, 2001, *Regolith Geology and Geomorphology*, Chichester, New York: Wiley.

Taylor, L. L., J. R. Leake, J. Quirk, K. Hardy, S. A. Banwart, and D. J. Beerling, 2009, Biological weathering and the long-term carbon cycle: integrating mycorrhizal evolution and function into the current paradigm, *Geobiology* **7**: 171–191.

Taylor, S. R. and S. M. McLennan, 1995, The geochemical evolution of the continental crust. *Reviews of Geophysics* **33**: 241–265.

Tesla, N., 1934, Radio power will revolutionize the world, *Modern Mechanics and Inventions* (July).

Terzhagi, R. D., 1962, Stability of steep slopes on hard unweathered rock, *Géotechnique* **12**: 251–270.

Tinkler, K. and E. Wohl, 1998, A primer on bedrock channels, in: K. Tinkler and E. Wohl (eds.), *Rivers Over Rock*, Geophysical Monograph **107**, American Geophysical Union, pp. 1–18.

Tinkler, K., J. W. Pengelly, W. G. Parkins, and G. Asselin, 1994, Postglacial recession of Niagara falls in relation to the great Lakes, *Quaternary Research* **42**: 20–29.

Toda, S., R. S. Stein, P. A. Reasenberg, and J. H. Dieterich, 1998, Stress transferred by the $Mw = 6.5$ Kobe, Japan, shock: effect on aftershocks and future earthquake probabilities, *Journal of Geophysical Research* **103**: 24 543–24 565.

Törnqvist, T. E., D. J. Wallace, J. E. A. Storms *et al.*, 2008, Mississippi Delta subsidence primarily caused by compaction of Holocene strata, *Nature Geoscience* **1**: 173–176.

Torres, R., 2002, A threshold condition for soil-water transport, *Hydrological Processes* **16**: 2703–2706.

Torres, R., W. E. Dietrich, K. Loague, D. R. Montgomery, and S. P. Anderson, 1998, Unsaturated zone processes and the hydrologic response of a steep, unchanneled catchment, *Water Resources Research* **34** (8): 1865–1879.

Trenberth, K. E. and J. M. Caron, 2001, Estimates of meridional atmosphere and ocean heat transports, *Journal of Climate* **14**, 3433–3443.

Trenhaile, A. S., 1987, *The Geomorphology of Rock Coasts*, Oxford: Clarendon Press.

Trudgill, B. D., 2002, Structural controls on drainage development in the Canyonlands grabens of southeast Utah, *AAPG Bulletin* **86** (6): 1095–1112.

Tucker, G. E. and K. X. Whipple, 2002, Topographic outcomes predicted by stream erosion models: sensitivity analysis and intermodel comparison, *Journal of Geophysical Research – Solid Earth* **107**, art. no. 2179.

Tulaczyk, S., B. Kamb, and H. Engelhardt, 2000a, Basal mechanics of Ice Stream B. I. Till mechanics, *Journal of Geophysical Research* **105**: 463–481.

Tulaczyk, S., B. Kamb, and H. Engelhardt, 2000b, Basal mechanics of Ice Stream B. II. Plastic-undrained-bed model, *Journal of Geophysical Research* **105**: 483–494.

Turcotte, D. L. and G. Schubert, 2002, *Geodynamics*, 2nd edition, Cambridge: Cambridge University Press.

Turnbull, J. C., S. J. Lehman, J. B. Miller, R. J. Sparks, J. R. Southon, and P. P. Tans, 2007, A new high precision $^{14}CO_2$ time series for North American continental air, *Journal of Geophysical Research* **112**: D11310, doi:10.1029/2006JD008184.

Tushingham, A. M. and W. R. Peltier, 1991, Ice-3G: a new global model of late Pleistocene deglaciation based upon geophysical predictions of post-glacial relative sea-level change, *Journal of Geophysical Research* **96**: 4497–4523.

Ungar, J. E. and P. K. Haff, 1987, Steady state saltation in air, *Sedimentology* **24**: 289–299.

US Geological Survey, 1996, *Perilous Beauty, the Hidden Dangers of Mt Rainier*, VHS video, 29 min, Reston, VA: US Geological Survey.

Valensise, G. and S. N. Ward, 1991, Long-term uplift of the Santa Cruz coastline in response to repeated earthquakes along the San Andreas fault, *Bulletin of the Seismological Society of America* **81**: 1694–1704.

Valvo, L. M., A. B. Murray, and A. Ashton, 2006, How does underlying geology affect coastline change? An initial modeling investigation, *Journal of Geophysical Research* **111**: F02025, doi:10.1029/2005JF000340.

van Dijk, A. I. J. M., L. A. Bruijnzeel, and C. J. Rosewell, 2002, Rainfall intensity-kinetic energy relationships: a critical literature appraisal, *Journal of Hydrology* **261**: 1–23.

Van Dyke, M., 1982, *An Album of Fluid Motion*, Stanford, CA: Parabolic Press.

Vanoni, V. A., 1941, Some experiments on the transportation of suspended load, *Transactions of the AGU 4/3–5/3*, pp. 608–621.

Vanoni, V. A., 1946, Transportation of suspended sediment by water, *Transactions of the American Society of Civil Engineers* **111**: 67–133.

Vanoni, V. A., P. C. Benedict, D. C. Bondurant, J. E. McKee, R. F. Piest, and J. Smallshaw, 1966, Sediment transportation mechanics: Initiation of motion, *Journal of the Hydraulic Division – American Society of Civil Engineers* **92**: 291–314.

Varnes, D. J., 1996, Landslide types and processes, in: A. K. Turner and R. L. Schuster, *Landslides: Investigation and Mitigation*, Transportation Research Board Special Report 247, National Research Council, Washington, DC: National Academy Press.

Vasconcelos, P. M., 1999, K-Ar and ^{40}Ar/^{39}Ar geochronology of weathering processes, *Annual Review of Earth Sciences* **27**: 183–229.

Veeh, H. H. and J. Chappell, 1970, Astronomical theory of climatic change: Support from New Guinea, *Science* **167**: 862–865.

Velbel, M. A., 1985, Geochemical mass balances and weathering rates in forested watersheds of the southern Blue Ridge, *American Journal of Science* **285**: 904–930.

Velbel, M. A., 1990, Influence of temperature and mineral surface characteristics on feldspar weathering, *Water Resources Research* **26**: 3049–3053.

Vermeesch, P. and N. Drake, 2008, Remotely sensed dune celerity and sand flux measurements of the world's fastest barchans (Bodélé, Chad), *Geophysical Research Letters* **35**: L24404, doi:C0.1029/2008GL035921.

Viles, H., 1988, *Biogeomorphology*, Oxford: Basil Blackwell.

Vitousek, P., O. Chadwick, P. Matson *et al.*, 2003, Erosion and the rejuvenation of weathering-derived nutrient supply in an old tropical landscape, *Ecosystems* **6**: 762–772.

von Blanckenburg, F., T. Hewawasam, and P. W. Kubik, 2004, Cosmogenic nuclide evidence for low weathering and denudation in the wet, tropical highlands of Sri Lanka, *Journal of Geophysical Research* **109**: F03008, doi:10.1029/2003JF000049.

Wahl, K. L., W. O. Thomas Jr., and R. M. Hirsch, 1995, *Stream-Gaging Program of the US Geological Survey*, US Geological Survey Circular 1123, Reston, VA: US Geological Survey.

Wahrhaftig, C., 1965, Stepped topography of the southern Sierra Nevada, California, *Geological Society of America Bulletin* **76**: 1165–1190.

Waitt, R. B., 1985, Case for periodic, colossal jokulhlaups from Pleistocene glacial Lake Missoula, *Geological Society of America Bulletin* **96** (10): 1271–1286.

Walcott, R. I., 1972, Late Quaternary vertical movements in eastern North America: Quantitative evidence of glacio-isostatic rebound, *Review of Geophysics and Space Physics* **10**: 849–884.

Waldbauer, J. R. and C. P. Chamberlain, 2005, Influence of uplift, weathering and base cation supply on past and future CO_2 levels, in: J. R. Ehleringer, T. E. Cerling, and M. D. Dearing (eds.), *A History of Atmospheric CO_2 and its Effects on Plants, Animals and Ecosystems*, New York: Springer-Verlag, pp. 166–184.

Walder, J. S. and J. E. Costa, 1996, Outburst floods from glacier-dammed lakes: the effect of mode of lake drainage on flood magnitude, *Earth Surface Processes and Landforms* **21** (8): 701–723.

Walder, J. S. and B. Hallet, 1985, A theoretical model of the fracture of rock due to freezing, *Geological Society of America Bulletin* **96** (3): 336–346, doi:10.1130/0016–7606 (1985)96 336:ATMOTF 2.0.CO;2.

Walder, J. S. and B. Hallet, 1986, The physical basis of frost weathering: toward a more fundamental and unified perspective, *Arctic and Alpine Research* **18** (1): 27–32.

Walder, J. S. and J. E. O'Connor, 1997, Methods for predicting peak discharge of floods caused by failure of natural and constructed earthen dams, *Water Resources Research* **33** (10): 2337–2348.

Walker, J. C. G., P. B. Hays, and J. F. Kasting, 1981, A negative feedback mechanism for long-term stabilization of Earth's surface temperature, *Journal of Geophysical Research* **86** (C10): 9776–9782.

Wallace, C. C. and B. R. Rosen, 2006, Diverse staghorn corals (*Acropora*) in high-latitude Eocene assemblages: implications for the evolution of modern diversity patterns of reef corals, *Proceedings Biological Sciences* **273** (1589): 975–982.

Wallace, R. E., 1970, Earthquake recurrence intervals of the San Andreas fault, *Geological Society of America Bulletin* **81**: 2875–2890.

Wallinga, J., 2002, Optically stimulated luminescence dating of fluvial deposits: a review, *Boreas* **31** (4): 303–322.

Ward, A. W. and R. Greeley, 1984, Evolution of the yardangs at Rogers Lake, California, *Geological Society of America Bulletin* **95**: 829–837.

Ward, D. J., R. S. Anderson, Z. S. Guido, and J. P. Briner, 2009, Numerical modeling of cosmogenic deglaciation records, Front Range and San Juan Mountains, Colorado, *Journal of Geophysical Research*, doi:10.1029/2008JF001057.

Ward, D., J. A. Spotila, G. S. Hancock, and J. M. Galbraith, 2005, New constraints on the late Cenozoic incision history of the New River, Virginia, *Geomorphology* **72** (1–4): 54–72.

Ward, J. V., 1998, River landscapes: biodiversity patterns, disturbance regimes, and aquatic conservation, *Biological Conservation* **83**: 269–278.

Ward, S. N. and S. D. B. Goes, 1993, How regularly do earthquakes recur? A synthetic seismicity model for the San Andreas fault, *Geophysical Research Letters* **20**: 2131–2134.

Ward, S. N. and G. Valensise, 1994, The Palos Verdes terraces, California: bathtub rings from a buried reverse fault, *Journal of Geophysical Research* **99** (B3): 4485–4494.

Washburn, A. L., 1965, Geomorphic and vegetational studies in the Mesters Vig district, Northeast Greenland: general introduction, *Meddelelser om Grønland* **166** (1), Kjøbenhavn: C.A. Reitzel.

Washburn, A. L., 1967, Instrumental observations of mass wasting in the Mesters Vig district, Northeast Greenland, *Meddelelser om Grønland* **166** (4), Kjøbenhavn: C.A. Reitzel.

Washburn, A. L., 1979, *Geocryology. A Survey of Periglacial Processes and Environments*, London: Edward Arnold.

Watts, A. B., 2001, *Isostasy and Flexure of the Lithosphere*, Cambridge: Cambridge University Press.

Waythomas, C. R, J. S. Walder, R. G. McGimsey, and C. A. Neal, 1996, A catastrophic flood caused by drainage of a caldera lake at Aniakchak volcano, Alaska, and implications for volcanic hazards assessment, *Geological Society of America Bulletin* **108**: 861–871.

Weaver, J. A., O. A. Saenko, P. U. Clark, and J. X. Mitrovica, 2003, Meltwater Pulse 1A from Antarctica as trigger of the Bølling-Allerød Warm Interval, *Science* **299** (5613): 1709–1713, doi:10.1126/science.1081002.

Webster, J. M., L. Wallace, E. Silver, *et al.*, 2004a, Drowned carbonate platforms in the Huon Gulf, Papua New Guinea, *Geochemistry, Geophysics, Geosystems* **5**: Q11008, doi:10.1029/2004GC000726.

Webster, J. M., L. Wallace, E. Silver, *et al.*, 2004b, Coralgal composition of drowned carbonate platforms in the Huon Gulf, Papua New Guinea: implications for lowstand reef development and drowning, *Marine Geology* **204** (1–2): 59–89.

Wendler, G., C. Stearns, G. Weidner, G. Dargaud, and T. Parish, 1997, On the extraordinary katabatic winds of Adélie Land, *Journal of Geophysical Research* **102**(D4): 4463–4474.

Werner, B. T., 1990, A steady-state model of wind-blown sand transport, *Journal of Geology* **98**: 1–17.

Werner, B. T., 1995, Eolian dunes: computer simulations and attractor interpretation, *Geology* **23**: 1107–1110.

Werner, B. T. and T. M. Fink, 1993, Beach cusps as self-organized patterns, *Science* **260** (5110): 968–971, doi:10.1126/science.260.5110.968.

Werner, B. T. and D. T. Gillespie, 1993, Fundamentally discrete stochastic model of wind ripple dynamics, *Physical Review Letters* **71**: 3230–3233.

Werner, B. T., P. K. Haff, R. P. Livi, and R. S. Anderson, 1986, The measurement of eolian ripple cross-sectional shapes, *Geology* **14**: 743–745.

Wessel, P. and B. H. Keating, 1994, Temporal variations of flexural deformation in Hawaii, *Journal of Geophysical Research* **99**: 2747–2756.

West, A. J., A. Galy, and M. Bickle, 2005, Tectonic and climatic controls on silicate weathering, *Earth and Planetary Science Letters* **235**: 211–228.

West, J. J. and L. J. Plug, 2008, Time-dependent morphology of thaw lakes and taliks in deep and shallow ground ice, *Journal of Geophysical Research* **113**: F01009, doi:10.1029/2006JF000696.

Wettlaufer, J. S. and J. G. Dash, 2000, Melting below zero, *Scientific American* **282** (2): 50–53.

Whipple, K. X., 2004, Bedrock rivers and the geomorphology of active orogens, *Annual Reviews of Earth and Planetary Science* **32**: 151–185.

Whipple, K. X., 1992, Predicting debris-flow runout and deposition on fans: the importance of the flow hydrograph, *IAHS Publication No.* **209**: 337–345.

Whipple, K. X. and T. Dunne, 1992, The influence of debris-flow rheology on fan morphology, Owens Valley, California, *Geological Society of America Bulletin* **104**: 887–900.

Whipple, K. X., G. S. Hancock, and R. S. Anderson, 2000, River incision into bedrock: mechanics and relative efficacy of plucking, abrasion, and cavitation, *Geological Society of America Bulletin* **112**: 490–503.

Whipple, K. X. and G. Tucker, 1999, Dynamics of the stream-power river incision model: implications for height limits of mountain ranges, landscape response timescales and research needs, *Journal of Geophysical Research* **104**: 17 661–17 674.

White, A. F., 2008, Quantitative approaches to characterizing natural chemical weathering rates, in: S. L. Brantley, J. D. Kubicki, and A. F. White (eds.), *Kinetics of Water–Rock Interaction*, New York and Heidelberg: Springer-Verlag, pp. 469–543.

White, A. F. and A. E. Blum, 1995, Effects of climate on chemical weathering in watersheds, *Geochimica et Cosmochimica Acta* **59**: 1729–1747.

White, A. F. and S. L. Brantley, 1995, Chemical weathering rates of silicate minerals: an overview, in: A. F. White and S. L. Brantley (eds.), *Chemical Weathering Rates of Silicate*

Minerals **31**: 1–22, Washington, DC: Mineralogical Society of America.

White, A. F. and S. L. Brantley, 2003, The effect of time on the weathering of silicate minerals: why do weathering rates differ in the laboratory and field?, *Chemical Geology* **202**: 479–50.

White, A. F., T. D. Bullen, M. S. Schulz, A. E. Blum, T. G. Huntington, and N. E. Peters, 2001, Differential rates of feldspar weathering in granitic regoliths, *Geochimica et Cosmochimica Acta* **65**: 847–869.

White, A. F., M. S. Schulz, D. V. Vivit, A. E. Blum, D. A. Stonestrom, and J. W. Harden, 2005, Chemical weathering rates of a soil chronosequence on granitic alluvium: III. Hydrochemical evolution and contemporary solute fluxes and rates, *Geochimica et Cosmochimica Acta* **69**: 1975–1996.

White, A. F., M. S. Schulz, D. V. Vivit, A. E. Blum, D. A. Stonestrom, and S. P. Anderson, 2008, Chemical weathering of a marine terrace chronosequence, Santa Cruz, California I: interpreting rates and controls based on soil concentration–depth profiles, *Geochimica et Cosmochimica Acta* **72**: 36–68.

Whiting, P. J. and W. E. Dietrich, 1990, Boundary shear stress and roughness over mobile alluvial beds, *American Society of Civil Engineers – Journal of Hydraulic Engineering* **116**: 1495–1511.

Whiting, P. J. and W. E. Dietrich, 1993a, Experimental studies of bed topography and flow patterns in large-amplitude meanders 1. Observations, *Water Resources Research* **29** (11): 3605–3614.

Whiting, P. J. and W. E. Dietrich, 1993b, Experimental studies of bed topography and flow patterns in large-amplitude meanders 2. Mechanisms, *Water Resources Research* **29** (11): 3615–3622.

Whittaker, A. C., P. A. Cowie, M. Attal, G. E. Tucker, and G. P. Roberts, 2007a, Bedrock channel adjustment to tectonic forcing: implications for predicting river incision rates. *Geology* **35** (2): 103–106, doi: 10.1130/G23106A.1.

Whittaker, A. C., P. A. Cowie, M. Attal, G. E. Tucker, and G. P. Roberts, 2007b, Contrasting transient and steady-state rivers crossing active normal faults: new field observations from the Central Apennines, Italy, *Basin Research*, doi:10.1111/j.1365–2117.2007.00337.x.

Wiberg, P. L. and J. D. Smith, 1987, Calculations of the critical shear stress for motion of uniform and heterogeneous sediments. *Water Resources Research* **23**: 1471–1480.

Wiberg, P. L. and J. D. Smith, 1989, Model for calculating bed load transport of sediment, *Journal of Hydraulic Engineering* **115** (1): 101–123.

Wilding, L. P. and H. Lin, 2006, Advancing the frontiers of soil science towards a geoscience, *Geoderma* **131**: 257–274.

Williams, P. J. and M. W. Smith, 1989, *The Frozen Earth – Fundamentals of Geocryology*, Cambridge: Cambridge University Press.

Wilshire, H. G., J. K. Nakata, and B. Hallet, 1981, Field observations of the December 1997 wind storm, San Joaquin Valley, California, in: T. L. Péwé (ed.), *Desert Dust: Origin, Characteristics, and Effect on Man*, Geological Society of America Special Paper 186, Boulder, CO: Geological Society of America, pp. 233–251.

Wobus, C. W., K. V. Hodges, and K. X. Whipple, 2003, Has focused denudation sustained active thrusting at the Himalayan topographic front?, *Geology* **31**: 861–864.

Wobus, C. W., J. W. Kean, G. E. Tucker, and R. S. Anderson, 2008b, Modeling the evolution of channel shape: balancing computational efficiency with hydraulic fidelity, *Journal of Geophysical Research* **113**: F02004, doi:10.1029/2007JF000914.

Wobus, C., M. Pringle, K. Whipple, and K. Hodges, 2008a, A Late Miocene acceleration of exhumation in the Himalayan crystalline core, *Earth and Planetary Science Letters* **269**: 1–10.

Wobus, C., G. Tucker, and R. S. Anderson, 2006, Self-formation of bedrock channels, *Geophysical Research Letters* **33**: L18408, doi:10.1029/2006GL027182.

Wohl, E. E. and H. Ikeda, 1997, Experimental simulation of channel incision into a cohesive substrate at varying gradients, *Geology* **25** (4): 295–298.

Wohl, E. E. and D. M. Merritt, 2001, Bedrock channel morphology, *Geological Society of America Bulletin* **113** (9): 1205–1212.

Wolkowinsky, A. and D. Granger, 2004, Early Pleistocene incision of the San Juan River, Utah, dated with ^{26}Al and ^{10}Be, *Geology* **32**: 749–752.

Wolman, M. G. and L. B. Leopold, 1957, *River Floodplains: Some Observations on their Formation*, US Geological Survey Professional Paper 282C, Reston, VA: US Geological Survey, pp. 87–109.

Wolman, M. G. and J. P. Miller, 1960, Magnitude and frequency of forces in geomorphic processes, *Journal of Geology* **68**: 54–74.

Wu, C. S., E. B. Thornton, and R. T. Guza, 1985, Waves and longshore currents: comparison of a numerical model with field data, *Journal of Geophysical Research* **90** (C3): 4951–4958.

Yalin, M. S., 1963, An expression for bed-load transport, *American Society of Civil Engineers, Journal of the Hydraulics Division*, **89**, HY3: 221–250.

Yalin, M. S., 1972, *Mechanics of Sediment Transport*, New York: Pergamon Press.

Yancey, T. E. and J. W. Lee, 1972, Major heavy mineral assemblages and heavy mineral provinces of the central California coast region, *Geological Society of America Bulletin* **83**: 2099–2103.

Yasso, W. E., 1965, Plan geometry of headland bay beaches, *Journal of Geology* **73**: 702–714.

Yatsu, E., 1988, *The Nature of Weathering*, Tokyo: Sozosha.

Yeats, R. S., K. Seih, and C. R. Allen, 1997, *The Geology of Earthquakes*, New York, Oxford: Oxford University Press.

Young, I. M. and J. W. Crawford, 2004, Interactions and self-organization in the soil-microbe complex, *Science* **304** (5677): 1634–1637.

Zachos, J. C., M. Pagani, L. Sloan, E. Thomas, and K. Billups, 2001, Trends, rhythms, and aberrations in global climate 65 Ma to Present, *Science* **292**: 686.

Zachos, J. C., N. J. Shackleton, J. S. Revenaugh, H. Pälike, and B. P. Flower, 2001, Climate response to orbital forcing across the Oligocene-Miocene boundary, *Science* **292** (5515): 274–278.

Zachos, J. C., M. W. Wara, S. M. Bohaty *et al.*, 2003, A transient rise in tropical sea surface temperature during the Paleocene-Eocene Thermal Maximum, *Science* **302**: 1551–1554.

Zandt, G., H. Gilbert, T. J. Owens, M. Ducea, J. Saleeby, and C. H. Jones, 2004, Active foundering of a continental arc root beneath the southern Sierra Nevada in California, *Nature* **431**: 41–46.

Zeitler, P. K., A. Herczeg, I. McDougall, and M. Honda, 1987, U-Th-He dating of Durango fluorapatite: a potential thermochronometer, *Geochimica et Cosmochimica Acta* **51**: 2865–2868.

Zeitler, P. K., A. S. Meltzer, P. O. Koons *et al.*, 2001, Erosion, Himalayan geodynamics, and the geomorphology of metamorphism, *GSA Today* **11**: 4–9.

INDEX